UNIVERSE
THE SOLAR SYSTEM

UNIVERSE
THE SOLAR SYSTEM

Roger A. Freedman

University of California, Santa Barbara

William J. Kaufmann III

Late of San Diego State University

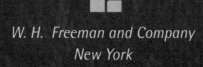

W. H. Freeman and Company
New York

ACQUISITIONS EDITOR: PATRICK FARACE

PUBLISHER: MICHELLE JULET

DIRECTOR OF MARKETING: JOHN BRITCH

EDITORIAL DEVELOPMENT: BARBARA BROOKS

PROJECT EDITOR: DIANE CIMINO DAVIS

MEDIA/SUPPLEMENTS EDITOR: CHARLES VAN WAGNER

ASSISTANT EDITOR: DANIELLE SWEARENGIN

COVER DESIGN: BLAKE LOGAN

TEXT DESIGN: VICKI TOMASELLI AND BLAKE LOGAN

ILLUSTRATION COORDINATOR: BILL PAGE

ILLUSTRATIONS: FINE LINE ILLUSTRATIONS

PHOTO RESEARCH: INGE KING AND VIKII WONG

PRODUCTION COORDINATOR: SUSAN WEIN

COMPOSITION: SHERIDAN SELLERS,
W. H. FREEMAN AND COMPANY ELECTRONIC PUBLISHING CENTER, AND TSI GRAPHICS

MANUFACTURING: R. R. DONNELLEY & SONS COMPANY

■ ■ ■

Library of Congress Cataloging-in-Publication Data

Freedman, Roger A.
    Universe : the solar system / Roger A. Freedman, William J. Kaufmann III.
        p. cm.
    Includes bibliographical references and index.
    ISBN 0-7167-4645-X
    1. Solar system. I. Kaufmann, William J. II. Title.
    QB501 .F74 2001
    523.2—dc21

                                            2001001695

Guest Essay photo credits:
p. 18, Don Fukuda/National Optical Astronomy Observatories;
p. 42, courtesy of James Randi; p. 177, courtesy of Geoff Marcy;
p. 418, courtesy of John N. Bahcall; p. 702, courtesy of Matthew Golombek

Printed in the United States of America
First printing 2001

To Lee Johnson Kaufmann and
Caroline Robillard-Freedman, strong survivors,
and to the memory of
S/Sgt. Ann Kazmierczak Freedman, WAC

Contents Overview

PART I Introducing Astronomy

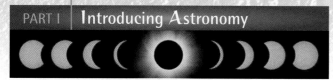

The total solar eclipse of August 11, 1999 (© 2000 by Fred Espenak, MrEclipse.com)

1 Astronomy and the Universe   1

   GUEST ESSAY: Why Astronomy?
   Sandra M. Faber   18

2 Knowing the Heavens   20

   GUEST ESSAY: Why Astrology Is Not Science
   James Randi   42

3 Eclipses and the Motion of the Moon   43

4 Gravitation and the Waltz of the Planets   62

5 The Nature of Light   92

6 Optics and Telescopes   122

PART II Planets and Moons

An artist's conception of the nucleus of a comet (STScI)

7 Our Solar System   150

   GUEST ESSAY: Alien Planets
   Geoff Marcy   177

8 Our Living Earth   179

9 Our Barren Moon   205

10 Sun-Scorched Mercury   225

11 Cloud-Covered Venus   240

12 Red Planet Mars   260

13 Jupiter: Lord of the Planets   283

14 The Galilean Satellites of Jupiter   302

15 The Spectacular Saturnian System   322

16 The Outer Worlds   342

17 Vagabonds of the Solar System   364

PART III The Sun and Beyond

An artist's conception of the explosion of a supernova (T. Goertel, STScI)

18 Our Star, the Sun   389

   GUEST ESSAY: Searching for Neutrinos Beyond the Textbooks
   John N. Bahcall   418

30 The Search for Extraterrestrial Life   689

   GUEST ESSAY: Exploring Mars
   Matthew Golombek   702

Contents

To the Instructor   xv

Instructors' Guided Tour of *Universe*   xix

Acknowledgments   xxii

To the Student: How to Get the Most
from *Universe*   xxv

A Guided Tour of *Universe*   xxvii

PART I | Introducing Astronomy

1 | Astronomy and the Universe   1

1-1   Astronomy and the Scientific Method   2

1-2   The Solar System   3

1-3   Stars and Stellar Evolution   4

1-4   Galaxies and Cosmology   6

1-5   Angles and Angular Measure   7

BOX 1-1   The Small-Angle Formula   8

1-6   Powers of Ten   8

BOX 1-2   Arithmetic with Powers-of-Ten
Notation   11

1-7   Astronomical Distances   11

BOX 1-3   Units of Length, Time, and Mass   12

1-8   The Adventure of Astronomy   14

Key Words and Ideas   15

Questions and Projects   15

GUEST ESSAY: Why Astronomy?
Sandra M. Faber   18

2 | Knowing the Heavens   20

2-1   Ancient Astronomy   21

2-2   Constellations   21

2-3   Motions of the Sky   23

2-4   The Celestial Sphere   26

BOX 2-1   Celestial Coordinates   28

2-5   The Seasons   29

2-6   Precession   32

2-7   Time   34

2-8   The Calendar   35

BOX 2-2   Sidereal Time   36

Key Words and Ideas   37

Questions and Projects   38

GUEST ESSAY: Why Astrology Is Not Science
James Randi   42

3 | Eclipses and the Motion of the Moon   43

3-1   Phases of the Moon   44

BOX 3-1   Phases and Shadows   46

3-2   The Moon's Rotation   47

3-3   Eclipses and the Line of Nodes   49

3-4   Lunar Eclipses   50

3-5   Solar Eclipses   52

3-6   Measuring the Earth   55

BOX 3-2   Predicting Solar Eclipses   56

Key Words and Ideas   58

Questions and Projects   59

p. 5

4 | Gravitation and the Waltz of the Planets   62

4-1   Geocentric Models   63

4-2   Copernicus and Heliocentric Models   66

BOX 4-1   Relating Synodic and Sidereal Periods   69

4-3   Galileo and the Telescope   70

4-4   Tycho Brahe's Observations   73

4-5   Kepler and the Orbits of the Planets   74

BOX 4-2   Using Kepler's Third Law   77

4-6   Newton's Laws of Motion   77

BOX 4-3   Newton's Laws in Everyday Life   79

4-7   Newton and Gravity   80

BOX 4-4   Newton's Form of Kepler's Third Law   83

4-8   Tides and the Moon   84

Key Words and Ideas   87

Questions and Projects   88

5 | The Nature of Light   92

5-1   The Speed of Light   93

5-2   The Wave Nature of Light   94

5-3   Blackbody Radiation   98

BOX 5-1   Temperatures and Temperature Scales   100

5-4   Wien's Law and the Stefan-Boltzmann Law   101

5-5   The Particle Nature of Light   102

BOX 5-2   Using the Laws of Blackbody Radiation   103

BOX 5-3   Photons at the Supermarket   105

5-6   Kirchhoff's Laws   105

BOX 5-4   Light Scattering   108

5-7   Atomic Structure   110

5-8   Spectral Lines and the Bohr Model   111

BOX 5-5   Atoms, the Periodic Table, and Isotopes   112

5-9   The Doppler Effect   115

BOX 5-6   Applications of the Doppler Effect   117

Key Words and Ideas   118

Questions and Projects   119

p. 144

6 | Optics and Telescopes   122

6-1   Refracting Telescopes   123

BOX 6-1   Magnification and Light-Gathering Power   127

6-2   Reflecting Telescopes   128

6-3   Angular Resolution   132

6-4   Charge-Coupled Devices (CCDs)   134

6-5   Spectrographs   136

6-6   Radio Telescopes   137

6-7   Telescopes in Space   140

Key Words and Ideas   146

Questions and Projects   147

PART II   Planets and Moons

7 | Our Solar System   150

7-1   Terrestrial and Jovian Planets   151

BOX 7-1   Average Density   154

7-2   Satellites of the Planets   154

7-3   The Evidence of Spectroscopy   155

7-4   Chemical Composition of the Planets   157

7-5   Asteroids and Comets   158

BOX 7-2   Kinetic Energy, Temperature, and Whether Planets Have Atmospheres   159

7-6   Abundances of the Elements   161

7-7   The Origin of the Solar System   163

7-8   The Origin of the Planets   165

7-9   Extrasolar Planets   169

Key Words and Ideas   172

Questions and Projects   173

GUEST ESSAY: Alien Planets
Geoff Marcy   177

8 | Our Living Earth   179

8-1   The Earth's Energy Sources   180

8-2   Earthquakes and the Earth's Interior   183

8-3   Plate Tectonics   186

8-4   The Earth's Magnetic Field   191

8-5   The Earth's Evolving Atmosphere   194

8-6   Humans and the Earth's Biosphere   198

Key Words and Ideas   201

Questions and Projects   202

9 | Our Barren Moon   205

9-1   The Moon's Airless Surface   206

9-2   Voyages to the Moon   210

9-3   The Moon's Interior   213

9-4   Moon Rocks   214

BOX 9-1   Calculating Tidal Forces   215

9-5   The Formation of the Moon   218

Key Words and Ideas   221

Questions and Projects   222

10 | Sun-Scorched Mercury   225

10-1   Mercury as Seen from Earth   226

10-2   Mercury's Curious Rotation   228

10-3   Mercury's Cratered Surface   231

10-4   The Interior Structure of Mercury   234

Key Words and Ideas   236

Questions and Projects   237

11 | Cloud-Covered Venus   240

11-1   Venus as Seen from Earth   241

11-2   Venus's Retrograde Rotation   243

11-3   Venus's Oppressive Atmosphere   244

11-4   Volcanoes on Venus   247

11-5   Climate Evolution on Venus   250

11-6   Geology on Venus   252

Key Words and Ideas   256

Questions and Projects   256

12 | Red Planet Mars   260

12-1   Mars as Seen from Earth   261

12-2   Speculations About Canals   263

12-3   Craters, Volcanoes, and Canyons   264

12-4   Dry Lakes and Polar Ice Caps   267

12-5   Climate Evolution on Mars   270

12-6   Landing on Mars   273

12-7   The Martian Seasons   276

12-8   The Moons of Mars   277

Key Words and Ideas   278

Questions and Projects   279

13 | Jupiter: Lord of the Planets   283

13-1   Jupiter as Seen from Earth   284

13-2   Jupiter's Clouds   286

13-3   Jupiter's Weather Systems   290

13-4   Probing Beneath Jupiter's Clouds   292

13-5   Jupiter's Rocky Core   295

13-6   The Interior Structure of Jupiter   295

13-7   Jupiter's Magnetosphere   296

Key Words and Ideas   298

Questions and Projects   298

p. 330

14 | The Galilean Satellites of Jupiter   302

14-1 The Satellites as Seen from Earth   303
14-2 Sizes, Masses, and Densities   305
14-3 Formation of the Galilean Satellites   306
14-4 Io's Active Volcanoes   306
14-5 Electric Currents in Io   310
14-6 Europa's Icy Crust   311
14-7 Cratered Ganymede and Callisto   313
14-8 Jupiter's Small Satellites and Ring   316
      Key Words and Ideas   318
      Questions and Projects   319

15 | The Spectacular Saturnian System   322

15-1 Saturn as Seen from Earth   323
15-2 Saturn's Icy Rings   326
15-3 The Structure of the Rings   327
15-4 Shepherd Satellites   328
15-5 Saturn's Atmosphere   330
15-6 The Interior Structure of Saturn   331
15-7 Saturn's Internal Heat   332
15-8 Titan   333
15-9 Saturn's Other Satellites   334
15-10 New Missions to Saturn   337
      Key Words and Ideas   338
      Questions and Projects   338

16 | The Outer Worlds   342

16-1 Discovering Uranus and Neptune   343
16-2 Weather and Seasons on Uranus   344
16-3 Cloud Patterns on Neptune   347
16-4 Inside Uranus and Neptune   349
16-5 Magnetic Fields of Uranus and Neptune   350
16-6 Rings of Uranus and Neptune   351
16-7 Uranus's Satellites   352
16-8 Triton   354
16-9 Discovering Pluto   356
16-10 Pluto and Charon   358
      Key Words and Ideas   360
      Questions and Projects   360

17 | Vagabonds of the Solar System   364

17-1 The Discovery of the Asteroids   365
17-2 Jupiter and the Asteroid Belt   367
17-3 The Nature of the Asteroids   368
17-4 Impacts on Earth   370
17-5 Classifying Meteorites   372
17-6 Meteorites and Our Origins   374
17-7 Comets   375
17-8 The Origin of Comets   379
17-9 Comets and Meteor Showers   381
      Key Words and Ideas   384
      Questions and Projects   385

PART III   The Sun and Beyond

18 | Our Star, the Sun   389

18-1   Thermonuclear Reactions   390
BOX 18-1 Converting Mass into Energy   393
18-2   A Model of the Sun   394

p. 412

18-3   Solar Seismology   397

18-4   Solar Neutrinos   398

18-5   The Photosphere   400

18-6   The Chromosphere   402

18-7   The Corona   403

18-8   Sunspots   405

18-9   The Sunspot Cycle   408

18-10 The Active Sun   411

    Key Words and Ideas   413

    Questions and Projects   414

GUEST ESSAY: Searching for Neutrinos Beyond the Textbooks
John N. Bahcall   418

30 | The Search for Extraterrestrial Life   689

30-1   Building Blocks of Life   690

30-2   Life in the Solar System   691

30-3   Meteorites from Mars   694

30-4   The Drake Equation   695

30-5   Radio Searches   696

30-6   Infrared Searches   697

    Key Words and Ideas   699

    Questions and Projects   699

GUEST ESSAY: Exploring Mars
Matthew Golombek   702

Appendices   A-1

1   The Planets: Orbital Data   A-1

2   The Planets: Physical Data   A-1

3   Satellites of the Planets   A-2

4   The Nearest Stars   A-4

5   The Visually Brightest Stars   A-5

6   Some Important Astronomical Quantities   A-6

7   Some Important Physical Constants   A-6

8   Some Useful Mathematics   A-6

Glossary   G-1

Answers to Selected Questions   Q-1

Index   I-1

Northern Hemisphere Star Charts   S-1

To the Instructor

From its first edition, Bill Kaufmann's *Universe* has been a tremendous help to my introductory astronomy students. They have benefited from Bill's compelling prose, his emphasis on how astronomers learn about the universe, and his clear exposition. I've benefited as well. The in-depth presentation in *Universe*—never "watered down," but always accessible to the beginning student—has made it possible for me to teach a more satisfying course. And so, it's been my special pleasure to carry on Bill's work by developing *Universe* into its new Sixth Edition.

A narrative of scientific discovery underscores each chapter

When I took over the stewardship of *Universe* five years ago, my goals were to continue its emphasis on the process of scientific discovery; to make it easier for students to grasp the concepts of astronomy; to help students develop the skills needed to solve quantitative problems; and to make the book as modern, lively, and interesting as I possibly could. Above all, I've worked to retain and enhance the strengths that have made *Universe* a standard in astronomy education, including its particular focus on *how scientists work and think*.

Many students who take an astronomy course will have no other exposure to the physical sciences in their college or university career. In a real sense, it is their last chance to gain an appreciation for science and the scientific method. Thus, it is essential that students learn not only *what* we know about our fascinating physical universe but also *how* we know it. The scientific way of knowing, based on observation and continual questioning, is at the heart of both the organization and the features of *Universe*.

Astronomy is an intensely visual and graphic subject, and *Universe* was the first astronomy textbook to be in full color throughout. Continuing this tradition of graphic innovation, new **flipbook** images created from sequences of animations appear in the upper right-hand corners of the odd-numbered pages in this edition. These images help motivate students to flip through the pages of the book, perhaps to run across an image or topic that piques their interest. The flipbook images also stimulate students to peruse the extensive collection of animations and videos on the *Universe* web site and CD-ROM. Perhaps most important of all, the flipbook images help make this book fun!

The flipbook images highlight the text's flexible organization, which has always suited my needs as an instructor. New to this edition, the split versions of *Universe* make it even easier for instructors to structure their courses as they see fit.

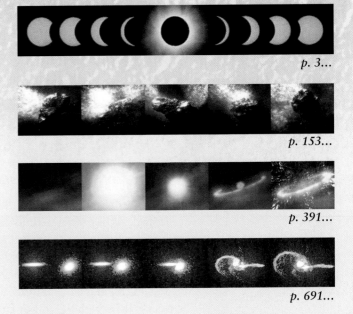

p. 3...

p. 153...

p. 391...

p. 691...

Three versions of UNIVERSE make it easy to customize course content

In addition to the complete, 30-chapter textbook, two briefer versions of the Sixth Edition are now available. These versions provide instructors with new options for shorter courses, as well as for courses that focus on either the solar system or stars and galaxies:

- *Universe* contains the full text of Chapters 1–30. (0-7167-4647-6)
- *Universe: The Solar System* contains the full text of Chapters 1–18 and 30. (0-7167-4645-X)
- *Universe: Stars and Galaxies* contains the full text of Chapters 1–7 and 18–30. (0-7167-4646-8)

All three versions include the fundamental material on positional astronomy, orbits, gravitation, light, and telescopes, as well as all Appendices, Glossary, Answers to Selected Questions, and Index. In addition, each version of *Universe* comes with the full CD-ROM, including *Starry Night*™ planetarium software.

UNIVERSE fully integrates the text with electronic media

Universe was the first textbook to use a web site and a CD-ROM. My vision for the Sixth Edition is that students will use the book, web site, and CD-ROM as aspects of a seamless learning experience. On almost every page, new **media icons** point students to specific, relevant information on the CD-ROM and the *Universe* web site:

 ANIMATIONS illustrating key concepts, including animated versions of selected text figures.

 VIDEOS from space missions as well as ground-based observations.

 IN-DEPTH DISCUSSIONS that supplement text topics.

 LINKS to external web sites where students can find additional information and images.

 INTERACTIVE EXERCISES test student understanding of selected images and diagrams in the text.

 ACTIVE INTEGRATED MEDIA MODULES are interactive learning tools that simplify repetitive calculations and allow students to gain quantitative understanding.

 Many end-of-chapter **OBSERVING PROJECTS** use the award-winning *Starry Night*™ planetarium software provided on the CD-ROM.

Review aids help students get the most out of the book

My students appreciate how the many features of *Universe* help them review what they've learned. As in previous editions, **chapter-opening questions and full-sentence section headings** help guide their reading. Chapters end with lists of **Key Words** referenced by page to where they are discussed in the chapter, as well as summaries of **Key Ideas**. Each chapter has an extensive collection of **Questions**, including **Problem-Solving Tips and Tools** to help students build their skills. **Observing Tips and Tools** provide students with practical hints to help them with the end-of-chapter **Observing Projects**.

A student-oriented introduction, **How to Get the Most from *Universe*,** follows this preface. I've included hints for students who may never have taken a college science course. Following this is a new **Guided Tour of *Universe*** that explains how best to use the book's text and media features, as well as a new self-quiz that helps students confirm their understanding of these features.

In the previous edition of *Universe*, I introduced **Caution paragraphs** that alert students to common misconceptions. I also added **Analogy paragraphs** and **"The Heavens on the Earth" boxes** that relate astronomical phenomena to aspects of everyday life. On the quantitative side, **"Tools of the Astronomer's Trade"** boxes have worked examples that show students how to solve "real" problems. **Links to other chapters** in blue underlined type help students see the connections among chapters and make it easier for them to locate material they need to review. I've worked hard to make these features even more useful in the Sixth Edition (see the Instructors' Guided Tour of *Universe* on page xix).

For this edition, I've added more than 300 entirely new end-of-chapter questions. In addition to traditional-style questions, many of the new questions ask students to interpret astronomical images or to evaluate how the mass media portray the ideas of astronomy. New **Web/CD-ROM Questions** ask them to analyze images or animations on the textbook CD-ROM or web site, or to do research on the World Wide Web. Other questions require them to use the *Starry Night™* planetarium software provided on the CD-ROM.

A thoroughly up-to-date text shows science in the making

I've brought the book up to date with the latest information from all aspects of the subject we love. In addition to dramatic new color images and enhanced explanatory diagrams, I've included the latest discoveries, including:

- Recent results from *Mars Global Surveyor, NEAR Shoemaker,* the Chandra X-ray Observatory, and the ESO's Very Large Telescope, as well as other missions and observatories.

- The latest information about brown dwarfs, subsurface water on Mars and Europa, ice on the Moon, and much more.

- New **guest essay** by Matthew Golombek (JPL) on the exploration of Mars.

I have endeavored to make this new edition even more accessible to students, while being careful to maintain the same intellectual level. One of the ways I have done this is by reorganizing several key chapters. Examples include:

- Chapter 4, Gravitation and the Waltz of the Planets, now presents tidal forces immediately after the discussion of Newton's laws and gravitation.

- Chapter 8, Our Living Earth, has been totally revamped to treat our planet's atmosphere, surface, and oceans as interacting systems.

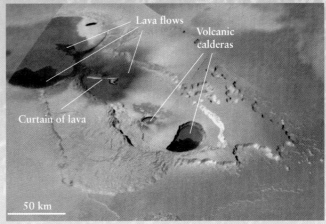

p. 307

p. 693

p. 411

• Chapter 12, Red Planet Mars, places greater emphasis on the joint evolution of the Martian ocean and atmosphere.

• Chapter 18, Our Star, the Sun, now discusses the Sun "from the inside out," beginning with how the Sun shines and ending with the solar atmosphere.

• Chapter 30, The Search for Extraterrestrial Life, has been expanded to include the search for extraterrestrial life of all kinds, from single cells to advanced civilizations. A new section describes the ongoing search for life within our solar system.

A constellation of supplements complements this textbook

Students can connect to the universe in two ways: via the web site (**www.whfreeman.com/universe6e**) and via the *Universe 6.0* CD-ROM packaged with every copy of the book. In addition, instructors can choose from a full range of print ancillaries, as well as software tools for course management, assessment, and presentation. For full details on *Universe* supplements for students and instructors, consult the Instructors' Guided Tour, which begins on page xix.

Instructors' Guided Tour of UNIVERSE

UNIVERSE WEB SITE at www.whfreeman.com/universe6e

For instructors, the site offers **Further information** about the topics in the book, **Links** to additional resources, and **a forum** for sharing ideas about using the book.

For students, the web site serves as a study guide. Its features include:

- **Chapter Objectives:** Learning objectives help students formulate study strategies.

- **Q & A:** The self-quizzing feature helps students prepare for exams.

- **Interactive Exercises:** "Drag & drop" exercises that help students understand the vocabulary of astronomy in the context of illustrations from the text.

- **Key Terms:** Flashcard exercises based on vocabulary and definitions.

- *Starry Night*™ **Observing Exercises:** In-depth projects that students can perform using the *Starry Night*™ software provided on the student CD-ROM. These complement the shorter *Starry Night*™ Observing Projects in the textbook.

- **Current Events in Astronomy:** Up-to-date links to information and resources.

- **Animations and Videos,** both original and from sources such as NASA and ESA, are updated regularly.

- **Web Links** lead to World Wide Web sites that contain supplemental material about astronomy.

- **Active Integrated Media Modules (AIMM):** Interactive modules that take students deeper into key topics from the text, including how to determine distances to the stars and beyond, the Doppler effect, the nearest stars, and relativistic redshift.

- **Looking Deeper:** In-depth boxes, thoroughly updated, extend topics discussed in the text, including the search for gravitational waves, how to interpret the shapes of spectral lines, and discoveries about the inner core of the Earth.

UNIVERSE 6.0 CD-ROM featuring STARRY NIGHT™

Packaged with every copy of the textbook, the *Universe 6.0* CD-ROM contains all the content that appears on the web site (except for the Web Links and Current Events in Astronomy). It also includes the easy-to-use, award-winning *Starry Night*™ planetarium software (for both Macintosh and Windows), the ideal electronic accompaniment for any astronomy course. *Starry Night*™ observational activities are provided in each chapter of *Universe*. Others are available on the *Universe* web site.

- **Freeman Observing Guide (ISBN 0-7167-4437-6)**, *William J. F. Wilson and T. Alan Clark, University of Calgary*—This printed supplement has 15 new, comprehensive lab activities for *Starry Night*™ that provide even more opportunities for students to explore the cosmos.

Presentation Tools

INSTRUCTOR'S CD-ROM

Supplied to all instructors who adopt *Universe*, this CD-ROM provides instructors with the tools to create in-class presentations:

- **All text images in JPEG, PICT, and PowerPoint format** for use with any standard presentation software.

- **Presentation Manager Pro software** lets you build classroom presentations using images and video from the *Universe* CD-ROM, *plus* your own digital material (including video) imported from the World Wide Web or other sources.

STARRY NIGHT™ PRO VERSION CD-ROM, FROM SPACE.COM

Available to qualified instructors, the deluxe version of *Starry Night*™ features over 19 million celestial objects (including the complete Hubble Guide Star Catalog), realistically rendered 3D planets, astrophotographs of all 110 Messier objects, a Datamaker application for adding your own custom astronomical databases, QuickTime movie-making abilities, and more.

THE FREEMAN ASTRONOMY IMAGE AND LECTURE GALLERY

- **Images and Diagrams:** This convenient database makes it easy to access electronic figures from the textbook and to use them in web pages or PowerPoint presentations. Registered users can browse, search, and download illustrations from Freeman astronomy titles. They can also create folders on a personalized home page for easy organization of the images.

- **PowerPoint Lecture Presentation:** Instructors adopting *Universe* have access to PowerPoint files for each chapter of the book. These include lecture notes written by Timothy Slater (Montana State University), as well as all the textbook images and their captions. The PowerPoint files can be downloaded and customized to suite the individual instructor's needs.

OVERHEAD TRANSPARENCY SET

100 full-color transparencies of key illustrations, photos, and tables from the text.

Course Management Tools

ASTRONOMY ONLINE, *Timothy Slater, Montana State University*

Available in WebCT, Blackboard, and Metatext formats, **Astronomy Online** combines text content from *Universe* with animations, videos, interactive exercises, images, and quizzing specifically created to accompany the book. It can serve either as the online component of a lecture-based course or as a complete online source for a distance learning course. Throughout **Astronomy Online**, the emphasis is on interactivity and active learning. Students are prompted to answer questions and make observations about the course materials. Upon completion of each module, they receive feedback on their responses. **Astronomy Online** can be used as is or tailored to meet instructors' individual needs. It can be accessed via the *Universe* web site.

ON-LINE COURSE MATERIALS IN WEBCT

Adopters using WebCT can use this service, which provides a fully loaded E-Pack that includes the instructor and student resources for *Universe*. The files can be used as is or customized to fit specific needs. Course outlines, pre-built quizzes, links, activities, and a whole array of materials are included.

INSTRUCTOR'S MANUAL AND RESOURCE GUIDE, *George A. Carlson, Citrus College*

The Manual includes worked-out solutions to all end-of-chapter questions in *Universe*. It also includes detailed chapter outlines as well as classroom-tested teaching strategies and hints. The extensive Resource Guide covers reading materials for students and instructors, audiovisual material, and discussion/paper topics.

Assessment Tools

TEST BANK, *William J. F. Wilson and T. Alan Clark, University of Calgary*

This set of more than 2300 multiple-choice questions is available both in print and on CD-ROM in both Macintosh and Windows formats. All questions are section-referenced and designated as "assignment" or "test" questions. The CD-ROM version makes it easy to add, edit, and re-sequence questions to suit your needs. The CD-ROM is also the access point for Diploma Online Testing.

DIPLOMA ONLINE TESTING, *from the Brownstone Research Group*

Diploma makes it easy to create and administer secure exams over either the Internet or a local area network. Questions (taken from the Test Bank CD-ROM) can incorporate multimedia and interactivity. The program lets you restrict tests to specific computers or time blocks. It includes an impressive suite of gradebook and result-analysis features.

ONLINE QUIZZING, *powered by Questionmark*

Accessed via the *Universe* web site, Questionmark's Perception enables instructors to quiz students online easily and securely using prewritten, multiple-choice questions (not from the Test Bank) for each text chapter. Students receive instant feedback and can take the quizzes multiple times. Instructors can go into a protected web site to view results by quiz, by student, or by question, or can get weekly results via e-mail.

SCIENTIFIC AMERICAN READER FOR UNIVERSE

This collection of 15 astronomy-related articles from *Scientific American* magazine is enhanced with introductions written by Roger Freedman. Each article is accompanied by five questions suitable for homework assignments.

Acknowledgments

I would like to thank my colleagues who carefully scrutinized the manuscript of this edition. This is a stronger and better textbook because of their conscientious efforts:

Robert R. J. Antonucci, *University of California, Santa Barbara*

David Bruning, *University of Wisconsin, Parkside*

Spencer L. Buckner, *Austin Peay State University*

Karen G. Castle, *Diablo Valley College*

Kim Coble, *University of California, Santa Barbara*

Malcolm Coe, *The University of Southampton*

David Dahl, *St. Olaf College*

Steve Danford, *University of North Carolina, Greensboro*

Mary V. Frohne, *Western Illinois University*

Pamela L. Gay, *University of Texas, Austin*

Robert M. Geller, *University of California, Santa Barbara*

Donna Hurlbut Gifford, *Pima Community College*

Owen Gingerich, *Harvard University*

Bernadette Londak Harris, *B.C. Open University*

Dave Kary, *Citrus College*

B. Alexander King III, *Austin Peay State University*

Arthur Kosowsky, *Rutgers University*

David Kriegler, *University of Nebraska, Omaha*

Andrew Lazarewicz, *Boston College*

Ntungwa Maasha, *Coastal Georgia Community College*

Margaret Mazzolini, *Swinburne Astronomy Online*

J. Ward Moody, *Brigham Young University*

Gerald H. Newsom, *The Ohio State University*

Erin O'Connor, *Allan Hancock College*

Fritz Osell, *Leeward Community College*

Robert L. Pompi, *Binghamton University*

Harrison B. Prosper, *Florida State University*

Jim F. Smeltzer, *Northwest Missouri State University*

Larry K. Smith, *Snow College*

Darryl Stanford, *City College of San Francisco*

Donald Terndrup, *The Ohio State University*

George F. Tucker, *The Sage Colleges*

Stephen Walton, *California State University, Northridge*

Joseph C. Wesney, *Greenwich High School*

A. B. Whiting, *U.S. Naval Academy*

My heartfelt thanks also go out to the following instructors, who responded to informal surveys for this edition or examined the fifth edition with particular care with an eye to our plans for revision. Their suggestions were tremendously helpful.

Grant Bazan, *Las Positas Community College*

Richard French, *Wellesley College*

Donald Gudehus, *Georgia State University*

Eric Harpell, *Las Positas Community College*

Paul Hintzen, *California State University, Long Beach*

Michael Kaufman, *San Jose State University*

Paul Marquard, *Casper College*

Donald H. Martins, *University of Alaska, Anchorage*

William Parke, *George Washington University*

Eric R. Peterson, *De Anza College*

Roger Romani, *Stanford University*

Michael Ruiz, *University of North Carolina, Asheville*

James Shea, *University of Wisconsin, Parkside*

Ron Stoner, *Bowling Green State University*

Carol Strong, *University of Alabama, Huntsville*

Colin Terry, *Ventura College*

David Weinberg, *The Ohio State University*

Lynda Williams, *San Francisco State University*

I would also like to thank the many people whose advice on previous editions has had an ongoing influence:

Robert Allen, *University of Wisconsin, La Crosse*; Robert R. J. Antonucci, *University of California, Santa Barbara*; Alice L. Argon, *Harvard-Smithsonian Center for Astrophysics*; Omer Blaes, *University of California, Santa Barbara*; David Van Blerkom, *University of Massachusetts*; John M. Burns, *Mt. San Antonio College*; Bel Campbell, *University of New Mexico*; George A. Carlson, *Citrus College*; Bruce W. Carney, *University of North Carolina*; Bradley W. Carroll, *Weber State University*; George L. Cassiday, *University of Utah*; John J. Cowan, *University of Oklahoma*; John E. Crawford, *McGill University*; Roger B. Culver, *Colorado State University*; Robert Dick, *Carleton University*; James N. Douglas, *University of Texas at Austin*; Robert J. Dukes, Jr., *College of Charleston*; Robert A. Egler, *North Carolina State University*; Debra Meloy Elmegreen, *Vassar College*; David S. Evans, *University of Texas, Austin*; George W. Ficken, Jr., *Cleveland State University*; Andrew Fraknoi, *Astronomical Society of the Pacific*; Juhan Frank, *Louisiana State University*; Steven Giddings, *University of California, Santa Barbara*; Owen Gingerich, *Harvard University*; Paul F. Goldsmith, *University of Massachusetts*; J. Richard Gott III, *Princeton University*; Austin F. Gulliver, *Brandon University*; Buford M. Guy, *Cleveland State Community College*; Carl Gwinn, *University of California, Santa Barbara*; Bruce Hanna, *Old Dominion University*; Charles L. Hartley, *Hartwick College*; Paul A. Heckert, *Western Carolina University*; Bill Herbst, *Wesleyan University*; Paul Hodge, *University of Washington*; Douglas P. Hube, *University of Alberta*; Icko Iben, Jr., *Pennsylvania State University*; Scott B. Johnson, *Idaho State University*; John K. Lawrence, *California State University, Northridge*; Marie E. Machacek, *Northeastern University*;

Laurence A. Marschall, *Gettysburg College;* Raymond C. McNeil, *Northern Kentucky University;* Dimitri Mihalas, *University of Illinois;* Robert M. O'Connell, *College of the Redwoods;* C. Robert O'Dell, *Rice University;* L. D. Opplinger, *Western Michigan University;* Stanton J. Peale, *University of California, Santa Barbara;* John R. Percy, *University of Toronto;* Robert L. Pompi, *State University of New York, Binghamton;* Carlton Pryor, *Rutgers University;* James L. Regas, *California State University, Chico;* Terry Retting, *University of Notre Dame;* Tina Riedinger, *University of Tennessee;* James A. Roberts, *University of North Texas;* Charles W. Rogers, *Southwestern Oklahoma State University;* Kenneth S. Rumstay, *Valdosta State College;* Richard Saenz, *California Polytechnic State University;* Thomas F. Scanlon, *Grossmont College;* Richard L. Sears, *University of Michigan;* Isaac Shlosman, *University of Kentucky;* Alan F. Sill, *Texas Tech University;* Caroline Simpson, *Florida International University;* Michael L. Sitko, *University of Cincinnati;* David B. Slavsky, *Loyola University of Chicago;* Joseph S. Tenn, *Sonoma State University;* Gordon B. Thomson, *Rutgers University;* Charles R. Tolbert, *University of Virginia;* Virginia Trimble, *University of California, Irvine;* Thomas Tsung, *Grossmont College;* Bruce A. Twarog, *University of Kansas;* George Wegner, *Dartmouth College;* Donat G. Wentzel, *University of Maryland;* Nicholas Wheeler, *Reed College;* Raymond E. White, *University of Arizona;* Louis Winkler, *The University of Pennsylvania;* and Robert L. Zimmerman, *University of Oregon.*

I'm particularly grateful to Matthew Golombek, whose new essay for this edition greatly enhances its coverage. I also thank Geoff Marcy and John Bahcall, who were kind enough to revise their essays in light of recent developments.

Many others have participated in the preparation of this book, and I thank them for their efforts. I am particularly grateful to Patrick Farace, my acquisitions editor, for keeping me entertained; to Barbara Brooks, my development editor, for keeping me encouraged; to Charlie Van Wagner, my media and supplements editor, for supervising the superb web site and CD-ROM; and to Kent Gardner, Annali Kiers, and their colleagues at Sumanas, for translating my rough ideas into elegant multimedia. Special thanks go to Liz Widdicombe, president; Louise B. Ketz, copy editor and indexer; Diane Cimino Davis, project editor; Susan Wein, production coordinator; Sheridan Sellers, W. H. Freeman and Company Electronic Publishing Center, and TSI Graphics, compositors; Inge King and Vikii Wong, photo researchers; John Britch, marketing manager; and the ever-charming Danielle Swearengin, assistant editor.

On a personal note, I would like to thank my father, Richard Freedman, for first cultivating my interest in space many years ago, and for his personal contributions to the exploration of the universe as an engineer for the Atlas and Centaur launch vehicle programs. Most of all, I thank my charming wife, Caroline, for putting up with my long nights slaving over the computer!

Although I have made a concerted effort to make this edition error free, some mistakes may have crept in unbidden. I would appreciate hearing from anyone who finds an error or wishes to comment on the text. You may e-mail or write me.

Roger A. Freedman
Department of Physics
University of California, Santa Barbara
Santa Barbara, CA 93106
airboy@physics.ucsb.edu

To the Student: How to Get the Most from UNIVERSE

If you're like most students just opening this textbook, you're enrolled in one of the few science courses you'll take in college. As you study astronomy, you'll probably do relatively little reading compared with a literature or history course—at least in terms of the number of pages. But your readings will be packed with information, much of it new to you and (I hope) exciting. You can't read this textbook like a novel and expect to learn much from it. Don't worry, though. I wrote this book with you in mind. In this section, I'll suggest how *Universe* can help you succeed in astronomy and take you on a guided tour of the book and media.

Apply these techniques to studying astronomy

- **Read before a lecture** You'll get the most out of your astronomy course if you read each chapter *before* hearing a lecture about its subject matter. That way, many of the topics will already be clear in your mind, and you'll understand the lecture better. You'll be able to spend more of your listening and note-taking time on the more challenging ideas presented in the lecture.

- **Take notes as you read and make use of office hours** Keep a notebook handy as you read, and write down the key points of each section so that you can review them later. If any parts of the section don't seem clear on first reading, make a note of them, too, including the page numbers. Once you've gone through the chapter, reread it with special emphasis on the ideas that gave you trouble the first time. If you're still unsure after the lecture, consult your instructor, either during office hours or after class. Bring your notes with you so your instructor can see which concepts are giving you trouble. Once your instructor has helped clarify things for you, revise your notes so you'll remember your new-found insights. You'll end up with a chapter summary in your own words. This will be a tremendous help when studying for exams!

- **Make use of your fellow students** Many students find it useful to form study groups for astronomy. You can hash out challenging topics with one another and have a good time while you're doing it. But make sure that you write up your homework by yourself, because the penalties for copying or plagiarizing other students' work can be severe in the extreme. Some students find individual assistance useful, too. If you think a tutor will be helpful, link up with one early. Getting a tutor late in the course, in the belief that you'll be able to catch up with what you missed earlier on, is almost always a pointless exercise.

- **Take advantage of the web site and CD-ROM** Take some time to explore the *Universe* web site (**www.whfreeman.com/universe6e**) and the CD-ROM that comes packaged with this book. On both of these you'll find review materials, animations, videos, interactive exercises, flashcards, and many other features keyed to chapters in *Universe*. All these features are designed to help you learn and enjoy astronomy, so make sure to take full advantage of them.

- **Try astronomy for yourself with your star charts and *Starry Night*™** At the back of this book you'll find a set of star charts for the each month of the year in the northern hemisphere. (For a set of southern hemisphere star charts, visit the *Universe* web site.) Star charts can get you started with your own observations of the universe. Hold the chart overhead in the same orientation as the compass points, with "southern horizon" toward the south and "western horizon" toward the west. (To save strain on your arms, you may want to photocopy these pages.) The CD-ROM packaged with this textbook includes the easy-to-use *Starry Night*™ planetarium program, developed in Canada, which you can use to view the sky on any date and time as seen from anywhere on Earth.

Before you study UNIVERSE, take this quiz

Universe has many features designed to help you succeed in your study of astronomy. To get the most from it, understanding these features and knowing how to use them are essential. To make sure that you do, read through **A Guided Tour of Universe**, which begins on the next page. Then take this brief quiz. If you can answer all the questions, you're ready to begin studying astronomy! (You can check your answers on page Q-1.)

1. Which specially labeled paragraphs alert you to common misconceptions and conceptual pitfalls?

2. Which specially labeled paragraphs draw analogies between ideas in astronomy and aspects of everyday life?

3. What is the significance of text that appears in blue underlined type?

4. In many chapters, in addition to the numbered sections, you will also find material set off in boxes. Which type of box provides extra help with solving mathematical problems? Which type relates astronomical principles to phenomena here on Earth?

5. Throughout the book you will encounter icons labeled "Web Link," "Animation," "Video," "AIMM," or "Looking Deeper." Where should you look to find the information to which these icons refer?

6. What is the *Starry Night*™ program? Where can you find a copy to install on your own computer? How much does it cost?

7. Many of the figures in this book are accompanied by the letters R I V U X G, with one of the letters highlighted. For instance, Figure 6-34*a* has R I **V** U X G, while Figure 6-34*d* has R I V U **X** G. What is the significance of the highlighted letter?

8. Where can you find self-tests and review material for each chapter of *Universe*?

9. Refer to the Appendices at the back of this book. On which page(s) of *Universe* would you look to find the following? (a) the value of the Stefan-Boltzmann constant; (b) the average orbital speed of Mars; (c) the average distance from the center of the planet Pluto to its moon, Charon; (d) the distance in light-years to the star Proxima Centauri.

10. Refer to the Index at the back of this book. On which page(s) of *Universe* would you find the following terms described? (a) spicule; (b) refraction; (c) tidal force; (d) aphelion.

11. Refer to the Answers to Selected Questions at the back of this book. What is the answer to Question 35 of Chapter 5? (*NOTE:* Your instructor may assign as homework some of the questions whose answers can be found in the Answers to Selected Questions. If so, your instructor will expect you to write out and explain your calculations to show how this answer is obtained.)

12. Where in this book can you find northern hemisphere star charts for each month of the year?

Here's the most important advice of all

I haven't mentioned the most important thing you should do when studying astronomy: Have fun! Of all the different kinds of scientists, astronomers are among the most excited about what they do and what they study. Let some of that excitement about the universe rub off on you, and you'll have a great time with this course and with this textbook.

In preparing this edition of *Universe,* I tried very hard to make it the kind of textbook that a student like you would find useful. I'm very interested in your comments and opinions! Please feel free to write me, or send me e-mail, and I will respond personally.

Best wishes for success in your studies!

Roger A. Freedman
Department of Physics
University of California, Santa Barbara
Santa Barbara, CA 93106
airboy@physics.ucsb.edu

A Guided Tour of UNIVERSE

This book is designed to help you get the most from your astronomy course

Our Star, the Sun 18

(Institute of Space and Astronomical Science, Yohkoh Project, SXT Group/NASA)

R A V U **X** G

The Sun is by far the brightest object in the sky. By earthly standards, the temperature of its glowing surface is remarkably high, about ... have found regions of the

Sun shines because at its core hundreds of millions of tons of hydrogen are converted to helium every second. We have found the by-products of this transmutation—

The Sun's size also helps us explain its tremendous energy output. Because the Sun is so large, the total number of square meters of radiating surface—that is, its surface area—is immense. Hence, the total amount of energy emitted by the Sun each second, called its **luminosity**, is very large indeed: about 3.9×10^{26} watts, or 3.9×10^{26} joules of energy emitted per second. (We discussed the relation between the Sun's surface temperature, radius, and luminosity in Box 5-2.) Astronomers denote the Sun's luminosity by the symbol $L_\odot$. A circle with a dot in the center is the astronomical symbol for the Sun and was also used by ancient astrologers.

Chapter openers A brief overview and question set guide you into each chapter's topics. Use the questions as a guide to your reading. You'll find the answer to each question within the corresponding section of the chapter.

Wavelength tabs In addition to using ordinary telescopes, astronomers rely heavily on special telescopes that are sensitive to nonvisible forms of light. To help you appreciate what astronomers learn from these different kinds of observations, wavelength tabs appear with all the images in this textbook. The highlighted letter on each tab shows whether the image was made with **R**adio waves, **I**nfrared radiation, **V**isible light, **U**ltraviolet light, **X** rays, or **G**amma rays. (We'll discuss these different kinds of light in Chapter 5.) The image of the Sun shown here was made with an X-ray telescope, as indicated by the highlighted **X** on the tab.

Key Words and Glossary Every new term that has a special importance in astronomy appears in **boldface type** when it's introduced and whenever it's discussed in depth. Be sure that you understand the meaning of each term and can explain it in your own words. Don't just memorize the definitions! The **Glossary**, which you'll find near the back of the book, gives brief definitions of all the **Key Words** in the book.

Links to other chapters References to topics from previous chapters appear in blue underlined type. The links help you find and review important ideas that apply to topics in different parts of the textbook. For example, the laws that govern the radiation from heated objects (discussed in Chapter 5) help us understand the light we receive from the Sun (Chapter 18).

Flipbook images You can create an animation— an eclipse, for example—when you flip through

the images in the upper corner of each right-hand page. These flipbook images show different animations in different sections of the book, so they help you locate information and have some fun in the process!

Confronting misconceptions You may think that the phases of the Moon are caused by the Earth's shadow falling on the Moon and that the Earth is closer to the Sun in summer than in winter. But in fact, these "commonsense" ideas are incorrect! (You'll learn the correct explanations in Chapters 3 and 4.) Paragraphs signaled by the caution icon alert you to conceptual pitfalls like these.

Bringing astronomy down to Earth To learn astronomical ideas, it can be helpful to relate them to your experience on Earth. Throughout the book you'll find analogy paragraphs that relate, for example, the motions of the planets to children on a merry-go-round and the bending of light in a telescope lens to a car driving on sand.

Box 18-1 | **Tools of the Astronomer's Trade**

Converting Mass into Energy

Figure 18-2 shows the steps involved in the thermonuclear fusion of hydrogen at the Sun's center. In these steps, four protons are converted into a single nucleus of ^{4}He, an isotope of helium with two protons and two neutrons. The reaction depicted in Figure 18-2*a* also produces a neutral, nearly massless particle called the *neutrino*. Neutrinos respond hardly at all to ordinary matter, so they travel almost unimpeded through the Sun's massive bulk. Hence, the energy that neutrinos carry is quickly lost into space. This loss is not great, however, because the neutrinos carry relatively little energy. (See Section 18-4 for more about these

4 hydrogen atoms	=	6.693×10^{-27} kg
−1 helium atom	=	-6.645×10^{-27} kg
Mass lost	=	0.048×10^{-27} kg

Thus, a small fraction (0.7%) of the mass of the hydrogen going into the nuclear reaction does not show up in the mass of the helium. This lost mass is converted into an amount of energy $E = mc^2$:

$$E = mc^2 = (0.048 \times 10^{-27} \text{ kg}) (3 \times 10^8 \text{ m/s})^2$$
$$= 4.3 \times 10^{-12} \text{ joule}$$

The Heavens on the Earth boxes show how the principles that astronomers use can explain everyday phenomena, from why diet soft drink cans float to why the daytime sky is blue.

Tools of the Astronomer's Trade boxes with highlighted, easy to follow, **Worked Examples** help you learn how to perform the calculations you'll probably be asked to work out on exams. Don't just read them; work the examples out on a piece of paper.

Take advantage of the end-of-chapter reviews

Key Words This list appears at the end of each chapter, along with a page number telling you where the word is discussed.

Key Ideas To get the most out of these brief summaries, use their outline format in conjunction with the notes you take while reading. Make sure you fully understand each description in the summary. If any point is unclear, go back and review the text.

KEY IDEAS

Hydrogen Burning in the Sun's Core: The Sun's energy is produced by hydrogen burning, a thermonuclear fusion process in which four hydrogen nuclei combine to produce a single helium nucleus.

• The energy released in a nuclear reaction corresponds to a slight reduction of mass according to Einstein's equation $E = mc^2$.

• Thermonuclear fusion occurs only at very high temperatures; for example, hydrogen burning occurs only at temperatures in excess of about 10^7 K. In the Sun, fusion occurs only in the dense, hot core.

Models of the Sun's Interior: A theoretical description of a star's interior can be calculated using the laws of physics.

• The ... Sun suggests that hydrogen burni... cente...

Questions and Projects Your instructor may assign items from these sections as homework. The questions help you review and apply text concepts. Some ask you to analyze images in the text or to evaluate how the mass media portray concepts in astronomy. *Web/CD-ROM Questions* challenge you to work with animations and physical concepts on the *Universe* CD-ROM or web site, or to research topics on the World Wide Web. Look for *Problem-Solving Tips and Tools* that provide guidance for the more advanced questions. (Near the end of the book, you'll find a section of *Answers to Selected Questions*.) *Observing Tips and Tools* accompany many *Observing Projects*, which make frequent use of the *Starry Night*™ planetarium software provided on the CD-ROM.

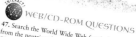

WEB/CD-ROM QUESTIONS

47. Search the World Wide Web for the latest information from the neutrino detectors at the Super-Kamiokande Observatory and the Sudbury Neutrino Observatory. What are the most recent results from these detectors? What is the current thinking about the solar neutrino problem? What is the status of a new detector called Borexino?

48. Search the World Wide Web for information about features in the solar atmosphere called *sigmoids*. What are they? What causes them? How might they provide a way to predict coronal mass ejections?

49. Determining the Lifetime of a Solar Granule. Access and view the video "Granules on the Sun's Surface" in Chapter 18 of the *Universe* Web site or CD-ROM. Your task is to determine the approximate lifetime of a solar granule on the photosphere. Select an area, then you can consistently repeat "start, stop, start, stop" until slowly and rhythmically of granules. While keeping your rhythm, move to a different area of the video and continue monitoring the appearance and disappearance of granules. When you are confident you have the timing right, move your eyes (or use a partner) to the clock shown in the video. Determine the length of time between the

Guest Essays Several chapters in the textbook end with essays by scientists at the cutting edge of astronomy. They include John Bahcall of the Institute for Advanced Study, writing on the physics of the Sun in an essay that follows Chapter 18, and Matthew Golombek of the Jet Propulsion Laboratory, who contributed the essay "Exploring Mars" at the end of Chapter 30. Their essays give you a sense of their excitement about astronomy. We hope you catch this excitement! For a full list of essays, see the Contents Overview on page vii.

MATTHEW GOLOMBEK

Exploring Mars

Dr. Matt Golombek, chief scientist of the *Mars Pathfinder* mission, is a research geologist at the Jet Propulsion Laboratory, California Institute of Technology, NASA's lead center for planetary exploration. He received his undergraduate degree from Rutgers University and his master's degree and Ph.D. from the University of Massachusetts, Amherst. Dr. Golombek was a postdoctoral fellow at the Lunar and Planetary Institute, Houston, before joining JPL.

Dr. Golombek has conducted research in the structural geology of the Earth, planets, and satellites. He is presently NASA's Mars Exploration Program Landing Site Scientist, leading the effort to select landing sites for Mars missions.

Dr. Golombek cowrote the book *Mars: Uncovering the Secrets of the Red Planet* for the National Geographic Society and has served as editor and associate editor of professional journals. He has received numerous awards, including the NASA Exceptional Scientific Achievement Medal. In his honor, asteroid 6456 was named Golombek.

The book, CD-ROM, and web site interact

Throughout each chapter, media icons point you to specific, relevant information on the CD-ROM and the *Universe* web site at **www.whfreeman.com/universe6e**

 Web Links, which lead to Internet sites with supplemental material about astronomy, are updated regularly.

 Animations and Videos, both original and from sources such as NASA and ESA.

 Interactive Exercises include "drag & drop" exercises that help you understand the vocabulary of astronomy in the context of illustrations from the text.

 Active Integrated Media Modules take you deeper into key topics from the text.

 Looking Deeper, in-depth coverage that extends topics discussed in the text, including the search for gravitational waves, how to interpret the shapes of spectral lines, and discoveries about the inner core of the Earth.

 Starry Night™ **Observing Exercises** are in-depth lab projects that you can perform using *Starry Night*™ software provided on the CD-ROM. These complement the shorter *Starry Night*™ Observing Projects in the textbook. To use *Starry Night*™, just follow the instructions on the CD-ROM to install the program on your computer.

The UNIVERSE web site also serves as a study guide

- **Chapter Objectives:** Learning objectives help you formulate study strategies.

- **Q & A:** The self-quizzing feature helps you prepare for exams.

- **Key Terms:** Flashcard exercises based on vocabulary and definitions.

- **Current Events in Astronomy:** Information about the latest astronomical developments.

UNIVERSE 6.0 CD-ROM featuring STARRY NIGHT™

 Packaged with every copy of the textbook, the CD-ROM contains all the content that appears on the web site—except for the Web Links and Current Events in Astronomy—and features the easy-to-use, award-winning *Starry Night*™ planetarium software.

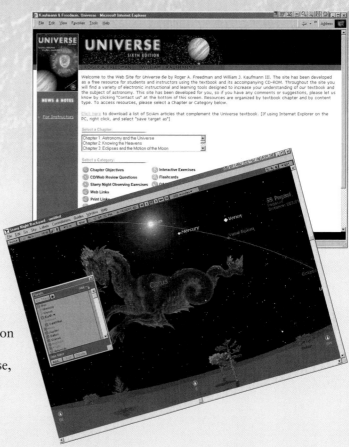

Astronomy and the Universe

Imagine yourself in the desert on a clear, dark, moonless night, far from the glare of city lights. As you gaze upward, you see a panorama that no poet's words can truly describe and that no artist's brush could truly capture. Literally thousands of stars are scattered from horizon to horizon. The delicate glow of the Milky Way—part of which is visible at the upper right of the photograph, taken from a remote location in northern Mexico—traces a luminous path across the sky. As you watch, the entire spectacle swings slowly overhead from east to west as the night progresses. If you are very lucky indeed, you may even see a rare celestial event such as the comet shown here.

For thousands of years people have looked up at the heavens and contemplated the universe. Like our ancestors, we find our thoughts turning to profound questions as we gaze at the stars. How was the universe created? Where did the Earth, Moon, and Sun come from? What are the planets and stars made of? And how do we fit in? What is our place in the cosmic scope of space and time?

To wonder about the universe is a particularly human endeavor. Our curiosity, our desire to explore and discover, and, most important, our ability to reason about what we have discovered are qualities that distinguish us from other animals. The study of the stars transcends all boundaries of culture, geography, and politics. In a literal sense, astronomy is a universal subject—its subject is the entire universe.

(Courtesy of D. L. Mammana)

R I **V** U X G

As you read the sections of this chapter, look for the answers to the following questions:

1-1 What methods do scientists use to expand our understanding of the universe?

1-2 What makes up our solar system?

1-3 What are the stars? Do they last forever?

1-4 What are galaxies? What do astronomers learn by studying them?

1-5 How does measuring angles help astronomers learn about objects in the sky?

1-6 What is powers-of-ten notation, and why is it useful in astronomy?

1-7 Why do astronomers measure distances in astronomical units, light-years, and parsecs?

1-8 How does studying the cosmos help us on Earth?

1-1 To understand the universe, astronomers use the laws of physics to construct testable theories and models

Astronomy has a rich heritage that dates back to the myths and legends of antiquity. Centuries ago, the heavens were thought to be populated with demons and heroes, gods and goddesses. Astronomical phenomena were explained as the result of supernatural forces and divine intervention.

The course of civilization was greatly affected by a profound realization: *The universe is comprehensible.* This awareness is one of the great gifts to come to us from ancient Greece. Greek astronomers discovered that by observing the heavens and carefully reasoning about what they saw, they could learn something about how the universe operates. For example, as we shall see in Chapter 3, they measured the size of the Earth and understood and predicted eclipses. Modern science is a direct descendant of the intellectual endeavors of these ancient Greek pioneers.

Like art, music, or any other human creative activity, science makes use of intuition and experience. But the approach used by scientists to explore physical reality differs from other forms of intellectual endeavor in that it is based fundamentally on *observation, logic,* and *skepticism.* This approach, called the **scientific method,** requires that our ideas about the world around us be consistent with what we actually observe.

The scientific method goes something like this: A scientist trying to understand some phenomenon proposes a **hypothesis,** which is a collection of ideas that seems to explain the phenomenon. It is in developing hypotheses that scientists are at their most creative, imaginative, and intuitive. But their hypotheses must always agree with existing observations and experiments, because a discrepancy with what is observed implies that the hypothesis is wrong. (The exception is if the scientist thinks that the existing results are wrong and can give compelling evidence to show that they are wrong.) The scientist then uses logic to work out the implications of the hypothesis and to make predictions that can be tested. A hypothesis is on firm ground only after it has accurately forecast the results of new experiments or observations. (In prac-

tice, scientists typically go through these steps in a less linear fashion than we have described.)

Scientists describe reality in terms of **models,** which are hypotheses that have withstood observational or experimental tests. A model tells us about the properties and behavior of some object or phenomenon. A familiar example is a model of the atom, which scientists picture as electrons orbiting a central nucleus. Another example, which we will encounter in Chapter 18, is a model that tells us about physical conditions (for example, temperature, pressure, density) in the interior of the Sun. A well-developed model uses mathematics—one of the most powerful tools for logical thinking—to make detailed predictions. For example, a successful model of the Sun's interior should describe in detail how the temperature changes as you look deeper into the Sun. For this reason, mathematics is one of the most important tools used by scientists.

A body of related hypotheses can be pieced together into a self-consistent description of nature called a **theory.** An example from Chapter 5 is the theory of electromagnetism, which in just four equations describes all electric and magnetic phenomena. Without models and theories there is no understanding and no science, only collections of facts.

 In everyday language the word "theory" is often used to mean an idea that looks good on paper, but has little to do with reality. In science, however, a good theory is one that explains reality very well. An excellent example is the theory of electromagnetism, which does a superb job of describing a host of electric and magnetic phenomena. For instance, it predicts that a wave made of oscillating electric and magnetic fields should travel through empty space at the speed of light. Experiment shows that this assertion is exactly correct, and that light itself is a wave of just this type. (We will learn more about the nature of light in Chapter 5.)

Skepticism is an essential part of the scientific method. New hypotheses must be able to withstand the close scrutiny of other scientists. The more radical the hypothesis, the more skepticism and critical evaluation it will receive from the scientific community, because the general rule in science is that extraordinary claims require extraordinary evidence. That is why scientists as a rule do not accept claims that people have been abducted by aliens and taken aboard UFOs. The evidence presented for these claims is flimsy, secondhand, and unverifiable.

At the same time, scientists must be open-minded. They must be willing to discard long-held ideas if these ideas fail to agree with new observations and experiments, provided the new data have survived evaluation. (If an alien spacecraft really did land on Earth, scientists would be the first to accept that aliens existed—provided they could take a careful look at the spacecraft and its occupants.) That is why scientific knowledge is always provisional. As you go through this book, you will encounter many instances where new observations have transformed our understanding of Earth, the planets, the Sun and stars, and indeed the very structure of the universe.

Theories that accurately describe the workings of physical reality have a significant effect on civilization. For example, basing his conclusions in part on observations of how the planets orbit the Sun, the seventeenth-century scientist Isaac Newton deduced a set of fundamental mathematical laws that describe how *all* objects move. These laws, which we will encounter in Chapter 4, work equally well on Earth as in the most distant corner of the universe. They represent our first complete, coherent description of the behavior of the physical universe. **Newtonian mechanics** had an immediate practical application in the construction of machines, buildings, and bridges. It is no coincidence that the Industrial Revolution followed hard on the heels of these theoretical and mathematical advances inspired by astronomy.

Newtonian mechanics and other physical theories have stood the test of time and been shown to have great and general validity. These proven principles are collectively referred to as the **laws of physics.** Astronomers use these laws to interpret and understand their observations of the universe. The laws governing light and its relationship to matter are of particular importance, because the only information we can gather about distant stars and galaxies is in the light that we receive from them. Using the physical laws that describe how objects absorb and emit light, astronomers have measured the temperature of the Sun and even learned what the Sun is made of. By analyzing starlight in the same way, they have discovered that our own Sun is a rather ordinary star, and that the observable universe may contain billions of stars just like the Sun.

An important part of science is the development of new tools for research and new techniques of observation. As an example, until fairly recently everything we knew about the distant universe was based on visible light. Astronomers would peer through telescopes to observe and analyze visible starlight. By the end of the nineteenth century, however, scientists had begun discovering forms of light invisible to the human eye: X rays, gamma rays, radio waves, microwaves, and ultraviolet and infrared radiation.

As we will see in Chapter 6, in recent years astronomers have constructed telescopes that can detect such nonvisible forms of light (Figure 1-1). These instruments give us views of the universe vastly different from anything our eyes can see. These new views have allowed us to see through the atmo-

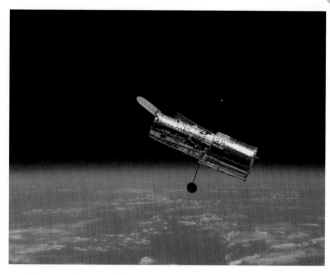

figure 1-1   R I **V** U X G

A Telescope in Space Because it orbits high above Earth in the vacuum of space, the Hubble Space Telescope (HST) can detect not only visible light but also ultraviolet and near-infrared light. These forms of nonvisible light are absorbed by our atmosphere and hence are difficult or impossible to detect with a telescope on the Earth's surface. This photo of HST was taken by the crew of the space shuttle *Discovery* after a servicing mission in 1997. (NASA)

spheres of distant planets, to study the thin but incredibly violent gas that surrounds our Sun, and even to observe new solar systems being formed around distant stars. Aided by high-technology telescopes, today's astronomers carry on the program of careful observation and logical analysis begun thousands of years ago by their ancient Greek predecessors.

1-2   By exploring the planets, astronomers uncover clues about the formation of the solar system

The science of astronomy is a way to let our intellects voyage across the cosmos. We can think of three stages in this voyage: from the Earth to the solar system, from the solar system to the stars, and from stars to galaxies and the grand scheme of the universe.

The star we call the Sun and all the celestial bodies that orbit the Sun—including Earth, the other eight planets, all their various moons, and smaller bodies such as asteroids and comets—make up the **solar system.** Since the 1960s a series of unmanned spacecraft has visited and explored all the planets except Pluto (Figure 1-2). Using the remote "eyes" of such spacecraft, we have flown over Mercury's cratered surface, peered beneath Venus's poisonous cloud cover, and discovered enormous canyons and extinct volcanoes on Mars. We have found active volcanoes on a moon of Jupiter, seen the rings of

figure 1-2 R I **V** U X G
The Sun and Planets to Scale
This montage of images from various spacecraft and ground-based telescopes shows the relative sizes of the nine planets and the Sun. The Sun is so large compared to the planets that only a portion of it fits into this illustration. The distances from the Sun to each planet are *not* shown to scale; the actual distance from the Sun to the Earth, for instance, is 12,000 times greater than the Earth's diameter. (Calvin J. Hamilton and NASA/JPL)

Saturn and Uranus up close, and looked down on the active atmosphere of Neptune.

Along with rocks brought back by the Apollo astronauts from the Moon (the only world beyond Earth yet visited by humans), this new information has revolutionized our understanding of the origin and evolution of the solar system. We have come to realize that many of the planets and their satellites were shaped by collisions with other objects. Craters on the Moon and on many other worlds are the relics of innumerable impacts by bits of interplanetary rock. The Moon may itself be the result of a catastrophic collision between the Earth and a planet-sized object shortly after the solar system was formed. Such a collision could have torn sufficient material from the primordial Earth to create the Moon.

The discoveries that we have made in our journeys across the solar system are directly relevant to the quality of human life on our own planet. Until recently, our understanding of geology, weather, and climate was based solely on data from the Earth. Since the advent of space exploration, however, we have been able to compare and contrast other worlds with our own. This new knowledge gives us valuable insight into our origins, the nature of our planetary home, and the limits of our natural resources.

1-3 | By studying stars and nebulae, astronomers discover how stars are born, grow old, and die

The nearest of all stars to Earth is the Sun. Although humans have used the Sun's warmth since the dawn of our species, it was not until the 1920s and 1930s that physicists figured out how the Sun shines. At the center of the Sun, thermonuclear reactions convert hydrogen (the Sun's primary constituent) into helium. This violent process releases a vast amount of energy, which eventually makes its way to the Sun's surface and escapes as light (Figure 1-3). By 1950 physicists could reproduce such thermonuclear reactions here on Earth in the form of a hydrogen bomb (Figure 1-4). While such thermonuclear weapons are capable of destroy-

ing life on our planet, peaceful applications of this same process may provide a clean source of energy sometime in the next several decades.

All the stars you can see in the nighttime sky also shine by thermonuclear reactions. These reactions consume the material of which stars are made, which means that stars cannot last forever. Rather, they must form, evolve, and eventually die.

Much of what we know about the life stories of stars comes from the study of huge clouds of interstellar gas, called **nebulae** (singular **nebula**), that are found scattered across the sky. Within some nebulae, such as the Orion Nebula shown in Figure 1-5, stars are born from the material of the nebula

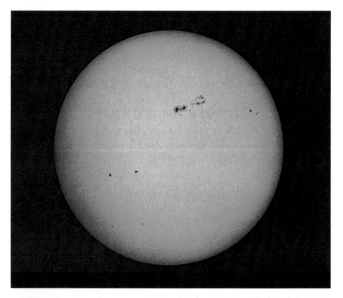

figure 1-3 R I **V** U X G
Our Star, the Sun The Sun is a typical star. Its diameter is about 1.39 million kilometers (roughly a million miles), and its surface temperature is about 5500°C (10,000°F). The Sun draws its energy from thermonuclear reactions occurring at its center, where the temperature is about 15 million degrees Celsius. (Celestron International)

figure 1-4 R I **V** U X G

A Thermonuclear Explosion A hydrogen bomb uses the same physical principle as the thermonuclear reactions at the Sun's center: the conversion of matter into energy by nuclear reactions. This thermonuclear detonation on October 31, 1952, had an energy output equivalent to 10.4 million tons of TNT. This is more than 500 times the output of the bombs exploded over Hiroshima and Nagasaki in 1945, but a mere ten-billionth of the amount of energy released by the Sun in one second. (Defense Nuclear Agency)

itself. Other nebulae reveal what happens when a star dies after millions or billions of years. Some stars that are far more massive than the Sun end their lives with a spectacular detonation called a **supernova** (plural **supernovae**) that blows the star apart. The Crab Nebula (Figure 1-6) is a striking example of a remnant left behind by a supernova.

Dying stars can produce some of the strangest objects in the sky. Some dead stars become **pulsars,** which emit pulses of radio waves, or **X-ray bursters,** which emit powerful bursts of X rays. And some stars end their lives as almost inconceivably dense objects called **black holes,** which are surrounded by gravity so powerful that nothing—not even light—can escape. Even though a black hole itself emits essentially no radiation, a number of black holes have been discovered by Earth-orbiting telescopes. This is done by detecting the X rays emitted by gases falling toward a black hole.

 Astronomers often use biological terms such as "birth" and "death" to describe stages in the evolution of inanimate objects like stars. Keep in mind that such terms are used only as *analogies*, which help us visualize these stages. They are not to be taken literally!

During their death throes, stars return the gas of which they are made to interstellar space. (Figure 1-6 shows these expelled gases expanding away from the site of a supernova explosion.) This gas contains heavy elements created during the star's lifetime by thermonuclear reactions in its interior.

figure 1-5 R I **V** U X G

The Orion Nebula—Birthplace of Stars This beautiful nebula is a stellar "nursery" where stars are formed out of the nebula's gas. Intense ultraviolet light from newborn stars excites the surrounding gas and causes it to glow. Many of the stars embedded in this nebula are less than a million years old, a brief interval in the lifetime of a typical star. The Orion Nebula is some 1500 light-years from Earth and is about 30 light-years across. (© Anglo-Australian Observatory; photograph by David Malin)

figure 1-6 R I **V** U X G

The Crab Nebula—Wreckage of an Exploded Star When a dying star exploded in a supernova, it left behind this elegant funeral shroud of glowing gases blasted violently into space. A thousand years after the explosion these gases are still moving outward at about 1800 km/s (roughly 4 million miles per hour). The Crab Nebula is 6500 light-years from Earth and about 10 light-years across. (The FORS Team, VLT, European Southern Observatory)

Interstellar space thus becomes enriched with newly manufactured atoms and molecules. The Sun and its planets were formed from interstellar material that was enriched in this way. This means that the atoms of iron and nickel that make up the body of the Earth, as well as the carbon in our bodies and the oxygen we breathe, were created deep inside ancient stars. By studying stars and their evolution, we are really studying our own origins.

1-4 By observing galaxies, astronomers learn about the origin and fate of the universe

Stars are not spread uniformly across the universe but are grouped together in huge assemblages called **galaxies.** Galaxies come in a wide range of shapes and sizes. A typical galaxy, like the Milky Way, of which our Sun is part, contains several hundred billion stars. Some galaxies are much smaller, containing only a few million stars. Others are monstrosities that devour neighboring galaxies in a process called "galactic cannibalism."

Our Milky Way Galaxy has arching spiral arms like those of the galaxy shown in Figure 1-7. These arms are particularly active sites of star formation. In recent years, astronomers have discovered a mysterious object at the center of the Milky Way with a mass millions of times greater than that of our Sun. It now seems certain that this curious object is an enormous black hole.

Some of the most intriguing galaxies appear to be in the throes of violent convulsions and are rapidly expelling matter. The centers of these strange galaxies, which may harbor even more massive black holes, are often powerful sources of X rays and radio waves.

Even more awesome sources of energy are found still deeper in space. At distances so great that their light takes billions of years to reach Earth, we find the mysterious **quasars.** Although quasars look like nearby stars (Figure 1-8), they are among the most distant and most luminous objects in the sky. A typical quasar shines with the brilliance of a hundred galaxies. Recent observations of quasars imply that they draw their energy from material falling into enormous black holes.

The motions of distant clusters of galaxies reveal that we live in an expanding universe. Extrapolating into the past, we learn that the universe must have been born from an incredibly dense state (perhaps infinitely dense) some 10 to 15 billion years ago. A variety of evidence indicates that at that moment—the beginning of time—the universe began with a cosmic explosion, known as the **Big Bang,** which occurred

ƒigure 1-7 R I **V** U X G

A Galaxy This spectacular galaxy, called M100, contains several hundred billion stars. M100 has a diameter of about 120,000 light-years and is located about 56 million light-years from Earth. Along this galaxy's spiral arms you can see a number of glowing clumps. Like the Orion Nebula in our own Milky Way galaxy (Figure 1-5), these are sites of active star formation. (© Anglo-Australian Observatory; photograph by David Malin)

ƒigure 1-8 R I **V** U X G

A Quasar The two bright starlike objects in this image look almost identical, but they are dramatically different. The object on the left is indeed a star that lies a few hundred light-years from Earth. But the "star" on the right is actually a quasar about 9 billion light-years away. To appear so bright even though they are so distant, quasars like this one must be some of the most luminous objects in the universe. The other objects in this image are galaxies like that in Figure 1-7. (Charles Steidel, California Institute of Technology; and NASA)

throughout all space. Events shortly after the Big Bang dictated the present nature of the universe.

Thanks to the combined efforts of astronomers and physicists, we are making steady advances in understanding these cosmic events. This understanding may reveal the origin of some of the most basic properties of physical reality. Studying the most remote galaxies may also answer questions about the ultimate fate of the universe. Will it continue expanding forever, or will it someday stop and collapse back in on itself?

 The work of unraveling the deepest mysteries of the universe requires specialized tools, including telescopes, spacecraft, and computers. But for many purposes the most useful device for studying the universe is the human brain itself. Our goal in this book is to help you use *your* brain to share in the excitement of scientific discovery.

In the remainder of this chapter we introduce some of the key concepts and mathematics that we will use in subsequent chapters. Study these carefully, for you will use them over and over again throughout your own study of astronomy.

1-5 Astronomers use angles to denote the positions and apparent sizes of objects in the sky

Whether they study planets, stars, galaxies, or the very origins of the universe, astronomers must know where to point their telescopes. For this reason, an important part of astronomy is keeping track of the positions of objects in the sky. *Angles* and a system of *angular measure* are essential tools for this aspect of astronomy.

An **angle** is the opening between two lines that meet at a point. **Angular measure** describes the size of an angle exactly. The basic unit of angular measure is the **degree**, designated by the symbol °. A full circle is divided into 360°, and a right angle measures 90° (Figure 1-9*a*). As Figure 1-9*b* shows, the angle between the two "pointer stars" in the Big Dipper—that is, the **angular distance** between these stars—is about 5°. (These two stars "point" to Polaris, or the North Star, as described in Chapter 2.) The angular distance between the stars that make up the top and bottom of the Southern Cross, which is visible from south of the equator, is about 6° (Figure 1-9*c*).

Astronomers also use angular measure to describe the apparent size of a celestial object—that is, what fraction of the sky that object seems to cover. For example, the angle covered by the diameter of the full moon is about ½° (Figure 1-9*a*). We therefore say that the **angular diameter** (or **angular size**) of the Moon is ½°. Alternatively, astronomers say that the Moon **subtends** an angle of ½°. Ten full moons could fit side by side between the two pointer stars in the Big Dipper.

The adult human hand held at arm's length provides a means of estimating angles, as Figure 1-10 shows. For example, your fist covers an angle of 10°, whereas the tip of your finger is about 1° wide. You can use various segments of your index finger extended to arm's length to estimate angles a few degrees across.

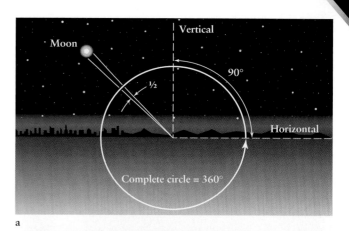

a

b

c

figure 1-9

Angles and Angular Measure (a) Angles are measured in degrees (°). There are 360° in a complete circle and 90° in a right angle. For example, the angle between the vertical direction (directly above you) and the horizontal direction (toward the horizon) is 90°. The angular diameter of the full moon in the sky is about ½°. (b) The Big Dipper is an easily recognized grouping of seven bright stars, visible from anywhere in the northern hemisphere. The angular distance between the two pointer stars at the front of the Big Dipper is about 5°, roughly ten times the angular diameter of the Moon. (c) The four bright stars that make up the Southern Cross can be seen from anywhere in the southern hemisphere. The angular distance between the stars at the top and bottom of the cross is about 6°.

figure 1-10

Estimating Angles with Your Hand Various parts of the adult human hand extended to arm's length can be used to estimate angular distances and angular sizes in the sky.

To talk about smaller angles, we subdivide the degree into 60 **arcminutes** (also called minutes of arc), which is commonly abbreviated as 60 arcmin or 60'. An arcminute is further sub-

divided into 60 **arcseconds** (or seconds of arc), usually written as 60 arcsec or 60". Thus,

$$1° = 60 \text{ arcmin} = 60'$$
$$1' = 60 \text{ arcsec} = 60''$$

For example, on January 1, 2001, the planet Saturn had an angular diameter of 19.7 arcsec as viewed from Earth. That is a convenient, precise statement of how big the planet appeared in Earth's sky on that date. (Because this angular diameter is so small, to the naked eye Saturn appears simply as a point of light. To see any detail on Saturn, such as the planet's rings, requires a telescope.)

If we know the angular size of an object as well as the distance to that object, we can determine the actual linear size of the object (measured in kilometers or miles, for example). Box 1-1 describes how this is done.

1-6 Powers-of-ten notation is a useful shorthand system for writing numbers

Astronomy is a subject of extremes. Astronomers investigate the largest structures in the universe, including galaxies and clusters of galaxies. But they must also study atoms and

---

box 1-1 Tools of the Astronomer's Trade

The Small-Angle Formula

You can estimate the angular sizes of objects in the sky with your hand and fingers (Figure 1-10). Using rather more sophisticated equipment, astronomers can measure angular sizes to a fraction of an arcsecond. Keep in mind, however, that *angular* size is not the same as *actual* size. As an example, if you extend your arm while looking at a full moon, you can completely cover the Moon with your thumb. That's because from your perspective, your thumb has a larger angular size (that is, it subtends a larger angle) than the Moon. But the actual size of your thumb (about 2 centimeters) is much less than the actual diameter of the Moon (more than 3000 kilometers).

The accompanying figure shows how the angular size of an object is related to its linear size. Part **a** of the figure shows that for a given angular size, the more distant the object, the larger its actual size. For example, your thumb held at arm's length just covers the full moon; the angular size of your thumb and the Moon are about the same, but the Moon is much farther away and is far larger in linear size. Part **b** shows that for a given linear

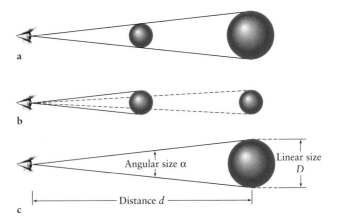

(a) Two objects that have the same angular size may have different linear sizes if they are at different distances from the observer. **(b)** For an object of a given linear size, the angular size is larger the closer the object is to the observer. **(c)** The small-angle formula relates the linear size D of an object to its angular size α and its distance d from the observer.

atomic nuclei, among the smallest objects in the universe, in order to explain how and why stars shine. They also study conditions ranging from the incredibly hot and dense centers of stars to the frigid near-vacuum of interstellar space. To describe such a wide range of phenomena, we need an equally wide range of both large and small numbers.

Astronomers avoid such confusing terms as "a million billion billion" by using a standard shorthand system called **powers-of-ten notation.** All the cumbersome zeros that accompany a large number are consolidated into one term consisting of 10 followed by an **exponent,** which is written as a superscript. The exponent indicates how many zeros you would need to write out the long form of the number. Thus,

$$10^0 = 1$$
$$10^1 = 10$$
$$10^2 = 100$$
$$10^3 = 1000$$
$$10^4 = 10{,}000$$

and so forth. The exponent also tells you how many tens must be multiplied together to give the desired number, which is why the exponent is also called the **power of ten.** For example, ten thousand can be written as 10^4 ("ten to the fourth" or "ten to the fourth power") because $10^4 = 10 \times 10 \times 10 \times 10 = 10{,}000$.

In powers-of-ten notation, numbers are written as a figure between one and ten multiplied by the appropriate power of ten. The approximate distance between Earth and the Sun, for example, can be written as 1.5×10^8 kilometers (or 1.5×10^8 km for short). Once you get used to it, this is more convenient than writing "150,000,000 kilometers" or "one hundred and fifty million kilometers." (The same number could also be written as 15×10^7 or 0.15×10^9, but the preferred form is *always* to have the first figure be between 1 and 10.)

Note that most electronic calculators use a shorthand for powers-of-ten notation. To enter the number 1.5×10^8, you first enter 1.5, then press a key labeled "EXP" or "EE," then enter the exponent 8. (The EXP or EE key takes care of the "$\times 10$" part of the expression.) The number will then appear on your calculator's display as "1.5 E 8," "1.5 8," or some variation of this; typically the "$\times 10$" is not displayed as such. There are some variations from one kind of calculator to another, so you should spend a few minutes reading over your calculator's instruction manual to make sure you know the correct procedure for working with numbers in powers-of-ten notation. You will be using this notation continually in your study of astronomy, so this is time well spent.

 Confusion can result from the way that calculators display powers-of-ten notation. Since 1.5×10^8 is displayed as "1.5 8" or "1.5 E 8," it is not uncommon to think that 1.5×10^8 is the same as 1.5^8. That is not

---

size, the angular size decreases the farther away the object. This is why a car looks smaller and smaller as it drives away from you.

We can put these relationships together into a single mathematical expression called the **small-angle formula.** Suppose that an object subtends an angle α (the Greek letter alpha) and is at a distance d from the observer, as in part **c** of the figure. If the angle α is small, as is almost always the case for objects in the sky, the linear size (D) of the object is given by the following expression:

The small-angle formula

$$D = \frac{\alpha d}{206{,}265}$$

D = linear size of an object
α = angular size of the object, in arcsec
d = distance to the object

The number 206,265 is required in the formula. Mathematically, it is equal to the number of arcseconds in a complete circle (that is, 360°) divided by the number 2π (the ratio of the circumference of a circle to that circle's radius).

EXAMPLE: On November 28, 2000, Jupiter was 609 million kilometers from Earth. Jupiter's angular diameter on that date was 48.6 arcseconds. Using the small-angle formula, we can calculate Jupiter's diameter as follows:

$$D = \frac{48.6 \times 609{,}000{,}000 \text{ km}}{206{,}265} = 143{,}000 \text{ km}$$

This answer agrees very well with an equatorial diameter of 142,984 km determined from spacecraft flybys.

EXAMPLE: Under excellent conditions, a telescope on Earth can see details with an angular size as small as 1 arcsec. We can use the small-angle formula to determine the greatest distance at which you could see details as small as 1.7 m (the height of a typical person) under these conditions. We rewrite the formula to solve for the distance d:

$$d = \frac{206{,}265 \, D}{\alpha} = \frac{206{,}265 \times 1.7 \text{ m}}{1} = 350{,}000 \text{ m} = 350 \text{ km}$$

This is much less than the distance to the Moon, which is 384,000 km. Thus, even the best telescope on Earth could not be used to see an astronaut walking on the surface of the Moon.

correct, however; 1.5^8 is equal to 1.5 multiplied by itself eight times, or 25.63, which is not even close to 150,000,000 $= 1.5 \times 10^8$. Another, not uncommon, mistake is to write 1.5×10^8 as 15^8. If you are inclined to do this, perhaps you are thinking that you can multiply 1.5 by 10, then tack on the exponent later. This also does not work; 15^8 is equal to 15 multiplied by itself eight times, or 2,562,890,625, which again is nowhere near 1.5×10^8. Reading over the manual for your calculator will help you to avoid these common errors.

You can use powers-of-ten notation for numbers that are less than one by using a minus sign in front of the exponent. A negative exponent tells you to *divide* by the appropriate number of tens. For example, 10^{-2} ("ten to the minus two") means to divide by 10 twice, so $10^{-2} = \frac{1}{10} \times \frac{1}{10} = \frac{1}{100} = 0.01$. This same idea tells us how to interpret other negative powers of ten:

$$10^0 = 1$$
$$10^{-1} = \frac{1}{10} = 0.1$$
$$10^{-2} = \frac{1}{10} \times \frac{1}{10} = \frac{1}{10^2} = 0.01$$
$$10^{-3} = \frac{1}{10} \times \frac{1}{10} \times \frac{1}{10} = \frac{1}{10^3} = 0.001$$
$$10^{-4} = \frac{1}{10} \times \frac{1}{10} \times \frac{1}{10} \times \frac{1}{10} = \frac{1}{10^4} = 0.0001$$

and so forth.

As these examples show, negative exponents tell you how many tenths must be multiplied together to give the desired number. For example, one ten–thousandth, or 0.0001, can be written as 10^{-4} ("ten to the minus four") because $10^{-4} = \frac{1}{10} \times \frac{1}{10} \times \frac{1}{10} \times \frac{1}{10} = 0.0001$.

A useful shortcut in converting a decimal to powers-of-ten notation is to notice where the decimal point is. For example,

the decimal point in 0.0001 is four places to the left of the "1," so the exponent is -4, that is, $0.0001 = 10^{-4}$.

You can also use powers-of-ten notation to express a number like 0.00245, which is not a multiple of $\frac{1}{10}$. For example, $0.00245 = 2.45 \times 0.001 = 2.45 \times 10^{-3}$. (Again, the standard for powers-of-ten notation is that the first figure is a number between one and ten.) This notation is particularly useful when dealing with very small numbers. A good example is the diameter of a hydrogen atom, which is much more convenient to state in powers-of-ten notation (1.1×10^{-10} meter, or 1.1×10^{-10} m) than as a decimal (0.00000000011 m) or a fraction (110 trillionths of a meter.)

Powers-of-ten notation lets us write familiar numerical terms in a compact way. For example,

$$\text{one hundred} = 100 = 10^2$$
$$\text{one thousand} = 1000 = 10^3$$
$$\text{one million} = 1{,}000{,}000 = 10^6$$
$$\text{one billion} = 1{,}000{,}000{,}000 = 10^9$$
$$\text{one trillion} = 1{,}000{,}000{,}000{,}000 = 10^{12}$$

and also

$$\text{one one-hundredth} = 0.01 = 10^{-2}$$
$$\text{one one-thousandth} = 0.001 = 10^{-3}$$
$$\text{one one-millionth} = 0.000001 = 10^{-6}$$
$$\text{one one-billionth} = 0.000000001 = 10^{-9}$$
$$\text{one one-trillionth} = 0.000000000001 = 10^{-12}$$

Because it bypasses all the awkward zeros, powers-of-ten notation is ideal for describing the size of objects as small as atoms or as big as galaxies (Figure 1-11). Box 1-2 explains

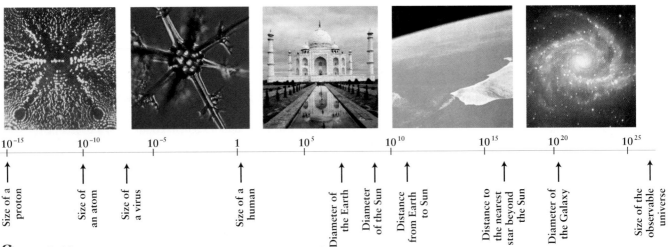

Size of a proton — Size of an atom — Size of a virus — Size of a human — Diameter of the Earth — Diameter of the Sun — Distance from Earth to Sun — Distance to the nearest star beyond the Sun — Diameter of the Galaxy — Size of the observable universe

10^{-15}　10^{-10}　10^{-5}　1　10^5　10^{10}　10^{15}　10^{20}　10^{25}

ƒigure 1-11

Examples of Powers-of-Ten Notation The scale gives the sizes of objects in meters, ranging from subatomic particles at the left to the entire observable universe on the right. The photograph at the left shows tungsten atoms, 10^{-10} meter in diameter. Second from left is the crystalline skeleton of a diatom (a single-celled organism), 10^{-4} meter (0.1 millimeter) in size. At the center is the Taj Mahal, about 60 meters tall and within reach of our unaided senses. On the right, looking across the Indian Ocean toward the South Pole, we see the curvature of the Earth, about 10^7 meters in diameter. At the far right is a galaxy, 10^{21} meters (100,000 light-years) in diameter. (Courtesy of Scientific American Books; NASA; and photograph by David Malin from the Anglo-Australian Observatory)

box 1-2 | Tools of the Astronomer's Trade

Arithmetic with Powers-of-Ten Notation

Using powers-of-ten notation makes it easy to multiply numbers. For example, suppose you want to multiply 100 by 1000. If you use ordinary notation, you have to write a lot of zeros:

$$100 \times 1000 = 100,000 \text{ (one hundred thousand)}$$

By converting these numbers to powers-of-ten notation, we can write this same multiplication more compactly as

$$10^2 \times 10^3 = 10^5$$

Because $2 + 3 = 5$, we are led to the following general rule for *multiplying* numbers expressed in terms of powers of ten: Simply *add* the exponents.

EXAMPLE:

$$10^4 \times 10^3 = 10^{4+3} = 10^7$$

To *divide* numbers expressed in terms of powers of ten, remember that $10^{-1} = \frac{1}{10}$, $10^{-2} = \frac{1}{100}$, and so on. The general rule for any exponent n is

$$10^{-n} = \frac{1}{10^n}$$

In other words, dividing by 10^n is the same as multiplying by 10^{-n}. To carry out a division, you first transform it into multiplication by changing the sign of the exponent, and then carry out the multiplication by adding the exponents.

EXAMPLE:

$$\frac{10^4}{10^6} = 10^4 \times 10^{-6} = 10^{4+(-6)} = 10^{4-6} = 10^{-2}$$

Usually a computation involves numbers like 3.0×10^{10}, that is, an ordinary number multiplied by a factor of 10 with an exponent. In such cases, to perform multiplication or division, you can treat the numbers separately from the factors of 10^n.

EXAMPLE: We can redo the first numerical example from Box 1-1 in a straightforward manner by using exponents:

$$D = \frac{48.6 \times 609,000,000 \text{ km}}{206,265}$$

$$= \frac{4.86 \times 10 \times 6.09 \times 10^8}{2.06265 \times 10^5} \text{ km}$$

$$= \frac{4.86 \times 6.09 \times 10^{1+8-5}}{2.06265} \text{ km} = 14.3 \times 10^4 \text{ km}$$

$$= 1.43 \times 10 \times 10^4 \text{ km} = 1.43 \times 10^5 \text{ km}$$

how powers-of-ten notation also makes it easy to multiply and divide numbers that are very large or very small.

1-7 Astronomical distances are often measured in astronomical units, parsecs, or light-years

Astronomers use many of the same units of measurement as do other scientists. They often measure lengths in meters (abbreviated m), masses in kilograms (kg), and time in seconds (s). (You can read more about these units of measurement, as well as techniques for converting between different sets of units, in Box 1-3.) Like other scientists, astronomers often find it useful to combine these units with powers of ten and create new units using prefixes. As an example, the number 1000 ($= 10^3$) is represented by the prefix "kilo," and so a distance of 1000 meters is the same as 1 kilometer (1 km).

Here are some of the most common prefixes, with examples of how they are used:

$$\text{one-billionth meter} = 10^{-9} \text{ m} = 1 \text{ nanometer}$$
$$\text{one-millionth second} = 10^{-6} \text{ s} = 1 \text{ microsecond}$$
$$\text{one-thousandth arcsecond} = 10^{-3} \text{ arcsec} = 1 \text{ milliarcsecond}$$
$$\text{one-hundredth meter} = 10^{-2} \text{ m} = 1 \text{ centimeter}$$
$$\text{one thousand meters} = 10^3 \text{ m} = 1 \text{ kilometer}$$
$$\text{one million tons} = 10^6 \text{ tons} = 1 \text{ megaton}$$

In principle, we could express all sizes and distances in astronomy using units based on the meter. Indeed, we will use kilometers to give the diameters of the Earth and Moon, as well as the Earth-Moon distance. But, while a kilometer (roughly equal to three-fifths of a mile) is an easy distance for humans to visualize, a megameter (10^6 m) is not. For this reason, astronomers have devised units of measure that are more appropriate for the tremendous distances between the planets and the far greater distances between the stars.

box 1-3 | Tools of the Astronomer's Trade

Units of Length, Time, and Mass

To understand and appreciate the universe, we need to describe phenomena not only on the large scales of galaxies but also on the submicroscopic scale of the atom. Astronomers generally use units that are best suited to the topic at hand. For example, interstellar distances are conveniently expressed in either light-years or parsecs, whereas the diameters of the planets are more comfortably presented in kilometers.

Most scientists prefer to use a version of the metric system called the International System of Units, abbreviated **SI** (after the French name Système International). In SI units, length is measured in meters (m), time is measured in seconds (s), and mass (a measure of the amount of material in an object) is measured in kilograms (kg). How are these basic units related to other measures?

When discussing objects on a human scale, sizes and distances are usually expressed in millimeters (mm), centimeters (cm), and kilometers (km). These units of length are related to the meter as follows:

$$1 \text{ millimeter} = 0.001 \text{ m} = 10^{-3} \text{ m}$$
$$1 \text{ centimeter} = 0.01 \text{ m} = 10^{-2} \text{ m}$$
$$1 \text{ kilometer} = 1000 \text{ m} = 10^{3} \text{ m}$$

Although the English system of inches (in.), feet (ft), and miles (mi) is much older than SI, today the English system is actually based on the SI system: The inch is defined to be exactly 2.54 cm. A useful set of conversions is

$$1 \text{ in.} = 2.54 \text{ cm}$$
$$1 \text{ ft} = 0.3048 \text{ m}$$
$$1 \text{ mi} = 1.609 \text{ km}$$

Each of these equalities can also be written as a fraction equal to 1. For example, you can write

$$\frac{0.3048 \text{ m}}{1 \text{ ft}} = 1$$

Fractions like this are useful for converting a quantity from one set of units to another. For example, the *Saturn V* rocket used to send astronauts to the Moon stands about 363 feet tall. How can we convert this height to meters? The trick is to remember that a quantity does not change if you multiply it by 1. Expressing the number 1 by the fraction (0.3048 m/1 ft), we can write the height of the rocket as

$$363 \text{ ft} \times 1 = 363 \text{ ft} \times \frac{0.3048 \text{ m}}{1 \text{ ft}} = 111 \frac{\cancel{\text{ft}} \times \text{m}}{\cancel{\text{ft}}} = 111 \text{ m}$$

EXAMPLE: The diameter of Mars is 6794 km. Let's try expressing this in miles.

 You can get into trouble if you are careless in applying the trick of taking the number whose units are to be converted and multiplying it by 1. For example, if we multiply the diameter by 1 expressed as (1.609 km)/(1 mi), we get

$$6794 \text{ km} \times 1 = 6794 \text{ km} \times \frac{1.609 \text{ km}}{1 \text{ mi}} = 10{,}930 \frac{\text{km}^2}{\text{mi}}$$

The unwanted units of km did not cancel, so this cannot be right. Furthermore, a mile is larger than a

When discussing distances across the solar system, astronomers use a unit of length called the **astronomical unit** (abbreviated AU). This is the average distance between Earth and the Sun:

$$1 \text{ AU} = 1.496 \times 10^{8} \text{ km} = 92.96 \text{ million miles}$$

Thus, the average distance between the Sun and Jupiter can be conveniently stated as 5.2 AU.

To talk about distances to the stars, astronomers use two different units of length. The **light-year** (abbreviated ly) is the distance that light travels in one year. This is a useful concept because the speed of light in empty space always has the same value, 3.00×10^{5} km/s (kilometers per second) or 1.86×10^{5} mi/s (miles per second).

In terms of kilometers or astronomical units, one light-year is given by

$$1 \text{ ly} = 9.46 \times 10^{12} \text{ km} = 63{,}240 \text{ AU}$$

This distance is roughly equal to 6 trillion miles.

Keep in mind that despite its name, the light-year is a unit of distance and not a unit of time. As an example, Proxima Centauri, the nearest star other than the Sun, is a distance of 4.2 light-years from Earth. This means that light takes 4.2 years to travel to us from Proxima Centauri.

Physicists often measure interstellar distances in light-years because the speed of light is one of nature's most impor-

kilometer, so the diameter expressed in miles should be a smaller number than when expressed in kilometers.

The correct approach is to write the number 1 so that the unwanted units *will* cancel. The number we are starting with is in kilometers, so we must write the number 1 with kilometers in the denominator ("downstairs" in the fraction). Thus, we express 1 as (1 mi)/(1.609 km):

$$6794 \text{ km} \times 1 = 6794 \text{ km} \times \frac{1 \text{ mi}}{1.609 \text{ km}}$$

$$= 4222 \text{ } \cancel{\text{km}} \times \frac{\text{mi}}{\cancel{\text{km}}} = 4222 \text{ mi}$$

Now the units of km cancel as they should, and the distance in miles is a smaller number than in kilometers (as it must be).

When discussing very small distances such as the size of an atom, astronomers often use the micrometer (μm) or the nanometer (nm). These are related to the meter as follows:

$$1 \text{ micrometer} = 1 \text{ } \mu\text{m} = 10^{-6} \text{ m}$$
$$1 \text{ nanometer} = 1 \text{ nm} = 10^{-9} \text{ m}$$

Thus, $1 \text{ } \mu\text{m} = 10^3 \text{ nm}$. (Note that the micrometer is often called the micron.)

The basic unit of time is the second (s). It is related to other units of time as follows:

$$1 \text{ minute (min)} = 60 \text{ s}$$
$$1 \text{ hour (h)} = 3600 \text{ s}$$
$$1 \text{ day (d)} = 86,400 \text{ s}$$
$$1 \text{ year (y)} = 3.156 \times 10^7 \text{ s}$$

In the SI system, speed is properly measured in meters per second (m/s). Quite commonly, however, speed is also expressed in km/s and mi/h:

$$1 \text{ km/s} = 10^3 \text{ m/s}$$
$$1 \text{ km/s} = 2237 \text{ mi/h}$$
$$1 \text{ mi/h} = 0.447 \text{ m/s}$$
$$1 \text{ mi/h} = 1.47 \text{ ft/s}$$

In addition to using kilograms, astronomers sometimes express mass in grams (g) and in solar masses ($M_\odot$), where the subscript $\odot$ is the symbol denoting the Sun. It is especially convenient to use solar masses when discussing the masses of stars and galaxies. These units are related to each other as follows:

$$1 \text{ kg} = 1000 \text{ g}$$
$$1 \text{ M}_\odot = 1.99 \times 10^{30} \text{ kg}$$

CAUTION! You may be wondering why we have not given a conversion between kilograms and pounds. The reason is that these units do not refer to the same physical quantity! A kilogram is a unit of *mass*, which is a measure of the amount of material in an object. By contrast, a pound is a unit of *weight*, which tells you how strongly gravity pulls on that object's material. Consider a person who weighs 110 pounds on Earth, corresponding to a mass of 50 kg. Gravity is only about one-sixth as strong on the Moon as it is on Earth, so on the Moon this person would weigh only one-sixth of 110 pounds, or about 18 pounds. But that person's mass of 50 kg is the same on the Moon; wherever you go in the universe, you take all of your material along with you. We will explore the relationship between mass and weight in Chapter 4.

tant numbers. But many astronomers prefer to use another unit of length, the *parsec*, because its definition is closely related to a method of measuring distances to the stars.

Imagine taking a journey far into space, beyond the orbits of the outer planets. As you look back toward the Sun, Earth's orbit subtends a smaller angle in the sky the farther you are from the Sun. As Figure 1-12 shows, the distance at which 1 AU subtends an angle of 1 arcsec is defined as 1 **parsec** (abbreviated pc):

$$1 \text{ pc} = 3.09 \times 10^{13} \text{ km} = 3.26 \text{ ly}$$

The distance to Proxima Centauri can be stated as 1.3 pc as well as 4.2 ly. Whether you choose to use parsecs or light-years is a matter of personal taste.

For even greater distances, astronomers commonly use **kiloparsecs** and **megaparsecs** (abbreviated kpc and Mpc). As we saw before, these prefixes simply mean "thousand" and "million," respectively:

$$1 \text{ kiloparsec} = 1 \text{ kpc} = 1000 \text{ pc} = 10^3 \text{ pc}$$
$$1 \text{ megaparsec} = 1 \text{ Mpc} = 1,000,000 \text{ pc} = 10^6 \text{ pc}$$

For example, the distance from Earth to the center of our Milky Way Galaxy is about 8 kpc, and the galaxy shown in Figure 1-7 is about 17 Mpc away.

Some astronomers prefer to talk about thousands or millions of light-years rather than kiloparsecs and megaparsecs. Once again, the choice is a matter of personal taste. As a general rule, astronomers use whatever yardsticks seem best

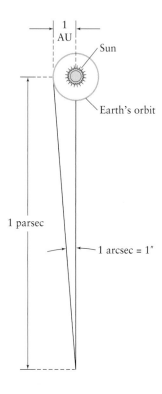

figure 1-12

A Parsec The parsec, a unit of length commonly used by astronomers, is equal to 3.26 light-years. The parsec is defined as the distance at which 1 AU perpendicular to the observer's line of sight subtends an angle of 1 arcsec.

suited for the issue at hand and do not restrict themselves to one system of measurement. For example, an astronomer might say that the supergiant star Antares has a diameter of 860 million kilometers and is located at a distance of 185 parsecs from Earth.

1-8 Astronomy is an adventure of the human mind

An underlying theme of this book is that the universe is rational. It is not a hodgepodge of unrelated things behaving in unpredictable ways. Rather, we find strong evidence that fundamental laws of physics govern the nature of the universe and the behavior of everything in it. These unifying concepts enable us to explore realms far removed from our earthly experience. Thus, a scientist can do experiments in a laboratory to determine the properties of light or the behavior of atoms and then use this knowledge to investigate the structure of the universe.

The discovery of fundamental laws of nature has had a profound influence on humanity. These laws have led to an immense number of practical applications that have fundamentally transformed commerce, medicine, entertainment, transportation, and other aspects of our lives. In particular,

space technology has given us instant contact with any point on the globe through communication satellites. Space technology has also made possible accurate weather forecasts from meteorological satellites, precise navigation to any point on Earth using signals from the satellites of the Global Positioning System (GPS), and long-term monitoring of Earth's climate and environment from orbit (Figure 1-13).

As important as the applications of science are, the pursuit of scientific knowledge for its own sake is no less important. We are fortunate to live in an age in which this pursuit is in full flower. Just as explorers such as Columbus and Magellan discovered the true size of our planet in the fifteenth and sixteenth centuries, astronomers of the twenty-first century are exploring the universe to an extent that is unparalleled in human history. Indeed, even the voyages into space imagined by such great science fiction writers as Jules Verne and H. G. Wells pale in comparison to today's reality. Over a few short

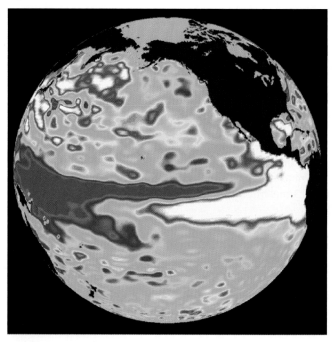

  figure 1-13

The Earth's Oceans from Space The *TOPEX/Poseidon* satellite (a joint project of the United States and France) measures variations in the height of the oceans. Areas of warm water bulge upward by several centimeters, so these measurements track changes in ocean temperature. In this map made from *TOPEX/Poseidon* data, elevated areas of warm water appear in white and red while depressed areas of cool water appear in blue and purple. Although the temperature variations are no more than a few degrees above or below normal, they have a profound effect on Earth's global climate. By monitoring these variations, scientists can give early warning of the potential for floods, droughts, and other severe weather. The savings in lives and property that result from these early warnings more than pay for the cost of the satellite. (NASA/JPL)

decades, humans have walked on the Moon, sent robot spacecraft to dig into the Martian soil and explore the satellites of Jupiter, and used the most powerful telescopes ever built to probe the limits of the observable universe. Never before has so much been revealed in so short a time.

As you proceed through this book, you will learn about the tools that scientists use to explore the natural world, as well as what they observe with these tools. But, most important, you will see how astronomers build from their observations an understanding of the universe in which we live. It is this search for understanding that makes science more than merely a collection of data and elevates it to one of the great adventures of the human mind. It is an adventure that will continue as long as there are mysteries in the universe—an adventure we hope you will come to appreciate and share.

KEY WORDS

Terms preceded by an asterisk (*) are discussed in the Boxes.

angle, p. 7
angular diameter (angular size), p. 7
angular distance, p. 7
angular measure, p. 7
arcminute (′, minute of arc), p. 8
arcsecond (″, second of arc), p. 8
astronomical unit (AU), p. 12
Big Bang, p. 6
black hole, p. 5
degree (°), p. 7

exponent, p. 9
galaxy, p. 6
hypothesis, p. 2
kiloparsec (kpc), p. 13
laws of physics, p. 3
light-year (ly), p. 12
megaparsec (Mpc), p. 13
model, p. 2
nebula (*plural* nebulae), p. 4
Newtonian mechanics, p. 3
parsec (pc), p. 13
power of ten, p. 9

powers-of-ten notation, p. 9
pulsar, p. 5
quasar, p. 6
scientific method, p. 2
*SI units, p. 12
*small-angle formula, p. 9
solar system, p. 3
subtend (an angle), p. 7
supernova (*plural* supernovae), p. 5
theory, p. 2
X-ray burster, p. 5

KEY IDEAS

Astronomy, Science, and the Nature of the Universe: The universe is comprehensible. The scientific method is a procedure for formulating hypotheses about the universe. These are tested by observation or experimentation in order to build consistent models or theories that accurately describe phenomena in nature.

• Observations of the heavens have helped scientists discover some of the fundamental laws of physics. The laws of physics are in turn used by astronomers to interpret their observations.

The Solar System: Exploration of the planets provides information about the origin and evolution of the solar system, as well as the history and resources of Earth.

Stars and Nebulae: Study of the stars and nebulae helps us learn about the origin and history of the Sun and the solar system.

Galaxies: Observations of galaxies tell us about the origin and history of the universe.

Angular Measure: Astronomers use angles to denote the positions and sizes of objects in the sky. The size of an angle is measured in degrees, arcminutes, and arcseconds.

Powers-of-ten notation is a convenient shorthand system for writing numbers. It allows very large and very small numbers to be expressed in a compact form.

Units of Distance: A variety of distance units are used by astronomers. These include the astronomical unit (the average distance from Earth to the Sun), the light-year (the distance that light travels in one year), and the parsec.

REVIEW QUESTIONS

1. Describe the role that skepticism plays in science.

2. What is the difference between a theory and a law of physics?

3. What caused the craters on the Moon?

4. What role do nebulae like the Orion Nebula play in the life stories of stars?

5. What is the difference between a solar system and a galaxy?

6. What are degrees, arcminutes, and arcseconds used for? What are the relationships among these units of measure?

7. How many arcseconds equal 1°?

8. With the aid of a diagram, explain what it means to say that the Moon subtends an angle of ½°.

9. What is an exponent? How are exponents used in powers-of-ten notation?

10. What are the advantages of using powers-of-ten notation?

11. Write the following numbers using powers-of-ten notation: (a) one hundred million, (b) forty thousand, (c) three one-thousandths, (d) twenty-two billion, (e) your age in years.

12. How is an astronomical unit (AU) defined? Give an example of a situation in which this unit of measure would be convenient to use.

13. What is the advantage to the astronomer of using the light-year as a unit of distance?

14. What is a parsec? How is it related to a kiloparsec and a megaparsec?

15. Give the word or phrase that corresponds to the following standard abbreviations: (a) km, (b) cm, (c) s, (d) km/s, (e) mi/h, (f) m, (g) m/s, (h) h, (i) y, (j) g, (k) kg. Which of these are units of speed? (*Hint:* You may have to refer to a dictionary. All of these abbreviations should be part of your working vocabulary.)

16. In the original (1977) *Star Wars* movie, Han Solo praises the speed of his spaceship by saying, "It's the ship that made the Kessel run in less than 12 parsecs!" Explain why this statement is obvious misinformation.

17. A reporter once described a light-year as "the time it takes light to reach us traveling at the speed of light." How would you correct this statement?

ADVANCED QUESTIONS

Questions preceded by an asterisk (*) involve topics discussed in the Boxes.

Problem-solving tips and tools

The small-angle formula, given in Box 1-1, relates the size of an astronomical object to the angle it subtends. Box 1-3 illustrates how to convert from one unit of measure to another. An object traveling at speed v for a time t covers a distance d given by $d = vt$; for example, a car traveling at 90 km/h (v) for 3 hours (t) covers a distance $d = (90 \text{ km/h})(3 \text{ h}) = 270$ km. Similarly, the time t required to cover a given distance d at speed v is $t = d/v$; for example, if $d = 270$ km and $v = 90$ km/h, then $t = (270 \text{ km})/(90 \text{ km/h}) = 3$ hours.

18. What is the meaning of the letters R I V U X G that appear under some of the figures in this chapter? Why in each case is one of the letters highlighted? (*Hint:* See the section "How to Use This Textbook" that precedes Chapter 1.)

19. A hydrogen atom has a radius of about 5×10^{-9} cm. The radius of the observable universe is about 15 billion light-years. How many times larger than a hydrogen atom is the observable universe? Use powers-of-ten notation.

20. The Sun's mass is 1.99×10^{30} kg, three-quarters of which is hydrogen. The mass of a hydrogen atom is 1.67×10^{-27} kg. How many hydrogen atoms does the Sun contain? Use powers-of-ten notation.

21. The diameter of the Sun is 1.4×10^{11} cm, and the distance to the nearest star, Proxima Centauri, is 4.2 ly. Suppose you want to build an exact scale model of the Sun and Proxima Centauri, and you are using a basketball 30 cm in diameter to represent the Sun. In your scale model, how far away would Proxima Centauri be from the Sun? Give your answer in kilometers, using powers-of-ten notation.

22. How many Suns would it take, laid side by side, to reach the nearest star? Use powers-of-ten notation. (*Hint:* See the preceding question.)

23. The average distance from the Earth to the Sun is 1.496×10^8 km. Express this distance (a) in light-years and (b) in parsecs. Use powers-of-ten notation. (c) Are light-years or parsecs useful units for describing distances of this size? Explain.

24. The speed of light is 3.00×10^8 m/s. How long does it take light to travel from the Sun to Earth? Give your answer in seconds, using powers-of-ten notation.

25. When *Voyager 2* sent back pictures of Neptune during its historic flyby of that planet in 1989, the spacecraft's radio signals traveled for 4 hours at the speed of light to reach Earth. How far away was the spacecraft? Give your answer in kilometers, using powers-of-ten notation. (*Hint:* See the preceding question.)

26. The star Epsilon Eridani is 3.22 pc from Earth. (a) What is the distance to Epsilon Eridani in kilometers? Use powers-of-ten notation. (b) How long does it take for light emanating from Epsilon Eridani to reach Earth? Give your answer in years. (*Hint:* You do not need to know the value of the speed of light.)

27. The age of the universe is about 15 billion years. What is this age in seconds? Use powers-of-ten notation.

*28. Explain where the number 206,265 in the small-angle formula comes from.

*29. At what distance would a person have to hold a nickel (which has a diameter of about 2.0 cm) in order for the nickel to subtend an angle of (a) 1°? (b) 1 arcmin? (c) 1 arcsec? Give your answers in meters.

Object	Distance (km)	Angular size (″)
Sun	1.5×10^8	1800
Saturn	1.5×10^9	16.5
Pluto	6.3×10^9	0.06

*30. The average distance to the Moon is 384,000 km, and the Moon subtends an angle of ½°. Use this information to calculate the diameter of the Moon in kilometers.

*31. Suppose your telescope can give you a clear view of objects and features that subtend angles of at least 2 arcsec. What is the diameter of the smallest crater you can see on the Moon? (*Hint:* See the preceding question.)

*32. A person with good vision can see details that subtend an angle of as small as 1 arcminute. If two dark lines on an eye chart are 2 millimeters apart, how far can such a person be from the chart and still be able to tell that there are two distinct lines? Give your answer in meters.

*33. On December 11, 2000, the planet Venus was at a distance of 0.951 AU from Earth. The diameter of Venus is 12,104 km. What was the angular size of Venus as seen from Earth on December 11, 2000? Give your answer in arcminutes.

DISCUSSION QUESTIONS

34. Scientists assume that "reality is rational." Discuss what this means and the thinking behind it.

35. How do astronomical observations differ from those of other sciences?

WEB/CD-ROM QUESTIONS

36. Use the links given in the *Universe* web site, Chapter 1, to learn about the Orion Nebula (Figure 1-5). Can the nebula be seen with the naked eye? Does the nebula stand alone, or is it part of a larger cloud of interstellar material? What has been learned by examining the Orion Nebula with telescopes sensitive to infrared light?

37. Use the links given in the *Universe* web site, Chapter 1, to learn more about the Crab Nebula (Figure 1-6). When did observers on Earth see the supernova that created this nebula? Does the nebula emit any radiation other than visible light? What kind of object is at the center of the nebula?

38. Use the links given in the *Universe* web site, Chapter 1, to learn more about *TOPEX/Poseidon* (see Figure 1-13) and other space missions that study the Earth. What are an "El Niño" and a "La Niña"? How can they be detected from space? How can the speed of winds over the oceans be measured by an orbiting satellite?

39. Access the AIMM (Active Integrated Media Module) called "Small-Angle Toolbox" in Chapter 1 of the *Universe* CD-ROM or web site. Use this to determine the diameters in kilometers of the Sun, Saturn, and Pluto given the following distances and angular sizes:

OBSERVING PROJECTS

40. On a dark, clear, moonless night, can you see the Milky Way from where you live ? If so, briefly describe its appearance. If not, what seems to be interfering with your ability to see the Milky Way?

41. Look up at the sky on a clear, cloud-free night. Is the Moon in the sky? If so, does it interfere with your ability to see the fainter stars? Why do you suppose astronomers prefer to schedule their observations on nights when the Moon is not in the sky?

42. Look up at the sky on a clear, cloud-free night and note the positions of a few prominent stars relative to such reference markers as rooftops, telephone poles, and treetops. Also note the location from where you make your observations. A few hours later, return to that location and again note the positions of the same bright stars that you observed earlier. How have their positions changed? From these changes, can you deduce the general direction in which the stars appear to be moving?

43. Use the CD-ROM that accompanies this book to install the *Starry Night* planetarium software on your computer. Use *Starry Night* to determine when the Moon is visible today during the day and when it is visible tonight. Determine which, if any, of the following planets are visible tonight: Mercury, Venus, Mars, Jupiter, and Saturn. *Hints:* (1) Feel free to experiment with *Starry Night*. You can always return to your starting screen by clicking on the "Home" button in the Control Panel at the top of the main window. To see the sky from your actual location on Earth, select **Set Home Location...** in the **Go** menu and click on the **Lookup...** button to find your city or town. (2) To change your viewing direction, move the mouse until the cursor changes into a little hand. Then, hold down the mouse button (on a Windows computer, the left button) as you move the mouse and you will move the sky. (3) Use the Control Panel at the top of the main window to change the time and date that is displayed as well as how rapidly time appears to change. (4) Use the **Find...** command in the **Edit** menu to locate specific planets or stars by name. (5) To get more information about any object in the sky, point the cursor at the object and double-click the mouse (on a Windows computer, click the right button).

You can find even more information about the program at the *Starry Night* web site.

SANDRA M. FABER

Why Astronomy?

Sandra M. Faber, professor of astronomy at the University of California, Santa Cruz, and astronomer at Lick Observatory, was intrigued by the origins of the universe when she entered Swarthmore College. She completed her doctorate in astronomy at Harvard University but did her dissertation at the Carnegie Institution's Department of Terrestrial Magnetism, where she was influenced by the distinguished astronomer Vera Rubin. Dr. Faber chaired the now legendary group of astronomers called the Seven Samurai, who surveyed the nearest 400 elliptical galaxies and discovered a new mass concentration, the Great Attractor. She is the recipient of many honors and awards for her research, including the Bok Prize from Harvard University in 1978. She was elected to the National Academy of Sciences in 1985 and in 1986 won the coveted Dannie Heineman Prize from the American Astronomical Society, given in recognition of a sustained body of especially influential astronomical research. She is also on the Board of Trustees of the Carnegie Institution.

As you study astronomy, it is appropriate for you to ask, "Why am I studying this subject? What good is it for human beings in general and for me in particular?" Astronomy, it must be admitted, does not offer the same practical benefits as other sciences. So how can astronomy be important to your life?

On the most basic level, I like to think of astronomy as providing the introductory chapters for the ultimate textbook on human history. Ordinary texts start with recorded history, going back some 3000 years. For events before that, we consult archeologists and anthropologists, who tell us about the early history of our species. For knowledge of the time before that, we consult paleontologists, biologists, and geologists about the origin and evolution of life and the evolution of our planet, altogether going back some five billion years. Astronomy tells us about the vast stretch of time before the origin of the Earth, the ten billion years or so that saw the formation of the Sun and solar system, the Milky Way Galaxy, and the origin of the universe in the Big Bang. Knowledge of astronomy is essential for a well-educated person's view of history.

Astronomy challenges our belief system and impels us to put our "philosophical house" in order. Take the origin of the world, for example. According to Genesis in the Bible, the world and all in it were created in six days by the hand of God. However, the ancient Egyptians believed that the Earth arose spontaneously from the infinite waters of the eternal universe, Nun, and Alaskan legends taught that the world was created by the conscious imaginings of a creator deity named Father Raven.

The modern astronomical story of the creation of the Earth differs from all these in that it claims to be buttressed by physics and by observational fact. The Sun, it is asserted, formed via gravitational collapse from a dense cloud of interstellar dust and gas about five billion years ago. The planets coalesced at the same time as condensates within the swirling solar nebula. This process took not weeks or days but several hundred thousand years. Confidence in this theory stems from our looking out into the Galaxy and seeing young stars actually form in this way.

At issue here, really, is the deep question of how we are to gain information about the nature of the physical world—whether by revelation and intuition or by logic and observation. Where science stops and faith begins is a thorny issue for all human beings, but particularly so for astronomers—and for astronomy students.

Astronomy cultivates our notions about cosmic time and cosmic evolution. It is all too easy, given the short span of human life, to overlook the fact that the universe is a dynamic place. When I first entered astronomy, I found myself handicapped by a conservative mind-set that assumed, subtly, that celestial objects were unchanging. This might seem strange for someone whose avowed interest was in learning how galaxies formed. Such are the vagaries of the human mind! Fortunately, I lost this bias as I grew older, mainly, I think, by observing evolution all around me. Many things that had seemed immutable to me as a child—social structures, customs, even the physical environment—turned out not to be so. With that realization, the veil fell from my eyes, and I was able to accept cosmic evolution as a core concept in my thinking about the universe.

Following closely on the concept of evolution is the concept of fragility—if something can change, it might even actually disappear someday. The most obvious example is the limited lifetime of our Sun. In another five billion years or so, the Sun will enter its death throes, during which it will swell up and brighten to 1000 times its present luminosity and, in the process, incinerate the Earth.

Five billion years is far enough in the future that neither you nor I feel any personal responsibility for preparing the human race to meet this challenge. However, there are many other cosmic catastrophes that will befall us in the meantime. The Earth will be hit by sizable pieces of space junk; eroded impact craters show this happens all the time—every few million years or so. Debris thrown up into the atmosphere from these events could be so extensive as to totally alter the climate for an extended period of time, possibly leading to mass extinctions. There will be enormous volcanic eruptions that will make Mount St. Helens look like a small firecracker. Another Ice Age, which will totally disrupt modern agriculture, is virtually certain within the next 20,000 years, unless we cook the Earth ourselves first by burning too much fossil fuel.

Closer to the human sphere, such common notions as the inevitability of human progress, the desirability of endless economic growth, the ability of the Earth to support its growing human population, are all based on limited experience and may not—and probably *will* not— prove viable in the long run. Consequently, we must rethink who we are as a species and what is our proper activity on Earth. Contemplating cosmic time puts us into the appropriate frame of mind for grappling with these issues. These are problems that are very long range and involve the whole human race yet are totally under human control and are vital to our long-term survival and well-being.

Astronomy is essential to developing a perspective on human existence and its relation to the cosmos. What question could be grander and at the same time more relevant to us? We can hardly contemplate human existence in general without taking at least a brief look at our own lives.

Earth is a mote in the vast cosmic sea, and our lives, in the evolutionary scheme of the universe, would seem at face value to be insignificant. Should that thought terrify us, depriving our lives of any meaning? Should we seek to provide meaning by introducing an intelligent creator? Or should we be fully comforted merely to know that this great universe has given birth to us in an entirely natural, perhaps even inevitable, chain of events?

Not long ago I visited the Southern Hemisphere for the first time to observe the sky from Chile. At that latitude, the center of the Milky Way passes overhead, where it makes a grand show, not the miserable, fuzzy patch that is visible, low on the horizon, from North America. Stepping out on the observatory catwalk one morning before dawn, I saw for the first time the galactic center in all its glory soaring overhead. At that instant, my perceptions underwent a profound alteration. One moment I was standing in an earthbound "room" with the Galaxy painted on the "ceiling." The next instant, the walls of this earthly room fell away, and I was floating freely in outer space, viewing the center of the Galaxy from an outpost on its periphery. The sense of depth was breathtaking: Sparkling star clusters and dense dust clouds were etched against the blazing inner bulge of the Galaxy many thousands of light-years distant. I saw the Galaxy as an immense, three-dimensional object in space for the first time. It became "real." At that moment, my sense of place in the universe underwent a fundamental and irreversible change. I remained a citizen of the Earth, but I also became a citizen of the Galaxy.

Many astronomers believe that the ultimate, proper concept of "home" for the human race is our universe. It seems more and more likely that there are a very large number of other universes. Virtually none of them would appear to resemble ours; they may have different numbers of space and time dimensions—perhaps even no space or time— and very different physical laws. The vast majority probably are incapable of harboring intelligent life as we know it.

The parallel with Earth is striking. Among the solar system planets, only Earth can support human life. Among the great number of planets that likely exist in our Galaxy, only a small fraction are apt to be such that we could call them home. The fraction of hospitable universes is likely to be smaller still. Our universe would seem, then, to be the ultimate instance of "home": a sanctuary in a vast sea of inhospitable universes.

I began this essay by talking about history and ended with questions that border on the ethical and religious. Astronomy is like that: It offers a modern-day version of Genesis—and of Apocalypse, too. Astronomy stimulates and challenges us on our deepest, most human levels. I hope that during this course you will be able to take time out to contemplate the broader implications of what you are studying. This is one of the rare opportunities in life to think about who you are and where you and the human race are going. Don't miss it.

Knowing the Heavens

(Luke Dodd/Science Photo Library/Photo Researchers)

R I **V** U X G

Hundreds of years ago, the peoples of Polynesia and Micronesia were among the most skilled navigators in the world. They were able to guide their wooden vessels accurately across vast stretches of the Pacific, far beyond sight of land, without compasses, and even by night. Their secret was to watch the positions of certain stars, such as Alpha and Beta Centauri (shown on the left in the photograph) and the Southern Cross (in the center). Each of these stars rises and sets at the same points on the horizon every night. Pacific navigators used these rising and setting points much as a modern sailor uses the points of a compass.

This is but one example of the many uses of *naked-eye astronomy*. You do this same sort of astronomy when you notice where the Moon is in the sky or when you trace the pattern of the Big Dipper.

Naked-eye astronomy cannot tell us what the Sun is made of or how far away the stars are. We need telescopes, and their attendant equipment, to answer those questions. But, as you will learn in this chapter, naked-eye astronomy is nonetheless tremendously important and useful. Before astronomers can train a telescope on a distant star, they must know where in the sky that star can be found. The cycle of day and night and the rhythm of the seasons are directly related to the apparent motions of the Sun across the sky. By studying naked-eye astronomy, you will learn how the Earth moves through space. In this way you will come to understand our true place in the cosmos.

As you read the sections of this chapter, look for the answers to the following questions.

2-1 What role did astronomy play in ancient civilizations?

2-2 Are the stars that make up a constellation actually close to one another?

2-3 Are the same stars visible every night of the year? What is so special about the North Star?

2-4 Are the same stars visible from any location on Earth?

2-5 What causes the seasons? Why are they opposite in the northern and southern hemispheres?

2-6 Has the same star always been the North Star?

2-7 Can we use the rising and setting of the Sun as the basis of our system of keeping time?

2-8 Why are there leap years?

2-1 Naked-eye astronomy had an important place in ancient civilizations

 Positional astronomy—the study of the positions of objects in the sky and how these positions change—has roots that extend far back in time. Four to five thousand years ago, the inhabitants of the British Isles erected stone structures, such as Stonehenge, that suggest a preoccupation with the motions of the sky. Alignments of these stones appear to show where the Sun rose and set at key times during the year. A similar structure in the New World is the Medicine Wheel, a circular ring of stones constructed by the Plains Indians atop a windswept plateau in Wyoming. Certain stones in this ring mark the rising points of the Sun and certain bright stars on the first day of summer.

Aztec, Mayan, and Incan architects in Central and South America also designed buildings with astronomical orientations. At the ruined city of Tiahuanaco in Bolivia, the walls of the Temple of the Sun were aligned north-south and east-west with an accuracy of better than one degree. The great Egyptian pyramids, built around 3000 B.C., are likewise oriented north-south and east-west with remarkable precision.

In addition to temples and tombs, some ancient buildings appear to have been dedicated expressly to astronomy. One of the best examples was built a thousand years ago in the Mayan city of Chichén Itzá, on the Yucatán Peninsula (Figure 2-1). The Caracol's cylindrical tower contains windows aligned with the northernmost and southernmost rising and setting points of both the Sun and the planet Venus. A similar four-story adobe building, probably constructed during the fourteenth century, is located at the Casa Grande site in Arizona.

These structures bear witness to an awareness of naked-eye astronomy by the peoples of many cultures. Many of the concepts of modern positional astronomy come to us from these ancients, including the idea of dividing the sky into constellations.

2-2 Eighty-eight constellations cover the entire sky

 Looking at the sky on a clear, dark night, you might think that you can see millions of stars. Actually, the unaided human eye can detect only about 6000 stars. Because half of the sky is below the horizon at any one time, you can see at most about 3000 stars. When ancient peoples looked at these thousands of stars, they imagined that groupings of stars traced out pictures in the sky.

ƒigure 2-1   R I V U X G

The Caracol at Chichén Itzá This ancient Mayan observatory in the Yucatán was built about A.D. 1000. Its architecture is based on alignments with important celestial events. Mayan astronomers developed a very accurate calendar and measured the motions of celestial bodies with great precision. Special significance was associated with the planet Venus, which inspired sacrificial rites and other ceremonies. (Courtesy of E. C. Krupp)

a   R I **V** U X G

Three Views of Orion Orion is a prominent constellation from December through March. During these months, Orion is easily seen high above the southern horizon from North America or Europe. From Australia or South America, it appears high above the northern horizon. **(a)** This time-exposure photograph of Orion shows that stars come in different colors, some red and others blue. You can see these colors with the naked eye; they are even more pronounced as seen through binoculars or a telescope. **(b)** A portion of a modern star atlas shows the names of some of the brightest stars in Orion, along with the distances in light-years (ly) to those stars. The yellow lines show the borders between Orion and its neighboring constellations (labeled in capitals), as defined by modern astronomers. **(c)** This fanciful drawing from a star atlas published in 1835 shows Orion the Hunter as well as other celestial creatures. (a: Luke Dodd/Science Photo Library/Photo Researchers; c: Courtesy of Janus Publications)

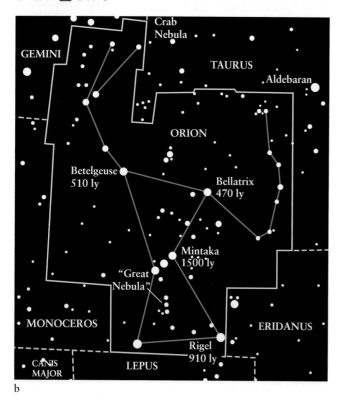

b

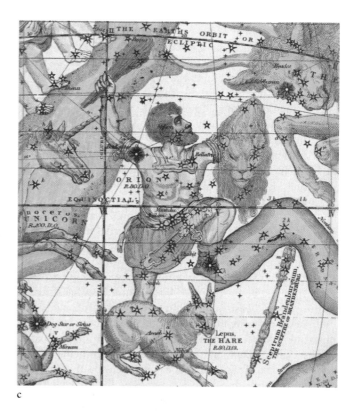

c

Astronomers still refer to these groupings, called **constellations** (from the Latin for "group of stars").

You may already be familiar with some of these pictures or patterns in the sky, such as the Big Dipper, which is actually part of the large constellation Ursa Major (the Great Bear). Many constellations, such as Orion in Figure 2-2, have names derived from the myths and legends of antiquity. Although some star groupings vaguely resemble the figures they are supposed to represent (see Figure 2-2*c*), most do not.

The term "constellation" has a broader definition in present-day astronomy. On modern star charts, the entire sky is divided into 88 regions, each of which is called a constellation. For example, the constellation Orion is now defined to be an irregular patch of sky whose borders are shown in Figure 2-2*b*. When astronomers refer to the "Great Nebula" M42 in Orion, they mean that as seen from Earth this nebula appears to be within Orion's patch of sky. Some constellations cover large areas of the sky (Ursa Major being one of the biggest) and others very small areas

(Crux, the Southern Cross, being the smallest). But because the modern constellations cover the entire sky, every star lies in one constellation or another.

 When you look at a constellation's star pattern, it is tempting to conclude that you are seeing a group of stars that are all relatively close together. In fact, most of these stars are nowhere near one another. As an example, Figure 2-2b shows the distances in light-years to four stars in Orion. Although Bellatrix (Arabic for "the Amazon") and Mintaka ("the belt") appear to be close to each other, Mintaka is actually about a thousand light-years farther away from us. They merely appear close to each other because they are in nearly the same direction as seen from Earth. The same illusion often appears when you see an airliner's lights at night. It is very difficult to tell how far away a single bright light is, which is why you can mistake an airliner a few kilometers away for a star trillions of times more distant.

 The star names shown in Figure 2-2b are from the Arabic. For example, Betelgeuse means "armpit," which makes sense when you look at the star atlas drawing in Figure 2-2c. Other types of names are also used for stars. For example, Betelgeuse is also known as α Orionis (α is the Greek letter alpha) and as HD 39801.

 A number of unscrupulous commercial firms offer to name a star for you for a fee. The money that they charge you for this "service" is real, but the star names are not; none of these names are recognized by astronomers. If you want to use astronomy to commemorate your name or the name of a friend or relative, consider making a donation to your local planetarium or science museum. The money will be put to much better use!

2-3 The appearance of the sky changes during the course of the night and from one night to the next

Go outdoors soon after dark, find a spot away from bright lights, and note the patterns of stars in the sky. Do the same a few hours later. You will find that the entire pattern of stars (as well as the Moon, if it is visible) has shifted its position. New constellations will have risen above the eastern horizon, and some will have disappeared below the western horizon. If you look again before dawn, you will see that the stars that were just rising in the east when the night began are now low in the western sky. This daily motion, or **diurnal motion,** of the stars is apparent in time-exposure photographs (Figure 2-3). We can understand this apparent motion of the sky by considering the rotation of the Earth.

At any given moment, it is daytime on the half of the Earth illuminated by the Sun and nighttime on the other half. The Earth rotates from west to east, making one complete rotation every 24 hours, which is why there is a daily cycle of day and night. Because of this rotation, stars appear to us to rise in the east and set in the west, as do the Sun and Moon.

Figure 2-4 helps to further explain diurnal motion. It shows two views of the Earth as seen from a point above the North Pole. At the instant shown in Figure 2-4a, it is day in Asia but night in most of North America and Europe. Figure 2-4b shows the Earth 4 hours later. Four hours is one-sixth of a complete 24-hour day, so the Earth has made one-sixth of a rotation between Figures 2-4a and 2-4b. Europe is now in the illuminated half of the Earth (the Sun has risen in Europe), while Alaska has moved from the illuminated to the dark half of the Earth (the Sun has set in Alaska). For a person in

ƒigure 2-3   R I **V** U X G
Diurnal Motion of the Night Sky This photograph is a time exposure of the night sky over the Kitt Peak National Observatory near Tucson, Arizona. The view is to the northwest. Because the Earth rotates from west to east, the sky appears to rotate from east to west. The stars thus appear as long trails in this photograph. (National Optical Astronomy Observatories)

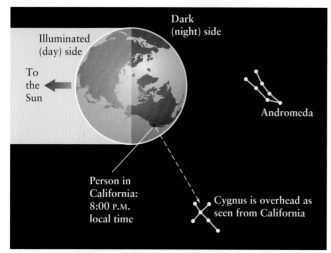

a Earth as seen from above the North Pole

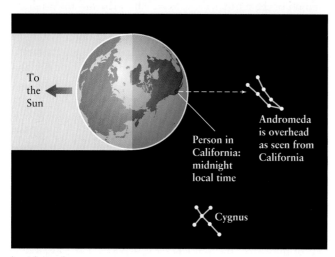

b 4 hours later

figure 2-4

Why Diurnal Motion Happens The diurnal (daily) motion of the stars, the Sun, and the Moon is a consequence of the Earth's rotation. **(a)** This drawing shows the Earth from a vantage point above the North Pole. It is day on the side of the Earth that is illuminated by the Sun, and night on the dark, unilluminated side. In this drawing, the local time in California is 8:00 P.M. and a person in California sees the constellation Cygnus directly overhead. **(b)** Four hours later, the Earth has rotated to the east by one-sixth of a complete rotation. From the perspective of a person on Earth, the entire sky appears to have rotated to the west by one-sixth of a complete rotation. It is now midnight in California, and the constellation directly over California is Andromeda.

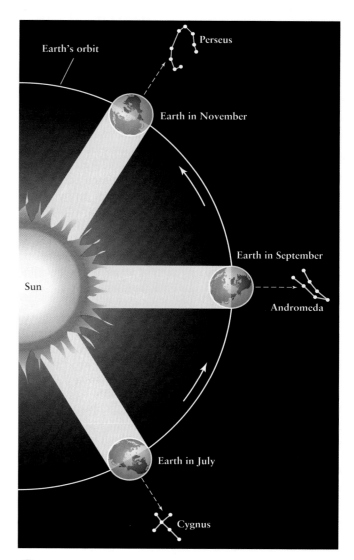

figure 2-5

Why the Night Sky Changes During the Year The particular stars that you see in the night sky are different at different times of the year. This is a consequence of the Earth's orbital motion; as we move around the Sun, the nighttime side of the Earth gradually turns toward different parts of the sky. This figure shows which constellation is overhead at midnight local time—when the Sun is on the opposite side of the Earth from your location—during different months for observers at midnorthern latitudes (including the United States). If you want to view the constellation Andromeda, the best time of the year to do it is in late September, when Andromeda is nearly overhead at midnight.

California, in Figure 2-4*a* the time is 8:00 P.M. and the constellation Cygnus (the Swan) is directly overhead. Four hours later, the constellation over California is Andromeda (named for a mythological princess). Because the Earth rotates from west to east, it appears to us on Earth that the entire sky is rotating around us in the opposite direction, from east to west.

In addition to the diurnal motion of the sky, the constellations visible in the night sky also change slowly over the

course of a year. This happens because the Earth orbits, or revolves around, the Sun (Figure 2-5). Over the course of a year, the Earth makes one complete orbit, and the darkened, nighttime side of the Earth gradually turns toward different parts of the heavens. For example, as seen from the northern hemisphere, at midnight in late July the constellation Cygnus is close to overhead; at midnight in late September the constellation Andromeda is close to overhead; and at midnight in

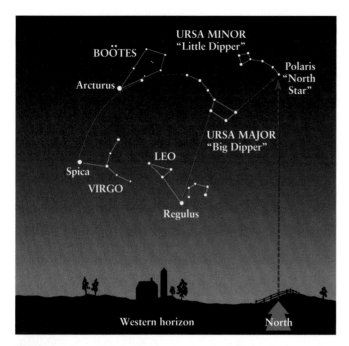

figure 2-6

The Big Dipper as a Guide The North Star can be seen from anywhere in the northern hemisphere on any night of the year. This star chart shows how the Big Dipper can be used to point out the North Star as well as the brightest stars in three other constellations. The chart shows the sky at around 2 A.M. (daylight saving time) on May 1, around midnight on June 1, or around 10 P.M. on July 1. Because of the Earth's rotation and orbital motion around the Sun, the view will be different on different dates and at different times, but the relative positions of the stars will be the same. The angular distance from Polaris to Spica is about 101°, so this chart covers a large fraction of the sky.

late November the constellation Perseus (commemorating a mythological hero) is close to overhead. If you follow a particular star on successive evenings, you will find that it rises approximately 4 minutes earlier each night, or 2 hours earlier each month.

Constellations can help you find your way around the sky. For example, if you live in the northern hemisphere, you can use the Big Dipper in Ursa Major to find the north direction by drawing a straight line through the two stars at the front of the Big Dipper's bowl (Figure 2-6). The first moderately bright star you come to is Polaris, also called the North Star because it is located almost directly over the Earth's north pole. If you draw a line from Polaris straight down to the horizon, you will find the north direction.

By drawing a line through the two stars at the rear of the bowl of the Big Dipper, you can find Leo (the Lion). As Figure 2-6 shows, that line points toward Regulus, the brightest star in the "sickle" tracing the lion's mane. By following the handle of the Big Dipper, you can locate the bright reddish star Arcturus in Boötes (the Shepherd) and the prominent bluish star Spica in Virgo (the Virgin). The saying "Follow the arc to Arcturus and speed to Spica" may help you remember these

stars, which are conspicuous in the evening sky during the spring and summer.

During winter in the northern hemisphere, you can see some of the brightest stars in the sky. Many of them are in the vicinity of the "winter triangle" (Figure 2-7), which connects bright stars in the constellations of Orion (the Hunter), Canis Major (the Large Dog), and Canis Minor (the Small Dog). The "winter triangle" is high above the eastern horizon during the middle of winter at midnight.

A similar feature, the "summer triangle," graces the summer sky in the northern hemisphere. This triangle connects the brightest stars in Lyra (the Harp), Cygnus (the Swan), and Aquila (the Eagle) (Figure 2-8). A conspicuous portion of the Milky Way forms a beautiful background for these constellations, which are nearly overhead during the middle of summer at midnight.

 A wonderful tool to help you find your way around the night sky is the planetarium program *Starry Night*, which you will find on the CD-ROM that accompanies this book. In addition, at the end of this book you will find a set of selected star charts for the evening

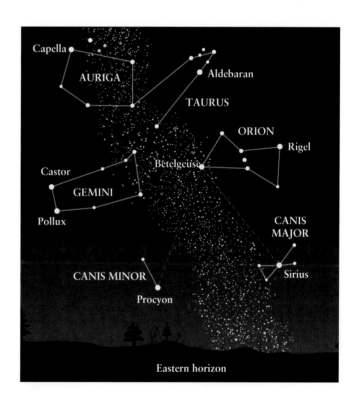

figure 2-7

The "Winter Triangle" This star chart shows the eastern sky as it appears on a late fall or early winter evening in the northern hemisphere (around midnight on November 1, around 10 P.M. on December 1, or around 8 P.M. on January 1). Three of the brightest stars in the sky make up the "winter triangle," which is about 26° on a side. In addition to the constellations involved in the triangle, the chart shows the prominent constellations Gemini (the Twins), Auriga (the Charioteer), and Taurus (the Bull).

figure 2-8

The "Summer Triangle" This star chart shows the eastern sky as it appears in the evening during spring and summer in the northern hemisphere (around 1 A.M.—daylight saving time—on June 1, around 11 P.M. on July 1, and around 9 P.M. on August 1). The angular distance from Deneb to Altair is about 38°. The constellations Sagitta (the Arrow) and Delphinus (the Dolphin) are much fainter than the three constellations that make up the triangle.

hours of all 12 months of the year. You may find stargazing an enjoyable experience, and *Starry Night* and the star charts will help you identify many well-known constellations.

Note that all the star charts in this section and at the end of this book are drawn for an observer in the northern hemisphere. If you live in the southern hemisphere, you can see constellations that are not visible from the northern hemisphere, and vice versa. In the next section we will see why this is so.

2-4 It is convenient to imagine that the stars are located on a celestial sphere

Many ancient societies believed that all the stars are the same distance from the Earth. They imagined the stars to be bits of fire imbedded into the inner surface of an immense hollow sphere, called the **celestial sphere**, with the Earth at its center. In this picture of the universe, the Earth was fixed and did not rotate. Instead, the entire celestial sphere rotated once a day around the Earth from east to west, thereby causing the diurnal motion of the sky. The picture of a rotating celestial sphere fit well with naked-eye observations, and for its time was a

useful model of how the universe works. (We discussed the role of models in science in Section 1-1).

We now know that this simple model of the universe is not correct. Diurnal motion is due to the rotation of the Earth, not the rest of the universe. Furthermore, as we learned when discussing the constellations in Section 2-2, it is impossible to tell with the naked eye how far away the stars are. Indeed, the brightest stars that you can see with the naked eye range from 4.2 to more than 1000 light-years away (see Figure 2-2), and telescopes allow us to see objects at distances of billions of light-years.

Thus, astronomers now recognize that the celestial sphere is an *imaginary* object that has no basis in physical reality. Nonetheless, the celestial sphere model remains a useful tool of positional astronomy. If we imagine, as did the ancients, that the Earth is stationary and that the celestial sphere rotates around us, it is relatively easy to specify the directions to different objects in the sky and to visualize the motions of these objects.

Figure 2-9 depicts the celestial sphere, with the Earth at its center. (A truly proportional drawing would show the celestial sphere as being millions of times larger than the Earth.) If we project the Earth's equator out into space, we obtain the

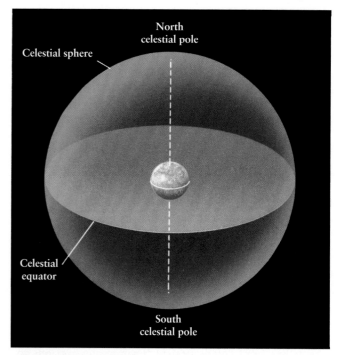

figure 2-9

The Celestial Sphere The celestial sphere is the apparent sphere of the sky. The view in this figure is from the outside of this (wholly imaginary) sphere. Because the Earth is at the center of the celestial sphere, however, our view is always of the *inside* of the sphere. The celestial equator and poles are the projections of the Earth's equator and axis of rotation out into space. The celestial poles are therefore located directly over the Earth's poles.

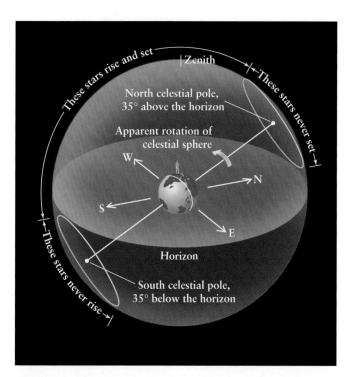

figure 2-10

The View from 35° North Latitude To an observer at 35° north latitude (roughly the latitude of Los Angeles, Atlanta, Tel Aviv, and Tokyo), the north celestial pole is always 35° above the horizon. Stars within 35° of the *north* celestial pole are always above the horizon on any night of the year. These stars are circumpolar; they trace out circles around the north celestial pole during the course of the night. Stars within 35° of the *south* celestial pole are always below the horizon and can never be seen from this latitude. Stars that lie between these two extremes rise in the east and set in the west.

figure 2-11   R I **V** U X G

Circumpolar Star Trails This long exposure, taken from Australia's Siding Spring Mountain and aimed at the south celestial pole, shows the rotation of the sky above the Anglo-Australian Telescope. During the exposure, someone carrying a flashlight walked along the catwalk on the outside of the telescope dome. Another flashlight made the wavy trail at ground level. (© Anglo-Australian Observatory, photograph by David Malin)

celestial equator. The celestial equator divides the sky into northern and southern hemispheres, just as the Earth's equator divides the Earth into two hemispheres.

If we project the Earth's north and south poles into space, we obtain the **north celestial pole** and the **south celestial pole.** The two celestial poles are where the Earth's axis of rotation (extended out into space) intersects the celestial sphere (see Figure 2-9). The star Polaris is less than 1° away from the north celestial pole, which is why it is called the North Star or the Pole Star.

The point in the sky directly overhead an observer anywhere on Earth is called that observer's **zenith.** The zenith and celestial sphere are shown in Figure 2-10 for an observer located at 35° north latitude (that is, at a location on the Earth's surface 35° north of the equator). The zenith is shown at the top of Figure 2-10, so the Earth and the celestial sphere appear "tipped" compared to Figure 2-9. At any time, an observer can see only half of the celestial sphere; the other half is below the horizon, hidden by the body of the Earth. The hidden half of the celestial sphere is darkly shaded in Figure 2-10.

For an observer anywhere in the northern hemisphere, including the observer in Figure 2-10, the north celestial pole is always above the horizon. As the Earth turns from west to east—or, equivalently, as the celestial sphere appears to turn from east to west—stars sufficiently near the north celestial pole revolve around the pole, never rising or setting. Such stars are called **circumpolar.** For example, as seen from North America, Polaris is a circumpolar star and can be seen at any time of night on any night of the year. Similarly, stars near the south celestial pole revolve around that pole but always remain below the horizon of an observer in the northern hemisphere. Hence, these stars can never be seen by the observer in Figure 2-10. Stars at intermediate positions on the celestial sphere rise in the east and set in the west, as in Figure 2-3.

For an observer at 35° south latitude (the latitude of Sydney, Australia), the roles of the north and south celestial poles are reversed: Objects close to the *south* celestial pole are circumpolar, that is, they revolve around that pole and never rise or set. The photograph in Figure 2-11 shows the circular trails of stars around the south celestial pole as seen from Australia.

Stars close to the *north* celestial pole are always below the horizon and can never be seen by an observer in the southern hemisphere. Hence, Australian astronomers never see the North Star but are able to see other stars that are forever hidden from North American observers.

Using the celestial equator and poles, we can define a coordinate system to specify the position of any star on the celestial sphere. As described in Box 2-1, the most commonly used coordinate system uses two angles, *right ascension* and *declination*, that are quite like longitude and latitude on Earth. These coordinates tell us in what direction we should look to see the star. To locate the star's true position in three-dimensional space, we must also know the distance to the star.

---

box 2-1 | Tools of the Astronomer's Trade

Celestial Coordinates

To denote the positions of objects in the sky, astronomers use a system based on *right ascension* and *declination*. Declination corresponds to latitude. As shown in the illustration, the **declination** of an object is its angular distance north or south of the celestial equator, measured along a circle passing through both celestial poles. It is measured in degrees, arcminutes, and arcseconds (see Section 1-5).

Right ascension corresponds to longitude. Astronomers measure right ascension from a specific point, called the vernal equinox, on the celestial equator. This point is one of two locations where the Sun crosses the celestial equator during its apparent annual motion, as we discuss in Section 2-5. In the Earth's northern hemisphere, spring officially begins when the Sun reaches the vernal equinox in late March. The **right ascension** of an object in the sky is the angular distance from the vernal equinox eastward along the celestial equator to the circle used in measuring its declination (see illustration). Astronomers do not measure this angular distance in degrees; instead, they use time units (hours, minutes, and seconds) corresponding to the time required for the celestial sphere to rotate through this angle. For example, suppose there is a star at your zenith right now with right ascension $6^h\ 0^m\ 0^s$. Two hours and 30 minutes from now, there will be a different object at your zenith with right ascension $8^h\ 30^m\ 0^s$.

 Right ascension and declination are very useful because they tell an astronomer precisely where in the sky an object is located. For example, the coordinates of the bright star Rigel for the year 2000 are R.A. = $5^h\ 14^m\ 32.2^s$, Decl. = $-8°\ 12'\ 06''$. (R.A. and Decl. are abbreviations for right ascension and declination.) A minus sign on the declination indicates that the star is south of the celestial equator; a plus sign (or no sign at all) indicates that an object is north of the celestial equator. As we discuss in Box 2-2, right ascension is especially useful for determining the best time to observe a particular object.

It is important to state the year for which a star's right ascension and declination are valid. This is so because of precession, which we discuss in Section 2-6.

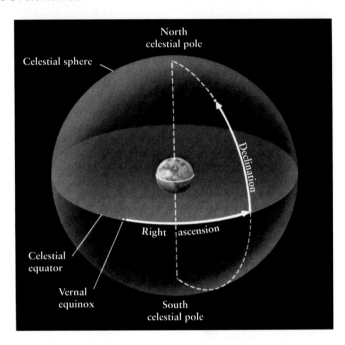

EXAMPLE: What are the coordinates of a star that lies exactly halfway between the vernal equinox and the south celestial pole? Since the circle used to measure this star's declination passes through the vernal equinox, this star's right ascension is R.A. = $0^h\ 0^m\ 0^s$. The angle between the celestial equator and south celestial pole is $90°\ 0'\ 0''$, so the declination of this star is Decl. = $-45°\ 0'\ 0''$. The declination is negative because the star is in the southern half of the celestial sphere.

EXAMPLE: At midnight local time you see a star with R.A. = $2^h\ 30^m\ 0^s$ at your zenith. When will you see a star at your zenith with R.A. = $21^h\ 0^m\ 0^s$? The time required for the sky to rotate through the angle between the stars is the difference in their right ascensions: $21^h\ 0^m\ 0^s - 2^h\ 30^m\ 0^s = 18^h\ 30^m\ 0^s$. So, the second star will be at your zenith 18½ hours after the first one, or at 6:30 P.M. the following evening.

2-5 The seasons are caused by the tilt of Earth's axis of rotation

In addition to rotating on its axis every 24 hours, the Earth revolves around the Sun in about 365¼ days. As we travel with the Earth around its orbit, we experience the annual cycle of seasons. But why *are* there seasons? Furthermore, the seasons are opposite in the northern and southern hemispheres. For example, February is midwinter in North America but midsummer in Australia. Why should this be?

The key fact that explains why we have seasons, and why they are different in different hemispheres, is that the Earth's axis of rotation is not perpendicular to the plane of the Earth's orbit. Instead, as Figure 2-12 shows, it is tilted about 23½° away from the perpendicular. The Earth maintains this tilt as it orbits the Sun, with the Earth's north pole pointing toward the north celestial pole. (This stability is a hallmark of all rotating objects. A top will not fall over as long as it is spinning, and the rotating wheels of a motorcycle help to keep the rider upright.)

During part of the year, when the Earth is in the part of its orbit shown on the left side of Figure 2-12, the northern hemisphere is tilted toward the Sun. As the Earth spins on its axis, a point in the northern hemisphere spends more than 12 hours in the sunlight. Thus, the days there are long and the nights are short, and it is summer in the northern hemisphere. The summer is hot not only because of the extended daylight hours but also because the Sun is high in the northern hemisphere's sky. As a result, sunlight strikes the ground at a nearly perpendicular angle that heats the ground efficiently (Figure 2-13a). During this same time of year in the southern hemisphere, the days are short and the nights are long, because a point in this hemisphere spends fewer than 12 hours a day in the sunlight. The Sun is low in the sky, so sunlight strikes the surface at a grazing angle that causes little heating (Figure 2-13b), and it is winter in the southern hemisphere.

Half a year later, the Earth is in the part of its orbit shown on the right side of Figure 2-12. Now the situation is reversed, with winter in the northern hemisphere (which is now tilted away from the Sun) and summer in the southern hemisphere. During spring and autumn, the two hemispheres receive roughly equal amounts of illumination from the Sun, and daytime and nighttime are of equal length everywhere on Earth.

CAUTION! A common misconception is that the seasons are caused by variations in the distance from the Earth to the Sun. According to this idea, the Earth is closer to the Sun in summer and farther away in winter. But in fact, the Earth's orbit around the Sun is very nearly circular, and the Earth-Sun distance varies only about 3% over the course of a year. (The Earth's orbit only *looks* elongated in Figure 2-12 because this illustration shows an oblique view.) We are slightly closer to the Sun in January than in July, but this small variation has little influence on the cycle of the seasons. Also, if the seasons were really caused by variations in the Earth-Sun distance, the seasons would be the same in both hemispheres!

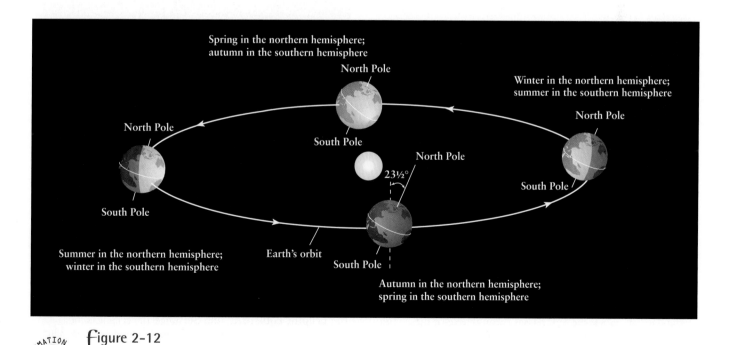

Figure 2-12

The Seasons The Earth's axis of rotation is inclined 23½° away from the perpendicular to the plane of the Earth's orbit. The Earth maintains this orientation (with its north pole aimed at the north celestial pole, near the star Polaris) throughout the year as it orbits the Sun. Consequently, the amount of solar illumination and the number of daylight hours at any location on Earth vary in a regular pattern throughout the year. This is the origin of the seasons.

a   b

figure 2-13

Solar Energy in Summer and Winter At different times of the year, sunlight strikes the ground at different angles. **(a)** In summer the Sun is high in the sky. The energy in a shaft of light falls onto a small, nearly circular area. This concentrated solar energy heats the ground effectively and makes the days warm. The days are also longest in summer, which further increases the heating. **(b)** In winter the Sun is low in the sky even at midday. The same shaft of light in (a) now falls on a larger, oval area. Because the sunlight is less concentrated and the days are short, little heating of the ground takes place. This accounts for the low temperatures in winter.

As the Earth orbits around the Sun, the Sun's position (as seen from the Earth) gradually shifts with respect to the background stars. As Figure 2-14 shows, the Sun therefore appears to trace out a circular path, called the **ecliptic,** on the celestial sphere. (This word suggests that the path traced out by the Sun has something to do with eclipses. We will discuss the connection in Chapter 3.) Because there are 365¼ days in a year and 360° in a circle, the Sun appears to move along the ecliptic at a rate of about 1° per day. This motion is from west to east, that is, in the direction opposite to the apparent motion of the celestial sphere.

ANALOGY It may help to envision the celestial sphere as a merry-go-round rotating clockwise, and the Sun as a restless child who is walking slowly around the merry-go-round's rim in the counterclockwise direction. During the time it takes the child to make a round trip, the merry-go-round rotates 365¼ times.

The ecliptic can also be thought of as the projection onto the celestial sphere of the plane of the Earth's orbit. This is *not* the same as the plane of the Earth's equator, thanks to the 23½° tilt of the Earth's rotation axis shown in Figure 2-12. Instead, the ecliptic and the celestial equator are inclined to each other by that same 23½° angle.

The ecliptic and the celestial equator intersect at only two points, which are exactly opposite each other on the celestial sphere. Each point is called an **equinox** (from the Latin for "equal night"), because when the Sun appears at either of these points, day and night are each about 12 hours long at all locations on Earth. The term "equinox" is also used to refer to the date on which the Sun passes through one of these special points on the ecliptic.

On about March 21 of each year, the Sun crosses northward across the celestial equator at the **vernal equinox.** This marks the beginning of spring in the northern hemisphere ("vernal" is from the Latin for "spring"). On about September 22 the Sun moves southward across the celestial equator at the **autumnal equinox,** marking the moment when fall

begins in the northern hemisphere. Since the seasons are opposite in the northern and southern hemispheres, for Australians and South Africans the vernal equinox actually marks the beginning of autumn. The names of the equinoxes come from astronomers of the past who lived north of the equator.

Between the vernal and autumnal equinoxes lie two other significant locations along the ecliptic. The point on the ecliptic farthest north of the celestial equator is called the **summer**

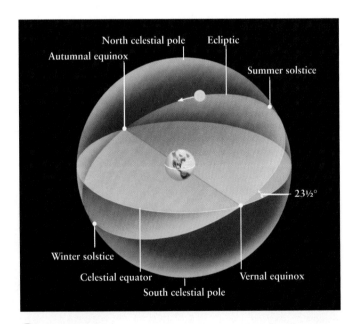

figure 2-14

The Ecliptic, Equinoxes, and Solstices The ecliptic is the apparent annual path of the Sun as projected onto the celestial sphere. Inclined to the celestial equator by 23½° because of the tilt of the Earth's axis of rotation, the ecliptic intersects the celestial equator at two points, called equinoxes. The northernmost point on the ecliptic is the summer solstice, and the corresponding southernmost point is the winter solstice. The Sun is shown in its approximate position for August 1.

solstice. "Solstice" is from the Latin for "solar standstill," and it is at the summer solstice that the Sun stops moving northward on the celestial sphere. At this point, the Sun is as far north of the celestial equator as it can get. It marks the location of the Sun at the moment summer begins in the northern hemisphere (about June 21). At the beginning of the northern hemisphere's winter, the Sun is farthest south of the celestial equator, at a point called the **winter solstice** (about December 21).

Because the Sun's position on the celestial sphere varies slowly over the course of a year, its daily path across the sky (due to the Earth's rotation) also varies with the seasons (Figure 2-15). On the first day of spring or the first day of fall, when the Sun is at one of the equinoxes, the Sun rises directly in the east and sets directly in the west.

When the northern hemisphere is tilted away from the Sun and it is winter in the northern hemisphere, the Sun rises in the southeast. Daylight lasts for fewer than 12 hours as the Sun skims low over the southern horizon and sets in the southwest. Northern hemisphere nights are longest when the Sun is at the winter solstice.

The closer you get to the north pole, the shorter the winter days and the longer the winter nights. In fact, anywhere within 23½° of the north pole (that is, north of latitude 90° − 23½° = 66½° N) the Sun is below the horizon for 24 continuous hours at least one day of the year. The circle around the Earth at 66½° north latitude is called the **Arctic Circle** (Figure 2-16*a*). The corresponding region around the south pole is bounded by the

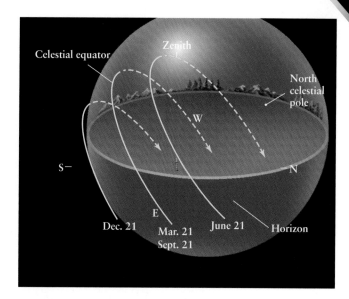

figure 2-15

The Sun's Daily Path Across the Sky Like Figure 2-10, this drawing is for an observer at 35° north latitude. On the first day of spring and the first day of fall, the Sun rises precisely in the east and sets precisely in the west. During summer in the northern hemisphere, the Sun rises in the northeast and sets in the northwest. In winter in the northern hemisphere, the Sun rises in the southeast and sets in the southwest.

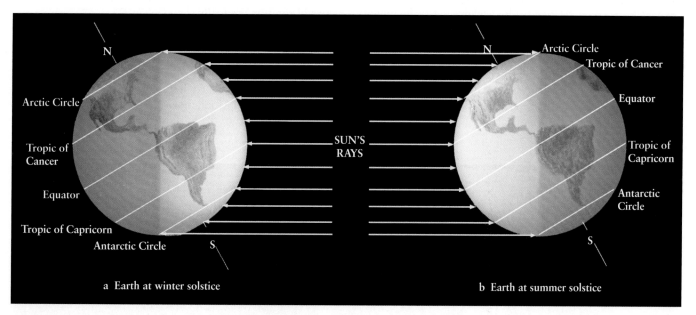

a Earth at winter solstice

b Earth at summer solstice

figure 2-16

Tropics and Circles (a) On the day when the Sun is at the winter solstice, the Sun does not rise at all at locations north of the Arctic Circle (66½° north latitude). On this same day the Sun is directly overhead at high noon anywhere on the Tropic of Capricorn (23½° south latitude), and is in the sky for 24 consecutive hours anywhere south of the Antarctic Circle (66½° south latitude). **(b)** The situation is reversed on the day when the Sun is at the summer solstice. The Sun does not set anywhere north of the Arctic Circle and does not rise anywhere south of the Antarctic Circle. The Sun is directly overhead at high noon anywhere on the Tropic of Cancer, at 23½° north latitude.

Antarctic Circle at 66½° south latitude. At the time of the winter solstice, explorers south of the Antarctic Circle enjoy "the midnight sun," or 24 hours of continuous daylight.

During summer in the northern hemisphere, when the northern hemisphere is tilted toward the Sun, the Sun rises in the northeast and sets in the northwest. The Sun is at its northernmost position at the summer solstice, giving the northern hemisphere the greatest number of daylight hours. At the summer solstice the Sun does not set at all north of the Arctic Circle, and does not rise at all south of the Antarctic Circle (Figure 2-16b).

The variations of the seasons are much less pronounced close to the equator. Between the **Tropic of Capricorn** at 23½° south latitude and the **Tropic of Cancer** at 23½° north latitude, the Sun is directly overhead—that is, at the zenith—at high noon at least one day a year. Outside of the tropics, the Sun is never directly overhead, but is always either south of the zenith (as seen from locations north of the Tropic of Cancer) or north of the zenith (as seen from south of the Tropic of Capricorn).

2-6 The Moon helps to cause precession, a slow, conical motion of Earth's axis of rotation

The Moon is by far the brightest and most obvious naked-eye object in the nighttime sky. Like the Sun, the Moon slowly changes its position relative to the background stars; unlike the Sun, the Moon makes a complete trip around the celestial sphere in only about four weeks, or about a month. (The word "month" comes from the same Old English root as the word "moon.") Ancient astronomers realized that this motion occurs because the Moon orbits the Earth in roughly four weeks. In one hour the Moon moves on the celestial sphere by about ½°, or roughly its own angular size.

The Moon's path on the celestial sphere is never far from the Sun's path (that is, the ecliptic). This is because the plane of the Moon's orbit around the Earth is inclined only slightly from the plane of the Earth's orbit around the Sun. The Moon's path varies somewhat from one month to the next, but always remains within a band called the **zodiac** that extends about 8° on either side of the ecliptic. Twelve famous constellations—Aries, Taurus, Gemini, Cancer, Leo, Virgo, Libra, Scorpius, Sagittarius, Capricornus, Aquarius, and Pisces—lie along the zodiac. The Moon is generally found in one of these 12 constellations. As it moves along its orbit, the Moon appears north of the celestial equator for about two weeks and then south of the celestial equator for about the next two weeks. We will learn more about the Moon's motion, as well as why the Moon goes through phases, in Chapter 3.

The Moon not only moves around the Earth but, in concert with the Sun, also causes a slow change in the Earth's rotation. This is because both the Sun and the Moon exert a gravitational pull on the Earth. We will learn much more about gravity in Chapter 4; for now, all we need is the idea that gravity is a universal attraction of matter for other matter.

The gravitational pull of the Sun and the Moon affects the Earth's rotation because the Earth is slightly fatter across the equator than it is from pole to pole: Its equatorial diameter is 43 kilometers (27 miles) larger than the diameter measured from pole to pole. The Earth is therefore said to have an "equatorial bulge." Because of the gravitational pull of the Moon and the Sun on this bulge, the orientation of the Earth's axis of rotation gradually changes.

The Earth behaves somewhat like a spinning top, as illustrated in Figure 2-17. If the top were not spinning, gravity would pull the top over on its side. When the top is spinning, gravity causes the top's axis of rotation to trace out a circle, producing a motion called **precession.**

As the Sun and Moon move along the zodiac, each spends half its time north of the Earth's equatorial bulge and half its

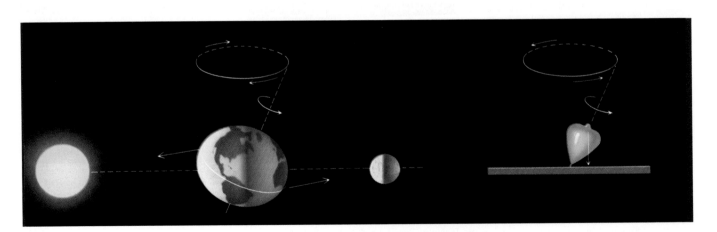

ƒigure 2-17

Precession The gravitational pull of the Moon and the Sun on the Earth's equatorial bulge causes the Earth to precess. As the Earth precesses, its axis of rotation slowly traces out a circle in the sky, like the shifting axis of a spinning top.

figure 2-18

Precession and the Path of the North Celestial Pole As the Earth precesses, the north celestial pole slowly traces out a circle among the northern constellations. At present, the north celestial pole is near the moderately bright star Polaris, which serves as the North Star. Twelve thousand years from now the bright star Vega will be the North Star.

time south of it. The gravitational pull of the Sun and Moon tugging on the equatorial bulge tries to twist the Earth's axis of rotation to be perpendicular to the plane of the ecliptic. But because the Earth is spinning, the combined actions of gravity and rotation cause the Earth's axis to trace out a circle in the sky, much like what happens to the toy top. As the axis precesses, it remains tilted about 23½° to the perpendicular.

As the Earth's axis of rotation slowly changes its orientation, the north and south celestial poles—which are the projections of that axis onto the celestial sphere—change their positions relative to the stars. At present, the north celestial pole lies within 1° of the star Polaris, which is why Polaris is the North Star. But 5000 years ago, the north celestial pole was closest to the star Thuban in the constellation of Draco (the Dragon). Thus, that star and not Polaris was the North Star. And 12,000 years from now, the North Star will be the bright star Vega in Lyra (the Harp). It takes 26,000 years for the north celestial pole to complete one full precessional circle around the sky (Figure 2-18). The south celestial pole executes a similar circle in the southern sky.

Precession also causes the Earth's equatorial plane to change its orientation. Because this plane defines the location of the celestial equator in the sky, the celestial equator precesses as well. The intersections of the celestial equator and the ecliptic define the equinoxes (see Figure 2-14), so these key locations in the sky also shift slowly from year to year. For this reason, the precession of the Earth is also called the **precession of the equinoxes.** The first person to detect the precession of the equinoxes, in the second century B.C., was the Greek astronomer Hipparchus, who compared his own observations with those of Babylonian astronomers three centuries earlier. Today, the vernal equinox is located in the constellation Pisces (the Fishes). Two thousand years ago, it was in Aries (the Ram). Around the year A.D. 2600, the vernal equinox will move into Aquarius (the Water Bearer).

Astrological terms like the "Age of Aquarius" involve boundaries in the sky that are not recognized by astronomers and are generally not even related to the positions of the constellations. Indeed, astrology is *not* a science at all, but merely a collection of superstitions and hokum. Its practitioners use some of the terminology of astronomy but reject the logical thinking that is at the heart of science. James Randi has more to say about astrology and other pseudosciences in his essay "Why Astrology Is Not Science" at the end of this chapter.

The astronomer's system of locating heavenly bodies by their right ascension and declination, discussed in Box 2-1, is tied to the positions of the celestial equator and the vernal equinox. Because of precession, these positions are changing, and thus the coordinates of stars in the sky are also constantly changing. These changes are very small and gradual, but they add up over the years. To cope with this difficulty, astronomers always make note of the date (called the **epoch**) for which a particular set of coordinates is precisely correct.

Consequently, star catalogs and star charts are periodically updated. Most current catalogs and star charts are prepared for the epoch 2000. The coordinates in these reference books, which are precise for January 1, 2000, will require very little correction over the next few decades.

2-7 Positional astronomy plays an important role in keeping track of time

Astronomers have traditionally been responsible for telling time. This is because we want the system of timekeeping used in everyday life to reflect the position of the Sun in the sky. Thousands of years ago, the sundial was invented to keep track of **apparent solar time.** To obtain more accurate measurements astronomers use the **meridian.** As Figure 2-19 shows, this is a north-south circle on the celestial sphere that passes through the zenith (the point directly overhead) and both celestial poles. *Local noon* is defined to be when the Sun crosses the **upper meridian,** which is the half of the meridian above the horizon. At *local midnight,* the Sun crosses the **lower meridian,** the half of the meridian below the horizon; this crossing cannot be observed directly.

The crossing of the meridian by any object in the sky is called a **meridian transit** of that object. If the crossing occurs above the horizon, it is an *upper* meridian transit. An **apparent solar day** is the interval between two successive upper meridian transits of the Sun as observed from any fixed spot on the Earth. Stated less formally, an apparent solar day is the time from one local noon to the next local noon, or from when the Sun is highest in the sky to when it is again highest in the sky.

Unfortunately, the Sun is not a good timekeeper. The length of an apparent solar day (as measured by a device such as an hourglass) varies from one time of year to another. There are two main reasons why this is so, both having to do with the way in which the Earth orbits the Sun.

The first reason is that the Earth's orbit is not a perfect circle but rather an ellipse, as Figure 2-20*a* shows in exaggerated

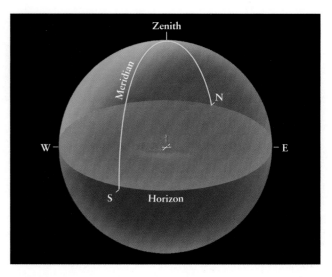

figure 2-19

The Meridian The meridian is a circle on the celestial sphere that passes through the observer's zenith (the point directly overhead) and the north and south points on the observer's horizon. The passing of celestial objects across the meridian can be used to measure time. The upper meridian is the part above the horizon, and the lower meridian (not shown) is the part below the horizon.

form. As we will learn in Chapter 4, the Earth moves more rapidly along its orbit when it is near the Sun than when it is farther away. Hence, the Sun appears to us to move more than 1° per day along the ecliptic in January, when the Earth is nearest the Sun, and less than 1° per day in July, when the Earth is farthest from the Sun. By itself, this effect would cause the apparent solar day to be longer in January than in July.

The second reason why the Sun is not a good timekeeper is the 23½° angle between the ecliptic and the celestial equator (see Figure 2-14). As Figure 2-20*b* shows, this causes a significant part of the Sun's apparent motion when near the equinoxes to be in a north-south direction. The net daily eastward progress in the sky is then somewhat foreshortened. At the summer and winter solstices, by contrast, the Sun's motion

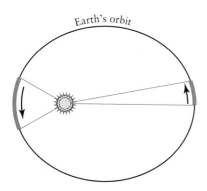

a A month's motion of the Earth along its orbit

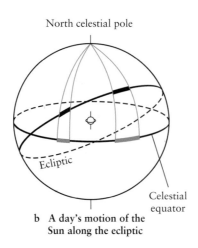

b A day's motion of the Sun along the ecliptic

figure 2-20

Why the Sun Is a Poor Timekeeper There are two main reasons that the Sun is a poor timekeeper. **(a)** The Earth's speed along its orbit varies during the year. It moves fastest when closest to the Sun in January and slowest when farthest from the Sun in July. Hence, the apparent speed of the Sun along the ecliptic is not constant. **(b)** Because of the tilt of the Earth's rotation axis, the ecliptic is inclined with respect to the celestial equator. Therefore, the projection of the Sun's daily progress along the ecliptic onto the celestial equator (shown in blue) varies during the year. This causes further variations in the length of the apparent solar day.

is parallel to the celestial equator. Thus, there is no comparable foreshortening around the beginning of summer or winter. This effect by itself would make the apparent solar day shorter in March and September than in June or December. Combining these effects with those due to the Earth's noncircular orbit, we find that the length of the apparent solar day varies in a complicated fashion over the course of a year.

To avoid these difficulties, astronomers invented an imaginary object called the **mean sun** that moves along the celestial equator at a uniform rate. (In science and mathematics, "mean" is a synonym for "average.") The mean sun is sometimes slightly ahead of the real Sun in the sky, sometimes behind. As a result, mean solar time and apparent solar time can differ by as much as a quarter of an hour at certain times of the year.

Because the mean sun moves at a constant rate, it serves as a fine timekeeper. A **mean solar day** is the interval between successive upper meridian transits of the mean sun. It is exactly 24 hours long, the average length of an apparent solar day. One 24-hour day as measured by your alarm clock or wristwatch is a mean solar day.

Time zones were invented for convenience in commerce, transportation, and communication. In a time zone, all clocks and watches are set to the mean solar time for a meridian of longitude that runs approximately through the center of the zone. Time zones around the world are generally centered on meridians of longitude at 15° intervals. In most cases, going from one time zone to the next requires you to change the time on your wristwatch by exactly one hour. The time zones for most of North America are shown in Figure 2-21.

In order to coordinate their observations with colleagues elsewhere around the globe, astronomers often keep track of time using Coordinated Universal Time, somewhat confusingly abbreviated UTC or UT. This is the time in a zone that includes Greenwich, England, a seaport just outside of London where the first internationally accepted time standard was kept. In former times UTC was known as Greenwich Mean Time. In North America, Eastern Standard Time (EST) is 5 hours different from UTC; 9:00 A.M. EST is 14:00 UTC. Coordinated Universal Time is also used by aviators and sailors, who regularly travel from one time zone to another.

Although it is natural to want our clocks and method of timekeeping to be related to the Sun, astronomers often use a system that is based on the apparent motion of the stars. This system, called **sidereal time,** is useful when aiming a telescope. Most observatories are therefore equipped with a clock that measures sidereal time, as discussed in Box 2-2.

2-8   Astronomical observations led to the development of the modern calendar

Just as the day is a natural unit of time based on the Earth's rotation, the year is a natural unit of time based on the Earth's revolution about the Sun. However, nature has not arranged things for our convenience. The year does not divide into

figure 2-21

Time Zones in North America For convenience, the Earth is divided into 24 time zones, generally centered on 15° intervals of longitude around the globe. There are four time zones across the continental United States, making for a 3-hour time difference between New York and California.

exactly 365 whole days. Ancient astronomers realized that the length of a year is approximately 365¼ days, so the Roman emperor Julius Caesar established the system of "leap years" to account for this extra quarter of a day. By adding an extra day to the calendar every four years, he hoped to ensure that seasonal astronomical events, such as the beginning of spring, would occur on the same date year after year.

Caesar's system would have been perfect if the year were exactly 365¼ days long and if there were no precession. Unfortunately, this is not the case. To be more accurate, astronomers now use several different types of years. For example, the **sidereal year** is defined to be the time required for the Sun to return to the same position with respect to the stars. It is equal to 365.2564 mean solar days, or $365^d\ 6^h\ 9^m\ 10^s$.

The sidereal year is the orbital period of the Earth around the Sun, but it is not the year on which we base our calendar. Like Caesar, most people want annual events—in particular, the first days of the seasons—to fall on the same date each year. For example, we want the first day of spring to occur on March 21. But spring begins when the Sun is at the vernal equinox, and the vernal equinox moves slowly against the background stars because of precession. Therefore, to set up a calendar we use the **tropical year,** which is equal to the time needed for the Sun to return to the vernal equinox. This period is equal to 365.2422 mean solar days, or $365^d\ 5^h\ 48^m\ 46^s$. Because of precession, the tropical year is 20 minutes and 24 seconds shorter than the sidereal year.

Caesar's assumption that the tropical year equals 365¼ days was off by 11 minutes and 14 seconds. This tiny error

box 2-2 | Tools of the Astronomer's Trade

Sidereal Time

If you want to observe a particular object in the heavens, the ideal time to do so is when the object is high in the sky, on or close to the upper meridian. This minimizes the distorting effects of the Earth's atmosphere, which increase as you view closer to the horizon. For astronomers who study the Sun, this means making observations at local noon, which is not too different from noon as determined using mean solar time. For astronomers who observe planets, stars, or galaxies, however, the optimum time to observe depends on the particular object to be studied. The problem is this: Given the location of a given object on the celestial sphere, when will that object be on the upper meridian?

To answer this question, astronomers use *sidereal time* rather than solar time. It is different from the time on your wristwatch. In fact, a sidereal clock and an ordinary clock even tick at different rates, because they are based on different astronomical objects. Ordinary clocks are related to the position of the Sun, while sidereal clocks are based on the position of the vernal equinox, the location from which right ascension is measured (see Box 2-1 for a discussion of right ascension).

Regardless of where the Sun is, midnight sidereal time at your location is defined to be when the vernal equinox crosses your upper meridian. (Like solar time, sidereal time depends on where you are on Earth.) A **sidereal day** is the time between two successive upper meridian passages of the vernal equinox. By contrast, an apparent solar day is the time between two successive upper meridian crossings of the Sun. The illustration shows why these two kinds of day are not equal. Because the Earth orbits the Sun, the Earth must make one complete rotation plus about 1° to get from one local solar noon to the next. This extra 1° of rotation corresponds to 4 minutes of time, which is the amount by which a solar day exceeds a sidereal day. To be precise:

$$1 \text{ sidereal day} = 23^h \ 56^m \ 4.091^s$$

where the hours, minutes, and seconds are in mean solar time.

One day according to your wristwatch is one mean solar day, which is exactly 24 hours of solar time long. A **sidereal clock** measures sidereal time in terms of sidereal hours, minutes, and seconds, where one sidereal day is divided into 24 sidereal hours. This explains why a sidereal clock ticks at a slightly different rate than your wristwatch. As a result, at some times of the year a sidereal clock will show a very different time than an ordinary clock. (On local noon on March 21, when the Sun is at the vernal equinox, a sidereal clock will say that it is midnight. Do you see why?)

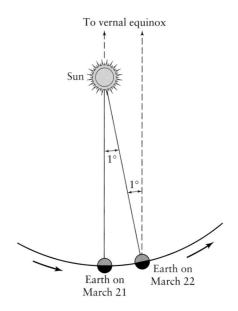

To vernal equinox

Sun

1°

1°

Earth on
March 22

Earth on
March 21

We can now answer the question in the opening paragraph. The vernal equinox, whose celestial coordinates are R.A. = $0^h \ 0^m \ 0^s$, Decl. = 0° 0′ 0″, crosses the upper meridian at midnight sidereal time (0:00). The autumnal equinox, which is on the opposite side of the celestial sphere at R.A. = $12^h \ 0^m \ 0^s$, Decl. = 0° 0′ 0″, crosses the upper meridian 12 sidereal hours later at noon sidereal time (12:00). As these examples illustrate, *any* object crosses the upper meridian when the sidereal time is equal to the object's right ascension. That is why astronomers measure right ascension in units of time rather than degrees, and why right ascension is always given in sidereal hours, minutes, and seconds.

EXAMPLE: Suppose you want to observe the bright star Regulus (Figure 2-6), which has epoch 2000 coordinates R.A. = $10^h \ 08^m \ 22.2^s$, Decl. = +11° 58′ 02″. What is the best time to do this? Regulus will be on the upper meridian and thus ideally suited for observation when the sidereal time at your location is equal to its right ascension, or about 10:08. As this example shows, it is useful to keep track of sidereal time, which is why most observatories are equipped with a sidereal clock.

While sidereal time is extremely useful in astronomy, mean solar time is still the best method of timekeeping for most earthbound purposes. All time measurements in this book are expressed in mean solar time unless otherwise stated.

adds up to about three days every four centuries. Although Caesar's astronomical advisers were aware of the discrepancy, they felt that it was too small to matter. However, by the sixteenth century the first day of spring was occurring on March 11.

The Roman Catholic Church became concerned because Easter kept shifting to progressively earlier dates. To straighten things out, Pope Gregory XIII instituted a calendar reform in 1582. He began by dropping ten days (October 4, 1582, was followed by October 15, 1582), which brought the first day of spring back to March 21. Next, he modified Caesar's system of leap years.

Caesar had added February 29 to every calendar year that is evenly divisible by four. Thus, for example, 1988, 1992, and 1996 were all leap years with 366 days. But we have seen that this system produces an error of about three days every four centuries. To solve the problem, Pope Gregory decreed that only the century years evenly divisible by 400 should be leap years. For example, the years 1700, 1800, and 1900 (which would have been leap years according to Caesar) were not leap years in the improved Gregorian system, but the year 2000, which can be divided evenly by 400, *was* a leap year.

We use the Gregorian system today. It assumes that the year is 365.2425 mean solar days long, which is very close to the true length of the tropical year. In fact, the error is only one day in every 3300 years. That won't cause any problems for a long time.

KEY WORDS

Terms preceded by an asterisk (*) are discussed in the Boxes.

Antarctic Circle, p. 32
apparent solar day, p. 34
apparent solar time, p. 34
Arctic Circle, p. 31
autumnal equinox, p. 30
celestial equator, p. 27
celestial sphere, p. 26
circumpolar, p. 27
constellation, p. 22
*declination, p. 28
diurnal motion, p. 23
ecliptic, p. 30

epoch, p. 33
equinox, p. 30
lower meridian, p. 34
mean solar day, p. 35
mean sun, p. 35
meridian, p. 34
meridian transit, p. 34
north celestial pole, p. 27
positional astronomy, p. 21
precession, p. 32
precession of the equinoxes, p. 33
*right ascension, p. 28
*sidereal clock, p. 36
*sidereal day, p. 36

sidereal time, p. 35
sidereal year, p. 35
south celestial pole, p. 27
summer solstice, pp. 30–31
time zone, p. 35
tropical year, p. 35
Tropic of Cancer, p. 32
Tropic of Capricorn, p. 32
upper meridian, p. 34
vernal equinox, p. 30
winter solstice, p. 31
zenith, p. 27
zodiac, p. 32

KEY IDEAS

Ideas preceded by an asterisk (*) are discussed in the Boxes.

Constellations and the Celestial Sphere: It is convenient to imagine the stars fixed to the celestial sphere with the Earth at its center.

• The surface of the celestial sphere is divided into 88 regions called constellations.

Diurnal (Daily) Motion of the Celestial Sphere: The celestial sphere appears to rotate around the Earth once in each 24-hour period. In fact, it is actually the Earth that is rotating.

• The poles and equator of the celestial sphere are determined by extending the axis of rotation and the equatorial plane of the Earth out to the celestial sphere.

• *The positions of objects on the celestial sphere are described by specifying their right ascension (in time units) and declination (in angular measure).

Seasons and the Tilt of the Earth's Axis: The Earth's axis of rotation is tilted at an angle of about 23½° from the perpendicular to the plane of the Earth's orbit.

• The seasons are caused by the tilt of the Earth's axis.

• Over the course of a year, the Sun appears to move around the celestial sphere along a path called the ecliptic. The ecliptic is inclined to the celestial equator by about 23½°.

• The ecliptic crosses the celestial equator at two points in the sky, the vernal and autumnal equinoxes. The northernmost point that the Sun reaches on the celestial sphere is the summer solstice, and the southernmost point is the winter solstice.

Precession: The orientation of the Earth's axis of rotation changes slowly, a phenomenon called precession.

• Precession is caused by the gravitational pull of the Sun and Moon on the Earth's equatorial bulge.

• Precession of the Earth's axis causes the positions of the equinoxes and celestial poles to shift slowly.

• *Because the system of right ascension and declination is tied to the position of the vernal equinox, the date (or epoch) of observation must be specified when giving the position of an object in the sky.

Timekeeping: Astronomers use several different means of keeping time.

• Apparent solar time is based on the apparent motion of the Sun across the celestial sphere, which varies over the course of the year.

• Mean solar time is based on the motion of an imaginary mean sun along the celestial equator, which produces a uniform mean solar day of 24 hours. Ordinary watches and clocks measure mean solar time.

• *Sidereal time is based on the apparent motion of the celestial sphere.

The Calendar: The tropical year is the period between two passages of the Sun across the vernal equinox. Leap year corrections are needed because the tropical year is not exactly 365 days. The sidereal year is the actual orbital period of the Earth.

REVIEW QUESTIONS

1. A fellow student tells you that only the stars in Figure 2-2*b* that are connected by blue lines are part of the constellation Orion. How would you respond?

2. How are constellations useful to astronomers? How many stars are not part of any constellation?

3. Why can't a person in Antarctica use the Big Dipper to find the north direction?

4. What is the celestial sphere? Why is this ancient concept still useful today?

5. Imagine that someone suggests sending a spacecraft to land on the surface of the celestial sphere. How would you respond to such a suggestion?

6. Is there any place on Earth where you could see the north celestial pole on the northern horizon? If so, where? Is there any place on Earth where you could see the north celestial pole on the western horizon? If so, where? Explain your answers.

7. What is the celestial equator? How is it related to the Earth's equator? How are the north and south celestial poles related to the Earth's axis of rotation? Where on Earth would you have to be for the celestial equator to pass through your zenith?

8. How many degrees is the angle from the horizon to the zenith? Does your answer depend on what point on the horizon you choose?

9. Using a diagram, explain why the tilt of the Earth's axis relative to the Earth's orbit causes the seasons as we orbit the Sun.

10. Give two reasons why it's warmer in summer than in winter.

11. What is the ecliptic? Why is it tilted with respect to the celestial equator? Does the Sun appear to move along the ecliptic, the celestial equator, or neither? By about how many degrees does the Sun appear to move on the celestial sphere each day?

12. Where on Earth do you have to be in order to see the north celestial pole directly overhead? What is the maximum possible elevation of the Sun above the horizon at that location? On what date can this maximum elevation be observed?

13. What are the vernal and the autumnal equinoxes? What are the summer and winter solstices? How are these four points related to the ecliptic and the celestial equator?

14. At what point on the horizon does the vernal equinox rise? Where on the horizon does it set? (*Hint:* See Figure 2-15.)

15. How does the daily path of the Sun across the sky change with the seasons? Why does it change?

16. Where on Earth do you have to be in order to see the Sun at the zenith? Will it be at the zenith every day? Explain.

17. What causes precession of the equinoxes? How long does it take for the vernal equinox to move 1° along the ecliptic?

18. What is the (fictitious) mean sun? What path does it follow on the celestial sphere? Why is it a better timekeeper than the actual Sun in the sky?

19. Why is it convenient to divide the Earth into time zones?

20. Why is the time given by a sundial not necessarily the same as the time on your wristwatch?

21. What is the difference between the sidereal year and the tropical year? Why are these two kinds of year slightly different in length? Why are calendars based on the tropical year?

22. When is the next leap year? Was 2000 a leap year? Will 2100 be a leap year?

ADVANCED QUESTIONS

Questions preceded by an asterisk (*) involve topics discussed in the Boxes.

Problem-solving tips and tools

To help you visualize the heavens, it is worth taking the time to become familiar with various types of star charts. These include the simple star charts at the end of this book, the monthly star charts published in such magazines as *Sky & Telescope* and *Astronomy*, and the more detailed maps of the heavens found in star atlases.

One of the best ways to understand the sky and its motions is to use the *Starry Night* computer program on the CD-ROM that accompanies this book. This easy-to-use program allows you to view the sky on any date and at any time, as seen from any point on Earth, and to animate the sky to visualize its diurnal and annual motions.

You may also find it useful to examine a planisphere, a device consisting of two rotatable disks. The bottom disk shows all the stars in the sky (for a particular latitude), and the top one is opaque with a transparent oval window through which only some of the stars can be seen. By rotating the top disk, you can immediately see which constellations are above the horizon at any time of the year. A planisphere is a convenient tool to carry with you when you are out observing the night sky.

23. On November 1 at 11:00 P.M. you look toward the eastern horizon and see the bright star Mintaka (shown in Figure 2-2b) rising. At approximately what time will Mintaka rise one week later, on November 8?

24. Figure 2-4 shows the situation on September 21, when Cygnus is highest in the sky at 8:00 P.M. local time and Andromeda is highest in the sky at midnight. But as Figure 2-5 shows, on July 21 Cygnus is highest in the sky at midnight. On July 21, at approximately what local time is Andromeda highest in the sky? Explain your reasoning.

25. Figure 2-5 shows which constellations are high in the sky (for observers in the northern hemisphere) in the months of July, September, and November. From this figure, would you be able to see Perseus at midnight on May 15? Draw a picture to justify your answer.

26. Figure 2-6 shows the appearance of Polaris, the Little Dipper, and the Big Dipper at midnight on June 1. Sketch how these objects would appear on this same date at (**a**) 9 P.M. and (**b**) 3 A.M. Include the horizon in your sketches, and indicate the north direction.

27. (**a**) Redraw Figure 2-10 for an observer at the North Pole. (*Hint:* The north celestial pole is directly above this observer.) (**b**) Redraw Figure 2-10 for an observer at the equator. (*Hint:* The celestial equator passes through this observer's zenith.) (**c**) Using Figure 2-10 and your drawings from (a) and (b), justify the following rule, long used by navigators: The latitude of an observer in the northern hemisphere is equal to the angle in the sky between that observer's horizon and the north celestial pole. (**d**) State the rule that corresponds to (c) for an observer in the southern hemisphere.

28. The time-exposure photograph in Figure 2-11 shows the trails made by individual stars as the celestial sphere appears to rotate around the Earth. (**a**) If you were standing in Australia next to the camera that took this picture and looking in the same direction that the camera was pointing, would you see the stars moving clockwise or counterclockwise? (**b**) For approximately what length of time was the camera shutter left open to take this photograph?

29. Is there any place on Earth where all the visible stars are circumpolar? If so, where? Is there any place on Earth where none of the visible stars is circumpolar? If so, where? Explain your answers.

30. In the northern hemisphere, houses are designed to have "southern exposure," that is, with the largest windows on the southern side of the house. But in the southern hemisphere houses are designed to have "northern exposure." Why are houses designed this way, and why is there a difference between the hemispheres?

31. Figure 2-15 shows the daily path of the Sun across the sky on March 21, June 21, September 21, and December 21 for an observer at 35° north latitude. Sketch a drawing of this kind for (**a**) an observer at 35° south latitude; (**b**) an observer at the equator; and (**c**) an observer at the North Pole.

32. Suppose that you live at a latitude of 40° N. What is the elevation of the Sun above the southern horizon at noon at the time of the winter solstice? Explain your reasoning. Include a drawing as part of your explanation.

33. The city of Mumbai (formerly Bombay) in India is 19° north of the equator. On how many days of the year, if any, is the Sun at the zenith at midday as seen from Mumbai? Explain your answer.

34. The Great Pyramid at Giza has a tunnel that points toward the north celestial pole. At the time the pyramid was built, around 2600 B.C., toward which star did it point? Toward which star does this same tunnel point today? (See Figure 2-18.)

35. Ancient records show that 2000 years ago, the stars of the Southern Cross were visible in the southern sky from Greece. Today, however, these stars cannot be seen from Greece. What accounts for this change?

36. Unlike western Europe, Imperial Russia did not use the revised calendar instituted by Pope Gregory XIII. Explain why the Russian Revolution, which started on November 7, 1917, according to the modern calendar is called "the

October revolution" in Russia. What was this date according to the Russian calendar at the time? Explain.

*37. What is the right ascension of a star that is on the meridian at midnight at the time of the autumnal equinox? Explain.

*38. The coordinates on the celestial sphere of the summer solstice are R.A. = 6^h 0^m 0^s, Decl. = +23° 27'. What are the right ascension and declination of the winter solstice? Explain your answer.

*39. Right ascension is measured in hours, minutes, and seconds. Because 24 hours of right ascension takes you all the way around the celestial equator, it follows that 24^h = 360°. What is the angle in the sky (measured in degrees) between a star with R.A. = 8^h 0^m 0^s, Decl. = 0° 0' 0" and a second star with R.A. = 11^h 20^m 0^s, Decl. = 0° 0' 0"? Explain your answer.

*40. At local noon on March 21, when the Sun is at the vernal equinox, a sidereal clock will say that it is midnight. Explain why.

*41. (a) What is the sidereal time when the vernal equinox rises? (b) On what date is the sidereal time nearly equal to the solar time? Explain.

*42. On a certain night, the first star in Advanced Question 39 passes through the zenith at 12:30 A.M. local time. At what time will the second star pass through the zenith? Explain your answer.

*43. How would the sidereal and solar days change (a) if the Earth's rate of rotation increased, (b) if the Earth's rate of rotation decreased, and (c) if the Earth's rotation were retrograde (that is, if the Earth rotated about its axis opposite to the direction in which it revolves about the Sun)?

44. The image of the Earth shown below was made by the *Galileo* spacecraft while en route to Jupiter. South America is at the center of the image and Antarctica is at the bottom of the image. (a) In which month of the year was this image made? Explain your reasoning. (b) When this image was made, was the Earth closer to perihelion or to aphelion? Explain your reasoning.

45. The photograph in Figure 2-11 was taken at the Anglo-Australian Observatory (AAO) on Siding Springs Mountain, Australia. Siding Springs is at longitude 149° 03' 58" east and latitude 31° 16' 37" south. (a) By making measurements on the photograph, find the approximate angular width and angular height of the photo. (b) How far (in degrees) from the south celestial pole can a star be and still be circumpolar as seen from the AAO?

DISCUSSION QUESTIONS

46. Examine a list of the 88 constellations. Are there any constellations whose names obviously date from modern times? Where are these constellations located? Why do you suppose they do not have archaic names?

47. Why is it useful to astronomers to have telescopes in both the southern hemisphere and the northern hemisphere? (See Figure 2-10.)

48. Describe how the seasons would be different if the Earth's axis of rotation, rather than having its present 23½° tilt, were tilted (a) by 0° or (b) by 90°.

49. In William Shakespeare's *Julius Caesar* (act 3, scene 1), Caesar says:

> *But I am constant as the northern star,*
> *Of whose true-fix'd and resting quality*
> *There is no fellow in the firmament.*

Translate Caesar's statement about the "northern star" into modern astronomical language. Is the northern star truly "constant"? Was the northern star the same in Shakespeare's time (1564–1616) as it is today?

WEB/CD-ROM QUESTIONS

50. Search the World Wide Web for information about the national flags of Australia, New Zealand, and Brazil and the state flag of Alaska. Which stars are depicted on these flags? Explain any similarities or differences among these flags.

51. Some people say that on the date that the Sun is at the vernal equinox, and only on this date, you can stand a raw egg on end. Others say that there is nothing special about the vernal equinox, and that with patience you can stand a raw egg on end on any day of the year. Search the World Wide Web for information about this story and for hints about how to stand an egg on end. Use these hints to try the experiment yourself on a day when the Sun is *not* at the

(NASA/JPL)   R I **V** U X G

vernal equinox. What do you conclude about the connection between eggs and equinoxes?

 52. Use the U.S. Naval Observatory web site to find the times of sunset and sunrise on (**a**) your next birthday and (**b**) the date this assignment is due. (**c**) Are the times the same for the two dates? Explain why or why not.

OBSERVING PROJECTS

Observing tips and tools

 Moonlight is so bright that it interferes with seeing the stars. For the best view of the constellations, do your observing when the Moon is below the horizon. You can find the times of moonrise and moonset in your local newspaper or on the World Wide Web. Each monthly issue of the magazines *Sky & Telescope* and *Astronomy* includes much additional observing information.

53. On a clear, cloud-free night, use the star charts at the end of this book to see how many constellations of the zodiac you can identify. Which ones were easy to find? Which were difficult? Are the zodiacal constellations the most prominent ones in the sky?

54. Examine the star charts that are published monthly in such popular astronomy magazines as *Sky & Telescope* and *Astronomy*. How do they differ from the star charts at the back of this book? On a clear, cloud-free night, use one of these star charts to locate the celestial equator and the ecliptic. Note the inclination of the Milky Way to the ecliptic and celestial equator. The Milky Way traces out the plane of our galaxy. What do your observations tell you about the orientation of the Earth and its orbit relative to the galaxy's plane?

55. Suppose you wake up before dawn and want to see which constellations are in the sky. Explain how the star charts at the end of this book can be quite useful, even though chart times are given only for the evening hours. Which chart most closely depicts the sky at 4:00 A.M. tomorrow morning? Set your alarm clock for 4:00 A.M. to see if you are correct.

 56. Use the *Starry Night* program to observe the diurnal motion of the sky. (**a**) First set *Starry Night* to display the sky as seen from where you live. Select **Set Home Location...** in the **Go** menu and click on the **Lookup** button to find your city or town. Then, using the hand cursor, center your field of view on the northern horizon (if you live in the northern hemisphere) or the southern horizon (if you live in the southern hemisphere). In the Control Panel at the top of the main window, set the time step to 3 minutes and click on the "Forward" button (a triangle that points to the right). To see the stars during the daytime, turn off daylight (select **Daylight** in the **Sky** menu). Do the stars appear to rotate clockwise or counterclockwise? Explain in terms of the Earth's rotation. Are any of the stars circumpolar? (**b**) Recenter your field of view on the southern horizon (if you live in the northern hemisphere) or the northern horizon (if you live in the southern hemisphere). Describe what you see. Are any of these stars circumpolar?

 57. Use the *Starry Night* program to observe the Sun's motion on the celestial sphere. Select **Atlas** in the **Go** menu to see the entire celestial sphere, including the part below the horizon. Center on the Sun by using the **Find...** command in the **Edit** menu. (**a**) Select **Auto Identify** in the **Constellations** menu to display the constellation at the center of the screen. In which constellation is the Sun located today? Is this the same as the astrological sign for today's date? Explain in terms of precession. (**b**) In the Control Panel at the top of the main window, set the time step to 3 days and click on the "Forward" button (a triangle that points to the right). Observe the Sun for a full year of simulated time. How does the Sun appear to move against the constellations? What path does it follow? Does it ever change direction?

JAMES RANDI

Why Astrology Is Not Science

James ("the Amazing") Randi works tirelessly to expose trickery, so that others can relish the greater wonder of science. As a magician, he has had his own television show and an enormous public following. As a lecturer, he addresses teachers, students, and others worldwide. His newsletter and column for *The Skeptic* are key resources for educators. His many books include *Flim-Flam!*, *The Faith Healers*, and *The Mask of Nostradamus*, about a legendary con man with secrets of his own.

Mr. Randi helped found the Committee for the Scientific Investigation of Claims of the Paranormal, and his $10,000 prize for "the performance of any paranormal event . . . under proper observing conditions" has gone unclaimed for more than 25 years. An amateur archeologist and astronomer as well, he lives in Florida with several untalented parrots and the occasional visiting magus.

I'm involved in the strange business of telling folks what they should already know. I meet audiences who believe in all sorts of impossible things, often despite their education and intelligence. My job is to explain how science differs from the unproven, illogical assumptions of pseudoscience—and why it matters. Perhaps my best example is the difference between astronomy and astrology.

Both astrology and astronomy arose from the wonders of the night sky, from the stars to comets, planets, the Sun, and the Moon. Surely, humans have long reasoned, there must be some meaning in their motions. Surely the Moon's effect on tides hints at hidden "causes" for strange events. *Judiciary* (literally "judging") astrology therefore attempted to foretell the future—our earthly future. To serve it, *horary* (literally "hourly") astrology carefully tracked the heavens.

It is the latter that has become astronomy. Thanks to its process of careful measurement and testing, we now understand more about the true nature of the starry universe than astrologers could ever have imagined. With the birth of a new science, astronomers had a logical framework based on physical causes and systematic observations.

Astrology remains a popular delusion. Far too many believe today that patterns in the sky govern our lives. They accept the vague tendencies and portents of seers who cast horoscopes. They shouldn't. Just a glance at the tenets of astrology provides ample evidence of its absurdity.

An individual is said to be born under a sign. To the astrologer, the Sun was located "in" that sign at the moment of birth. (Stars are not seen in the daytime, but no matter—a calculation tells where the Sun is.) Each sign takes its name from a constellation, a totally imaginary figure invented for our convenience in referring to stars. Different cultures have different mythical figures up there, and so different schools of astrology assign different meanings to the signs they use.

In the spirit of equal-opportunity swindling, astrologers divide up the year fairly, ignoring variations in the size of constellations. Since Libra is tiny, while Virgo is huge, they chop some of the sky off Virgo and add it—along with bits of Scorpio—to bring Libra up to size. The Sun could well be declared "in" Libra when it is actually outside that constellation.

It gets worse. Science constantly challenges itself and changes. The rules of astrology could not, although they were made up thousands of years ago, and since then the "fixed" stars have moved. In particular, precession of the equinoxes has shifted objects in the sky relative to our calendar. The constellations have changed but astrology has not. If you were born August 7, you are said to be a Leo, but the Sun that day was really in the same part of the sky as the constellation Cancer.

With a theory like this to back it up, we should not be surprised at the bottom line: *A pseudoscience does not work.* Test after test has checked its predictions, and the result is always the same. One such investigator is Shawn Carlson of the University of California, San Diego. As he put it in *Nature* magazine, astrology is "a hopeless cause." Johannes Kepler, the pioneering astronomer, himself cast horoscopes, but they are little remembered today. Owen Gingerich, a historian of science at Harvard, puts it well: Kepler was the astrologer who destroyed astrology.

Astronomy works, and it works very well indeed. That isn't easy. Because we humans tend to find what we want in any body of data, it takes science's careful process of observation, creative insight, and critical thinking to understand and predict changes in nature. As I write, a transit of Ganymede is due next Thursday at 21:47:20. At exactly that time, the satellite of Jupiter will cross in front of its planet as seen from Earth, and yet most of us will never know it. Still other moons of Jupiter may hold fresh clues to the formation of our entire solar system and the conditions for life elsewhere.

For most people, astronomy has too little fantasy or money in it, and they will never experience the beauty in its predictions. The dedicated labors of generations of scientists have enabled us to perform a genuine wonder.

Eclipses and the Motion of the Moon

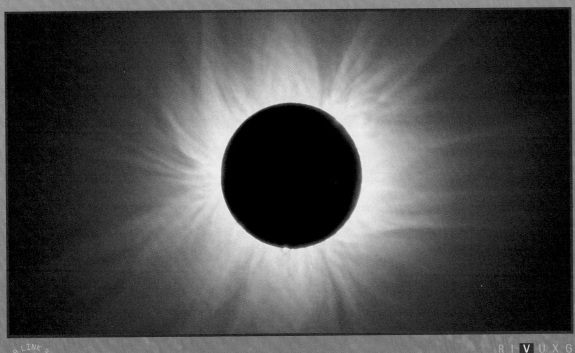

R I **V** U X G

(Fred Espenak, NASA/Goddard Space Flight Center; © 2000 by Fred Espenak, MrEclipse.com)

On August 11, 1999, millions of humans stood along a narrow corridor of land that extended from England through Europe and the Middle East to India. For a few brief minutes, they saw the rare cosmic spectacle shown here: a total solar eclipse. As the Moon covered the disk of the Sun, the stars and planets became visible in the darkening sky, and the solar corona—the Sun's thin outer atmosphere, which glows with an unearthly pearlescent light—was revealed.

Not everyone will ever see the Moon cover the Sun in this way. But anyone can find the Moon in the sky and observe how its appearance changes from night to night, from new moon to full moon and back again. The times when the Moon rises and sets also differ noticeably from one night to the next.

In this chapter our subject is how the Moon moves as seen from the Earth. We will explore why the Moon goes through a regular cycle of phases, and how the Moon's orbit around the Earth leads to solar eclipses as well as lunar eclipses. We will also see how ancient astronomers used their observations of the Moon to determine the size and shape of the Earth, as well as other features of the solar system. Thus, the Moon—which has always loomed large in the minds of poets, lovers, and dreamers—has also played a key role in the development of our modern picture of the universe.

As you read the sections of this chapter, look for the answers to the following questions.

3-1   Why does the Moon go through phases?

3-2   Is there such a thing as the "dark side of the Moon"?

3-3   What is the difference between a lunar eclipse and a solar eclipse?

3-4   How often do lunar eclipses happen? When one is taking place, where do you have to be to see it?

3-5   How often do solar eclipses happen? Why are they visible only from certain special locations on Earth?

3-6   How did ancient astronomers deduce the sizes of the Earth, the Moon, and the Sun?

3-1   The phases of the Moon are caused by its orbital motion

As seen from Earth, both the Sun and the Moon appear to move from west to east on the celestial sphere—that is, relative to the background of stars. They move at very different rates, however. The Sun takes one year to make a complete trip around the imaginary celestial sphere along the path we call the *ecliptic* (Section 2-5). By comparison, the Moon takes only about four weeks. In the past, these similar motions led people to believe that both the Sun and the Moon orbit around the Earth. We now know that only the Moon orbits the Earth, while the Earth-Moon system as a whole (Figure 3-1) orbits the Sun. (In Chapter 4 we will learn how this was discovered.)

One key difference between the Sun and the Moon is the nature of the light that we receive from them. The Sun emits its own light. So do the stars, which are objects like the Sun but much farther away, and so does an ordinary lightbulb. By contrast, the light that we see from the Moon is reflected light. This is sunlight that has struck the Moon's surface, bounced off, and ended up in our eyes here on Earth.

You probably associate *reflection* with shiny objects like a mirror or the surface of a still lake. In science, however, the term refers to light bouncing off any object. You see most objects around you by reflected light. When you look at your hand, for example, you are seeing light from the Sun (or from a light fixture) that has been reflected from the skin of your hand and into your eye. In the same way, moonlight is really sunlight that has been reflected by the Moon's surface.

Figure 3-1 shows both the Moon and the Earth as seen from a spacecraft. When this picture was recorded, the Sun was far off to the right. Hence, only the right-hand hemispheres of both worlds were illuminated by the Sun; the left-hand hemispheres were in darkness and are not visible in the picture. In the same way, when we view the Moon from the Earth, we see only the half of the Moon that faces the Sun and

is illuminated. However, not all of the illuminated half of the Moon is necessarily facing us. As the Moon moves around the Earth, from one night to the next we see different amounts of the illuminated half of the Moon. These different appearances of the Moon are called **lunar phases.**

ƒigure 3-1   R I V U X G

The Earth and the Moon This picture of the Earth and the Moon was taken in 1992 by the *Galileo* spacecraft on its way toward Jupiter. The Sun, which provides the illumination for both the Earth and the Moon, was far to the right and out of the camera's field of view when this photograph was taken. (NASA/JPL)

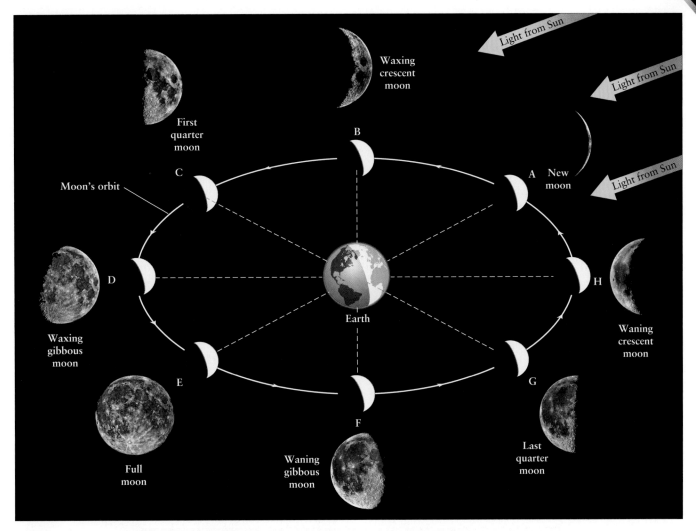

 figure 3-2

Why the Moon Goes Through Phases This figure shows the Moon at eight positions on its orbit, along with photographs of what the Moon looks like at each position as seen from Earth. The changes in phase occur because light from the Sun illuminates one half of the Moon, and as the Moon orbits the Earth we see varying amounts of the Moon's illuminated half. It takes about 29½ days for the Moon to go through a complete cycle of phases. (Photographs from Lick Observatory)

Figure 3-2 shows the relationship between the lunar phase visible from Earth and the position of the Moon in its orbit. For example, when the Moon is at position A, we see it in roughly the same direction in the sky as the Sun. Hence, the dark hemisphere of the Moon faces the Earth. This phase, in which the Moon is not visible, is called **new moon.**

As the Moon continues around its orbit from position A in Figure 3-2, more of its illuminated half becomes exposed to our view. The result, shown at position B, is a phase called **waxing crescent moon** ("waxing" is a synonym for "increasing"). About a week after new moon, the Moon is at position C; we then see half of the Moon's illuminated hemisphere and half of the dark hemisphere. This phase is called **first quarter moon.**

Despite the name, a first quarter moon appears to be *half* illuminated, not one-quarter illuminated! The name means that this phase is one-quarter of the way through the complete cycle of lunar phases.

About four days later, the Moon reaches position D in Figure 3-2. Still more of the illuminated hemisphere can now be seen from Earth, giving us the phase called **waxing gibbous moon** ("gibbous" is another word for "swollen"). When you look at the Moon in this phase, as in the waxing crescent and first quarter phases, the illuminated part of the Moon is on the right (toward the west). Two weeks after new moon, when the Moon stands opposite the Sun in the sky (position E), we see the fully illuminated hemisphere. This phase is called **full moon.**

box 3-1 | **The Heavens on the Earth**

Phases and Shadows

Figure 3-2 shows how the relative positions of the Earth, Moon, and Sun explain the phases of the Moon. You can visualize lunar phases more clearly by doing a simple experiment here on Earth. All you need are a small round object, such as an orange or a baseball, and a bright source of light, such as a street lamp or the Sun.

In this experiment, you play the role of an observer on the Earth looking at the Moon, and the round object plays the role of the Moon. The light source plays the role of the Sun. Hold the object in your right hand with your right arm stretched straight out in front of you, with the object directly between you and the light source (position A in the accompanying illustration). In this orientation, the illuminated half of the object faces away from you, like the Moon when it is its new phase (position A in Figure 3-2). By analogy, the round object in your hand is in its "new" phase when you hold it between your eyes and the light source.

Now, slowly turn your body to the left so that the object in your hand "orbits" around you (toward positions C, E, and G in the illustration). As you turn, more and more of the illuminated side of the "moon" in your hand will become visible, and it will appear to go through the same cycle of phases—waxing crescent, first quarter, and waxing gibbous—as does the real Moon. When you have rotated

through half a turn so that the light source is directly behind you, you will be looking face on at the illuminated side of the object in your hand. This corresponds to a full moon (position E in Figure 3-2). Make sure your body does not cast a shadow on the "moon" in your hand—that would correspond to a lunar eclipse!

As you continue turning to the left, more of the unilluminated half of the object will become visible as its phase moves through waning gibbous, last quarter, and waning crescent. When your body has rotated back to the same orientation that you were in originally, the unilluminated half of your hand-held "moon" will again be facing toward you, and its phase will again be new. If you continued to rotate, the object in your hand would repeat the cycle of "phases," just as the Moon does as it orbits around the Earth.

The experiment works best when there is just one light source around. If there are several light sources, such as in a room with several lamps turned on, the different sources will create multiple shadows, and it will be difficult to see the phases of your hand-held "moon." If you do the experiment outdoors using sunlight, you may find that it is best to perform it in the early morning or late afternoon when shadows are most pronounced and the Sun's rays are nearly horizontal.

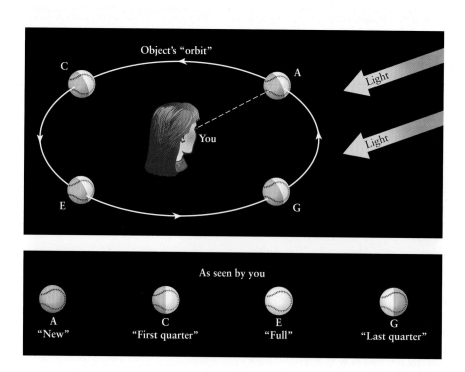

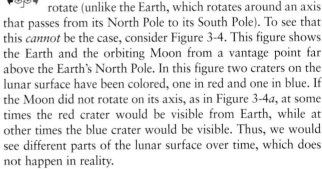

figure 3-3   R I **V** U X G

The Moon During the Day The Moon can be seen during the daytime as well as at night. The time of day or night when it is visible depends on its phase. When the Moon is full, as shown in this photograph, it is in the part of the sky directly opposite the Sun (see position E in Figure 3-2). Thus, a full moon rises on the eastern horizon when the Sun is setting in the west, and sets on the western horizon when the Sun is rising in the east. (Art Wolfe)

Over the following two weeks, we see less and less of the Moon's illuminated hemisphere as it continues along its orbit. The Moon is said to be *waning*, or decreasing in illumination as seen from Earth. While the Moon is waning, its illuminated side is on the left (toward the east). The phases are called **waning gibbous moon** (position F), **last quarter moon** (position G), and **waning crescent moon** (position H). The Moon takes about four weeks to complete one orbit around the Earth, so it likewise takes about four weeks for a complete cycle of phases from new moon to full moon and back to new moon.

Figure 3-2 also explains why the Moon is often visible in the daytime, as shown in Figure 3-3. From any location on Earth, about half of the Moon's orbit is visible at any time. For example, if it is midnight at your location, you are in the middle of the dark side of the Earth that faces away from the Sun. At that time you can easily see the Moon at positions C, D, E, F, or G. If it is midday at your location, you are in the middle of the Earth's illuminated side, and the Moon will be easily visible if it is at positions A, B, C, G, or H. (The Moon is so bright that it can be seen even against the bright blue sky.) You can see that the Moon is prominent in the midnight sky for about half of its orbit, and prominent in the midday sky for the other half.

CAUTION! A very common misconception about lunar phases is that they are caused by the shadow of the *Earth* falling on the Moon. As Figure 3-2 shows, this is not the case at all. Instead, phases are simply the result of our seeing the illuminated half of the Moon at different angles as the Moon moves around its orbit. To help you better visualize how this works, Box 3-1 describes how you can simulate the cycle shown in Figure 3-2 using ordinary objects on Earth. (As we will learn in Section 3-3, the Earth's shadow does indeed fall on the Moon on rare occasions. When this happens, we see a lunar eclipse.)

3-2   The Moon's rotation always keeps the same face toward the Earth

The phase of the Moon is continuously changing. But one constant aspect of the Moon is that it always keeps essentially the same hemisphere, or face, toward the Earth. Thus, you will always see the same craters and mountains on the Moon, no matter when you look at it; the only difference will be the angle at which these surface features are illuminated by the Sun. (You can verify this by carefully examining the photographs of the Moon in Figure 3-2.)

CAUTION! Why is it that we only ever see one face of the Moon? You might think that it is because the Moon does not rotate (unlike the Earth, which rotates around an axis that passes from its North Pole to its South Pole). To see that this *cannot* be the case, consider Figure 3-4. This figure shows the Earth and the orbiting Moon from a vantage point far above the Earth's North Pole. In this figure two craters on the lunar surface have been colored, one in red and one in blue. If the Moon did not rotate on its axis, as in Figure 3-4a, at some times the red crater would be visible from Earth, while at other times the blue crater would be visible. Thus, we would see different parts of the lunar surface over time, which does not happen in reality.

In fact, the Moon always keeps the same face toward us because it *is* rotating, but in a very special way: It takes exactly as long to rotate on its axis as it does to make one orbit around the Earth. This situation is called **synchronous rotation**. As Figure 3-4b shows, this keeps the crater shown in red always facing the Earth, so that we always see the same face

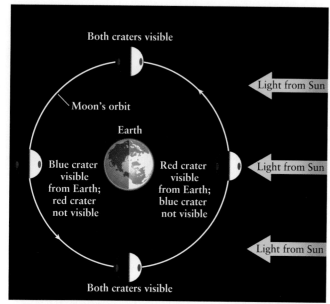

a If the Moon did not rotate, we could see all sides of the Moon

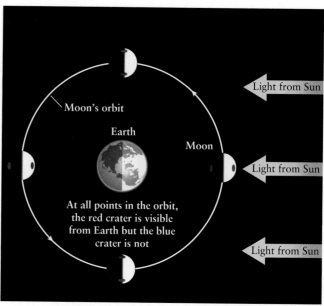

b In fact the Moon does rotate and we see only one face of the Moon

figure 3-4

The Moon's Rotation These diagrams show the Moon at four points in its orbit as viewed from high above the Earth's north pole. **(a)** If the Moon did not rotate, then at various times the red crater would be visible from Earth while at other times the blue crater would be visible. Over a complete orbit, the entire surface of the Moon would be visible. **(b)** In reality, like the Earth, the Moon rotates on its north-south axis. Because it makes one rotation in exactly the same time that it makes one orbit around the Earth, we only see one face of the Moon.

of the Moon. In Chapter 4 we will learn why the Moon's rotation and orbital motion are in step with each other.

An astronaut standing at the spot shown in red in Figure 3-4*b* would spend two weeks (half of a lunar orbit) in darkness, or lunar nighttime, and the next two weeks in sunlight, or lunar daytime. Thus, as seen from the Moon, the Sun rises and sets, and no part of the Moon is perpetually in darkness. This means that there really is no "dark side of the Moon." The side of the Moon that constantly faces away from the Earth is properly called the *far* side. The Sun rises and sets on the far side just as on the side toward the Earth. Hence, the blue crater on the far side of the Moon in Figure 3-4*b* is in sunlight for half of each lunar orbit.

The time for a complete lunar "day"—the same as the time that it takes the Moon to rotate once on its axis—is about four weeks. (Remember that it takes the same time for one complete lunar orbit.) It also takes about four weeks for the Moon to complete one cycle of its phases as seen from Earth. This regular cycle of phases inspired our ancestors to invent the concept of a month. For historical reasons, none of which has much to do with the heavens, the calendar we use today has months of differing lengths. Astronomers find it useful to define two other types of months, depending on whether the Moon's motion is measured relative to the stars

or to the Sun. Neither corresponds exactly to the familiar months of the calendar.

The **sidereal month** is the time it takes the Moon to complete one full orbit of the Earth, as measured with respect to the stars. This true orbital period is equal to about 27.32 days. The **synodic month**, or **lunar month**, is the time it takes the Moon to complete one cycle of phases (that is, from new moon to new moon or from full moon to full moon) and thus is measured with respect to the Sun rather than the stars. The length of the "day" on the Moon is a synodic month, not a sidereal month.

The synodic month is longer than the sidereal month because the Earth is orbiting the Sun while the Moon goes through its phases. As Figure 3-5 shows, the Moon must travel *more* than 360° along its orbit to complete a cycle of phases (for example, from one new moon to the next). Because of this extra distance, the synodic month is equal to about 29.53 days, about two days longer than the sidereal month.

Both the sidereal month and synodic month vary somewhat from one orbit to another, the latter by as much as half a day. The reason is that the Sun's gravity sometimes causes the Moon to speed up or slow down slightly in its orbit, depending on the relative positions of the Sun, Moon, and Earth. Furthermore, the Moon's orbit changes slightly from one month to the next.

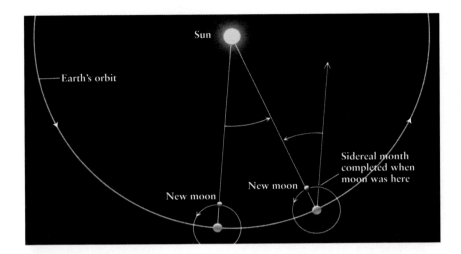

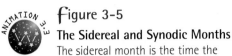 **figure 3-5**

The Sidereal and Synodic Months
The sidereal month is the time the Moon takes to complete one full revolution around the Earth with respect to the background stars. However, because the Earth is constantly moving along its orbit about the Sun, the Moon must travel through slightly more than 360° to get from one new moon to the next. Thus, the synodic month—the time from one new moon to the next—is longer than the sidereal month.

3-3 Eclipses occur only when the Sun and Moon are both on the line of nodes

From time to time the Sun, Earth, and Moon all happen to lie along a straight line. When this occurs, the shadow of the Earth can fall on the Moon or the shadow of the Moon can fall on the Earth. Such phenomena are called **eclipses.** They are perhaps the most dramatic astronomical events that can be seen with the naked eye.

A **lunar eclipse** occurs when the Moon passes through the Earth's shadow. This occurs when the Sun, Earth, and Moon are in a straight line, with the Earth between the Sun and Moon so that the Moon is at full phase (position E in Figure 3-2). At this point in the Moon's orbit, the face of the Moon seen from Earth would normally be fully illuminated by the Sun. Instead, it appears quite dim because the Earth casts a shadow on the Moon.

A **solar eclipse** occurs when the Earth passes through the Moon's shadow. As seen from Earth, the Moon moves in front of the Sun. Once again, this can happen only when the Sun, Moon, and Earth are in a straight line. However, for a solar eclipse to occur, the Moon must be between the Earth and the Sun. Therefore, a solar eclipse can occur only at new moon (position A in Figure 3-2).

 Both new moon and full moon occur at intervals of 29½ days. Hence, you might expect that there would be a solar eclipse every 29½ days, followed by a lunar eclipse about two weeks (half a lunar orbit) later. But in fact, there are only a few solar eclipses and lunar eclipses per year. Solar and lunar eclipses are so infrequent because the plane of the Moon's orbit and the plane of the Earth's orbit are not exactly aligned, as Figure 3-6 shows. The angle between the plane of the Earth's orbit and the plane of the Moon's orbit is about 5°. Because of this tilt, new moon and full moon usually occur when the Moon is either above or below the plane of the Earth's orbit. When the Moon is not in the plane of the Earth's orbit, the Sun, Moon, and Earth cannot align perfectly, and an eclipse cannot occur.

In order for the Sun, Earth, and Moon to be lined up for an eclipse, the Moon must lie in the same plane as the Earth's orbit around the Sun. This is called the **plane of the ecliptic** because its projection onto the celestial sphere is the same as the path that the Sun appears to us to follow among the stars—that is, the ecliptic (see Figure 2-14). Thus, whenever an eclipse occurs, the Moon appears from Earth to be on the ecliptic (which is how the ecliptic gets its name).

The planes of the Earth's orbit and the Moon's orbit intersect along a line called the **line of nodes,** shown in Figure 3-6. The line of nodes passes through the Earth and is pointed in a particular direction in space. Eclipses can occur only if the line of nodes is pointed toward the Sun—that is, if the Sun lies on or near the line of nodes—and if, at the same time, the Moon lies on or very near the line of nodes. Only then do the Sun,

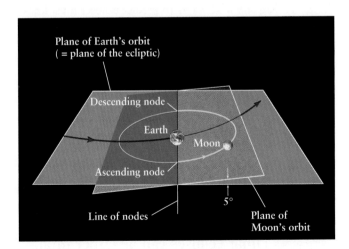

figure 3-6

The Inclination of the Moon's Orbit This drawing shows the Moon's orbit around the Earth (in yellow) and part of the Earth's orbit around the Sun (in red). The plane of the Moon's orbit (shown in brown) is tilted by about 5° with respect to the plane of the Earth's orbit, also called the plane of the ecliptic (shown in blue). These two planes intersect along a line called the line of nodes.

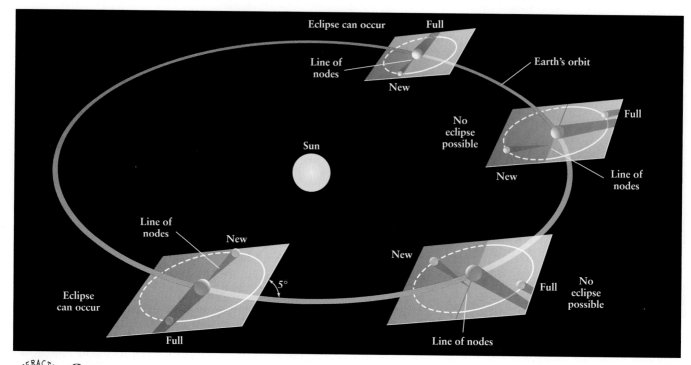

figure 3-7

Conditions for Eclipses Eclipses can take place only if the Sun and Moon are both very near or on the line of nodes. Only then can the Sun, Earth, and Moon all lie along a straight line. A solar eclipse occurs only if the Moon is very near the line of nodes at new moon; a lunar eclipse occurs only if the Moon is very near the line of nodes at full moon. If the Sun and Moon are not near the line of nodes, the Moon's shadow cannot fall on the Earth and the Earth's shadow cannot fall on the Moon.

Earth, and Moon lie in a line straight enough for an eclipse to occur (Figure 3-7).

Anyone who wants to predict eclipses must know the orientation of the line of nodes. But the line of nodes is gradually shifting because of the gravitational pull of the Sun on the Moon. As a result, the line of nodes rotates slowly westward. Astronomers calculate such details to fix the dates and times of upcoming eclipses.

There are at least two—but never more than five—solar eclipses each year. The last year in which five solar eclipses occurred was 1935. The least number of eclipses possible (two solar, zero lunar) happened in 1969. Lunar eclipses occur just about as frequently as solar eclipses, but the maximum possible number of eclipses (lunar and solar combined) in a single year is seven.

3-4 Lunar eclipses can be either total, partial, or penumbral, depending on the alignment of the Sun, Earth, and Moon

The character of a lunar eclipse depends on exactly how the Moon travels through the Earth's shadow. As Figure 3-8 shows, the shadow of the Earth has two distinct parts. In the **umbra**, the darkest part of the shadow, no portion of the

Sun's surface can be seen. A portion of the Sun's surface is visible in the **penumbra**, which therefore is not quite as dark. Most people notice a lunar eclipse only if the Moon passes into the Earth's umbra. As this umbral phase of the eclipse begins, a bite seems to be taken out of the Moon.

The inset in Figure 3-8 shows the different ways in which the Moon can pass into the Earth's shadow. When the Moon passes through only the Earth's penumbra (Path 1), we see a **penumbral eclipse**. During a penumbral eclipse, the Earth blocks only part of the Sun's light and so none of the lunar surface is completely shaded. Because the Moon still looks full but only a little dimmer than usual, penumbral eclipses are easy to miss. If the Moon travels completely into the umbra (Path 2), a **total lunar eclipse** occurs. If only part of the Moon passes through the umbra (Path 3), we see a **partial lunar eclipse**.

If you were on the Moon during a total lunar eclipse, the Sun would be hidden behind the Earth. But some sunlight would be visible through the thin ring of atmosphere around the Earth, just as you can see sunlight through a person's hair if they stand with their head between your eyes and the Sun. As a result, a small amount of light reaches the Moon during a total lunar eclipse, and so the Moon does not completely disappear from the sky as seen from Earth. Most of the light that passes through the Earth's atmosphere is red, and thus the eclipsed Moon glows faintly in reddish hues (Figure 3-9).

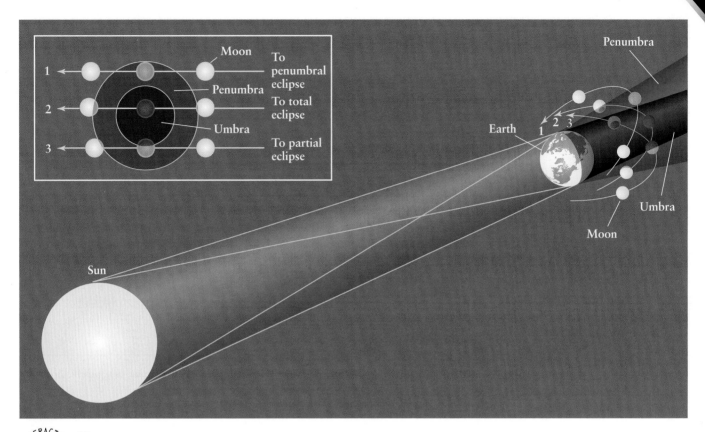

![INTERACTIVE EXERCISE 3.3] **figure 3-8**

Three Types of Lunar Eclipse People on the nighttime side of the Earth see a lunar eclipse when the Moon moves through the Earth's shadow. In the umbra, the darkest part of the shadow, the Sun is completely covered by the Earth. The penumbra is less dark because only part of the Sun is covered by the Earth. The three paths show the motion of the Moon if the lunar eclipse is penumbral (Path 1), total (Path 2), or partial (Path 3). The inset shows these same paths, along with the umbra and penumbra, as viewed from the Earth.

figure 3-9   R I **V** U X G

A Total Lunar Eclipse This sequence of nine photographs was taken over a 3-hour period during the lunar eclipse of January 20, 2000. The sequence, which runs from right to left, shows the Moon moving through the Earth's umbra. During the total phase of the eclipse (shown in the center), the Moon has a distinct reddish color. (Fred Espenak, NASA/Goddard Space Flight Center; ©2000 Fred Espenak, MrEclipse.com)

table 3-1	Lunar Eclipses, 2001–2005		
Date	Type	Where visible	Duration of totality (h = hours, m = minutes)
2001 January 9	Total	Eastern Americas, Europe, Africa, Asia	1h 2m
2001 July 5	Partial	Eastern Africa, Asia, Australia, Pacific	—
2001 December 30	Penumbral	Eastern Asia, Australia, Americas	—
2002 May 26	Penumbral	Eastern Asia, Australia, Pacific, western Americas	—
2002 June 24	Penumbral	South America, Europe, Africa, central Asia, Australia	—
2002 November 20	Penumbral	Americas, Europe, Africa, eastern Asia	—
2003 May 16	Total	Central Pacific, Americas, Europe, Africa	53m
2003 November 9	Total	Americas, Europe, Africa, central Asia	24m
2004 May 4	Total	South America, Europe, Africa, Asia, Australia	1h 16m
2004 October 28	Total	Americas, Europe, Africa, central Asia	1h 21m
2005 April 24	Penumbral	Eastern Asia, Australia, Pacific, Americas	—
2005 October 17	Partial	Asia, Australia, Pacific, North America	—

Eclipse predictions by Fred Espenak, NASA/Goddard Space Flight Center. All dates are given in standard astronomical format: year, month, day.

Lunar eclipses occur at full moon, when the Moon is directly opposite the Sun in the sky. Hence, a lunar eclipse can be seen at any place on Earth where the Sun is below the horizon (that is, where it is nighttime). A lunar eclipse has the maximum possible duration if the Moon travels directly through the center of the umbra. The Moon's speed through the Earth's shadow is roughly 1 kilometer per second (3600 kilometers per hour, or 2280 miles per hour), which means that **totality**—the period when the Moon is completely within the Earth's umbra—can last for as long as 1 hour and 42 minutes.

 On average, two or three lunar eclipses occur in a year. Table 3-1 lists all twelve lunar eclipses from 2001 to 2005. As the table shows, about half of all lunar eclipses are penumbral.

3-5 Solar eclipses can be either total, partial, or annular, depending on the alignment of the Sun, Earth, and Moon

As seen from Earth, the angular diameter of the Moon is almost exactly the same as the angular diameter of the far larger but more distant Sun—about 0.5°. Thanks to this coincidence of nature, the Moon just "fits" over the Sun during a **total solar eclipse.**

A total solar eclipse is a dramatic event. The sky begins to darken, the air temperature falls, and winds increase as the Moon gradually covers more and more of the Sun's disk (Figure 3-10). All nature responds: Birds go to roost, flowers close their petals, and crickets begin to chirp as if evening had arrived. As the last few rays of sunlight peek out from behind the edge of the Moon and the eclipse becomes total,

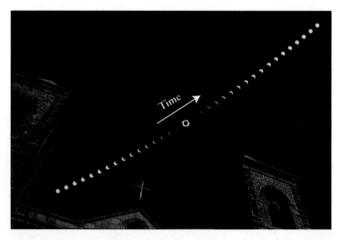

figure 3-10 R I **V** U X G

A Total Solar Eclipse This multiple-exposure photograph shows the stages of the total solar eclipse of July 11, 1991, as seen from La Paz, Baja California. The individual exposures were taken at 5-minute intervals as the Sun and Moon moved across the sky from left to right. Notice how the Moon progressively covers then uncovers the Sun. At the center of the photograph, the Sun is totally eclipsed and the solar corona can be seen surrounding the Moon. (Akira Fujii, Hiroyuki Tomioka, and Yonematsu Shiono)

the landscape around you is bathed in an eerie gray or, less frequently, in shimmering bands of light and dark. Finally, for a few minutes the Moon completely blocks out the dazzling solar disk and not much else. The **solar corona**—the Sun's thin, hot outer atmosphere, which is normally too dim to be seen—blazes forth in the darkened daytime sky (see the photo that opens this chapter). It is an awe-inspiring sight.

If you are fortunate enough to see a solar eclipse, keep in mind that the only time when it is safe to look at the Sun is during **totality,** when the solar disk is blocked by the Moon and only the solar corona is visible. Viewing this magnificent spectacle cannot harm you in any way. But you must *never* look directly at the Sun when even a portion of its intensely brilliant disk is exposed. *If you look directly at the Sun at any time without a special filter approved for solar viewing, you will suffer permanent eye damage or blindness.*

To see the remarkable spectacle of a total solar eclipse, you must be inside the darkest part of the Moon's shadow, also called the umbra, where the Moon completely blocks the Sun. Because the Sun and the Moon have nearly the same

angular diameter as seen from Earth, only the tip of the Moon's umbra reaches the Earth's surface (Figure 3-11). As the Earth rotates, the tip of the umbra traces an **eclipse path** across the Earth's surface. Only those locations within the eclipse path are treated to the spectacle of a total solar eclipse. The inset in Figure 3-11 shows the dark spot on the Earth's surface produced by the Moon's umbra.

Immediately surrounding the Moon's umbra is the region of partial shadow called the penumbra. As seen from this area, the Sun's surface appears only partially covered by the Moon. During a solar eclipse, the Moon's penumbra covers a large portion of the Earth's surface, and anyone standing inside the penumbra sees a **partial solar eclipse.** Such eclipses are much less interesting events than total solar eclipses, which is why astronomy enthusiasts strive to be inside the eclipse path. If you are within the eclipse path, you will see a partial eclipse before and after the brief period of totality (see Figure 3-10).

The width of the eclipse path depends primarily on the Earth–Moon distance during totality. The eclipse path is widest if the Moon happens to be at **perigee**, the point in its orbit nearest the Earth. In this case the width of the eclipse

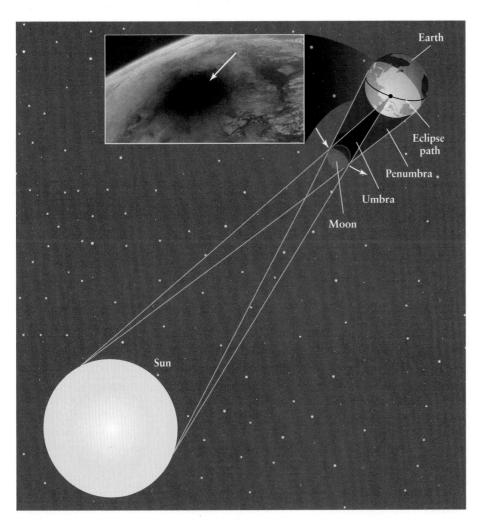

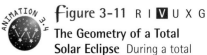

figure 3-11 R I **V** U X G

The Geometry of a Total Solar Eclipse During a total solar eclipse, the tip of the Moon's umbra reaches the Earth's surface. As the Earth and Moon move along their orbits, this tip traces an eclipse path across the Earth's surface. People within the eclipse path see a total solar eclipse as the tip moves over them. Anyone within the penumbra sees only a partial eclipse. The inset photograph was taken from the *Mir* space station during the August 11, 1999, total solar eclipse (the same eclipse shown in the photo that opens this chapter). The tip of the umbra appears as a black spot on the Earth's surface. At the time the photograph was taken, this spot was 105 km (65 mi) wide and was crossing the English Channel at 3000 km/h (1900 mi/h). (Photograph by Jean-Pierre Haigneré, Centre National d'Etudes Spatiales, France/GSFS)

path can be as great as 270 kilometers (170 miles). In most eclipses, however, the path is much narrower.

In some eclipses the Moon's umbra does not reach all the way to the Earth's surface. This can happen if the Moon is at or near **apogee,** its farthest position from Earth. In this case, the Moon appears too small to cover the Sun completely. The result is a third type of solar eclipse, called an **annular eclipse.** During an annular eclipse, a thin ring of the Sun is seen around the edge of the Moon (Figure 3-12). The length of the Moon's umbra is nearly 5000 kilometers (3100 miles) less than the average distance between the Moon and the Earth's surface. Thus, the Moon's shadow often fails to reach the Earth even when the Sun, Moon, and Earth are properly aligned for an eclipse. Hence, annular eclipses are slightly more common—as well as far less dramatic—than total eclipses.

Even during a total eclipse, most people along the eclipse path observe totality for only a few moments. The Earth's rotation, coupled with the orbital motion of the Moon, causes the umbra to race eastward along the eclipse path at speeds in excess of 1700 kilometers per hour (1060 miles per hour). Because of the umbra's high speed, totality never lasts for more than 7½ minutes. In a typical total solar eclipse, the Sun-Moon-Earth alignment and the Earth-Moon distance are such that totality lasts much less than this maximum.

 The details of solar eclipses are calculated well in advance. They are published in such reference books as the *Astronomical Almanac* and are available on the World Wide Web. Figure 3-13 shows the eclipse paths for all total solar eclipses from 1997 to 2020. Table 3-2 lists all the total, annular, and partial eclipses from 2001 to 2005, including the maximum duration of totality for total eclipses.

ꜰigure 3-12   R I **V** U X G

An Annular Solar Eclipse This composite of six photographs taken at sunrise in Costa Rica shows the progress of an annular eclipse of the Sun on December 24, 1973. (Five photographs were made of the Sun, plus one of the hills and sky.) Note that at mideclipse the limb, or outer edge, of the Sun is visible around the Moon. (Courtesy of Dennis di Cicco)

ꜰable 3-2	**Solar Eclipses, 2001–2005**		
Date	**Type**	**Where visible**	**Notes**
2001 June 21	Total	Eastern South America, southern Atlantic, Africa	Maximum duration of totality 4 m 57 s
2001 December 14	Annular	North and Central America, northwest South America	—
2002 June 10	Annular	Eastern Asia, Australia, western North America	—
2002 December 4	Total	Southern Africa, Antarctica, Indonesia, Australia	Maximum duration of totality 2 m 4 s
2003 May 31	Annular	Europe, Asia, northwest North America	—
2003 November 23	Total	Australia, New Zealand, Antarctica, southern South America	Maximum duration of totality 1 m 57 s
2004 April 19	Partial	Antarctica, southern Africa	74% eclipsed
2004 October 14	Partial	Northeast Asia, Hawaii, Alaska	93% eclipsed
2005 April 8	Annular and Total	New Zealand, North and South America	Annular along part of path; maximum duration of totality 0 m 42 s
2005 October 3	Annular	Europe, Africa, southern Asia	—

Eclipse predictions by Fred Espenak, NASA/Goddard Space Flight Center. All dates are given in standard astronomical format: year, month, day.

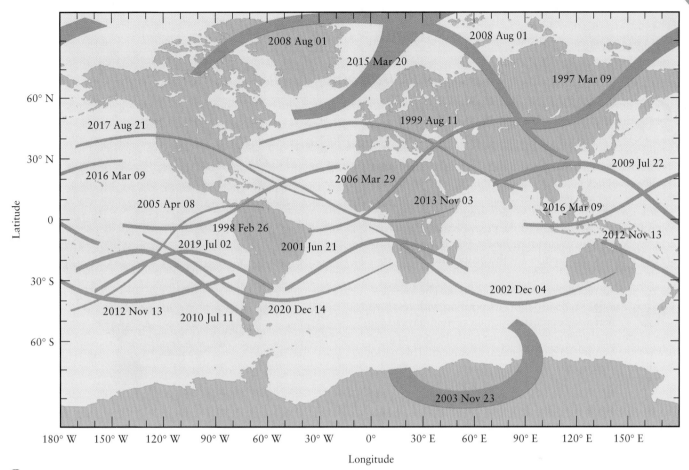

figure 3-13

Eclipse Paths for Total Eclipses, 1997–2020 This map shows the eclipse paths for all 18 total solar eclipses occurring from 1997 through 2020. In each eclipse, the Moon's shadow travels along the eclipse path in a generally eastward direction across the Earth's surface. (Courtesy of Fred Espenak, NASA/Goddard Space Flight Center)

Ancient astronomers achieved a limited ability to predict eclipses. In those times, religious and political leaders who were able to predict such awe-inspiring events as eclipses must have made a tremendous impression on their followers. One of three priceless manuscripts to survive the devastating Spanish Conquest shows that the Mayan astronomers of Mexico and Guatemala had a fairly reliable method for predicting eclipses. The great Greek astronomer Thales of Miletus is said to have predicted the famous eclipse of 585 B.C., which occurred during the middle of a war. The sight was so unnerving that the soldiers put down their arms and declared peace.

In retrospect, it seems that what ancient astronomers actually produced were eclipse "warnings" of various degrees of reliability rather than true predictions. Working with historical records, these astronomers generally sought to discover cycles and regularities from which future eclipses could be anticipated. To see how you might produce eclipse warnings yourself, read Box 3-2.

3-6 Ancient astronomers measured the size of the Earth and attempted to determine distances to the Sun and Moon

The prediction of eclipses was not the only problem attacked by ancient astronomers. More than 2000 years ago, centuries before sailors of Columbus's era crossed the oceans, Greek astronomers were fully aware that the Earth is not flat. They had come to this conclusion using a combination of observation and logical deduction, much like modern scientists. The Greeks noted that during lunar eclipses, when the Moon passes through the Earth's shadow, the edge of the shadow is always circular. Because a sphere is the only shape that always casts a circular shadow from any angle, they concluded that the Earth is spherical.

Around 200 B.C., the Greek astronomer Eratosthenes devised a way to measure the circumference of the spherical

box 3-2 Tools of the Astronomer's Trade

Predicting Solar Eclipses

Suppose that you observe a solar eclipse in your hometown and want to figure out when you and your neighbors might see another eclipse. How would you begin?

First, remember that a solar eclipse can occur only if the line of nodes points toward the Sun at the same time that there is a new moon (see Figure 3-7). Second, you must know that it takes 29.53 days (one synodic month) to go from one new moon to the next. Because solar eclipses occur only during new moon, you must wait several whole lunar months for the proper alignment to occur again.

 However, there is a complication: The line of nodes gradually shifts its position with respect to the background stars. It takes 346.6 days to move from one alignment of the line of nodes pointing toward the Sun to the next identical alignment. This period is called the **eclipse year.**

Therefore, to predict when you will see another solar eclipse, you need to know how many whole lunar months equal some whole number of eclipse years. This will tell you how long you will have to wait for the next virtually identical alignment of the Sun, the Moon, and the line of nodes. By trial and error, you find that 223 lunar months is the same length of time as 19 eclipse years, because

$$223 \times 29.53 \text{ days} = 19 \times 346.6 \text{ days} = 6585 \text{ days}$$

This calculation is accurate to within a few hours. A more accurate calculation gives an interval, called the **saros,** that is about one-third day longer, or 6585.3 days (18 years, 11.3 days). Eclipses separated by the saros interval are said to form an *eclipse series*.

You might think that you and your neighbors would simply have to wait one full saros interval to go from one solar eclipse to the next. However, because of the extra one-third day, the Earth will have rotated by an extra 120°

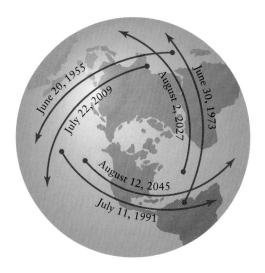

(one-third of a complete rotation) when the next solar eclipse of a particular series occurs. The eclipse path will thus be one-third of the way around the world from you. Therefore, you must wait three full saros intervals (54 years, 34 days) before the eclipse path comes back around to your part of the Earth. The illustration shows a series of six solar eclipse paths, each separated from the next by one saros interval.

There is evidence that ancient Babylonian astronomers knew about the saros interval. However, the discovery of the saros is more likely to have come from lunar eclipses than solar eclipses. If you are far from the eclipse path, there is a good chance that you could fail to notice a solar eclipse. Even if half the Sun is covered by the Moon, the remaining solar surface provides enough sunlight for the outdoor illumination not to be greatly diminished. By contrast, anyone on the nighttime side of the Earth can see an eclipse of the Moon unless clouds block the view.

Earth. It was known that on the date of the summer solstice (the first day of summer; see Section 2-5) in the town of Syene in Egypt, near present-day Aswan, the Sun shone directly down the vertical shafts of water wells. Hence, at local noon on that day the Sun was at the zenith (see Section 2-4) as seen from Syene. Eratosthenes knew that the Sun never appeared at the zenith at his home in the Egyptian city of Alexandria, which is on the Mediterranean Sea almost due north of Syene. Rather, on the summer solstice in Alexandria, the posi-

tion of the Sun at local noon was about 7° south of the zenith (Figure 3-14). This angle is about one-fiftieth of a complete circle, so he concluded that the distance from Alexandria to Syene must be about one-fiftieth of the Earth's circumference.

In Eratosthenes's day, the distance from Alexandria to Syene was said to be 5000 stades. Therefore, Eratosthenes found the Earth's circumference to be

$$50 \times 5000 \text{ stades} = 250,000 \text{ stades}$$

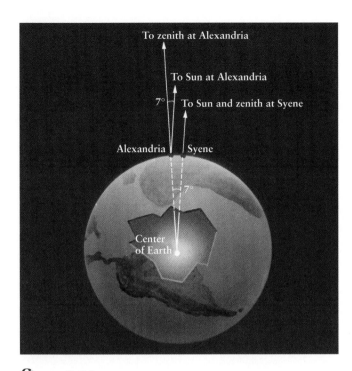

figure 3-14

Eratosthenes's Method of Determining the Diameter of the Earth Around 200 B.C., Eratosthenes noticed that the Sun is about 7° south of the zenith at Alexandria when it is directly overhead at Syene. This angle is about one-fiftieth of a circle, so the distance between Alexandria and Syene must be about one-fiftieth of the Earth's circumference.

Unfortunately, no one today is sure of the exact length of the Greek unit called the stade. One guess is that the stade was about one-sixth of a kilometer, which would mean that Eratosthenes obtained a circumference for the Earth of about 42,000 kilometers. This is remarkably close to the modern value of 40,000 kilometers.

Eratosthenes was only one of several brilliant astronomers to emerge from the so-called Alexandrian school, which by his time had a distinguished tradition. One of the first Alexandrian astronomers, Aristarchus of Samos, had proposed a method of determining the relative distances to the Sun and Moon, perhaps as long ago as 280 B.C.

Aristarchus knew that the Sun, Moon, and Earth form a right triangle at the moment of first or last quarter moon, with the right angle at the location of the Moon (Figure 3-15). He estimated that, as seen from Earth, the angle between the Moon and the Sun at first and last quarters is 87°, or 3° less than a right angle. Using the rules of geometry, Aristarchus concluded that the Sun is about 20 times farther from us than is the Moon. We now know that Aristarchus erred in measuring angles and that the average distance to the Sun is about 390 times larger than the average distance to the Moon. It is nevertheless impressive that people were trying to measure distances across the solar system more than 2000 years ago.

Aristarchus also made an equally bold attempt to determine the relative sizes of the Earth, Moon, and Sun. From his observations of how long the Moon takes to move through the Earth's shadow during a lunar eclipse, Aristarchus estimated the diameter of the Earth to be about 3 times larger

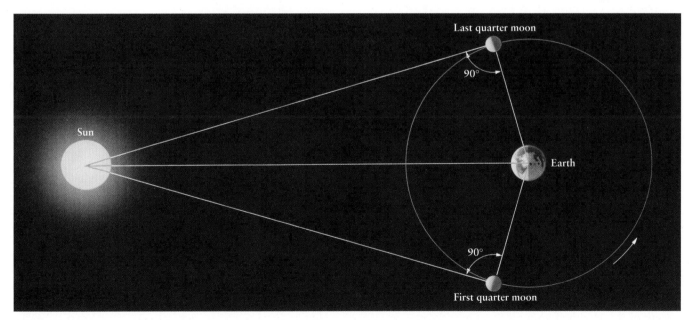

figure 3-15

Aristarchus's Method of Determining Distances to the Sun and Moon Aristarchus knew that the Sun, Moon, and Earth form a right triangle at first and last quarter phases. Using geometrical arguments, he calculated the relative lengths of the sides of these triangles, thereby obtaining the distances to the Sun and Moon.

than the diameter of the Moon. To determine the diameter of the Sun, Aristarchus simply pointed out that the Sun and the Moon have the same angular size in the sky. Therefore, their diameters must be in the same proportion as their distances (see part *a* of the figure in Box 1-1). In other words, because Aristarchus thought the Sun to be 20 times farther from the Earth than the Moon, he concluded that the Sun must be 20 times larger than the Moon. Once Eratosthenes had measured the Earth's circumference, astronomers of the Alexandrian school could estimate the diameters of the Sun and Moon as well as their distances from Earth.

Table 3-3 summarizes some ancient and modern measurements of the sizes of Earth, the Moon, and the Sun and the distances between them. Some of these ancient measurements are far from the modern values. Yet the achievements of our ancestors still stand as impressive applications of observation and reasoning and important steps toward the development of the scientific method.

table 3-3 — Comparison of Ancient and Modern Astronomical Measurements

	Ancient measure (km)	Modern measure (km)
Earth's diameter	13,000	12,756
Moon's diameter	4,300	3,476
Sun's diameter	9×10^4	1.39×10^6
Earth-Moon distance	4×10^5	3.84×10^5
Earth-Sun distance	10^7	1.50×10^8

KEY WORDS

Terms preceded by an asterisk (*) are discussed in the Boxes.

annular eclipse, p. 54

apogee, p. 54

eclipse, p. 49

eclipse path, p. 53

*eclipse year, p. 56

first quarter moon, p. 45

full moon, p. 45

last quarter moon, p. 47

line of nodes, p. 49

lunar eclipse, p. 49

lunar month, p. 48

lunar phases, p. 44

new moon, p. 45

partial lunar eclipse, p. 50

partial solar eclipse, p. 53

penumbra (*plural* penumbrae), p. 50

penumbral eclipse, p. 50

perigee, p. 53

plane of the ecliptic, p. 49

*saros, p. 56

sidereal month, p. 48

solar corona, p. 53

solar eclipse, p. 49

synchronous rotation, p. 47

synodic month, p. 48

totality (lunar eclipse), p. 52

totality (solar eclipse), p. 53

total lunar eclipse, p. 50

total solar eclipse, p. 52

umbra (*plural* umbrae), p. 50

waning crescent moon, p. 47

waning gibbous moon, p. 47

waxing crescent moon, p. 45

waxing gibbous moon, p. 45

KEY IDEAS

Lunar Phases: The phases of the Moon occur because light from the Moon is actually reflected sunlight. As the relative positions of the Earth, the Moon, and the Sun change, we see more or less of the illuminated half of the Moon.

Length of the Month: Two types of months are used in describing the motion of the Moon.

• With respect to the stars, the Moon completes one orbit around the Earth in a sidereal month, averaging 27.32 days.

• The Moon completes one cycle of phases (one orbit around the Earth with respect to the Sun) in a synodic month (also called a lunar month), averaging 29.53 days.

The Moon's Orbit: The plane of the Moon's orbit is tilted by about 5° from the plane of the Earth's orbit, or ecliptic.

• The line of nodes is the line where the planes of the Moon's orbit and the Earth's orbit intersect. The gravitational pull of the Sun gradually shifts the orientation of the line of nodes with respect to the stars.

Conditions for Eclipses: During a lunar eclipse, the Moon moves through the Earth's shadow. During a solar eclipse, the Earth passes through the Moon's shadow.

• Lunar eclipses occur at full moon, while solar eclipses occur at new moon.

• Either type of eclipse can occur only when the Sun and Moon are both on or very near the line of nodes. If this condition is not met, the Earth's shadow cannot fall on the Moon and the Moon's shadow cannot fall on the Earth.

Umbra and Penumbra: The shadow of an object has two parts: the umbra, within which the light source is completely blocked, and the penumbra, where the light source is only partially blocked.

Lunar Eclipses: Depending on the relative positions of the Sun, Moon, and Earth, lunar eclipses may be total (the Moon passes completely into the Earth's umbra), partial (only part of the Moon passes into the Earth's umbra), or penumbral (the Moon passes only into the Earth's penumbra).

Solar Eclipses: Solar eclipses may be total, partial, or annular.

• During a total solar eclipse, the Moon's umbra traces out an eclipse path over the Earth's surface as the Earth rotates. Observers outside the eclipse path but within the penumbra see only a partial solar eclipse.

• During an annular eclipse, the umbra falls short of the Earth, and the outer edge of the Sun's disk is visible around the Moon at mideclipse.

The Moon and Ancient Astronomers: Ancient astronomers such as Aristarchus and Eratosthenes made great progress in determining the sizes and relative distances of the Earth, the Moon, and the Sun.

REVIEW QUESTIONS

Questions preceded by an asterisk (*) involve topics discussed in the Boxes.

1. (a) Explain why the Moon exhibits phases. (b) A common misconception about the Moon's phases is that they are caused by the Earth's shadow. Use Figure 3-2 to explain why this not correct.

2. Suppose it is the first day of autumn in the northern hemisphere. What is the phase of the Moon if the Moon is located at (a) the vernal equinox? (b) the summer solstice? (c) the autumnal equinox? (d) the winter solstice? (*Hint:* Make a drawing showing the relative positions of the Sun, Earth, and Moon. Compare with Figure 3-2.)

3. How would the sequence and timing of lunar phases be affected if the Moon moved around its orbit (a) in the same direction, but at twice the speed; (b) at the same speed, but in the opposite direction? Explain your answers.

4. Is the far side of the Moon (the side that can never be seen from Earth) the same as the dark side of the Moon? Explain.

5. What is the difference between a sidereal month and a synodic month? Which is longer? Why?

6. (a) If you lived on the Moon, would you see the Sun rise and set, or would it always be in the same place in the sky? Explain. (b) Would you see the Earth rise and set, or would it always be in the same place in the sky? Explain.

7. Astronomers sometimes refer to lunar phases in terms of the age of the Moon. This is the time that has elapsed since new moon phase. Thus, the age of a full moon is half of a 29½-day synodic period, or approximately 15 days. Find the approximate age of (a) a waxing crescent moon; (b) a last quarter moon; (c) a waning gibbous moon.

8. What is the difference between the umbra and the penumbra of a shadow?

9. Why doesn't a lunar eclipse occur at every full moon and a solar eclipse at every new moon?

10. What is the line of nodes? Why is it important to the subject of eclipses?

11. What is a penumbral eclipse of the Moon? Why do you suppose that it is easy to overlook such an eclipse?

12. The maximum duration of totality of a lunar eclipse is 1 hour, 42 minutes. But none of the lunar eclipses listed in Table 3-1 lasts this long. Why is this?

13. Can one ever observe an annular eclipse of the Moon? Why or why not?

14. If you were looking at the Earth from the side of the Moon that faces the Earth, what would you see during (a) a total lunar eclipse? (b) a total solar eclipse? Explain your answers.

15. If there is a total eclipse of the Sun in April, can there be a lunar eclipse three months later in July? Why or why not?

16. Which type of eclipse—lunar or solar—do you think most people on Earth have seen? Why?

17. How is an annular eclipse of the Sun different from a total eclipse of the Sun? What causes this difference?

*18. What is the saros? How did ancient astronomers use it to predict eclipses?

19. How did Eratosthenes measure the size of the Earth?

20. How did Aristarchus try to estimate the distance from the Earth to the Sun and Moon?

ADVANCED QUESTIONS

Problem-solving tips and tools

To estimate the average angular speed of the Moon along its orbit (that is, how many degrees around its orbit the Moon travels per day), divide 360° by the length of a sidereal month. It is helpful to know that the saros interval of 6585.3 days equals 18 years and 11⅓ days if the interval includes four leap years, but is 18 years and 10⅓ days if it includes five leap years.

21. The dividing line between the illuminated and unilluminated halves of the Moon is called the *terminator.* The terminator appears curved when there is a crescent or gibbous moon, but appears straight when there is a first quarter or last quarter moon (see Figure 3-2). Describe how you could use these facts to explain to a friend why lunar phases cannot be caused by the Earth's shadow falling on the Moon.

22. This photograph of the Earth was taken by the crew of the *Apollo 8* spacecraft as they orbited the Moon. A portion of the lunar surface is visible at the right-hand side of the photo. In this photo, the Earth is oriented with its North Pole approximately at the top. When this photo was taken, was the Moon waxing or waning as seen from Earth? Explain your answer with a diagram.

(NASA/JSC)　　　　　　　　　R I **V** U X G

23. (a) The Moon moves noticeably over the space of a single night. To show this, calculate how long it takes the Moon to move through an angle equal to its own angular diameter (½°) against the background of stars. Give your answer in hours. **(b)** Through what angle (in degrees) does the Moon move during a 12-hour night? Can you notice an angle of this size? (*Hint:* See Figure 1-10.)

24. During an occultation, or "covering up," of Jupiter by the Moon, an astronomer notices that it takes the Moon's edge 90 seconds to cover Jupiter's disk completely. If the Moon's motion is assumed to be uniform and the occultation was "central" (that is, center over center), find the angular diameter of Jupiter. (*Hint:* Assume that Jupiter does not appear to move in the sky during this brief 90-second interval. You will need to convert the Moon's angular speed from degrees per day to arcseconds per second.)

25. How many more sidereal months than synodic months are there in a year? Explain.

26. Suppose the Earth moved a little faster around the Sun, so that it took a bit less than one year to make a complete orbit. If the speed of the Moon's orbit around the Earth were unchanged, would the length of the sidereal month be the same, longer, or shorter than it is now? What about the synodic month? Explain your answers.

27. If the Moon revolved about the Earth in the same orbit but in the opposite direction, would the synodic month be longer or shorter than the sidereal month? Explain your reasoning.

28. A "blue moon" is the second full moon within the same calendar month. There is usually only one full moon within a calendar month, which is why the phrase "once in a blue moon" means "hardly ever." Why are blue moons so rare? Are there any months of the year in which it would be impossible to have two full moons? Explain your answer.

29. The total lunar eclipse of November 29, 1993, was visible from North America. The duration of totality was 48 minutes. Was this total eclipse also visible from India, on the opposite side of the Earth? Explain your reasoning.

30. You are watching a lunar eclipse from some place on the Earth's night side. Will you see the Moon enter the Earth's shadow from the east or from the west? Explain your reasoning.

31. During a total solar eclipse, the Moon's umbra moves in a generally eastward direction across the Earth's surface. Use a drawing like Figure 3-11 to explain why the motion is eastward, not westward.

32. A total solar eclipse was visible from southern Germany on August 11, 1999. (See the photo that opens this chapter.) Draw what the eclipse would have looked like as seen from Switzerland, to the south of the path of totality. Explain the reasoning behind your drawing.

33. Just as the distance from the Earth to the Moon varies somewhat as the Moon orbits the Earth, the distance from the Sun to the Earth changes as the Earth orbits the Sun. The Earth is closest to the Sun at its *perihelion*; it is farthest from the Sun at its *aphelion*. In order for a total solar eclipse to have the maximum duration of totality, should the Earth be at perihelion or aphelion? Assume that the Earth-Moon distance is the same in both situations. As part of your explanation, draw two pictures like Figure 3-11, one with the Earth relatively close to the Sun and one with the Earth relatively far from the Sun.

34. (a) Suppose the diameter of the Moon were doubled, but the orbit of the Moon remained the same. Would total solar eclipses be more common, less common, or just as common as they are now? Explain. **(b)** Suppose the diameter of the Moon were halved, but the orbit of the Moon remained the same. Explain why there would be *no* total solar eclipses.

*35. On June 21, 2001, residents of southern Africa were treated to a total solar eclipse. **(a)** On what date and over what part of the world will the next total eclipse of that series occur? Explain. **(b)** On what date might you next expect a total eclipse of that series to be visible from southern Africa? Explain.

DISCUSSION QUESTIONS

36. Describe the cycle of lunar phases that would be observed if the Moon moved around the Earth in an orbit perpendicular to the plane of the Earth's orbit. Would it be possible for both solar and lunar eclipses to occur under these circumstances? Explain your reasoning.

37. How would a lunar eclipse look if the Earth had no atmosphere? Explain your reasoning.

38. Why do you suppose that total solar eclipse paths fall more frequently on oceans than on land? (You may find it useful to look at Figure 3-13.)

39. In his 1885 novel *King Solomon's Mines*, H. Rider Haggard described a total solar eclipse that was seen in both South Africa and in the British Isles. Is such an eclipse possible? Why or why not?

40. Examine Figure 3-13, which shows all of the total solar eclipses from 1997 to 2020. What are the chances that you might be able to travel to one of the eclipse paths? Do you think you might go through your entire life without ever seeing a total eclipse of the Sun?

WEB/CD-ROM QUESTIONS

41. Search the World Wide Web for information about the next total solar eclipse. Through which major cities, if any, does the path of totality pass? What is the maximum duration of totality? At what location is this maximum duration observed? Will this eclipse be visible (even as a partial eclipse) from your location? Draw a picture showing the Sun, Earth, and Moon when the totality is at its maximum duration, and indicate your location on the drawing of the Earth.

42. Search the World Wide Web for information about the next total lunar eclipse. Will the total phase of the eclipse be visible from your location? If not, will the penumbral phase be visible? Draw a picture showing the Sun, Earth, and Moon when the totality is at its maximum duration, and indicate your location on the drawing of the Earth.

43. Access the animation "The Moon's Phases" in Chapter 3 of the *Universe* web site or CD-ROM. This shows the Earth-Moon system

as seen from a vantage point looking down onto the North Pole. **(a)** Describe where you would be on the diagram if you are on the equator and the time is 6:00 P.M. **(b)** If it is 6:00 P.M. and you are standing on Earth's equator, would a third-quarter moon be visible? Why or why not? If it would be visible, describe its appearance.

44. Access the animation "A Solar Eclipse Viewed from the Moon" in Chapter 3 of the *Universe* web site or CD-ROM. This shows the solar eclipse of August 11, 1999, as viewed from the Moon. Using a diagram, explain why the stars and the Moon's shadow move in the directions shown in this animation.

OBSERVING PROJECTS

45. Observe the Moon on each clear night over the course of a month. On each night, note the Moon's location among the constellations and record that location on a star chart that also shows the ecliptic. After a few weeks, your observations will begin to trace the Moon's orbit. Identify the orientation of the line of nodes by marking the points where the Moon's orbit and the ecliptic intersect. On what dates is the Sun near the nodes marked on your star chart? Compare these dates with the dates of the next solar and lunar eclipses.

46. It is quite possible that a lunar eclipse will occur while you are taking this course. Look up the date of the next lunar eclipse in Table 3-1, in the current issue of a reference such as the *Astronomical Almanac* or *Astronomical Phenomena*, or on the World Wide Web. Then make arrangements to observe this lunar eclipse. You can observe the eclipse with the naked eye, but binoculars or a small telescope will enhance your viewing experience. If the eclipse is partial or total, note the times at which the Moon enters and exits the Earth's umbra. If the eclipse is penumbral, can you see any changes in the Moon's brightness as the eclipse progresses?

47. Use the *Starry Night* program to observe the motion of the Moon. **(a)** Display the entire celestial sphere, including the part below the horizon (select **Atlas** in the **Go** menu). Center on the Moon by using the **Find...** command in the **Edit** menu. In the Control Panel at the top of the main window, set the time step to 1 day and click on the "Forward" button (a triangle that ponts to the right). How does the Moon appear to move against the background of stars? Does it ever change direction? **(b)** Determine how many days elapse between successive times when the Moon is on the ecliptic. (If you don't see a green line representing the ecliptic, select **The Ecliptic** in the **Guides** menu.) Then move forward in time to a date when the Moon is on the ecliptic *and* either full or new. What type of eclipse will occur on that date? Confirm your answer by comparing with Tables 3-1 and 3-2 or with lists of eclipses on the World Wide Web.

Gravitation and the Waltz of the Planets

In July 1969 humans first journeyed from the Earth to the surface of the Moon. By studying the rocks that the astronauts collected on the Moon during this and later missions, geologists learned how the surface of our nearest celestial neighbor has evolved over the past several billion years. These discoveries revolutionized our understanding of the very nature of the Moon.

But several other revolutions in human understanding had to take place before these discoveries could be made. One revolution overthrew the ancient idea that the Earth is an immovable object at the center of the universe, around which moved the Sun, the Moon, and the planets. In this chapter we will learn how Nicolaus Copernicus, Galileo Galilei, Tycho Brahe, and Johannes Kepler helped us understand that the Earth is itself one of several planets orbiting the Sun.

We will learn, too, about Isaac Newton's revolutionary discovery of why the planets move in the way that they do. This was just one aspect of Newton's immense body of work, which included formulating the fundamental laws of physics and developing a precise mathematical description of the force of gravity—the force that holds the planets in their orbits.

Newton's laws apply on Earth as well as in the heavens. They also govern the flight of spacecraft, which made missions to the Moon possible. That is why astronaut William Anders radioed the following while en route from the Moon to the Earth: "I think Isaac Newton is doing most of the driving right now."

(Michael Collins, *Apollo 11*, NASA)

R I **V** U X G

As you read the sections of this chapter, look for the answers to the following questions.

4-1 How did ancient astronomers explain the motions of the planets?

4-2 Why did Copernicus think that the Earth and the other planets go around the Sun?

4-3 What did Galileo see in his telescope that confirmed that the planets orbit the Sun?

4-4 How did Tycho Brahe attempt to test the ideas of Copernicus?

4-5 What paths do the planets follow as they move around the Sun?

4-6 What fundamental laws of nature explain the motions of objects on Earth as well as the motions of the planets?

4-7 Why don't the planets fall into the Sun?

4-8 What keeps the same face of the Moon always pointed toward the Earth?

4-1 Ancient astronomers invented geocentric models to explain planetary motions

Since the dawn of civilization, scholars have attempted to explain the nature of the universe. The ancient Greeks were the first to use the principle that still guides scientists today: The universe can be described and understood logically. For example, more than 2500 years ago Pythagoras and his followers put forth the idea that nature can be described with mathematics. About 200 years later, Aristotle asserted that the universe is governed by physical laws.

As they attempted to create a model (see Section 1-1) of how the universe works, most Greeks assumed that the Sun, the Moon, the stars, and the planets revolve about a stationary Earth. A model of this kind, in which the Earth is at the center of the universe, is called a **geocentric model**. Similar ideas were held by the scholars of ancient China.

Today we recognize that the stars are not merely points of light on an immense celestial sphere. But this is just how the ancient Greeks regarded the stars in their geocentric model of the universe. To explain the diurnal motions of the stars, they assumed that the entire celestial sphere rotated around the stationary Earth once a day. The Sun and Moon both participated in this daily rotation of the sky, which explained their rising and setting motions. To explain why the Sun and Moon both move slowly with respect to the stars, the ancient Greeks imagined that both of these objects orbit around the Earth.

ANALOGY Imagine a merry-go-round that rotates clockwise as seen from above, as in Figure 4-1a. As it rotates, two children walk slowly counterclockwise at different speeds around the merry-go-round's platform. Thus, the children rotate along with the merry-go-round and also change

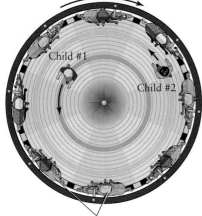

Merry-go-round rotates clockwise

Child #1

Child #2

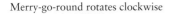

Wooden horses fixed on merry-go-round

a   A rotating merry-go-round

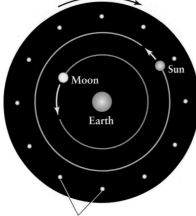

Celestial sphere rotates to the west

Sun

Moon

Earth

Stars fixed on celestial sphere

b   The Greek geocentric cosmogony

Figure 4-1

A Merry-Go-Round Analogy

(a) Two children walk at different speeds around a rotating merry-go-round with its wooden horses. (b) In an analogous way, the ancient Greeks imagined that the Sun and Moon move at different speeds around the rotating celestial sphere with its fixed stars. Thus, the Sun and Moon move from east to west across the sky every day and also move slowly eastward from one night to the next relative to the background of stars.

their positions with respect to the merry-go-round's wooden horses. This scene is analogous to the way the ancient Greeks pictured the motions of the stars, Sun, and Moon. In their model, the celestial sphere rotated to the west around a stationary Earth (Figure 4-1b). The stars rotate along with the celestial sphere just as the wooden horses rotate along with the merry-go-round in Figure 4-1a. The Sun and Moon are analogous to the two children; they both turn westward with the celestial sphere, making one complete turn each day, and also move slowly eastward at different speeds with respect to the stars.

The geocentric model of the heavens also had to explain the motions of the planets. The ancient Greeks and other cultures of that time knew of five planets: Mercury, Venus, Mars, Jupiter, and Saturn. These planets are quite obvious because they are bright objects in the night sky. For example, when Venus is at its maximum brilliancy, it is 16 times brighter than the brightest star. (By contrast, Uranus, Neptune, and Pluto are quite dim and were not discovered until after the invention of the telescope.)

Like the Sun and Moon, all of the planets rise in the east and set in the west once a day. And like the Sun and Moon, from night to night the planets slowly move on the celestial sphere, that is, with respect to the background of stars. However, the character of this motion on the celestial sphere is quite different for the planets. Both the Sun and the Moon always move from west to east on the celestial sphere, that is, opposite the direction in which the celestial sphere appears to rotate. Furthermore, the Sun and the Moon each move at relatively constant speeds around the celestial sphere. (The Moon's speed is faster than that of the Sun, because it travels all the way around the celestial sphere in about a month while the Sun takes an entire year). By contrast, each of the planets appears to wander back and forth on the celestial sphere, moving from west to east at most times but at other times from east to west. (The name *planet* is well deserved; it comes from a Greek word meaning "wanderer.") As a planet wanders, its speed on the celestial sphere can vary substantially.

You can see this distinctive behavior in Figure 4-2, which shows the motion of Mars with respect to the background of stars during 2005 and 2006. In the language of the merry-go-round analogy in Figure 4-1, the Greeks imagined the planets as children walking around the rotating merry-go-round but who keep changing their minds about which direction to walk!

As seen from Earth, the planets wander primarily across the 12 constellations of the zodiac. (Figure 4-2 shows that during 2005–2006, Mars moves through Pisces, Aries, and Taurus.) As we learned in Section 2-5, these constellations encircle the sky in a continuous band centered on the ecliptic, which is the apparent path of the Sun across the celestial sphere.

If you observe a planet as it travels across the zodiac from night to night, you will find that the planet usually moves

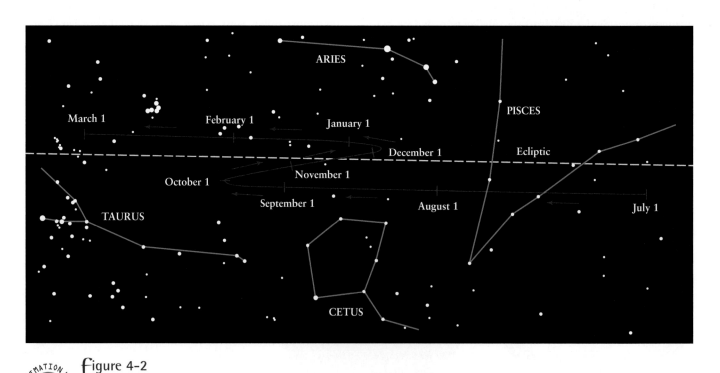

figure 4-2

The Path of Mars in 2005–2006 From July 2005 through February 2006, Mars will move across the constellations Pisces, Aries, and Taurus. Mars's motion will be direct (from west to east, or from right to left in this figure) most of the time but will be retrograde (from east to west, or from left to right in this figure) during October and November 2005. Notice that the speed of Mars relative to the stars is not constant: The planet moves faster and travels farther across the sky from August 1 to September 1 than it does from September 1 to October 1.

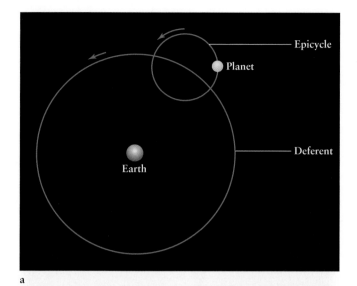

a

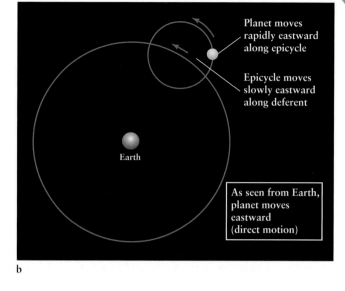

Planet moves
rapidly eastward
along epicycle

Epicycle moves
slowly eastward
along deferent

As seen from Earth,
planet moves
eastward
(direct motion)

b

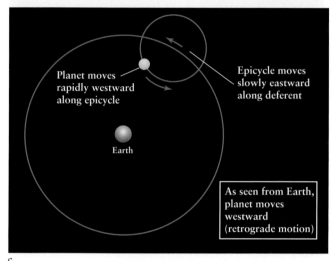

Planet moves
rapidly westward
along epicycle

Epicycle moves
slowly eastward
along deferent

As seen from Earth,
planet moves
westward
(retrograde motion)

c

figure 4-3

A Geocentric Explanation of Retrograde Motion
(a) The ancient Greeks imagined that each planet moves along an epicycle, which in turn moves along a deferent centered approximately on the Earth. The planet moves along the epicycle more rapidly than the epicycle moves along the deferent. **(b)** At most times the eastward motion of the planet on the epicycle adds to the eastward motion of the epicycle on the deferent. Then the planet moves eastward in direct motion as seen from Earth. **(c)** When the planet is on the inside of the deferent, its motion along the epicycle is westward. Because this motion is faster than the eastward motion of the epicycle on the deferent, the planet appears from Earth to be moving westward in retrograde motion.

slowly eastward against the background stars. This eastward progress is called **direct motion.** Occasionally, however, the planet will seem to stop and then back up for several weeks or months. This occasional westward movement is called **retrograde motion.** Both direct and retrograde motions are much slower than the apparent daily rotation of the sky caused by the Earth's rotation. Hence, they are best detected by mapping the position of a planet against the background stars from night to night over a long period. Figure 4-2 is a map of just this sort. Mars undergoes retrograde motion about every 22½ months; all the other planets go through retrograde motion, but at different intervals.

Explaining the nonuniform motions of the five planets was one of the main challenges facing the astronomers of antiquity. The Greeks developed many theories to account for retrograde motion and the loops that the planets trace out against the background stars. One of the most successful and enduring models was originated by Hipparchus in the second century B.C. and expanded upon by Ptolemy, the last of the great Greek astronomers, during the second century A.D. Fig-

ure 4-3a sketches the basic concept, usually called the **Ptolemaic system.** Each planet is assumed to move in a small circle called an **epicycle,** whose center in turn moves in a larger circle, called a **deferent,** which is centered approximately on the Earth. Both the epicycle and deferent rotate in the same direction, shown as counterclockwise in Figure 4-3a.

As viewed from Earth, the epicycle moves eastward along the deferent. Most of the time the eastward motion of the planet on its epicycle adds to the eastward motion of the epicycle on the deferent (Figure 4-3b). Then the planet is seen to be in direct (eastward) motion against the background stars. However, when the planet is on the part of its epicycle nearest Earth, the motion of the planet along the epicycle is opposite to the motion of the epicycle along the deferent. The planet therefore appears to slow down and halt its usual eastward movement among the constellations, and actually goes backward in retrograde (westward) motion for a few weeks or months (Figure 4-3c). Thus, the concept of epicycles and deferents enabled Greek astronomers to explain the retrograde loops of the planets.

Using the wealth of astronomical data in the library at Alexandria, including records of planetary positions for hundreds of years, Ptolemy deduced the sizes and rotation rates of the epicycles and deferents needed to reproduce the recorded paths of the planets. After years of tedious work, Ptolemy assembled his calculations into 13 volumes, collectively called the *Almagest*. His work was used to predict the positions and paths of the Sun, Moon, and planets with unprecedented accuracy. In fact, the *Almagest* was so successful that it became the astronomer's bible, and for more than 1000 years, the Ptolemaic system endured as a useful description of the workings of the heavens.

A major problem posed by the Ptolemaic system was a philosophical one: It treated each planet independent of the others. There was no rule in the *Almagest* that related the size and rotation speed of one planet's epicycle and deferent to the corresponding sizes and speeds for other planets. This problem made the Ptolemaic system very unsatisfying to many Islamic and European astronomers of the Middle Ages. They felt that a correct model of the universe should be based on a simple set of underlying principles that applied to all of the planets.

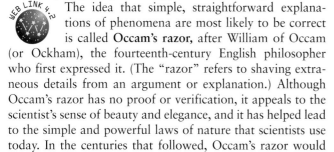 The idea that simple, straightforward explanations of phenomena are most likely to be correct is called **Occam's razor,** after William of Occam (or Ockham), the fourteenth-century English philosopher who first expressed it. (The "razor" refers to shaving extraneous details from an argument or explanation.) Although Occam's razor has no proof or verification, it appeals to the scientist's sense of beauty and elegance, and it has helped lead to the simple and powerful laws of nature that scientists use today. In the centuries that followed, Occam's razor would help motivate a new and revolutionary view of the universe.

4-2 Nicolaus Copernicus devised the first comprehensive heliocentric model

During the first half of the sixteenth century, a Polish lawyer, physician, canon of the church, and gifted mathematician named Nicolaus Copernicus (Figure 4-4) began to construct a new model of the universe. His model, which placed the Sun at the center, explained the motions of the planets in a more natural way than the Ptolemaic system. As we will see, it also helped lay the foundations of modern physical science. But Copernicus was not the first to conceive a Sun-centered model of planetary motion.

Imagine riding on a fast racehorse. As you pass a slowly walking pedestrian, he appears to move backward, even though he is traveling in the same direction as you and your horse. In the third century B.C., this sort of simple observation inspired the Greek astronomer Aristarchus to suggest a straightforward explanation of retrograde motion.

In Aristarchus's **heliocentric** (Sun-centered) **model,** all the planets, including Earth, revolve about the Sun. Different planets take different lengths of time to complete an orbit, so from time to time one planet will overtake another, just as a fast-moving horse overtakes a person on foot. When the Earth

figure 4-4

Nicolaus Copernicus (1473–1543) Copernicus was the first person to work out the details of a heliocentric system in which the planets, including the Earth, orbit the Sun. (E. Lessing/Magnum)

overtakes Mars, for example, Mars appears to move backward in retrograde motion, as Figure 4-5 shows. Thus, in the heliocentric picture, the occasional retrograde motion of a planet is merely the result of the Earth's motion.

As we saw in Section 3-6, Aristarchus demonstrated that the Sun is bigger than the Earth (see Table 3-3). This made it sensible to imagine the Earth orbiting the larger Sun. He also imagined that the Earth rotated on its axis once a day, which explained the daily rising and setting of the Sun, Moon, and planets and the diurnal motions of the stars. To explain why the apparent motions of the planets never take them far from the ecliptic, Aristarchus proposed that the orbits of the Earth and all the planets must lie in nearly the same plane. (Recall from Section 2-5 that the ecliptic is the projection onto the celestial sphere of the plane of the Earth's orbit.)

This heliocentric model is conceptually much simpler than an Earth-centered system, such as that of Ptolemy, with all its "circles upon circles." In Aristarchus's day, however, the idea of an orbiting, rotating Earth seemed inconceivable, given the Earth's apparent stillness and immobility. Nearly 2000 years would pass before a heliocentric model found broad acceptance.

In the years after 1500, Copernicus came to realize that a heliocentric model has several advantages beyond providing a natural explanation of retrograde motion. In the Ptolemaic system, the arrangement of the planets—that is, which are close to the Earth and which are far away—was chosen

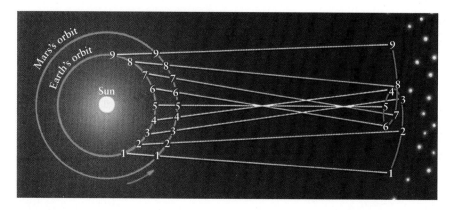

ANIMATION 4.4

figure 4-5

A Heliocentric Explanation of Retrograde Motion In the heliocentric model of Aristarchus, the Earth and the other planets orbit the Sun. The Earth travels around the Sun more rapidly than Mars. Consequently, as the Earth overtakes and passes this slower-moving planet, Mars appears for a few months (from points 4 through 6) to fall behind and move backward with respect to the background of stars.

in large part by guesswork. But using a heliocentric model, Copernicus could determine the arrangement of the planets without ambiguity.

Copernicus realized that because Mercury and Venus are always observed fairly near the Sun in the sky, their orbits must be smaller than the Earth's. Planets in such orbits are called **inferior planets** (Figure 4-6). The other visible planets—Mars, Jupiter, and Saturn—are sometimes seen on the side of the celestial sphere opposite the Sun, so these planets appear high above the horizon at midnight (when the Sun is far below the horizon). When this happens, the Earth must lie between the Sun and these planets. Copernicus therefore concluded that the orbits of Mars, Jupiter, and Saturn must be larger than the Earth's orbit. Hence, these planets are called **superior planets.**

Three additional planets, Uranus, Neptune, and Pluto, were discovered after the telescope was invented (and after the death of Copernicus). All three can be seen at times in the midnight sky, so these are also superior planets with orbits larger than the Earth's.

The heliocentric model also explains why planets appear in different parts of the sky on different dates. Both inferior planets (Mercury and Venus) go through cycles: The planet is seen in the west after sunset for several weeks or months, then for several weeks or months in the east before sunrise, and then in the west after sunset again.

Figure 4-6 shows the reason for this cycle. When Mercury or Venus is visible after sunset, it is near **greatest eastern elongation.** (The angle between the Sun and a planet as viewed from Earth is called the planet's **elongation.**) The planet's position in the sky is as far east of the Sun as possible, so it appears above the western horizon after sunset (that is, to the east of the Sun) and is often called an "evening star." At **greatest western elongation,** Mercury or Venus is as far west of the Sun as it can possibly be. It then rises before the Sun, gracing the predawn sky as a "morning star" in the east. When Mercury or Venus is at **inferior conjunction,** it is between us and the Sun, and it is moving from the evening sky into the morning sky. At **superior conjunction,** when the planet is on the opposite side of the Sun, it is moving back into the evening sky.

A superior planet such as Mars, whose orbit is larger than the Earth's, is best seen in the night sky when it is at **opposi-**

tion. At this point the planet is in the part of the sky opposite the Sun and is highest in the sky at midnight. This is also when the planet appears brightest, because it is closest to us. But when a superior planet like Mars is located behind the Sun at **conjunction,** it is above the horizon during the daytime and thus is not well placed for nighttime viewing.

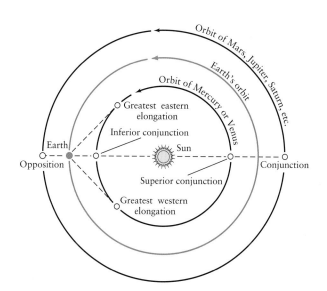

INTERACTIVE EXERCISE 4.7

figure 4-6

Planetary Orbits and Configurations When and where in the sky a planet can be seen from Earth depends on the size of its orbit and its location on that orbit. The inferior planets Mercury and Venus, whose orbits are smaller than the Earth's, are visible after sunset when at or near greatest eastern elongation, and before sunrise when at or near greatest western elongation. The best view of a superior planet (one whose orbit is larger than the Earth's) happens when it is at opposition, so that the planet is highest in the sky at midnight. When an inferior planet is at inferior conjunction or superior conjunction, or when a superior planet is at conjunction, it is only up during the daytime and cannot be seen at night. (Note that in this figure you are looking down onto the solar system from a point far above the Earth's northern hemisphere.)

table 4-1	Synodic and Sidereal Periods of the Planets	
Planet	Synodic period	Sidereal period
Mercury	116 days	88 days
Venus	584 days	225 days
Earth	—	1.0 year
Mars	780 days	1.9 years
Jupiter	399 days	11.9 years
Saturn	378 days	29.5 years
Uranus	370 days	84.0 years
Neptune	368 days	164.8 years
Pluto	367 days	248.5 years

The Ptolemaic system has no simple rules relating the motion of one planet to another. But Copernicus showed that there are such rules in a heliocentric model. In particular, he found a correspondence between the time a planet takes to complete one orbit—that is, its **period**—and the size of the orbit.

Determining the period of a planet takes some care, because the Earth, from which we must make the observations, is also moving. Realizing this, Copernicus was careful to distinguish between two different periods of each planet. The **synodic period** is the time that elapses between two successive identical configurations as seen from Earth—from one opposition to the next, for example, or from one conjunction to the next. The **sidereal period** is the true orbital period of a planet, the time it takes the planet to complete one full orbit of the Sun relative to the stars.

The synodic period of a planet can be determined by observing the sky, but the sidereal period has to be found by calculation. Copernicus figured out how to do this (Box 4-1). Table 4-1 shows the results for all of the planets.

To find a relationship between the sidereal period of a planet and the size of its orbit, Copernicus still had to determine the relative distances of the planets from the Sun. He devised a straightforward geometric method of determining the relative distances of the planets from the Sun using trigonometry. His answers turned out to be remarkably close to the modern values, as shown in Table 4-2.

The distances in Table 4-2 are given in terms of the astronomical unit, which is the average distance from Earth to the Sun (Section 1-7). Copernicus did not know the precise value of this distance, so he could only determine the *relative* sizes of the orbits of the planets. One method used by modern astronomers to determine the astronomical unit is to measure the Earth-Venus distance very accurately using radar. At the same time, they measure the angle in the sky between Venus and the Sun, and then calculate the Earth-Sun distance using

table 4-2	Average Distances of the Planets from the Sun	
Planet	Copernican value (AU*)	Modern value (AU)
Mercury	0.38	0.39
Venus	0.72	0.72
Earth	1.00	1.00
Mars	1.52	1.52
Jupiter	5.22	5.20
Saturn	9.07	9.54
Uranus	—	19.19
Neptune	—	30.06
Pluto	—	39.53

*1 AU = 1 astronomical unit = average distance from the Earth to the Sun.

box 4-1 | Tools of the Astronomer's Trade

Relating Synodic and Sidereal Periods

We can derive a mathematical formula that relates a planet's sidereal period (the time required for the planet to complete one orbit) to its synodic period (the time between two successive identical configurations). To start with, let's consider an inferior planet (Mercury or Venus) orbiting the Sun as shown in the diagram. Let P be the planet's sidereal period, S the planet's synodic period, and E the sidereal period of Earth or sidereal year (see Section 2-8), which Copernicus knew to be nearly 365¼ days.

The rate at which the Earth moves around its orbit is the number of degrees around the orbit divided by the time to complete the orbit, or $360°/E$ (equal to a little less than 1° per day). Similarly, the rate at which the inferior planet moves along its orbit is $360°/P$.

During a given time interval, the angular distance that the Earth moves around its orbit is its rate, $(360°/E)$, multiplied by the length of the time interval. Thus, during a time S, or one synodic period of the inferior planet, the Earth covers an angular distance of $(360°/E)S$ around its orbit. In that same time, the inferior planet covers an angular distance of $(360°/P)S$. Note, however, that the inferior planet has gained one full lap on the Earth, and hence has covered 360° more than the Earth has (see the diagram). Thus, $(360°/P)S = (360°/E)S + 360°$. Dividing each term of this equation by $360°\ S$ gives

For an inferior planet:

$$\frac{1}{P} = \frac{1}{E} + \frac{1}{S}$$

P = inferior planet's sidereal period
E = Earth's sidereal period = 1 year
S = inferior planet's synodic period

A similar analysis for a superior planet (for example, Mars, Jupiter, or Saturn) yields

For a superior planet:

$$\frac{1}{P} = \frac{1}{E} - \frac{1}{S}$$

P = superior planet's sidereal period
E = Earth's sidereal period = 1 year
S = superior planet's synodic period

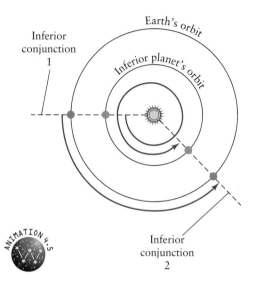

Inferior conjunction 1

Earth's orbit

Inferior planet's orbit

Inferior conjunction 2

ANIMATION 4.5

Using these formulas, we can calculate a planet's sidereal period P from its synodic period S. Often astronomers express P, E, and S in terms of years, by which they mean Earth years of approximately 365.26 days.

EXAMPLE: Jupiter has an observed synodic period S of 398.9 days, or 1.092 years. Because Jupiter is a superior planet and because $E = 1$ year exactly, we determine Jupiter's sidereal period using the second of the two equations above:

$$\frac{1}{P} = \frac{1}{1} - \frac{1}{1.092} = 0.08425,$$

so

$$P = \frac{1}{0.08425} = 11.87 \text{ years}$$

It takes 11.87 years for Jupiter to complete one full orbit of the Sun. The much shorter synodic period of 1.092 years is the time from one opposition to the next, or the time that elapses from when the Earth overtakes Jupiter to when it next overtakes Jupiter. The reason why this is so much shorter than the sidereal period is that Jupiter moves quite slowly around its orbit. The Earth overtakes it a little less often than once per Earth orbit, that is, at intervals of a little bit more than a year.

trigonometry. In this way, the astronomical unit is found to be 1.496×10^8 km (92.96 million miles). Once the astronomical unit is known, the average distances from the Sun to each of the planets in kilometers can be determined from Table 4-2. A table of these distances is given in Appendix 2.

By comparing Tables 4-1 and 4-2, you can see the unifying relationship between planetary orbits in the Copernican model: The farther a planet is from the Sun, the longer it takes to travel around its orbit (that is, the longer its sidereal period). That is not simply because a planet must travel farther to complete a larger orbit; in addition, a planet moves more slowly in a larger orbit. For example, Mercury, with its small orbit, moves at an average speed of 47.9 km/s (107,000 mi/h). Saturn travels around its large orbit much more slowly, at an average speed of 9.64 km/s (21,600 mi/h). The older Ptolemaic model offers no such simple relations between the motions of different planets.

At first, Copernicus assumed that Earth travels around the Sun along a circular path. He found that perfectly circular orbits could not accurately describe the paths of the other planets, so he had to add an epicycle to each planet. (This was *not* to explain retrograde motion, which Copernicus realized was because of the differences in orbital speeds of different planets, as shown in Figure 4-4. Rather, the small epicycles helped Copernicus account for slight variations in each planet's speed along its orbit.) Even though he clung to the old notion that orbits must be made up of circles, Copernicus had shown that a heliocentric model could explain the motions of the planets. He compiled his ideas and calculations into a book entitled *De revolutionibus orbium coelestium* (On the Revolutions of the Celestial Spheres), which was published in 1543, the year of his death.

For several decades after Copernicus, most astronomers saw little reason to change their allegiance from the older geocentric model of Ptolemy. The predictions that the Copernican model makes for the apparent positions of the planets are, on average, no better or worse than those of the Ptolemaic model. The test of Occam's razor does not really favor either model, because both use a combination of circles to describe each planet's motion.

More concrete evidence was needed to convince scholars to abandon the old, comfortable idea of a stationary Earth at the center of the universe. Obtaining this evidence required a quantum leap in the technology of astronomical observation—the invention of the telescope.

4-3 Galileo's discoveries with a telescope strongly supported a heliocentric model

When Dutch opticians invented the telescope during the first decade of the seventeenth century, astronomy was changed forever. The scholar who used this new tool to amass convincing evidence that the planets orbit the Sun, not the Earth, was the Italian mathematician and physical scientist Galileo Galilei (Figure 4-7).

WEB LINK 4.4

ƒigure 4-7

Galileo Galilei (1564–1642) Galileo was one of the first people to use a telescope to observe the heavens. He discovered craters on the Moon, sunspots on the Sun, the phases of Venus, and four moons orbiting Jupiter. His observations strongly suggested that the Earth orbits the Sun, not vice versa. (Art Resource)

While Galileo did not invent the telescope, he was the first to point one of these new devices toward the sky and to publish his observations. Beginning in 1610, he saw sights of which no one had ever dreamed. He discovered mountains on the Moon, sunspots on the Sun, and the rings of Saturn, and he was the first to see that the Milky Way is not a featureless band of light but rather "a mass of innumerable stars."

One of Galileo's most important discoveries with the telescope was that Venus exhibits phases like those of the Moon (Figure 4-8). Galileo also noticed that the apparent size of Venus as seen through his telescope was related to the planet's phase. Venus appears small at gibbous phase and largest at crescent phase. There is also a correlation between the phases of Venus and the planet's angular distance from the Sun.

Figure 4-9 shows that these relationships are entirely compatible with a heliocentric model in which the Earth and Venus both go around the Sun. They are also completely *incompatible* with the Ptolemaic system, in which the Sun and Venus both orbit the Earth. To explain why Venus is never seen very far from the Sun, the Ptolemaic model had to assume that the deferents of Venus and of the Sun move together in lockstep, with the epicycle of Venus centered on a

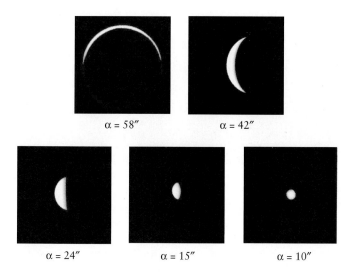

Galileo also found more unexpected evidence for the ideas of Copernicus. In 1610 Galileo discovered four moons, now called the Galilean satellites, orbiting Jupiter (Figure 4-11). He realized that they were orbiting Jupiter because they appeared to move back and forth from one side of the planet to the other. Figure 4-12 shows confirming observations made by Jesuit observers in 1620. Astronomers soon realized that the larger the orbit of one of the moons around Jupiter, the slower that moon moves and the longer it takes that moon to travel around its orbit. These are the same relationships that Copernicus deduced for the motions of the planets around the Sun. Thus, the moons of Jupiter behave like a Copernican system in miniature.

Galileo's telescopic observations constituted the first fundamentally new astronomical data in almost 2000 years. Contradicting prevailing opinion, his discoveries strongly suggested a heliocentric structure of the universe. At the time, the Roman Catholic Church attacked Galileo's ideas, because they could not be reconciled with certain passages in the Bible or with the writings of Aristotle and Plato. Galileo was condemned to spend his latter years under house arrest "for

figure 4-8   R I **V** U X G

The Phases of Venus This series of photographs shows how the appearance of Venus changes as it moves along its orbit. The number below each view is the angular diameter α of the planet in arcseconds. Venus has the largest angular diameter when it is a crescent, and the smallest angular diameter when it is gibbous (nearly full). (New Mexico State University Observatory)

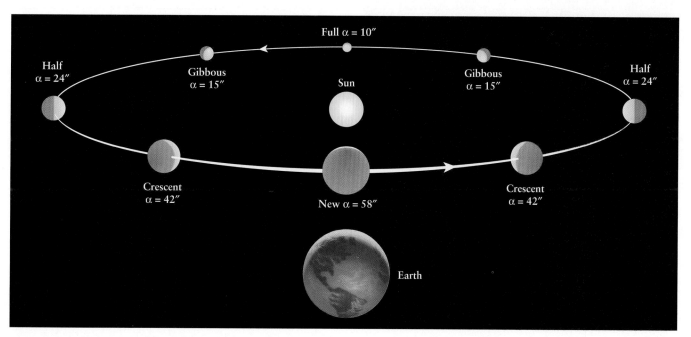

figure 4-9

The Changing Appearance of Venus Explained in a Heliocentric Model A heliocentric model, in which the Earth and Venus both orbit the Sun, provides a natural explanation for the changing appearance of Venus shown in Figure 4-8. When Venus is on the opposite side of the Sun from the Earth, it appears full and has a small angular size. When Venus is on the same side of the Sun as the Earth, we see it in a "new" phase and with a larger angular size. When Venus is at the greatest angle to either side of the Sun (at greatest eastern or western elongation), it appears half illuminated and has an intermediate angular size.

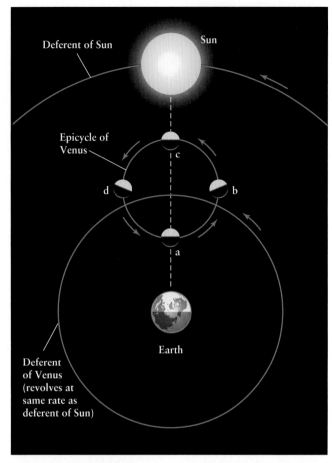

figure 4-10

The Appearance of Venus in the Ptolemaic Model In the geocentric Ptolemaic model, the deferents of Venus and the Sun rotate together, with the epicycle of Venus centered on a line (shown dashed) that connects the Sun and the Earth. In this model an Earth observer would never see Venus as more than half illuminated. (At positions *a* and *c*, Venus appears in a "new" phase; at positions *b* and *d*, it appears as a crescent. Compare with Figure 3-2, which shows the phases of the Moon.) Because Galileo saw Venus in nearly fully illuminated phases, he concluded that the Ptolemaic model must be incorrect.

vehement suspicion of heresy." Nevertheless, there was no turning back.

Galileo's observations showed convincingly that the Ptolemaic model was entirely wrong and that a heliocentric model is the more nearly correct one. Galileo did not, however, prove that the particular heliocentric model proposed by Copernicus is *the* correct one. What was needed was a heliocentric description of planetary motion that makes accurate predictions for the apparent positions of planets in the sky.

As it happened, the details of such a model were being worked out at about the same time that Galileo was making his telescopic discoveries. To construct such a model, it was essential to have an extensive set of data on planetary positions against which the predictions of the model could be tested.

figure 4-11   R I **V** U X G

Jupiter and Its Largest Moons This photograph, taken by an amateur astronomer with a small telescope, shows the four Galilean satellites alongside an overexposed image of Jupiter. Each satellite is bright enough to be seen with the unaided eye, were it not overwhelmed by the glare of Jupiter. (Courtesy of C. Holmes)

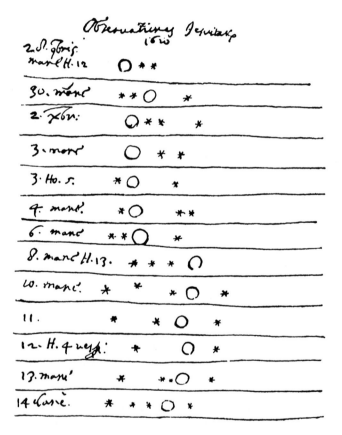

figure 4-12

Early Observations of Jupiter's Moons In 1610 Galileo discovered four "stars" that move back and forth across Jupiter from one night to the next. He concluded that these are four moons that orbit Jupiter, much as our Moon orbits the Earth. This drawing shows notations made by Jesuit observers on successive nights in 1620. The circle represents Jupiter and the stars its moons. Compare the drawing numbered 13 with the photograph in Figure 4-11. (Yerkes Observatory)

The story of how these data came to be collected begins nearly 40 years before Galileo's observations, when a young Danish astronomer pondered the nature of a new star in the heavens.

4-4 Tycho Brahe's astronomical observations disproved ancient ideas about the heavens

On November 11, 1572, a bright star suddenly appeared in the constellation of Cassiopeia. At first, it was even brighter than Venus, but then it began to grow dim. After 18 months, it faded from view. Modern astronomers recognize this event as a supernova explosion, the violent death of a massive star.

In the sixteenth century, however, the vast majority of scholars held with the ancient teachings of Aristotle and Plato, who had argued that the heavens are permanent and unalterable. Consequently, the "new star" of 1572 could not really be a star at all, because the heavens do not change; it must instead be some sort of bright object quite near the Earth, perhaps not much farther away than the clouds overhead.

The 25-year-old Danish astronomer Tycho Brahe (1546–1601) realized that straightforward observations might reveal the distance to the new star. It is common experience that when you walk from one place to another, nearby objects appear to change position against the background of more distant objects. This phenomenon, whereby the apparent position of an object changes because of the motion of the observer, is called **parallax**. If the new star was nearby, then its position should shift against the background stars over the course of a single night because Earth's rotation changes our viewpoint. Figure 4-13 shows this predicted shift. (Actually, Tycho believed that the heavens rotate about the Earth, as in the Ptolemaic model, but the net effect is the same.)

Tycho's careful observations failed to disclose any parallax. The farther away an object is, the less it appears to shift against the background as we change our viewpoint, and the smaller the parallax. Hence, the new star had to be quite far away, farther from Earth than anyone had imagined. Tycho also attempted to measure the parallax of a bright comet that appeared in 1577 and, again, found it too small to measure. Thus, the comet also had to be far beyond the Earth. With these observations, Tycho showed that the heavens are by no means pristine and unchanging. This discovery flew in the face of nearly 2000 years of astronomical thought.

Tycho's observations of the supernova of 1572 attracted the attention of the king of Denmark, who was so impressed that he financed the construction of two magnificent observatories for Tycho on the island of Hven, just off the Danish coast. The bequest for these two observatories—Uraniborg ("heavenly castle") and Stjerneborg ("star castle")—allowed Tycho to design and have built a set of astronomical instruments vastly superior in quality to any earlier instruments (Figure 4-14). With this state-of-the-art equipment, Tycho proceeded to measure the positions of stars and planets with unprecedented accuracy. In addition, he and his assistants were careful to make several observations of the same star or planet

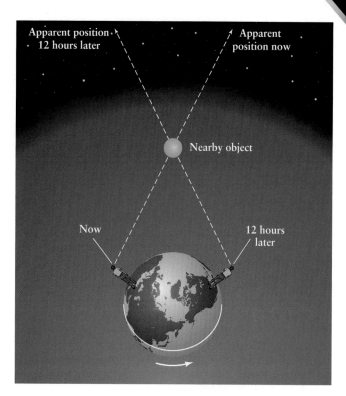

ƒigure 4-13

The Parallax of a Nearby Object Tycho Brahe argued that if an object is near the Earth, its position relative to the background stars should change over the course of a night. Tycho failed to measure such changes for a supernova in 1572 and a comet in 1577. He therefore concluded that these objects were far from the Earth.

with different instruments, in order to identify any errors that might be caused by the instruments themselves. This painstaking approach to mapping the heavens revolutionized the practice of astronomy and is used by astronomers today.

A key goal of Tycho's observations during this period was to test the ideas Copernicus had proposed decades earlier about the Earth going around the Sun. Tycho argued that if the Earth was in motion, then nearby stars should appear to shift their positions with respect to background stars as we orbit the Sun. Tycho failed to detect any such parallax, and he concluded that the Earth was at rest and the Copernican system was wrong.

On this point Tycho was in error, for nearby stars do in fact shift their positions as he had suggested. But even the nearest stars are so far away that the shifts in their positions are less than an arcsecond, too small to be seen with the naked eye. Tycho would have needed a telescope to detect the parallax that he was looking for, but the telescope was not invented until after his death in 1601. Indeed, the first accurate determination of stellar parallax was not made until 1838.

Although he remained convinced that the Earth was at the center of the universe, Tycho nonetheless made a tremendous contribution toward putting the heliocentric model on a solid foundation. From 1576 to 1597, he used his instruments to

figure 4-14

Tycho Brahe (1546–1601) Observing Tycho Brahe measured the positions of stars and planets with greater accuracy than ever before. This contemporary illustration shows Tycho with some of the measuring apparatus at Uraniborg, one of the two observatories that he built under the patronage of Frederik II of Denmark. (This magnificent observatory lacked a telescope, which had not yet been invented.) The data that Tycho collected were crucial to the development of astronomy in the years after his death. (Photo Researchers, Inc.)

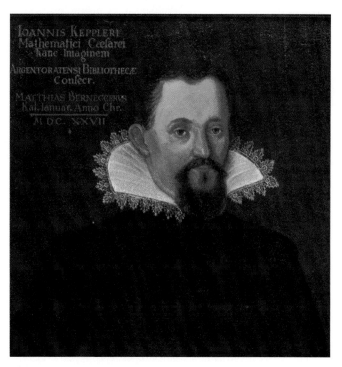

figure 4-15

Johannes Kepler (1571–1630) By analyzing Tycho Brahe's detailed records of planetary positions, Kepler developed three general principles, called Kepler's laws, that describe how the planets move about the Sun. Kepler was the first to realize that the orbits of the planets are ellipses and not circles. (E. Lessing/Magnum)

make comprehensive measurements of the positions of the planets with an accuracy of 1 arcminute. This is as well as can be done with the naked eye and was far superior to any earlier measurements. Within the reams of data that Tycho compiled lay the truth about the motions of the planets. The person who would extract this truth was a German mathematician who became Tycho's assistant in 1600, a year before the great astronomer's death. His name was Johannes Kepler (Figure 4-15).

4-5 Johannes Kepler proposed elliptical paths for the planets about the Sun

The task that Kepler took on at the beginning of the seventeenth century was to find a model of planetary motion that agreed completely with Tycho's extensive and very accurate observations of planetary positions. To do this, Kepler found that he had to break with an ancient prejudice about planetary motions.

Astronomers had long assumed that heavenly objects move in circles, which were considered the most perfect and harmonious of all geometric shapes. They believed that if a perfect God resided in heaven along with the stars and planets, then the motions of these bodies must be perfect too. Against this context, Kepler dared to try to explain planetary motions with noncircular curves. In particular, he found that he had the best success with a particular kind of curve called an **ellipse**.

You can draw an ellipse by using a loop of string, two thumbtacks, and a pencil, as shown in Figure 4-16a. Each thumbtack in the figure is at a **focus** (plural **foci**) of the ellipse; an ellipse has two foci. The longest diameter of an ellipse, called the **major axis**, passes through both foci. Half of that distance is called the **semimajor axis** and is usually designated by the letter *a*. A circle is a special case of an ellipse in which the two foci are at the same point (this corresponds to using only a single thumbtack in Figure 4-16a). The semimajor axis of a circle is equal to its radius.

By assuming that planetary orbits were ellipses, Kepler found, to his delight, that he could make his theoretical calculations match precisely to Tycho's observations. This impor-

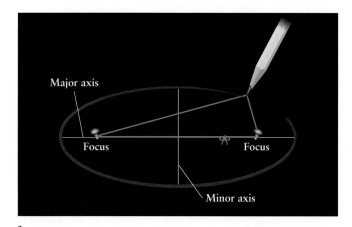

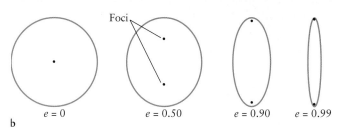

a

b

figure 4-16

Ellipses (a) To draw an ellipse, use two thumbtacks to secure the ends of a piece of string, then use a pencil to pull the string taut. If you move the pencil while keeping the string taut, the pencil traces out an ellipse. The thumbtacks are located at the two foci of the ellipse. The major axis is the greatest distance across the ellipse; the semimajor axis is half of this distance. The minor axis is perpendicular to the major axis and passes through the center of the ellipse. **(b)** A series of ellipses with the same major axis but different eccentricities. An ellipse can have any eccentricity from $e = 0$ (a circle) to just under $e = 1$ (virtually a straight line).

tant discovery, first published in 1609, is now called **Kepler's first law:**

The orbit of a planet about the Sun is an ellipse with the Sun at one focus.

There is no object at the other focus of a planet's elliptical orbit. This "empty focus" has geometrical significance, because it helps to define the shape of the ellipse, but plays no other role. The semimajor axis *a* of a planet's orbit is the average distance between the planet and the Sun.

Ellipses come in different shapes, depending on the elongation of the ellipse. The shape of an ellipse is described by its **eccentricity,** designated by the letter *e*. The value of *e* can range from 0 (a circle) to just under 1 (nearly a straight line). The greater the eccentricity, the more elongated the ellipse. Figure 4-16*b* shows a few examples of ellipses with different eccentricities. Because a circle is a special case of an ellipse, it is possible to have a perfectly circular orbit. But all of the planets have orbits that are at least slightly elliptical, with eccentricities that range from 0.007 for Venus (the most nearly circular of any planetary orbit) to 0.248 for Pluto (the most elongated orbit).

Once he knew the shape of a planet's orbit, Kepler was ready to describe exactly *how* it moves on that orbit. As a planet travels in an elliptical orbit, its distance from the Sun varies. Kepler realized that the speed of a planet also varies along its orbit. A planet moves most rapidly when it is nearest the Sun, at a point on its orbit called **perihelion.** Conversely, a planet moves most slowly when it is farthest from the Sun, at a point called **aphelion** (Figure 4-17).

After much trial and error, Kepler found a way to describe just how a planet's speed varies as it moves along its orbit. This discovery, also published in 1609, is illustrated in Figure 4-17. Suppose that it takes 30 days for a planet to go from point *A* to point *B*. During that time, an imaginary line joining the Sun and the planet sweeps out a nearly triangular

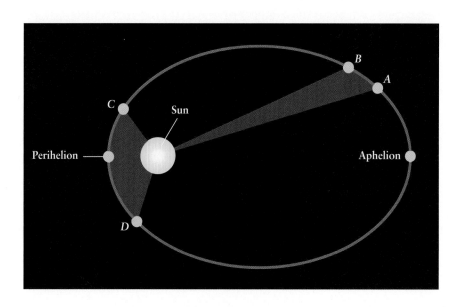

figure 4-17

Kepler's First and Second Laws
According to Kepler's first law, a planet travels around the Sun along an elliptical orbit with the Sun at one focus. According to his second law, a planet moves fastest when closest to the Sun (at perihelion) and slowest when farthest from the Sun (at aphelion). As the planet moves, an imaginary line joining the planet and the Sun sweeps out equal areas in equal intervals of time (from *A* to *B* or from *C* to *D*). Kepler found that by using these laws, his calculations gave a perfect fit to the apparent motions of the planets.

area. Kepler discovered that a line joining the Sun and the planet also sweeps out exactly the same area during any other 30-day interval. In other words, if the planet also takes 30 days to go from point *C* to point *D*, then the two shaded segments in Figure 4-17 are equal in area. **Kepler's second law** can be stated thus:

A line joining a planet and the Sun sweeps out equal areas in equal intervals of time.

This relationship is also called the **law of equal areas**. In the idealized case of a circular orbit, a planet would have to move at a constant speed around the orbit in order to satisfy Kepler's second law.

ANALOGY A good analogy for Kepler's second law is a twirling ice skater. If the skater pulls her arms straight in to her body, she spins faster; if she lets her arms extend away from her body, her rate of spin decreases. In the same way, a planet in an elliptical orbit travels at a higher speed when it moves closer to the Sun (toward perihelion) and travels at a lower speed when it moves away from the Sun (toward aphelion).

Kepler's second law describes how the speed of a given planet changes as it orbits the Sun. Kepler also deduced from Tycho's data a relationship that can be used to compare the motions of *different* planets. Published in 1618 and now called **Kepler's third law**, it states a relationship between the size of a planet's orbit and the time the planet takes to go once around the Sun:

The square of the sidereal period of a planet is directly proportional to the cube of the semimajor axis of the orbit.

Kepler's third law says that the larger a planet's orbit—that is, the larger the semimajor axis, or average distance from the

planet to the Sun—the longer the sidereal period, which is the time it takes the planet to complete an orbit. From Kepler's third law one can show that the larger the semimajor axis, the slower the average speed at which the planet moves around its orbit. (By contrast, Kepler's *second* law describes how the speed of a given planet is sometimes faster and sometimes slower than its average speed.) This qualitative relationship between orbital size and orbital speed is just what Aristarchus and Copernicus used to explain retrograde motion, as we saw in Section 4-2. Kepler's great contribution was to make this relationship a quantitative one.

It is useful to restate Kepler's third law as an equation. If a planet's sidereal period *P* is measured in years and the length of its semimajor axis *a* is measured in astronomical units (AU), where 1 AU is the average distance from the Earth to the Sun (Section 1-7), then Kepler's third law is

Kepler's third law

$$P^2 = a^3$$

P = planet's sidereal period, in years
a = planet's semimajor axis, in AU

If you know either the sidereal period of a planet or the semimajor axis of its orbit, you can find the other quantity using this equation. Box 4-2 gives some examples of how this is done.

We can verify Kepler's third law for all of the planets, including those that were discovered after Kepler's death, using data from Tables 4-1 and 4-2. If Kepler's third law is correct, for each planet the numerical values of P^2 and a^3 should be equal. This is indeed true to very high accuracy, as Table 4-3 shows.

Kepler's laws are a landmark in the history of astronomy. They made it possible to calculate the motions of the planets with far better accuracy than any geocentric model ever had, and they helped to justify the idea of a heliocentric model.

table 4-3	A Demonstration of Kepler's Third Law		
Planet	**Sidereal period** P **(years)**	**Semimajor axis** a **(AU)**	P^2	a^3
Mercury	0.24	0.39	0.06	0.06
Venus	0.61	0.72	0.37	0.37
Earth	1.00	1.00	1.00	1.00
Mars	1.88	1.52	3.53	3.51
Jupiter	11.86	5.20	140.7	140.6
Saturn	29.46	9.54	867.9	868.3
Uranus	84.01	19.19	7,058	7,067
Neptune	64.79	30.06	27,160	27,160
Pluto	248.54	39.53	61,770	61,770

Box 4-2 | **T**ools of the Astronomer's Trade

Using Kepler's Third Law

Kepler's third law relates the sidereal period P of an object orbiting the Sun to the semimajor axis a of its orbit:

$$P^2 = a^3$$

You must keep two essential points in mind when working with this equation:

1. The period P *must* be measured in years, and the semimajor axis a *must* be measured in astronomical units (AU). Otherwise you will get nonsensical results.

2. This equation applies *only* to the special case of an object, like a planet, that orbits the Sun. If you want to analyze the orbit of the Moon around the Earth, of a spacecraft around Mars, or of a planet around a distant star, you must use a different, generalized form of Kepler's third law. We discuss this alternative equation in Section 4-7 and Box 4-4.

EXAMPLE: The average distance from Venus to the Sun—that is, the semimajor axis a of Venus's orbit—is 0.72 AU. To determine the sidereal period of Venus, we first cube the semimajor axis (multiply it by itself twice):

$$a^3 = (0.72)^3 = 0.72 \times 0.72 \times 0.72 = 0.373$$

According to Kepler's third law this is also equal to P^2, the square of the sidereal period. So, to find P, we have to "undo" the square, that is, take the square root. Using a calculator, we find

$$P = \sqrt{P^2} = \sqrt{0.373} = 0.61$$

The sidereal period of Venus is 0.61 years, or a bit more than seven Earth months.

EXAMPLE: Consider a small asteroid (a rocky body a few tens of kilometers across) that takes eight years to complete one orbit around the Sun. We can use Kepler's third law to determine the semimajor axis a of the asteroid's orbit. We first square the period:

$$P^2 = 8^2 = 8 \times 8 = 64$$

From Kepler's third law, this is also equal to a^3. To determine a, we must take the *cube root* of a^3, that is, find the number whose cube is 64. If your calculator has a cube root function, denoted by the symbol $\sqrt[3]{}$, you can use it to find that the cube root of 64 is 4: $\sqrt[3]{64} = 4$. Otherwise, you can determine by trial and error that the cube of 4 is 64:

$$4^3 = 4 \times 4 \times 4 = 64$$

Because the cube of 4 is 64, it follows that the cube root of 64 is 4 (taking the cube root "undoes" the cube).

With either technique you find that the orbit of this asteroid has semimajor axis $a = 4$ AU. This is intermediate between the orbits of Mars and Jupiter. Many asteroids are known with semimajor axes in this range, forming a region in the solar system called the asteroid belt.

Kepler's laws also pass the test of Occam's razor, for they are simpler in every way than the schemes of Ptolemy or Copernicus, both of which used a complicated combination of circles.

But the significance of Kepler's laws goes beyond understanding planetary orbits. These same laws are also obeyed by spacecraft orbiting the Earth, by two stars revolving about each other in a binary star system, and even by galaxies in their orbits about each other. Throughout this book, we shall use Kepler's laws in a wide range of situations.

What Kepler was not able to explain is *why* planets move in accordance with his three laws. The first person who was able to provide such an explanation was the Englishman Isaac Newton, born on Christmas Day of 1642, a dozen years after the death of Kepler and the same year that Galileo died. While Galileo and Kepler revolutionized our understanding of planetary motions, Newton's contribution was far greater: He deduced the basic laws that govern all motions on Earth as well as in the heavens.

4-6 Isaac Newton formulated three laws that describe fundamental properties of physical reality

Until the mid-seventeenth century, virtually all attempts to describe the heavens mathematically were *empirical*, or based directly on data and observations. From Ptolemy to Kepler, astronomers would adjust their ideas and calculations by trial and error until they ended up with answers that agreed with observation.

Isaac Newton (Figure 4-18) introduced a new approach. He began with three quite general statements, now called **Newton's laws of motion.** These laws, deduced from experimental observation, apply to all forces and all bodies. Newton then showed that Kepler's three laws follow logically from these laws of motion and from a formula for the force of gravity that he derived from observation.

In other words, Kepler's laws are not just an empirical description of the motions of the planets, but a direct consequence of the fundamental laws of physical matter. Using this deeper insight into the nature of motions in the heavens, Newton and his successors were able to describe accurately not just the orbits of the planets but also the orbits of the Moon and comets.

Newton's laws of motion describe objects on Earth as well as in the heavens. Thus, we can understand each of these laws by considering the motions of objects around us. We begin with **Newton's first law of motion,** or **law of inertia:**

A body remains at rest, or moves in a straight line at a constant speed, unless acted upon by a net outside force.

figure 4-18

Isaac Newton (1642–1727) Using mathematical techniques that he devised, Isaac Newton formulated the law of universal gravitation and demonstrated that the planets orbit the Sun according to simple mechanical rules. (National Portrait Gallery, London)

By **force** we mean any push or pull that acts on the body. An *outside* force is one that is exerted on the body by something other than the body itself. The net, or total, outside force is the combined effect of all of the individual outside forces that act on the body.

Right now, you are demonstrating the first part of Newton's first law. As you sit in your chair reading this, there are two outside forces acting on you: The force of gravity pulls you downward, and the chair pushes up on you. These two forces are of equal strength but of opposite direction, so their effects cancel—there is no *net* outside force. Hence, your body remains at rest as stated in Newton's first law. If you try to lift yourself out of your chair by grabbing your knees and pulling up, you will remain at rest because this force is not an outside force.

The second part of Newton's first law, about objects in motion, may seem to go against common sense. If you want to make this book move across the floor in a straight line at a constant speed, you must continually push on it. You might therefore think that there *is* a net outside force, the force of your push. But another force also acts on the book—the force of friction as the book rubs across the floor. As you push the book across the floor, the force of your push exactly balances the force of friction, so again there is no net outside force. The effect is to make the book move in a straight line at constant speed, just as Newton's first law says. If you stop pushing, there will be nothing to balance the effects of friction. Then there will be a net outside force and the book will slow to a stop.

Newton's first law tells us that a force must be acting on the planets. A planet moving through empty space encounters no friction, so it would tend to fly off into space along a straight line if there were no other outside force acting on it. Because this does not happen, Newton concluded that there must be a force that acts continuously on the planets to keep them in their elliptical orbits.

Newton's second law describes how the motion of an object changes if there is a net outside force acting on it. To appreciate Newton's second law, we must first understand three quantities that describe motion—speed, velocity, and acceleration.

Speed is a measure of how fast an object is moving. Speed and direction of motion together constitute an object's **velocity.** Compared with a car driving north at 100 km/h (62 mi/h), a car driving east at 100 km/h has the same speed but a different velocity.

Acceleration is the rate at which velocity changes. Because velocity involves both speed and direction, acceleration can result from changes in either. Contrary to popular use of the term, acceleration does not simply mean speeding up. A car is accelerating if it is speeding up, and it is also accelerating if it is slowing down or turning (that is, changing the direction in which it is moving).

You can verify these statements about acceleration if you think about the sensations of riding in a car. If the car is moving with a constant velocity (in a straight line at a constant speed), you feel the same as if the car were not moving at all.

Box 4-3 **The Heavens on the Earth**

Newton's Laws in Everyday Life

In our study of astronomy, we use Newton's three laws of motion to help us understand the motions of objects in the heavens. But you can see applications of Newton's laws every day in the world around you. By considering these everyday applications, we can gain insight into how Newton's laws apply to celestial events that are far removed from ordinary human experience.

Newton's *first* law, or principle of inertia, says that an object at rest naturally tends to remain at rest and that an object in motion naturally tends to remain in motion. This explains the sensations that you feel when riding in an automobile. When you are waiting at a red light, your car and your body are both at rest. When the light turns green and you press on the gas pedal, the car accelerates forward but your body attempts to stay where it was. Hence, the seat of the accelerating car pushes forward into your body, and it feels as though you are being pushed back in your seat.

Once the car is up to cruising speed, your body wants to keep moving in a straight line at this cruising speed. If the car makes a sharp turn to the left, the right side of the car will move toward you. Thus, you will feel as though you are being thrown to the car's right side (the side on the outside of the turn). If you bring the car to a sudden stop by pressing on the brakes, your body will continue moving forward until the seat belt stops you. In this case, it feels as though you are being thrown toward the front of the car.

Newton's *second* law states that the net outside force on an object equals the product of the object's mass and its acceleration. You can accelerate a crumpled-up piece of paper to a pretty good speed by throwing it with a moderate force. But if you try to throw a backpack full of books by using the same force, the acceleration will be

much less because the backpack has much more mass than the crumpled paper. Because of the reduced acceleration during the throw, the backpack will leave your hand moving at only a slow speed.

Automobile airbags are based on the relationship between force and acceleration. It takes a large force to bring a fast-moving object suddenly to rest because this requires a large acceleration. In a collision, the driver of a car not equipped with airbags is jerked to a sudden stop and the large forces that act can cause major injuries. But if the car has airbags that deploy in an accident, the driver's body will slow down more gradually as it contacts the airbag, and the driver's acceleration will be less. (Remember that *acceleration* can refer to slowing down as well as to speeding up.) Hence, the force on the driver and the chance of injury will both be greatly reduced.

Newton's *third* law, the principle of action and reaction, explains how a car can accelerate at all. It is not correct to say that the engine pushes the car forward, because Newton's second law tells us that it takes a force acting from outside the car to make the car accelerate. Rather, the engine makes the wheels and tires turn, and the tires push backward on the ground. (You can see this backward force acting when a car drives through mud and sprays mud backward from the spinning tires.) From Newton's third law, the ground must exert an equally large forward force on the car, and this is the force that pushes the car forward.

You use the same principles when you walk: You push backward on the ground with your foot, and the ground pushes forward on you. Icy pavement or a freshly waxed floor have greatly reduced friction. In these situations, your feet and the surface under you can exert only weak forces on each other, and it is much harder to walk.

But you can feel it when the car accelerates in any way: You feel thrown back in your seat if the car speeds up, thrown forward if the car slows down, and thrown sideways if the car changes direction in a tight turn. In Box 4-3 we discuss the reasons for these sensations, along with other applications of Newton's laws to everyday life.

An apple falling from a tree is a good example of acceleration that involves only an increase in speed. Initially, at the moment the stem breaks, the apple's speed is zero. After 1 second, its downward speed is 9.8 meters per second, or 9.8 m/s (32 feet per second, or 32 ft/s). After 2 seconds, the apple's speed is twice this, or 19.6 m/s. After 3 seconds, the speed is 29.4 m/s. Because the apple's speed increases by 9.8 m/s for each second of free fall, the rate of acceleration is

9.8 meters per second per second, or 9.8 m/s^2 (32 ft/s^2). Thus, the Earth's gravity gives the apple a constant acceleration of 9.8 m/s^2 downward, toward the center of the Earth.

A planet revolving about the Sun along a perfectly circular orbit is an example of acceleration that involves change of direction only. As the planet moves along its orbit, its speed remains constant. Nevertheless, the planet is continuously being accelerated because its direction of motion is continuously changing.

Newton's second law of motion says that the acceleration of an object is proportional to the net outside force acting on the object. In other words, the harder you push on an object, the greater the resulting acceleration. This law can be succinctly stated as an equation. If a net outside force F acts on an object of mass m, the object will experience an acceleration a such that

Newton's second law

$$F = ma$$

F = net outside force on an object
m = mass of object
a = acceleration of object

The **mass** of an object is a measure of the total amount of material in the object. It is usually expressed in kilograms (kg) or grams (g). For example, the mass of the Sun is 2×10^{30} kg, the mass of a hydrogen atom is 1.7×10^{-27} kg, and the mass of an average adult is 75 kg. The Sun, a hydrogen atom, and a person have these masses regardless of where they happen to be in the universe.

It is important not to confuse the concepts of mass and weight. **Weight** is the force of gravity that acts on a body and, like any force, is usually expressed in pounds or newtons (1 newton = 0.225 pound).

We can use Newton's second law to relate mass and weight. We have seen that the acceleration caused by the Earth's gravity is 9.8 m/s^2. When a 50-kg swimmer falls from a diving board, the only outside force acting on her as she falls is her weight. Thus, from Newton's second law ($F = ma$), her weight is equal to her mass multiplied by the acceleration due to gravity:

$$50 \text{ kg} \times 9.8 \text{ m/s}^2 = 490 \text{ newtons} = 110 \text{ pounds}$$

Note that this answer is correct only when the swimmer is on Earth. She would weigh less on the Moon, where the pull of gravity is weaker, and more on Jupiter, where the gravitational pull is stronger. Floating deep in space, she would have no weight at all; she would be "weightless." Nevertheless, in all these circumstances, she would always have exactly the same mass, because mass is an inherent property of matter unaffected by details of the environment. Whenever we describe the properties of planets, stars, or galaxies, we speak of their masses, never of their weights.

We have seen that a planet is continually accelerating as it orbits the Sun. From Newton's second law, this means that there must be a net outside force that acts continually on each of the planets. As we will see in the next section, this force is the gravitational attraction of the Sun.

The last of Newton's general laws of motion, called **Newton's third law of motion**, is the famous statement about action and reaction:

Whenever one body exerts a force on a second body, the second body exerts an equal and opposite force on the first body.

For example, if you weigh 110 pounds, when you are standing up you are pressing down on the floor with a force of 110 pounds. Newton's third law tells us that the floor is also pushing up against your feet with an equal force of 110 pounds.

Newton realized that because the Sun is exerting a force on each planet to keep it in orbit, each planet must also be exerting an equal and opposite force on the Sun. However, the planets are much less massive than the Sun (for example, the Earth has only 1/300,000 of the Sun's mass). Therefore, although the Sun's force on a planet is the same as the planet's force on the Sun, the planet's much smaller mass gives it a much larger acceleration, according to Newton's second law. This is why the planets circle the Sun instead of vice versa. Thus, Newton's laws reveal the reason for our heliocentric solar system.

4-7 Newton's description of gravity accounts for Kepler's laws and explains the motions of the planets

Tie a ball to one end of a piece of string, hold the other end of the string in your hand, and whirl the ball around in a circle. As the ball "orbits" your hand, it is continuously accelerating because its velocity is changing. (Even if its speed is constant, its direction of motion is changing.) In accordance with Newton's second law, this can happen only if the ball is continuously acted on by an outside force—the pull of the string. The pull is directed along the string toward your hand. In the same way, Newton saw, the force that keeps a planet in orbit around the Sun is a pull that always acts toward the Sun. That pull is **gravity**, or **gravitational force**.

Newton's discovery about the forces that act on planets led him to suspect that the force of gravity pulling a falling apple straight down to the ground is fundamentally the same as the force on a planet that is always directed straight at the Sun. In other words, gravity is the force that shapes the orbits of the planets. What is more, he was able to determine how the force of gravity depends on the distance between the Sun and the planet. His result was a law of gravitation that could apply to the motion of distant planets as well as to the flight of a football on Earth. Using this law, Newton achieved the remarkable goal of deducing Kepler's laws from fundamental principles of nature.

To see how Newton reasoned, think again about a ball attached to a string. If you use a short string, so that the ball orbits in a small circle, and whirl the ball around your hand at a high speed, you will find that you have to pull fairly hard on the string (Figure 4-19a). But if you use a longer string, so that the ball moves in a larger orbit, and if you make the ball orbit your hand at a slow speed, you only have to exert a light tug on the string (Figure 4-19b). The orbits of the planets behave in the same way; the larger the size of the orbit, the slower the planet's speed. By analogy to the force of the string on the orbiting ball, Newton concluded that the force that attracts a planet toward the Sun must decrease with increasing distance between the Sun and the planet.

Using his own three laws and Kepler's three laws, Newton succeeded in formulating a general statement that describes the nature of the gravitational force. Newton's **law of universal gravitation** is as follows:

Two bodies attract each other with a force that is directly proportional to the mass of each body and inversely proportional to the square of the distance between them.

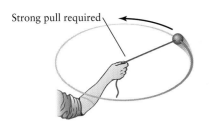

Strong pull required

a **Ball moves at a high speed in a small circle**

Only a weak pull required

b **Ball moves at a low speed in a large circle**

figure 4-19

A Ball on a String (a) To make a ball on a string move at high speed around a small circle, you have to exert a substantial pull on the string. Similarly, a planet that orbits close to the Sun moves at high speed and requires a substantial gravitational force from the Sun. **(b)** If you lengthen the string and make the same ball move at low speed around a large circle, much less pull is required. In the same way, a planet in a large orbit moves at low speed and requires less gravitational force to stay in orbit. This example shows that the Sun's gravitational pull must decrease with increased distance.

This law states that *any* two objects exert gravitational pulls on each other. Normally, you notice only the gravitational force that the Earth exerts on you, otherwise known as your weight. In fact, you feel gravitational attractions to *all* the objects around you. For example, this book is exerting a gravitational force on you as you read it. But because the force exerted on you by this book is proportional to the book's mass, which is very small compared to the Earth's mass, the force is too small to notice. (It can actually be measured with sensitive equipment.)

Consider two 1-kg objects separated by a distance of 1 meter. Newton's law of universal gravitation says that the force is directly proportional to the mass, so if we double the mass of one object to 2 kg, the force between the objects will double. If we double both masses so that we have two 2-kg objects separated by 1 meter, the force will be $2 \times 2 = 4$ times what it was originally (the force is directly proportional to the mass of *each* object). If we go back to two 1-kg masses, but double their separation to 2 meters, the force will be only one-quarter its original value. This is because the force is inversely proportional to the distance: If we double the distance, the force is multiplied by a factor of

$$\frac{1}{2^2} = \frac{1}{4}$$

Newton's law of universal gravitation can be stated more succinctly as an equation. If two objects have masses m_1 and m_2 and are separated by a distance r, then the gravitational force F between these two objects is given by the equation

Newton's law of universal gravitation

$$F = G\left(\frac{m_1 m_2}{r^2}\right)$$

F = gravitational force between two objects
m_1 = mass of first object
m_2 = mass of second object
r = distance between objects
G = universal constant of gravitation

If the masses are measured in kilograms and the distance between them in meters, then the force is measured in new-tons. In this formula, G is a number called the **universal constant of gravitation.** Laboratory experiments have yielded a value for G of

$$G = 6.67 \times 10^{-11} \text{ newton m}^2/\text{kg}^2$$

We can use Newton's law of universal gravitation to calculate the force with which any two bodies attract each other. For example, to compute the gravitational force that the Sun exerts on the Earth, we substitute values for the Earth's mass ($m_1 = 5.98 \times 10^{24}$ kg), the Sun's mass ($m_2 = 1.99 \times 10^{30}$ kg), the distance between them ($r = 1$ AU $= 1.5 \times 10^{11}$ m), and the value of G into Newton's equation. We get

$$F_{\text{Sun-Earth}} = 6.67 \times 10^{-11} \left[\frac{(5.98 \times 10^{24}) \times (1.99 \times 10^{30})}{(1.50 \times 10^{11})^2} \right]$$

$$= 3.53 \times 10^{22} \text{ newtons}$$

If we calculate the force that the Earth exerts on the Sun, we get exactly the same result. (Mathematically, we just let m_1 be the Sun's mass and m_2 be the Earth's mass instead of the other way around. The product of the two numbers is the same, so the force is the same.) This is in accordance with Newton's third law: Any two objects exert *equal* gravitational forces on each other.

Because there is a gravitational force between any two objects, Newton concluded that gravity is also the force that keeps the Moon in orbit around the Earth. It is also the force that keeps artificial satellites in orbit. But if the force of gravity attracts two objects to each other, why don't satellites immediately fall to Earth? Why doesn't the Moon fall into the Earth? And, for that matter, why don't the planets fall into the Sun?

To see the answer, imagine (as Newton did) dropping a ball from a great height above the Earth's surface, as in Figure 4-20. After you drop the ball, it, of course, falls straight down (path A in Figure 4-20). But if you *throw* the ball horizontally, it travels some distance across the Earth's surface before hitting the ground (path B). If you throw the ball harder, it travels a greater distance (path C). If you could throw at just the right speed, the curvature of the

WEB LINK 4.8 **figure 4-20**

An Explanation of Orbits If a ball is dropped from a great height above the Earth's surface, it falls straight down (A). If the ball is thrown with some horizontal speed, it follows a curved path before hitting the ground (B, C). If thrown with just the right speed (E), the ball goes into circular orbit; the ball's path curves but it never gets any closer to the Earth's surface. If the ball is thrown with a speed that is slightly less (D) or slightly more (F) than the speed for a circular orbit, the ball's orbit is an ellipse.

ball's path will exactly match the curvature of the Earth's surface (path E). Although the Earth's gravity is making the ball fall, the Earth's surface is falling away under the ball at the same rate. Hence, the ball does not get any closer to the surface, and the ball is in circular orbit. So the ball in path E is in fact falling, but it is falling *around* the Earth rather than *toward* the Earth.

A spacecraft is launched into orbit in just this way—by throwing it fast enough. The thrust of a rocket is used to give the spacecraft the necessary orbital speed. Once the spacecraft is in orbit, the rocket turns off and the spacecraft falls continually around the Earth.

CAUTION, An astronaut on board an orbiting spacecraft feels "weightless," but this is *not* because she is "beyond the pull of gravity." The astronaut is herself an independent satellite of the Earth, and the Earth's gravitational pull is what holds her in orbit. She feels "weightless" because

she and her spacecraft are falling *together* around the Earth, so there is nothing pushing her against any of the spacecraft walls. You feel the same "weightless" sensation whenever you are falling, such as when you jump off a diving board or ride the free-fall ride at an amusement park.

If the ball in Figure 4-20 is thrown with a slightly slower speed than that required for a circular orbit, its orbit will be an ellipse (path D). An elliptical orbit also results if instead the ball is thrown a bit too fast (path F). In this way, spacecraft can be placed into any desired orbit around the Earth by adjusting the thrust of the rockets.

Just as the ball in Figure 4-20 will not fall to Earth if given enough speed, the Moon does not fall to Earth and the planets do not fall into the Sun. The planets acquired their initial speeds around the Sun when the solar system first formed 4.6 billion years ago. Figure 4-20 shows that a circular orbit is a very special case, so it is no surprise that the orbits of the planets are not precisely circular.

CAUTION, Orbiting satellites do sometimes fall out of orbit and crash back to Earth. When this happens, however, the real culprit is not gravity but air resistance. A satellite in a relatively low orbit is actually flying through the tenuous outer wisps of the Earth's atmosphere. The resistance of the atmosphere slows the satellite and changes a circular orbit like E in Figure 4-20 to an elliptical one like D. As the satellite sinks to lower altitude, it encounters more air resistance and sinks even lower. Eventually, it either strikes the Earth or burns up in flight due to air friction. By contrast, the Moon and planets orbit in the near-vacuum of interplanetary space. Hence, they are unaffected by this kind of air resistance, and their orbits are much more long-lasting.

Using his three laws of motion and his law of gravity, Newton found that he could prove Kepler's three laws mathematically. Kepler's first law, concerning the elliptical shape of planetary orbits, proved to be a direct consequence of the $1/r^2$ factor in the law of universal gravitation. (Had the nature of gravity in our universe been different, so that this factor was given by a different function such as $1/r$ or $1/r^3$, elliptical orbits would not have been possible.) The law of equal areas, or Kepler's second law, turns out to be a consequence of the Sun's gravitational force on a planet being directed straight toward the Sun.

Newton also demonstrated that Kepler's third law follows logically from his law of gravity. Specifically, he proved that if two objects with masses m_1 and m_2 orbit each other, the period P of their orbit and the semimajor axis a of their orbit (that is, the average distance between the two objects) are related by the equation

$$P^2 = \left[\frac{4\pi^2}{G(m_1 + m_2)} \right] a^3$$

This equation is called **Newton's form of Kepler's third law.** It is valid whenever two objects orbit each other because of their mutual gravitational attraction. It is invaluable in the

Box 4-4

Newton's Form of Kepler's Third Law

Kepler's original statement of his third law, $P^2 = a^3$, is valid only for objects that orbit the Sun. (Box 4-2 shows how to use this equation.) But Newton's form of Kepler's third law is much more general: It can be used in *any* situation where two bodies of masses m_1 and m_2 orbit each other. For example, Newton's form is the equation to use for a moon orbiting a planet or a satellite orbiting the Earth. This equation is

Newton's form of Kepler's third law

$$P^2 = \left[\frac{4\pi^2}{G(m_1 + m_2)} \right] a^3$$

P = sidereal period of orbit, in seconds
a = semimajor axis of orbit, in meters
m_1 = mass of first object, in kilograms
m_2 = mass of second object, in kilograms
G = universal constant of gravitation = 6.67×10^{-11}

Notice that P, a, m_1, and m_2 *must* be expressed in these particular units. If you fail to use the correct units, your answer will be incorrect.

EXAMPLE: Io is one of the four large moons of Jupiter discovered by Galileo. It orbits at a distance of 421,600 km from the center of Jupiter and has an orbital period of 1.77 days. Because this is not an orbit around the Sun, we must use Newton's form of Kepler's third law. We can use

this form to determine the combined mass of Jupiter (m_1) and Io (m_2). We first rewrite the equation in the form

$$m_1 + m_2 = \frac{4\pi^2 a^3}{G P^2}$$

To use this equation, we have to convert the distance a from kilometers to meters and convert the period P from days to seconds. There are 1000 meters in 1 kilometer and 86,400 seconds in 1 day, so

$$a = (421{,}600 \text{ km}) \times \frac{1000 \text{ m}}{1 \text{ km}} = 4.216 \times 10^8 \text{ m}$$

$$P = (1.77 \text{ days}) \times \frac{86{,}400 \text{ s}}{1 \text{ day}} = 1.529 \times 10^5 \text{ s}$$

We can now put these values and the value of G into the above equation:

$$m_1 + m_2 = \frac{4\pi^2 (4.216 \times 10^8)^3}{(6.67 \times 10^{-11})(1.529 \times 10^5)^2} = 1.90 \times 10^{27} \text{ kg}$$

Io is very much smaller than Jupiter, so its mass is only a small fraction of the mass of Jupiter. Thus, $m_1 + m_2$ is very nearly the mass of Jupiter alone. We conclude that Jupiter has a mass of 1.90×10^{27} kg, or about 300 times the mass of Earth. This technique can be used to determine the mass of any object that has a second, much smaller object orbiting around it. Astronomers use this technique to find the masses of stars, black holes, and entire galaxies of stars.

---

study of binary star systems, in which two stars orbit each other. If the orbital period P and semimajor axis a of the two stars in a binary system are known, astronomers can use this formula to calculate the sum $m_1 + m_2$ of the masses of the two stars. Within our own solar system, Newton's form of Kepler's third law makes it possible to learn about the masses of planets. By measuring the period and semimajor axis for a satellite, astronomers can determine the sum of the masses of the planet and the satellite. (The satellite can be a moon of the planet or a spacecraft that we place in orbit around the planet. Newton's laws apply in either case.) Box 4-4 gives some examples using Newton's form of Kepler's third law.

Newton also discovered new features of orbits around the Sun. For example, his equations soon led him to conclude that the orbit of an object around the Sun need not be an ellipse. It could be any one of a family of curves called conic sections.

A **conic section** is any curve that you get by cutting a cone with a plane, as shown in Figure 4-21. You can get circles and ellipses by slicing all the way through the cone. You can also get two types of open curves called **parabolas** and **hyperbolas.** If you were to throw the ball in Figure 4-20 with a fast enough speed, it would follow a parabolic or hyperbolic orbit and would fly off into space, never to return. Comets hurtling toward the Sun from the depths of space sometimes follow hyperbolic orbits.

Newton's ideas turned out to be applicable to an incredibly wide range of situations. Using his laws of motion, Newton himself proved that the Earth's axis of rotation must precess because of the gravitational pull of the Moon and the Sun on the Earth's equatorial bulge (see Figure 2-17). In fact, all the details of the orbits of the planets and their satellites could be explained mathematically with a body of knowledge built on Newton's work that is today called **Newtonian mechanics.**

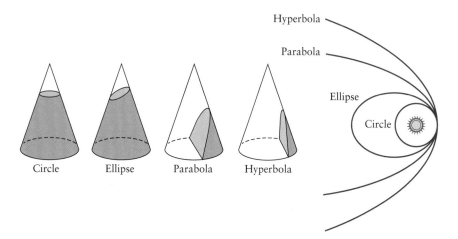

Hyperbola

Parabola

Ellipse

Circle

Circle   Ellipse   Parabola   Hyperbola

figure 4-21
Conic Sections A conic section is any one of a family of curves obtained by slicing a cone with a plane. The orbit of one body about another can be any one of these curves: a circle, an ellipse, a parabola, or a hyperbola.

Not only could Newtonian mechanics explain a variety of known phenomena in detail, but it could also predict new phenomena. For example, one of Newton's friends, Edmund Halley, was intrigued by three similar historical records of a comet that had been sighted at intervals of 76 years. Assuming these records to be accounts of the same comet, Halley used Newton's methods to work out the details of the comet's orbit and predicted its return in 1758. It was first sighted on Christmas night of 1757, a fitting memorial to Newton's birthday, and to this day the comet bears Halley's name (Figure 4-22).

Perhaps the most dramatic success of Newton's ideas was their role in the discovery of the eighth planet from the Sun. The seventh planet, Uranus, had been discovered accidentally by William Herschel in 1781 during a telescopic survey of the sky. Fifty years later, however, it was clear that Uranus was not following its predicted orbit. Two mathematicians, John Couch Adams in England and Urbain Le Verrier in France,

independently calculated that the gravitational pull of a yet unknown, more distant planet could explain the deviations of Uranus from its orbit. They each predicted that the planet would be found at a certain location in the constellation of Aquarius. A brief telescopic search on September 23, 1846, revealed the planet Neptune within 1° of the calculated position. Before it was sighted with a telescope, Neptune was actually predicted with pencil and paper.

Because it has been so successful in explaining and predicting many important phenomena, Newtonian mechanics has become the cornerstone of modern physical science. Even today, as we send astronauts into Earth orbit and spacecraft to the outer planets, Newton's equations are used to calculate the orbits and trajectories of these spacecraft.

In the twentieth century, scientists found that Newton's laws do not apply in all situations. A new theory called *quantum mechanics* had to be developed to explain the behavior of matter on the very smallest of scales, such as within the atom and within the atomic nucleus. Albert Einstein developed the *theory of relativity* to explain what happens at very high speeds approaching the speed of light and in places where gravitational forces are very strong. For many purposes in astronomy, however, Newton's laws are as useful today as when Newton formulated them more than three centuries ago.

4-8   Gravitational forces between the Earth and Moon produce tides

We have seen how Newtonian mechanics explains why the Moon stays in orbit around the Earth. It also explains why there are ocean tides, as well as why the Moon always keeps the same face toward the Earth. Both of these are consequences of *tidal forces*—an aspect of gravity that deforms planets and reshapes galaxies.

Tidal forces are differences in the gravitational pull at different points in an object. As an illustration, imagine that three billiard balls are lined up in space at some distance from a planet, as in Figure 4-23a. According to Newton's law of universal gravitation, the force of attraction between two

figure 4-22   R I **V** U X G
Comet Halley This most famous of all comets orbits the Sun with an average period of about 76 years. During the twentieth century, the comet passed near the Sun in 1910 and again in 1986. It will next be prominent in the sky in 2061. This photograph shows how the comet looked in 1986.
(David Malin, Anglo-Australian Observatory/Royal Observatory, Edinburgh)

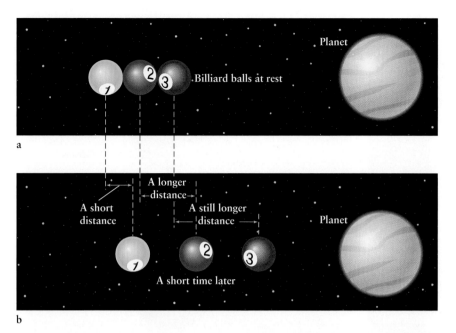

a

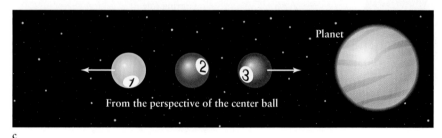

b

c

ANIMATION 4.7

Figure 4-23

The Origin of Tidal Forces

(a) Imagine three identical billiard balls placed some distance from a planet and released. **(b)** The closer a ball is to the planet, the more gravitational force the planet exerts on it. Thus, a short time after the balls are released, the blue 2-ball has moved farther toward the planet than the yellow 1-ball, and the red 3-ball has moved farther still. **(c)** From the perspective of the 2-ball in the center, it appears that forces have pushed the 1-ball away from the planet and pulled the 3-ball toward the planet. These forces are called tidal forces.

objects is greater the closer the two objects are to each other. Thus, the planet exerts more force on the 3-ball (in red) than on the 2-ball (in blue), and, in turn, the planet exerts more force on the 2-ball than on the 1-ball (in yellow). Now, imagine that the three balls are released and allowed to fall toward the planet. Figure 4-23*b* shows the situation a short time later. Because of the differences in gravitational pull, a short time later the 3-ball will have moved farther than the 2-ball, which will in turn have moved farther than the 1-ball. But now imagine that same motion from the perspective of the 2-ball. From this perspective, it appears as though the 3-ball is pulled toward the planet while the 1-ball is pushed away from the planet (Figure 4-23*c*). These apparent pushes and pulls are called tidal forces.

The Moon has a similar effect on the Earth as the planet in Figure 4-23 has on the three billiard balls. The arrows in Figure 4-24*a* indicate the strength and direction of the gravitational force of the Moon at several locations on the Earth. The side of the Earth closest to the Moon feels a greater gravitational pull than the Earth's center does, and the side of the Earth that faces away from the Moon feels less gravitational pull than does the Earth's center. Like the billiard balls in Figure 4-23, the result is that there are tidal forces acting on the Earth (Figure 4-24*b*). These forces try to elongate the

Earth along a line connecting the centers of the Earth and the Moon and try to squeeze the Earth inward in the direction perpendicular to that axis. The net effect is that the Moon's gravity tries to deform the Earth into a football shape.

Because the body of the Earth is largely rigid, it cannot deform very much in response to the Moon's tidal forces. But the water in the oceans can and does deform into a football shape, as Figure 4-25*a* shows. As the Earth rotates, a point on its surface goes from where the water is shallow to where the water is deep and back again. This is the origin of low and high ocean tides. (In this simplified description we have assumed that the Earth is completely covered with water. The full story of the tides is much more complex, because the shapes of the continents and the effects of winds must also be taken into account.)

Actually, the Sun also exerts tidal forces on the Earth's oceans. (The tidal effects of the Sun are about half as great as those of the Moon.) When the Sun, Moon, and the Earth are aligned, which happens at either new moon or full moon, the tidal effects of the Sun and Moon reinforce each other and the tidal distortion of the oceans is greatest. This produces large shifts in water level called **spring tides** (Figure 4-25*b*). At first quarter and last quarter, when the Sun and Moon form a right angle with the Earth, the tidal effects of the Sun and Moon

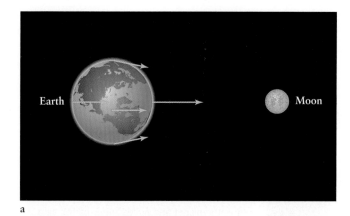

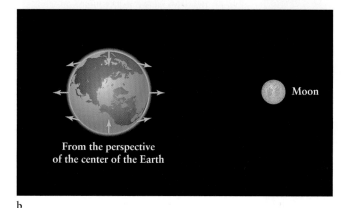

figure 4-24

Tidal Forces on the Earth **(a)** These arrows indicate the strength and direction of the Moon's gravitational pull at selected points on the Earth. **(b)** These arrows indicate the strength and direction of the tidal forces acting on the Earth. At any location, the tidal force equals the Moon's gravitational pull at that point minus the gravitational pull of the Moon at the center of the Earth. These tidal forces tend to deform the Earth into a nonspherical shape.

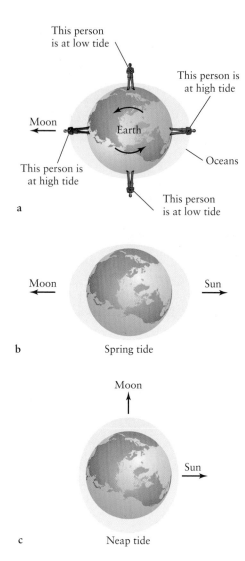

figure 4-25

High and Low Tides **(a)** The gravitational forces of the Moon and Sun deform the oceans. As the Earth rotates underneath the oceans, a given location experiences alternating low tides and high tides. **(b)** The greatest deformation (spring tides) occurs when the Sun, Earth, and Moon are aligned and the tidal effects of the Sun and Moon reinforce each other. **(c)** The least deformation (neap tides) occurs when the Sun, Earth, and Moon form a right angle. In this situation the tidal effects of the Sun and Moon partially cancel each other.

partially cancel each other. Hence, the tidal distortion of the oceans is the least pronounced, producing smaller tidal shifts called **neap tides** (Figure 4-25*c*).

Note that spring tides have nothing to do with the season of the year called spring. Instead, the name refers to the way that the ocean level "springs up" to a greater than normal height. Spring tides occur whenever there is a new moon or full moon, no matter what the season of the year.

Just as the Moon exerts tidal forces on the Earth, the Earth exerts tidal forces on the Moon. Soon after the Moon formed some 4.6 billion years ago, it was molten throughout its volume. The Earth's tidal forces deformed the molten Moon into a slightly elongated shape, with the long axis of the Moon pointed toward the Earth. The Moon retained this shape and orientation when it cooled and solidified. To keep its long axis

pointed toward the Earth, the Moon spins once on its axis as it makes one orbit around the Earth—that is, it is in synchronous rotation (Section 3-2). Hence, the same side of the Moon always faces the Earth, and this is the side that we see. For the same reason, most of the satellites in the solar system are in synchronous rotation, and thus always keep the same side facing their planet.

Tidal forces are also important on scales much larger than the solar system. Figure 4-26 shows two spiral galaxies, like

IC 2163

NGC 2207

Stars and gas pulled out of IC 2163 by tidal forces from NGC 2207

figure 4-26 R I **V** U X G

Tidal Forces on a Galaxy For millions of years the galaxies NGC 2207 and IC 2163 have been moving ponderously past each other. The larger galaxy's tremendous tidal forces have drawn a streamer of material a hundred thousand light-years long out of IC 2163. If you lived on a planet orbiting a star within this streamer, you would have a magnificent view of both galaxies. NGC 2207 and IC 2163 are respectively 143,000 light-years and 101,000 light-years in diameter. Both galaxies are 114 million light-years away in the constellation Canis Major. (NASA and the Hubble Heritage Team, AURA/STScI)

the one in Figure 1-7, undergoing a near-collision. During the millions of years that this close encounter has been taking place, the tidal forces of the larger galaxy have pulled an immense streamer of stars and interstellar gas out of the smaller galaxy.

Many galaxies, including our own Milky Way Galaxy, show signs of having been disturbed at some time by tidal interactions with other galaxies. By their effect on the interstellar gas from which stars are formed, tidal interactions can actually trigger the birth of new stars. Our own Sun and solar system may have been formed as a result of tidal interactions of this kind. Hence we may owe our very existence to tidal forces. In this and many other ways, the laws of motion and of universal gravitation shape our universe and our destinies.

KEY WORDS

acceleration, p. 78

aphelion, p. 75

conic section, p. 83

conjunction, p. 67

deferent, p. 65

direct motion, p. 65

eccentricity, p. 75

ellipse, p. 74

elongation, p. 67

epicycle, p. 65

focus (of an ellipse; *plural* foci), p. 74

force, p. 78

geocentric model, p. 63

gravitational force, p. 80

gravity, p. 80

greatest eastern elongation, p. 67

greatest western elongation, p. 67

heliocentric model, p. 66

hyperbola, p. 83

inferior conjunction, p. 67

inferior planet, p. 67

Kepler's first law, p. 75

Kepler's second law, p. 76

Kepler's third law, p. 76

law of equal areas, p. 76

law of inertia, p. 78

law of universal gravitation, p. 80

major axis (of an ellipse), p. 74

mass, p. 80

neap tides, p. 86

Newtonian mechanics, p. 83

Newton's first law of motion, p. 78

Newton's form of Kepler's third law, p. 82

Newton's second law of motion, p. 79

Newton's third law of motion, p. 80

Occam's razor, p. 66

opposition, p. 67

parabola, p. 83

parallax, p. 73

perihelion, p. 75

period (of a planet), p. 68

Ptolemaic system, p. 65

retrograde motion, p. 65

semimajor axis (of an ellipse), p. 74

sidereal period, p. 68

speed, p. 78

spring tides, p. 85

superior conjunction, p. 67

superior planet, p. 67

synodic period, p. 68

tidal forces, p. 84

universal constant of gravitation, p. 81

velocity, p. 78

weight, p. 80

KEY IDEAS

Apparent Motions of the Planets: Like the Sun and Moon, the planets move on the celestial sphere with respect to the background of stars. Most of the time a planet moves eastward in direct motion, in the same direction as the Sun and the Moon, but from time to time it moves westward in retrograde motion.

Geocentric Model: Ancient astronomers believed the Earth to be at the center of the universe. They invented a complex system of epicycles and deferents to explain the direct and retrograde motions of the planets on the celestial sphere.

Heliocentric Model: Copernicus's heliocentric (Sun-centered) theory simplified the general explanation of planetary motions.

• In a heliocentric system, the Earth is one of the planets orbiting the Sun.

• A planet undergoes retrograde motion as seen from Earth when the Earth and the planet pass each other.

• The sidereal period of a planet, its true orbital period, is measured with respect to the stars. Its synodic period is measured with respect to the Earth and the Sun (for example, from one opposition to the next).

Evidence for the Heliocentric Model: The invention of the telescope led Galileo to new discoveries that supported a heliocentric model. These included his observations of the phases of Venus and of the motions of four moons around Jupiter.

Elliptical Orbits and Kepler's Laws: Copernicus thought that the orbits of the planets were combinations of circles. Using data collected by Tycho Brahe, Kepler deduced three laws of planetary motion: (1) the orbits are in fact ellipses; (2) a planet's speed varies as it moves around its elliptical orbit; and (3) the orbital period of a planet is related to the size of its orbit.

Newton's Laws of Motion: Isaac Newton developed three principles, called the laws of motion, that apply to the motions of objects on Earth as well as in space. These are (1) the law of inertia, (2) the relationship between the net outside force on an object and the object's acceleration, and (3) the principle of action and reaction. These laws and Newton's law of universal gravitation can be used to deduce Kepler's laws. They lead to extremely accurate descriptions of planetary motions.

• The mass of an object is a measure of the amount of matter in the object. Its weight is a measure of the force with which the gravity of some other object pulls on it.

• In general, the path of one object about another, such as that of a planet or comet about the Sun, is one of the curves called conic sections: circle, ellipse, parabola, or hyperbola.

Tidal Forces: Tidal forces are caused by differences in the gravitational pull that one object exerts on different parts of a second object.

• The tidal forces of the Moon and Sun produce tides in the Earth's oceans.

• The tidal forces of the Earth have locked the Moon into synchronous rotation.

REVIEW QUESTIONS

1. In what direction does a planet move relative to the stars when it is in direct motion? When it is in retrograde motion? How do these compare with the direction in which we see the Sun move relative to the stars?

2. (a) In what direction does a planet move relative to the horizon over the course of one night? (b) The answer to (a) is the same whether the planet is in direct motion or retrograde motion. What does this tell you about the speed at which planets move on the celestial sphere?

3. What is an epicycle? How is it important in Ptolemy's explanation of the retrograde motions of the planets?

4. What is the significance of Occam's razor as a tool for analyzing theories?

5. How did Copernicus explain the retrograde motion of the planets?

6. At what configuration (for example, superior conjunction, greatest eastern elongation, and so on) would it be best to observe Mercury or Venus with an Earth-based telescope? At what configuration would it be best to observe Mars, Jupiter, or Saturn? Explain your answers.

7. Which planets can never be seen at opposition? Which planets can never be seen at inferior conjunction? Explain your answers.

8. What is the difference between the synodic period and the sidereal period of a planet?

9. What observations did Galileo make that reinforced the heliocentric model? Why did these observations contradict the older model of Ptolemy? Why could these observations not have been made before Galileo's time?

10. What did Tycho Brahe's observations of a supernova and a comet suggest about the ancient Greek model of the universe?

11. What observations did Tycho Brahe make in an attempt to test the heliocentric model? What were his results? Explain why modern astronomers get different results.

12. What are Kepler's three laws? Why are they important?

13. What are the foci of an ellipse? If the Sun is at one focus of a planet's orbit, what is at the other focus?

14. A line joining the Sun and an asteroid is found to sweep out 6.3 AU^2 of space during 2004. How much area is swept out during 2005? Over a period of five years?

15. The orbit of a spacecraft about the Sun has a perihelion distance of 0.1 AU and an aphelion distance of 0.4 AU. What is the semimajor axis of the spacecraft's orbit? What is its orbital period?

16. A comet with a period of 125 years moves in a highly elongated orbit about the Sun. At perihelion, the comet comes very close to the Sun's surface. What is the comet's average distance from the Sun? What is the farthest it can get from the Sun?

17. What are Newton's three laws? Give an everyday example of each law.

18. How much force do you have to exert on a 4-kg brick to give it an acceleration of 3 m/s^2? If you double this force, what is the brick's acceleration? Explain.

19. What is your weight in pounds and in newtons? What is your mass in kilograms?

20. What is the difference between weight and mass?

21. Suppose that the Earth were moved to a distance of 0.25 AU from the Sun. How much stronger or weaker would the Sun's gravitational pull be on the Earth? Explain.

22. The mass of the Moon is 7.35×10^{22} kg, while that of the Earth is 5.98×10^{24} kg. The average distance from the center of the Moon to the center of the Earth is 384,400 km. What is the size of the gravitational force that the Earth exerts on the Moon? What is the size of the gravitational force that the Moon exerts on the Earth? How do your answers compare with the force between the Sun and the Earth calculated in the text?

23. How far would you have to go from Earth to be completely beyond the pull of its gravity? Explain.

24. What are conic sections? In what way are they related to the orbits of planets in the solar system?

25. Why was the discovery of Neptune an important confirmation of Newton's law of universal gravitation?

26. What is a tidal force? How do tidal forces produce tides in the Earth's oceans?

27. What is the difference between spring tides and neap tides?

Questions preceded by an asterisk (*) involve topics discussed in the Boxes.

Problem-solving tips and tools

Box 4-1 explains sidereal and synodic periods in detail. The semimajor axis of an ellipse is half the length of the long, or major, axis of the ellipse. For data about the planets and their satellites, see Appendices 1, 2, and 3 at the back of this book. If you want to calculate the gravitational force that you feel on the surface of a planet, the distance r to use is the planet's radius (the distance between you and the center of the planet). Boxes 4-2 and 4-4 show how to use Kepler's third law in its original form and in Newton's form.

28. Figure 4-2 shows the retrograde motion of Mars as seen from Earth. Sketch a similar figure that shows how Earth would appear to move against the background of stars during this same time period as seen by an observer on Mars.

*29. The synodic period of Mercury (an inferior planet) is 115.88 days. Calculate its sidereal period in days.

*30. Table 4-1 shows that the synodic period is *greater* than the sidereal period for the inferior planets Mercury and Venus, and that the synodic period is *less* than the sidereal period for all the superior planets. Draw diagrams like the one in Box 4-1 to explain why this is so.

*31. A general rule for superior planets is that the greater the average distance from the planet to the Sun, the more frequently that planet will be at opposition. Explain how this rule comes about.

32. In 2000, Mercury was at greatest western elongation on March 28, July 27, and November 15. It was at greatest eastern elongation on February 15, June 9, and October 6. Does Mercury take longer to go from eastern to western elongation, or vice versa? Why do you suppose this is the case?

33. A certain asteroid is 2 AU from the Sun at perihelion and 6 AU from the Sun at aphelion. (a) Find the semimajor axis of the asteroid's orbit. (b) Find the sidereal period of the orbit.

34. A comet orbits the Sun with a sidereal period of 27.0 years. (a) Find the semimajor axis of the orbit. (b) At aphelion, the comet is 17.5 AU from the Sun. How far is it from the Sun at perihelion?

35. One trajectory that can be used to send spacecraft from the Earth to Venus is an elliptical orbit that has the Sun at

one focus, its aphelion at the Earth, and its perihelion at Venus. The spacecraft is launched from Earth and coasts along this ellipse until it reaches Venus, when a rocket is fired either to put the spacecraft into orbit around Venus or to cause it to land on Venus. (**a**) Find the semimajor axis of the ellipse. (*Hint:* Draw a picture showing the Sun and the orbits of the Earth, Venus, and the spacecraft. Treat the orbits of the Earth and Venus as circles.) (**b**) Calculate how long (in days) such a one-way trip to Venus would take.

36. Suppose that you traveled to a planet with 3 times the mass and 3 times the diameter of the Earth. Would you weigh more or less on that planet than on Earth? By what factor?

37. The mass of Saturn is approximately 100 times that of Earth, and the semimajor axis of Saturn's orbit is approximately 10 AU. To this approximation, how does the gravitational force that the Sun exerts on Saturn compare to the gravitational force that the Sun exerts on the Earth? How do the accelerations of Saturn and the Earth compare?

38. On Earth, a 50-kg astronaut weighs 490 newtons. What would she weigh if she landed on Jupiter's moon Europa? What fraction is this of her weight on Earth? See Appendix 3 for relevant data about Europa.

39. A satellite is said to be in a "geosynchronous" orbit if it appears always to remain over the exact same spot on Earth. (**a**) What is the period of this orbit? (**b**) At what distance from the center of the Earth must such a satellite be placed into orbit? (**c**) Explain why the orbit must be in the plane of the Earth's equator.

40. Imagine a planet like the Earth orbiting a star with 4 times the mass of the Sun. If the semimajor axis of the planet's orbit is 1 AU, what would be the planet's sidereal period? (*Hint:* Use Newton's form of Kepler's third law. Compared with the case of the Earth orbiting the Sun, by what factor has the quantity $m_1 + m_2$ changed? Has a changed? By what factor must P^2 change?)

41. The photograph that opens this chapter shows the lunar module *Eagle* in orbit around the Moon after completing the first successful lunar landing in July 1969. (The photograph was taken from the command module *Columbia*, in which the astronauts returned to Earth.) The spacecraft orbited 111 km above the surface of the Moon. Calculate the period of the spacecraft's orbit. See Appendix 3 for relevant data about the Moon.

*42. In Box 4-4 we analyzed the orbit of Jupiter's moon Io. Look up information about the orbits of Jupiter's three other large moons (Europa, Ganymede, and Callisto) in the appendices. Demonstrate that these data are in agreement with Newton's form of Kepler's third law.

*43. Suppose a newly discovered asteroid is in a circular orbit with synodic period 1.25 years. The asteroid lies between the orbits of Mars and Jupiter. (**a**) Find the sidereal period of the orbit. (**b**) Find the distance from the asteroid to the Sun.

44. The average distance from the Moon to the center of the Earth is 384,400 km, and the diameter of the Earth is 12,756 km. Calculate the gravitational force that the Moon exerts (**a**) on a 1-kg rock at the point on the Earth's surface closest to the Moon and (**b**) on a 1-kg rock at the point on the Earth's surface farthest from the Moon. (**c**) Find the difference between the two forces you calculated in parts (a) and (b). This difference is the tidal force pulling these two rocks away from each other, like the 1-ball and 3-ball in Figure 4-23. Explain why tidal forces cause only a very small deformation of the Earth.

DISCUSSION QUESTIONS

45. Which planet would you expect to exhibit the greatest variation in apparent brightness as seen from Earth? Which planet would you expect to exhibit the greatest variation in angular diameter? Explain your answers.

46. Use two thumbtacks, a loop of string, and a pencil to draw several ellipses. Describe how the shape of an ellipse varies as the distance between the thumbtacks changes.

WEB/CD-ROM QUESTIONS

47. (**a**) Search the World Wide Web for information about Galileo. What were his contributions to physics? Which of Galileo's new ideas were later used by Newton to construct his laws of motion? (**b**) Search the World Wide Web for information about Kepler. Before he realized that the planets move on elliptical paths, what other models of planetary motion did he consider? What was Kepler's idea of "the music of the spheres?" (**c**) Search the World Wide Web for information about Newton. What were some of the contributions that he made to physics other than developing his laws of motion? What contributions did he make to mathematics?

48. Monitoring the Retrograde Motion of Mars. Watching Mars night after night reveals that it changes its position with respect to the background stars. To track its motion, access and view the animation "The Path of Mars in 2004–2005" in Chapter 4 of the *Universe* web site or CD-ROM. (**a**) Through which two constellations does Mars move? (**b**) On approximately what date does Mars stop its direct (west-to-east) motion and begin its retrograde motion? (*Hint:* Use the "Stop" function on your animation controls.) (**c**) Over how many days does Mars move retrograde?

OBSERVING PROJECTS

49. It is quite probable that within a few weeks of your reading this chapter one of the planets will be near opposition or greatest eastern elongation, making it readily visible in the evening sky. Select a planet that is at or near such a configuration by searching the World Wide Web or by consulting a reference book, such as the current issue of the *Astronomical Almanac* or the pamphlet entitled *Astronomical Phenomena* (both published by the U.S. government). At that configuration, would you expect the planet to be moving rapidly or slowly from night to night against the background stars? Verify your expectations by observing the planet once a week for a month, recording your observations on a star chart.

50. If Jupiter happens to be visible in the evening sky, observe the planet with a small telescope on five consecutive clear nights. Record the positions of the four Galilean satellites by making nightly drawings, just as the Jesuit priests did in 1620 (see Figure 4-12). From your drawings, can you tell which moon orbits closest to Jupiter and which orbits farthest? Was there a night when you could see only three of the moons? What do you suppose happened to the fourth moon on that night?

51. If Venus happens to be visible in the evening sky, observe the planet with a small telescope once a week for a month. On each night, make a drawing of the crescent that you see. From your drawings, can you determine if the planet is nearer or farther from the Earth than the Sun is?

Do your drawings show any changes in the shape of the crescent from one week to the next? If so, can you deduce if Venus is coming toward us or moving away from us?

52. Use the *Starry Night* program to observe the moons of Jupiter. Turn off daylight (select **Daylight** in the **Sky** menu) and show the entire celestial sphere (select **Atlas** in the **Go** menu). Center on the planet Jupiter (use **Find...** in the **Edit** menu), then use the controls at the right-hand end of the Control Panel to zoom in or out until you can see the planet and its four Galilean satellites. (**a**) Step forward in increments of 1 hour and draw the positions of the moons at each time step. (**b**) From your drawings, can you tell which moon orbits closest to Jupiter and which orbits farthest away? Explain your reasoning. (**c**) Are there times when only three of the satellites are visible? What happens to the fourth moon at those times?

53. Use the *Starry Night* program to observe the changing appearance of Mercury. Turn off daylight (select **Daylight** in the **Sky** menu) and show the entire celestial sphere (select **Atlas** in the **Go** menu). Center on the planet Mercury (use **Find...** in the **Edit** menu), then use the controls at the right-hand end of the Control Panel to zoom in or out until you can clearly see details on the planet's surface. (**a**) Step forward in increments of one day and record the changes in Mercury's phase and apparent size. (**b**) Explain why the phase and apparent size change in this way.

The Nature of Light

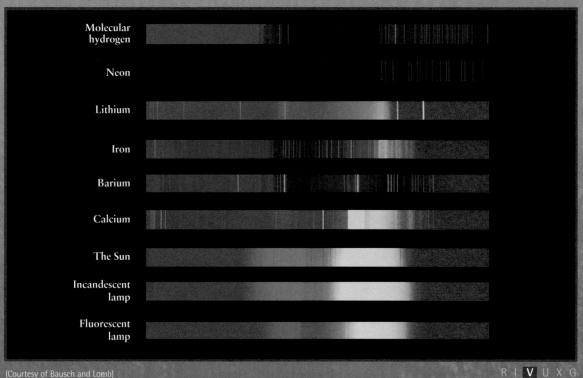

Molecular hydrogen
Neon
Lithium
Iron
Barium
Calcium
The Sun
Incandescent lamp
Fluorescent lamp

(Courtesy of Bausch and Lomb)

R I **V** U X G

In the early 1800s, the French philosopher Auguste Comte argued that because the stars are so far away, humanity would never know their nature and composition. But the means to learn about the stars was already there for anyone to see—starlight. Just a few years after Comte's bold pronouncement, scientists began analyzing starlight to learn the very things that he had deemed unknowable.

We now know that atoms of each chemical element emit and absorb light at a unique set of wavelengths characteristic of that element alone. Some of these atomic "fingerprints" are shown in the accompanying photo, along with the spectra of light from the Sun and from ordinary lamps. By looking for these wavelengths in the light from celestial objects, astronomers have learned that the same chemical elements found on Earth are also found in nearby planets, distant stars, and remote galaxies.

In this chapter we learn about the basic properties of light. Light is a form of energy with the properties of both waves and particles. Because the light emitted by an object depends upon the object's temperature, we can determine the surface temperatures of stars. By studying the structure of atoms, we will learn why each element emits and absorbs light only at specific wavelengths, and see how astronomers determine what the atmospheres of planets and stars are made of. The motion of a light source also affects wavelengths, permitting us to deduce how fast stars and other objects are approaching or receding. These are but a few of the reasons why understanding light is a prerequisite to understanding the universe.

As you read the sections of this chapter, look for the answers to the following questions.

5-1 How fast does light travel? How can this speed be measured?

5-2 Why do we think light is a wave? What kind of wave is it?

5-3 How is the light from an ordinary lightbulb different from the light emitted by a neon sign?

5-4 How can astronomers measure the temperatures of the Sun and stars?

5-5 What is a photon? How does an understanding of photons help explain why ultraviolet light causes sunburns?

5-6 How can astronomers tell what distant celestial objects are made of?

5-7 What are atoms made of?

5-8 How does the structure of atoms explain what kind of light those atoms can emit or absorb?

5-9 How can we tell if a star is approaching us or receding from us?

5-1 Light travels through empty space at a speed of 300,000 km/s

Galileo Galilei and Isaac Newton were among the first to ask basic questions about light. Does light travel instantaneously from one place to another, or does it move with a measurable speed? Whatever the nature of light, it does seem to travel swiftly from a source to our eyes. We see a distant event before we hear the accompanying sound. (For example, we see a flash of lightning before we hear the thunderclap.)

In the early 1600s, Galileo tried to measure the speed of light. He and an assistant stood at night on two hilltops a known distance apart, each holding a shuttered lantern. First, Galileo opened the shutter of his lantern; as soon as his assistant saw the flash of light, he opened his own. Galileo used his pulse as a timer to try to measure the time between opening his lantern and seeing the light from his assistant's lantern. From the distance and time, he hoped to compute the speed at which the light had traveled to the distant hilltop and back.

Galileo found that the measured time failed to increase noticeably, no matter how distant the assistant was stationed. Galileo therefore concluded that the speed of light is too high to be measured by slow human reactions. Thus, he was unable to tell whether or not light travels instantaneously.

The first evidence that light does *not* travel instantaneously was presented in 1676 by Olaus Rømer, a Danish astronomer. Rømer had been studying the orbits of the moons of Jupiter by carefully timing the moments when they passed into or out of Jupiter's shadow. To Rømer's surprise, the timing of these eclipses of Jupiter's moons seemed to depend on the relative positions of Jupiter and the Earth. When the Earth was far from Jupiter (that is, near conjunction; see Figure 4-6), the eclipses occurred several minutes later than when the Earth was close to Jupiter (near opposition).

Rømer realized that this puzzling effect could be explained if light needs time to travel from Jupiter to the Earth. When the Earth is closest to Jupiter, the image of a satellite disappearing into Jupiter's shadow arrives at our telescopes a little sooner than it does when Jupiter and the Earth are farther apart (Figure 5-1). The range of variation in the times at which such eclipses are observed is about 16.6 minutes, which Rømer interpreted as the length of time required for light to travel across the diameter of the Earth's orbit (a distance of 2 AU). The size of the Earth's orbit was not accurately known in Rømer's day, and he never actually calculated the speed of light. Today, using the modern value of 150 million kilometers for the astronomical unit, Rømer's method yields a value for the speed of light equal to roughly 300,000 km/s (186,000 mi/s).

Almost two centuries after Rømer, the speed of light was measured very precisely in an experiment carried out on Earth.

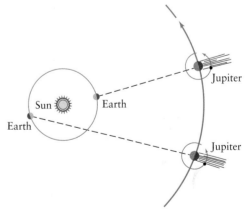

figure 5-1

Rømer's Proof that Light Does Not Travel Instantaneously
When the Earth is near Jupiter, eclipses of Jupiter's moons occur slightly earlier than anticipated. When the Earth-Jupiter distance is large, the eclipses occur slightly later than anticipated. Rømer correctly attributed this effect to variations in the time required for light to travel from Jupiter to the Earth.

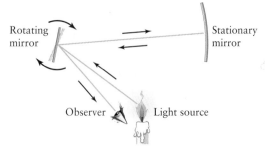

figure 5-2

The Fizeau-Foucault Method of Measuring the Speed of Light Light from a light source is reflected off a rotating mirror to a stationary mirror and from there back to the rotating mirror. The ray that reaches the observer is deflected away from the path of the initial beam because the rotating mirror has moved slightly while the light was making the round trip. The speed of light is calculated from the deflection angle and the dimensions of the apparatus.

In 1850, the French physicists Armand-Hippolyte Fizeau and Jean Foucault built the apparatus sketched in Figure 5-2. Light from a light source reflects from a rotating mirror toward a stationary mirror 20 meters away. The rotating mirror moves slightly while the light is making the round trip, so the returning light ray is deflected away from the source by a small angle. By measuring this angle and knowing the dimensions of their apparatus, Fizeau and Foucault could deduce the speed of light. Once again, the answer was very nearly 300,000 km/s.

The speed of light in a vacuum is usually designated by the letter c (from the Latin *celeritas*, meaning "speed"). The modern accepted value is $c = 299,792.458$ km/s (186,282.397 mi/s). In most calculations you can use

$$c = 3.0 \times 10^5 \text{ km/s} = 3.0 \times 10^8 \text{ m/s}$$

The most convenient set of units to use for c is different in different situations. The value in kilometers per second (km/s) is often most useful when comparing c to the speeds of objects in space, while the value in meters per second (m/s) is preferred when doing calculations involving the wave nature of light (which we will discuss in Section 5-2).

Note that the quantity c is the speed of light *in a vacuum*. Light travels more slowly through air, water, glass, or any other transparent substance than it does in a vacuum. In our study of astronomy, however, we will almost always consider light traveling through the vacuum (or near vacuum) of space.

The speed of light in empty space is one of the most important numbers in modern physical science. This value appears in many equations that describe atoms, gravity, electricity, and magnetism. According to Einstein's special theory of relativity, nothing can travel faster than the speed of light.

5-2 Light is electromagnetic radiation and is characterized by its wavelength

Light is energy. This fact is apparent to anyone who has felt the warmth of the sunshine on a summer's day. But what exactly is light? How is it produced? What is it made of? How does it move through space? Scholars have struggled with these questions throughout history.

The first major breakthrough in understanding light came from a simple experiment performed by Isaac Newton around 1670. Newton was familiar with what he called the "celebrated Phenomenon of Colours," in which a beam of sunlight passing through a glass prism spreads out into the colors of the rainbow (Figure 5-3). This rainbow is called a **spectrum** (plural **spectra**).

 Until Newton's time, it was thought that a prism somehow added colors to white light. To test this idea, Newton placed a second prism so that just one color of the spectrum passed through it (Figure 5-4). According to the old theory, this should have caused a further

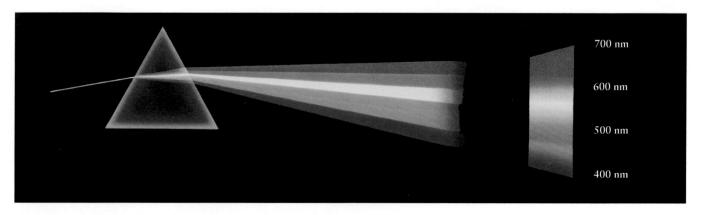

figure 5-3

A Prism and a Spectrum When a beam of sunlight passes through a glass prism, the light is broken into a rainbow-colored band called a spectrum. The numbers on the right side of the spectrum indicate the wavelengths of different colors of light.

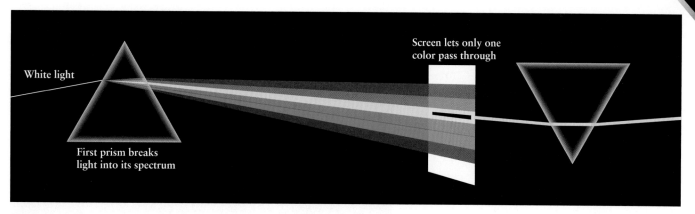

Figure 5-4

Newton's Experiment on the Nature of Light In a crucial experiment, Newton took sunlight that had passed through a prism and sent it through a second prism. Between the two prisms was a screen with a hole in it that allowed only one color of the spectrum to pass through. This same color emerged from the second prism. Newton's experiment proved that prisms do not add color to light but merely bend different colors through different angles. It also proved that white light, such as sunlight, is actually a combination of all the colors that appear in its spectrum.

change in the color of the light. But Newton found that each color of the spectrum was unchanged by the second prism; red remained red, blue remained blue, and so on. He concluded that a prism merely separates colors and does not add color. Hence, the spectrum produced by the first prism shows that sunlight is a mixture of all the colors of the rainbow.

Newton suggested that light is composed of particles too small to detect individually. (The modern theory of quantum mechanics, developed in the early 1900s, takes a similar viewpoint and postulates that light is composed of tiny packets of energy. We will explore these ideas in Section 5-5.) In 1678, however, the Dutch physicist and astronomer Christiaan Huygens proposed a rival explanation. He suggested that light travels in the form of waves rather than particles. Scientists now realize that light possesses both particlelike and wavelike characteristics.

Around 1801, Thomas Young in England carried out an experiment that convincingly demonstrated the wavelike aspect of light. He passed a beam of light through two thin, parallel slits in an opaque screen, as shown in Figure 5-5a. On a white surface some distance beyond the slits, the light formed a pattern of alternating bright and dark bands. Young reasoned that if a beam of light was a stream of particles (as Newton had suggested), the two beams of light from the slits should simply form bright images of the slits on the white surface. The pattern of bright and dark bands he observed is just what would be expected, however, if light had wavelike properties. An analogy with water waves demonstrates why.

ANALOGY Imagine ocean waves pounding against a reef or breakwater that has two openings (Figure 5-5b). A pattern of ripples is formed on the other side of the barrier as the waves come through the two openings and interfere with each other. At certain points, wave crests arrive simultaneously from the two openings. These reinforce each

other and produce high waves. At other points, a crest from one opening meets a trough from the other opening. These cancel each other out, leaving areas of still water. This process of combining two waves also takes place in Young's double-slit experiment: The bright bands are regions where waves from the two slits reinforce each other, while the dark bands appear where waves from the two slits cancel each other.

The discovery of the wave nature of light posed some obvious questions. What exactly is "waving" in light? That is, what is it about light that goes up and down like water waves on the ocean? Because we can see light from the Sun, planets, and stars, light waves must be able to travel across empty space. Hence, whatever is "waving" cannot be any material substance. What, then, is it?

The answer came from a seemingly unlikely source—a comprehensive theory that described electricity and magnetism. Numerous experiments during the first half of the nineteenth century demonstrated an intimate connection between electric and magnetic forces. A central idea to emerge from these experiments is the concept of a *field,* an immaterial yet measurable disturbance of any region of space in which electric or magnetic forces are felt. Thus, an electric charge is surrounded by an electric field, and a magnet is surrounded by a magnetic field. Experiments in the early 1800s demonstrated that moving an electric charge produces a magnetic field; conversely, moving a magnet gives rise to an electric field.

In the 1860s, the Scottish mathematician and physicist James Clerk Maxwell succeeded in describing all the basic properties of electricity and magnetism in four equations. This mathematical achievement demonstrated that electric and magnetic forces are really two aspects of the same phenomenon, which we now call **electromagnetism.**

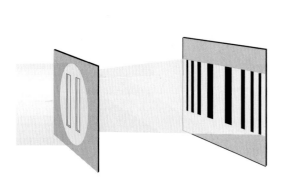

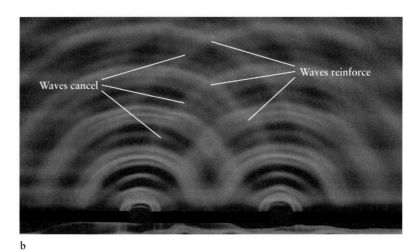

Waves cancel

Waves reinforce

a

b

figure 5-5

Young's Double-Slit Experiment **(a)** Thomas Young's classic double-slit experiment can easily be repeated in the modern laboratory by shining light from a laser onto two closely spaced parallel slits. Alternating dark and light bands appear on a screen beyond the slits. **(b)** The intensity of light on the screen in (a) is analogous to the height of water waves that pass through a barrier with two openings. (The photograph shows this experiment with water waves in a small tank.) In certain locations, wave crests from both openings reinforce each other to produce extra high waves. At other locations a crest from one opening meets a trough from the other. The crest and trough cancel each other, producing still water. (© Eric Schrempp/Photo Researchers)

By combining his four equations, Maxwell showed that electric and magnetic fields should travel through space in the form of waves at a speed of 3.0×10^5 km/s—a value exactly equal to the best available value for the speed of light. Maxwell's suggestion that these waves do exist and are observed as light was soon confirmed by experiments. Because of its electric and magnetic properties, light is also called **electromagnetic radiation.**

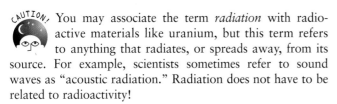 You may associate the term *radiation* with radioactive materials like uranium, but this term refers to anything that radiates, or spreads away, from its source. For example, scientists sometimes refer to sound waves as "acoustic radiation." Radiation does not have to be related to radioactivity!

Electromagnetic radiation consists of oscillating electric and magnetic fields, as shown in Figure 5-6. The distance between two successive wave crests is called the **wavelength** of the light, usually designated by the Greek letter λ (lambda). No matter what the wavelength, electromagnetic radiation always travels at the same speed $c = 3.0 \times 10^5$ km/s = 3.0×10^8 m/s in a vacuum.

More than a century elapsed between Newton's experiments with a prism and the confirmation of the wave nature of light. One reason for this delay is that **visible light,** the light to which the human eye is sensitive, has extremely short wavelengths—less than a thousandth of a millimeter—that are not easily detectable. To express such tiny distances conve-

niently, scientists use a unit of length called the **nanometer** (abbreviated nm), where 1 nm = 10^{-9} m. Experiments demonstrated that visible light has wavelengths covering the range from about 400 nm for violet light to about 700 nm for red light. Intermediate colors of the rainbow like yellow (550 nm) have intermediate wavelengths, as shown in Figure 5-7. (Some astronomers prefer to measure wavelengths in *angstroms.* One angstrom, abbreviated Å, is one-tenth of a nanometer: 1 Å = 0.1 nm = 10^{-10} m. In these units, the wavelengths of visible light extend from about 4000 Å to about 7000 Å. We will not use these units in this book, however.)

Maxwell's equations place no restrictions on the wavelength of electromagnetic radiation. Hence, electromagnetic waves could and should exist with wavelengths both longer and shorter than the 400–700 nm range of visible light. Consequently, researchers began to look for *invisible* forms of light. These are forms of electromagnetic radiation to which the cells of the human retina do not respond.

The first kind of invisible radiation to be discovered actually preceded Maxwell's work by more than a half century. Around 1800 the British astronomer William Herschel passed sunlight through a prism and held a thermometer just beyond the red end of the visible spectrum. The thermometer registered a temperature increase, indicating that it was being exposed to an invisible form of energy. This invisible energy, now called **infrared radiation,** was later realized to be electromagnetic radiation with wavelengths somewhat longer than those of visible light.

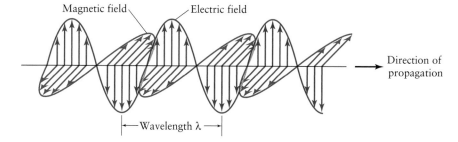

figure 5-6

Electromagnetic Radiation All forms of light consist of oscillating electric and magnetic fields that move through space at a speed of 3.0×10^5 km/s $= 3.0 \times 10^8$ m/s. This figure shows a "snapshot" of these fields at one instant. The distance between two successive crests, called the wavelength of the light, is usually designated by the Greek letter λ (lambda).

In experiments with electric sparks in 1888, the German physicist Heinrich Hertz succeeded in producing electromagnetic radiation with even longer wavelengths of a few centimeters or more. These are now known as **radio waves.** In 1895 another German physicist, Wilhelm Röntgen, invented a machine that produces electromagnetic radiation with wavelengths shorter than 10 nm, now known as **X rays.** The X-ray machines in modern medical and dental offices are direct descendants of Röntgen's invention. Over the years radiation has been discovered with many other wavelengths.

Thus, visible light occupies only a tiny fraction of the full range of possible wavelengths, collectively called the **electromagnetic spectrum.** As Figure 5-7 shows, the electromagnetic spectrum stretches from the longest-wavelength radio waves to the shortest-wavelength gamma rays.

On the long-wavelength side of the visible spectrum, infrared radiation covers the range from about 700 nm to 1 mm. Astronomers interested in infrared radiation often express wavelength in *micrometers* or *microns,* abbreviated μm, where 1 μm $= 10^{-3}$ mm $= 10^{-6}$ m. **Microwaves** have wavelengths from roughly 1 mm to 10 cm, while radio waves have even longer wavelengths.

At wavelengths shorter than those of visible light, **ultraviolet radiation** extends from about 400 nm down to 10 nm. Next are X rays, which have wavelengths between about 10 and 0.01 nm, and beyond them at even shorter wavelengths are **gamma rays.** Note that these rough boundaries are simply arbitrary divisions in the electromagnetic spectrum.

Astronomers who work with radio telescopes often prefer to speak of *frequency* rather than wavelength. The **frequency** of a wave is the number of wave crests that pass a given point in one second. Equivalently, it is the number of complete *cycles* of the wave that pass per second (a complete cycle is from one crest to the next). Frequency is usually denoted by the Greek letter ν (nu). The unit of frequency is the cycle per second, also called the hertz (abbreviated Hz) in honor of Heinrich Hertz, the physicist who first produced radio waves. For example, if 500 crests of a wave pass you in one second, the frequency of the wave is 500 cycles per second or 500 Hz.

In working with frequencies, it is often convenient to use the prefix *mega-* (meaning "million," or 10^6, and abbreviated M) or *kilo-* (meaning "thousand," or 10^3, and abbreviated k). For example, AM radio stations broadcast at frequencies between 535 and 1605 kHz (kilohertz), while FM radio

stations broadcast at frequencies in the range from 88 to 108 MHz (megahertz).

The relationship between the frequency and wavelength of an electromagnetic wave is a simple one. Because light moves at a constant speed $c = 3 \times 10^8$ m/s, if the wavelength (distance from one crest to the next) is made shorter, the frequency must increase (more of those closely spaced crests pass

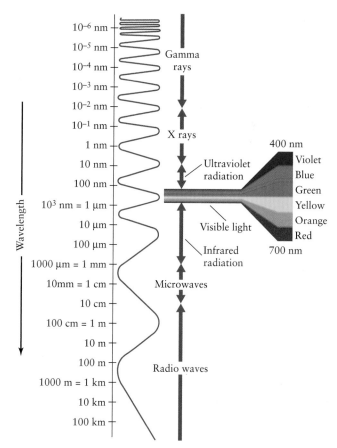

figure 5-7

The Electromagnetic Spectrum The full array of all types of electromagnetic radiation is called the electromagnetic spectrum. It extends from the longest-wavelength radio waves to the shortest-wavelength gamma rays. Visible light occupies only a tiny portion of the full electromagnetic spectrum.

you each second). Mathematically, the frequency ν of light is related to its wavelength λ by

Frequency and wavelength of an electromagnetic wave

$$\nu = \frac{c}{\lambda}$$

ν = frequency of an electromagnetic wave (in Hz)

c = speed of light = 3×10^8 m/s

λ = wavelength of the wave (in meters)

For example, hydrogen atoms in space emit radio waves with a wavelength of 21.12 cm. To calculate the frequency of this radiation, we must first express the wavelength in meters rather than centimeters: λ = 0.2112 m. Then we can use the above formula to find the frequency ν:

$$\nu = \frac{c}{\lambda} = \frac{3 \times 10^8 \text{ m/s}}{0.2112 \text{ m/s}} = 1.42 \times 10^9 \text{ Hz} = 1420 \text{ MHz}$$

Visible light has a much shorter wavelength and higher frequency than radio waves. You can use the above formula to show that for yellow-orange light of wavelength 600 nm, the frequency is 5×10^{14} Hz or 500 *million* megahertz!

5-3 A dense object emits electromagnetic radiation according to its temperature

To learn about objects in the heavens, astronomers study the character of the electromagnetic radiation coming from those objects. Such studies can be very revealing because different kinds of electromagnetic radiation are typically produced in different ways. As an example, on Earth the most common way to generate radio waves is to make an electric current oscillate back and forth (as is done in the broadcast antenna of a radio station). By contrast, X rays for medical and dental purposes are usually produced by bombarding atoms with fast-moving particles extracted from within other atoms. Our own Sun emits radio waves from near its glowing surface and X rays from its corona (see the photo that opens Chapter 3). Hence, these observations indicate the presence of electric currents near the Sun's surface, and of fast-moving particles in the Sun's outermost regions. (We will discuss the Sun at length in Chapter 18.)

The simplest and most common way to produce electromagnetic radiation, either on or off the Earth, is to heat an object. The hot filament of wire inside an ordinary lightbulb emits white light, and a neon sign has a characteristic red glow because neon gas within the tube is heated by an electric current. In like fashion, almost all the visible light that we receive from space comes from hot objects like the Sun and the stars. The kind and amount of light emitted by a hot object tell us not only how hot it is but also about other properties of the object.

We can tell whether the hot object is made of relatively dense or relatively thin material. Consider the difference between a lightbulb and a neon sign. The dense, solid filament of a lightbulb makes white light, which is a mixture of all different visible wavelengths, while the thin neon gas produces light of a rather definite red color and, hence, a rather definite wavelength. For now we will concentrate our attention on the light produced by dense objects. (We will return to gases in Section 5-6.) Even though the Sun and stars are gaseous, not solid, it turns out that they emit light with many of the same properties as light emitted by a hot, glowing, solid object.

Imagine a welder or blacksmith heating a bar of iron. As the bar becomes hot, it begins to glow deep red, as shown in Figure 5-8a. (You can see this same glow from the coils of a toaster, or from an electric range turned on "high.") As the temperature rises further, the bar begins to give off a brighter, reddish-orange light (Figure 5-8b). At still higher temperatures, it shines with a brilliant yellowish-white light (Figure 5-8c). If the bar could be prevented from melting and vaporizing, at extremely high temperatures it would emit a dazzling blue-white light.

As this example shows, the amount of energy emitted by the hot, dense object and the dominant wavelength of the emitted radiation both depend on the temperature of the object. The hotter the object, the more energy it emits and the shorter the wavelength at which most of the energy is

a

b

c

Figure 5-8

Heating a Bar of Iron This sequence of drawings shows how the appearance of a heated bar of iron changes with temperature. As the temperature increases, the bar glows more brightly because it radiates more energy. The color of the bar also changes because the dominant wavelength of light emitted by the bar decreases as the temperature goes up.

figure 5-9 R **I** V U X G

An Infrared Boy and His Dog A camera sensitive to infrared radiation took this picture of a boy and his dog. The different colors in the image represent regions of different temperature; white areas are the warmest and emit the most infrared light, while black areas (including the dog's cold nose) are at the lowest temperatures and emit the least radiation. (R. P. Clark and M. Goff)

If something is hot, its atoms are moving at high speeds; if it is cold, its atoms are moving slowly. Scientists usually prefer to use the Kelvin temperature scale, on which temperature is measured in **kelvins** (K) upward from **absolute zero.** This is the coldest possible temperature, at which atoms move as slowly as possible (they can never quite stop completely). On the more familiar Celsius and Fahrenheit temperature scales, absolute zero (0 K) is −273°C and −460°F. Ordinary room temperature is 293 K, 20°C, or 68°F. Box 5-1 discusses the relationships among the Kelvin, Celsius, and Fahrenheit temperature scales.

Figure 5-10 depicts quantitatively how the radiation from a dense object depends on its Kelvin temperature. Each curve in this figure shows the intensity of light emitted at each wavelength by a dense object at a given temperature. In other words, the curves show the spectrum of light emitted by such an object. At any temperature, a hot, dense object emits at all wavelengths, so its spectrum is a smooth, continuous curve with no gaps in it.

emitted. Colder objects emit relatively little energy, and this emission is primarily at long wavelengths.

These observations explain why you can't see in the dark. The temperatures of people, animals, and furniture are rather less than even that of the iron bar in Figure 5-8a. So, while these objects emit radiation even in a darkened room, most of this emission is at wavelengths greater than those of red light, in the infrared part of the spectrum (see Figure 5-7). Your eye is not sensitive to infrared, and you thus cannot see ordinary objects in a darkened room. But you can detect this radiation by using a camera that is sensitive to infrared light (Figure 5-9).

To better understand the relationship between the temperature of a dense object and the radiation it emits, it is helpful to know just what "temperature" means. The temperature of a substance is directly related to the average speed of the tiny **atoms**—the building blocks that come in distinct forms for each distinct chemical element—that make up the substance. (Typical atoms are about 10^{-10} m = 0.1 nm in diameter, or about 1/5000 as large as a typical wavelength of visible light.)

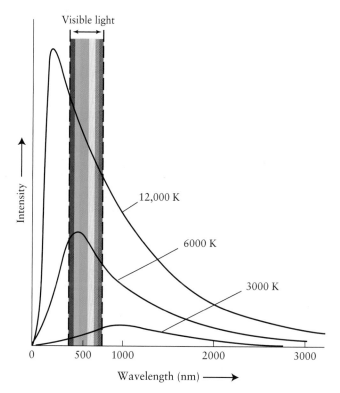

figure 5-10

Blackbody Curves Each of these curves shows the intensity of light at every wavelength that is emitted by a blackbody (an idealized case of a dense object) at a particular temperature. The higher the temperature, the shorter the wavelength of maximum emission (where the curve peaks) and the greater the amount of light emitted at every wavelength. The rainbow-colored band shows the range of visible wavelengths. The vertical scale has been compressed so that all three curves can be seen; the peak intensity for the 12,000-K curve is actually about 1000 times greater than the peak intensity for the 3000-K curve.

Box 5-1 Tools of the Astronomer's Trade

Temperatures and Temperature Scales

Three temperature scales are in common use. Throughout most of the world, temperatures are expressed in **degrees Celsius** (°C). The Celsius temperature scale is based on the behavior of water, which freezes at 0°C and boils at 100°C at sea level on Earth. This scale is named in honor of the Swedish astronomer Anders Celsius, who proposed it in 1742.

Astronomers usually prefer the Kelvin temperature scale. This is named after the nineteenth-century British physicist Lord Kelvin, who made many important contributions to our understanding of heat and temperature. Absolute zero, the temperature at which atomic motion is at the absolute minimum, is −273°C in the Celsius scale but 0 K in the Kelvin scale. Because it is impossible for atomic motion to be any less than the minimum, nothing can be colder than 0 K; hence, there are no negative temperatures on the Kelvin scale. Note that we do *not* use degree (°) with the Kelvin temperature scale.

A temperature expressed in kelvins is always equal to the temperature in degrees Celsius plus 273. On the Kelvin scale, water freezes at 273 K and boils at 373 K. Water must be heated through a change of 100 K or 100°C to go from its freezing point to its boiling point. Thus, the "size" of a kelvin is the same as the "size" of a Celsius degree. When considering temperature changes, measurements in kelvins and Celsius degrees are the same.

The now-archaic Fahrenheit scale, which expresses temperature in **degrees Fahrenheit** (°F), is used only in the United States. When the German physicist Gabriel Fahrenheit introduced this scale in the early 1700s, he intended 100°F to represent the temperature of a healthy human body. On the Fahrenheit scale, water freezes at 32°F and boils at 212°F. There are 180 Fahrenheit degrees between the freezing and boiling points of water, so a degree Fahrenheit is only 100/180 = 5/9 as large as either a Celsius degree or a kelvin.

Two simple equations allow you to convert a temperature from the Celsius scale to the Fahrenheit scale and from Fahrenheit to Celsius:

$$T_F = \frac{9}{5} T_c + 32$$

$$T_C = \frac{5}{9} (T_F - 32)$$

T_F = temperature in degrees Fahrenheit

T_C = temperature in degrees Celsius

EXAMPLE: A typical room temperature is 68°F. We can convert this to the Celsius scale using the second equation:

$$T_C = \frac{5}{9}(68 - 32) = 20°C$$

To convert this to the Kelvin scale, we simply add 273 to the Celsius temperature. Thus,

$$68°F = 20°C = 293 \text{ K}$$

The diagram displays the relationships among these three temperature scales.

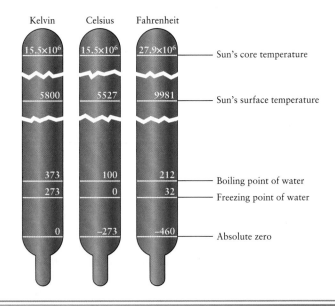

The shape of the spectrum depends on temperature, however. An object at relatively low temperature (say, 3000 K) has a low curve, indicating a low intensity of radiation. The **wavelength of maximum emission,** at which the curve has its peak and the emission of energy is strongest, is at a long wavelength. The higher the temperature, the higher the curve (indicating greater intensity) and the shorter the wavelength of maximum emission.

Figure 5-10 shows that for a dense object at a temperature of 3000 K, the wavelength of maximum emission is around 1000 nm (1 μm). Because this is an infrared wavelength well outside the visible range, you might think that you cannot see the radiation from an object at this temperature. In fact, the glow from such an object *is* visible; the curve shows that this object emits plenty of light within the visible range, as well as at even shorter wavelengths. The 3000-K curve is quite a bit higher at the red end of the visible spectrum than at the violet end, so a dense object at this temperature will appear red in color. Similarly, the 12,000-K curve has its wavelength of maximum emission in the ultraviolet part of the spectrum, at a wavelength shorter than visible light. But such a hot, dense object also emits copious amounts of visible light (much more than at 6000 K or 3000 K, for which the curves are lower) and thus will have a very visible glow. The curve for this temperature is higher for blue light than for red light, and so the color of a dense object at 12,000 K is a brilliant blue or blue-white. These conclusions agree with the color changes of a heated rod shown in Figure 5-8. The same principles apply to stars: A star that looks blue has a high surface temperature, while a red star has a relatively cool surface.

The curves in Figure 5-10 are drawn for an idealized type of dense object called a **blackbody.** A perfect blackbody does not reflect any light at all; instead, it absorbs all radiation falling on it. Because it reflects no electromagnetic radiation, the radiation that it does emit is entirely the result of its temperature. Ordinary objects, like tables, textbooks, and people, are not perfect blackbodies; they reflect light, which is why they are visible. A star such as the Sun, however, behaves very much like a perfect blackbody, because it absorbs almost completely any radiation falling on it from outside. The light emitted by a blackbody is called **blackbody radiation,** and the curves in Figure 5-10 are often called **blackbody curves.**

Despite its name, a blackbody does not necessarily look black. The Sun, for instance, does not look black because its temperature is high (around 5800 K), and so it glows brightly. But a room-temperature (around 300 K) blackbody would appear very black indeed. Even if it were as large as the Sun, it would emit only about 1/100,000 as much energy. (Its blackbody curve is far too low to graph in Figure 5-10.) Furthermore, most of this radiation would be at wavelengths that are too long for our eyes to perceive.

Figure 5-11 illustrates how accurately a star like the Sun can be modeled as a blackbody. The graph shows how the intensity of sunlight varies with wavelength, as measured from above the Earth's atmosphere. (This is necessary because the Earth's atmosphere absorbs certain wavelengths.) The peak of the curve is at a wavelength of about 500 nm, near the middle of the visible spectrum.

Figure 5-11 also shows the blackbody curve for a temperature of 5800 K. Note how closely the observed intensity curve for the Sun matches the blackbody curve. This is a strong indication that the temperature of the Sun's glowing surface is about 5800 K—a temperature that we can measure across a distance of 150 million kilometers! The close correlation between blackbody curves and the observed intensity curves for most stars is a key reason why astronomers are interested in the physics of blackbody radiation.

Blackbody radiation depends only on the temperature of the object emitting the radiation, not on the chemical composition of the object. The light emitted by molten gold at

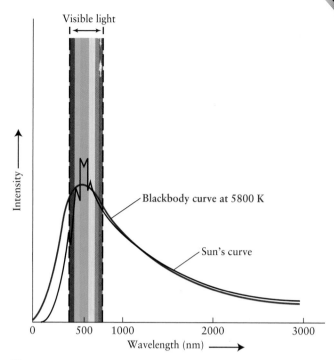

figure 5-11

The Sun as a Blackbody This graph compares the intensity of sunlight over a wide range of wavelengths (black curve) with the intensity of radiation coming from a blackbody at a temperature of 5800 K (red curve). The measurements of the Sun's intensity were made above the Earth's atmosphere. The Sun mimics a blackbody remarkably well. Note that the Sun's spectrum peaks in the visible spectrum. This is not surprising, because the human eye evolved to take advantage of the most plentiful light available.

2000 K is very nearly the same as that emitted by molten lead at 2000 K. Therefore, it might seem that analyzing the light from the Sun or from a star can tell astronomers the object's temperature but not what the star is made of. As Figure 5-11 shows, however, the intensity curve for the Sun (a typical star) is not precisely that of a blackbody. We will see later in this chapter that the *differences* between a star's spectrum and that of a blackbody allow us to determine the chemical composition of the star.

5-4 Wien's law and the Stefan-Boltzmann law are useful tools for analyzing glowing objects like stars

The mathematical formula that describes the blackbody curves in Figure 5-10 is a rather complicated one. But there are two simpler formulas for blackbody radiation that prove to be very useful in many branches of astronomy. They are used by astronomers who investigate the stars as well as by those who study the planets (which are dense, relatively cool objects that emit infrared radiation). One of these formulas relates the temperature of a blackbody to its wavelength of maximum emission, and the other relates the temperature to the amount of energy that the blackbody emits.

Figure 5-10 shows that the higher the temperature (T) of a blackbody, the shorter its wavelength of maximum emission (λ_{max}). In 1893 the German physicist Wilhelm Wien used ideas about both heat and electromagnetism to make this relationship quantitative. The formula that he derived, which today is called **Wien's law,** is

Wien's law for a blackbody

$$\lambda_{max} = \frac{0.0029}{T}$$

λ_{max} = wavelength of maximum emission of the object (in meters)

T = temperature of object (in kelvins)

 According to Wien's law, the wavelength of maximum emission of a blackbody is inversely proportional to its temperature in kelvins. In other words, if the temperature of the blackbody doubles, its wavelength of maximum emission is halved, and vice versa. For example, Figure 5-10 shows blackbody curves for temperatures of 3000 K, 6000 K, and 12,000 K. From Wien's law, a blackbody with a temperature of 6000 K has a wavelength of maximum emission λ_{max} = (0.0029)/(6000) = 4.8 × 10^{-7} m = 480 nm, in the visible part of the electromagnetic spectrum. At 12,000 K, or twice the temperature, the blackbody has a wavelength of maximum emission half as great, or λ_{max} = 240 nm; this is in the ultraviolet. At 3000 K, just half our original temperature, the value of λ_{max} is twice the original value—960 nm, which is an infrared wavelength. You can see that these wavelengths agree with the peaks of the curves in Figure 5-10.

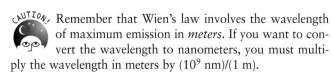 Remember that Wien's law involves the wavelength of maximum emission in *meters*. If you want to convert the wavelength to nanometers, you must multiply the wavelength in meters by (10^9 nm)/(1 m).

Wien's law is very useful for determining the surface temperatures of stars. It is not necessary to know how far away the star is, how large it is, or how much energy it radiates into space. All we need to know is the dominant wavelength of the star's electromagnetic radiation.

The other useful formula for the radiation from a blackbody involves the total amount of energy the blackbody radiates at all wavelengths. (By contrast, the curves in Figure 5-10 show how much energy a blackbody radiates at each individual wavelength.)

Energy is usually measured in **joules** (J), named after the nineteenth-century English physicist James Joule. A joule is the amount of energy contained in the motion of a 2-kilogram mass moving at a speed of 1 meter per second. The joule is a convenient unit of energy because it is closely related to the familiar **watt** (W): 1 watt is 1 joule per second, or 1 W = 1 J/s = 1 J s^{-1}. (The superscript −1 means you are dividing by that quantity.) For example, a 100-watt lightbulb uses energy at a rate

of 100 joules per second, or 100 J/s. The energy content of food is also often measured in joules; in most of the world, diet soft drinks are labeled as "low joule" rather than "low calorie."

The amount of energy emitted by a blackbody depends both on its temperature and on its surface area. This makes sense: A large burning log radiates much more heat than a burning match, even though the temperatures are the same. To consider the effects of temperature alone, it is convenient to look at the amount of energy emitted from each square meter of an object's surface in a second. This quantity is called the **energy flux** (F). Flux means "rate of flow," and thus F is a measure of how rapidly energy is flowing out of the object. It is measured in joules per square meter per second, usually written as J/m^2/s or J m^{-2} s^{-1}. Alternatively, because 1 watt equals 1 joule per second, we can express flux in watts per square meter (W/m^2, or W m^{-2}).

In 1879 the Austrian physicist Josef Stefan reported experimental evidence that the flux from a blackbody is proportional to the fourth power of the object's temperature (measured in kelvins). Five years after Stefan announced his law, another Austrian physicist, Ludwig Boltzmann, showed how it could be derived mathematically from basic assumptions about atoms and molecules. For this reason, Stefan's law is commonly known as the **Stefan-Boltzmann law.** Written as an equation, the Stefan-Boltzmann law is

Stefan-Boltzmann law for a blackbody

$$F = \sigma T^4$$

F = energy flux, in joules per square meter of surface per second

σ = a constant = 5.67 × 10^{-8} W m^{-2} K^{-4}

T = object's temperature, in kelvins

The value of the constant σ (the Greek letter sigma) is known from laboratory experiments.

The Stefan-Boltzmann law says that if you double the temperature of an object (for example, from 300 K to 600 K), then the energy emitted from the object's surface each second increases by a factor of 2^4 = 16. If you increase the temperature by a factor of 10 (for example, from 300 to 3000 K), the rate of energy emission increases by a factor of 10^4 = 10,000. Thus, a chunk of iron at room temperature (around 300 K) emits very little electromagnetic radiation (and essentially no visible light), but an iron bar heated to 3000 K glows quite intensely.

Box 5-2 gives several examples of applying Wien's law and the Stefan-Boltzmann law to typical astronomical problems.

5-5 Light has properties of both waves and particles

At the end of the nineteenth century, physicists mounted a valiant effort to explain all the characteristics of blackbody radiation. To this end they constructed theories based on

Using the Laws of Blackbody Radiation

The Sun and stars behave like nearly perfect blackbodies. Wien's law and the Stefan-Boltzmann law can therefore be used to relate the surface temperature of the Sun or a distant star to the energy flux and wavelength of maximum emission of its radiation. The following examples show how to do this.

EXAMPLE: The surface temperature of the Sun can be determined using Wien's law. The Sun emits energy over a wide range of wavelengths, but the maximum intensity of sunlight is at a wavelength of roughly 500 nm = 5.0×10^{-7} m. From Wien's law, we find the Sun's surface temperature $T_{\odot}$ to be

$$T_{\odot} = \frac{0.0029}{\lambda_{max}} = \frac{0.0029}{5.0 \times 10^{-7}} = 5800 \text{ K}$$

This is about the same temperature as an iron welding arc. (The symbol $\odot$ is the standard astronomical symbol for the Sun.)

EXAMPLE: We can also find the Sun's surface temperature using the Stefan-Boltzmann law. Using detectors above the Earth's atmosphere, astronomers have measured the average flux of solar energy arriving at Earth; this value, called the **solar constant,** is equal to 1370 W m^{-2}. However, the quantity F in the Stefan-Boltzmann law refers to the flux measured at the Sun's surface, not at the Earth.

To determine the value of F, we first imagine a huge sphere of radius 1 AU with the Sun at its center, as shown in the figure. Each square meter of that sphere receives 1370 watts of power from the Sun, so the total energy radiated by the Sun per second is equal to the solar constant multiplied by the sphere's surface area. The result, called the **luminosity** of the Sun and denoted by the symbol $L_{\odot}$, is $L_{\odot} = 3.90 \times 10^{26}$ W. That is, in 1 second the Sun radiates 3.90×10^{26} joules of energy into space. Because we know the size of the Sun, we can compute the energy flux (energy emitted per square meter per second) at its surface. The radius of the Sun is $R_{\odot} = 6.96 \times 10^8$ m, and the Sun's surface area is $4\pi R_{\odot}^2$. Therefore, its energy flux is the luminosity (total energy emitted by the Sun per second) divided by the surface area (the number of square meters of surface):

$$F_{\odot} = \frac{L_{\odot}}{4\pi R_{\odot}^2} = \frac{3.90 \times 10^{26} \text{ W}}{4\pi(6.96 \times 10^8 \text{ m})^2} = 6.41 \times 10^7 \text{ W m}^{-2}$$

Notice that the solar constant of 1370 W m^{-2} is very much less than this. By the time the Sun's radiation reaches Earth, it is spread over a greatly increased area.

Once we have the Sun's energy flux $F_{\odot}$, we can use the Stefan-Boltzmann law to find the Sun's surface temperature $T_{\odot}$:

$$T_{\odot}^4 = \frac{F_{\odot}}{\sigma} = 1.13 \times 10^{15} \text{ K}^4$$

Taking the fourth root (the square root of the square root) of this value, we find the surface temperature of the Sun to be $T_{\odot} = 5800$ K. This result agrees with the value we computed in the previous example using Wien's law.

EXAMPLE: Sirius, the brightest star in the night sky, has a surface temperature of about 10,000 K. We can use Wien's law to calculate the wavelength (λ_{max}) at which Sirius emits most intensely:

$$\lambda_{max} = \frac{0.0029}{T} = \frac{0.0029}{10,000} = 2.9 \times 10^{-7} \text{ m} = 290 \text{ nm}$$

Sirius therefore emits light most intensely in the ultraviolet. In the visible part of the spectrum, it emits more blue light than red light (see the curve for 12,000 K in Figure 5-10), so Sirius has a distinct blue color.

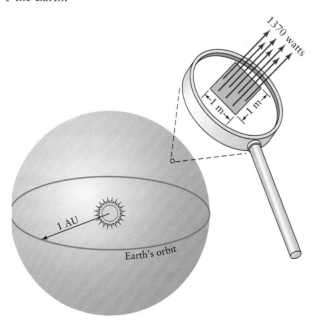

(continued on following page)

box 5-2 *(continued)*

EXAMPLE: We can use the Stefan-Boltzmann law to compare the energy flux from Sirius with that from the Sun. For the Sun, the Stefan-Boltzmann law is $F_\odot = \sigma T_\odot^4$, and for Sirius we can likewise write $F_* = \sigma T_*^4$, where the subscripts $\odot$ and $*$ refer to the Sun and Sirius, respectively. If we divide one equation by the other, the Stefan-Boltzmann con°stants cancel out, giving us an expression in terms of ratios:

$$\frac{F_*}{F_\odot} = \frac{T_*^4}{F_\odot^4}$$

In the preceding examples, we found that the temperatures of Sirius and the Sun are 10,000 K and 5800 K, respectively. Thus, we get

$$\frac{F_*}{F_\odot} = \frac{(10{,}000 \text{ K})^4}{(5800 \text{ K})^4} = \left(\frac{10{,}000}{5800}\right)^4 = 8.8$$

In other words, each square meter of Sirius's surface emits 8.8 times more energy per second than a square meter of the Sun's surface.

Maxwell's description of light as electromagnetic waves. But all such theories failed to explain the characteristic shapes of blackbody curves shown in Figure 5-10.

In 1900, however, the German physicist Max Planck discovered that he could derive a formula that correctly described blackbody curves if he assumed that electromagnetic energy is emitted in discrete, particlelike packets. He further had to assume that the energy of each such packet, today called a **photon,** is related to the wavelength of light: The greater the wavelength, the lower the energy of a photon associated with that wavelength. Thus, a photon of red light (wavelength $\lambda = 700$ nm) has less energy than a photon of violet light ($\lambda = 400$ nm). In this picture, light has a dual personality; it behaves as a stream of particlelike photons, but each photon has wavelike properties. In this sense, the best answer to the question "Is light a wave or a stream of particles?" is "Yes!"

It was soon realized that Planck's photon hypothesis explains more than just the detailed shape of blackbody curves. For example, it explains why only ultraviolet light causes suntans and sunburns. The reason is that tanning or burning involves a chemical reaction in the skin. High-energy, short-wavelength ultraviolet photons can trigger these reactions, but the lower-energy, longer-wavelength photons of visible light cannot. Similarly, normal photographic film is sensitive to visible light but not to infrared light; a long-wavelength infrared photon does not have enough energy to cause the chemical change that occurs when film is exposed to the higher-energy photons of visible light. Another phenomenon explained by the photon hypothesis is the **photoelectric effect.** In this effect, a metal plate is illuminated by a light beam. If ultraviolet light is used, tiny negatively charged particles called **electrons** are emitted from the metal plate. (We will see in Section 5-7 that the electron is one of the basic particles of the atom.) But if visible light is used, no matter how bright, no electrons are emitted.

In 1905 the great German-born physicist Albert Einstein explained this behavior by noting that a certain minimum amount of energy is required to remove an electron from the metal plate. The energy of a short-wavelength ultraviolet photon is greater than this minimum value, so an electron that absorbs a photon of ultraviolet light will have enough energy to escape from the plate. But an electron that absorbs a photon of visible light, with its longer wavelength and lower energy, does not gain enough energy to escape and so remains within the metal. Einstein and Planck both won Nobel prizes for their contributions to understanding the nature of light.

The relationship between the energy E of a single photon and the wavelength of the electromagnetic radiation can be expressed in a simple equation:

Energy of a photon (in terms of wavelength)

$$E = \frac{hc}{\lambda}$$

E = energy of a photon

h = Planck's constant

c = speed of light

λ = wavelength of light

The value of the constant h in this equation, now called *Planck's constant,* has been shown in laboratory experiments to be

$$h = 6.625 \times 10^{-34} \text{ J s}$$

The units of h are joules multiplied by seconds, called "joule-seconds" and abbreviated J s.

Because the value of h is so tiny, a single photon carries a very small amount of energy. For example, a photon of red light with wavelength 633 nm has an energy of only 3.14×10^{-19} J (Box 5-3). This is why we ordinarily do not notice that light comes in the form of photons; even a dim light source emits so many photons per second that it seems to be radiating a continuous stream of energy.

box 5-3 | **The Heavens on the Earth**

Photons at the Supermarket

A beam of light can be regarded as a stream of tiny packets of energy called photons. The Planck relationships $E = hc/\lambda$ and $E = h\nu$ can be used to relate the energy E carried by a photon to its wavelength λ and frequency ν.

As an example, the laser bar-code scanners used at stores and supermarkets emit orange-red light of wavelength 633 nm. To calculate the energy of a single photon of this light, we must first express the wavelength in meters. A nanometer (nm) is equal to 10^{-9} m, so the wavelength is

$$\lambda = (633 \text{ nm})\left(\frac{10^{-9} \text{ m}}{1 \text{ nm}}\right)$$

$$= 633 \times 10^{-9} \text{ m}$$

$$= 6.33 \times 10^{-7} \text{ m}$$

Then, using the Planck formula $E = hc/\lambda$, we find that the energy of a single photon is

$$E = \frac{hc}{\lambda} = \frac{(6.625 \times 10^{-34} \text{ J s})(3 \times 10^8 \text{ m s}^{-1})}{6.33 \times 10^{-7} \text{ m}} = 3.14 \times 10^{-19} \text{ J}$$

This is a very small amount of energy. The laser in a typical bar-code scanner emits 10^{-3} joule of light energy per second, so the number of photons emitted per second is

$$\frac{10^{-3} \text{ joule per second}}{3.14 \times 10^{-19} \text{ joule per photon}} = 3.2 \times 10^{15} \text{ photons per second}$$

This is such a large number that the laser beam seems like a continuous flow of energy rather than a stream of little energy packets.

The energies of photons are sometimes expressed in terms of a small unit of energy called the **electron volt** (eV). One electron volt is equal to 1.602×10^{-19} J, so a 633-nm photon has an energy of 1.96 eV. If energy is expressed in electron volts, Planck's constant is best expressed in electron volts multiplied by seconds, abbreviated eV s:

$$h = 4.135 \times 10^{-15} \text{ eV s}$$

Because the frequency ν of light is related to the wavelength λ by $\nu = c/\lambda$, we can rewrite the equation for the energy of a photon as

Energy of a photon (in terms of frequency)

$$E = h\nu$$

E = energy of a photon

h = Planck's constant

ν = frequency of light

The equations $E = hc/\lambda$ and $E = h\nu$ are together called **Planck's law.** Both equations express a relationship between a particlelike property of light (the energy E of a photon) and a wavelike property (the wavelength λ or frequency ν).

The photon picture of light is essential for understanding the detailed shapes of blackbody curves. As we will see, it also helps to explain how and why the spectra of the Sun and stars differ from those of perfect blackbodies.

5-6 Each chemical element produces its own unique set of spectral lines

In 1814 the German master optician Joseph von Fraunhofer repeated Newton's classic experiment of shining a beam of sunlight through a prism (see Figure 5-3). But this time Fraunhofer subjected the resulting rainbow-colored spectrum to intense magnification. To his surprise, he discovered that the solar spectrum contains hundreds of fine, dark lines, now called **spectral lines.** By contrast, if the light from a perfect blackbody were sent through a prism, it would produce a smooth, continuous spectrum with no dark lines. Fraunhofer counted more than 600 dark lines in the Sun's spectrum; today we know of more than 30,000. The photograph of the Sun's spectrum in Figure 5-12 shows hundreds of these spectral lines.

Half a century later, chemists discovered that they could produce spectral lines in the laboratory and use these spectral lines to analyze what kinds of atoms different substances are made of. Chemists had long known that many substances emit distinctive colors when sprinkled into a flame. To facilitate study of these colors, around 1857 the German chemist Robert Bunsen invented a gas burner (today called a Bunsen burner) that produces a clean flame with no color of its own. Bunsen's colleague, the Prussian-born physicist Gustav Kirchhoff, suggested that the colored light produced by substances in a flame might best be studied by passing the light through a prism (Figure 5-13). The two scientists promptly discovered that the spectrum from a flame consists of a pattern of thin, bright

figure 5-12 R I **V** U X G

The Sun's Spectrum Numerous dark spectral lines are seen in this photograph of the Sun's spectrum. The spectrum is spread out so much that it had to be cut into segments to fit on this page. (National Optical Astronomy Observatories)

spectral lines against a dark background. The same kind of spectrum is produced by heated gases such as neon or argon.

Kirchhoff and Bunsen then found that each chemical element produces its own unique pattern of spectral lines. Thus was born in 1859 the technique of **spectral analysis,** the identification of chemical substances by their unique patterns of spectral lines.

A chemical **element** is a fundamental substance that cannot be broken down into more basic chemicals. Some examples are hydrogen, oxygen, carbon, iron, gold, and silver. After Kirchhoff and Bunsen had recorded the prominent spectral lines of all the then-known elements, they soon began to discover other spectral lines in the spectra of vaporized mineral samples. In this way they discovered elements whose presence had never before been suspected. In 1860 Kirchhoff and Bunsen found a new line in the blue portion of the spectrum of a sample of mineral water. After isolating the previously unknown element responsible for making the line, they named it cesium (from the Latin *caesium,* "gray-blue"). The next year, a new line in the red portion of the spectrum of a mineral sample led them to discover the element rubidium (Latin *rubidium,* "red").

Spectral analysis even allowed the discovery of new elements outside Earth. During the solar eclipse of 1868, astronomers found a new spectral line in light coming from the hot gases at the upper surface of the Sun while the main body of the Sun was hidden by the Moon. This line was attributed to a new element that was named helium (from the Greek *helios,* "sun"). Helium was not discovered on Earth until 1895, when it was found in gases obtained from a uranium mineral.

The spectrum of the Sun, with its dark spectral lines superimposed on a bright background (see Figure 5-12), may seem to

be unrelated to the spectra of bright lines against a dark background produced by substances in a flame (see Figure 5-13). But by the early 1860s Kirchhoff's experiments had revealed a direct connection between these two types of spectra. His conclusions are summarized in three important statements about spectra that are today called **Kirchhoff's laws.** These laws, which are illustrated in Figure 5-14, are as follows:

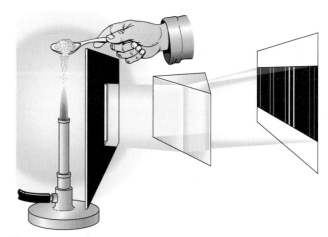

figure 5-13

The Kirchhoff-Bunsen Experiment In the mid-1850s, Gustav Kirchhoff and Robert Bunsen discovered that when a chemical substance is heated and vaporized, the spectrum of the emitted light exhibits a series of bright spectral lines. They also found that each chemical element produces its own characteristic pattern of spectral lines. (In an actual laboratory experiment, lenses would be needed to focus the image of the slit onto the screen.)

Law 1 A hot opaque body, such as a perfect blackbody, or a hot, dense gas produces a **continuous spectrum**—a complete rainbow of colors without any spectral lines.

Law 2 A hot, transparent gas produces an **emission line spectrum**—a series of bright spectral lines against a dark background.

Law 3 A cool, transparent gas in front of a source of a continuous spectrum produces an **absorption line spectrum**—a series of dark spectral lines among the colors of the continuous spectrum. Furthermore, the dark lines in the absorption spectrum of a particular gas occur at exactly the *same* wavelengths as the bright lines in the emission spectrum of that same gas.

Kirchhoff's laws imply that if a beam of white light is passed through a gas, the atoms of the gas somehow extract light of very specific wavelengths from the white light. Hence, an observer who looks straight through the gas at the white-light source (the blackbody in Figure 5-14) will receive light whose spectrum has dark absorption lines superimposed on the continuous spectrum of the white light. The gas atoms then radiate light of precisely these same wavelengths in all directions. An observer at an oblique angle (that is, one who is not sighting directly through the cloud toward the blackbody) will receive only this light radiated by the gas cloud; the spectrum of this light is bright emission lines on a dark background.

 Figure 5-14 shows that light can either pass through a cloud of gas or be absorbed by the gas. But there is also a third possibility: The light can simply bounce off the atoms or molecules that make up the gas, a phenomenon called **light scattering**. In other words, photons passing through a gas cloud can miss the gas atoms altogether, be swallowed whole by the atoms (absorption), or bounce off the atoms like billiard balls colliding (scattering). Box 5-4 describes how light scattering explains the blue color of the sky and the red color of sunsets.

 Whether an emission line spectrum or an absorption line spectrum is observed from a gas cloud depends on the relative temperatures of the gas cloud and its background. Absorption lines are seen if the background is hotter than the gas, and emission lines are seen if the background is cooler.

For example, if sodium is placed in the flame of a Bunsen burner in a darkened room, the flame will emit a characteristic

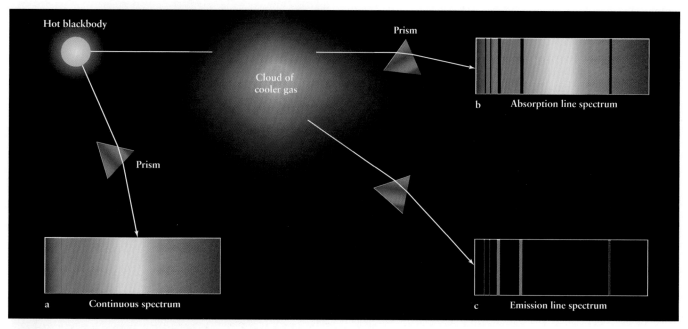

Figure 5-14

Continuous, Absorption Line, and Emission Line Spectra A hot, glowing object such as a blackbody emits a continuous spectrum of light (spectrum *a*). If this light is passed through a cloud of a cooler gas, the cloud absorbs light of certain specific wavelengths. Hence, the light from the hot, glowing object that passes directly through the cloud is depleted in these wavelengths, and the spectrum of this light has dark absorption lines (spectrum *b*). The cloud does not retain all the light energy that it absorbs but radiates it outward in all directions. The spectrum of this reradiated light contains bright emission lines (spectrum *c*). The dark absorption lines in spectrum *b* have exactly the same wavelengths as the bright emission lines in spectrum *c*. The specific wavelengths observed depend on the chemical composition of the cloud.

Box 5-4   The Heavens on the Earth

Light Scattering

Light scattering is the process whereby photons bounce off particles in their path. These particles can be atoms, molecules, or clumps of molecules. You are reading these words using photons from the Sun or a lamp that bounced off the page—that is, were scattered by the particles that make up the page.

An important fact about light scattering is that very small particles—ones that are smaller than a wavelength of visible light—are quite effective at scattering short-wavelength photons of blue light, but less effective at scattering long-wavelength photons of red light. This fact explains a number of phenomena that you can see here on Earth.

The light that comes from the daytime sky is sunlight that has been scattered by the molecules that make up our atmosphere (see part *a* of the accompanying figure). Air molecules are less than 1 nm across, far smaller than the wavelength of visible light, so they scatter blue light more than red light—which is why the sky looks blue. Smoke particles are also quite small, which explains why the smoke from a cigarette or a fire has a bluish color.

Distant mountains often appear blue thanks to sunlight's being scattered from the atmosphere between the mountains and your eyes. (The Blue Ridge Mountains, which extend from Pennsylvania to Georgia, and Australia's Blue Mountains derive their names from this effect.) Sunglasses often have a red or orange tint, which blocks out blue light. This cuts down on the amount of scattered light from the sky reaching your eyes and allows you to see distant objects more clearly.

Light scattering also explains why sunsets are red. The light from the Sun contains photons of all visible wavelengths, but as this light passes through our atmosphere the blue photons are scattered away from the straight-line path from the Sun to your eye. Red photons undergo relatively little scattering, so the Sun always looks a bit redder than it really is. When you look toward the setting sun, the sunlight that reaches your eye has had to pass through a relatively thick layer of atmosphere (part *b* of the accompanying figure). Hence, a large fraction of the blue light from the Sun has been scattered, and the Sun appears quite red.

The same effect also applies to sunrises, but sunrises seldom look as red as sunsets do. The reason is that dust is lifted into the atmosphere during the day by the wind (which is typically stronger in the daytime than at night), and dust particles in the atmosphere help to scatter even more blue light.

If the small particles that scatter light are sufficiently concentrated, there will be almost as much scattering of red light as of blue light, and the scattered light will appear white. This explains the white color of clouds, fog, and haze, in which the scattering particles are ice crystals or water droplets. Whole milk looks white because of light scattering from tiny fat globules; nonfat milk has only a very few of these globules and so has a slight bluish cast.

Light scattering has many applications to astronomy. For example, it explains why very distant stars in our Galaxy appear surprisingly red. The reason is that there is a very thin gas in the space between the stars, and that gas scatters blue photons. By studying how much scattering takes place, astronomers have learned about the tenuous material that fills interstellar space.

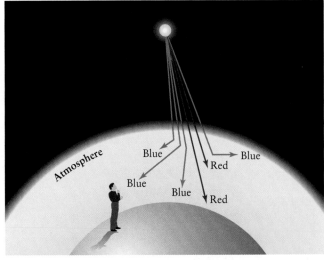

a

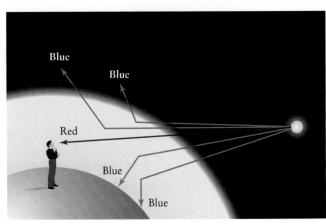

b

orange-yellow glow. (This same glow is produced if we use ordinary table salt, which is a compound of sodium and chlorine.) If we pass the light from the flame through a prism, it displays an emission line spectrum with two closely spaced spectral lines at wavelengths of 588.99 and 589.59 nm, in the orange-yellow part of the spectrum. We now turn on a lightbulb whose filament is hotter than the flame and shine the bulb's white light through the flame. The spectrum of this light after it passes through the flame's sodium vapor is the continuous spectrum from the lightbulb, but with two closely spaced *dark* lines at 588.99 and 589.59 nm. Thus, the chemical composition of the gas is revealed by either bright emission lines or dark absorption lines.

Spectroscopy is the systematic study of spectra and spectral lines. Spectral lines are tremendously important in astronomy, because they provide reliable evidence about the chemical composition of distant objects. As an example, the spectrum of the Sun shown in Figure 5-12 is an absorption line spectrum. The continuous spectrum comes from the hot surface of the Sun, which acts like a blackbody. The dark absorption lines are caused by this light passing through a cooler gas; this gas is the atmosphere that surrounds the Sun. Therefore, by identifying the spectral lines present in the solar spectrum, we can determine the chemical composition of the Sun's atmosphere.

Figure 5-15 shows both a portion of the Sun's absorption line spectrum and the emission line spectrum of iron vapor over the same wavelength range. This pattern of bright spectral lines in the lower spectrum is iron's own distinctive "fingerprint," which no other substance can imitate. Because some absorption lines in the Sun's spectrum coincide with the iron lines, some vaporized iron must exist in the Sun's atmosphere.

Spectroscopy can also help us analyze gas clouds in space, such as the nebula NGC 2363 shown in Figure 5-16. Such glowing clouds have emission line spectra, because we see them against the black background of space. The dominant red color of NGC 2363 is due to an emission line at a wavelength near 656 nm. This is one of the characteristic wavelengths emitted by hydrogen gas, so we can conclude that this nebula contains hydrogen. More detailed analyses of this kind show that hydrogen is the most common element in gaseous nebulae, and indeed in the universe as a whole.

What is truly remarkable about spectroscopy is that it can determine chemical composition at any distance. The 656-nm red light produced by a sample of heated hydrogen gas on

 figure 5-16   R I **V** U X G

The Nebula NGC 2363 The glowing gas cloud in this Hubble Space Telescope image lies within a galaxy some 10 million light-years away in the constellation Camelopardis (the Camel). Hot stars within the nebula emit high-energy, ultraviolet photons, which are absorbed by the surrounding gas and heat the gas to high temperature. This heated gas produces light with an emission line spectrum. The particular wavelength of red light emitted by the nebula is 656 nm, characteristic of hydrogen gas. (L. Drissen, J.-R. Roy, and C. Robert/Département de Physique and Observatoire du Mont Mégantic, Université Laval; and NASA)

Earth is the same as that observed coming from NGC 2363 in Figure 5-16, located about 10 million light-years away. By using the basic principles outlined by Kirchhoff, astronomers have the tools to make chemical assays of objects that are almost inconceivably distant.

To make full use of Kirchhoff's laws, it is helpful to understand why they work. Why does an atom absorb light of only particular wavelengths? And why does it then emit light of only these same wavelengths? Maxwell's theory of electromagnetism

figure 5-15   R I **V** U X G

Iron in the Sun The upper part of this figure is a portion of the Sun's absorption line spectrum from 420 to 430 nm. Numerous dark spectral lines are visible. The lower part of the figure is a corresponding portion of the emission line spectrum of vaporized iron. Several bright spectral lines are seen against a black background. The iron lines coincide with some of the solar lines, which proves that there is some iron (albeit a very tiny amount) in the Sun's atmosphere. (Carnegie Observatories)

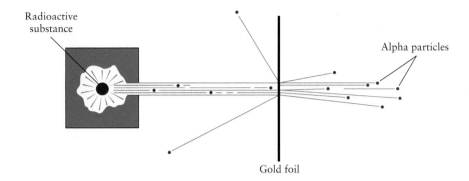

Radioactive substance

Alpha particles

Gold foil

figure 5-17

Rutherford's Experiment Alpha particles from a radioactive source are directed at a thin metal foil. Most alpha particles pass through the foil with very little deflection. Occasionally, however, an alpha particle will recoil, indicating that it has collided with the massive nucleus of an atom. This experiment provided the first evidence that the nuclei of atoms are relatively massive and compact.

(Section 5-2) could not answer these questions. The answers did not come until early in the twentieth century, when scientists began to discover the structure and properties of atoms.

5-7 An atom consists of a small, dense nucleus surrounded by electrons

The first important clue about the internal structure of atoms came from an experiment conducted in 1910 by Ernest Rutherford, a gifted chemist and physicist from New Zealand. Rutherford and his colleagues at the University of Manchester in England had been investigating the recently discovered phenomenon of radioactivity. Certain radioactive elements, such as uranium and radium, were known to emit particles of various types. One type, the alpha particle, has about the same mass as a helium atom and is emitted from some radioactive substances with considerable speed.

In one series of experiments, Rutherford and his colleagues were using alpha particles as projectiles to probe the structure of solid matter. They directed a beam of these particles at a thin sheet of metal (Figure 5-17). Almost all the alpha particles passed through the metal sheet with little or no deflection from their straight-line paths. To the surprise of the experimenters, however, an occasional alpha particle bounced back from the metal sheet as though it had struck something quite dense. Rutherford later remarked, "It was almost as incredible as if you fired a fifteen-inch shell at a piece of tissue paper and it came back and hit you."

Rutherford concluded from this experiment that most of the mass of an atom is concentrated in a compact, massive lump of matter that occupies only a small part of the atom's volume. Most of the alpha particles pass freely through the nearly empty space that makes up most of the atom, but a few particles happen to strike the dense mass at the center of the atom and bounce back.

Rutherford proposed a new model for the structure of an atom, shown in Figure 5-18. According to this model, a massive, positively charged **nucleus** at the center of the atom is orbited by tiny, negatively charged electrons. Rutherford concluded that at least 99.98% of the mass of an atom must be concentrated in its nucleus, whose diameter is only about

10^{-14} m. (The diameter of a typical atom is far larger, about 10^{-10} m.)

ANALOGY To appreciate just how tiny the nucleus is, imagine expanding an atom by a factor of 10^{12}, so that it was 100 meters across, about the size of a sports stadium. To this scale, the nucleus would be just a centimeter across—no larger than your thumbnail.

We know today that the nucleus of an atom contains two types of particles, **protons** and **neutrons**. A proton has a positive electric charge, equal and opposite to that of an electron. As its name suggests, a neutron has no electric charge—it is electrically neutral. As an example, an alpha particle (such as those Rutherford's team used) is actually a nucleus of the

figure 5-18

Rutherford's Model of the Atom Electrons orbit the atom's nucleus, which contains most of the atom's mass. The nucleus contains two types of particles, protons and neutrons.

helium atom, with two protons and two neutrons. Protons and neutrons are held together in a nucleus by the so-called strong nuclear force, whose great strength overcomes the electric repulsion between the positively charged protons. A proton and a neutron have almost the same mass, 1.7×10^{-27} kg, and each has about 2000 times as much mass as an electron (9.1×10^{-31} kg). In an ordinary atom there are as many positive protons as there are negative electrons, so the atom has no net electric charge. Because the mass of the electron is so small, the mass of an atom is not much greater than the mass of its nucleus. That is why an alpha particle has nearly the same mass as an atom of helium.

While the solar system is held together by gravitational forces, atoms are held together by electrical forces. The electric forces attracting the positively charged protons and the negatively charged electrons keep the atom from coming apart. Box 5-5 describes more about the connection between the structure of atoms and the chemical and physical properties of substances made of those atoms.

Rutherford's experiments clarified the structure of the atom, but they did not explain how these tiny particles within the atom give rise to spectral lines. The task of reconciling Rutherford's atomic model with Kirchhoff's laws of spectral analysis was undertaken by the young Danish physicist Niels Bohr, who joined Rutherford's group at Manchester in 1911.

5-8   Spectral lines are produced when an electron jumps from one energy level to another within an atom

WEB LINK 5.7 Niels Bohr began his study of the connection between atomic spectra and atomic structure by trying to understand the structure of hydrogen, the simplest and lightest of the elements. (As we discussed in Section 5-6, hydrogen is also the most common element in the universe.) When Bohr was done, he had not only found a way to explain this atom's spectrum but had also found a justification for Kirchhoff's laws in terms of atomic physics.

The most common type of hydrogen atom consists of a single electron and a single proton. Hydrogen has a simple

visible-light spectrum consisting of a pattern of lines that begins at a wavelength of 656.3 nm and ends at 364.6 nm. The first spectral line is called H_α (H-alpha), the second spectral line is called H_β (H-beta), the third is H_γ (H-gamma), and so forth. The closer you get to the short-wavelength end of the spectrum at 364.6 nm, the more spectral lines you see.

The regularity in this spectral pattern was described mathematically in 1885 by Johann Jakob Balmer, a Swiss schoolteacher. The spectral lines of hydrogen at visible wavelengths are today called **Balmer lines,** and the entire pattern from H_α onward is called the **Balmer series.** More than two dozen Balmer lines are seen in the spectrum of the star shown in Figure 5-19. Stars in general, including the Sun, have Balmer absorption lines in their spectra, which shows they have atmospheres that contain hydrogen.

Using trial and error, Balmer discovered a formula from which the wavelengths (λ) of hydrogen's spectral lines can be calculated. Balmer's formula is usually written

$$\frac{1}{\lambda} = R\left(\frac{1}{4} - \frac{1}{n^2}\right)$$

where n can be any integer (whole number) greater than 2. Here R is the *Rydberg constant* ($R = 1.097 \times 10^7$ m^{-1}), named in honor of the Swedish spectroscopist Johannes Rydberg. To get the wavelength λ_α of the spectral line H_α, you first put $n = 3$ into Balmer's formula:

$$\frac{1}{\lambda_\alpha} = (1.097 \times 10^7 \text{ m}^{-1})\left(\frac{1}{4} - \frac{1}{3^2}\right) = 1.524 \times 10^6 \text{ m}^{-1}$$

Then take the reciprocal:

$$\lambda_\alpha = \frac{1}{1.524 \times 10^6 \text{ m}^{-1}} = 6.563 \times 10^{-7} \text{ m} = 656.3 \text{ nm}$$

To get the wavelength of H_β, use $n = 4$, and to get the wavelength of H_γ, use $n = 5$. If you use $n = \infty$ (the symbol ∞ stands for infinity), you get the short-wavelength end of the hydrogen spectrum at 364.6 nm. (Note that 1 divided by infinity equals zero.)

Bohr realized that to fully understand the structure of the hydrogen atom, he had to be able to derive Balmer's formula

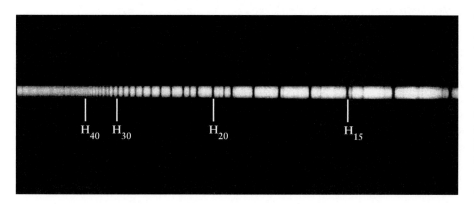

Figure 5-19   R I V **U** X G

Balmer Lines in the Spectrum of a Star
This portion of the spectrum of the star HD 193182 shows nearly two dozen Balmer lines, from H_{13} through H_{40}. (Numbers are used beyond the first few Balmer lines, so you do not have to memorize the entire Greek alphabet!) The series converges at 364.6 nm, just to the left of H_{40}. This star's spectrum also contains the first 12 Balmer lines (H_α through H_{12}), but they are located beyond the right edge of this photograph. (Carnegie Observatories)

box 5-5 | Tools of the Astronomer's Trade

Atoms, the Periodic Table, and Isotopes

Each different chemical element is made of a specific type of atom. Each specific atom has a characteristic number of protons in its nucleus. For example, a hydrogen atom has 1 proton in its nucleus, an oxygen atom has 8 protons in its nucleus, and so on.

The number of protons in an atom's nucleus is the **atomic number** for that particular element. The chemical elements are most conveniently listed in the form of a **periodic table** (shown in the figure). Elements are arranged in the periodic table in order of increasing atomic number. With only a few exceptions, this sequence also corresponds to increasing average mass of the atoms of the elements. Thus, hydrogen (symbol H), with atomic number 1, is the lightest element. Iron (symbol Fe) has atomic number 26 and is a relatively heavy element.

All the elements listed in a single vertical column of the periodic table have similar chemical properties. For example, the elements in the far right column are all gases under the conditions of temperature and pressure found at the Earth's surface, and they are all very reluctant to react chemically with other elements.

In addition to nearly 100 naturally occurring elements, the periodic table includes a number of artificially produced elements. All these elements are heavier than uranium (symbol U) and are highly radioactive, which means that they decay into lighter elements within a short time of being created in laboratory experiments. Scientists have succeeded in creating only a few atoms of elements 110 through 118 in this way, and these "superheavy" elements have not yet been given official names.

The number of protons in the nucleus of an atom determines which element that atom is. Nevertheless, the same element may have different numbers of neutrons in its nucleus. For example, oxygen (O) has atomic number 8, so every oxygen nucleus has exactly 8 protons. But oxygen nuclei can have 8, 9, or 10 neutrons. These three slightly different kinds of oxygen are called **isotopes.** The isotope with 8 neutrons is by far the most abundant variety. It is written as ^{16}O, or oxygen-16. The rarer isotopes with 9 and 10 neutrons are designated as ^{17}O and ^{18}O, respectively.

The superscript that precedes the chemical symbol for an element equals the total number of protons and neutrons in a nucleus of that particular isotope. For example, a nucleus of the most common isotope of iron, ^{56}Fe or iron-56, contains a total of 56 protons and neutrons. From the periodic table, the atomic number of iron is 26, so every iron atom has 26 protons in its nucleus. Therefore, the number of neutrons in an iron-56 nucleus is $56 - 26 = 30$. (Most nuclei have more neutrons than protons, especially in the case of the heaviest elements.)

It is extremely difficult to distinguish chemically between the various isotopes of a particular element. Ordinary

---

using the laws of physics. He first made the rather wild assumption that the electron in a hydrogen atom can orbit the nucleus only in certain specific orbits. (This was a significant break with the ideas of Newton, in whose mechanics any orbit should be possible.) Figure 5-20 shows the four smallest of these **Bohr orbits**, labeled by the numbers $n = 1$, $n = 2$, $n = 3$, and so on.

Although confined to one of these allowed orbits while circling the nucleus, an electron can jump from one Bohr orbit to another. For an electron to do this, the hydrogen atom must gain or lose a specific amount of energy. The atom must absorb energy for the electron to go from an inner to an outer orbit; the atom must release energy for the electron to go from an outer to an inner orbit. As an example, Figure 5-21 shows an electron jumping between the $n = 2$ and $n = 3$ orbits of a hydrogen atom as the atom absorbs or emits an H_α photon.

When the electron jumps from one orbit to another, the energy of the photon that is emitted or absorbed equals the difference in energy between these two orbits. This energy difference, and hence the photon energy, is the same whether the jump is from a low orbit to a high orbit (Figure 5-21*a*) or from the high orbit back to the low one (Figure 5-21*b*). According

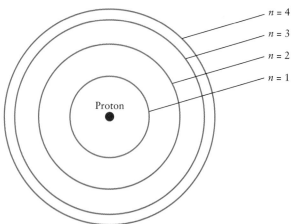

figure 5-20

The Bohr Model of the Hydrogen Atom According to the Bohr model of the hydrogen atom, an electron circles the nucleus (a proton) only in allowed orbits $n = 1, 2, 3$, and so forth. The first four Bohr orbits are shown here. This figure is not drawn to scale; in the Bohr model, the $n = 2, 3$, and 4 orbits are respectively 4, 9, and 16 times larger than the $n = 1$ orbit.

chemical reactions involve only the electrons that orbit the atom, never the neutrons buried in its nucleus. But there are small differences in the wavelengths of the spectral lines for different isotopes of the same element. For example, the spectral line wavelengths of the hydrogen isotope ^{2}H are about 0.03% greater than the wavelengths for the most common hydrogen isotope, ^{1}H. Thus, different isotopes can be distinguished by careful spectroscopic analysis.

Isotopes are important in astronomy for a variety of reasons. By measuring the relative amounts of different isotopes of a given element in a Moon rock or meteorite, the age of that sample can be determined. The mixture of isotopes left behind when a star explodes into a supernova (see Section 1-3) tells astronomers about the processes that led to the explosion. And knowing the properties of different isotopes of hydrogen and helium is crucial to understanding the nuclear reactions that make the Sun shine. Look for these and other applications of the idea of isotopes in later chapters.

Periodic Table of the Elements

1 H Hydrogen																	2 He Helium
3 Li Lithium	4 Be Beryllium											5 B Boron	6 C Carbon	7 N Nitrogen	8 O Oxygen	9 F Fluorine	10 Ne Neon
11 Na Sodium	12 Mg Magnesium											13 Al Aluminum	14 Si Silicon	15 P Phosphorus	16 S Sulfur	17 Cl Chlorine	18 Ar Argon
19 K Potassium	20 Ca Calcium	21 Sc Scandium	22 Ti Titanium	23 V Vanadium	24 Cr Chromium	25 Mn Manganese	26 Fe Iron	27 Co Cobalt	28 Ni Nickel	29 Cu Copper	30 Zn Zinc	31 Ga Gallium	32 Ge Germanium	33 As Arsenic	34 Se Selenium	35 Br Bromine	36 Kr Kryton
37 Rb Rubidium	38 Sr Strontium	39 Y Yttrium	40 Zr Zirconium	41 Nb Niobium	42 Mo Molybdenum	43 Tc Technetium	44 Ru Ruthenium	45 Rh Rhodium	46 Pd Palladium	47 Ag Silver	48 Cd Cadmium	49 In Indium	50 Sn Tin	51 Sb Antimony	52 Te Tellurium	53 I Iodine	54 Xe Xenon
55 Cs Cesium	56 Ba Barium	71 Lu Lutetium	72 Hf Hafnium	73 Ta Tantaium	74 W Tungsten	75 Re Rhenium	76 Os Osmium	77 Ir Iridium	78 Pt Platinum	79 Au Gold	80 Hg Mercury	81 Tl Thallium	82 Pb Lead	83 Bi Bismuth	84 Po Polonium	85 At Astatine	86 Rn Radon
87 Fr Francium	88 Ra Radium	103 Lr Lawrencium	104 Rf Rutherfordium	105 Db Dubnium	106 Sg Seaborgium	107 Bh Bohrium	108 Hs Hassium	109 Mt Meitnerium	110	111	112	113	114	115	116	117	118

57 La Lanthanum	58 Ce Cerium	59 Pr Praseodymium	60 Nd Neodymium	61 Pm Promethium	62 Sm Samarium	63 Eu Europium	64 Gd Gadolinium	65 Tb Terbium	66 Dy Dysprosium	67 Ho Holmium	68 Er Erbium	69 Tm Thulium	70 Yb Ytterbium
89 Ac Actinium	90 Th Thorium	91 Pa Protactinium	92 U Uranium	93 Np Neptunium	94 Pu Plutonium	95 Am Americium	96 Cm Curium	97 Bk Berkelium	98 Cf Californium	99 Es Einsteinium	100 Fm Fermium	101 Md Mendelevium	102 No Nobelium

to Planck and Einstein, if two photons have the same energy E, the relationship $E = hc/\lambda$ tells us that they must also have the same wavelength λ. It follows that if an atom can emit photons of a given energy and wavelength, it can also absorb photons of precisely the same energy and wavelength. Thus, Bohr's picture explains Kirchhoff's observation that atoms emit and absorb the same wavelengths of light.

The Bohr picture also helps us visualize what happens to produce an emission line spectrum. When a gas is heated, its atoms move around rapidly and can collide forcefully with each other. These energetic collisions excite the atoms' electrons into high orbits. The electrons then cascade back down to the inner-most possible orbit, emitting photons whose energies are equal to the energy differences between different Bohr orbits. In this

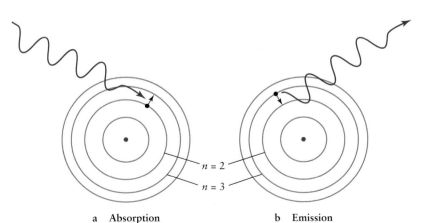

a Absorption   b Emission

figure 5-21
The Absorption and Emission of an H$_\alpha$ Photon This schematic diagram, drawn according to the Bohr model of the atom, shows what happens when a hydrogen atom absorbs or emits a photon whose wavelength is 656.3 nm. **(a)** The photon is absorbed by the atom, causing the electron to jump from the $n = 2$ orbit up to $n = 3$. **(b)** The photon is emitted by the atom as the electron falls from the $n = 3$ orbit down to the $n = 2$ orbit.

ANIMATION 5.2

fashion, a hot gas produces an emission line spectrum with a variety of different wavelengths.

To produce an absorption line spectrum, begin with a relatively cool gas, so that the electrons in most of the atoms are in inner, low-energy orbits. If a beam of light with a continuous spectrum is shone through the gas, most wavelengths will pass through undisturbed. Only those photons will be absorbed whose energies are just right to excite an electron to an allowed outer orbit. Hence, only certain wavelengths will be absorbed, and dark lines will appear in the spectrum at those wavelengths.

Using his picture of allowed orbits and the formula $E = hc/\lambda$, Bohr was able to prove mathematically that the wavelength λ of the photon emitted or absorbed as an electron jumps between an inner orbit N and an outer orbit n is

Bohr formula for hydrogen wavelengths

$$\frac{1}{\lambda} = R\left(\frac{1}{N^2} - \frac{1}{n^2}\right)$$

N = number of inner orbit

n = number of outer orbit

R = Rydberg constant = 1.097×10^7 m^{-1}

λ = wavelength (in meters) of emitted or absorbed photon

If Bohr let $N = 2$ in this formula, he got back the formula that Balmer discovered by trial and error. Hence, Bohr deduced the meaning of the Balmer series: All the Balmer lines are produced by electrons jumping between the second Bohr orbit ($N = 2$) and higher orbits ($n = 3, 4, 5,$ and so on).

Bohr's formula also correctly predicts the wavelengths of other series of spectral lines that occur at nonvisible wavelengths. Using $N = 1$ gives the **Lyman series,** which is entirely in the ultraviolet. All the spectral lines in this series involve electron transitions between the lowest Bohr orbit and all higher orbits ($n = 2, 3, 4,$ and so on). This pattern of spectral lines begins with L_α (Lyman alpha) at 122 nm and converges on L_∞ at 91.2 nm. Using $N = 3$ gives a series of infrared wavelengths called the **Paschen series.** This series, which involves transitions between the third Bohr orbit and all higher orbits, begins with P_α (Paschen alpha) at 1875 nm and converges on P_∞ at 822 nm. Additional series exist at still longer wavelengths. Figure 5-22 shows some examples of electron transitions in the Bohr atom for these series, along with the wavelength for each transition.

Today's view of the atom owes much to the Bohr model, but is different in certain ways. The modern picture is based on **quantum mechanics,** a branch of physics dealing with photons and subatomic particles that was developed during the 1920s. As a result of this work, physicists no longer picture electrons as moving in specific orbits about the nucleus. Instead, electrons are now known to have both particle and wave properties and are said to occupy only certain **energy levels** in the atom.

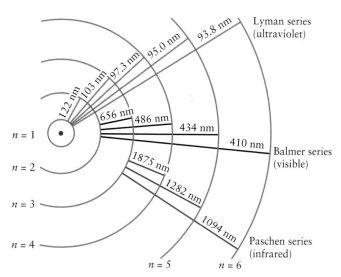

 Figure 5-22

Electron Transitions in the Hydrogen Atom
Electron transitions between the lowest orbit ($n = 1$) and higher orbits ($n = 2, 3, 4, \ldots$) in the hydrogen atom give rise to spectral lines in the ultraviolet called the Lyman series. Transitions between the second orbit ($n = 2$) and higher orbits ($n = 3, 4, 5, \ldots$) cause visible spectral lines in the Balmer series, and transitions between the third orbit ($n = 3$) and higher orbits ($n = 4, 5, 6, \ldots$) create the Paschen series of spectral lines in the infrared. The same wavelengths occur when the electron drops from a high orbit to a low one, emitting a photon, and when the electron jumps from a low orbit to a high one, absorbing a photon. The orbits are not shown to scale.

An extremely useful way of displaying the structure of an atom is with an **energy-level diagram.** Figure 5-23 shows such a diagram for hydrogen. The lowest energy level, called the **ground state,** corresponds to the $n = 1$ Bohr orbit. Higher energy levels, called **excited states,** correspond to successively larger Bohr orbits.

An electron can jump from the ground state up to the $n = 2$ level only if the atom absorbs a Lyman-alpha photon with a wavelength of 122 nm. Such a photon has energy $E = hc/\lambda = 10.2$ eV (electron volts; see Section 5-5). That's why the energy level of $n = 2$ is shown in Figure 5-23 as having an energy 10.2 eV above that of the ground state (which is usually assigned a value of 0 eV). Similarly, the $n = 3$ level is 12.1 eV above the ground state, and so forth. Electrons can make transitions to higher energy levels by absorbing a photon or in a collision between atoms; they can make transitions to lower energy levels by emitting a photon.

On the energy-level diagram for hydrogen, the $n = \infty$ level has an energy of 13.6 eV. (This corresponds to an infinitely large orbit in the Bohr model.) If the electron is initially in the ground state and the atom absorbs a photon of any energy greater than 13.6 eV, the electron will be removed completely from the atom. This process is called **ionization.** A 13.6-eV photon has a wavelength of 91.2 nm, equal to the shortest wavelength in the ultraviolet Lyman series (L_∞). So any photon with a wavelength of 91.2 nm or less can ionize hydrogen. (The Planck formula $E = hc/\lambda$ tells us that the higher the photon energy, the shorter the wavelength.)

As an example, the gaseous nebula NGC 2363 shown in Figure 5-16 has hot stars in its neighborhood that produce copious amounts of ultraviolet photons with wavelengths less than 91.2 nm. Hydrogen atoms in the nebula that absorb these photons become ionized and lose their electrons. When the electrons recombine with the nuclei, they cascade down the energy levels to the ground state and emit visible light in the process. This is what makes the nebula glow.

LOOKING DEEPER 5.1 The same basic principles that explain the hydrogen spectrum also apply to the atoms of other elements. Electrons in each kind of atom can be only in certain energy levels, so only photons of certain wavelengths can be emitted or absorbed. Because each kind of atom has its own unique arrangement of electron levels, the pattern of spectral lines is likewise unique to that particular type of atom (see the photograph that opens this chapter). These patterns are in general much more complicated than for the hydrogen atom. Hence, there is no simple relationship analogous to the Bohr formula that applies to the spectra of all atoms.

The idea of energy levels explains the emission line spectra and absorption line spectra of gases. But what about the continuous spectra produced by dense objects like the filament of a lightbulb or the coils of a toaster? These objects are made of atoms, so why don't they emit light with an emission line spectrum characteristic of the particular atoms of which they are made?

The reason is directly related to the difference between a gas on the one hand and a liquid or solid on the other. In a gas, atoms are widely separated and can emit photons without interference from other atoms. But in a liquid or a solid, atoms are so close that they almost touch, and thus these atoms interact strongly with each other. These interactions interfere with the process of emitting photons. As a result, the pattern of distinctive bright spectral lines that the atoms would emit in isolation becomes "smeared out" into a continuous spectrum.

ANALOGY Think of atoms like tuning forks. If you strike a single tuning fork, it produces a sound wave with a single clear frequency and wavelength, just as an isolated atom emits light of definite wavelengths. But if you shake a box packed full of tuning forks, you will hear a clanging noise that is a mixture of sounds of all different frequencies and wavelengths. This is directly analogous to the continuous spectrum of light emitted by a dense object with closely packed atoms.

With the work of such people as Planck, Einstein, Rutherford, and Bohr, the interchange between astronomy and physics came full circle. Modern physics was born when Newton set out to understand the motions of the planets. Two and a half centuries later, physicists in their laboratories probed the properties of light and the structure of atoms. Their labors had immediate applications in astronomy. Armed with this new understanding of light and matter, astronomers were able to probe in detail the chemical and physical properties of planets, stars, and galaxies.

5-9 The wavelength of a spectral line is affected by the relative motion between the source and the observer

In addition to telling us about temperature and chemical composition, the spectrum of a planet, star, or galaxy can also reveal something about that object's motion through space. This idea dates from 1842, when Christian Doppler, a professor of mathematics in Prague, pointed out that the observed wavelength of light must be affected by motion. Figure 5-24 shows why. In this figure, a light source is moving from right to left, and the circles represent the crests of waves emitted from the moving source at various positions. Each successive wave crest is emitted from a position slightly closer to the observer on the left, so she sees a shorter wavelength—the distance from one crest to the next—than she would if the source were stationary. All the lines in the spectrum of an approaching source are shifted toward the

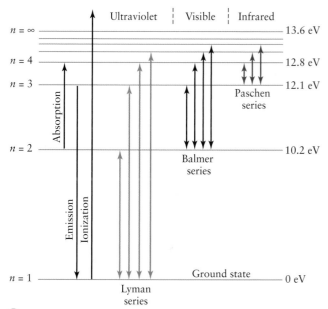

Figure 5-23

Energy-Level Diagram of Hydrogen A convenient way to display the structure of the hydrogen atom is in a diagram like this, which shows the allowed energy levels. The diagram shows a number of possible electron jumps, or transitions, between energy levels. An upward transition occurs when the atom absorbs a photon; a downward transition occurs when the atom emits a photon. (Compare with Figure 5-22.)

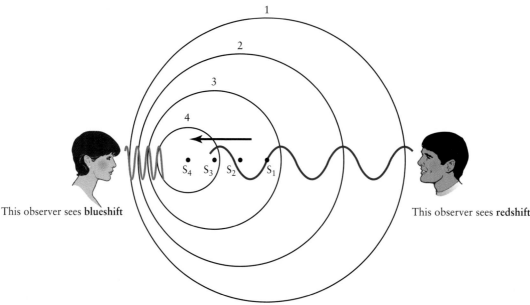

figure 5-24

The Doppler Effect The wavelength of light is affected by motion between the light source and an observer. The light source shown here is moving toward the left. Circles 1 through 4 indicate the crests of light waves that were emitted by the source when it was at points S_1 through S_4, respectively. The waves are crowded together in front of the moving source but are spread out behind it. Consequently, wavelengths appear shortened (blueshifted) if the source is moving toward the observer. They appear lengthened (redshifted) if the source is moving away from the observer. Motion perpendicular to an observer's line of sight does not affect wavelength.

short-wavelength (blue) end of the spectrum. This phenomenon is called a **blueshift.**

The source is receding from the observer on the right in Figure 5-24. The wave crests that reach him are stretched apart, so that he sees a longer wavelength than he would if the source were stationary. All the lines in the spectrum of a receding source are shifted toward the longer-wavelength (red) end of the spectrum, producing a **redshift.** In general, the effect of relative motion on wavelength is called the **Doppler effect.**

You have probably noticed a similar Doppler effect for sound waves. When a police car is approaching, the sound waves from its siren have a shorter wavelength and higher frequency than if the siren were at rest, and hence you hear a higher pitch. After the police car passes you and is moving away, you hear a lower pitch from the siren because the sound waves have a longer wavelength and a lower frequency.

Suppose that λ_0 is the wavelength of a particular spectral line from a light source that is not moving. It is the wavelength that you might look up in a reference book or determine in a laboratory experiment for this spectral line. If the source is moving, this particular spectral line is shifted to a different wavelength λ. The size of the wavelength shift is usually written as $\Delta\lambda$, where $\Delta\lambda = \lambda - \lambda_0$. Thus, $\Delta\lambda$ is the difference between the

wavelength listed in reference books and the wavelength that you actually observe in the spectrum of a star or galaxy.

Doppler proved that the wavelength shift ($\Delta\lambda$) is governed by the following simple equation:

Doppler shift equation

$$\frac{\Delta\lambda}{\lambda_0} = \frac{v}{c}$$

$\Delta\lambda$ = wavelength shift

λ_0 = wavelength if source is not moving

v = velocity of the source measured along the line of sight

c = speed of light = 3.0×10^5 km/s

The capital Greek letter Δ, or "delta," is commonly used to denote a change in the value of a quantity. Thus, $\Delta\lambda$ is the change in the wavelength λ due to the Doppler effect. It is *not* equal to a quantity Δ multiplied by a second quantity λ!

The velocity determined from the Doppler effect is called **radial velocity,** because v is the component of the star's motion parallel to our line of sight, or along the "radius" drawn from Earth to the star. Of course, a sizable fraction of a star's motion may be perpendicular to our line of sight. The speed

box 5-6 | **Tools of the Astronomer's Trade**

Applications of the Doppler Effect

Doppler's formula relates the radial velocity of an astronomical object to the wavelength shift of its spectral lines. Here are two examples that show how to use this remarkably powerful formula.

EXAMPLE: As measured in the laboratory, the prominent H_α spectral line of hydrogen has a wavelength $\lambda_0 = 656.285$ nm. But in the spectrum of the star Vega, the brightest star in the constellation Lyra (the Harp), this line has a wavelength $\lambda = 656.255$ nm. The wavelength shift is

$$\Delta\lambda = \lambda - \lambda_0 = 656.255 \text{ nm} - 656.285 \text{ nm} = -0.030 \text{ nm}$$

The negative value means that we see the light from Vega shifted to shorter wavelengths—that is, there is a blueshift. (Note that the shift is very tiny and can be measured only using specialized equipment.) From Doppler's formula, the star's radial velocity is:

$$v = c\frac{\Delta\lambda}{\lambda_0} = (3.0 \times 10^5 \text{ km/s})\left(\frac{-0.030 \text{ nm}}{656.285 \text{ nm}}\right) = -14 \text{ km/s}$$

The minus sign indicates that the star is coming toward us at 14 km/s.

By plotting the motions of different stars toward and away from us, astronomers have been able to learn how the Milky Way Galaxy (of which our Sun is a part) is rotating. From this knowledge, and aided by Newton's universal law of gravitation (Section 4-7), they have made the surprising discovery that the Milky Way contains roughly 10 times more matter than had once been thought! The nature of this unseen *dark matter* is still a subject of debate.

EXAMPLE: In the radio region of the electromagnetic spectrum, hydrogen atoms emit and absorb photons with a wavelength of 21.12 cm, giving rise to a spectral feature commonly called the *21-centimeter line*. The galaxy NGC 3840 in the constellation Leo (the Lion) is receding from us at a speed of 7370 km/s, or about 2.5% of the speed of light. Using the Doppler formula, we can predict the wavelength at which we expect to detect the 21-cm line from this galaxy. The wavelength shift is

$$\Delta\lambda = \lambda_0\left(\frac{v}{c}\right) = (21.12 \text{ cm})\left(\frac{7370 \text{ km/s}}{3.0 \times 10^5 \text{ km/s}}\right) = (0.52 \text{ cm})$$

Therefore, we will detect the 21-cm line of hydrogen from this galaxy at a wavelength of

$$\lambda = \lambda_0 + \Delta\lambda = 21.12 \text{ cm} + 0.52 \text{ cm} = 21.64 \text{ cm}$$

The 21-cm line has been redshifted to a longer wavelength because the galaxy is receding from us. In fact, most galaxies are receding from us. This observation is one of the key pieces of evidence that the universe is expanding outward from a Big Bang that took place some 10 to 15 billion years ago.

---

of this transverse movement across the sky does not affect wavelengths if the speed is small compared with c. Box 5-6 includes two examples of calculations with radial velocity using the Doppler formula.

CAUTION! The redshifts and blueshifts of stars visible to the naked eye, or even through a small telescope, are only a small fraction of a nanometer. These tiny wavelength changes are far too small to detect visually. (Astronomers were able to detect the tiny Doppler shifts of starlight only after they had developed highly sensitive equipment for measuring wavelengths. This was done around 1890, a half-century after Doppler's original proposal.) So if you see a star with a red color, it means that the star really is red; it does *not* mean that it is moving rapidly away from us.

The Doppler effect is an important tool in astronomy because it uncovers basic information about the motions of planets, stars, and galaxies. For example, the rotation of the planet Venus was deduced from the Doppler shift of radar waves reflected from its surface. Small Doppler shifts in the spectrum of sunlight have shown that the entire Sun is vibrating like an immense gong. The back-and-forth Doppler shifting of the spectral lines of certain stars reveals that these stars are being orbited by unseen companions; from this astronomers have discovered planets around other stars and massive objects that may be black holes. Astronomers also use the Doppler effect along with Kepler's third law to measure the masses of galaxies. These are but a few examples of how Doppler's discovery has empowered astronomers in their quest to understand the universe.

In this chapter we have glimpsed how much can be learned by analyzing light from the heavens. To analyze this light, however, it is first necessary to collect as much of it as possible, because most light sources in space are very dim. Collecting the faint light from distant objects is a key purpose of telescopes. In the next chapter we will describe both how telescopes work and how they are used.

KEY WORDS

Terms preceded by an asterisk (*) are discussed in the Boxes.

absolute zero, p. 99

absorption line spectrum, p. 107

atom, p. 99

*atomic number, p. 112

Balmer line, p. 111

Balmer series, p. 111

blackbody, p. 101

blackbody curves, p. 101

blackbody radiation, p. 101

blueshift, p. 116

Bohr orbits, p. 112

continuous spectrum, p. 107

*degrees Celsius, p. 100

*degrees Fahrenheit, p. 100

Doppler effect, p. 116

electromagnetic radiation, p. 96

electromagnetic spectrum, p. 97

electromagnetism, p. 95

electron, p. 104

electron volt, p. 105

element, p. 106

emission line spectrum, p. 107

energy flux, p. 102

energy level, p. 114

energy-level diagram, p. 114

excited state, p. 114

frequency, p. 97

gamma rays, p. 97

ground state, p. 114

infrared radiation, p. 96

ionization, p. 115

*isotope, p. 112

joule, p. 102

kelvin, p. 99

Kirchhoff's laws, p. 106

light scattering, p. 107

*luminosity, p. 103

Lyman series, p. 114

microwaves, p. 97

nanometer, p. 96

neutron, p. 110

nucleus, p. 110

Paschen series, p. 114

*periodic table, p. 112

photoelectric effect, p. 104

photon, p. 104

Planck's law, p. 105

proton, p. 110

quantum mechanics, p. 114

radial velocity, p. 116

radio waves, p. 97

redshift, p. 116

*solar constant, p. 103

spectral analysis, p. 106

spectral line, p. 105

spectroscopy, p. 109

spectrum (*plural* spectra), p. 94

Stefan-Boltzmann law, p. 102

ultraviolet radiation, p. 97

visible light, p. 96

watt, p. 102

wavelength, p. 96

wavelength of maximum emission, p. 100

Wien's law, p. 102

X rays, p. 97

KEY IDEAS

The Nature of Light: Light is electromagnetic radiation. It has wavelike properties described by its wavelength λ and frequency ν, and travels through empty space at the constant speed $c = 3.0 \times 10^8$ m/s $= 3.0 \times 10^5$ km/s.

Blackbody Radiation: A blackbody is a hypothetical object that is a perfect absorber of electromagnetic radiation at all wavelengths. Stars closely approximate the behavior of blackbodies, as do other hot, dense objects.

• The intensities of radiation emitted at various wavelengths by a blackbody at a given temperature are shown by a blackbody curve.

• Wien's law states that the dominant wavelength at which a blackbody emits electromagnetic radiation is inversely proportional to the Kelvin temperature of the object: λ_{max} (in meters) $= (0.0029)/T$.

• The Stefan-Boltzmann law states that a blackbody radiates electromagnetic waves with a total energy flux F directly proportional to the fourth power of the Kelvin temperature T of the object: $F = \sigma T^4$.

Photons: An explanation of blackbody curves led to the discovery that light has particlelike properties. The particles of light are called photons.

• Planck's law relates the energy E of a photon to its frequency ν or wavelength λ: $E = h\nu = hc/\lambda$, where h is Planck's constant.

Kirchhoff's Laws: Kirchhoff's three laws of spectral analysis describe conditions under which different kinds of spectra are produced.

• A hot, dense object such as a blackbody emits a continuous spectrum covering all wavelengths.

• A hot, transparent gas produces a spectrum that contains bright (emission) lines.

• A cool, transparent gas in front of a light source that itself has a continuous spectrum produces dark (absorption) lines in the continuous spectrum.

Atomic Structure: An atom has a small dense nucleus composed of protons and neutrons. The nucleus is surrounded by electrons that occupy only certain orbits or energy levels.

- When an electron jumps from one energy level to another, it emits or absorbs a photon of appropriate energy (and hence of a specific wavelength).

- The spectral lines of a particular element correspond to the various electron transitions between energy levels in atoms of that element.

- Bohr's model of the atom correctly predicts the wavelengths of hydrogen's spectral lines.

The Doppler Shift: The Doppler shift enables us to determine the radial velocity of a light source from the displacement of its spectral lines.

- The spectral lines of an approaching light source are shifted toward short wavelengths (a blueshift); the spectral lines of a receding light source are shifted toward long wavelengths (a redshift).

- The size of a wavelength shift is proportional to the radial velocity between the light source and the observer.

REVIEW QUESTIONS

1. How long does it take light to travel from the Moon to the Earth, a distance of 384,000 km?

2. Approximately how many times around the world could a beam of light travel in one second?

3. (a) Describe an experiment in which light behaves like a wave. (b) Describe an experiment in which light behaves like a particle.

4. What is meant by the frequency of light? How is frequency related to wavelength?

5. A light source emits infrared radiation at a wavelength of 1060 nm. What is the frequency of this radiation?

6. A cellular phone is actually a radio transmitter and receiver. You receive an incoming call in the form of a radio wave of frequency 893.07 MHz. What is the wavelength (in meters) of this wave?

7. Using Wien's law and the Stefan-Boltzmann law, explain the color and intensity changes that are observed as the temperature of a hot, glowing object increases.

8. What is a blackbody? In what way is a blackbody black? If a blackbody is black, how can it emit light? If you were to shine a flashlight beam on a perfect blackbody, what would happen to the light?

9. Explain why astronomers are interested in blackbody radiation.

10. Why do astronomers find it convenient to use the Kelvin temperature scale in their work rather than the Celsius or Fahrenheit scale?

11. The bright star Aldebaran in the constellation Taurus (the Bull) has a surface temperature of 3850 K. What is its wavelength of maximum emission in nanometers? What color is this star?

12. The bright star Hadar in the constellation Centaurus (the Centaur) emits the greatest intensity of radiation at a wavelength of 121 nm. What is the surface temperature of the star? What color is this star?

13. If you double the Kelvin temperature of a hot piece of steel, how much more energy will it radiate per second?

14. How is the energy of a photon related to its wavelength? What kind of photons carry the most energy? What kind of photons carry the least energy?

15. How do we know that atoms have massive, compact nuclei?

16. Describe the spectrum of hydrogen at visible wavelengths and explain how Bohr's model of the atom accounts for the Balmer lines.

17. Why do different elements display different patterns of lines in their spectra?

18. What is the Doppler effect? Why is it important to astronomers?

19. If you see a blue star, what does its color tell you about how the star is moving through space? Explain your answer.

ADVANCED QUESTIONS

Questions preceded by an asterisk (*) involve topics discussed in the Boxes.

Problem-solving tips and tools

You can find formulas in Box 5-1 for converting between temperature scales. Box 5-2 discusses how a star's radius, luminosity, and surface temperature are related. Box 5-3 shows how to use Planck's law to calculate the energy of a photon. To learn how to do calculations using the Doppler effect, see Box 5-6.

20. What is the temperature of the Sun's surface in degrees Fahrenheit?

21. Your normal body temperature is 98.6°F. What kind of radiation do you predominantly emit? At what wavelength (in nm) do you emit the most radiation?

22. What wavelength of electromagnetic radiation is emitted with greatest intensity by this book? To what region of the electromagnetic spectrum does this wavelength correspond?

*23. The bright star Deneb in the constellation Cygnus (the Swan) has a surface temperature of 8700 K. How much more energy is emitted each second from each square meter of Deneb's surface than from each square meter of the Sun's surface?

*24. Jupiter's moon Io has an active volcano named Pele whose temperature can be as high as 320°C. (a) What is the wavelength of maximum emission for the volcano at this temperature? In what part of the electromagnetic spectrum is this? (b) The average temperature of Io's surface is −150°C. Compared with a square meter of surface at this temperature, how much more energy is emitted per second from each square meter of Pele's surface?

*25. The bright star Sirius in the constellation of Canis Major (the Large Dog) has a radius of 1.67 $R_\odot$ and a luminosity of 25 $L_\odot$. (a) What is the energy flux at the surface of Sirius? (b) What is the star's surface temperature?

26. Black holes are objects whose gravity is so strong that not even an object moving at the speed of light can escape from their surface. Hence, black holes do not themselves emit light. But it is possible to detect radiation from material falling *toward* a black hole. Calculations suggest that as this matter falls, it is compressed and heated to temperatures around 10^6 K. Calculate the wavelength of maximum emission for this temperature. In what part of the electromagnetic spectrum does this wavelength lie?

27. In Figure 5-12 you can see two distinct dark lines at the boundary between the orange and yellow parts of the Sun's spectrum (in the center of the third colored band from the top of the figure). The wavelengths of these dark lines are 588.99 and 589.59 nm. What do you conclude from this about the chemical composition of the Sun's atmosphere?

28. Since the 1970s, instruments on board balloons and spacecraft have detected 511-keV photons coming from the direction of the center of our Galaxy. (The prefix k means *kilo*, or thousand, so 1 keV = 10^3 eV.) What is the wavelength of these photons? To what part of the electromagnetic spectrum do these photons belong?

29. (a) Calculate the wavelength of P_8, the fourth wavelength in the Paschen series. (b) Draw a schematic diagram of the hydrogen atom and indicate the electron transition that gives rise to this spectral line. (c) In what part of the electromagnetic spectrum does this wavelength lie?

30. (a) Calculate the wavelength of H_{13}, the spectral line for an electron transition between the $n = 14$ and $n = 2$ orbits of hydrogen. (b) In what part of the electromagnetic spectrum does this wavelength lie? Use this to explain why Figure 5-19 is labeled R I V **U** X G.

31. (a) Can a hydrogen atom in the ground state absorb an H-alpha (H_α) photon? Explain why or why not. (b) Can a hydrogen atom in the $n = 2$ state absorb a Lyman-alpha (L_α) photon? Explain why or why not.

32. The spaces between the galaxies are filled with a very cold, very thin gas of hydrogen atoms. Ultraviolet radiation with any wavelength shorter than 91.2 nm cannot pass through this gas; instead, it is absorbed. Explain why.

33. An imaginary atom has just 3 energy levels: 0 eV, 1 eV, and 3 eV. Draw an energy-level diagram for this atom. Show all possible transitions between these energy levels. For each transition, determine the photon energy and the photon wavelength. Which transitions involve the emission or absorption of visible light?

34. The wavelength of H_β in the spectrum of the star Megrez in the Big Dipper (part of the constellation Ursa Major, the Great Bear) is 486.112 nm. Laboratory measurements demonstrate that the normal wavelength of this spectral line is 486.133 nm. Is the star coming toward us or moving away from us? At what speed?

35. The nebula NGC 2363 shown in Figure 5-16 is located within the galaxy NGC 2366 in the constellation Camelopardis (the Camel). This galaxy and the nebula within it are moving away from us at 252 km/s. At what wavelength does the red H_α line of hydrogen (which causes the color of the nebula) appear in the nebula's spectrum?

36. You are given a traffic ticket for going through a red light (wavelength 700 nm). You tell the police officer that because you were approaching the light, the Doppler effect caused a blueshift that made the light appear green (wavelength 500 nm). How fast would you have had to be going for this to be true? Would the speeding ticket be justified? Explain.

DISCUSSION QUESTIONS

37. The equation that relates the frequency, wavelength, and speed of a light wave, $\nu = c/\lambda$, can be rewritten as $c = \nu\lambda$. A friend who has studied mathematics but not much astronomy or physics might look at this equation and say: "This equation tells me that the higher the frequency ν, the greater the wave speed c. Since visible light has a higher frequency than radio waves, this means that visible light goes faster than radio waves." How would you respond to your friend?

38. Why do you suppose that ultraviolet light can cause skin cancer but ordinary visible light does not?

39. The accompanying visible-light image shows the star cluster NGC 3293 in the constellation Carina (the Ship's Keel). What can you say about the surface temperatures of most of the bright stars in this cluster? In what part of the electromagnetic spectrum do these stars emit most intensely? Are your eyes sensitive to this type of radiation? If not, how is it possible to see these stars at all? There is at least one bright star in this cluster with a distinctly

(David Malin/Anglo-Australian Observatory)   R I **V** U X G

different color from the others; what can you conclude about its surface temperature?

40. (a) If you could see ultraviolet radiation, how might the night sky appear different? Would ordinary objects appear different in the daytime? **(b)** What differences might there be in the appearance of the night sky and in the appearance of ordinary objects in the daytime if you could see infrared radiation?

41. The human eye is most sensitive over the same wavelength range at which the Sun emits the greatest intensity of radiation. Suppose creatures were to evolve on a planet orbiting a star somewhat hotter than the Sun. To what wavelengths would their vision most likely be sensitive?

 ## WEB/CD-ROM QUESTIONS

42. Search the World Wide Web for information about rainbows. Why do rainbows form? Why do they appear as circular arcs? Why can you see different colors?

43. Measuring Stellar Temperatures. Access the Active Integrated Media Module "Blackbody Curves" in Chapter 5 of the *Universe* web site or CD-ROM. **(a)** Use the module to determine the range of temperatures over which a star's peak wavelength is in the visible spectrum. **(b)** Determine if any of the following stars have a peak wavelength in the visible spectrum: Rigel, $T = 14,000$ K; Deneb, $T = 9500$ K; Arcturus, $T = 4500$ K; Vega, $T = 11,500$ K; Betelgeuse, $T = 3100$ K.

OBSERVING PROJECTS

44. Turn on an electric stove or toaster oven and carefully observe the heating elements as they warm up. Relate your observations to Wien's law and the Stefan-Boltzmann law.

45. Obtain a glass prism (or a diffraction grating, which is probably more readily available and is discussed in the next chapter) and look through it at various light sources, such as an ordinary incandescent light, a neon sign, and a mercury vapor street lamp. **Do not look at the sun! Looking directly at the Sun causes permanent eye damage or blindness.** Do you have any trouble seeing spectra? What do you have to do to see a spectrum? Describe the differences in the spectra of the various light sources you observed.

 46. Use the *Starry Night* program to examine some distant celestial objects. First turn off daylight (select **Daylight** in the **Sky** menu) and show the entire celestial sphere (select **Atlas** in the **Go** menu). Then search for objects (i), (ii), and (iii) listed below (select **Find...** in the **Edit** menu). For each object, zoom in until you can see it in detail. For each object, state whether it has a continuous spectrum, an absorption line spectrum, or an emission line spectrum, and explain your reasoning. (i) The Trifid Nebula in Sagittarius. (*Hint:* See Figure 5-16.) (ii) M31, the great galaxy in the constellation Andromeda. (*Hint:* The light coming from this galaxy is the combined light of hundreds of billions of individual stars.) (iii) The Moon. (*Hint:* Recall from Section 3-1 that moonlight is simply reflected sunlight.)

Optics and Telescopes

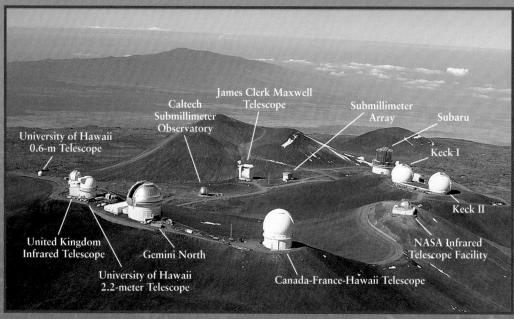

University of Hawaii
0.6-m Telescope

Caltech
Submillimeter
Observatory

James Clerk Maxwell
Telescope

Submillimeter
Array

Subaru

Keck I

Keck II

United Kingdom
Infrared Telescope

Gemini North

University of Hawaii
2.2-meter Telescope

Canada-France-Hawaii Telescope

NASA Infrared
Telescope Facility

(Richard J. Wainscoat, University of Hawaii)

R I **V** U X G

Just a few kilometers from the resorts of Hawaii's Kona Coast, in a place where few tourists ever go, is one of the best locations on Earth to view the heavens. This is the summit of Mauna Kea, an extinct volcano that reaches some 4200 m (14,000 ft) above the waters of the Pacific. Astronomers prize Mauna Kea because the skies there are unusually clear, still, and dark. And so it is that Mauna Kea has become the home of nearly a dozen powerful telescopes (see the accompanying photo).

Large telescopes such as those atop Mauna Kea can make images that are far brighter and sharper than the images formed by our eyes. They can also produce finely detailed spectra that reveal the chemical composition of nearby planets as well as distant galaxies.

A number of modern telescopes can even compensate for the "twinkling" of starlight, which occurs when light passes through turbulent air. Some telescopes deal with this problem by avoiding it altogether; the most famous example is the Hubble Space Telescope, which orbits high above the blurring effects of the atmosphere.

Special telescopes on Earth and in orbit explore the electromagnetic spectrum at wavelengths outside the visible range. Radio telescopes have mapped out the structure of our Milky Way Galaxy; with infrared telescopes, astronomers have peered at stars in the process of formation; and X-ray telescopes may have located black holes in orbit around ordinary stars. The telescope, in all its variations, is by far the most useful tool that astronomers have for collecting data about the universe.

As you read the sections of this chapter, look for the answers to the following questions.

6-1 Why is it important that telescopes be large?

6-2 Why do most modern telescopes use a large mirror rather than a large lens?

6-3 Why are observatories in such remote locations?

6-4 Do astronomers use ordinary photographic film to take pictures of the sky? Do they actually look through large telescopes?

6-5 How do astronomers use telescopes to measure the spectra of distant objects?

6-6 Why do astronomers need telescopes that detect radio waves and other nonvisible forms of light?

6-7 Why is it useful to put telescopes in orbit?

6-1 A refracting telescope uses a lens to concentrate incoming light at a focus

The **optical telescope**—that is, a telescope designed for use with visible light—was invented in the Netherlands in the early seventeenth century. Soon after, Galileo used one of these new inventions for his groundbreaking astronomical observations (see Section 4-3). These first telescopes used *lenses* to make distant objects appear larger and brighter. Telescopes of this same basic design are used today by many amateur astronomers. To understand telescopes of this kind, we need to understand how lenses work.

All lenses, including those used in telescopes, make use of the same physical principle: *Light travels at a slower speed in a dense substance*. Thus, although the speed of light in a vacuum is 3.0×10^8 m/s, its speed in glass is less than 2×10^8 m/s. Just as a woman's walking pace slows suddenly when she walks from a boardwalk onto a sandy beach, so light slows abruptly as it enters a piece of glass. Upon exiting the glass, light resumes its original speed, just as a woman stepping back onto a boardwalk easily resumes her original pace.

A material through which light travels is called a **medium** (plural **media**). As a beam of light passes from one transparent medium into another—say, from air into glass, or from glass back into air—the direction of the light can change. This phenomenon, called **refraction**, is caused by the change in speed of light.

ANALOGY Imagine driving a car from a smooth pavement onto a sandy beach (Figure 6-1*a*). If the car approaches the beach head-on, it slows down when it enters the sand but keeps moving straight ahead. If the car approaches the beach at an angle, however, one of the front wheels will be slowed by the sand before the other is, and the car will veer from its original direction. In the same way, a beam of light changes direction when it enters a piece of glass at an angle (Figure 6-1*b*).

Figure 6-2*a* shows the refraction of a beam of light passing through a piece of flat glass. As the beam enters the

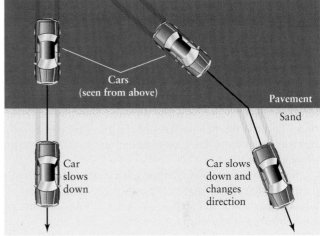

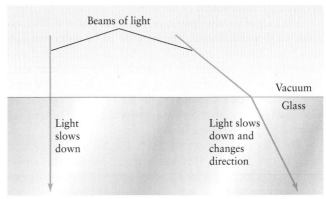

b

figure 6-1

Refraction (a) When a car drives from smooth pavement into soft sand, it slows down. If it enters the sand at an angle, the front wheel on one side feels the drag of the sand before the other wheel, causing the car to veer to the side and change direction. (b) Similarly, light slows down when it passes from a vacuum into glass. The direction of the light beam changes if it enters the glass at an angle.

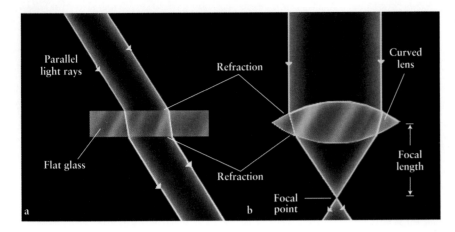

ƒigure 6-2

Refraction and Lenses (a) Refraction is the change in direction of a light ray when it passes into or out of a transparent medium such as glass. When light rays pass through a flat piece of glass, the two refractions bend the rays in opposite directions. There is no overall change in the direction in which the light travels. **(b)** If the glass is in the shape of a convex lens, parallel light rays converge to a focus at a special point called the focal point. The distance from the lens to the focal point is called the focal length of the lens.

upper surface of the glass, refraction takes place, and the beam is bent to a direction more nearly perpendicular to the surface of the glass. As the beam exits from the glass back into the surrounding air, a second refraction takes place, and the beam bends in the opposite sense. (The amount of bending depends on the speed of light in the glass, so different kinds of glass produce slightly different amounts of refraction.) Because the two surfaces of the glass are parallel, the beam emerges from the glass traveling in the same direction in which it entered.

Something more useful happens if the glass is curved into a convex shape (one that is fatter in the middle than at the edges), like the lens in Figure 6-2*b*. When a beam of light rays passes through the lens, refraction causes all the rays to converge at a point called the **focus**. If the light rays entering the lens are all parallel, the focus occurs at a special point called the **focal point**. The distance from the lens to the focal point is called the **focal length** of the lens.

The case of parallel light rays, shown in Figure 6-2*b*, is not merely a theoretical ideal. The stars are so far away that light rays from them are essentially parallel, as Figure 6-3 shows. Consequently, a lens always focuses light from an astronomical object to the focal point. If the object has a very small angular size, like a distant star, all the light entering the lens from that object converges onto the focal point. The resulting image is just a single bright dot.

But if the object is *extended*—that is, has a relatively large angular size, like the Moon or a planet—then light coming from each point on the object is brought to a focus at its own individual point. The result is an extended image that lies in the **focal plane** of the lens (Figure 6-4), which is a plane that includes the focal point. You can use an ordinary magnifying glass in this way to make an image of the Sun on the ground.

To use a lens to make a photograph of an astronomical object, you would place a piece of film in the focal plane. An ordinary camera works in the same way for taking photographs of relatively close objects here on Earth. But if you instead want to view the image visually, you would add a second lens to magnify the image formed in the focal plane. Such an arrangement of two lenses is called a **refracting telescope,**

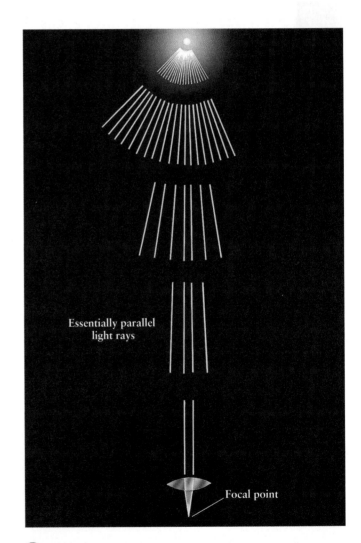

ƒigure 6-3

Light Rays from Distant Objects Are Parallel Light rays travel away in all directions from an ordinary light source. If a lens is located very far from the light source, only a few of the light rays will enter the lens, and these rays will be essentially parallel. This is why we drew parallel rays entering the lens in Figure 6-2*b*.

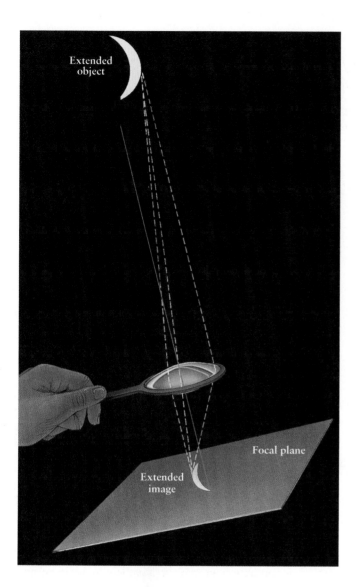

figure 6-4

A Lens Creates an Extended Image of an Extended Object

Light coming from each point on an extended object passes through a lens and produces an image of that point. All of these tiny images put together make an extended image of the entire object. The image of a very distant object is formed in a plane called the focal plane. The distance from the lens to the focal plane is called the focal length.

Extended object

Focal plane

Extended image

or **refractor** (Figure 6-5). The large-diameter, long-focal-length lens at the front of the telescope, called the **objective lens,** forms the image; the smaller, shorter-focal-length lens at the rear of the telescope, called the **eyepiece lens,** magnifies the image for the observer.

In addition to the focal length, the other important dimension of the objective lens of a refractor is the diameter. Compared with a small-diameter lens, a large-diameter lens captures more light, produces brighter images, and allows astronomers to detect fainter objects. (For the same reason, the iris of your eye opens when you go into a darkened room to allow you to see dimly lit objects.)

The **light-gathering power** of a telescope is directly proportional to the area of the objective lens, which in turn is proportional to the square of the lens diameter (Figure 6-6). Thus, if you double the diameter of the lens, the light-gathering power increases by a factor of $2^2 = 2 \times 2 = 4$. Box 6-1 describes how to compare the light-gathering power of different telescopes.

Because light-gathering power is so important for seeing faint objects, the lens diameter is almost always given in describing a telescope. For example, the Lick telescope on Mount Hamilton in California is a 90-cm refractor, which

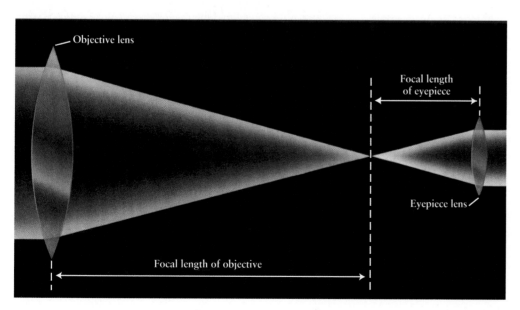

Objective lens

Focal length of eyepiece

Eyepiece lens

Focal length of objective

figure 6-5

A Refracting Telescope

A refracting telescope consists of a large-diameter objective lens with a long focal length and a small eyepiece lens of short focal length. The eyepiece lens magnifies the image formed by the objective lens in its focal plane (shown as a dashed line).

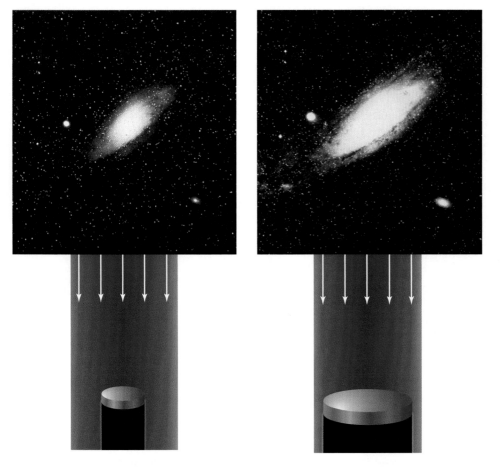

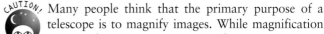 figure 6-6   R I **V** U X G

Light-Gathering Power These two photographs of the galaxy M31 in Andromeda were taken using the same exposure time and at the same magnification, but with two different telescopes with objective lenses of different diameters. The right-hand photograph is brighter and shows more detail because it was made using the larger-diameter lens, which intercepts more starlight than a small-diameter lens. This same principle applies to telescopes that use curved mirrors rather than lenses to collect light (see Section 6-2). (Association of Universities for Research in Astronomy)

means that it is a refracting telescope whose objective lens is 90 cm in diameter. By comparison, Galileo's telescope of 1610 was a 3-cm refractor. The Lick telescope has an objective lens 30 times larger in diameter, and so has $30 \times 30 = 900$ times the light-gathering power of Galileo's instrument.

In addition to their light-gathering power, telescopes are useful because they magnify distant objects. As an example, the angular diameter of the Moon as viewed with the naked eye is about 0.5°. But when Galileo viewed the Moon through his telescope, its apparent angular diameter was 10°, large enough so that he could identify craters and mountain ranges. The **magnification,** or **magnifying power,** of a telescope is the ratio of an object's angular diameter seen through the telescope to its naked-eye angular diameter. Thus, the magnification of Galileo's telescope was 10°/0.5° = 20 times, usually written as 20×.

 The magnification of a refracting telescope depends on the focal lengths of both of its lenses:

$$\text{Magnification} = \frac{\text{focal length of objective lens}}{\text{focal length of eyepiece lens}}$$

This formula shows that using a long-focal-length objective lens with a short-focal-length eyepiece gives a large magnification. Box 6-1 illustrates how this formula is used.

CAUTION! Many people think that the primary purpose of a telescope is to magnify images. While magnification is certainly important, it is *not* the most important aspect of a telescope. The reason is that there is a limit to how sharp any astronomical image can be. For telescopes that view the heavens from Earth's surface, this limit is usually due to the blurring caused by having to view through Earth's atmosphere. (We will discuss this phenomenon in more detail in Section 6-3.) Magnifying a blurred image may make it look bigger but will not make it any clearer. Thus, beyond a certain point, there is nothing to be gained by further magnification. Astronomers put much more store in the light-gathering power of a telescope than in its magnification. Greater light-gathering power means brighter images, which makes it easier to see faint details.

If you were to build a telescope like that in Figure 6-5 using only the instructions given so far, you would probably be disappointed with the results. The problem is that a lens bends different colors of light through different angles, just as a prism does (recall Figure 5-3). As a result, different colors do not focus at the same point, and stars viewed through a telescope that uses a simple lens are surrounded by fuzzy, rainbow-colored halos. Figure 6-7a shows this optical defect, called **chromatic aberration.**

box 6-1 | Tools of the Astronomer's Trade

Magnification and Light-Gathering Power

The magnification of a telescope is equal to the focal length of the objective divided by the focal length of the eyepiece. Telescopic eyepieces are usually interchangeable, so the magnification of a telescope can be changed by using eyepieces of different focal lengths.

EXAMPLE: A small refracting telescope has an objective of focal length 120 cm. If the eyepiece has a focal length of 4.0 cm, the magnification of the telescope is

$$\text{Magnification} = \frac{120}{4.0} = 30 \text{ (usually written as 30×)}$$

As viewed through this telescope, a large lunar crater that subtends an angle of 1 arcminute to the naked eye will appear to subtend an angle 30 times greater, or 30 arcminutes (one-half of a degree). This makes the details of the crater much easier to see.

If a 2.0-cm-focal-length eyepiece is used instead, the magnification will be 120/2.0 = 60×. The shorter the focal length of the eyepiece, the greater the magnification.

The light-gathering power of a telescope depends on the diameter of the objective lens; it does not depend on the focal length. The light-gathering power is proportional to the square of the diameter.

EXAMPLE: A fully dark adapted human eye has a pupil diameter of about 5 mm. By comparison, a small telescope whose objective lens is 5 cm in diameter has 10 times the diameter and $10^2 = 100$ times the light-gathering power of the eye. (Recall that there are 10 mm in 1 cm.) Hence, this telescope allows you to see objects 100 times fainter than you can see without a telescope.

The same relationships apply to reflecting telescopes, discussed in Section 6-2. Each of the two Keck telescopes on Mauna Kea in Hawaii (see the chapter opening photograph) uses a concave mirror 10 m in diameter to bring starlight to a focus. Because there are 1000 mm in one meter, this diameter can also be expressed as

$$10 \text{ m} \times \frac{1000 \text{ mm}}{1 \text{ m}} = 10,000 \text{ mm}$$

Thus, the light-gathering power of either of the Keck telescopes is greater than that of the human eye by a factor of

$$\frac{(10,000 \text{ mm})^2}{(5 \text{ mm})^2} = (2000)^2 = 4 \times 10^6, \text{ or } 4,000,000$$

When it comes to light-gathering power, the bigger the telescope, the better!

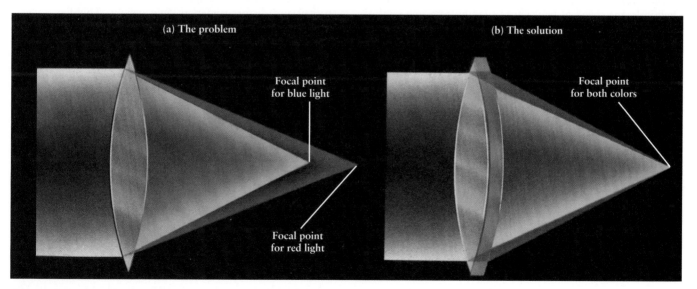

(a) The problem

Focal point for blue light

Focal point for red light

(b) The solution

Focal point for both colors

Figure 6-7

Chromatic Aberration (a) A single lens suffers from a defect called chromatic aberration, in which different colors of light are brought to a focus at different distances from the lens. (b) This problem can be corrected by adding a second lens made from a different kind of glass.

a

b

figure 6-8   R I ☑ U X G

A Large Refracting Telescope **(a)** This giant refractor, built at the end of the 1800s, is housed at Yerkes Observatory near Chicago. The objective lens is at the upper end of the telescope tube, and the eyepiece is at the lower end. The telescope tube is 19.5 m (64 ft) long; it has to be this long because the focal length of the objective is just under 19.5 m (see Figure 6-5). The entire observatory floor can be raised to bring the eyepiece within reach. **(b)** This historical photograph shows the astronomer George van Biesbrock with the objective lens of the Yerkes refractor. This lens, the largest ever made, is 102 cm (40 in.) in diameter. (Yerkes Observatory)

One way to correct for chromatic aberration is to use an objective lens that is not just a single piece of glass. Different types of glass can be manufactured by adding small amounts of chemicals to the glass when it is molten. Because of these chemicals, the speed of light varies slightly from one kind of glass to another, and the refractive properties vary as well. If a thin lens is mounted just behind the main objective lens of a telescope, as shown in Figure 6-7b, and if the telescope designer carefully chooses two different kinds of glass for these two lenses, different colors of light can be brought to a focus at the same point.

Chromatic aberration is only the most severe of a host of optical problems that must be solved in designing a high-quality refracting telescope. Master opticians of the nineteenth century devoted their careers to solving these problems, and several magnificent refractors were constructed in the late 1800s. Figure 6-8 shows the world's largest refracting telescope, completed in 1897.

Unfortunately, there are several negative aspects of refractors that not even the finest optician can overcome. First, because faint light must readily pass through the objective lens, the glass from which the lens is made must be totally free of defects, such as the bubbles that frequently form when molten glass is poured into a mold. Such defect-free glass is extremely expensive. Second, glass is opaque to certain kinds of light. Ultraviolet light is absorbed almost completely, and even visible light is dimmed substantially as it

passes through the thick slab of glass that makes up the objective lens. Third, it is impossible to produce a large lens that is entirely free of chromatic aberration. Fourth, because the lens can be supported only around its edges, it tends to sag and distort under its own weight. This has adverse effects on the image clarity.

For these reasons and more, few major refractors have been built since the beginning of the twentieth century. Instead, astronomers have avoided all of the limitations of refractors by building telescopes that use a mirror instead of a lens to form an image.

6-2   A reflecting telescope uses a mirror to concentrate incoming light at a focus

Almost all modern telescopes form an image using the principle of **reflection**. To understand reflection, imagine drawing a dashed line perpendicular to the surface of a flat mirror at the point where a light ray strikes the mirror (Figure 6-9). The angle i between the *incident* (arriving) light ray and the perpendicular is always equal to the angle r between the *reflected* ray and the perpendicular.

In 1663, the Scottish mathematician James Gregory first proposed a telescope using reflection from a concave mirror—

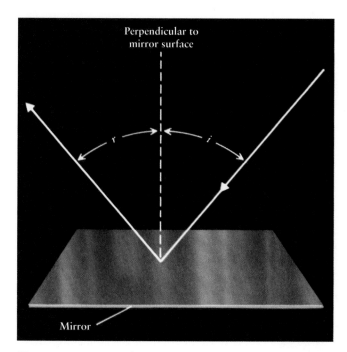

figure 6-9

Reflection by a Flat Mirror The angle at which a beam of light approaches a mirror, called the angle of incidence (i), is always equal to the angle at which the beam is reflected from the mirror, called the angle of reflection (r). That is, $i = r$.

one that is fatter at the edges than at the middle. Such a mirror makes parallel light rays converge to a focus (Figure 6-10). The distance between the reflecting surface and the focus is the focal length of the mirror. A telescope that uses a curved mirror to make an image of a distant object is called a **reflecting telescope**, or **reflector**. Using terminology similar to that used for refractors, the mirror that forms the image is called the **objective mirror** or **primary mirror**.

To make a reflector, an optician grinds and polishes a large slab of glass into the appropriate concave shape. The glass is then coated with silver, aluminum, or a similar highly reflective substance. Because light reflects off the surface of the glass rather than passing through it, defects within the glass—which would have very negative consequences for the objective lens of a refracting telescope—have no effect on the optical quality of a reflecting telescope.

Another advantage of reflectors is that they do not suffer from the chromatic aberration that plagues refractors. This is because reflection is not affected by the wavelength of the incoming light, so all wavelengths are reflected to the same focus. (A small amount of chromatic aberration may arise if the image is viewed using an eyepiece lens.) Furthermore, the mirror can be fully supported by braces on its back, so that a large, heavy mirror can be mounted without much danger of breakage or surface distortion.

Although a reflecting telescope has many advantages over a refractor, the arrangement shown in Figure 6-10 is not ideal.

One problem is that the focal point is in front of the objective mirror. If you try to view the image formed at the focal point, your head will block part or all of the light from reaching the mirror.

To get around this problem, in 1668 Isaac Newton simply placed a small, flat mirror at a 45° angle in front of the focal point, as sketched in Figure 6-11*a*. This secondary mirror deflects the light rays to one side, where Newton placed an eyepiece lens to magnify the image. A reflecting telescope with this optical design is appropriately called a **Newtonian reflector**. The magnifying power of a Newtonian reflector is calculated in the same way as for a refractor: The focal length of the objective mirror is divided by the focal length of the eyepiece (see Box 6-1).

Later astronomers modified Newton's original design. The objective mirrors of some modern reflectors are so large that the astronomer could actually sit in an "observing cage" at the undeflected focal point directly in front of the objective mirror. (In practice, riding in this cage on a winter's night is a remarkably cold and uncomfortable experience.) This arrangement is called a **prime focus** (Figure 6-11*b*). It usually provides the highest-quality image, because there is no need for a secondary mirror (which might have imperfections).

Another popular optical design, called a **Cassegrain focus** after the French contemporary of Newton who first proposed it, also has a convenient, accessible focal point. A hole is drilled directly through the center of the primary mirror, and a convex secondary mirror placed in front of the original focal point reflects the light rays back through the hole (Figure 6-11*c*).

A fourth design is useful when there is optical equipment too heavy or bulky to mount directly on the telescope. Instead, a series of mirrors channels the light rays away from the telescope to a remote focal point where the equipment is located.

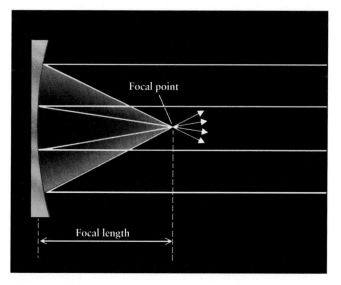

figure 6-10

Reflection by a Concave Mirror A concave mirror causes parallel light rays to converge to a focus at the focal point. The distance between the mirror and the focus is the focal length of the mirror.

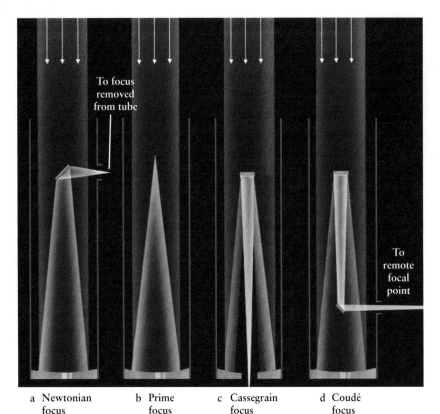

figure 6-11

Designs for Reflecting Telescopes Four of the most popular optical designs for reflecting telescopes are shown here. **(a)** The Newtonian focus is used today on some small reflecting telescopes for amateur astronomers. **(b)** Prime focus is used only on some large telescopes; an observer or instrument is placed directly at the focal point, within the barrel of the telescope. **(c)** The Cassegrain focus is used on reflecting telescopes of all sizes, from 90-mm (3.5-in.) reflectors used by amateur astronomers to the giant 10-m Keck telescopes on Mauna Kea (see the chapter opening figure). **(d)** The coudé focus is useful when large and heavy optical apparatus is to be used at the focal point. Light reflects off the objective mirror to a secondary mirror, then back down to an angled tertiary mirror. With this arrangement, the heavy apparatus at the focal point does not have to move when the telescope is repositioned.

Image labels: To focus removed from tube; To remote focal point

a Newtonian focus   b Prime focus   c Cassegrain focus   d Coudé focus

This design is called a **coudé focus,** from a French word meaning "bent like an elbow" (Figure 6-11*d*).

CAUTION! You might think that the secondary mirror in the Newtonian, Cassegrain, and coudé designs shown in Figure 6-11 would cause a black spot or hole in the center of the telescope image. But this does not happen. The reason is that light from every part of the object lands on every part of the primary, objective mirror. Hence, any portion of the mirror can itself produce an image of the distant object, as Figure 6-12 shows. The only effect of the secondary mirror is that it prevents part of the light from reaching the objective mirror, which reduces somewhat the light-gathering power of the telescope.

A reflecting telescope must be designed to minimize a defect called **spherical aberration.** At issue is the precise shape of a mirror's concave surface. A spherical surface is easy to grind and polish, but different parts of a spherical mirror have slightly different focal lengths (Figure 6-13*a*). This results in a fuzzy image.

One common way to eliminate spherical aberration is to polish the mirror's surface to a parabolic shape, because a parabola reflects parallel light rays to a common focus (Figure 6-13*b*). Unfortunately, the astronomer then no longer has a wide-angle view. Furthermore, unlike spherical mirrors, parabolic mirrors suffer from a defect called **coma,** wherein star images far from the center of the field of view are elongated to look like tiny teardrops. A different approach is to use a spherical mirror, thus minimizing coma, and to place a thin correcting lens at the front of the telescope to eliminate spherical aberration (Figure 6-13*c*). This approach is only used on relatively small reflecting telescopes for amateur astronomers.

At the time of this writing, there are eleven optical reflectors around the world with primary mirrors more than 8 meters (26.2 feet) in diameter. These are listed in Table 6-1. The largest full-capability optical telescopes are the twin 10-meter reflectors of the W. M. Keck Observatory, located atop the dormant volcano Mauna Kea in Hawaii (see the photograph that opens this chapter). Figure 6-14 shows the objective mirror of the Keck I telescope. Even larger is the 11-meter Hobby-Eberly Telescope, located in the Davis Mountains of western Texas. It cannot be pointed to all portions of the sky, but it cost only one-seventh as much as the $94 million price of either Keck telescope.

A matching pair of 8.1-meter telescopes, called Gemini, have recently gone into operation atop Mauna Kea and Cerro Pachón, Chile. These twins will allow astronomers to observe both the northern and southern parts of the celestial sphere with essentially the same state-of-the-art instrument. Two other "twins" are the side-by-side 8.4-m objective mirrors of the Large Binocular Telescope in Arizona. Combining the light from these two mirrors gives double the light-gathering power, equivalent to a single 11.8-m mirror. In the following section we will learn about the special characteristics of the other telescopes listed in Table 6-1.

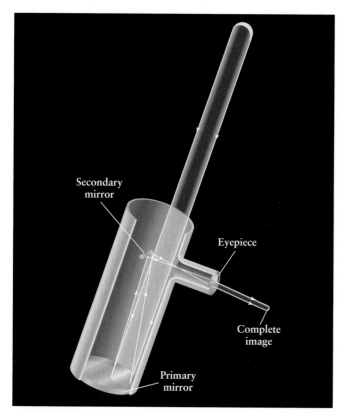

figure 6-12

The Secondary Mirror Does Not Cause a Hole in the Image
This illustration shows how even a small portion of the primary (objective) mirror of a reflecting telescope can make a complete image of the Moon. Thus, the secondary mirror does not cause a black spot or hole in the image. (It does, however, make the image a bit dimmer by reducing the total amount of light that reaches the primary mirror.)

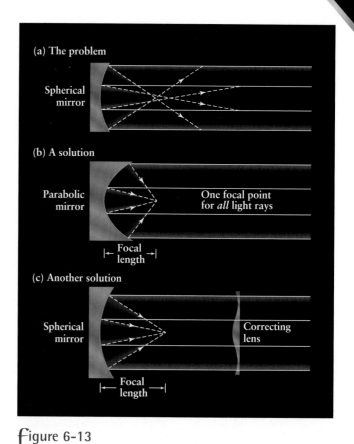

(a) The problem

Spherical mirror

(b) A solution

Parabolic mirror

One focal point for *all* light rays

Focal length

(c) Another solution

Spherical mirror

Correcting lens

Focal length

figure 6-13

Spherical Aberration (a) Different parts of a spherically concave mirror reflect light to slightly different focal points. This effect, called spherical aberration, causes image blurring. This difficulty can be corrected by either (b) using a parabolic mirror or (c) using a correcting lens in front of the mirror.

table 6-1	The World's Largest Optical Telescopes		
Telescope	**Location**	**Year completed**	**Mirror diameter (m)**
Hobby-Eberly Telescope	McDonald Observatory, Texas	1998	11.0
Keck II	Mauna Kea, Hawaii	1996	10.0
Keck I	Mauna Kea, Hawaii	1993	10.0
Large Binocular Telescope	Mount Graham, Arizona	2002	Two 8.4
Subaru	Mauna Kea, Hawaii	1999	8.3
VLT UT1–Antu	Cerro Paranal, Chile	1998	8.2
VLT UT 2–Kueyen	Cerro Paranal, Chile	1999	8.2
VLT UT 3–Melipal	Cerro Paranal, Chile	2000	8.2
VLT UT 4–Yepun	Cerro Paranal, Chile	2000	8.2
Gemini North	Mauna Kea, Hawaii	1999	8.1
Gemini South	Cerro Pachón, Chile	2000	8.1

f igure 6-14 R I **V** U X G

The Keck I Telescope on Mauna Kea This photograph is a view down the Keck I telescope, looking toward the 10-meter objective mirror. (Note the astronomers standing on platforms to either side of the telescope.) The objective mirror is actually an arrangement of 36 hexagonal mirrors, each of which is 1.8 m (6 ft) across. Sensors and mechanical actuators on the back of each hexagonal mirror keep the arrangement in perfect alignment. The entire system is much lighter than a single 10-meter mirror would be. The hole in the center of the mirror arrangement is for the Cassegrain focus (see Figure 6-11c); the 1.4-m secondary mirror is housed in the dark hexagonal structure above the center of the photograph. (Roger Ressmeyer-Starlight)

In addition to the "giants" listed in Table 6-1, several other reflectors have objective mirrors between 3 and 6 meters in diameter, and dozens of smaller but still powerful telescopes have mirrors in the range of 1 to 3 meters. There are thousands of professional astronomers, each of whom has several ongoing research projects, and thus the demand for all of these telescopes is high. On any night of the year, nearly every research telescope in the world is being used to explore the universe.

6-3 Telescope images are degraded by the blurring effects of the atmosphere and by light pollution

In addition to providing a brighter image, a large telescope also helps achieve a second major goal: It produces star images that are sharp and crisp. A quantity called **angular resolution** gauges how well fine details can be seen. Poor angular resolution causes star images to be fuzzy and blurred together.

To determine the angular resolution of a telescope, pick out two adjacent stars whose separate images are just barely discernible (Figure 6-15). The angle θ (the Greek letter theta) between these stars is the telescope's angular resolution; the *smaller* that angle, the finer the details that can be seen and the sharper the image.

When you are asked to read the letters on an eye chart, what's being measured is the angular resolution of your eye. If you have 20/20 vision, the angular resolution θ of your eye is about 1 arcminute, or 60 arcseconds. (You may want to review the definitions of these angular measures in Section 1-5.) Hence, with the naked eye it is impossible to distinguish two stars less than 1 arcminute apart or to see details on the Moon with an angular size smaller than this. All the planets have angular sizes (as seen from Earth) of 1 arcminute or less, which is why they appear as featureless points of light to the naked eye.

One factor limiting angular resolution is **diffraction**, which is the tendency of light waves to spread out when they are confined to a small area like the lens or mirror of a telescope. (A rough analogy is the way water exiting a garden hose sprays out in a wider angle when you cover part of the end of the hose with your thumb.) As a result of diffraction, a narrow beam of light tends to spread out within a telescope's optics, thus blurring the image. If diffraction were the only limit, the angular resolution of a telescope would be given by the formula

Angular resolution

$$\theta = 2.5 \times 10^5 \frac{\lambda}{D}$$

θ = diffraction-limited angular resolution of a telescope, in arcseconds

λ = wavelength of light, in meters

D = diameter of telescope objective, in meters

For a given wavelength of light, using a telescope with an objective of *larger* diameter D *reduces* the amount of diffraction and makes the angular resolution θ *smaller* (and hence better). For example, with visible light with wavelength 600 nm, or 6×10^{-7} m, the diffraction-limited resolution of a 20-cm (8-in.) telescope of the sort used by amateur astronomers would be θ = (2.5 × 10⁵) (6 × 10⁻⁷ m)/(0.20 m) = 0.75 arcsec. With the 10-meter Keck telescope in Figure 6-14, the angular resolution would be 0.015 arcsec, or 50 times better.

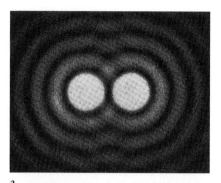

a

b

figure 6-15   R I **V** U X G

Angular Resolution The angular resolution of a telescope indicates the sharpness of the telescope's images. **(a)** This telescope view shows two sources of light whose angular separation is greater than the angular resolution. The two sources can easily be distinguished. **(b)** The light sources have been moved together so that their angular separation is equal to the angular resolution. In this case it is just barely possible to tell that there are two sources. If the sources were moved any closer together, the telescope image would show them as a single source.

In practice, however, ordinary optical telescopes cannot achieve such exceptionally fine angular resolution. The problem is that turbulence in the air causes star images to jiggle around and twinkle. Even through the largest telescopes, a star still looks like a tiny blob rather than a pinpoint of light. A measure of the limit that atmospheric turbulence places on a telescope's resolution is called the **seeing disk.** This disk is the angular diameter of a star's image broadened by turbulence. The size of the seeing disk varies from one observatory site to another and from one night to another. At the observatories on Kitt Peak in Arizona and Cerro Tololo in Chile, the seeing disk is typically around 1 arcsec. Some of the very best conditions in the world can be found at the observatories atop Mauna Kea in Hawaii, where the seeing disk is often as small as 0.5 arcsec. This is one reason why so many telescopes have been built there (see the photograph that opens this chapter).

In many cases the angular resolution of a telescope will be even worse than the limit imposed by the seeing disk. This will be the case if the objective mirror deforms even slightly due to variations in air temperature or flexing of the telescope mount. To combat this, most of the large telescopes listed in Table 6-1 are equipped with an **active optics** system. Such a system adjusts the mirror shape every few seconds to help keep the telescope in optimum focus and properly aimed at its target.

Changing the mirror shape is also at the heart of a more refined technique called **adaptive optics.** The goal of this technique is to compensate for atmospheric turbulence, so that the angular resolution can be smaller than the size of the seeing disk and can even approach the theoretical limit set by diffraction. Turbulence causes the image of a star to "dance" around erratically. Optical sensors monitor this dancing motion 10 to 100 times per second, and a powerful computer rapidly calculates the mirror shape needed to compensate. Fast-acting mechanical devices called *actuators* then deform the mirror accordingly, at a much faster rate than in an active optics system. (In some adaptive-optics systems, the actuators deform a small secondary mirror rather than the large objective mirror.)

Figure 6-16 shows the dramatic improvement in angular resolution possible with adaptive optics. Images made with adaptive optics are nearly as sharp as if the telescope were in the vacuum of space, where there is no atmospheric distortion whatsoever and the only limit on angular resolution is

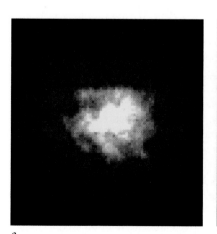

a   b

figure 6-16   R **I** V U X G

Using Adaptive Optics to "Unblur" Telescope Images Both these images of the star κ (Kappa) Pegasi (in the constellation Pegasus, the Winged Horse) were made with the same 3.5-m telescope. **(a)** Without adaptive optics, the star's image is blurred. Here, the star's image is blurred across a seeing disk of more than 1 arcsecond by turbulence in the atmosphere. **(b)** This image was made with an adaptive optics system turned on. The angular resolution is much finer, revealing that κ Pegasi is actually a binary star (two stars orbiting each other). The angular separation between the stars is only 0.3 arcsec. (R. Q. Fugate, USAF Research Laboratory, Starfire Optical Range)

Figure 6-17 R I **V** U X G

The Very Large Telescope This photograph shows the four domes of the European Southern Observatory's Very Large Telescope (VLT). The VLT is atop a 2640-m (8660-ft) mountain in the Atacama desert of Chile, where the skies are very dark and cloudless and the seeing is excellent. The four 8.2-m telescopes bear the names of celestial objects in the local Mapuche language: Antu (Sun), Kueyen (Moon), Melipal (Southern Cross), and Yepun (Venus, the evening star). (European Southern Observatory)

diffraction. Several of the telescopes listed in Table 6-1, including the Japanese telescope Subaru (the Japanese term for the Pleiades star cluster), are designed to make use of adaptive optics. At present, however, adaptive optics can be used only with rather bright objects.

Perhaps the finest angular resolution will be obtained with the Very Large Telescope (VLT), a project of the multinational European Southern Observatory. The VLT is actually four reflecting telescopes located in adjacent buildings at Cerro Paranal, Chile (Figure 6-17). Each telescope has a relatively flexible 8.2-meter objective mirror equipped with 214 active-optics actuators for controlling the mirror's shape. The four reflectors can be used individually to look at different objects, but by 2006 it will be possible to observe the same object simultaneously with all four telescopes and to combine the light received by all four. This will give the same light-gathering power as a single 16.4-m (53.8-ft) mirror. But because the reflectors are spread out over a distance of 200 m (670 ft, or more than two football fields), the angular resolution will be that of a 200-m telescope. In this way it will be possible to achieve the amazing resolution of 0.001 arcsec, which corresponds to being able to distinguish the two headlights on a car located on the Moon!

Light from city street lamps and from buildings also degrades telescope images. This **light pollution** illuminates the sky, making it more difficult to see the stars. You can appreciate the problem if you have ever looked at the night sky from a major city. Only a few of the very brightest stars can be seen, as against the thousands that can be seen with the naked eye in the desert or the mountains. To avoid light pollution, observatories are built in remote locations far from any city lights.

Unfortunately, the expansion of cities has brought light pollution to observatories that in former times had none. As an example, the growth of Tucson, Arizona, has had deleterious effects on observations at the nearby Kitt Peak National Observatory. Efforts have been made to have cities adopt light fixtures that provide safe illumination for their citizens but produce little light pollution. These efforts have met with only mixed success.

One factor over which astronomers have absolutely no control is the weather. Optical telescopes cannot see through clouds, so it is important to build observatories where the weather is usually good. One advantage of mountaintop observatories such as Mauna Kea or Cerro Paranal is that most clouds form at altitudes below the observatory, giving astronomers a better chance of having clear skies.

In many ways the best location for a telescope is in orbit around Earth, where it is unaffected by weather, light pollution, or atmospheric turbulence. We will discuss orbiting telescopes in Section 6-7.

6-4 An electronic device is commonly used to record the image at a telescope's focus

Telescopes provide astronomers with detailed pictures of distant objects. The task of recording these pictures is called **imaging.**

Astronomical imaging really began in the nineteenth century with the invention of photography. It was soon realized that this new invention was a boon to astronomy. By taking long exposures with a camera mounted at the focus of a telescope, an astronomer can record features too faint to be seen by simply looking through the telescope. Such long exposures can reveal details in galaxies, star clusters, and nebulae that would not be visible to an astronomer looking through a telescope. Indeed, most large, modern telescopes do not have eyepieces at all.

Unfortunately, photographic film is not a very efficient light detector. Only about 1 out of every 50 photons striking photographic film triggers the chemical reaction needed to produce an image. Thus, roughly 98% of the light falling onto photographic film is wasted.

The most sensitive light detector currently available to astronomers is the **charge-coupled device (CCD).** At the heart of a CCD is a semiconductor wafer divided into an array of small light-sensitive squares called picture elements or, more commonly, **pixels** (Figure 6-18). For example, some state-of-the-art CCDs for astronomy have more than 16 million pixels arranged in 4096 rows by 4096 columns. They have about a thousand times more pixels per square centimeter than on a typical computer screen, which means that a CCD of this type can record

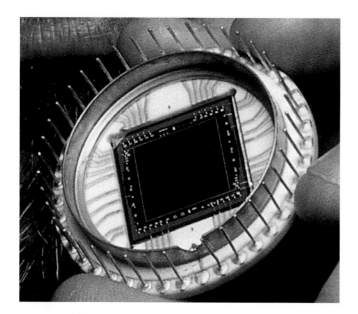

Figure 6-18 R I **V** U X G

A Charge-Coupled Device (CCD) This tiny silicon square records images in its 16,777,216 light-sensitive pixels. At the end of each exposure, additional circuits control the transfer and readout of the data to a waiting computer. Somewhat less elaborate CCDs are used by amateur astronomers and in consumer electronics. (Institute for Astronomy, University of Hawaii, and Roger Ressmeyer ©1993 Corbis)

very fine image details. CCDs with smaller numbers of pixels are used in digital cameras, scanners, and fax machines.

When an image from a telescope is focused on the CCD, an electric charge builds up in each pixel in proportion to the number of photons falling on that pixel. When the exposure is finished, the amount of charge on each pixel is read into a computer, where the resulting image can be stored in digital form and either viewed on a monitor or printed out. Compared with photographic film, CCDs are some 35 times more sensitive to light (they commonly respond to 70% of the light falling on them, versus 2% for film), can record much finer details, and respond more uniformly to light of different colors. Figure 6-19 shows the dramatic difference between photographic and CCD images. The great sensitivity of CCDs also makes them useful for **photometry**, which is the measurement of the brightnesses of stars and other astronomical objects.

In the modern world of CCD astronomy, astronomers need no longer spend the night in the unheated dome of a telescope. Instead, they operate the telescope electronically from a separate control room, where the electronic CCD images can be viewed on a computer monitor. The control room need not even be adjacent to the telescope. Although the Keck telescopes (see Figure 6-14) are at an altitude of 4200 m (13,800 feet), astronomers can now make observations from a facility elsewhere on the island of Hawaii that is much closer to sea level. This saves the laborious drive to the summit of Mauna Kea and eliminates the need for astronomers to acclimate to the high altitude.

a

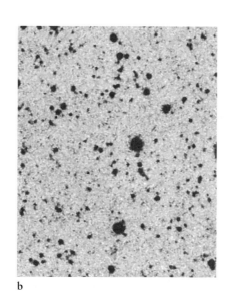

b

c

Figure 6-19 R I **V** U X G

Imaging: Photographic Film Versus CCD **(a)** This is a negative print (black stars and white sky) of a portion of the sky as imaged with a 4-meter telescope and photographic film. **(b)** This negative image of the same area as in (a) was made using the same telescope, but with the photographic film replaced by a CCD. You can see dozens of stars and galaxies that did not appear at all in the photograph. **(c)** To produce this color positive view of the same region, a series of CCD images were made using different color filters. These were then combined using a computer image-processing program. (Patrick Seitzer, National Optical Astronomy Observatories)

Most of the images that you will see in this book were made with CCDs. Because of their extraordinary sensitivity and their ability to be used in conjunction with computers, CCDs have attained a role of central importance in astronomy.

6-5 Spectrographs record the spectra of astronomical objects

We saw in Section 5-6 how the spectrum of an astronomical object provides a tremendous amount of information about that object. This is why measuring spectra, or **spectroscopy,** is one of the most important uses of telescopes. (The giant 11-meter Hobby-Eberly Telescope described in Section 6-2 is designed solely for measuring the spectra of distant, faint objects; it will never be used for imaging.) An essential tool of spectroscopy is the **spectrograph,** a device that records spectra on a CCD or photographic film. This optical device is mounted at the focus of a telescope.

An older design for a spectrograph, shown in Figure 6-20, uses a prism to form the spectrum of a planet, star, or galaxy. A slit, two lenses, and the prism are arranged to focus the spectrum onto a small piece of photographic film. Next, the exposed portion of the photographic plate is covered, and light from a hot gas (usually an element like iron or argon) is focused on the spectrograph slit. This result is a *comparison spectrum* above and below the spectrum of the star or galaxy (Figure 6-21). The wavelengths of the bright spectral lines of the comparison spectrum are already known from laboratory experiments and can therefore serve as reference markers.

There are drawbacks to this old-fashioned spectrograph. A prism does not disperse the colors of the rainbow evenly: Blue and violet portions of the spectrum are spread out more than the red portion. In addition, because the blue and violet wavelengths must pass through more glass than the red wavelengths (examine Figure 5-3), light is absorbed unevenly across the spectrum. Indeed, a glass prism is opaque to near-ultraviolet light.

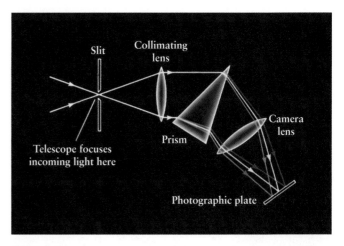

ƒigure 6-20

A Prism Spectrograph This optical device uses a prism to break up the light from a source into a spectrum. The collimating lens directs incoming light rays so that they enter the prism parallel to one another. A second lens, called a camera lens, then focuses the spectrum onto a photographic plate.

A better device for breaking starlight into the colors of the rainbow is a **diffraction grating,** or **grating** for short. This is a piece of glass on which thousands of closely spaced parallel lines have been cut. Some of the finest diffraction gratings have more than 10,000 lines per centimeter, which are usually cut by drawing a diamond back and forth across the glass. The spacing of the lines must be very regular, and the best results are obtained when the grooves are beveled. When light is shone on a diffraction grating, a spectrum is produced by the way in which light waves leaving different parts of the grating interfere with each other. (This same effect produces the rainbow of colors you see reflected from a compact disc or CD-ROM. Information is stored on the disc in a series of closely spaced pits, which act as a diffraction grating.) Figure 6-22 shows the design of a modern grating spectrograph.

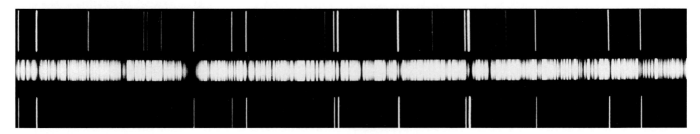

ƒigure 6-21   R I **V** U X G

A Spectrogram The photographic record of a spectrum is called a spectrogram. The spectrogram running across the middle of this figure shows the spectrum of a star. Note the many dark absorption lines, which are characteristic of the light from stars (see Section 5-6). Above and below the star's spectrum are emission lines produced by an iron arc at the observatory. These emission lines serve as a comparison spectrum. (Palomar Observatory)

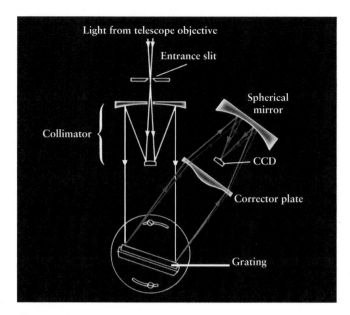

figure 6-22

A Grating Spectrograph This optical device uses a diffraction grating to break up the light from a source into a spectrum. The collimator ensures that light rays striking the grating are parallel. A corrector lens and mirror then focus the spectrum onto a CCD, which is far more sensitive than a photographic plate.

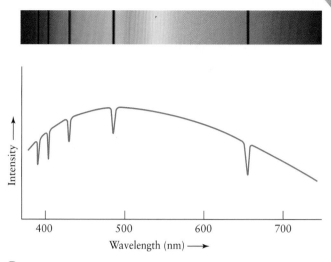

figure 6-23

Two Representations of a Spectrum This drawing compares two ways of displaying a spectrum. When a photographic plate is placed at the focus of a spectrograph, a rainbow-colored spectrum is obtained. The dark absorption lines in this example are the Balmer lines of hydrogen (see Section 5-8). When a CCD is placed at the focus of a spectrograph, a computer program can be used to convert the CCD's data into a graph of intensity versus wavelength. Note that absorption lines appear as dips on such a graph.

In Figure 6-22 a CCD, instead of a photographic plate, is placed at the focus of the spectrograph. When the exposure is finished, electronic equipment measures the charge that has accumulated in each pixel. These data are used to graph light intensity against wavelength. Dark absorption lines in the spectrum appear as depressions or valleys on the graph, while bright emission lines appear as peaks. Figure 6-23 compares two ways of exhibiting a spectrum in which several absorption lines appear. Later in this book we shall see spectra presented in both these ways.

6-6 A radio telescope uses a large concave dish to reflect radio waves to a focus

For thousands of years, all the information that astronomers gathered about the universe was based on ordinary visible light. In the twentieth century, however, astronomers first began to explore the nonvisible electromagnetic radiation coming from astronomical objects. In this way they have discovered aspects of the cosmos that are forever hidden to optical telescopes.

Astronomers have used ultraviolet light to map the outer regions of the Sun and the clouds of Venus, and used infrared radiation to see new stars and perhaps new planetary systems in the process of formation. By detecting radio waves from Jupiter and Saturn, they have mapped the intense magnetic fields that surround those giant planets; by detecting curious bursts of X rays from space, they have learned about the utterly alien conditions in the vicinity of a black hole. It is no exaggeration to say that today's astronomers learn as much about the universe using telescopes for nonvisible wavelengths as they do using visible light.

Radio waves were the first part of the electromagnetic spectrum beyond the visible to be exploited for astronomy. This happened as a result of a research project seemingly unrelated to astronomy. In the early 1930s, Karl Jansky, a young electrical engineer at Bell Telephone Laboratories, was trying to locate what was causing interference with the then-new transatlantic radio link. By 1932 he realized that one kind of radio noise is strongest when the constellation Sagittarius is high in the sky. The center of our Galaxy is located in the direction of Sagittarius, and Jansky concluded that he was detecting radio waves from an astronomical source.

At first only Grote Reber, a radio engineer living in Illinois, took up Jansky's research. In 1936 Reber built in his backyard the first **radio telescope,** a radio-wave detector dedicated to astronomy. He modeled his design after an ordinary reflecting telescope, with a parabolic metal "dish" (reflecting antenna) measuring 31 ft (10 m) in diameter and a radio receiver at the focal point of the dish.

Reber spent the years from 1938 to 1944 mapping radio emissions from the sky at wavelengths of 1.9 m and 0.63 m. He found radio waves coming from the entire Milky Way, with the greatest emission from the center of the Galaxy. These results, together with the development of improved radio technology during World War II, encouraged the growth

ƒigure 6-24 R I V U X G

A Radio Telescope The dish of this radio telescope, silhouetted by the setting Sun, is 64 m (210 ft) in diameter. It is located at the Parkes Observatory in New South Wales, Australia. (Roger Ressmeyer-Starlight)

of radio astronomy and the construction of new radio telescopes around the world. Radio observatories are as common today as major optical observatories.

Like Reber's prototype, a typical modern radio telescope has a large parabolic dish (Figure 6-24). A small antenna tuned to the desired frequency is located at the focus, and the incoming signal is relayed to amplifiers and recording instruments, typically located in a room at the base of the telescope's pier.

 The radio telescope in Figure 6-24 looks like a radar dish but is used in a different way. In radar, the dish is used to send out a narrow beam of radio waves. If this beam encounters an object like an airplane, some of the radio waves will be reflected back to the radar dish and detected by a receiver at the focus of the dish. Thus, a radar dish looks for radio waves *reflected* by distant objects. By contrast, a radio telescope is designed to receive radio waves *emitted* by objects in space.

Many radio telescope dishes, like the one in Figure 6-24, have visible gaps in them like a wire mesh. This does not affect their reflecting power because the holes are much smaller than the wavelengths of the radio waves. The same idea is used in the design of microwave ovens. The glass window in the oven door would allow the microwaves to leak out, so the window is covered by a metal screen with small holes. These holes are much smaller than the 12.2-cm (4.8-in.) wavelength of the microwaves, so the screen reflects the microwaves back into the oven.

One great drawback of early radio telescopes was their very poor angular resolution. Recall from Section 6-3 that angular resolution is the smallest angular separation between two stars that can just barely be distinguished as separate objects. For radio telescopes, the limitation on angular resolution is not atmospheric turbulence, as it is for optical telescopes. Rather, the problem is that angular resolution is

directly proportional to the wavelength being observed (see the formula for the angular resolution θ in Section 6-3). The longer the wavelength, the larger (and hence worse) the angular resolution and the fuzzier the picture. As an example, a 1-m radio telescope detecting radio waves of 5-cm wavelength has 100,000 times poorer angular resolution than a 1-m optical telescope. Because radio radiation has very long wavelengths, astronomers thought that radio telescopes could produce only blurry, indistinct images.

 A very large radio telescope can produce a somewhat sharper radio image, because as the diameter of the telescope increases, the angular resolution decreases. In other words, the bigger the dish, the better the resolution. For this reason, most modern radio telescopes have dishes more than 30 m (100 ft) in diameter. This is also useful for increasing light-gathering power, because radio signals from astronomical objects are typically very weak in comparison with the intensity of visible light. But even the largest single radio dish in existence, the 305-m (1000-ft) Arecibo radio telescope in Puerto Rico, cannot come close to the resolution of the best optical instruments.

A very clever technique makes it possible to produce radio images with excellent resolution. Unlike ordinary light, radio signals can be carried over electrical wires. Consequently, two radio telescopes observing the same astronomical object can be hooked together, even if they are separated by many kilometers. This technique is called **interferometry**, because the incoming radio signals are made to "interfere," or blend together. This makes the combined signal sharp and clear. The effective angular resolution of two such radio telescopes is equivalent to that of one gigantic dish with a diameter equal to the **baseline**, or distance between the two telescopes.

One of the largest arrangements of radio telescopes for interferometry is the Very Large Array (VLA), located in the desert near Socorro, New Mexico (Figure 6-25). The VLA consists of 27 parabolic dishes, each 25 m (82 ft) in diameter. These 27 telescopes are arranged along the arms of a gigantic **Y** that covers an area 27 km (17 mi) in diameter. By pointing all 27 telescopes at the same object and combining the 27 radio signals, this system can produce radio views of the sky with an angular resolution comparable to that of the very best optical telescopes. (The Very Large Telescope, shown in Figure 6-17, will combine the images from its four optical reflectors to do interferometry with *visible* light. In this way the VLT may achieve even better angular resolution with visible light than the VLA does using radio waves.)

 Dramatically better angular resolution can be obtained by combining the signals from radio telescopes at different observatories thousands of kilometers apart. This technique is called **very-long-baseline interferometry (VLBI)**. VLBI is used by a system called the Very Long Baseline Array (VLBA), which consists of ten 25-meter dishes at different locations between Hawaii and the Caribbean. Although the ten dishes are not physically connected, they are all used to observe the same object at the same time. The data from each telescope are recorded electronically and processed later. By carefully synchronizing the ten recorded

figure 6-25 R I **V** U X G

The Very Large Array (VLA) The 27 radio telescopes of the VLA are arranged along the arms of a **Y** in central New Mexico. (Only 13 of the 27 telescopes can be seen in this photograph.) The north arm of the array is 19 km long; the southwest and southeast arms are each 21 km long. By spreading the telescopes out along the legs and combining the signals received, the VLA can give the same angular resolution as a single dish many kilometers in radius. (National Radio Astronomy Observatory)

signals, they can be combined just as if the telescopes had been linked together during the observation. With VLBA, features smaller than 0.001 arcsec can be distinguished at radio wavelengths. This angular resolution is 100 times better than a large optical telescope with adaptive optics.

Even better angular resolutions can be obtained by adding radio telescopes in space; the baseline is then the distance from the VLBA to the orbiting telescope. The first such space radio telescope, the Japanese HALCA spacecraft, went into operation in 1997. Orbiting at a maximum distance of 21,400 km (13,300 mi) above the Earth's surface, HALCA gives a baseline 3 times longer—and thus an angular resolution 3 times better—than can be obtained with Earthbound telescopes alone.

Figure 6-26 shows how optical and radio images of the same object can give different and complementary kinds of information. The visible-light image of Saturn (Figure 6-26a) shows clouds in the planet's atmosphere and the structure of the rings. Like the visible light from the Moon, the light used to make this image is just reflected sunlight. By contrast, the radio image of Saturn (Figure 6-26b) is a record of waves emitted by electrically charged particles outside the planet. These particles emit because of their motion in Saturn's magnetic field, so this radio image provides clues about the conditions in the interior of Saturn where the magnetic field is produced. This information could never be obtained from a visible-light image such as Figure 6-26a.

Figure 6-26b is also an example of how radio astronomers often use "false color" to display their views of astronomical objects. In this figure the most intense radio emission is shown in red, the least intense in blue. Intermediate colors of the rainbow represent intermediate levels of radio intensity. Black indicates no detectable radio emission. Astronomers working at other nonvisible wavelength ranges also frequently use false-color techniques to display images obtained from their instruments. You will find such false-color images throughout this book.

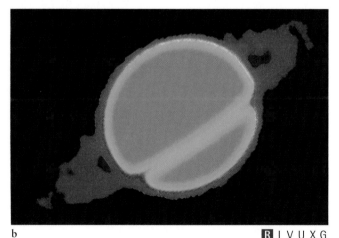

a   R I **V** U X G   b   **R** I V U X G

figure 6-26

Optical and Radio Views of Saturn **(a)** This picture was taken by a spacecraft 18 million kilometers from Saturn. The view was produced by sunlight reflecting from the planet's cloudtops and rings. **(b)** This VLA image shows radio emission from Saturn at a wavelength of 2 cm. This emission is caused by electrically charged particles moving within Saturn's strong magnetic field and acting as tiny radio transmitters. In this image, color represents the intensity of radio emission, ranging from blue for the weakest to red for the strongest. Note the blue radio "shadow" caused by Saturn's rings where they lie in front of the planet. (a: NASA; b: National Radio Astronomy Observatory)

6-7 Telescopes in orbit around the Earth detect radiation that does not penetrate the atmosphere

The many successes of radio astronomy show the value of observations at nonvisible wavelengths. But the Earth's atmosphere is opaque to many wavelengths. Other than visible light and radio waves, very little radiation from space manages to penetrate the air we breathe. To overcome this, astronomers have placed a variety of telescopes in orbit.

Figure 6-27 shows the transparency of the Earth's atmosphere to different wavelengths of electromagnetic radiation. The atmosphere is most transparent in two wavelength regions, the **optical window** (which includes the entire visible spectrum) and the **radio window** (which includes part, but not all, of the radio spectrum). There are also several relatively transparent regions at infrared wavelengths between 1 and 40 μm. Infrared radiation within these wavelength intervals can penetrate the Earth's atmosphere somewhat and can be detected with ground-based telescopes. This wavelength range is called the *near-infrared*, because it lies just beyond the red end of the visible spectrum.

Water vapor is the main absorber of infrared radiation from space, which is why infrared observatories are located at sites with exceptionally low humidity. The site must also be at high altitude to get above as much of the atmosphere's water vapor as possible. One site that meets both criteria is the summit of Mauna Kea in Hawaii, shown in the photograph that opens this chapter. (The complete lack of vegetation on the summit attests to its extreme dryness.) Some of the telescopes on Mauna Kea are designed exclusively for detecting infrared radiation. Others, such as the two

Keck telescopes, are used for both visible and near-infrared observations.

 Even at the elevation of Mauna Kea, water vapor in the atmosphere restricts the kinds of infrared observations that astronomers can make. This situation can be improved by carrying telescopes on board high-altitude balloons or aircraft. But the ultimate solution is to place a telescope in Earth orbit and radio its data back to astronomers on the ground. The first such orbiting infrared observatory, the Infrared Astronomical Satellite (IRAS), was launched in 1983. During its nine-month mission, IRAS used its 57-cm (22-in.) telescope to map almost the entire sky at wavelengths from 12 to 100 μm.

The IRAS data revealed the presence of dust disks around nearby stars. Planets are thought to coalesce from disks of this kind, so this was the first concrete (if indirect) evidence that there might be planets orbiting other stars. The dust that IRAS detected is warm enough to emit infrared radiation but too cold to emit much visible light, so it remained undetected by ordinary optical telescopes. IRAS also discovered distant, ultraluminous galaxies that emit almost all their radiation at infrared wavelengths.

In 1995 the Infrared Space Observatory (ISO), a more advanced 60-cm reflector with better light detectors, was launched into orbit by the European Space Agency (Figure 6-28). During its 2½-year mission, ISO made a number of groundbreaking observations of very distant galaxies and of the thin, cold material between the stars of our own Galaxy. Like IRAS, ISO had to be cooled by liquid helium to temperatures just a few degrees above absolute zero. Had this not been done, the infrared blackbody radiation from the telescope itself would have outshone the infrared radiation from astronomical objects. The ISO mission came to an end when the last of the helium evaporated into space.

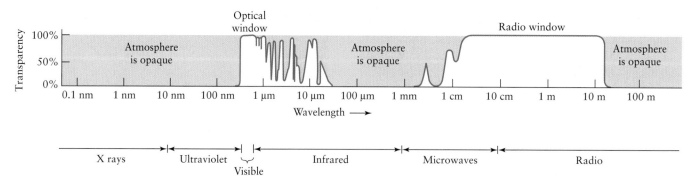

figure 6-27

The Transparency of the Earth's Atmosphere This graph shows the percentage of radiation that can penetrate the Earth's atmosphere at different wavelengths. At wavelengths where the curve of transparency is high, such as in the optical and radio windows, the atmosphere is relatively transparent; at wavelengths where the curve is low, the atmosphere is opaque. Absorption by oxygen and nitrogen makes the atmosphere completely opaque at wavelengths less than about 290 nm. The opaque region between the optical and radio windows is due to absorption by water vapor and carbon dioxide. At wavelengths longer than about 20 m, radiation is reflected back into space by ionized gases in the upper atmosphere.

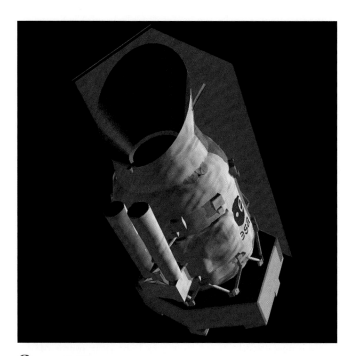

figure 6-28

The Infrared Space Observatory (ISO) Between 1995 and 1998, this spacecraft observed the sky at wavelengths between 2.5 and 240 μm. It provided astronomers with more than 26,000 infrared images and other data about planets, comets, gas and dust around stars and in interstellar space, galaxies, and the large-scale distribution of matter in the universe. (European Space Agency)

The next great orbiting infrared observatory is SIRTF (Space Infrared Telescope Facility), an 85-cm infrared telescope that will survey the infrared sky with unprecedented resolution. SIRTF will be used to study brown dwarfs (exotic objects larger than planets but smaller than stars), search for more direct evidence of planets forming around other stars, and probe the interiors of galaxies billions of light-years away. As of this writing NASA plans to launch SIRTF in 2002 on a mission lasting from 2½ to 5 years.

Astronomers are also very interested in observing at ultraviolet wavelengths. These observations can reveal a great deal about hot stars, ionized clouds of gas between the stars, and the Sun's high-temperature corona (see Section 3-5), all of which emit copious amounts of ultraviolet light. The spectrum of ultraviolet sunlight reflected from a planet can also reveal the composition of the planet's atmosphere. However, Earth's atmosphere is opaque to ultraviolet light except for the narrow *near-ultraviolet* range, which extends from about 400 nm (the violet end of the visible spectrum) down to 300 nm.

To see shorter-wavelength *far-ultraviolet* light, astronomers must again make their observations from space. The first ultraviolet telescope was placed in orbit in 1962, and several others have since followed it into space. Small rockets have also lifted ultraviolet cameras briefly above the Earth's atmosphere. Figure 6-29 shows an ultraviolet view of the constellation Orion, along with infrared and visible views.

The Far Ultraviolet Spectroscopic Explorer (FUSE), which went into orbit in 1999, specializes in measuring spectra at wavelengths from 90 to 120 nm.

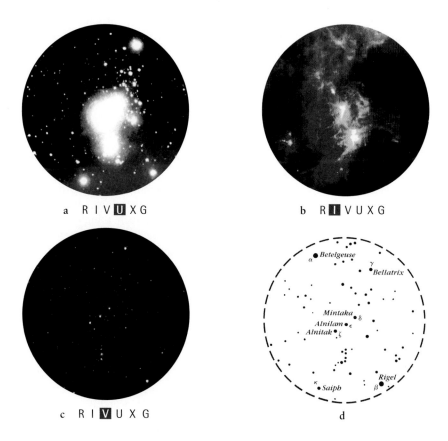

a R I V **U** X G

b R **I** V U X G

c R **I** **V** U X G

d

figure 6-29

Orion Seen at Ultraviolet, Infrared, and Visible Wavelengths (a) An ultraviolet view of the constellation of Orion was obtained during a brief rocket flight in 1975. This 100-s exposure covers the wavelength range 125–200 nm. (b) The false-color view from the Infrared Astronomical Satellite displays emission at different wavelengths in different colors: red for 100-μm radiation, green for 60-μm radiation, and blue for 12-μm radiation. Compare these images with (c) an ordinary optical photograph and (d) a star chart of Orion. (a: G. R. Carruthers, Naval Research Laboratory; b: NASA; c: R. C. Mitchell, Central Washington University)

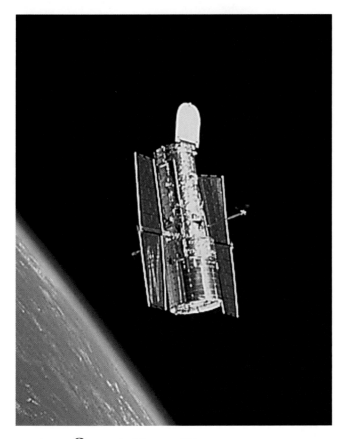

figure 6-30   R I **V** U X G

The Hubble Space Telescope (HST) This photograph of the HST in orbit was taken from the space shuttle *Discovery* during a servicing mission in 1999. Light enters the telescope through the open aperture cover on the top. The large, dark-colored solar panels on either side of the telescope provide electric power. The telescope is 13.1 m (43.0 ft) long and has a mass of 11,600 kg (13 tons). (NASA/Space Telescope Science Institute)

Highly ionized oxygen atoms, which can exist only in an extremely high-temperature gas, have a characteristic spectral line in this range. By looking for this spectral line in various parts of the sky, FUSE has confirmed that our Milky Way Galaxy (Section 1-4) is surrounded by an immense "halo" of gas at temperatures in excess of 200,000 K. Only an ultraviolet telescope could have detected this "halo," which is thought to have been produced by exploding stars called supernovae (Section 1-3).

Infrared and ultraviolet satellites give excellent views of the heavens at selected wavelengths. But since the 1940s astronomers had dreamed of having one large telescope that could be operated at any wavelength from the near-infrared through the visible range and out into the ultraviolet. This is the mission of the Hubble Space Telescope (HST), which was placed in a 600-km-high orbit by the space shuttle *Discovery* in 1990 (Figure 6-30). HST has a 2.4-meter (7.9-ft) objective mirror and

was designed to observe at wavelengths from 115 nm to 1 μm. Like most ground-based telescopes, HST uses a CCD to record images. (In fact, the development of HST helped drive advances in CCD technology.) The images are then radioed back to Earth in digital form.

The great promise of HST was that from its vantage point high above the atmosphere, its angular resolution would be limited only by diffraction. But soon after HST was placed in orbit, astronomers discovered that a manufacturing error had caused the telescope's 2.4-m primary mirror to suffer from spherical aberration. The mirror should have been able to concentrate 70% of a star's light into an image with an angular diameter of 0.1 arcsec. Instead, only 20% of the light was focused into this small area. The remainder was smeared out over an area about 1 arcsec wide, giving images little better than those achieved at major ground-based observatories.

On an interim basis, astronomers used only the 20% of incoming starlight that was properly focused and, with computer processing, discarded the remaining poorly focused 80%. This was practical only for brighter objects on which astronomers could afford to waste light. But many of the observing projects scheduled for HST involved extremely dim galaxies and nebulae.

These problems were resolved by a second space shuttle mission in 1993. Astronauts installed a set of small secondary mirrors whose curvature exactly compensated for the error in curvature of the primary mirror. Once these were in place, HST was able to make truly sharp images of extremely faint objects (Figure 6-31). Astronomers have used the repaired HST to make discoveries about the nature of planets, the evolution of stars, the inner workings of galaxies, and the expansion of the universe. You will see many HST images in later chapters.

The success of HST has inspired plans for its sequel, the Next Generation Space Telescope (NGST). This 8-m reflector, the largest ever placed in orbit, will observe at visible and infrared wavelengths with more than 10 times the light-gathering power of HST. Planned for a 2007 launch, NGST will study faint objects such as planetary systems forming around other stars and galaxies near the limit of the observable universe.

Space telescopes have also made it possible to explore objects whose temperatures reach the almost inconceivable values of 10^6 to 10^8 K. Atoms in such a high-temperature gas move so fast that when they collide, they emit X-ray photons of very high energy and very short wavelengths less than 10 nm. X-ray telescopes designed to detect these photons must be placed in orbit, since Earth's atmosphere is totally opaque at these wavelengths.

X-ray telescopes work on a very different principle from the X-ray devices used in medicine and dentistry. If you have your foot "X-rayed" to check for a broken bone, a piece of photographic film sensitive to X rays is placed under your foot and an X-ray beam is directed at your foot from above. The radiation penetrates through soft

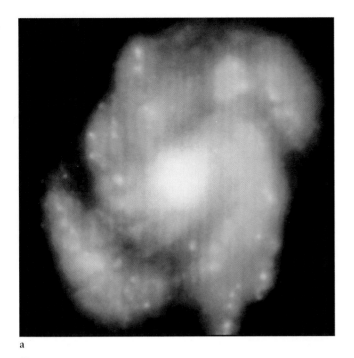

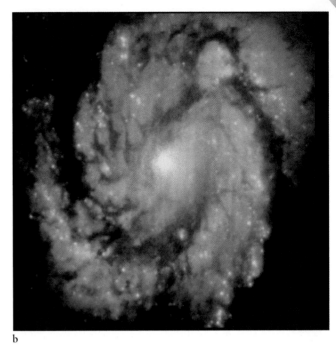

a
b

ƒigure 6-31   R I ☒ U X G

HST Images: Before and After Repair These Hubble Space Telescope images show the central region of the galaxy M100 in the constellation Coma Berenices (Berenice's Hair), some 70 million light-years from Earth. Each image shows an area about 25 arcseconds wide. **(a)** Spherical aberration in the original, uncorrected HST gave this blurred image. **(b)** This image, made after a 1993 space shuttle mission to install corrective optics, shows details that are less than a tenth of an arcsecond across. (Space Telescope Science Institute)

tissue but not through bone, so the bones cast an "X-ray shadow" on the film. A fracture will show as a break in the shadow. X-ray telescopes, by contrast, do *not* send beams of X rays toward astronomical objects in an attempt to see inside them. Rather, these telescopes detect X rays that the objects emit on their own.

Astronomers got their first quick look at the X-ray sky from brief rocket flights during the late 1940s. These observations confirmed that the Sun's corona (see Section 3-5) is a source of X rays, and must therefore be at a temperature of millions of kelvins. In 1962 a rocket experiment revealed that objects beyond the solar system also emit X rays.

Since 1970, a series of increasingly sensitive and sophisticated X-ray observatories have been placed in orbit, including NASA's Einstein Observatory, the European Space Agency's Exosat, and the German-British-American ROSAT. These telescopes have shown that other stars also have high-temperature coronae, and have found hot, X-ray-emitting gas clouds so immense that hundreds of galaxies fit inside them. They also discovered unusual stars that emit X rays in erratic bursts. These bursts are now thought to be coming from heated gas swirling around a small but massive object—possibly a black hole.

X-ray astronomy took a quantum leap forward in 1999 with the launch of NASA's Chandra X-ray Observatory and the European Space Agency's XMM-Newton. Named for the Indian-American astrophysicist Subrahmanyan Chandrasekhar, Chandra can view the X-ray sky with an angular resolution of 0.5 arcsec (Figure 6-32*a*). This is comparable to the best ground-based optical telescopes, and more than a thousand times better than the resolution of the first orbiting X-ray telescope. Chandra can also measure X-ray spectra 100 times more precisely than any previous spacecraft, and can detect variations in X-ray emissions on time scales as short as 16 microseconds. This latter capability is essential for understanding how X-ray bursts are produced around black holes.

XMM-Newton (for *X-ray Multi-mirror Mission*) is actually three X-ray telescopes that all point in the same direction (Figure 6-32*b*). Their combined light-gathering power is 5 times greater than that of Chandra, which makes XMM-Newton able to observe fainter objects. (For reasons of economy, the mirrors were not ground as precisely as those on Chandra, so the angular resolution of XMM-Newton is only about 6 arcseconds.) It also carries a small but highly capable telescope for ultraviolet and visible observations. Hot X-ray sources are usually accompanied by

a

b

figure 6-32

WEB LINK 6.16

Two Orbiting X-Ray Observatories **(a)** X rays are absorbed by ordinary mirrors like those used in optical reflectors, but they can be reflected if they graze the mirror surface at a very shallow angle. In the Chandra X-ray Observatory, X rays (shown in red) enter the telescope from the left and graze the inner, polished surface of a gently curved cylinder (shown in blue). The reflected rays come together in the focal plane (shown in yellow) 10 m (33 ft) behind the mirror. **(b)** XMM-Newton is about the same size as Chandra and forms images in the same way. This artist's rendering shows the large apertures for the three X-ray telescopes. (a: NASA/Chandra X-ray Observatory Center/Smithsonian Astrophysical Observatory; b: D. Ducros/European Space Agency)

cooler material that radiates at these longer wavelengths, so XMM-Newton will be able to observe these hot and cool regions simultaneously.

Gamma rays, the shortest-wavelength photons of all, help us to understand phenomena even more energetic than those that produce X rays. As an example, when a massive star explodes into a supernova, it produces radioactive atomic nuclei that are strewn across interstellar space. Observing the gamma rays emitted by these nuclei helps astronomers understand the nature of supernova explosions.

Like X rays, gamma rays do not penetrate Earth's atmosphere, so space telescopes are required. The most powerful gamma-ray telescope placed in orbit to date is the Compton Gamma Ray Observatory (CGRO), shown in Figure 6-33. One particularly important task for CGRO was the study of gamma-ray bursts, which are brief, unpredictable, and very intense flashes of gamma rays that occur unpredictably from all parts of the sky. Data from CGRO and other orbiting observatories have shown that the sources of these gamma-ray bursts are billions of light-years away. For these bursts to be visible across such great distances, their sources must be among the most energetic objects in the universe.

The advantages and benefits of Earth-orbiting observatories cannot be overemphasized. We are no longer limited to the narrow ranges of whatever wavelengths manage to leak through our shimmering, hazy atmosphere (Figure 6-34). For the first time, we are really *seeing* the universe.

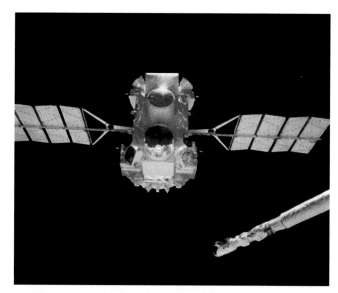

figure 6-33 R I **V** U X G

WEB LINK 6.17

The Compton Gamma Ray Observatory (CGRO) This photograph shows the Compton Observatory being deployed from the space shuttle *Atlantis* in 1991. Named in honor of Arthur Holly Compton, an American scientist who made important discoveries about gamma rays, CGRO carried four different gamma-ray detectors. When its mission ended in 2000, CGRO disintegrated as it re-entered our atmosphere. (NASA)

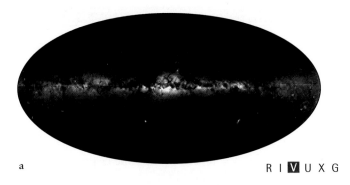

a   R **V** U X G

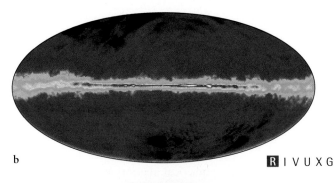

b   **R** I V U X G

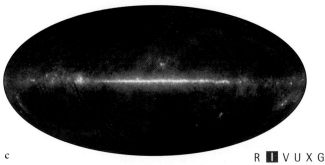

c   R **I** V U X G

d   R I V U **X** G

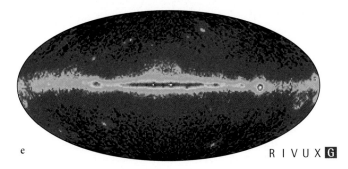

e   R I V U X **G**

 figure 6-34

The Entire Sky at Five Wavelength Ranges These five views show the entire sky at visible, radio, infrared, X-ray, and gamma-ray wavelengths. The entire celestial sphere is mapped onto an oval, with the Milky Way stretching horizontally across the center. The black crescents in the infrared and X-ray images are where data are missing. (GFSC/NASA)

(a) In the visible view the constellation Orion is at the right, Sagittarius in the middle, and Cygnus toward the left. Many of the dark areas along the Milky Way are locations where interstellar dust is sufficiently thick to block visible light.

(b) The radio view shows the sky at a wavelength of 21 cm. This wavelength is emitted by hydrogen atoms in interstellar space. The brightest regions (shown in red) are in the plane of the Milky Way, where the hydrogen is most concentrated.

(c) The infrared view from IRAS shows emission at 100 μm, 60 μm, and 12 μm. Most of the emission is from dust particles in the plane of the Milky Way that have been warmed by starlight.

(d) The X-ray view from ROSAT shows wavelengths of 0.8 nm (blue), 1.7 nm (green), and 5.0 nm (red), corresponding to photon energies of 1500, 750, and 250 eV. Extremely high temperature gas emits these X rays. The white regions, which emit strongly at all X-ray wavelengths, are remnants of supernovae.

(e) The gamma-ray view from the Compton Gamma Ray Observatory includes all wavelengths less than about 1.2×10^{-5} nm (photon energies greater than 10^8 eV). The diffuse radiation from the Milky Way is emitted when fast-moving subatomic particles collide with the nuclei of atoms in interstellar gas clouds. The bright spots above and below the Milky Way are distant, extremely energetic galaxies.

KEY WORDS

active optics, p. 133

adaptive optics, p. 133

angular resolution, p. 132

baseline, p. 138

Cassegrain focus, p. 129

charge-coupled device (CCD), p. 134

chromatic aberration, p. 126

coma, p. 130

coudé focus, p. 130

diffraction, p. 132

diffraction grating, p. 136

eyepiece lens, p. 125

focal length, p. 124

focal plane, p. 124

focal point, p. 124

focus (of a lens or mirror), p. 124

grating, p. 136

imaging, p. 134

interferometry, p. 138

light-gathering power, p. 125

light pollution, p. 134

magnification (magnifying power), p. 126

medium (*plural* media), p. 123

Newtonian reflector, p. 129

objective lens, p. 125

objective mirror (primary mirror), p. 129

optical telescope, p. 123

optical window (in the Earth's atmosphere), p. 140

photometry, p. 135

pixel, p. 134

prime focus, p. 129

radio telescope, p. 137

radio window (in the Earth's atmosphere), p. 140

reflecting telescope (reflector), p. 129

reflection, p. 128

refracting telescope (refractor), pp. 124–125

refraction, p. 123

seeing disk, p. 133

spectrograph, p. 136

spectroscopy, p. 136

spherical aberration, p. 130

very-long-baseline interferometry (VLBI), p. 138

KEY IDEAS

Refracting Telescopes: Refracting telescopes, or refractors, produce images by bending light rays as they pass through glass lenses.

• Chromatic aberration is an optical defect whereby light of different wavelengths is bent in different amounts by a lens.

• Glass impurities, chromatic aberration, opacity to certain wavelengths, and structural difficulties make it inadvisable to build extremely large refractors.

Reflecting Telescopes: Reflecting telescopes, or reflectors, produce images by reflecting light rays to a focus point from curved mirrors.

• Reflectors are not subject to most of the problems that limit the useful size of refractors.

Angular Resolution: A telescope's angular resolution, which indicates ability to see fine details, is limited by two key factors.

• Diffraction is an intrinsic property of light waves. Its effects can be minimized by using a larger objective lens or mirror.

• The blurring effects of atmospheric turbulence can be minimized by placing the telescope atop a tall mountain with very smooth air. They can be dramatically reduced by the use of adaptive optics and can be eliminated entirely by placing the telescope in orbit.

Charge-Coupled Devices: Sensitive light detectors called charge-coupled devices (CCDs) are often used at a telescope's focus to record faint images.

Spectrographs: A spectrograph uses a diffraction grating and lenses to form the spectrum of an astronomical object.

Radio Telescopes: Radio telescopes use large reflecting antennas or dishes to focus radio waves.

• Very large dishes provide reasonably sharp radio images. Higher resolution is achieved with interferometry techniques that link smaller dishes together.

Transparency of the Earth's Atmosphere: The Earth's atmosphere absorbs much of the radiation that arrives from space.

• The atmosphere is transparent chiefly in two wavelength ranges known as the optical window and the radio window. A few wavelengths in the near-infrared also reach the ground.

Telescopes in Space: For observations at wavelengths to which the Earth's atmosphere is opaque, astronomers depend on telescopes carried above the atmosphere by rockets or spacecraft.

• Satellite-based observatories provide new information about the universe and permit coordinated observation of the sky at all wavelengths.

REVIEW QUESTIONS

1. Describe refraction and reflection. Explain how these processes enable astronomers to build telescopes.

2. With the aid of a diagram, describe a refracting telescope. Which dimensions of the telescope determine its light-gathering power? Which dimensions determine the magnification?

3. What is the purpose of a telescope eyepiece? What aspect of the eyepiece determines the magnification of the image? In what circumstances would the eyepiece not be used?

4. Do most professional astronomers actually look through their telescopes? Why or why not?

5. What is chromatic aberration? For what kinds of telescopes does it occur? How can it be corrected?

6. Quite often advertisements appear for telescopes that extol their magnifying power. Is this a good criterion for evaluating telescopes? Explain your answer.

7. With the aid of a diagram, describe a reflecting telescope. Describe four different ways in which an astronomer can access the focal plane.

8. Explain some of the advantages of reflecting telescopes over refracting telescopes.

9. What is spherical aberration? How can it be corrected?

10. What kind of telescope would you use if you wanted to take a color photograph entirely free of chromatic aberration? Explain your answer.

11. Explain why a Cassegrain reflector can be substantially shorter than a refractor of the same focal length.

12. No major observatory has a Newtonian reflector as its primary instrument, whereas Newtonian reflectors are extremely popular among amateur astronomers. Explain why this is so.

13. What is active optics? What is adaptive optics? Why are they useful? Would either of these be a good feature to include on a telescope to be placed in orbit?

14. Compare an optical reflecting telescope and a radio telescope. What do they have in common? How are they different?

15. Why can radio astronomers make observations at any time during the day, whereas optical astronomers are mostly limited to observing at night? (*Hint:* Does your radio work any better or worse in the daytime than at night?)

16. What are the optical window and the radio window? Why isn't there an X-ray window or an ultraviolet window?

17. How are the images made by an X-ray telescope different from those made by a medical X-ray machine?

18. Why must astronomers use satellites and Earth-orbiting observatories to study the heavens at X-ray and gamma-ray wavelengths?

ADVANCED QUESTIONS

Problem-solving tips and tools

You may find it useful to review the small-angle formula discussed in Box 1-1. The area of a circle is proportional to the square of its diameter. Data on the planets can be found in the appendices at the end of this book. Section 5-2 discusses the relationship between frequency and wavelength. Box 6-1 gives examples of how to calculate magnifying power and light-gathering power.

19. Ordinary photographs made with a telephoto lens make distant objects appear close. How does the focal length of a telephoto lens compare with that of a normal lens? Explain your reasoning.

20. Show by means of a diagram why the image formed by a simple refracting telescope is upside down.

21. The observing cage in which an astronomer sits at the prime focus of the 5-m telescope on Palomar Mountain is about 1 m in diameter. Calculate what fraction of the incoming starlight is blocked by the cage.

22. Compare the light-gathering power of the Subaru 8.3-m telescope with that of the Hubble Space Telescope (HST), which has a 2.4-m objective mirror. What advantages does HST have over Subaru?

23. Suppose your Newtonian reflector has an objective mirror 20 cm (8 in.) in diameter with a focal length of 2 m. What magnification do you get with eyepieces whose focal lengths are (a) 9 mm, (b) 20 mm, and (c) 55 mm? What is the telescope's diffraction-limited angular resolution when used with orange light of wavelength 600 nm?

24. The Hobby-Eberly Telescope (HET) has a spherical mirror (the least expensive shape to grind). Consequently, the telescope has spherical aberration. Explain why this doesn't affect the usefulness of HET for spectroscopy.

25. The four largest moons of Jupiter are roughly the same size as our Moon and are about 628 million (6.28×10^8) kilometers from Earth at opposition. What is the size in kilometers of the smallest surface features that the Hubble Space Telescope (resolution of 0.1 arcsec) can detect? How does this compare with the smallest features that can be

seen on the Moon with the unaided human eye (resolution of 1 arcmin)?

26. Figure 6-31*b* shows the center of the galaxy M100. **(a)** If the angular resolution of this Hubble Space Telescope image is 0.1 arcsec, what is the diameter in light-years of the smallest detail that can be discerned in the image? Use the small-angle formula and the information given in the caption to Figure 6-31. **(b)** At what distance would a U.S. dime (diameter 1.8 cm) have an angular size of 0.1 arcsec? Give your answer in kilometers.

27. Can the Hubble Space Telescope distinguish any features on Pluto? Justify your answer using calculations.

28. The Russian Space Agency plans to place a radio telescope into an even higher orbit than the Japanese HALCA telescope. Using this telescope in concert with the VLBA, baselines as long as 77,000 km may be obtainable. Astronomers want to use this combination to study radio emission at a frequency of 1665 MHz from distant objects called quasars. **(a)** What is the wavelength of this emission? **(b)** Taking the baseline to be the effective diameter of this radio-telescope array, what angular resolution can be achieved?

29. The mission of the Submillimeter Wave Astronomy Satellite (SWAS), launched in 1998, is to investigate interstellar clouds within which stars form. One of the frequencies at which it observes these clouds is 557 GHz (1 GHz = 1 gigahertz = 10^9 Hz), characteristic of the emission from interstellar water molecules. **(a)** What is the wavelength (in meters) of this emission? In what part of the electromagnetic spectrum is this? **(b)** Why is it necessary to use a satellite for these observations? **(c)** SWAS has an angular resolution of 4 arcminutes. What is the diameter of its primary mirror?

30. To search for ionized oxygen gas surrounding our Milky Way Galaxy, astronomers aimed the ultraviolet telescope of the FUSE spacecraft at a distant galaxy far beyond the Milky Way. They then looked for an ultraviolet spectral line of ionized oxygen in that galaxy's spectrum. Were they looking for an emission line or an absorption line? Explain.

31. A sufficiently thick interstellar cloud of cool gas can absorb low-energy X rays but is transparent to high-energy X rays and gamma rays. Explain why both part *b* and part *d* of Figure 6-34 reveal the presence of cool gas in the Milky Way. Could you infer the presence of this gas from the visible-light image in Figure 6-34*a*? Explain.

DISCUSSION QUESTIONS

32. If you were in charge of selecting a site for a new observatory, what factors would you consider important?

33. Discuss the advantages and disadvantages of using a small telescope in Earth orbit versus a large telescope on a mountaintop.

WEB/CD-ROM QUESTIONS

34. Several telescope manufacturers build telescopes with a design called a "Schmidt-Cassegrain." These use a correcting lens in an arrangement like that shown in Figure 6-13*b*. Consult advertisements on the World Wide Web to see the appearance of these telescopes and find out their cost. Why do you suppose they are very popular among amateur astronomers?

35. Projects are under way to build large optical reflectors in South Africa (SALT, the Southern African Large Telescope) and in the Canary Islands (GTC, the Gran Telescopio Canarias). Search for current information about these telescopes on the World Wide Web. What will be the sizes of the primary mirrors? Are the telescopes designed for imaging, for spectroscopy, or both? Will they observe only at visible wavelengths? In what ways do they complement or surpass existing telescopes?

36. Three of the telescopes shown in the photo that opens this chapter— the James Clerk Maxwell Telescope (JCMT), the Caltech Submillimeter Observatory (CSO), and the Submillimeter Array (SMA)—are designed to detect radiation with wavelengths close to 1 mm. Search for current information about JCMT, CSO, and SMA on the World Wide Web. What kinds of celestial objects emit radiation at these wavelengths? What can astronomers see using JCMT, CSO, and SMA that cannot be observed at other wavelengths? Why is it important that they be at high altitude? How large are the primary mirrors used in JCMT, CSO, and SMA? What are the differences among the three telescopes? Which can be used in the daytime? What recent discoveries have been made using JCMT, CSO, or SMA?

37. The Hubble Space Telescope was originally intended to have a 15-year lifetime. At the time of this writing, NASA has decided to extend this to 20 years. Consult the Space Telescope Science Institute web site to learn about plans for HST's final years of operation. Are future space shuttle missions planned to service HST? If so, what changes will be made to HST on such missions? What will become of HST at the end of its mission lifetime?

38. NASA and the European Space Agency (ESA) have plans to launch a number of advanced space telescopes. These include Constellation-X, the Far Infrared and Submillimeter Telescope (FIRST), the Gamma-ray Large Area Space Telescope (GLAST), and the Space Interferometry Mission (SIM). Search the World Wide Web for information about at least two of these. What are the scientific goals of these projects? What is unique about each telescope? What advantages would they have over existing ground-based or orbiting telescopes? What kind of orbit will each telescope be in? When will each be launched and placed in operation?

39. Telescope Magnification. Access the Active Integrated Media Module "Telescope Magnification" in Chapter 6 of the *Universe* web site or CD-ROM. A common telescope found in department stores is a 3-inch (76-mm) diameter refractor that boasts a magnification of 300 times. Use the magnification calculator to determine the magnifications that are achieved by using each of the following commonly found eyepieces on that telescope: Eyepiece A with focal length 40 mm; Eyepiece B with focal length 25 mm; Eyepiece C with focal length 12 mm; and Eyepiece D with focal length 2.5 mm.

OBSERVING PROJECTS

40. Obtain a telescope during the daytime along with several eyepieces of various focal lengths. If you can determine the telescope's focal length, calculate the magnifying powers of the eyepieces. Focus the telescope on some familiar object, such as a distant lamppost or tree. **DO NOT FOCUS ON THE SUN! Looking directly at the Sun can cause blindness.** Describe the image you see through the telescope. Is it upside down? How does the image move as you slowly and gently shift the telescope left and right or up and down? Examine the eyepieces, noting their focal lengths. By changing the eyepieces, examine the distant object under different magnifications. How do the field of view and the quality of the image change as you go from low power to high power?

41. On a clear night, view the Moon, a planet, and a star through a telescope using eyepieces of various focal lengths and known magnifying powers. (To determine the locations in the sky of the Moon and planets, you may want to use the *Starry Night* program on the CD-ROM that comes with this book. You may also want to consult such magazines as *Sky & Telescope* and *Astronomy* or their web sites.) In what way does the image seem to degrade as you view with increasingly higher magnification? Do you see any chromatic aberration? If so, with which object and which eyepiece is it most noticeable?

42. Many towns and cities have amateur astronomy clubs. If you are so inclined, attend a "star party" hosted by your local club. People who bring their telescopes to such gatherings are delighted to show you their instruments and take you on a telescopic tour of the heavens. Such an experience can lead to a very enjoyable, lifelong hobby.

 43. The field of view of a typical small telescope for amateur astronomers is about 30 arcminutes (30´). Use the *Starry Night* program to simulate the view that such a telescope provides of various celestial objects. First turn off daylight (select **Daylight** in the **Sky** menu) and show the entire celestial sphere (select **Atlas** in the **Go** menu). Then search for objects (i), (ii), and (iii) listed below (select **Find...** in the **Edit** menu). For each object, use the controls at the right-hand end of the Control Panel to adjust the field of view to approximately 30´. Describe how much detail you can see on each object. (i) Jupiter. (ii) Saturn. (iii) M31, the great galaxy in the constellation Andromeda.

Our Solar System

(GSFC/NASA)

R I **V** U X G

for a good view of the planets, go to a dark location far from city lights. But for even better views, leave the Earth altogether and view the sky from space—like this view from the *Clementine* spacecraft, which shows (from left to right) Saturn, Mars, and Mercury, as well as the Sun and the Moon. What are the planets like? Are they larger or smaller than the Earth? What are they made of? And where did they come from?

Fifty years ago, astronomers knew precious little about the solar system. Even the best telescopes provided images of the planets that were frustratingly hazy and indistinct. Of asteroids, comets, and the satellites of the planets, we knew even less.

Today, our knowledge of the solar system has grown many thousandfold. Spacecraft such as *Clementine* have flown at close range past all the

planets save Pluto, showing us details that astronomers of an earlier generation could only dream about. We have landed spacecraft on the Moon, Venus, and Mars and dropped a probe into the immense atmosphere of Jupiter. This is truly the golden age of solar system exploration.

In this chapter we paint in broad outline our present understanding of the solar system. We will see that the planets come in a variety of sizes and chemical compositions. There is also rich variety among the moons of the planets and among smaller bodies we call asteroids and comets. By studying this diverse cast of characters, we will come to understand how the solar system formed some 4.6×10^9 years ago. We will also learn about planets beyond our solar system and how even more planets may be forming around other stars.

As you read the sections of this chapter, look for the answers to the following questions.

7-1 Are all the other planets similar to Earth, or are they very different?

7-2 Do other planets have moons like Earth's Moon?

7-3 How do astronomers know what the other planets are made of?

7-4 Are all the planets made of basically the same material?

7-5 What is the difference between an asteroid and a comet?

7-6 Why are some elements (like gold) quite rare, while others (like carbon) are more common?

7-7 How do astronomers think the solar system formed?

7-8 Did all of the planets form in the same way?

7-9 Are there planets orbiting other stars? How do astronomers search for such planets?

7-1 There are two broad categories of planets: Earthlike and Jupiterlike

Each of the nine planets that orbit the Sun is unique. Only Earth has liquid water and an atmosphere breathable by humans; only Venus has a perpetual cloud layer made of sulfuric acid droplets; only Jupiter has immense storm systems that persist for centuries; and only Pluto always keeps the same face toward its moon. But there are also striking similarities among planets. Volcanoes exist not only on Earth but also on Venus and Mars; there are rings around Jupiter, Saturn, Uranus, and Neptune; and craters on Mercury, Venus, Earth, and Mars show that all of these planets have been bombarded by interplanetary debris.

How can we make sense of the many similarities and differences among the planets? An important step is to organize our knowledge of the planets in a systematic way. There are two useful ways to do this. First, we can contrast the orbits of different planets around the Sun; and second, we can compare the planets' physical properties such as diameter, mass, average density, and chemical composition.

The planets fall naturally into two classes according to the sizes of their orbits. As Figure 7-1 shows, the orbits of the four inner planets (Mercury, Venus, Earth, and Mars) are crowded in close to the Sun. In contrast, the orbits of the next four planets (Jupiter, Saturn, Uranus, and Neptune) are widely spaced at great distances from the Sun. Table 7-1 lists the orbital characteristics of all the planets.

 Note that while Figure 7-1 shows the orbits of the planets, it does not show the planets themselves. The reason is simple: If Jupiter, the largest of the planets, were to be drawn to the same scale as the rest of this figure, it would be a dot just 0.0002 cm across—about $\frac{1}{300}$ of the width of a human hair and far too small to see without a microscope. The planets themselves are *very* small compared to the distances between the planets. Indeed, while an airliner traveling at 1000 km/h (620 mi/h) can fly around the Earth in less than two days, at this speed it would take 17 *years* to fly from the Earth to the Sun. The solar system is a very large and very empty place!

Most of the planets have orbits that are nearly circular. As we learned in Section 4-5, Kepler discovered in the sixteenth century that these orbits are actually ellipses. Astronomers denote the elongation of an ellipse by its *eccentricity* (see Figure 4-16b). The eccentricity of a circle is zero, and indeed most planets have orbital eccentricities that are very close to zero. The exceptions are Mercury and Pluto. In fact, Pluto's noncircular orbit sometimes takes it nearer the Sun than its neighbor, Neptune. (Happily, the orbits of Pluto and Neptune are such that the two planets will never collide.)

If you could observe the solar system from a point several astronomical units (AU) above the Earth's North Pole, you would see that all the planets orbit the Sun in the same counterclockwise direction. Furthermore, the orbits of the planets all lie in nearly the same plane. In other words, the orbits of the planets are inclined at only slight angles to the plane of the ecliptic, which is the plane of the Earth's orbit around the Sun. (Again, however, Pluto is an exception; the plane of its orbit is tilted at about 17° to the ecliptic, as Figure 7-1 shows.) Furthermore, the plane of the Sun's equator is very closely aligned with the orbital planes of the planets. As we will learn, these near-alignments are not a coincidence. They provide important clues about the origin of the planets.

When we compare the physical properties of the planets, we again find that they fall naturally into two classes—the four inner planets and the four outer ones—with Pluto as an exception. The four inner planets are called **terrestrial planets** because they resemble the Earth (in Latin, *terra*). They all have hard, rocky surfaces with mountains, craters, valleys, and volcanoes. You could stand on the surface of any one of them, although you would need a protective spacesuit on Mercury, Venus, or Mars. The outer four planets are called **Jovian planets** because they resemble Jupiter. (Jove was another name for the Roman god Jupiter.) An attempt to land a spacecraft on the surface of any of the Jovian planets would be futile, because the materials of which these planets are made are mostly gaseous or liquid. The visible "surface" features of a Jovian planet are actually cloud formations in the planet's atmosphere. The photographs in Table 7-1 show the distinctive appearances of the two classes of planets.

The most apparent difference between the terrestrial and Jovian planets is their diameters. You can compute the

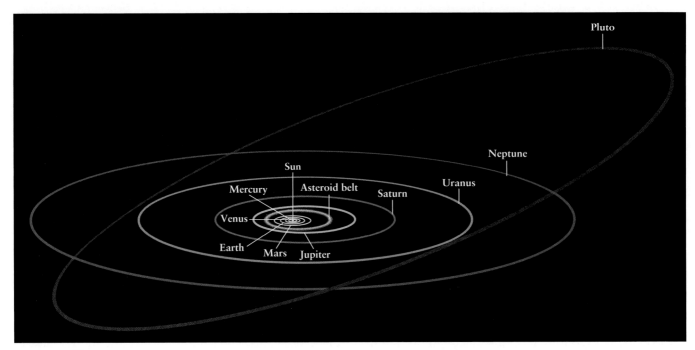

figure 7-1

The Solar System to Scale This scale drawing shows a perspective view of the orbits of the planets around the Sun. The four inner planets are crowded in close to the Sun, while the five outer planets orbit the Sun at much greater distances. Many smaller objects called asteroids orbit within a belt between the orbits of Mars and Jupiter. On the scale of this drawing, the planets themselves would be much smaller than the diameter of a human hair and too small to see.

diameter of a planet if you know its angular diameter as seen from Earth and its distance from Earth. For example, on December 11, 2000, Venus was 1.42×10^8 km from Earth and had an angular diameter of 17.6 arcsec. Using the small-angle formula from Box 1-1, we can calculate the diameter of Venus to be 12,100 km (7520 mi). Similar calculations demonstrate that the Earth, with its diameter of about 12,756 km (7926 mi), is the largest of the four inner, terrestrial planets. In sharp contrast, the four outer, Jovian planets are much larger than the terrestrial planets. First place goes to Jupiter, whose equa-

torial diameter is more than 11 times that of the Earth. Pluto, always the exception, is even smaller than the inner planets, despite being the outermost planet. Its diameter is less than one-fifth that of the Earth. Figure 7-2 shows the Sun and the planets drawn to the same scale. The diameters of the planets are given in Table 7-1.

The masses of the terrestrial and Jovian planets are also dramatically different. If a planet has a satellite, you can calculate the planet's mass from the satellite's period and semi-major axis by using Newton's form of Kepler's third law (see

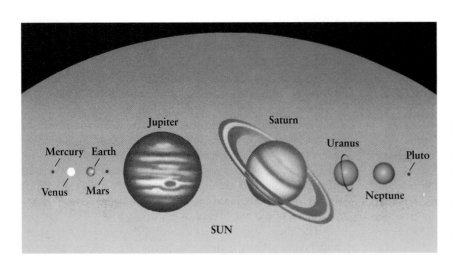

figure 7-2

The Sun and the Planets to Scale This drawing shows the nine planets in front of the disk of the Sun, with all ten bodies drawn to the same scale. The four terrestrial planets (Mercury, Venus, Earth, and Mars) have orbits nearest the Sun. They are small and dense, and made of rocky materials. The Jovian planets are the next four planets from the Sun (Jupiter, Saturn, Uranus, and Neptune). They are large and of low density and are composed primarily of hydrogen and helium. Pluto, an exceptional case, is a mixture of ice and rock.

Table 7-1	Characteristics of the Planets

	The Inner Planets			
	Mercury	**Venus**	**Earth**	**Mars**
Average distance from Sun (10^6 km)	57.91	108.2	149.60	227.93
Average distance from Sun (AU)	0.3871	0.7233	1.0000	1.5236
Orbital period (years)	0.2408	0.6152	1.0000	1.8808
Orbital eccentricity	0.206	0.007	0.017	0.093
Inclination of orbit to the ecliptic	7.00°	3.39°	0.00°	1.85°
Equatorial diameter (km)	4880	12,104	12,756	6794
Equatorial diameter (Earth = 1)	0.383	0.949	1.000	0.533
Mass (kg)	3.302×10^{23}	4.868×10^{24}	5.974×10^{24}	6.418×10^{23}
Mass (Earth = 1)	0.0553	0.8150	1.0000	0.1074
Average density (kg/m³)	5430	5243	5515	3934

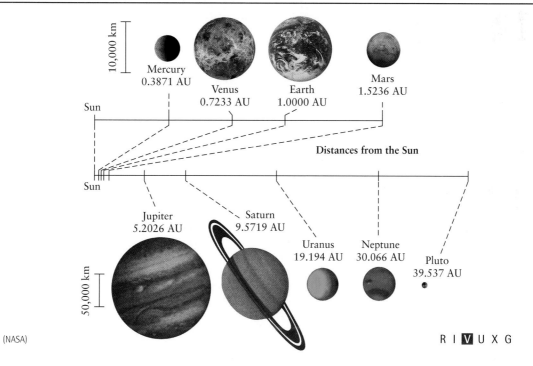

10,000 km

Mercury
0.3871 AU

Venus
0.7233 AU

Earth
1.0000 AU

Mars
1.5236 AU

Sun

Distances from the Sun

Sun

Jupiter
5.2026 AU

Saturn
9.5719 AU

Uranus
19.194 AU

Neptune
30.066 AU

Pluto
39.537 AU

50,000 km

(NASA)

R I **V** U X G

	The Outer Planets				
	Jupiter	**Saturn**	**Uranus**	**Neptune**	**Pluto**
Average distance from Sun (10^6 km)	778.30	1431.9	2877.4	4497.8	5914.7
Average distance from Sun (AU)	5.2026	9.5719	19.194	30.066	39.537
Orbital period (years)	11.856	29.369	84.099	164.86	248.60
Orbital eccentricity	0.048	0.053	0.043	0.010	0.250
Inclination of orbit to the ecliptic	1.30°	2.48°	0.77°	1.77°	17.12°
Equatorial diameter (km)	142,984	120,536	51,118	49,528	2300
Equatorial diameter (Earth = 1)	11.209	9.449	4.007	3.883	0.180
Mass (kg)	1.899×10^{27}	5.685×10^{26}	8.682×10^{25}	1.024×10^{26}	1.31×10^{22}
Mass (Earth = 1)	317.8	95.16	14.53	17.15	0.002
Average density (kg/m³)	1326	687	1318	1638	2000

Box 7-1 The Heavens on the Earth

Average Density

Average density—the mass of an object divided by that object's volume—is a useful quantity for describing the differences between planets in our solar system. This same quantity has many applications here on Earth.

A rock tossed into a lake sinks to the bottom, while an air bubble produced at the bottom of a lake (for example, by the air tanks of a scuba diver) rises to the top. These are examples of a general principle: An object sinks in a fluid if its average density is greater than that of the fluid, but rises if its average density is less than that of the fluid. The average density of water is 1000 kg/m^3, which is why a typical rock (with an average density of about 3000 kg/m^3) sinks, while an air bubble (average density of about 1.2 kg/m^3) rises.

At many summer barbecues, cans of soft drinks are kept cold by putting them in a container full of ice. When the ice melts, the cans of diet soda always rise to the top, while the cans of regular soda sink to the bottom. Why is this? The average density of a can of diet soda—which includes water, flavoring, artificial sweetener, and the trapped gas that makes the drink fizzy—is slightly less than the density of water, and so the can floats. A can of regular soda contains sugar instead of artificial sweetener, and the sugar is a bit heavier than the sweetener. The extra weight is just enough to make the average density of a can of regular soda slightly more than that of water, making the can sink. (You can test these statements by putting unopened cans of diet soda and regular soda in a sink or bathtub full of water.)

The concept of average density provides geologists with important clues about the early history of the Earth. The average density of surface rocks on Earth, about 3000 kg/m^3, is less than the Earth's average density of 5515 kg/m^3. The simplest explanation is that in the ancient past, the Earth was completely molten throughout its volume, so that low-density materials rose to the surface and high-density materials sank deep into the Earth's interior in a process called *chemical differentiation*. This series of events also suggests that the Earth's core must be made of relatively dense materials, such as iron and nickel. A tremendous amount of other geological evidence has convinced scientists that this picture is correct.

Section 4-7 and Box 4-4). Astronomers have also measured the mass of each planet except Pluto by sending a spacecraft to pass near the planet. The planet's gravitational pull (which is proportional to its mass) deflects the spacecraft's path, and the amount of deflection tells us the planet's mass. Using these techniques, astronomers have found that the four Jovian planets have masses that are tens or hundreds of times greater than the mass of any of the terrestrial planets. Again, first place goes to Jupiter, whose mass is 318 times greater than the Earth's.

Once we know the diameter and mass of a planet, we can learn something about what that planet is made of. The trick is to calculate the planet's **average density,** or mass divided by volume, measured in kilograms per cubic meter (kg/m^3). The average density of any substance depends in part on that substance's composition. For example, air near sea level on Earth has an average density of 1.2 kg/m^3, water's average density is 1000 kg/m^3, and a piece of concrete has an average density of 2000 kg/m^3. Box 7-1 describes some applications of the idea of average density to everyday phenomena on Earth.

The four inner, terrestrial planets have very high average densities (see Table 7-1); the average density of the Earth, for example, is 5515 kg/m^3. By contrast, a typical rock found on the Earth's surface has a lower average density, about 3000 kg/m^3. Thus, the Earth must contain a large amount of material that is denser than rock. This information provides our first clue that terrestrial planets have dense iron cores.

In sharp contrast, the outer, Jovian planets have quite low densities. Saturn has an average density less than that of water. This information strongly suggests that the giant outer planets are composed primarily of light elements such as hydrogen and helium. All four Jovian planets probably have large cores of mixed rock and highly compressed water buried beneath low-density outer layers tens of thousands of kilometers thick.

Pluto is again an oddity. Although it is even smaller than the dense inner planets, its average density is closer to that of the giant outer planets. Because its average density of about 1900 kg/m^3 is intermediate between the densities of ice (that is, frozen water) and of rock, Pluto is probably composed of a mixture of these two substances.

7-2 Seven large satellites are almost as big as the terrestrial planets

All the planets except Mercury and Venus have satellites. More than 60 satellites are known (Jupiter, Saturn, and Uranus each have at least 16), and dozens of other small ones probably remain to be discovered. The known satellites fall into two distinct categories. Seven giant satellites are larger than Pluto

table 7-2	The Seven Giant Satellites						
	Moon	**Io**	**Europa**	**Ganymede**	**Callisto**	**Titan**	**Triton**
Parent planet	Earth	Jupiter	Jupiter	Jupiter	Jupiter	Saturn	Neptune
Diameter (km)	3476	3642	3130	5268	4806	5150	2706
Mass (kg)	7.35×10^{22}	8.93×10^{22}	4.80×10^{22}	1.48×10^{23}	1.08×10^{23}	1.34×10^{23}	2.15×10^{22}
Average density (kg/m^3)	3340	3530	2970	1940	1850	1880	2050
Substantial atmosphere?	No	No	No	No	No	Yes	No

| Moon | Io | Europa | Ganymede | Callisto | Titan | Triton |

(JPL/NASA)

R I **V** U X G

and roughly as big as the planet Mercury. (Table 7-2 shows these satellites to the same scale.) Like the terrestrial planets, all seven giant satellites have hard surfaces of either rock or ice. All the other satellites also have solid surfaces but are much smaller, with diameters less than 2000 km.

Interplanetary spacecraft have made many surprising and fascinating discoveries about these giant satellites. We now know that Jupiter's satellite Io is the most geologically active world in the solar system, with numerous geyserlike volcanoes that continually belch forth sulfur-rich compounds. The fractured surface of Europa, another of Jupiter's large satellites, suggests that a worldwide ocean of liquid water may lie beneath its icy surface. Saturn's largest satellite, Titan, has an atmosphere nearly twice as dense as the Earth's atmosphere, with a perpetual haze layer that gives it a featureless appearance (see the photographs in Table 7-2).

7-3 Spectroscopy reveals the chemical composition of the planets

The average densities of the planets and satellites give us a crude measure of their **chemical compositions**—that is, what substances they are made of. For example, the low average density of the Moon (3344 kg/m^3) compared with the Earth (5515 kg/m^3)

tells us that the Moon contains relatively little of dense metals like iron. But to truly understand the nature of the planets and satellites, we need to know their chemical compositions in much greater detail than we can learn from average density alone.

The most accurate way to determine chemical composition is by directly analyzing samples taken from a planet's atmosphere and soil. Unfortunately, we have such direct information only for the Earth and the three worlds on which spacecraft have landed—Venus, the Moon, and Mars. In all other cases, astronomers must analyze sunlight reflected from the distant planets and their satellites. To do that, astronomers bring to bear one of their most powerful tools, **spectroscopy**, the systematic study of spectra and spectral lines. (We discussed spectroscopy in Sections 5-6 and 6-5.)

Spectroscopy is a sensitive probe of the composition of a planet's atmosphere. If a planet has an atmosphere, then sunlight reflected from that planet must have passed through its atmosphere. During this passage, some of the wavelengths of sunlight will have been absorbed. Hence, the spectrum of this reflected sunlight will have dark absorption lines. Astronomers look at the particular wavelengths absorbed and the amount of light absorbed at those wavelengths. Both of these depend on the kinds of chemicals present in the planet's atmosphere and the abundance of those chemicals.

For example, astronomers have used spectroscopy to analyze the atmosphere of Saturn's largest satellite, Titan (see

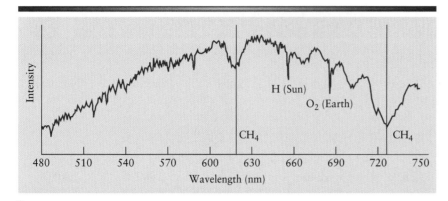

a

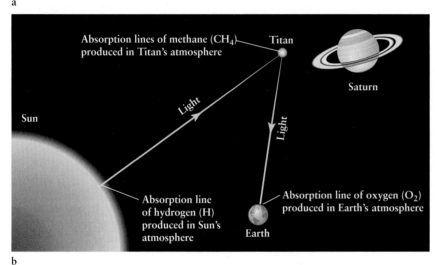

b

WEB LINK 7.4

Figure 7-3

The Spectrum of Saturn's Satellite Titan

(a) This graph shows the spectrum of sunlight reflected from Titan, the largest satellite of Saturn and the only satellite in the solar system with a substantial atmosphere. The dips in the curve are due to absorption by atoms of hydrogen (H), molecules of oxygen (O_2), and molecules of methane (CH_4). Of these gases, only methane is actually present in Titan's atmosphere. **(b)** The spectrum in (a) was made by light that was emitted from the Sun's glowing surface, passed through the Sun's atmosphere, was reflected by Titan's atmosphere, and passed through the Earth's atmosphere before reaching a telescope. The absorption lines of hydrogen and oxygen are produced in the atmospheres of the Sun and Earth, respectively. Only the methane dips are caused by absorption in Titan's atmosphere.

Table 7-2 for a photograph of Titan). The graph in Figure 7-3*a* shows the spectrum of visible sunlight reflected from Titan. (We first saw this method of displaying spectra in Figure 6-23.) The dips in this curve of intensity versus wavelength represent absorption lines. However, not all of these absorption lines are produced in the atmosphere of Titan (Figure 7-3*b*). Before reaching Titan, light from the Sun's glowing surface must pass through the Sun's hydrogen-rich atmosphere. This produces the hydrogen absorption line in Figure 7-3*a* at a wavelength of 656 nm. After being reflected from Titan, the light must pass through the Earth's atmosphere before reaching the telescope; this is where the oxygen absorption line in Figure 7-3*a* is produced. Only the two dips near 620 nm and 730 nm are caused by gases in Titan's atmosphere.

These two absorption lines are caused not by individual atoms in the atmosphere of Titan but by atoms combined to form **molecules.** For example, two hydrogen atoms can combine with an atom of oxygen to form a molecule of water. Water's chemical formula, H_2O, describes the atoms in a water molecule. In a similar way, two oxygen atoms can bond with a carbon atom to produce a molecule of carbon dioxide, whose formula is CO_2. Molecules, like atoms, also produce unique patterns of lines in the spectra of astronomical objects. The absorption lines in Figure 7-3*a* indicate the presence in Titan's atmosphere of molecules of methane, or CH_4 (a molecule made of one carbon atom and four hydrogen atoms).

This shows that Titan is a curious place indeed, because on Earth, methane is the primary ingredient in natural gas! The spectra of all of the planetary atmospheres in the solar system display absorption lines of molecules of various types.

In addition to visible-light measurements such as those in Figure 7-3*a*, it is very useful to study the *infrared* and *ultraviolet* spectra of planetary atmospheres. Many molecules have much stronger spectral lines in these nonvisible wavelength bands than in the visible. As an example, the ultraviolet spectrum of Titan shows that nitrogen molecules (N_2) are the dominant constituent of Titan's atmosphere. Furthermore, Titan's infrared spectrum includes spectral lines of a variety of molecules of carbon and hydrogen, indicating that Titan's atmosphere has a very complex chemistry. None of these molecules could have been detected by visible light alone.

Spectroscopy can also provide useful information about the solid surfaces of planets and satellites without atmospheres. When light shines on a solid surface, some wavelengths are absorbed while others are reflected. (For example, a plant leaf absorbs red and violet light but reflects green light—which is why leaves look green.) Unlike a gas, a solid illuminated by sunlight does not produce sharp, definite spectral lines. Instead, only broad absorption features appear in the spectrum. By comparing such a spectrum with the spectra of samples of different substances on Earth, astronomers can infer the chemical composition of the surface of a planet or satellite.

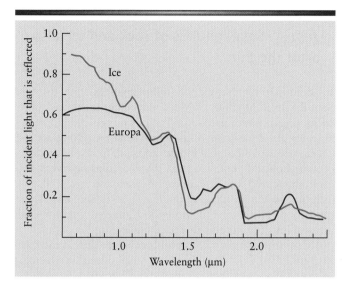

figure 7-4

The Spectrum of Jupiter's Moon Europa Infrared
light from the Sun that is reflected from the surface
of Europa, one of Jupiter's moons, has almost exactly the same
spectrum as sunlight reflected from ordinary water ice. This
shows that the surface of Europa is made predominantly of ice
and not rock.

As an example, Figure 7-4 shows the infrared spectrum of
light reflected from the surface of Jupiter's satellite Europa
(Table 7-2 includes a photograph of Europa). Because this
spectrum is so close to that of ordinary ice (that is, frozen
water), astronomers conclude that ice is the dominant constituent of Europa's surface. However, this technique tells us
little about what the material is like just below the surface.
There is no substitute for sending a spacecraft to a planet and
examining its surface directly.

7-4 Hydrogen and helium are abundant on the Jovian planets, whereas the terrestrial planets are composed mostly of heavy elements

Spectroscopic observations from Earth and spacecraft show
that the outer layers of the Jovian planets are composed primarily of the lightest gases, hydrogen and helium. In contrast,
chemical analysis of soil samples from Venus, Earth, and Mars
demonstrate that the terrestrial planets are made mostly of
heavy elements, such as iron, silicon, magnesium, sulfur, and
nickel. Spacecraft images such as Figures 7-5 and 7-6 only hint
at these striking differences in chemical composition.

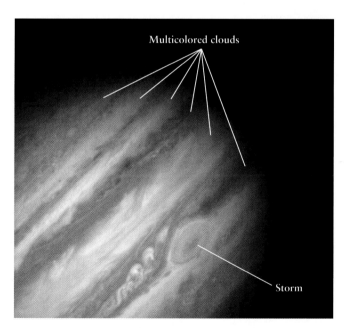

figure 7-5   R I **V** U X G

A Jovian Planet This detailed view of Jupiter's
cloudtops was made by the Hubble Space Telescope in
1999. Jupiter is composed mostly of the lightest elements, hydrogen
and helium. Hydrogen and helium are colorless; the colors in the
atmosphere are caused by trace amounts of other substances. The
giant storm at lower right, called the Great Red Spot, has been raging
for more than 300 years. (Space Telescope Science Institute/JPL/NASA)

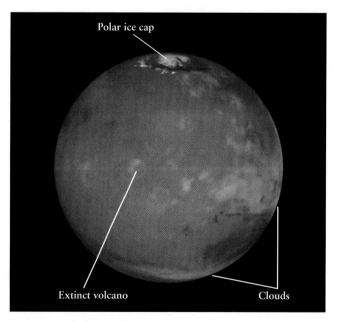

figure 7-6   R I **V** U X G

A Terrestrial Planet Mars is composed mostly of
heavy elements such as iron, silicon, magnesium, and
sulfur. The planet's red surface can be seen clearly in this Hubble
Space Telescope image because the Martian atmosphere is thin and
nearly cloudless. Olympus Mons, the extinct volcano to the left of
center, is nearly 3 times the height of Mount Everest. (Space
Telescope Science Institute/JPL/NASA)

Temperature plays a major role in determining whether the materials of which planets are made exist as solids, liquids, or gases. Hydrogen and helium are gaseous except at extremely low temperatures and extraordinarily high pressures. By contrast, rock-forming compounds such as iron and silicon are solids except at temperatures exceeding 1000 K. (You may want to review the discussion of temperature scales in Box 5-1.) Between these two extremes are substances such as water (H_2O), carbon dioxide (CO_2), methane (CH_4), and ammonia (NH_3). At low temperatures, typically below 100 to 300 K, these common chemicals solidify into solids called **ices**. (In astronomy, frozen water is just one kind of "ice.") At somewhat higher temperatures, they can exist as liquids or gases. For example, clouds of ammonia ice crystals are found in the cold upper atmosphere of Jupiter, but within Jupiter's warmer interior, ammonia exists primarily as a liquid.

As you might expect, a planet's surface temperature is related to its distance from the Sun. The four inner planets are quite warm. For example, midday temperatures on Mercury may climb to 700 K ($= 427°C = 801°F$), and during midsummer on Mars, it is sometimes as warm as 290 K ($= 17°C = 63°F$). The outer planets, which receive much less solar radiation, are cooler. Typical temperatures range from about 125 K ($= -148°C = -234°F$) in Jupiter's upper atmosphere to about 40 K ($= -233°C = -387°F$) on Pluto.

The higher surface temperatures of the terrestrial planets help to explain the following observation: The atmospheres of the terrestrial planets contain virtually no hydrogen or helium. Instead, the atmospheres of Venus, Earth, and Mars are composed of heavier molecules such as nitrogen (N_2, 14 times more massive than a hydrogen molecule), oxygen (O_2, 16 times more massive), and carbon dioxide (22 times more massive). To understand the connection between surface temperature and the absence of hydrogen and helium, we need to know a few basic facts about gases.

The temperature of a gas is directly related to the speeds at which the atoms or molecules of the gas move: The higher the gas temperature, the greater the speed of its atoms or molecules. Furthermore, for a given temperature, lightweight atoms and molecules move more rapidly than heavy ones. On the four inner, terrestrial planets, where atmospheric temperatures are high, low-mass hydrogen molecules and helium atoms move so swiftly that they can escape from the relatively weak gravity of these planets. Hence, the atmospheres that surround the terrestrial planets are composed primarily of more massive, slower-moving molecules such as CO_2, N_2, O_2, and water vapor (H_2O). On the four Jovian planets, low temperatures and relatively strong gravity prevent even lightweight hydrogen and helium gases from escaping into space, and so their atmospheres are much more extensive. The combined mass of Jupiter's atmosphere, for example, is about a million (10^6) times greater than that of the Earth's atmosphere. This is comparable to the mass of the entire Earth! Box 7-2 describes more about the ability of a planet's gravity to retain gases.

7-5 Small chunks of rock and ice also orbit the Sun

In addition to the nine planets, many smaller objects orbit the Sun. Between the orbits of Mars and Jupiter are thousands of rocky objects called **asteroids**. There is no sharp dividing line between planets and asteroids, which is why asteroids are also called **minor planets**. The largest asteroid, Ceres, has a diameter of about 900 km. The next largest, Pallas and Vesta, are each about 500 km in diameter. Still smaller ones, like the asteroid shown in close-up in Figure 7-7, are increasingly numerous. There are thousands of kilometer-sized asteroids and millions of asteroids that are boulder-sized or smaller.

Most (though not all) asteroids orbit the Sun at distances of 2 to 3.5 AU. This region of the solar system between the orbits of Mars and Jupiter is called the **asteroid belt**.

One common misconception about asteroids is that they are the remnants of an ancient planet that somehow broke apart or exploded, like the fictional planet Krypton in the comic book adventures of Superman. In fact, the asteroids were probably never part of any planet-sized body. The early solar system is thought to have been filled with asteroidlike objects, most of which coalesced to form the planets. The "leftover" objects that missed out on this process make up our present-day population of asteroids.

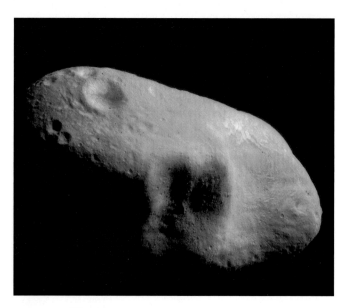

figure 7-7 R I **V** U X G

An Asteroid The asteroid shown in this image, 433 Eros, is only 33 km (21 mi) long and 13 km (8 mi) wide. Because Eros is so small, its gravity is too weak to have pulled it into a spherical shape. This image was taken in March 2000 by *NEAR Shoemaker*, the first spacecraft to go into orbit around an asteroid. (*NEAR* Project, Johns Hopkins University, Applied Physics Laboratory, and NASA)

box 7-2 | **T**ools of the Astronomer's Trade

Kinetic Energy, Temperature, and Whether Planets Have Atmospheres

A moving object possesses energy as a result of its motion. The faster it moves, the more energy it has. Energy of this type is called **kinetic energy**. If an object of mass m is moving with a speed v, its kinetic energy E_k is given by

Kinetic energy

$$E_k = \frac{1}{2} mv^2$$

E_k = kinetic energy of an object

m = mass of object

v = speed of object

This expression for kinetic energy is valid for all objects, both big and small, from atoms and molecules to planets and stars, as long as their speed is slow in relation to the speed of light. If the mass is expressed in kilograms and the speed in meters per second, the kinetic energy is expressed in joules (J).

EXAMPLE: An automobile of mass 1000 kg driving at a typical freeway speed of 30 m/s (= 108 km/h = 67 mi/h) has a kinetic energy of

$$E_k = \frac{1}{2}(1000 \times 30^2) = 450,000 \text{ J} = 4.5 \times 10^5 \text{ J}$$

Consider a gas, such as the atmosphere of a star or planet. Some of the gas atoms or molecules will be moving slowly, with little kinetic energy, while others will be moving faster and have more kinetic energy. The temperature of the gas is a direct measure of the *average* amount of kinetic energy per atom or molecule. The hotter the gas, the faster atoms or molecules move, on average, and the greater the average kinetic energy of an atom or molecule.

If the gas temperature is sufficiently high, typically several thousand kelvins, molecules move so fast that when they collide with one another, the energy of the collision can break the molecules apart into their constituent atoms. Thus, the Sun's atmosphere, where the temperature is 5800 K, consists primarily of individual hydrogen atoms rather than hydrogen molecules. By contrast, the hydrogen atoms in the Earth's atmosphere (temperature 290 K) are combined with oxygen atoms into molecules of water vapor (H_2O).

The physics of gases tells us that in a gas of temperature T (in kelvins), the average kinetic energy of an atom or molecule is

Kinetic energy of a gas atom or molecule

$$E_k = \frac{3}{2} kT$$

E_k = average kinetic energy of a gas atom or molecule, in joules

$k = 1.38 \times 10^{-23}$ J/K

T = temperature of gas, in kelvins

The quantity k is called the Boltzmann constant. Note that the higher the gas temperature, the greater the average kinetic energy of an atom or molecule of the gas. This average kinetic energy becomes zero at absolute zero, or $T = 0$, the temperature at which molecular motion is at a minimum.

At a given temperature, all kinds of atoms and molecules will have the same average kinetic energy. But the average *speed* of a given kind of atom or molecule depends on its mass. To see this, note that the average kinetic energy of a gas atom or molecule can be written in two equivalent ways:

$$E_k = \frac{1}{2} mv^2 = \frac{3}{2} kT$$

where v represents the average speed of an atom or molecule in a gas with temperature T. Rearranging this equation, we obtain

Average speed of a gas atom or molecule

$$v = \sqrt{\frac{3kT}{m}}$$

v = average speed of a gas atom or molecule, in m/s

$k = 1.38 \times 10^{-23}$ J/K

T = temperature of gas, in kelvins

m = mass of the atom or molecule, in kilograms

For a given gas temperature, the greater the mass of a given type of gas atom or molecule, the slower its average speed. (The value of v given by this equation is actually slightly higher than the average speed of the atoms or molecules in the gas, but it is close enough for our purposes here. If you are studying physics, you may know that v is actually the root-mean-square speed.)

(*continued on following page*)

box 7-2 *(continued)*

EXAMPLE: Suppose you want to know the average speed of the oxygen molecules that you breathe at a room temperature of 20°C (= 68°F = 293 K). From a reference book, you can find that the mass of an oxygen atom is 2.66×10^{-26} kg. The mass of an oxygen molecule (O_2) is twice the mass of an oxygen atom, or $2(2.66 \times 10^{-26}$ kg) $= 5.32 \times 10^{-26}$ kg. Thus, the average speed is

$$v = \sqrt{\frac{3(1.38 \times 10^{-23})(293)}{5.32 \times 10^{-26}}} = 478 \text{ m/s} = 0.478 \text{ km/s}$$

This is more than 1700 kilometers per hour (1100 miles per hour). Hence, atoms and molecules move rapidly in even a moderate-temperature gas.

In some situations, atoms and molecules in a gas may be moving so fast that they can overcome the attractive force of a planet's gravity and escape into interplanetary space. The minimum speed that an object at a planet's surface must have in order to permanently leave the planet is called the planet's **escape speed**. The escape speed for a planet of mass M and radius R is given by

$$v_{\text{escape}} = \sqrt{\frac{2GM}{R}}$$

where $G = 6.67 \times 10^{-11}$ N m²/kg² is the universal constant of gravitation.

The accompanying table gives the escape speed for various objects in the solar system. For example, to get to another planet, a spacecraft must leave Earth with a speed greater than 11.2 km/s (25,100 mi/h).

A good rule of thumb is that a planet can retain a gas if the escape speed is at least 6 times greater than the average speed of the molecules in the gas. (Some molecules are moving slower than average, and others are moving faster, but very few are moving more than 6 times faster than average.) In such a case, very few molecules will be moving fast enough to escape from the planet's gravity.

EXAMPLE: Consider the Earth's atmosphere. We saw that the average speed of oxygen molecules is 0.478 km/s at room temperature. The escape speed from the Earth (11.2 km/s) is much more than 6 times the average speed of the oxygen molecules, so the Earth has no trouble keeping oxygen in its atmosphere.

A similar calculation for hydrogen molecules (H_2) gives a different result, however. At 293 K, the average speed of a hydrogen molecule is 1.9 km/s. Six times this speed is 11.4 km/s, which is slightly higher than the escape speed from the Earth. Thus, the Earth does not retain hydrogen in its atmosphere. Any hydrogen released into the air slowly leaks away into space.

Planet	Escape speed (km/s)
Mercury	4.3
Venus	10.4
Earth	11.2
Moon	2.4
Mars	5.0
Jupiter	59.5
Saturn	35.5
Uranus	21.3
Neptune	23.5
Pluto	1.3

Quite far from the Sun, beyond the orbit of Neptune, are chunks of very dirty ice called **comets**. Many comets have highly elongated orbits that occasionally bring them close to the Sun. When this happens, the Sun's radiation vaporizes some of the comet's ices, producing long flowing tails of gas and dust particles (Figure 7-8).

 Science-fiction movies and television programs sometimes show comets tearing across the night sky like a rocket. That would be a pretty impressive sight—but that is not what comets look like. Like the planets, comets orbit the Sun. And like the planets, comets move hardly at all against the background of stars over the course of a single night (see Section 4-1). If you are lucky enough to see a bright comet, it will not zoom dramatically from horizon to horizon. Instead, it will seem to hang majestically among the stars, so you can admire it at your leisure.

Both asteroids and comets are thought to be debris left over from the formation of the solar system. In the inner regions of the solar system, rocky fragments have been able to endure continuous exposure to the Sun's heat, but any ice originally present would have evaporated. Far from the Sun, chunks of ice have survived for billions of years. Thus, debris in the solar system naturally divides into two families (asteroids and comets), which can be arranged according to dis-

figure 7-8 R I **V** U X G

A Comet This photograph shows Comet Hale-Bopp as it appeared in April 1997. The solid part of a comet like this is a chunk of dirty ice a few tens of kilometers in diameter. When a comet passes near the Sun, solar radiation vaporizes some of the icy material. The released material forms a bluish tail of gas and a white tail of dust, both of which can extend for tens of millions of kilometers. (Richard J. Wainscoat, University of Hawaii)

tance from the Sun, just like the two categories of planets (terrestrial and Jovian).

These ideas show that to understand the solar system, we must understand how it formed and how it evolved over time. To do this, we need to understand the origin of the basic ingredients of the solar system—the particular mix of chemical elements out of which planets, satellites, asteroids, comets, and the Sun itself are made. The answers to many of these questions lie in the stars, for this is where most of the chemical elements are manufactured.

7-6 The relative abundances of the elements are the result of cosmic processes

Some chemical elements are very common in our solar system, but others are very rare. Hydrogen is by far the most abundant substance in the solar system: It makes up nearly three-quarters of the combined mass of the Sun and planets. (We have seen that Jupiter, whose mass is greater than that of all the other planets combined, is composed mostly of hydrogen.) Helium is the second most abundant element. Together, hydrogen and helium account for about 98% of the mass of all the material in the solar system. All of the other chemical elements combined make up the remaining 2%.

The dominance of hydrogen and helium is not merely a characteristic of our local part of the universe. By analyzing the spectra of stars and galaxies, astronomers have found the same pattern of chemical abundances out to the farthest dis-

tance attainable by the most powerful telescopes. It is fair to say that the vast majority of the atoms in the universe are hydrogen and helium atoms. The elements that make up the bulk of the Earth (mostly iron, oxygen, and silicon) are relatively rare in the universe as a whole, as are the elements of which living organisms are made (carbon, oxygen, nitrogen, and phosphorus, among others).

There is a good reason for this overwhelming abundance of hydrogen and helium. A wealth of evidence has led astronomers to conclude that the universe began some 10 to 15 billion years ago with a violent event called the Big Bang (see Section 1-4). Only the lightest elements—hydrogen, helium, and tiny amounts of lithium and beryllium—emerged from the enormously high temperatures following this cosmic event. All the heavier elements were manufactured by stars later, either by nuclear fusion reactions deep in their interiors or by the violent explosions that mark the end of massive stars. Were it not for these processes that take place only in stars, there would be no heavy elements in the universe, no planets like Earth, no living creatures, and no one to contemplate the nature of the cosmos.

Because our solar system contains heavy elements, it must be that at least some of its material was once inside other stars. But how did this material become available to help build our solar system? The answer is that near the ends of their lives, stars cast much of their matter back out into space. For most stars this process is a comparatively gentle one, in which a star's outer layers are gradually expelled. Figure 7-9 shows a star which is losing material in this fashion. This ejected

figure 7-9 R I **V** U X G

A Mass-Loss Star The star Antares is shedding material from its outer layers that forms a cloud around the star. Although this cloud is very thin, we can see it because some of the ejected material has condensed into tiny grains of dust that reflect the star's light. Dust particles in the air around you reflect light in the same way, which is why you can see them within a shaft of sunlight in a darkened room. Antares lies some 600 light-years from Earth in the constellation Scorpio. (David Malin/ Anglo-Australian Observatory)

material appears as the cloudy region, or **nebulosity** (from *nubes,* Latin for "cloud"), that surrounds the star and is illuminated by it. A few stars eject matter much more dramatically at the very end their lives, in a spectacular detonation called a *supernova,* which blows the star apart (see Figure 1-6).

No matter how it escapes, the ejected material contains heavy elements dredged up from the star's interior, where they were formed. This material becomes part of the **interstellar medium,** a tenuous collection of gas and dust that pervades the spaces between the stars. As different stars die, they increasingly enrich the interstellar medium with heavy elements. Observations show that new stars form as condensations in the interstellar medium (Figure 7-10). Thus, these new stars have an adequate supply of heavy elements from which to develop a system of planets, satellites, comets, and asteroids. Our own solar system must have formed from enriched material in just this way.

Stars create different heavy elements in different amounts. For example, carbon, oxygen, silicon, and iron are readily produced in the interiors of massive stars, whereas gold is created only under special circumstances. Consequently, gold is rare in our solar system and in the universe as a whole, while carbon is relatively abundant (although still much less abundant than hydrogen or helium).

A convenient way to express the relative abundances of the various elements is to say how many atoms of a particular element are found for every trillion (10^{12}) hydrogen atoms. For example, for every 10^{12} hydrogen atoms in space, there are about 70 billion (70×10^9, or 7×10^{10}) helium atoms. From spectral analysis of stars and chemical analysis of Earth rocks, Moon rocks, and meteorites, scientists have determined the relative abundances of the elements in our part of the Galaxy today. Table 7-3 lists the ten most abundant elements, in order of their **atomic number.** An element's atomic number is the number of protons in the nucleus of an atom of that element. It is also equal to the number of electrons orbiting the nucleus (see Box 5-5). In general, the greater the atomic number of an atom, the greater its mass.

figure 7-10   R I **V** U X G
A Dusty Region of Star Formation Unlike Figure 7-9, which depicts an old star that is ejecting material into space, this image shows young stars in the constellation Orion (the Hunter) that have only recently formed from a cloud of gas and dust. The bluish, wispy appearance of the cloud (called NGC 1973-1975-1977) is caused by starlight reflecting off interstellar dust grains within the cloud. The grains are made of heavy elements produced by earlier generations of stars. (David Malin/Anglo-Australian Observatory)

In addition to these ten very common elements, five elements are moderately abundant: sodium, aluminum, argon, calcium, and nickel. These elements have abundances in the range of 10^6 to 10^7 relative to the standard 10^{12} hydrogen atoms. Most of the other elements are much rarer. For exam-

table 7-3	Abundances of the Most Common Elements		
Atomic number	Element	Symbol	Relative abundance
1	hydrogen	H	1×10^{12}
2	helium	He	7×10^{10}
6	carbon	C	4×10^8
7	nitrogen	N	9×10^7
8	oxygen	O	7×10^8
10	neon	Ne	1×10^8
12	magnesium	Mg	4×10^7
14	silicon	Si	4×10^7
16	sulfur	S	2×10^7
26	iron	Fe	3×10^7

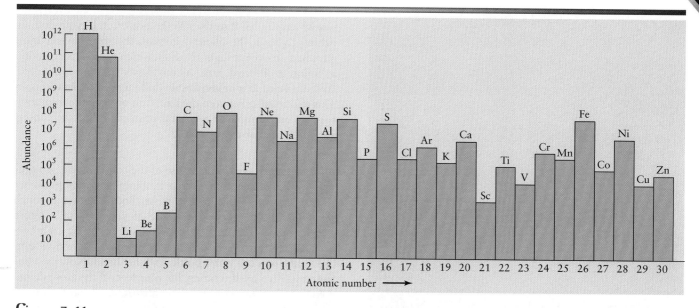

figure 7-11

Abundances of the Lighter Elements This graph shows the abundances of the 30 lightest elements (listed in order of increasing atomic number) compared to a value of 10^{12} for hydrogen. Notice that the vertical scale is not a linear one; each division on the scale corresponds to a tenfold increase in abundance. All elements heavier than zinc (Zn) have abundances less than 1000 atoms per 10^{12} atoms of hydrogen.

ple, for every 10^{12} hydrogen atoms in the solar system, there are only six atoms of gold. Figure 7-11, an extension of Table 7-3, shows the relative abundances of the 30 lightest elements.

Some 4.6 billion years ago, a collection of hydrogen, helium, and heavy elements came together to form the Sun and all the objects that orbit around it. All of those heavy elements, including the carbon atoms in your body and the oxygen atoms that you breathe, were created and cast off by stars that lived and died during the first 5 to 10 billion years of the universe's existence. Thus, by studying the abundances of the elements, we are led to a remarkable insight: We are literally made of star dust.

7-7 The Sun and planets formed from a solar nebula

Processes in the Big Bang and within ancient stars produced the ingredients to make our solar system. But given these ingredients, how did the Sun and planets form? In other words, where did the solar system come from?

This question has tantalized astronomers for centuries. While we do not yet have a wholly complete answer, we think we have identified the most likely series of events that led to the present-day system of Sun and planets.

A key piece of evidence about the origin of the solar system is that all the planets orbit the Sun in the same direction and in nearly the same plane, as we saw in Section 7-1. In the eighteenth century, the German philosopher Immanuel

Kant and the French scientist Pierre-Simon de Laplace independently suggested that this state of affairs could not be a coincidence. They proposed that our entire solar system—the Sun as well as all of its planets and satellites—formed from a vast, rotating cloud of gas and dust called the **solar nebula** (Figure 7-12a).

The consensus among today's astronomers is that Kant and Laplace were exactly right. In the modern version of their theory, at the outset the solar nebula was similar in character to the nebulosity shown in Figure 7-10 and had a mass somewhat greater than that of our present-day Sun.

Each part of the nebula exerted a gravitational attraction on the other parts, and these mutual gravitational pulls tended to make the nebula contract. As it contracted, the greatest concentration of matter occurred at the center of the nebula, forming a relatively dense region called the **protosun**. As its name suggests, this part of the solar nebula eventually developed into the Sun. The planets formed from the much sparser material in the outer regions of the solar nebula. Indeed, the mass of all the planets together is only 0.1% of the Sun's mass.

When you drop a ball, the gravitational attraction of the Earth makes the ball fall faster and faster as it falls; in the same way, material falling inward toward the protosun would have gained speed. As this fast-moving material ran into the protosun, the energy of the collision was converted into thermal energy, causing the temperature deep inside the solar nebula to climb. This process, in which the gravitational energy of a contracting gas cloud is converted into thermal energy, is called **Kelvin-Helmholtz contraction**, after the nineteenth-century physicists who first described it.

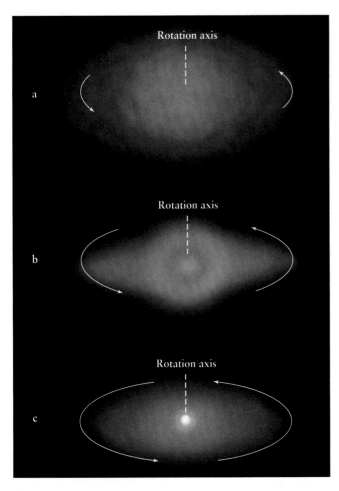

figure 7-12

The Birth of the Solar System This sequence of drawings shows three stages in the formation of the solar system. **(a)** A slowly rotating cloud of interstellar gas and dust begins to contract because of its own gravity. **(b)** A central condensation forms as the cloud flattens and spins more rapidly around its rotation axis. **(c)** Roughly 10^5 years after the cloud begins to contract, a glowing protosun has formed at its center. Surrounding this is a flattened disk of gas and dust, out of which the planets will form.

Temperatures within the newly created protosun soon climbed to several thousand kelvins. While the protosun's surface temperature stayed roughly constant, the temperature inside the protosun increased even more by means of further contraction. Eventually, after perhaps 10^7 (10 million) years had passed since the solar nebula first began to contract, the central temperature of the protosun reached a few million kelvins (that is, a few times 10^6 K), and nuclear reactions that convert hydrogen into helium began in the protosun's interior. When this happened, the contraction stopped and a true star was born. Nuclear reactions in the interior of the present-day Sun are the source of all the energy that the Sun radiates into space.

If the solar nebula had not been rotating at all, everything would have fallen directly into the protosun, leaving nothing behind to form the planets. Instead, the solar nebula must have had an overall slight rotation, which caused its evolution to follow a different path. As the slowly rotating nebula collapsed inward, it would naturally have tended to rotate faster. This relationship between the size of an object and its rotation speed is an example of a general principle called the **conservation of angular momentum.**

ANALOGY Figure skaters make use of the conservation of angular momentum. When a spinning skater pulls her arms and legs in close to her body, the rate at which she spins automatically increases. (If you are not a figure skater, you can demonstrate this yourself by sitting on an office chair. Sit with your arms outstretched and hold a weight, like a shoe or a full water bottle, in either hand. Now use your feet to start yourself rotating, lift your feet off the ground, and then pull your arms inward. Your rotation will speed up quite noticeably.)

As the solar nebula began to rotate more rapidly, it also tended to flatten out (Figure 7-12*b*). From the perspective of a particle rotating along with the nebula, it felt as though there were a force pushing the particle away from the nebula's axis of rotation. (Likewise, passengers on a spinning carnival ride seem to feel a force pushing them outward and away from the ride's axis of rotation.) This apparent force was directed opposite to the inward pull of gravity, and so it tended to slow the contraction of material toward the nebula's rotation axis. But there was no such effect opposing contraction in a direction parallel to the rotation axis. Some 10^5 (100,000) years after the solar nebula first began to contract, it had developed the structure shown in Figure 7-12*c*, with a rotating, flattened disk surrounding the protosun. The planets formed from the material in this disk, which explains why their orbits all lie in essentially the same plane and why they all orbit the Sun in the same direction.

There were no humans to observe these processes taking place during the formation of the solar system. But Earth astronomers have seen disks of material surrounding other stars that formed only recently. These are called **protoplanetary disks,** or **proplyds,** because it is thought that planets can eventually form from these disks around other stars. Thus, proplyds are planetary systems that are still "under construction." By studying these proplyds, astronomers are able to examine what our solar nebula may have been like some 4.6×10^9 years ago.

WEB LINK 7.13 Figure 7-13 shows a number of proplyds in the Orion Nebula, a region of active star formation. A star is visible at the center of each proplyd, which reinforces the idea that our Sun began to shine before the planets were fully formed. A study of 110 stars in the Orion Nebula detected proplyds around 56 of them, which suggests that systems of planets may form around a substantial fraction of stars. Later in this chapter we will see direct evidence for planets that have formed around stars other than the Sun.

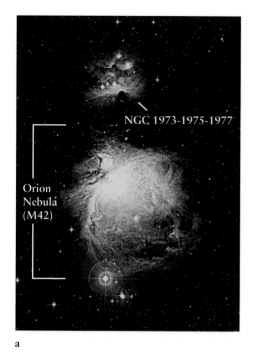

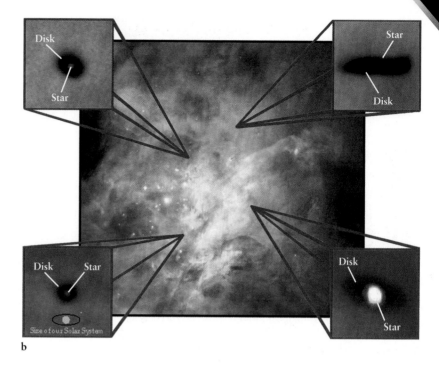

a   b

VIDEO 7.1

figure 7-13   R I **V** U X G

Protoplanetary Disks (Proplyds) in the Orion Nebula **(a)** The Orion Nebula is a star-forming
region located some 1500 light-years from Earth. To the naked eye, it is the middle star in Orion's
"sword" (see Figure 2-2). The smaller, bluish nebula is the object shown in close-up in Figure 7-13. **(b)** This
view of the center of the Orion Nebula is a mosaic of Hubble Space Telescope images. The four insets are
extreme close-ups of four proplyds, or protoplanetary disks, that lie within the nebula. False colors are used in
the insets to show brightness variations within each proplyd. The inset at the lower left shows the size of our
own solar system for comparison. A young, recently formed star is at the center of each proplyd. (The proplyd
at upper right is seen nearly edge-on.) The material in each proplyd may, with time, form into planets.
(a: Anglo-Australian Observatory image by David Malin; b: C. R. O'Dell and S. K. Wong, Rice University; NASA)

7-8 The planets formed by the accretion of planetesimals and the accumulation of gases in the solar nebula

We have seen how the solar nebula would have contracted to form a young Sun with a protoplanetary disk, or proplyd, rotating around it. But how did this disk coalesce into planets? And how did the formation of the inner, high-density, terrestrial planets differ from that of the outer, low-density, Jovian planets?

To understand how the planets, asteroids, and comets formed, we must look at the conditions that prevailed within the solar nebula. The density of material in the nebula was rather low, as was the pressure of the nebula's gas. If the pressure is sufficiently low, a substance cannot remain in the liquid state, but must end up as either a solid or a gas. For a given pressure, what determines whether a certain substance is a solid or a gas is its **condensation temperature.** If the temperature of a substance is above its condensation temperature,

the substance is a gas; if the temperature is below the condensation temperature, the substance solidifies into tiny specks of dust or snowflakes. You can often see similar behavior on a cold morning. The air temperature can be above the condensation temperature of water, while the cold windows of parked cars may have temperatures below the condensation temperature. Thus, water molecules in the air remain as a gas (water vapor) but form solid ice particles on the car windows.

Substances such as water (H_2O), methane (CH_4), and ammonia (NH_3) have low condensation temperatures, ranging from 100 to 300 K. Rock-forming substances have much higher condensation temperatures, in the range from 1300 to 1600 K. The gas cloud from which the solar system formed had an initial temperature near 50 K, so all of these substances could have existed in solid form. Thus, the solar nebula would have been populated by an abundance of small ice particles and dust grains like the one in Figure 7-14. (Recall that "ice" can refer to frozen CO_2, CH_4, or NH_3, as well as frozen water.) But hydrogen and helium, the most abundant elements in the solar nebula, have condensation temperatures so near absolute zero that these substances always existed as gases

figure 7-14

A Grain of Cosmic Dust This greatly enlarged image shows a tiny dust grain from interplanetary space that entered Earth's upper atmosphere and was collected by a high-flying aircraft. Dust grains of this sort are abundant in star-forming regions like that shown in Figure 7-13, and were almost certainly abundant in the nebula from which our solar system formed. These tiny grains served as the building blocks of the planets. The grain shown here is about 20 µm (0.02 mm) long. (Donald Brownlee, University of Washington)

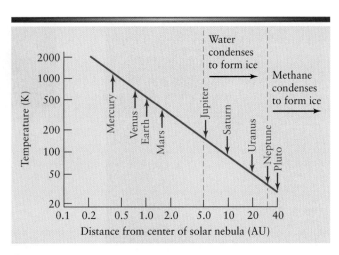

figure 7-15

Temperature Distribution in the Solar Nebula This graph shows how temperatures probably varied across the solar nebula as the planets were forming. For the inner planets, temperatures ranged from roughly 1200 K at Mercury to 500 K at the orbit of Mars. From the position of Jupiter outward, temperatures were everywhere less than 200 K. Beyond 5 AU from the center of the nebula, temperatures were low enough for water to condense and form ice; beyond 30 AU, methane (CH_4) could also condense.

during the creation of the solar system. You can best visualize the initial state of the solar nebula as a thin gas of hydrogen and helium strewn with tiny dust particles.

This state of affairs changed as the central part of the solar nebula underwent Kelvin-Helmholtz contraction to form the protosun. During this phase, the protosun was actually quite a bit more luminous than the present-day Sun, and it heated the inner part of the solar nebula to temperatures above 2000 K. Meanwhile, temperatures in the outermost regions of the solar nebula remained below 50 K.

Figure 7-15 shows the probable temperature distribution in the solar nebula at this stage in the formation of our solar system. In the inner regions of the solar nebula, water, methane, and ammonia were vaporized by the high temperatures. Only materials with high condensation temperatures could have remained solid. Of these materials, iron, silicon, magnesium, and sulfur were particularly abundant, followed closely by aluminum, calcium, and nickel. (These elements were also present in the form of oxides, which are chemical compounds containing oxygen. These elements also have high condensation temperatures.) In contrast, ice particles and ice-coated dust grains were able to survive in the cooler, outer portions of the solar nebula.

Small chunks would have formed in the inner part of the solar nebula from collisions between neighboring dust grains. Initially electric forces (chemical bonds) held them together. Over a few million years, these chunks coalesced into roughly 10^9 asteroidlike objects called **planetesimals,** with diameters

of a kilometer or so. During the next stage, the gravitational attraction between the planetesimals caused them to collide and coalesce into still-larger objects called **protoplanets,** which were roughly the size and mass of our Moon. This accumulation of material to form larger and larger objects is called **accretion.** During the final stage, these Moon-sized protoplanets collided to form the terrestrial planets. This final episode must have involved some truly spectacular, world-shattering collisions.

Important clues to this stage in the evolution of the inner solar system come from studies of **meteorites,** bits of rocky solar system debris that have collided with the Earth. Many of these are fragments of planetesimals that were never incorporated into the planets. Most meteorites contain not only dust grains but also **chondrules,** which are small, glassy, roughly spherical blobs (Figure 7-16). (The "ch" in "chondrule" is pronounced as in "chord.") Liquids tend to form spherical drops (like drops of water), so the shape of chondrules shows that they were once molten.

Attempts to produce chondrules in the laboratory show that the blobs must have melted suddenly, then solidified over the space of only an hour or so. This means they could not have been melted by the temperature of the inner solar nebula, which would have remained high for hundreds of thousands of years after the formation of the protosun. Instead, chondrules must have been produced by sudden, high-energy events that quickly melted material and then permitted it to cool. It is not clear what these events were, though the sheer number of

Figure 7-16 R I **V** U X G

Chondrules This is not a collection of pebbles; it is a cross-section of the interior of a meteorite. The "pebbles" are individual chondrules ranging in size from a few millimeters to a few tenths of a millimeter. Many of the chondrules are coated with dust grains like the one shown in Figure 7-14. (Courtesy of John A. Wood and the Smithsonian Astrophysical Observatory)

chondrules suggests that they happened throughout the inner solar nebula over a period of more than 10^6 years. The words "cloud" and "nebula" may suggest material moving in gentle orbits around the protosun; the existence of chondrules suggests that conditions in the inner solar nebula were at times quite violent.

Astronomers use computer simulations to learn how the inner planets could have formed from planetesimals. A computer is programmed to simulate a large number of particles circling a newborn sun along orbits dictated by Newtonian mechanics. As the simulation proceeds, the particles coalesce to form larger objects, which in turn collide to form planets. By performing a variety of simulations, each beginning with somewhat different numbers of planetesimals in different orbits, it is possible to see what kinds of planetary systems would have been created under different initial conditions. Such studies demonstrate that a wide range of initial conditions ultimately lead to basically the same result in the inner solar system: Accretion continues for roughly 10^8 (100 million) years and typically forms four or five terrestrial planets with orbits between 0.3 and 1.6 AU from the Sun.

Figure 7-17 shows one particular computer simulation. The calculations began with 100 planetesimals, each having a mass of 1.2×10^{23} kg. This choice ensures that the total mass (1.2×10^{25} kg) equals the combined mass of the four terrestrial planets (Mercury through Mars) plus their satellites. The initial orbits of planetesimals are inclined to each other by angles of less than 5° to simulate a thin layer of particles orbiting the protosun.

After an elapsed time simulating 30 million (3×10^7) years, the 100 original planetesimals have coalesced into 22 protoplanets. After 79 million (7.9×10^7) years, 11 larger protoplanets remain. Nearly another 100 million (10^8) years elapse before the total number of growing protoplanets is reduced to six. Figure 7-17c shows four planets following nearly circular orbits after a total elapsed time of 441 million (4.41×10^8) years. In this particular simulation, the fourth planet from the Sun ends up being the most massive; in our own solar system, it is the third planet (Earth) that is the most massive of the terrestrial planets. Note that the four planets in the simulation end up in orbits that are nearly circular, just like the orbits of most of the planets in our solar system.

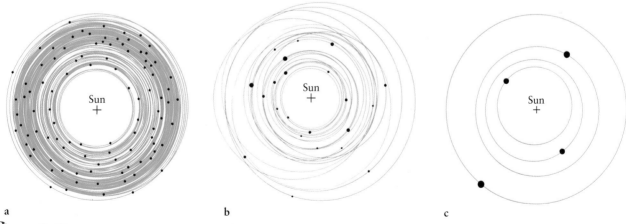

a   b   c

Figure 7-17

Accretion of the Terrestrial Planets These three drawings show the results of a computer simulation of the formation of the inner planets. **(a)** The simulation begins with 100 planetesimals orbiting the Sun. **(b)** After 30 million years, these planetesimals have coalesced into 22 protoplanets. **(c)** This final view shows four planets orbiting the Sun after an elapsed time of 441 million years. In fact, in this simulation the inner planets were essentially formed after 150 million years. (Adapted from George W. Wetherill)

As we have seen, the material from which the inner protoplanets accreted was rich in substances with high condensation temperatures. This material remained largely in solid form despite the high temperatures in the inner solar nebula. But as the protoplanets grew, they were heated by the violent impacts of planetesimals as well as the decay of radioactive elements such as uranium, and this heat caused melting. Thus, the terrestrial planets began their existence as spheres of at least partially molten rock. Material was free to move within these molten spheres, so the denser, iron-rich minerals sank to the centers of the planets while the less dense silicon-rich minerals floated to their surfaces. This process is called **chemical differentiation** (see Box 7-1). In this way the terrestrial planets developed their dense iron cores.

Like the inner planets, the outer planets may have begun to form by the accretion of planetesimals. The key difference is that ices as well as rocky grains were able to survive in the colder outer regions of the solar nebula. The elements of which ices are made are much more abundant than those that form rocky grains. Thus, much more solid material would have been available to form planetesimals in the outer solar nebula than in the inner part. As a result, objects several times larger than any of the terrestrial planets formed in the outer solar nebula. Each such object, made up of a mixture of ices and rock, could have become the core of a Jovian planet and served as a "seed" around which the rest of the planet eventually grew.

Thanks to the lower temperatures in the outer solar system, gas atoms (principally hydrogen and helium) were moving relatively slowly and so could more easily be captured by the strong gravity of these massive cores (see Box 7-2). Thus, the core of a Jovian protoplanet began to capture an envelope of gas as it continued to grow by accretion. Calculations suggest that both rock and gas slowly accumulated for about a million years, until the masses of the core and the envelope became equal. From that critical moment on, the envelope pulled in all the gas it could get, dramatically increasing the protoplanet's mass and size.

This runaway growth of the protoplanet continued until all the available gas was used up. The result was a huge planet with an enormously thick, hydrogen-rich envelope surrounding a rocky core with 5 to 10 times the mass of the Earth. This scenario occurred at four different distances from the Sun, thus creating the four Jovian planets. Figure 7-18 summarizes this story of the formation of the solar system.

Like the terrestrial planets, the Jovian planets were quite a bit hotter during their formation than they are today. A heated gas expands, so these gas-rich planets must also have been much larger than at present. As each planet cooled and contracted, it would have formed a disk like a solar nebula in miniature (see Figure 7-12). The satellites of the Jovian planets are thought to have formed from ice particles and dust grains within these disks.

The Jovian planets may also explain why there are asteroids and comets. Jupiter, the most massive of the planets, was almost certainly the first to form. Jupiter's tremendous mass would have exerted strong gravitational forces on any nearby planetesimals. Some would have been sent crashing into the terrestrial planets or into the Sun, while others would have

a

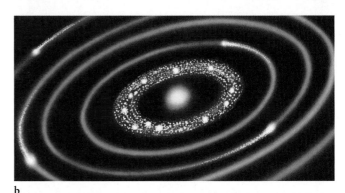

b

c

Figure 7-18

Formation of the Planets These sketches show the major stages in the birth of the solar system. **(a)** The solar nebula in its initial stages. **(b)** Terrestrial planets accrete from rocky material in the warm inner regions of the solar nebula. Meanwhile, the huge, gaseous Jovian planets form in the cold outer regions. **(c)** Planetary formation is nearly complete after some 10^8 (100 million) years.

been ejected completely from the solar system. A relatively few rocky planetesimals survived to produce the present-day population of asteroids. Comets, by contrast, began as icy planetesimals beyond the orbit of Jupiter. The strong gravitational forces from all of the Jovian planets pushed these into orbits even further from the Sun. Only a miniscule fraction of these wander close enough to the Sun to produce a visible tail (see Figure 7-8).

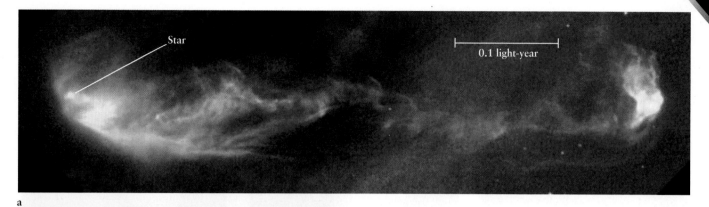

a

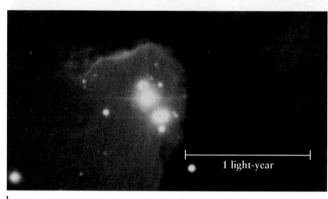

b

Figure 7-19 R I **V** U X G

Jets and Winds from Young Stars **(a)** This young star in the southern constellation Vela (the Sails) is ejecting matter at speeds of hundreds of kilometers per second, forming an immense jet. The jet is about 0.2 light-year (12,500 AU) in length. **(b)** An outpouring of particles and radiation from the surfaces of these young stars has carved out a cavity in the surrounding dusty material. The stars lie within the Trifid Nebula in the constellation Sagittarius (the Archer). (a: C. Burrows, STScI; J. Hester, Arizona State University; J. Morse, STScI; and NASA; b: David Malin/Anglo-Australian Observatory)

While the planets, satellites, asteroids, and comets were forming, the protosun was also evolving into a full-fledged star with nuclear reactions occurring in its core (see Section 1-3). The time required for this to occur was about 10^8 years, roughly the same as that required for the formation of the terrestrial planets. Before the onset of nuclear reactions, however, the young Sun probably expelled a substantial portion of its mass into space. Magnetic fields within the solar nebula would have funneled a portion of the nebula's mass into oppositely directed **jets** along the rotation axis of the nebula. Figure 7-19a shows one such jet emanating from a young star.

In addition, instabilities within the young Sun would have caused it to eject its tenuous outermost layers into space. This brief but intense burst of mass loss, observed in many young stars across the sky, is called a **T Tauri wind,** after the star in the constellation Taurus (the Bull) where it was first identified (see Figure 7-19b). Each of the proplyds in Figure 7-13 has a T Tauri star at its center. (The present-day Sun also loses mass from its outer layers in the form of high-speed electrons and protons, a flow called the **solar wind.** But this is miniscule compared with a T Tauri wind, which causes a star to lose mass 10^6 to 10^7 times faster than in the solar wind.)

With the passage of time, the combined effects of jets, the T Tauri wind, and accretion onto the planets would have swept the solar system nearly clean of gas and dust. With no more interplanetary material to gather up, the planets would have stabilized at roughly their present-day sizes, and the formation of the solar system would have been complete.

7-9 Astronomers have discovered planets orbiting other stars

If planets formed around our Sun, have they formed around other stars? That is, are there **extrasolar planets** orbiting stars other than the Sun? Our model for the formation of the planets would seem to suggest so. This model is based on the laws of physics and chemistry, which to the best of our understanding are the same throughout the universe. The discovery of a set of planets orbiting another star, with terrestrial planets in orbit close to the star and Jovian planets orbiting further away, would be a tremendous vindication of our theory of solar system formation. It would also be one of the most remarkable scientific discoveries of all time, for it would tell us that our planetary system is not unique in the universe. Because at least one planet in our solar system—the Earth—has the ability to support life, perhaps other planetary systems could also harbor living organisms.

Since 1995 evidence has accumulated for the existence of dozens of planets orbiting stars other than the Sun. However, none of these is a terrestrial planet. Most of them are more massive than Jupiter, and some are in eccentric, noncircular orbits quite unlike planetary orbits in our solar system. There is even some controversy as to whether these objects orbiting other stars are really planets at all. To understand this controversy, we must look at the process that astronomers go through to search for extrasolar planets.

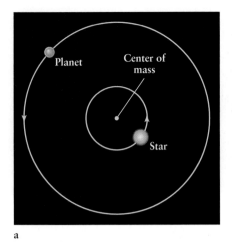
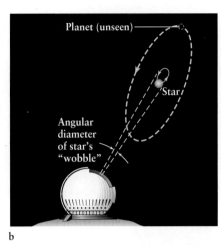
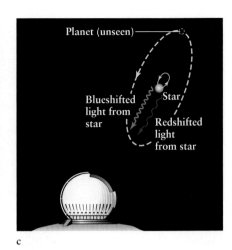

a　　　　　　　　　　　b　　　　　　　　　　　c

ƒigure 7-20

Detecting Planets Orbiting Other Stars **(a)** A planet and its star both orbit around their common center of mass, always staying on opposite sides of this point. Even if the planet cannot be seen, its presence can be inferred if the star's motion can be detected. **(b)** The astrometric method of detecting the unseen planet involves making direct measurements of the star's orbital motion. **(c)** In the radial velocity method, astronomers measure the Doppler shift of the star's spectrum as it moves alternately toward and away from the Earth. The amount of Doppler shift can be used to determine the size of the star's orbit, which in turn allows a determination of the unseen planet's orbit.

It is very difficult to make direct observations of planets orbiting other stars. The problem is that planets are small and dim compared with stars; at visible wavelengths, the Sun is 10^9 times brighter than Jupiter and 10^{10} times brighter than the Earth. A hypothetical planet orbiting a distant star, even a planet 10 times larger than Jupiter, would be lost in the star's glare as seen through even the largest telescope on Earth.

Instead, indirect methods must be used to search for extrasolar planets. One very powerful method is to search for stars that appear to "wobble." If a star has a planet, it is not quite correct to say that the planet orbits the star. Rather, both the planet and the star move in elliptical orbits around a point called the **center of mass**. Imagine the planet and the star as sitting at opposite ends of a very long seesaw; the center of mass is the point where you would have to place the fulcrum in order to make the seesaw balance. Because of the star's much greater mass, the center of mass is much closer to the star than to the planet. Thus, while the planet may move in an orbit that is hundreds of millions of kilometers across, the star will move in a much smaller orbit (Figure 7-20a).

As an example, the Sun and Jupiter both orbit their common center of mass with an orbital period of 5.2 years. (Jupiter has more mass than the other eight planets put together, so it is a reasonable approximation to consider the Sun's wobble as being due to Jupiter alone.) Jupiter's orbit has a semimajor axis of 7.78×10^8 km, while the Sun's orbit has a much smaller semimajor axis of 742,000 km. The Sun's radius is 696,000 km, so the Sun slowly wobbles around a point not far outside its surface. If astronomers elsewhere in the Galaxy could detect the Sun's wobbling motion, they could tell that there was a large planet (Jupiter) orbiting the Sun. They could

even determine the planet's mass and the size of its orbit, even though the planet itself was unseen.

Detecting the wobble of other stars is not an easy task. One approach to the problem, called the **astrometric method**, involves making very precise measurements of a star's position in the sky relative to other stars. The goal is to find stars whose positions change in a cyclic way (Figure 7-20b). The measurements must be made with very high accuracy (0.001 arcsec or better) and, ideally, over a long enough time to span an entire orbital period of the star's motion.

A different approach to the problem is the **radial velocity method** (Figure 7-20c). This is based on the Doppler effect, which we described in Section 5-9. A wobbling star will alternately move away from and toward the Earth. This will cause the dark absorption lines in the star's spectrum (see Figures 5-12 and 5-18) to change their wavelengths in a periodic fashion. When the star is moving away from us, its spectrum will undergo a redshift to longer wavelengths. When the star is approaching, there will be a blueshift of the spectrum to shorter wavelengths. These wavelength shifts are very small because the star's motion around its orbit is quite slow. As an example, the Sun moves around its small orbit at only 12.5 m/s (45 km/h, or 28 mi/h). If the Sun was moving directly toward an observer at this speed, the hydrogen absorption line at a wavelength of 656 nm in the Sun's spectrum would be shifted by only 2.6×10^{-5} nm, or about 1 part in 25 million. Detecting these tiny shifts requires extraordinarily careful measurements and painstaking data analysis.

In 1995, Michel Mayor and Didier Queloz of the Geneva Observatory in Switzerland announced that they had used the radial velocity method to discover a planet orbiting the

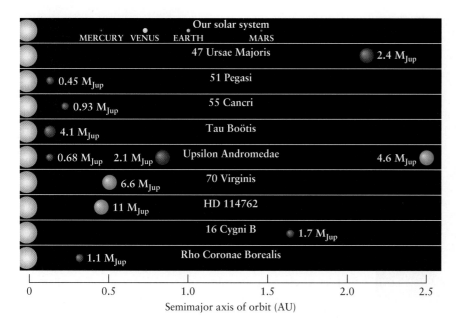

figure 7-21

A Selection of Extrasolar Planets

This figure summarizes what we know about planets orbiting nine other stars. The star name is given at the middle of each line. Each planet is shown at its average distance from its star (equal to the semimajor axis of its orbit). Comparison with our own solar system (at the top of the figure) shows how closely these extrasolar planets orbit their stars. The planets are *not* shown to the same scale as their stars, and the relative sizes of the extrasolar planets are estimates only. The mass of each planet—actually a lower limit—is given as a multiple of Jupiter's mass (M_{Jup}). (Adapted from G. Marcy and R. P. Butler)

star 51 Pegasi, which is 48 light-years from Earth in the constellation Pegasus (the Winged Horse). Their results were soon confirmed by the team of Geoff Marcy of San Francisco State University and Paul Butler of the University of California, Berkeley, using observations made at the University of California's Lick Observatory. For the first time, solid evidence had been found for a planet orbiting a star like our own Sun. Marcy and Butler, along with other astronomers, have since discovered more than 30 planets orbiting other stars by means of the radial velocity method. Figure 7-21 depicts a selection of these discoveries.

The extrasolar planets discovered by the radial velocity method all have masses comparable with or larger than that of Saturn, and thus are presumably Jovian planets made primarily of hydrogen and helium. (This cannot be confirmed directly, because the spectra of the planets are too faint to be detected with present technology.) According to the picture we presented in Section 7-8, such Jovian planets would be expected to orbit relatively far from their stars, where temperatures were low enough to allow the buildup of a massive envelope of hydrogen and helium gas. But as Figure 7-21 shows, many extrasolar planets are in fact found orbiting very *close* to their stars. For example, the planet orbiting 51 Pegasi has a mass at least 0.45 times as great as that of Jupiter but orbits only 0.051 AU from its star with an orbital period of just 4.2 days. In our own solar system, this orbit would lie well inside the orbit of Mercury!

Another surprising result is that about a third of the extrasolar planets found so far have very eccentric orbits. As an example, the planet around 16 Cygni B has an orbital eccentricity of 0.67; its distance from the star varies between 0.6 AU and 2.7 AU. This is quite unlike planetary orbits in our own solar system, where no planet has an orbital eccentricity greater than 0.25.

Do these observations mean that our picture of how planets form is incorrect? If Jupiterlike extrasolar planets such as

that orbiting 51 Pegasi formed close to their stars, the mechanism of their formation must have been very different from that which operated in our solar system. But another possibility is that extrasolar planets actually formed at large distances from their stars and have migrated inward since their formation. If enough gas and dust remains in a disk around a star after its planets form, interactions between the disk material and an orbiting planet will cause the planet to lose energy and to spiral inward toward the star around which it orbits. The planet's inward migration can eventually stop because of subtle gravitational effects from the disk or from the star. Gravitational interactions between the planet and the disk, or between planets in the same planetary system, could also have forced an extrasolar planet into a highly eccentric, noncircular orbit.

Yet a third possibility is that some of these extrasolar planets are not planets at all! The problem is that the radial velocity method cannot give precise values for the masses of planets, only lower limits. (An exact determination of the mass would require knowing how the plane of the planet's orbit is inclined to our line of sight. Unfortunately, this angle is not known because the planets themselves are unseen.) Hence, the actual masses of some of the "planets" shown in Figure 7-21 may be much larger than the values shown in that figure. It is thought that planets more massive than about 13 times the mass of Jupiter are actually **brown dwarfs,** objects that are like the Sun but insufficiently massive to sustain nuclear reactions in their cores (Figure 7-22). Like stars, but unlike true planets, brown dwarfs are thought to form directly from gas by gravitational contraction rather than by accretion of planetesimals.

Happily, some extrasolar planets appear to be of a sort that could fit into our own solar system. The best example is the planet orbiting 47 Ursae Majoris, a star very similar to the Sun. This planet is in a nearly circular orbit with a radius of 2.1 AU, intermediate between the orbits of Mars and Jupiter, and has a mass of at least 2.4 times that of Jupiter. Some

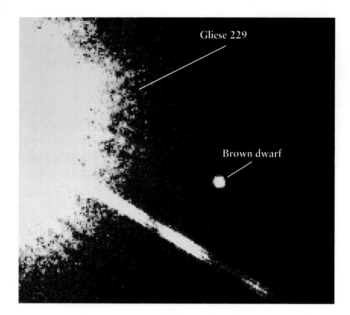

Gliese 229

Brown dwarf

figure 7-22   R **I** V U X G

A Brown Dwarf This false-color infrared image shows a brown dwarf that orbits the star Gliese 229 at an average distance of 32 AU. The brown dwarf is estimated to have a mass of 30 to 55 times the mass of Jupiter (between 0.030 and 0.055 of the Sun's mass) but about the same radius as Jupiter. Its surface temperature is about 1000 K. The spike running diagonally is an illusion caused by the telescope optics. Gliese 229 and its brown dwarf are 19 ly from Earth in the constellation Lepus (the Hare). (S. Kulkarni, Caltech; D. Golimowski, Johns Hopkins University; and NASA)

simulations of planet formation produce planets with orbital radii and masses in this range. Evidence for another solar system-like case has been presented by George Gatewood of the Allegheny Observatory. Applying the astrometric method to the star Lalande 21185, he finds that the wobble of that star indicates the presence of *two* planets, one with 0.9 Jupiter masses at a distance of 2.2 AU from the star and another with 1.1 Jupiter masses at 11 AU from the star. (These masses are actual values, not lower limits, because the astrometric method makes it possible to determine the inclinations of the planets' orbits.) There is also evidence for a system of three planets orbiting Upsilon Andromedae (see Figure 7-21).

The study of extrasolar planets is still in its infancy. We know little about how they formed or how their formation may have differed from that of our own solar system. We do not yet know what extrasolar planets look like or what they are made of. No Earth-sized extrasolar planets have yet been discovered, because their masses are too small to be detected by either the radial velocity method or the astrometric method using current telescope technology. Hence, we do not yet know whether solar systems like our own exist around other stars.

Eventually, however, advances in adaptive optics (see Section 6-3) may make it possible to record images of extrasolar planets the size of Jupiter. Orbiting telescopes are now being planned by both NASA and the European Space Agency to search for terrestrial planets around other stars. The next few decades may tell whether our own solar system is an exception or just one of a host of similar systems throughout the Galaxy.

KEY WORDS

Terms preceded by an asterisk (*) are discussed in the Boxes.

accretion, p. 166

asteroid, p. 158

asteroid belt, p. 158

astrometric method (for detecting extrasolar planets), p. 170

atomic number, p. 162

average density, p. 154

brown dwarf, p. 171

center of mass, p. 170

chemical composition, p. 155

chemical differentiation, p. 168

chondrule, p. 166

comet, p. 160

condensation temperature, p. 165

conservation of angular momentum, p. 164

*escape speed, p. 160

extrasolar planet, p. 169

ices, p. 158

interstellar medium, p. 162

jets, p. 169

Jovian planet, p. 151

Kelvin-Helmholtz contraction, p. 163

*kinetic energy, p. 159

meteorite, p. 166

minor planet, p. 158

molecule, p. 156

nebulosity, p. 162

planetesimal, p. 166

protoplanet, p. 166

protoplanetary disk (proplyd), p. 164

protosun, p. 163

radial velocity method (for detecting extrasolar planets), p. 170

solar nebula, p. 163

solar wind, p. 169

spectroscopy, p. 155

T Tauri wind, p. 169

terrestrial planet, p. 151

KEY IDEAS

Properties of the Planets: All of the planets orbit the Sun in the same direction and in almost the same plane. Most of the planets have nearly circular orbits.

• The four inner planets are called terrestrial planets. They are relatively small (with diameters of 5000 to 13,000 km), have high average densities (4000 to 5500 kg/m³), and are composed primarily of rock.

• The four giant outer planets are called Jovian planets. They have large diameters (50,000 to 143,000 km) and low average densities (700 to 1700 kg/m³) and are composed primarily of hydrogen and helium.

• Pluto appears to be a special case. It is smaller than any of the terrestrial planets and has an intermediate average density of about 1900 kg/m³, suggesting that it is composed of a mixture of ice and rock.

Satellites and Small Bodies in the Solar System: Besides the planets, the solar system includes satellites of the planets, asteroids, and comets.

• Seven large planetary satellites (one of which is the Moon) are comparable in size to the planet Mercury. The remaining satellites of the solar system are much smaller.

• Asteroids are small, rocky objects, while comets are small objects made of dirty ice. Both are remnants left over from the formation of the planets.

Spectroscopy and the Composition of the Planets: Spectroscopy, the study of spectra, provides information about the chemical composition of objects in the solar system.

• By mass, 98% of the ordinary matter in the universe is hydrogen and helium, the two lightest elements. These elements were probably formed shortly after the universe was created. The heavier elements were produced much later, in the centers of stars, and were cast into space when the stars died.

• The atoms of various elements can combine to form many different kinds of molecules.

• The basic planet-forming substances can be classified as gases, ices, or rock, depending on their condensation temperatures. The terrestrial planets are composed primarily of rock, whereas the Jovian planets are composed largely of gas.

Formation of the Solar System: The solar system formed from a disk-shaped cloud of hydrogen and helium that also contained ice and dust particles.

• The four inner planets formed through the accretion of dust particles into planetesimals, then into larger protoplanets.

• The four outer planets probably began as rocky protoplanetary cores, similar in character to the terrestrial planets. Gas then accreted onto these cores in a runaway fashion.

• The Sun formed by gravitational contraction of the center of the nebula. After about 10^8 years, temperatures at the protosun's center became high enough to ignite nuclear reactions that convert hydrogen into helium, thus forming a true star.

Extrasolar Planets: Astronomers have discovered planets orbiting other stars. The planets themselves are not visible; their presence is detected by the "wobble" of the stars around which they orbit.

• Most of the extrasolar planets discovered to date are quite massive and have orbits that are very different from planets in our solar system.

REVIEW QUESTIONS

1. What are the characteristics of a terrestrial planet?

2. What are the characteristics of a Jovian planet?

3. In what ways does Pluto not fit the usual classification of either terrestrial or Jovian planets?

4. In what ways are the largest satellites similar to the terrestrial planets? In what ways are they different? Which satellites are these?

5. What is an asteroid? What is a comet? In what ways are these minor members of the solar system like or unlike the planets?

6. What is meant by the average density of a planet? What does the average density of a planet tell us?

7. If hydrogen and helium account for 98% of the mass of all the material in the universe, why aren't the Earth and Moon composed primarily of these two gases?

8. What is meant by a substance's condensation temperature? What role did condensation temperatures play in the formation of the planets?

9. What is a planetesimal? How did planetesimals give rise to the planets?

10. What is meant by accretion?

11. What is a chondrule? How do we know they were not formed by the ambient heat of the solar nebula?

12. Why did the terrestrial planets form close to the Sun while the Jovian planets formed far from the Sun?

13. Why is the combined mass of all the asteroids so small?

14. Explain how our current understanding of the formation of the solar system can account for the following characteristics of the solar system: (a) All planetary orbits lie in nearly the same plane. (b) All planetary orbits are nearly circular. (c) The planets orbit the Sun in the same direction that the Sun itself rotates.

15. Explain why most of the satellites of Jupiter orbit that planet in the same direction that Jupiter rotates.

16. What techniques are used to detect planets orbiting other stars? Why are these techniques unable to detect planets like Earth?

17. Summarize the differences between the planets of our solar system and those found orbiting other stars.

18. A 1999 news story about the discovery of three planets orbiting the star Upsilon Andromedae (see Figure 7-21) stated that "the newly discovered galaxy, with three large planets orbiting a star known as Upsilon Andromedae, is 44 light-years away from Earth." What is wrong with this statement?

ADVANCED QUESTIONS

Questions preceded by an asterisk (*) involve topics discussed in the Boxes.

| Problem-solving tips and tools |

The volume of a sphere of radius r is $\frac{4}{3}\pi r^3$, and the volume of a disk of radius r and thickness t is $\pi r^2 t$. The average density of an object is its mass divided by its volume. To calculate escape speeds, you will need to review Box 7-2; to compute the mass of Mars or 70 Virginis, you may have to review Box 4-4. Be sure to use the same system of units (meters, seconds, kilograms) in all your calculations involving escape speeds, orbital speeds, and masses. Appendix 6 gives conversion factors between different sets of units. Box 5-1 has formulas relating various temperature scales, and Box 1-1 explains the relationship between the angular size of an object and its actual size.

19. Mars has two small satellites, Phobos and Deimos. Phobos circles Mars once every 0.31891 day at an average altitude of 5980 km above the planet's surface. The diameter of Mars is 6794 km. Using this information, calculate the mass and average density of Mars.

20. Figure 7-3 shows the spectrum of Saturn's largest satellite, Titan. Can you think of a way that astronomers can tell which absorption lines are due to Titan's atmosphere and which are due to the atmospheres of the Sun and Earth?

21. Table 7-3 shows that carbon, nitrogen, and oxygen are the most abundant elements (after hydrogen and helium). In our solar system, the atoms of these elements are found primarily in the molecules CH_4 (methane), NH_3 (ammonia), and H_2O (water). Why do you suppose this is?

*22. A hypothetical spherical asteroid 2 km in diameter, and composed of rock with an average density of 2500 kg/m³, strikes the Earth with a speed of 25 km/s. (a) What is the kinetic energy of the asteroid at the moment of impact? (b) How does this energy compare with that released by a 20-kiloton nuclear weapon, like the device that destroyed Hiroshima, Japan, on August 6, 1945? (*Hint*: 1 kiloton of TNT releases 4.2×10^{12} joules of energy.)

*23. (a) Find the speed required to escape from the surface of the asteroid described in Question 22. (b) A typical jogging speed is 3 m/s. What would happen to an astronaut who decided to go for a jog on this asteroid?

*24. A hydrogen atom has a mass of 1.673×10^{-27} kg, and the temperature of the Sun's surface is 5800 K. What is the average speed of hydrogen atoms at the Sun's surface?

*25. The Sun's mass is 1.989×10^{30} kg, and its radius is 6.96×10^8 m. (a) What is the escape speed from the Sun's surface? (b) Using your answer to Question 24, explain why the Sun has lost very little hydrogen over its entire 4.6-billion-year history.

*26. Suppose a spacecraft landed on Jupiter's moon Europa (see Table 7-2), which moves around Jupiter in an orbit of radius 670,900 km. After collecting samples from the satellite's surface, the spacecraft prepares to return to Earth. (a) Calculate the escape speed from Europa. (b) Calculate the escape speed from Jupiter at the distance of Europa's orbit. (c) In order to begin its homeward journey, the spacecraft must leave Europa with a speed greater than either your answer to (a) or your answer to (b). Explain why.

27. (a) If the Earth had retained hydrogen and helium in the same proportion to the heavier elements that exist elsewhere in the universe, what would its mass be? Give your answer as a multiple of the Earth's actual mass, given in Table 7-1. Explain your reasoning. (b) How does your answer to (a) compare with the masses of the Jovian planets? What does your answer imply about how large the cores of the Jovian planets must be?

28. Imagine a planet with roughly the same mass as Earth but located 50 AU from the Sun. (a) What do you think this planet would be made of? Explain your reasoning. (b) On the basis of this speculation, assume a reasonable density for this planet and calculate its diameter. How many times bigger or smaller than Earth would it be?

29. The distance from the asteroid 433 Eros (Figure 7-7) to the Sun varies between 1.13 and 1.78 AU. (a) Find the period of Eros's orbit. (b) Does Eros lie in the asteroid belt? How can you tell?

30. Suppose you were to use the Hubble Space Telescope to monitor one of the protoplanetary disks shown in Figure 7-13*b*. Over the course of ten years, would you expect to see planets forming within the disk? Why or why not?

31. The protoplanetary disk, or proplyd, at the upper right of Figure 7-13*b* is seen edge-on. The diameter of the disk is about 700 AU. (a) Make measurements on this image to

determine the thickness of the disk in AU. (b) Explain why the disk will continue to flatten as time goes by.

32. The accompanying infrared image shows IRAS 04302+2247, a young star that is still surrounded by a disk of gas and dust. The scale bar at the lower right of the image shows that at the distance of IRAS 04302+2247, an angular size of 2 arcseconds corresponds to a linear size of 280 AU. Find the distance to IRAS 04302+2247.

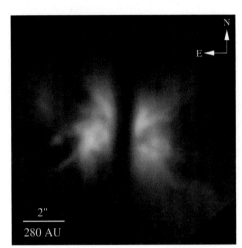

2"

280 AU

(Courtesy of D. Padgett; W. Brandner,   R **I** V U X G
IPAC/Caltech; K. Stapelfeldt, JPL; and NASA)

33. The image accompanying Question 32 shows a dark, opaque disk of material surrounding the young star IRAS 04302+2247. The disk is edge-on to our line of sight, so it appears as a dark band running vertically across this image (north to south in the sky). The material to the left and right of this band is still falling onto the disk. (a) Make measurements on this image to determine the diameter of the disk in AU. Use the scale bar at the lower right of this image. (b) If the thickness of the disk is 50 AU, find its volume in cubic meters. (c) The total mass of the disk is perhaps 2×10^{28} kg (0.01 of the mass of the Sun). How many atoms are in the disk? Assume that the disk is all hydrogen. A single hydrogen atom has a mass of 1.673×10^{-27} kg. (d) Find the number of atoms per cubic meter in the disk. By comparison, the air that you breathe contains about 5.4×10^{25} atoms per cubic meter. Would you describe the disk material as thick or thin by Earth standards?

34. Discuss why the "planet" that orbits the star HD 114762 (see Figure 7-21) could possibly be a brown dwarf, while the planet that orbits 51 Pegasi probably is not.

*35. The planet discovered orbiting the star 70 Virginis, 59 light-years from Earth, moves in an orbit with semimajor axis 0.47 AU and eccentricity 0.40. The period of the orbit is 116.6 days. Find the mass of 70 Virginis. Compare your answer with the mass of the Sun.

*36. Because of the presence of Jupiter, the Sun moves in a small orbit of radius 742,000 km with a period of

11.86 years. (a) Calculate the Sun's orbital speed in meters per second. (b) An astronomer on a hypothetical planet orbiting the star Vega, 25 light-years from the Sun, wants to use the astrometric method to search for planets orbiting the Sun. What would be the angular diameter of the Sun's orbit as seen by this alien astronomer? Would the Sun's motion be discernible if the alien astronomer could measure positions to an accuracy of 0.001 arcsec? (c) Repeat part (b), but now let the astronomer be located on a hypothetical planet in the Pleiades star cluster, 360 light-years from the Sun. Would the Sun's motion be discernible to this astronomer?

DISCUSSION QUESTIONS

37. Propose an explanation why the Jovian planets are orbited by terrestrial-like satellites.

38. Suppose that a planetary system is now forming around some protostar in the sky. In what ways might this planetary system turn out to be similar to or different from our own solar system? Explain your reasoning.

39. Suppose astronomers discovered a planetary system in which the planets orbit a star along randomly inclined orbits. How might a theory for the formation of that planetary system differ from that for our own?

WEB/CD-ROM QUESTIONS

40. Search the World Wide Web for information about recent observations of protoplanetary disks. What insights have astronomers gained from these observations? Is there any evidence that planets have formed within these disks?

41. In 2000, extrasolar planets with masses comparable to that of Saturn were first detected around the stars HD 16141 (also called 79 Ceti) and HD 46375. (All extrasolar planets discovered prior to these are more massive than Jupiter.) Search the World Wide Web for information about these "lightweight" planets. Do these planets move around their stars in the same kind of orbit as Saturn follows around the Sun? Why do you suppose this is? How does the discovery of these planets reinforce the model of planet formation described in Section 7-8?

42. As of this writing, not even the most powerful telescopes have been able to actually *see* planets orbiting other stars. Nonetheless, astronomers have evidence of having "seen" an extrasolar planet come between its star and us. Search the World Wide Web for information about this planet, which orbits the star HD 209458. If the planet is too small and faint to be seen, how can astronomers tell that it came between HD 209458 and us? What have they learned about the planet from this event that cannot be learned with the radial velocity method?

43. Determining Terrestrial Planet Orbital Periods. Access the animation "Planetary Orbits" in Chapter 6 of the *Universe* web site or CD-ROM. Focus on the motions of the inner planets at the last half of the animation. Using the stop and start buttons, determine how many days it takes Mars, Venus, and Mercury to orbit the Sun once if Earth takes approximately 365 days.

OBSERVING PROJECTS

44. There are many young stars still embedded in the clouds of gas and dust from which they formed. In the winter evening sky, for example, is the famous Orion Nebula (Figure 7-13*a*). In the summer night sky, the Lagoon, Omega, and Trifid nebulae are found in the Milky Way. Examine some of these nebulae with a telescope. Describe their appearance. Can you guess which stars in your field of view are actually associated with the nebulosity? You can use the *Starry Night* program that comes with this book to help you find some of these nebulae. The table below gives their coordinates (right ascension and declination; see Box 2-1) for the year 2000.

Nebula	Right ascension	Declination
Lagoon	18ʰ 03.8ᵐ	−24° 23′
Omega	18 20.8	−16 11
Trifid	18 02.3	−23 02
Orion	5 35.4	−5 27

45. Use the *Starry Night* program to observe magnified images of at least four of the planets. First turn off daylight (select **Daylight** in the **Sky Display** menu) and show the entire celestial sphere (select **Atlas** in the **Go** menu). Select **Planet List** in the **Window** menu, and then double-click on the name of the planet you wish to view. Use the controls at the right-hand end of the Control Panel to zoom in until you can see a detailed view of the planet. Describe each planet's appearance. From what you observe, is there any way of knowing whether you are looking at a planet's surface or simply a cloud cover?

GEOFF MARCY

Alien Planets

When Geoff Marcy sat down to write this essay, 50 planets were known to orbit other stars. He leads the team that has found 32 of them. Dr. Marcy became interested in astronomy at age 14, when his parents bought him a small reflecting telescope. Since receiving his Ph.D. from the University of California, Santa Cruz, he has studied stars similar to our Sun. He helped show that magnetic regions on their surfaces cause dark "star spots" and stellar flares. He also showed that brown dwarfs—stars too small to burn hydrogen—rarely orbit other stars. His work may soon reveal whether a planetary system like our own is common or a quirk of the cosmos. Dr. Marcy is an astronomer at the University of California, Berkeley.

Astronomers have now surveyed 1000 nearby stars in the search for orbiting planets. My team is using the 10-meter Keck telescope in Hawaii to measure the Doppler effect in stars that wobble because of planets orbiting around them. So far, we and other teams have found 50 planets, some more massive than Jupiter and some probably less massive than Saturn.

Two systems of multiple planets orbiting a single star have been found, and many more such planetary systems will be discovered in the next few years. We hope to compare the architecture of these planetary systems to that of our own solar system. Is our solar system a common, garden-variety type? Do all planetary systems have giant planets orbiting far away from the host star, with small, rocky planets orbiting close in?

Remarkably, the planetary systems observed so far seem very different from our own solar system. In some, a giant planet orbits very close to the star. More surprising is that almost all alien planets that orbit beyond 0.1 AU from their star reside in highly elliptical orbits that resemble eccentric comet orbits. Why are these planets in such elliptical orbits, so different from the nearly circular orbits of planets in our own solar system? Astronomers are baffled by this difference.

One possibility is that planets usually form as close-knit families around a star. Over time, the planets may gravitationally pull on one another, yanking themselves out of their original circular orbits. These perturbed planets may then venture near other planets, yanking on them, too. Soon, all the planets may pass close to other planets, gravitationally slinging some of them completely out of the planetary system. These far-flung planets are destined to roam the darkness of galactic space, cooling to frigid temperatures that are inhospitable to life.

The extrasolar planets cast a mystery back on the Earth. Why is our solar system immune to this chaotic episode that scatters planets into wacky orbits? Imagine the disaster if our Earth were suddenly thrust into a highly elliptical orbit. During half the year, we would be roasting too close to the Sun, with oceans vaporized into an enormous steam bath. The other half of the year our orbit would carry us too far from the Sun, causing all the oceans to freeze over. Life on Earth would be severely challenged to survive in such an elongated orbit.

Why are we humans so lucky as to live in a stable planetary system in which circular orbits ensure that we receive the same warming light intensity from our Sun, all year round? Perhaps the mystery is explained by our very existence. After all, intelligent life might not have evolved here if the Earth suffered from wild temperature swings that would make liquid water rare. We humans, and intelligent species in general, may flourish mostly on worlds that maintain nearly constant temperature. If so, our solar system and the Earth might be a bizarre, rare galactic quirk that just happens to be suitable for life. We wouldn't be here otherwise. If this picture is correct, our Earth is indeed a precious oasis in our Milky Way Galaxy.

In the next decade, we will discover more planetary systems by using the Doppler method. We are working night and day to find Jovian planets that orbit 5 AU from their star, similar to the orbital radius of our Jupiter. Their orbits, circular or elliptical, will shed light on the most intriguing questions of our time: How common are configurations like our Solar System, and how common is advanced life in the universe?

Our Living Earth

 (GSFC/NASA)

R I V U X G

When astronauts first journeyed from the Earth to the Moon between 1968 and 1972, they often reported that our planet is the most beautiful sight visible from space. Indeed, no other world in the solar system has the Earth's contrast between blue oceans and white clouds. No other world but Earth has the inviting green color of vegetation. Compared with the Moon, whose surface has been ravaged by billions of years of impacts by interplanetary debris, the Earth seems an inviting and tranquil place.

Yet the appearance of tranquility is deceiving. The seemingly solid surface of the planet is in a state of slow but constant motion, driven by the flow of molten rock within the Earth's interior. Sunlight provides the Earth with the energy to create huge storms, like the hurricane off the western coast of Mexico apparent in this image.

Unseen in this image are immense clouds of subatomic particles that wander around the outside of the planet in two giant belts, held in thrall by the Earth's magnetic field. And equally unseen are trace gases in the Earth's atmosphere whose abundance may determine the future of our climate, and on which may depend the survival of entire species.

In this chapter our goal is to learn about the Earth's components—its dynamic oceans and atmosphere, its ever-changing surface, and its hot, active interior—and how they interact to make up our planetary home. In later chapters we will use this knowledge as a point of reference for studying the Moon and the other planets.

As you read the sections of this chapter, look for the answers to the following questions.

8-1 What is the greenhouse effect? How does it affect the average temperature of the Earth?

8-2 Is the Earth completely solid inside? How can scientists tell?

8-3 How is it possible for entire continents to move across the face of the Earth?

8-4 Where does the Earth's magnetic field come from?

8-5 Why is Earth the only planet with an oxygen-rich atmosphere?

8-6 What are global warming and the "ozone hole"? Why should they concern us?

8-1 The Earth's atmosphere, oceans, and surface are extraordinarily active

 The crew of an alien spacecraft exploring our solar system might be tempted to overlook the Earth altogether. Although the largest of the terrestrial planets, with a mass greater than that of Mercury, Venus, and Mars put together, Earth is far smaller than any of the giant Jovian planets (see Table 7-1 and Table 8-1). But a closer inspection would reveal the Earth to be unique among all the planets that orbit the Sun.

Unlike the arid surfaces of Venus and Mars, the Earth is very wet. Indeed, nearly 71% of the Earth's surface is covered with water (Figure 8-1). Water is also locked into the chemical structure of many Earth rocks, including those found in the driest deserts. Furthermore, the Earth's liquid water is in constant motion. The oceans ebb and flow with the tide, streams and rivers flow down to the sea, and storms whip lake waters into a frenzy.

The Earth's atmosphere is also in a state of perpetual activity. Winds blow at all altitudes, with speeds and directions that change from one hour to the next. The most dynamic activity occurs when water evaporates into the atmosphere to form clouds, then returns to the surface as precipitation (Figure 8-2). While other worlds of the solar system have atmospheres, only the Earth's contains the oxygen that animals (including humans) need to sustain life.

 The combined effects of water and wind cause erosion of mountains and beaches. But these are by no means the only forces reshaping the face of our planet. All across the Earth we find evidence that the surface has been twisted, deformed, and folded (Figure 8-3). What is more, new material is continually being added to the Earth's surface as lava pours forth from volcanoes and from immense cracks in the ocean floors. These processes and others work together to renew and refresh our planet's exterior. Thus, although the Earth is some 4.6 billion (4.6×10^9) years old, much of its surface is less than 100 million (10^8) years old.

What powers all this activity in the Earth's oceans, atmosphere, and surface? There are three energy sources: the Sun, the tidal effects of the Moon, and the Earth's own internal heat.

The Sun is the principal source of energy for the atmosphere. The Earth's surface is warmed by sunlight, which in turn warms the air next to the surface. Hot air is less dense than cool air and so tends to rise (see the discussion of density in Box 7-1). As the air rises, it transfers heat to its surroundings. As a result, the rising air cools down and becomes denser. It then sinks downward to be heated again, and the process starts over. This up-and-down motion is called **convection,** and the overall pattern of circulation is called a **convection current.** (You can see convection currents in action by heating water on a stove, as Figure 8-4 shows.) In Section 8-5 we will see how the Earth's rotation modifies the up-and-down motion of the atmosphere to produce a pattern of global circulation.

Solar energy also powers atmospheric activity by evaporating water from the surface. The energy in the water vapor is released when it condenses to form water droplets, like those that make up clouds. In a typical thunderstorm (see Figure 8-2), some 5×10^8 kg of water vapor is lifted to great heights. The amount of energy released when this water condenses is as much as a city of 100,000 people uses in a month!

One complication with solar energy is that the Earth does not absorb all the radiation that it receives from the Sun. About 39% of it is reflected back into space by clouds, snow, ice, and sand. The fraction of incoming sunlight that a planet reflects is called its **albedo** (from the Latin for "whiteness"); thus the Earth's albedo is about 0.39.

Another important effect is that the Earth *emits* radiation into space because of its temperature, in accordance with the laws that describe heated dense objects (Section 5-3). Wien's law tells us that the wavelength at which such an object emits most strongly (λ_{max}) is inversely proportional to its temperature (T) on the Kelvin scale (see Section 5-4 and Box 5-2). For example, the Sun's surface temperature is about 5800 K, and sunlight has its greatest intensity at a wavelength λ_{max} of 500 nm, in the middle of the visible spectrum. The Earth's average surface temperature is far lower than the Sun's, only about 280 K (9°C), so it emits most strongly at longer wave-

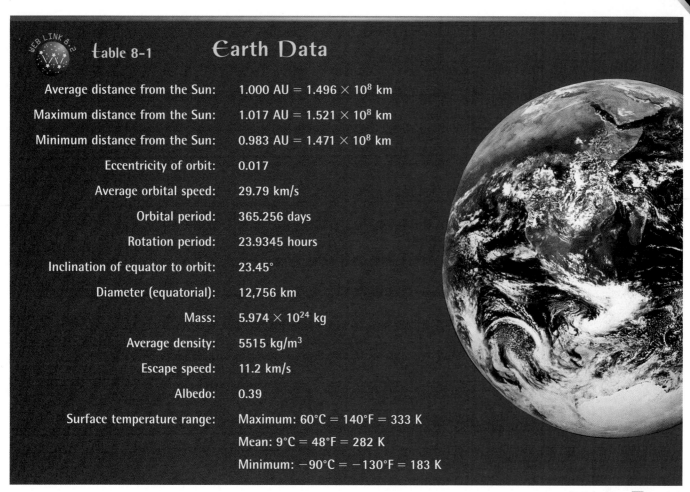

WEB LINK 8.2

table 8-1   Earth Data

Average distance from the Sun:	$1.000 \text{ AU} = 1.496 \times 10^8$ km
Maximum distance from the Sun:	$1.017 \text{ AU} = 1.521 \times 10^8$ km
Minimum distance from the Sun:	$0.983 \text{ AU} = 1.471 \times 10^8$ km
Eccentricity of orbit:	0.017
Average orbital speed:	29.79 km/s
Orbital period:	365.256 days
Rotation period:	23.9345 hours
Inclination of equator to orbit:	23.45°
Diameter (equatorial):	12,756 km
Mass:	5.974×10^{24} kg
Average density:	5515 kg/m³
Escape speed:	11.2 km/s
Albedo:	0.39
Surface temperature range:	Maximum: 60°C = 140°F = 333 K
	Mean: 9°C = 48°F = 282 K
	Minimum: −90°C = −130°F = 183 K

(NASA)   R I **V** U X G

lengths in the infrared portion of the electromagnetic spectrum. Temperature also determines the amount of radiation that the Earth emits: The higher the temperature, the more energy it radiates.

The Earth's average surface temperature is nearly constant, which means that on the whole it is neither gaining nor losing energy. Thus, the rate at which the Earth emits energy must equal the rate at which it receives energy from the Sun (not including the solar energy that is reflected back into space). Since the rate of emission depends on temperature, we can use this condition to calculate what the Earth's average surface temperature should be. The result is a very

figure 8-1   R I **V** U X G

The Earth's Dynamic Oceans Nearly three-quarters of the Earth's surface is covered with water. In contrast, there is no liquid water at all on Mercury, Venus, Mars, or the Moon. (Farley Lewis/Photo Researchers)

Our Living Earth | 181

figure 8-2 R I ∨ U X G

The Earth's Dynamic Atmosphere This space shuttle image shows thunderstorm clouds over the African nation of Zaire. The tops of these clouds frequently reach altitudes of 10,000 m (33,000 ft) or higher. At any given time, nearly 2000 thunderstorms are in progress over the Earth's surface. (JSC/NASA)

figure 8-3 R I ∨ U X G

The Earth's Dynamic Surface These tan-colored ridges, called hogbacks, were once layers of sediment at the bottom of an ancient body of water. Forces within the Earth folded this terrain and rotated the layers into a vertical orientation. The layers were revealed when wind and rain eroded away the surrounding material. These hogback ridges are in the Rocky Mountains of Colorado. (Tom Till/DRK)

chilly 246 K ($-27°C = -17°F$). Yet the Earth's actual average surface temperature is 282 K ($9°C = 48°F$)! What is wrong with our model? Why is the Earth warmer than we would expect?

The explanation for this discrepancy is called the **greenhouse effect:** The Earth's atmosphere prevents some of the Earth's surface radiation from escaping into space. Certain gases in our atmosphere called **greenhouse gases,** among them water vapor and carbon dioxide, are transparent to visible light but not to infrared radiation. Consequently, visible sunlight has no trouble entering our atmosphere and warming the surface. But the infrared radiation coming from the heated surface is partially trapped by the atmosphere, thus raising the temperatures of both the atmosphere and the surface. As the surface and atmosphere become hotter, they both emit more infrared radiation, part of which is able to escape into space. The temperature levels off when the amount of infrared energy that escapes just balances the amount of solar energy reaching the surface (Figure 8-5). The net result is that our atmosphere is some 36°C (65°F) warmer than it would be without the greenhouse effect.

⚠️*CAUTION!* Despite its name, the greenhouse effect is *not* how a greenhouse—a building with glass walls and roof and used for growing plants—stays warm inside. Plants that grow outside are heated by the Sun, and in turn heat the air

Hot, less dense water rises

Water cools, becomes more dense, sinks

figure 8-4

Convection in the Kitchen Water being heated in a pot moves up and down in convection currents. The flames heat the water at the bottom of the pot, making it expand and lowering its average density. This low-density water rises and transfers heat to its cooler surroundings. The water that began at the bottom thus cools down, becomes denser, and sinks back to the bottom to repeat the process.

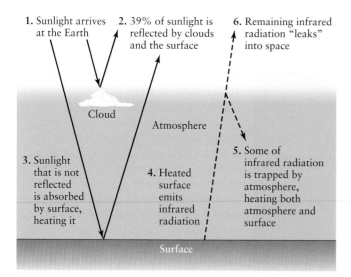

figure 8-5

The Greenhouse Effect A portion of the sunlight reaching the Earth penetrates through the atmosphere and heats the planet's surface. The heated surface emits infrared radiation, much of which is absorbed by water vapor and carbon dioxide. The trapped radiation helps raise the average temperatures of the surface and atmosphere. Some infrared radiation does penetrate the atmosphere and leaks into space. In a state of equilibrium, the rate at which the Earth loses energy to space in this way is equal to the rate at which it absorbs energy from the Sun.

1. Sunlight arrives at the Earth

2. 39% of sunlight is reflected by clouds and the surface

6. Remaining infrared radiation "leaks" into space

Cloud

Atmosphere

3. Sunlight that is not reflected is absorbed by surface, heating it

4. Heated surface emits infrared radiation

5. Some of infrared radiation is trapped by atmosphere, heating both atmosphere and surface

Surface

around them. This hot air rises and transfers its heat to higher-altitude air by convection, thus keeping the plants from getting very warm. Plants in a greenhouse are also heated by the Sun, but the roof keeps the warm air around the plants from rising very far. Hence, the heat is trapped and the plants stay quite warm. The same effect explains why the inside of a car with the windows rolled up gets uncomfortably hot on a sunny day. If you roll the windows down by just a few centimeters, hot air can escape and the inside feels noticeably cooler.

Solar energy also helps to power the oceans. Warm water from near the equator moves toward the poles, while cold polar water returns toward the equator. As we saw in Section 4-8, however, the back-and-forth motion of the tides is due to the tidal forces of the Moon and Sun. Sometimes these two influences can reinforce each other, as when a storm (caused by solar energy) reaches a coastline at high tide (caused by the tidal forces), producing waves strong enough to seriously erode beaches and sea cliffs.

Neither solar energy nor tidal forces can explain the re-shaping of the Earth's surface suggested by Figure 8-3. Rather, all this geologic activity is powered by heat flowing from the interior of the Earth itself. The Earth formed by collisions among planetesimals (see Section 7-8), and these collisions heated the body of the Earth. The interior remains hot to this day. An additional source of heat is the decay of radioactive

elements such as uranium, thorium, and potassium deep inside the Earth

The heat flow from the Earth's interior to its surface is miniscule—just 1/20,000 as great as the flow of energy we receive from the Sun—but it has a profound effect on the face of our planet. To understand the connection between geologic activity on the Earth's surface and heat coming from its interior, we must first learn something about the interior structure of our dynamic Earth.

8-2 Studies of earthquakes reveal the Earth's layered interior structure

The kinds of rocks found on and near the Earth's surface provide an important clue about our planet's interior. The densities of typical surface rocks are around 3000 kg/m^3, but the average density of the Earth as a whole (that is, its mass divided by its volume) is 5515 kg/m^3. The interior of the Earth must therefore be composed of a substance much denser than the crust. But what is this substance? How did it come to be that the Earth's interior is more dense than its crust? And is the Earth's interior solid like the crust or molten like lava?

Iron (chemical symbol Fe) is a good candidate for the substance that makes up most of the Earth's interior. This is so for two reasons. First, iron atoms are quite massive (a typical iron atom has 56 times the mass of a hydrogen atom), and second, iron is rather abundant. (Table 7-3 shows that it is the tenth most abundant element in the universe.) Other elements such as lead and uranium have more massive atoms, but these elements are quite rare. Hence, the solar nebula could not have had enough of these massive atoms to create the Earth's dense interior. Furthermore, iron is common in meteoroids that strike the Earth, which suggests that it was abundant in the planetesimals from which the Earth formed.

The Earth was almost certainly molten throughout its volume soon after its formation, about 4.6×10^9 years ago. Energy released by the violent impacts of numerous meteoroids and asteroids and by the decay of radioactive isotopes likely melted the solid material collected from the earlier planetesimals. Gravity caused abundant, dense iron to sink toward the Earth's center, forcing less dense material to the surface. Figure 8-6a shows this process of *chemical differentiation*. (We discussed chemical differentiation in Section 7-8.) The result was a planet with the layered structure shown in Figure 8-6b—a central **core** composed of almost pure iron, surrounded by a **mantle** of dense, iron-rich minerals. The mantle, in turn, is surrounded by a thin **crust** of relatively light silicon-rich minerals. We live on the surface of this crust.

How do we know that this layered structure is correct? The challenge is that the Earth's interior is as inaccessible as the most distant galaxies in space. The deepest wells go down only a few kilometers, barely penetrating the surface of our planet. Despite these difficulties, geologists have learned basic properties of the Earth's interior by studying earthquakes and the seismic waves that they produce.

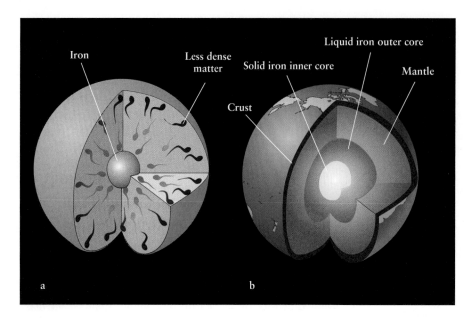

figure 8-6

Chemical Differentiation and the Earth's Internal Structure **(a)** When the Earth first formed, it was probably molten throughout its volume. Dense materials such as iron (shown in orange) sank toward the center, while low-density materials (shown in blue) rose toward the surface. **(b)** The present-day Earth is no longer completely molten inside. A dense, solid iron core is surrounded by a less dense liquid core and an even less dense mantle. The crust, which includes the continents and ocean floors, is the least dense of all; it floats atop the mantle, like the skin that forms on the surface of a cup of cocoa as it cools.

Over the centuries, stresses build up in the Earth's crust. Occasionally, these stresses are relieved with a sudden motion called an **earthquake.** Most earthquakes occur deep within the Earth's crust. The point on the Earth's surface directly over an earthquake's location is called the **epicenter.**

Earthquakes produce three different kinds of **seismic waves,** which travel around or through the Earth in different ways and at different speeds. Geologists use sensitive instruments called **seismographs** to detect and record these vibratory motions. The first type of wave, which is analogous to water waves on the ocean, causes the rolling motion that people feel around an epicenter. These are called **surface waves** because they travel only over the Earth's surface. The two other kinds of waves, called **P waves** (for "primary") and **S waves** (for

"secondary"), travel through the interior of the Earth. P waves are called *longitudinal* waves because their oscillations are parallel to the direction of wave motion, like a spring that is alternately pushed and pulled. In contrast, S waves are called *transverse* waves because their vibrations are perpendicular to the direction in which the waves move. S waves are analogous to waves produced by a person shaking a rope up and down (Figure 8-7).

What makes seismic waves useful for learning about the Earth's interior is that they do not travel in straight lines. Instead, the paths that they follow through the body of the Earth are bent because of the varying density and composition of the Earth's interior. We saw in Section 6-1 that light waves behave in a very similar way. Just as light waves bend, or refract, when they pass from air into glass or vice versa

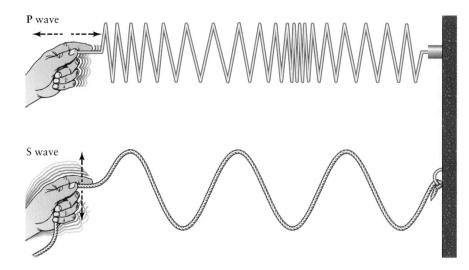

figure 8-7

Seismic Waves Earthquakes produce two kinds of waves that travel through the body of our planet. One kind, called P waves, are longitudinal waves. They are analogous to those produced by pushing a spring in and out. The other kind, S waves, are transverse waves analogous to the waves produced by shaking a rope up and down.

([Figure 6-2](#)), seismic waves refract as they pass through different parts of the Earth's interior. By studying how the paths of these waves bend, geologists can map out the general interior structure of the Earth.

One key observation about seismic waves and how they bend has to do with the differences between S and P waves. When an earthquake occurs, seismographs relatively close to the epicenter record both S and P waves, but those on the opposite side of Earth record only P waves. The absence of S waves was first explained in 1906 by British geologist Richard Dixon Oldham, who noted that transverse vibrations such as S waves cannot travel far through liquids. Oldham therefore concluded that our planet has a molten core. Furthermore, there is a region in which neither S waves nor P waves from an earthquake can be detected (Figure 8-8). This "shadow zone" results from the specific way in which P waves are refracted at the boundary between the solid mantle and the molten core. By measuring the size of the shadow zone, geologists have concluded that the radius of the molten core is about 3500 km (2200 mi), about 55% of our planet's overall radius but about double the radius of the Moon (1738 km = 1080 mi).

As the quality and sensitivity of seismographs improved, geologists discovered faint traces of P waves in an earthquake's shadow zone. In 1936, the Danish seismologist Inge Lehmann explained that some of the P waves passing through the Earth are deflected into the shadow zone by a small, solid **inner core** at the center of our planet. The radius of this inner core is about 1300 km (800 mi).

The interior of our planet therefore has a curious structure—a liquid **outer core** sandwiched between a solid inner core and a solid mantle. Table 8-2 summarizes this structure. To understand this arrangement, we must look at how temperature and pressure inside the Earth affect the melting point of rock.

Both temperature and pressure increase with increasing depth below the Earth's surface. The temperature of the Earth's interior rises steadily from about 9°C on the surface to nearly 5000°C at our planet's center (Figure 8-9).

The Earth's outermost layer, the crust, is only about 5 to 35 km thick. It is composed of rocks for which the **melting point,** or temperature at which the rock changes from solid to liquid, is far higher than the temperatures actually found in the crust. Thus, the crust is solid.

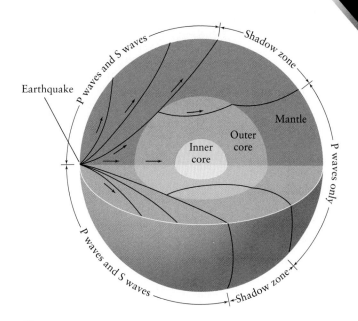

figure 8-8

The Earth's Internal Structure and the Paths of Seismic Waves Seismic waves follow curved paths because of differences in the density and composition of the material in the Earth's interior. For the most part, the paths curve only gradually, because the density and composition change only gradually. Sharp bends occur only where there is an abrupt change from one kind of material to another, such as at the boundary between the outer core and the mantle. Only P waves can pass through the Earth's liquid outer core.

The Earth's mantle, which extends to a depth of about 2900 km (1800 mi), is largely composed of substances rich in iron and magnesium. On the Earth's surface, specimens of these substances have melting points slightly over 1000°C. However, the melting point of a substance depends on the pressure to which it is subjected—the higher the pressure, the higher the melting point. As Figure 8-9 shows, the actual temperatures throughout the mantle are less than the melting point of all the substances of which the mantle is made. Hence, the mantle is primarily solid. The upper levels of the

table 8-2	The Earth's Internal Structure		
Region	Depth below surface (km)	Distance from center (km)	Average density (kg/m³)
Crust (solid)	0–5 (under oceans) 0–35 (under continents)	6343–6378	3500
Mantle (solid)	from bottom of crust to 2900	3500–6343	3500–5500
Outer core (liquid)	2900–5100	1300–3500	10,000–12,000
Inner core (solid)	5100–6400	0–1300	13,000

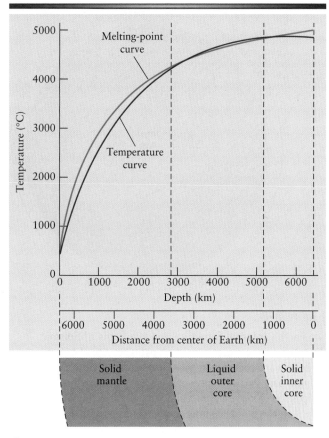

Figure 8-9

Temperature and Melting Point of Rock Inside the Earth The temperature (red curve) rises steadily from the Earth's surface to its center. The melting point of the Earth's material (blue curve) is also shown on this graph. Where the temperature is below the melting point, as in the mantle and inner core, the material is solid; where the temperature is above the melting point, as in the outer core, the material is liquid. (Adapted from F. Press and R. Siever)

mantle are, however, able to flow slowly and are therefore referred to as being **plastic.**

It may seem strange to think of a plastic material as one that is able to flow. Normally we think of plastic objects as being hard and solid, like the plastic out of which audio and video cassettes are made. But to make these objects, the plastic material is heated so that it can flow into a mold, then cooled so that it solidifies. Strictly, the material is only "plastic" when it is able to flow.

At the boundary between the mantle and the outer core, there is an abrupt change in chemical composition, from iron-rich materials to almost-pure iron with a small admixture of nickel. Because this nearly pure iron has a lower melting point than the iron-rich materials, the melting-point curve

in Figure 8-9 dips below the temperature curve as it crosses from the mantle to the outer core, and remains below it down to a depth of about 5100 km (3200 mi). Hence, from depths of about 2900 to 5100 km, the core is liquid.

At depths greater than about 5100 km, the pressure is more than 10^{11} newtons per square meter. This is about 10^6 times ordinary atmospheric pressure, or about 10^4 tons per square inch. Because the melting point of the iron-nickel mixture under this pressure is higher than the actual temperature (see Figure 8-9), the Earth's inner core is solid.

Until the 1980s, geologists knew little more about the inner core than that it is solid and dense. Since then, evidence has accumulated that suggests the inner core is a single **crystal** of iron—that is, iron atoms arranged in orderly rows, like carbon atoms are arranged within a diamond. If so, the inner core is the largest crystalline object anywhere in the solar system. Furthermore, there are strong indications that the inner core is rotating at a slightly faster rate than the rest of the Earth! These remarkable discoveries suggest that even more surprises may lurk deep within the Earth's interior.

Heat naturally flows from where the temperature is high to where it is low. Figure 8-9 thus explains why there is a heat flow from the center of the Earth outward. We will see in the next section how this heat flow acts as the "engine" that powers our planet's geologic activity.

8-3 Plate tectonics produces earthquakes, mountain ranges, and volcanoes that shape the Earth's surface

One of the most important geological discoveries of the twentieth century was the realization that the Earth's crust is constantly changing. We have learned that the crust is divided into huge **plates** whose motions produce earthquakes, volcanoes, mountain ranges, and oceanic trenches. This picture of the Earth's crust has come to be the central unifying theory of geology, much as the theory of evolution has become the centerpiece of modern biology.

Anyone who carefully examines a map of the Earth might come up with the idea of moving continents. South America, for example, would fit snugly against Africa were it not for the Atlantic Ocean. As Figure 8-10 shows, the fit between land masses on either side of the Atlantic Ocean is quite remarkable. This observation inspired the German meteorologist Alfred Wegener to advocate "continental drift"—the idea that the continents on either side of the Atlantic Ocean have simply drifted apart. After much research, in 1915 Wegener published the theory that there had originally been a single gigantic supercontinent, which he called Pangaea (meaning "all lands"), that began to break up and drift apart some 200 million years ago. Other geologists refined this theory, arguing that Pangaea must have first split into two smaller supercontinents, which they called Laurasia and Gondwanaland, separated by what they called the Tethys Sea. Gondwanaland later split into Africa and

South America, with Laurasia dividing to become North America and Eurasia. According to this theory, the Mediterranean Sea is a surviving remnant of the ancient Tethys Sea (Figure 8-11).

Initially, most geologists ridiculed Wegener's ideas. Although it was generally accepted that the continents do "float" on the denser, somewhat plastic mantle beneath them, few geologists could accept the idea that entire continents could move around the Earth at speeds as great as several centimeters per year. The "continental drifters" could not explain what forces could be shoving the massive continents around. Beginning in the mid-1950s, however, geologists found evidence that material is being forced upward to the crust from deep within the Earth.

Bruce C. Heezen of Columbia University and his colleagues began discovering long mountain ranges on the ocean floors, such as the Mid-Atlantic Ridge, which stretches all the way from Iceland to Antarctica (Figure 8-12). During the 1960s, Harry Hess of Princeton University and Robert Dietz of the University of California again carefully examined the floor of the Atlantic Ocean. They concluded that rock from the Earth's

a 200 million years ago

b 180 million years ago

c Today

Figure 8-11

The Breakup of the Supercontinent Pangaea **(a)** The shapes of the continents led Alfred Wegener to conclude that 200 million (2×10^8) years ago, the continents were merged into a single supercontinent, which he called Pangaea. **(b)** Pangaea first split into two smaller land masses, Laurasia and Gondwanaland. **(c)** Over millions of years, the continents moved to their present-day locations. This picture of continental drift, considered in Wegener's day to be a wild speculation, has been confirmed by a variety of evidence. For example, nearly identical rock formations are found in locations that today are thousands of kilometers apart but would have been side by side on Pangaea.

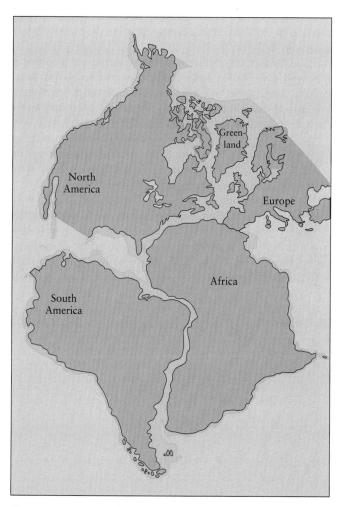

Figure 8-10

Fitting the Continents Together Africa, Europe, Greenland, and North and South America fit together remarkably well. The fit is especially convincing if the edges of the continental shelves (shown in yellow) are used, rather than today's shorelines. This strongly suggests that these continents were in fact joined together at some point in the past. (Adapted from P. M. Hurley)

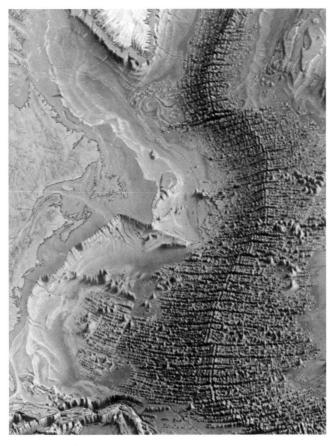

figure 8-12

The Mid-Atlantic Ridge This artist's rendition shows the floor of the North Atlantic Ocean. The immense mountain range in the middle of the ocean floor is called the Mid-Atlantic Ridge. It is caused by lava seeping up from the Earth's interior along a rift that extends from Iceland to Antarctica. (Courtesy of M. Tharp and B. C. Heezen)

mantle is being melted and then forced upward along the Mid-Atlantic Ridge, which is in essence a long chain of underwater volcanoes.

The upwelling of new material from the mantle to the crust forces the existing crusts apart, causing **seafloor spreading.** For example, the floor of the Atlantic Ocean to the east of the Mid-Atlantic Ridge is moving eastward and the floor to the west is moving westward. By explaining what fills in the gap between continents as they move apart, seafloor spreading helps to fill out the theories of continental drift. Because of the seafloor spreading from the Mid-Atlantic Ridge, South America and Africa are moving apart at a speed of roughly 3 cm per year. Working backward, these two continents would have been next to each other some 200 million years ago—just as Wegener suggested (see Figure 8-10).

In the early 1960s, geologists began to find additional evidence supporting the existence of large, moving plates. Thus was born the modern theory of crustal motion, which came to be known as **plate tectonics** (from the Greek *tekton*, meaning "builder").

Geologists today realize that earthquakes tend to occur at the boundaries of the Earth's crustal plates, where the plates are colliding, separating, or rubbing against each other. The boundaries of the plates therefore stand out clearly when the epicenters of earthquakes are plotted on a map (Figure 8-13).

The vast majority of volcanoes also occur at plate boundaries. Figure 8-14 shows a volcanic eruption at the boundary between the Pacific and Eurasian plates, near the upper left corner of Figure 8-13.

What makes the plates move? The answer is twofold. First, heat flows outward from the Earth's hot core to its cool crust, and second, this heat flows through the mantle by convection, the same process that takes place in the Earth's atmosphere or in a pot of boiling water (recall Figure 8-4).

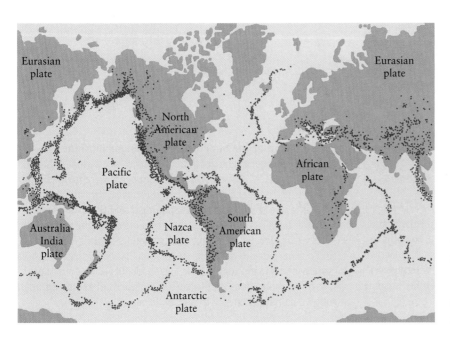

figure 8-13

The Earth's Major Plates The boundaries of the Earth's major plates are the scenes of violent seismic and geologic activity. Most earthquakes occur where plates separate, collide, or rub together. Plate boundaries are therefore easily identified simply by plotting earthquake epicenters (shown here as dots) on a map.

figure 8-14 R I **V** U X G

An Erupting Volcano On September 30, 1994, a volcano erupted at the boundary of the Pacific and Eurasian plates on Russia's Kamchatka Peninsula. This space shuttle photograph shows an immense cloud of ash belching forth from this volcano. The cloud reached an altitude of 18 km (60,000 ft) above sea level and was carried by the winds for at least 1000 km from the volcano. (*STS-68*, NASA)

The upper levels of the mantle that are hot and soft enough to permit an oozing, plastic flow are called the **asthenosphere** (from the Greek *asthenia*, meaning "weakness"). Atop the asthenosphere is a rigid layer, called the **lithosphere** (from the Greek word for "rock.") The lithosphere is divided into plates that ride along the convection currents of the asthenosphere. The crust is simply the uppermost layer of the lithosphere.

Figure 8-15 shows how convection causes plate movement. Molten subsurface rock seeps upward along **oceanic rifts**, where plates are separating. The Mid-Atlantic Ridge, shown in Figure 8-12, is an oceanic rift. Where plates collide, cool crustal material from one of the plates sinks back down into the mantle along a **subduction zone.** One such subduction zone is found along the west coast of South America, where the oceanic Nazca plate is being subducted into the mantle under the continental South American plate at a relatively speedy 10 centimeters per year. As the material from the subducted plate sinks, it pulls the rest of its plate along with it, thus helping to keep the plates in motion. New material is added to the crust from the mantle at the oceanic rifts and is "recycled" back into the mantle at the subduction zones. In this way the total amount of crust remains essentially the same.

The boundaries between plates are the sites of some of the most impressive geological activity on our planet. Great mountain ranges, such as the Sierras and Cascades along the western coast of North America and the Andes along South America's west coast, are thrust up by ongoing collisions between continental plates and the plates of the ocean floor. Subduction zones, where old crust is pushed back down into the mantle, are typically the locations of deep oceanic trenches, such as the Peru-Chile Trench off the west coast of

Rift   Ocean floor   Trench   Mountain range   Lithosphere   Continent

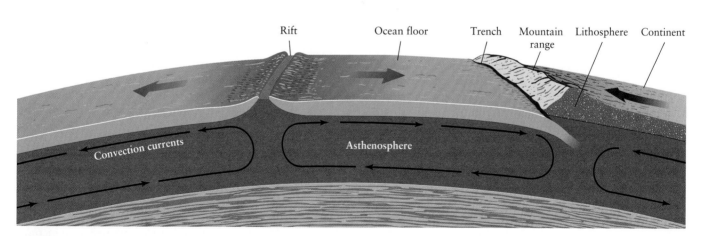

Convection currents   Asthenosphere

figure 8-15

The Mechanism of Plate Tectonics Convection currents in the soft upper layer of the mantle (called the asthenosphere) are responsible for pushing around rigid, low-density crustal plates. New crust forms in oceanic rifts, where lava oozes upward between separating plates. Mountain ranges and deep oceanic trenches are formed where plates collide.

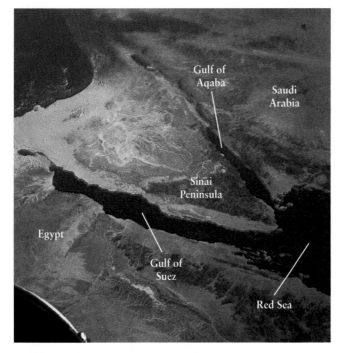

figure 8-16 R I **V** U X G

The Separation of Two Plates The plates that carry Africa and Arabia are moving apart, leaving a great rift that has been flooded to form the Red Sea. This view from orbit shows the northern Red Sea, which splits into the Gulf of Suez and the Gulf of Aqaba. (*Gemini 12*, NASA)

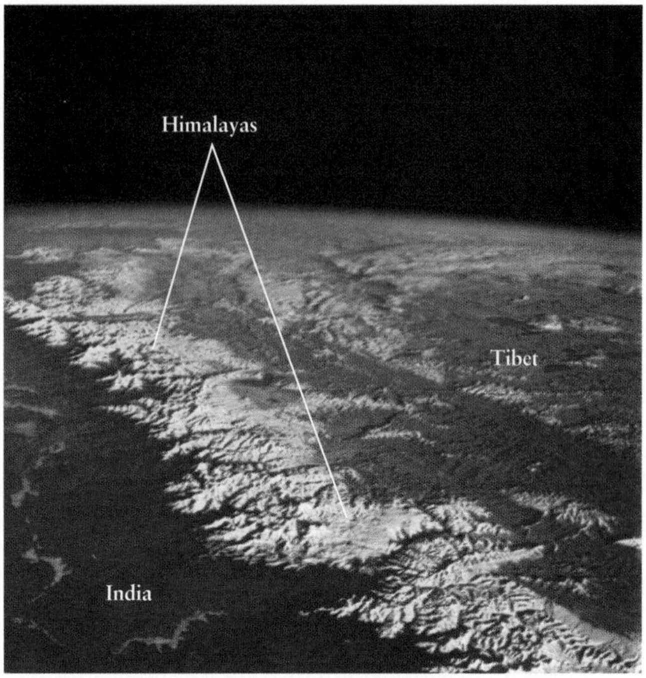

figure 8-17 R I **V** U X G

The Collision of Two Plates The plates that carry India and China are colliding. Instead of one plate being subducted beneath the other, both plates are pushed upward, forming the Himalayas. Mount Everest is one of the snow-covered peaks near the center of this photograph taken from orbit. (*Apollo 7*, NASA)

South America. Figures 8-16 and 8-17 show two well-known geographic features that resulted from tectonic activity at plate boundaries.

Plate tectonics also helps to explain geology on the scale of individual rocks and minerals. These are composed of chemical elements (see the periodic table in Box 5-5), but individual elements are rarely found in a pure state. The exceptions include diamonds, which are crystals of carbon, as well as nuggets of gold, silver, and copper. Somewhat more common are solids composed of a particular chemical combination of elements called a *compound*. An example is feldspar, a crystalline material made up of potassium, aluminum, silicon, and oxygen atoms. Diamond and feldspar are both **minerals**—naturally occurring solids composed of a single element or compound. Most common of all are **rocks**, which are solid parts of the Earth's crust that are composed of one or more minerals. For example, granite contains both feldspar and quartz (a mineral containing silicon and oxygen).

Geologic processes create three major categories of rocks. **Igneous rocks** result when minerals cool from a molten state. (Molten rock is called **magma** when it is buried below the surface and **lava** when it flows out upon the surface, as in a volcanic eruption.) The ocean floor is made predominantly of a type of igneous rock called basalt

(Figure 8-18a). This is just what we would expect if the ocean floor is produced by material welling up from the mantle, as predicted by the theory of seafloor spreading.

Sedimentary rocks are produced by the action of wind, water, or ice. For example, winds can pile up layer upon layer of sand grains. Other minerals present amid the sand can gradually cement the grains together to produce sandstone (Figure 8-18b). Minerals that precipitate out of the oceans can cover the ocean floor with layers of a sedimentary rock called limestone. The motion of tectonic plates can move such rocks to places far from where they were formed. This explains why sedimentary rock from the floor of an ancient ocean can now be found in central Alaska and as hogback ridges in Colorado (Figure 8-3).

Sometimes igneous or sedimentary rocks become buried far beneath the surface, where they are subjected to enormous pressure and high temperatures. These severe conditions change the structure of the rocks, producing **metamorphic rock** (Figure 8-18c). The presence of metamorphic rocks at the Earth's surface tells us that tectonic activity sometimes lifts up material from deep within the crust. This can happen when two plates collide, as shown in Figure 8-17. The boundaries of the colliding plates are thrust upward to form a new chain of mountains, and buried metamorphic rocks are brought to the surface.

a   b   c

ƒigure 8-18   R I **V** U X G

Igneous, Sedimentary, and Metamorphic Rocks **(a)** Igneous rocks are created when molten materials solidify. The example shown here is basalt, which is a fine-grained mixture of feldspar along with other, iron-rich minerals that give the rock its dark color. **(b)** Sedimentary rocks are typically formed when loose particles of soil or sand are fused into rock by the presence of other minerals, which act as a cement. Sandstone, shown here, is made primarily of grains of feldspar and quartz. **(c)** Metamorphic rocks are produced when igneous or sedimentary rocks are subjected to high temperatures and pressures deep within the Earth's crust. Marble (left) is formed from sedimentary limestone; schist (right) is formed from igneous rock. (W. J. Kaufmann III; specimens courtesy Mineral Museum, California Division of Mines)

Plate tectonic theory offers insight into geology on the largest of scales, that of an entire supercontinent. In recent years, geologists have uncovered evidence that points to a whole succession of supercontinents that once broke apart and then reassembled. Pangaea is only the most recent supercontinent in this cycle, which repeats about every 500 million (5×10^8) years. As a result, intense episodes of mountain building have occurred at roughly 500-million-year intervals.

Apparently, a supercontinent sows the seeds of its own destruction because it blocks the flow of heat from the Earth's interior. As soon as a supercontinent forms, temperatures beneath it rise, much as they do under a book lying on an electric blanket. As heat accumulates, the lithosphere domes upward and cracks. Molten rock from the overheated asthenosphere wells up to fill the resulting fractures, which continue to widen as pieces of the fragmenting supercontinent move apart.

It can take a very long time for the heat trapped under a supercontinent to escape. Although Pangaea broke apart some 200 million years ago (see Figure 8-11), the mantle under its former location is still hot and is still trying to rise upward. As a result, Africa—which lies close to the center of this mass of rising material—sits several tens of meters higher than the other continents.

The changes wrought by plate tectonics are very slow on the scale of a human lifetime, but they are very rapid in comparison with the age of the Earth. For example, the period over which Pangaea broke into Laurasia and Gondwanaland was only about 0.4% of the Earth's age of 4.6×10^9 years. (To put this in perspective, 0.4% of a human lifetime is about 4 months.) The lesson of plate tectonics is that the seemingly permanent face of the Earth is in fact dynamic and ever-changing.

8-4   The Earth's magnetic field produces a magnetosphere that traps particles from the solar wind

One of the simplest tools for probing the Earth's interior is an ordinary compass. The needle of a compass points north because it aligns with the Earth's magnetic field. The mere existence of this field is further evidence that the interior of our planet is partially molten.

 Magnetism arises whenever electrically charged particles are in motion. For example, a loop of wire carrying an electric current generates a magnetic field in the space around it. The magnetic field that surrounds an ordinary bar magnet (Figure 8-19a) is created by the motions of negatively charged electrons within the iron atoms of which the magnet is made. The Earth's magnetic field is similar to that of a bar magnet, as Figure 8-19b shows. The consensus among geologists is that this magnetic field is caused by electric currents flowing in the outer, liquid portions of the Earth's iron core. As material deep in the liquid core cools and solidifies to join the solid portion of the core, it releases the energy needed to stir up the motions of the remaining liquid material. (In Section 8-1 we described a similar process in the Earth's atmosphere: When water vapor in the air cools and forms liquid drops, it releases the energy that powers thunderstorms.) Our planet's rotation helps to sustain these currents, which produce a magnetic field that dominates space for tens of thousands of kilometers in all directions.

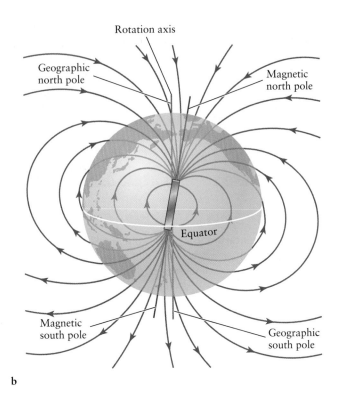

R I **V** U X G   b

ƒigure 8-19

The Magnetic Fields of a Bar Magnet and of the Earth **(a)** The magnetic field of a bar magnet is revealed by iron filings on a piece of paper. The pattern of the filings show the magnetic field lines, which appear to stream from one of the magnet's poles to the other. **(b)** The Earth's magnetic field is generated by electric currents in the liquid outer core. But the effect is much the same as if there were a giant bar magnet inside the Earth, tilted by 11.5° from the planet's rotation axis. Because of this tilt, the magnetic north and south poles (where the magnetic axis passes through the Earth's surface) are not at the same locations as the true, or geographic, poles (where the rotation axis passes through the surface). A compass needle points toward the north magnetic pole, not the true North Pole. (a: Jules Bucher/Photo Researchers)

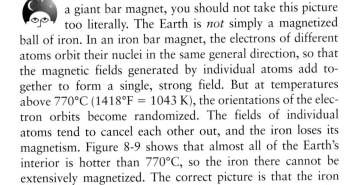 While the Earth's magnetic field is similar to that of a giant bar magnet, you should not take this picture too literally. The Earth is *not* simply a magnetized ball of iron. In an iron bar magnet, the electrons of different atoms orbit their nuclei in the same general direction, so that the magnetic fields generated by individual atoms add together to form a single, strong field. But at temperatures above 770°C (1418°F = 1043 K), the orientations of the electron orbits become randomized. The fields of individual atoms tend to cancel each other out, and the iron loses its magnetism. Figure 8-9 shows that almost all of the Earth's interior is hotter than 770°C, so the iron there cannot be extensively magnetized. The correct picture is that the iron carries electric currents, and it is these currents that create the Earth's magnetic field.

Studies of ancient rocks reinforce the idea that our planet's magnetism is due to fluid material in motion. When iron-bearing lava cools and solidifies to form igneous rock, it becomes magnetized in the direction of the Earth's magnetic field. By analyzing samples of igneous rock of different ages from around the world, geologists have found that the Earth's magnetic field actually flips over and reverses direction on an irregular schedule ranging from tens of thousands to hundreds of thousands of years. As an example, lava that solidified 30,000 years ago is magnetized in the opposite direction to lava that has solidified recently. Therefore, 30,000 years ago a compass needle would have pointed south, not north! If the Earth were a permanent magnet, like the small magnets used to attach notes to refrigerators, it would be hard to imagine how its magnetic field could spontaneously reverse direction. But computer simulations show that fields produced by moving fluids in the Earth's outer core can indeed change direction from time to time.

The Earth's magnetic field interacts dramatically with charged particles from the Sun. This *solar wind* is a flow of mostly protons and electrons that streams constantly outward from the Sun's upper atmosphere. Near the Earth, the particles in the solar wind move at speeds of roughly 450 km/s, or about a million miles per hour. Because this is considerably

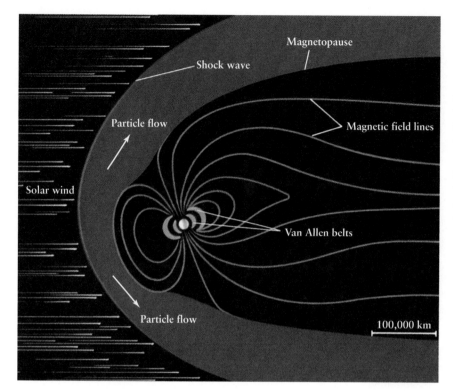

figure 8-20

The Earth's Magnetosphere The Earth's magnetic field carves out a cavity in the solar wind, shown here in cross section. A shock wave marks the boundary where the supersonic solar wind is abruptly slowed to subsonic speeds. Most of the particles of the solar wind are deflected around the Earth in a turbulent region (colored purple in this drawing). The Earth's magnetic field also traps some charged particles in two huge, doughnut-shaped rings called the Van Allen belts (in blue). This figure shows only a slice through the Van Allen belts.

faster than sound waves can travel in the very thin gas between the planets, the solar wind is said to be *supersonic*. (Because the gas between the planets is so thin, interplanetary sound waves carry too little energy to be heard by astronauts.)

If the Earth had no magnetic field, we would be continually bombarded by the solar wind. But our planet does have a magnetic field, and the forces that this field can exert on charged particles are strong enough to deflect them away from us. The region of space around a planet in which the motion of charged particles is dominated by the planet's magnetic field is called the planet's **magnetosphere.** Figure 8-20 is a scale drawing of the Earth's magnetosphere, which was discovered in the late 1950s by the first satellites placed in orbit.

When the supersonic particles in the solar wind first encounter the Earth's magnetic field, they abruptly slow to subsonic speeds. The boundary where this sudden decrease in velocity occurs is called a **shock wave.** Still closer to the Earth lies another boundary, called the **magnetopause,** where the outward magnetic pressure of the Earth's field is exactly counterbalanced by the impinging pressure of the solar wind. Most of the particles of the solar wind are deflected around the magnetopause, just as water is deflected to either side of the bow of a ship.

Some charged particles of the solar wind manage to leak through the magnetopause. When they do, they are trapped by the Earth's magnetic field in two huge, doughnut-shaped rings around Earth called the **Van Allen belts.** These belts were discovered in 1958 during the flight of the first successful U.S. Earth-orbiting satellite. They are named after the physicist James Van Allen, who insisted that the satel-

lite carry a Geiger counter to detect charged particles. The inner Van Allen belt, which extends over altitudes of about 2000 to 5000 km, contains mostly protons. The outer Van Allen belt, about 6000 km thick, is centered at an altitude of about 16,000 km above the Earth's surface and contains mostly electrons.

Sometimes the magnetosphere becomes overloaded with particles. The particles then leak through the magnetic fields at their weakest points and cascade down into the Earth's upper atmosphere, usually in a ring-shaped pattern (Figure 8-21a). As these high-speed charged particles collide with atoms in the upper atmosphere, they excite the atoms to high energy levels. The atoms then emit visible light as they drop down to their ground states, like the excited gas atoms in a neon light (see Section 5-8). The result is a beautiful, shimmering display called the **northern lights (aurora borealis)** or **southern lights (aurora australis),** depending on the hemisphere in which the phenomenon is observed. Figures 8-21b and 8-21c show the aurora borealis as seen from orbit and from the Earth's surface.

Occasionally a violent event on the Sun's surface called a **coronal mass ejection** sends a burst of protons and electrons toward Earth. The resulting auroral display can be exceptionally bright and can often be seen over a wide range of latitudes. Such events also disturb radio transmissions and can damage communication satellites and transmission lines.

It is remarkable that the Earth's magnetosphere, including its vast belts of charged particles, was entirely unknown until a few decades ago. Such discoveries remind us of how little we truly understand and how much remains to be learned even about our own planet.

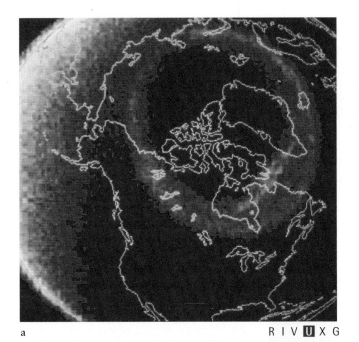

a   R I **U** X G

b   R I **V** U X G

c   R I **V** U X G

figure 8-21

The Aurora An increased flow of charged particles from the Sun can overload the Van Allen belts and cascade toward the Earth, producing aurorae. **(a)** This false-color ultraviolet image from the *Dynamics Explorer 1* spacecraft shows a glowing oval of auroral emission about 4500 km in diameter centered on the north magnetic pole. **(b)** This space shuttle photograph shows the aurora australis over Antarctica. **(c)** This view from Alaska shows aurorae at their typical altitudes of 100 to 400 km. (a: Courtesy of L. A. Frank and J. D. Craven, University of Iowa; b: JSC/NASA; c: Courtesy of S.-I. Akasofu, Geophysical Institute, University of Alaska)

8-5 The Earth's atmosphere has changed substantially over our planet's history

Both plate tectonics and the Earth's magnetic field paint a picture of a planet with a dynamic, evolving surface and interior. Our atmosphere, too, has changed and evolved substantially over the Earth's 4.6-billion-year history. As we will see, this evolution explains why we have an atmosphere totally unlike that of any other world in the solar system.

When the Earth first formed by accretion of planetesimals (see Section 7-8), gases were probably trapped within the Earth's interior in the same proportions that they were present in the solar nebula. But since the early Earth was hot enough to be molten throughout its volume, most of these trapped gases were released. As we discussed in Section 7-4 and Box 7-2, the Earth's gravity was too weak to prevent hydrogen and helium—the most two common kinds of atoms in the universe, but also the least massive—from leaking away into space. The atmosphere that remained still contained substantial amounts of hydrogen, but in the form of relatively massive molecules of water vapor (H_2O) made by combining two atoms of hydrogen with one atom of oxygen, the third most common element in the universe (see Table 7-3). In fact, water vapor was probably the dominant constituent of the early atmosphere.

Water vapor is one of the greenhouse gases that we discussed in Section 8-1. It traps infrared radiation from the Earth's surface, and its presence in the early atmosphere helped to sustain the high temperature of the surface. But if water vapor had been the *only* constituent of the atmosphere, you would not be reading these words! The reason is that as the molten Earth inevitably cooled, water vapor in the atmosphere condensed into raindrops and fell to Earth to form the oceans. As water vapor was lost from the atmosphere, its contribution to the greenhouse effect weakened, which would have made the surface temperature drop even lower. A further cooling factor is that the young Sun was only about 70% as bright as it is today. The net result is that the entire Earth should have been frozen over! Ice is so reflective that the Earth would have remained frozen even as the Sun aged and became more luminous. Liquid water is absolutely essential to all living creatures, so life as we know it would probably never have evolved on such an icy Earth.

In fact, the first living organisms appeared on Earth within 400 million years after the planet first formed. Hence, the Earth could not have been frozen over for very long, if at all. What saved our planet from a perpetual deep freeze was the presence of carbon, the fourth most common of the elements. When combined with oxygen, carbon forms carbon dioxide (CO_2), a greenhouse gas that remains gaseous at low temperatures. Carbon dioxide would have been released into the atmosphere by volcanic activity, a process called **outgassing** (see Figure 8-14). It would also have been added to the atmosphere by carbon-rich meteors striking the planet's surface. Once CO_2 was in the atmosphere, its greenhouse effect raised the planet's temperature and melted the ice. Some of the water evaporated into the atmosphere, further enhancing the Earth's greenhouse effect.

Fewer than one in 2500 of the molecules in today's atmosphere is carbon dioxide. But CO_2 must have been thousands of times more abundant in the early Earth's atmosphere in order to melt the global ice sheath. Unlike hydrogen or helium, this excess CO_2 was not lost into space; instead, it has been trapped in rocks. Carbon dioxide dissolves in rainwater and falls into the oceans, where it combines with other substances to form a class of minerals called *carbonates*. (Limestone and marble are examples of carbonate-bearing rock.) These form sediments on the ocean floor, which are eventually recycled into the crust by subduction. As an example, marble (Figure 8-18c) is a metamorphic rock formed deep within the crust from limestone, a carbonate-rich sedimentary rock.

This process removed most of the CO_2 from the atmosphere within the first billion (10^9) years after the formation of the Earth. Although this weakened the greenhouse effect, temperatures remained warm as the Sun's brightness increased. The small amounts of atmospheric CO_2 that remain today are the result of a balance between volcanic activity (which releases CO_2 into the atmosphere) and the formation of carbonates

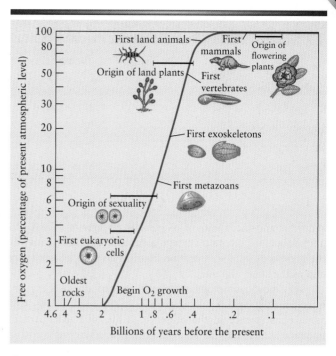

figure 8-22

The Growth of Atmospheric Oxygen This graph shows how the amount of oxygen in the atmosphere (expressed as a percentage of its present-day value) has evolved with time. Note that the atmosphere contained essentially no oxygen until about 2 billion years ago. (Adapted from Preston Cloud, "The Biosphere," *Scientific American*, September 1983, p. 176)

(which removes CO_2). (As we will see in Section 8-6, human activity is upsetting this balance.)

The appearance of life on Earth set into a motion a radical transformation of the atmosphere. Early single-cell organisms converted energy from sunlight into chemical energy using **photosynthesis,** a chemical process that consumes CO_2 and water and releases oxygen (O_2). Oxygen molecules are very reactive, so originally most of the O_2 produced by photosynthesis combined with other substances to form minerals called oxides. But as life proliferated, the amount of photosynthesis increased dramatically. Eventually so much oxygen was being produced that it could not all be absorbed to form oxides, and O_2 began to accumulate in the atmosphere. Figure 8-22 shows how the amount of O_2 in the atmosphere has increased over the history of the Earth.

About two billion (2×10^9) years ago, a new type of life evolved to take advantage of the newly abundant oxygen. These new organisms produce energy by consuming oxygen and releasing carbon dioxide—a process called **respiration** that is used by all modern animals, including humans. Such organisms thrived because photosynthetic plants continued to add even more oxygen to the atmosphere.

Several hundred million years ago the number of oxygen molecules in the atmosphere stabilized at 21% of the total, the same as the present-day value. This value represents a balance between the release of oxygen from plants by photo-

table 8-3	Chemical Compositions of Three Planetary Atmospheres		
	Venus	Earth	Mars
Nitrogen (N_2)	3.5%	78.08%	2.7%
Oxygen (O_2)	almost zero	20.95%	almost zero
Carbon dioxide (CO_2)	96.5%	0.035%	95.3%
Water vapor (H_2O)	0.003%	about 1%	0.03%
Other gases	almost zero	almost zero	2%

synthesis and the intake of oxygen by respiration. Thus, the abundance of oxygen is regulated almost exclusively by the presence of life—a situation that has no parallel anywhere else in the solar system.

The most numerous molecules in our atmosphere are nitrogen (N_2), which make up 78% of the total. These, too, are a consequence of the presence of life on Earth. Certain bacteria extract oxygen from minerals called nitrates, and in the process liberate nitrogen into the atmosphere. The amount of atmospheric nitrogen is kept in check by lightning. The energy in a lightning flash causes atmospheric nitrogen and oxygen to combine into nitrogen oxides, which dissolve in rainwater, fall into the oceans, and form nitrates.

Table 8-3 shows the dramatic differences between the Earth's atmosphere and those of Venus and Mars. The greater intensity of sunlight on Venus caused higher temperatures, which boiled any liquid water and evaporated CO_2 out of the rocks. The atmosphere thus became far denser than our own and rich in greenhouse gases. The resulting greenhouse effect raised temperatures on Venus to their present value of about 460°C (733 K = 855°F).

Just the opposite happened on Mars, where sunlight is less than half as intense as on Earth. The lower temperatures drove CO_2 from the atmosphere into Martian rocks, and froze all of the water beneath the planet's surface. The atmosphere that remains on Mars has a similar composition to that of Venus but is less than 1/10,000 as dense. On neither Venus or Mars was life able to blossom and transform the atmosphere as it did here on Earth.

While life has shaped our atmosphere's chemical composition, the structure of the atmosphere is controlled by the influence of sunlight. Scientists describe the structure of any atmosphere in terms of two properties, temperature and atmospheric pressure, which vary with altitude.

Atmospheric pressure at any height in the atmosphere is caused by the weight of all the air above that height. The average atmospheric pressure at sea level is defined to be one **atmosphere** (1 atm), equal to 1.01×10^5 N/m² or 14.7 pounds per square inch. As you go up in the atmosphere, there is less air above you to weigh down on you. Hence, atmospheric pressure decreases smoothly with increasing altitude.

Unlike atmospheric pressure, temperature varies with altitude in a complex way. Figure 8-23 shows that temperature decreases with increasing altitude in some layers of the atmosphere, but in other layers actually *increases* with increasing altitude. These differences result from the individual ways in which each layer is heated.

The lowest layer, called the **troposphere**, extends from the surface to an average altitude of 12 km (roughly 7.5 miles, or 39,000 ft). It is heated only indirectly by the Sun. Sunlight warms the Earth's surface, which heats the lower part of the troposphere. By contrast, the upper part of the troposphere remains at cooler temperatures. This vertical temperature variation causes convection currents that move up and down through the troposphere (see Section 8-1). All of the Earth's weather is a consequence of this convection.

Convection on a grand scale is caused by the temperature difference between the Earth's equator and its poles. If the Earth did not rotate, heated air near the equator would rise

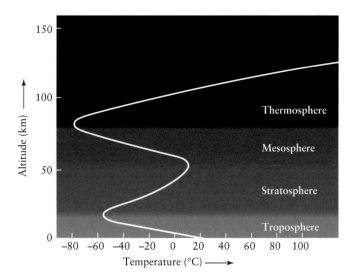

figure 8-23

Temperature Profile of the Earth's Atmosphere This graph shows how the temperature in the Earth's atmosphere varies with altitude. In the troposphere and mesosphere, temperature decreases with increasing altitude; in the stratosphere and thermosphere, temperature actually increases with increasing altitude.

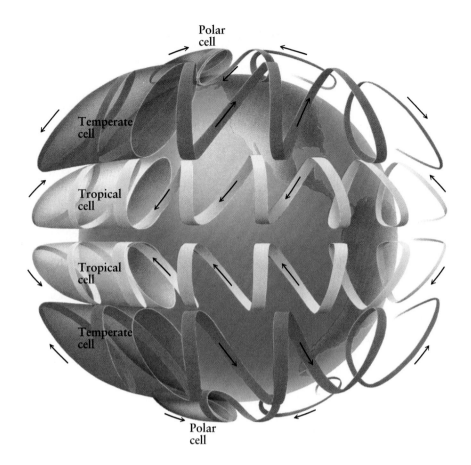

Polar cell

Temperate cell

Tropical cell

Tropical cell

Temperate cell

Polar cell

ƒigure 8-24

Circulation Patterns in the Earth's Atmosphere The dominant circulation in our atmosphere consists of six convection cells, three in the northern hemisphere and three in the southern hemisphere. In the northern temperate region (including the continental United States), the prevailing winds at the surface are from the southwest toward the northeast. Farther south, within the northern tropical region (for example, Hawaii), the prevailing surface winds are from northeast to southwest.

upward and flow at high altitude toward the poles. There it would cool and sink to lower altitudes, at which it would flow back to the equator. However, the Earth's rotation breaks up this simple convection pattern into a series of **convection cells.** In these cells, air flows east and west as well as vertically and in a north-south direction. The structure of these cells explains why the prevailing winds blow in different directions at different altitudes (Figure 8-24).

Almost all the oxygen in the troposphere is in the form of O_2, a molecule made of two oxygen atoms. But in the **stratosphere,** which extends from about 12 to 50 km (about 7.5 to 31 mi) above the surface, an appreciable amount of oxygen is in the form of **ozone,** a molecule made of *three* oxygen atoms (O_3). Unlike O_2, ozone is very efficient at absorbing ultraviolet radiation from the Sun, which means that the stratosphere can directly absorb solar energy. The result is that the temperature actually increases as you move upward in the stratosphere. Convection requires that temperature decrease, not increase, with increasing altitude, so there are essentially no convection currents in the stratosphere.

Above the stratosphere lies the **mesosphere.** Very little ozone is found here, so solar ultraviolet radiation is not absorbed within the mesosphere, and atmospheric temperature again declines with increasing altitude.

The temperature of the mesosphere reaches a minimum of about −75°C (= −103°F = 198 K) at an altitude of about 80 km (50 mi). This minimum marks the bottom of the **ther-**

mosphere, in which temperature once again rises with increasing altitude. This is not due to the presence of ozone, because in this very low-density region oxygen and nitrogen are found as individual atoms rather than in molecules. Instead, the thermosphere is heated because these isolated atoms absorb very-short-wavelength solar ultraviolet radiation (which oxygen and nitrogen molecules cannot).

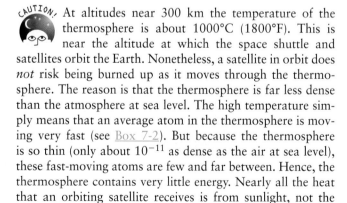 At altitudes near 300 km the temperature of the thermosphere is about 1000°C (1800°F). This is near the altitude at which the space shuttle and satellites orbit the Earth. Nonetheless, a satellite in orbit does *not* risk being burned up as it moves through the thermosphere. The reason is that the thermosphere is far less dense than the atmosphere at sea level. The high temperature simply means that an average atom in the thermosphere is moving very fast (see Box 7-2). But because the thermosphere is so thin (only about 10^{-11} as dense as the air at sea level), these fast-moving atoms are few and far between. Hence, the thermosphere contains very little energy. Nearly all the heat that an orbiting satellite receives is from sunlight, not the thermosphere.

 There is no definite upper limit to the Earth's atmosphere. At sufficiently high altitudes, the thermosphere merges into the very thin gas between the planets, called the interplanetary medium.

8-6 A burgeoning human population is profoundly altering the Earth's biosphere

One of the extraordinary characteristics of the Earth is that it is covered with life. Living organisms thrive in a wide range of environments, from ocean floors to mountaintops and from frigid polar caps to blistering deserts.

Living organisms subsist in a relatively thin layer enveloping the Earth called the **biosphere,** which includes the oceans, the lowest few kilometers of the troposphere, and the crust to a depth of almost 3 kilometers. Figure 8-25 is a portrait of the Earth's biosphere based on NASA satellite data. The biosphere, which has taken billions of years to evolve to its present state, is a delicate, highly complex system in which plants and animals depend on each other for their mutual survival.

The state of the biosphere depends crucially on the temperature of the oceans and atmosphere. Even small temperature changes can have dramatic consequences. An example that recurs every three to seven years is the El Niño phenomenon, in which temperatures at the surface of the equatorial Pacific Ocean rise by 2 to 3°C. (Figure 1-13 shows ocean temperatures during the 1997–1998 El Niño.) Ordinarily water from the cold depths of the ocean is able to well upward, bringing with it nutrients that are used by microscopic marine organisms called phytoplankton that live near the surface (see Figure 8-25). But during an El Niño, the warm surface water suppresses this upwelling, and the phytoplankton starve. This wreaks havoc on organisms such as mollusks that feed on phytoplankton, on the fish that feed on the mollusks, and on the birds and mammals who eat the fish. During the 1982–1983 El Niño, a quarter of the adult sea lions off the Peruvian coast starved, along with all of their pups.

Many different factors can change the surface temperature of our planet. One is that the amount of energy radiated by the Sun can vary up or down by a few tenths of a percent. (Reduced solar brightness may explain the period from 1450 to 1850, when European winters were substantially colder than today.) Other factors are the gravitational influences of the Moon and the other planets.

Thanks to these influences, the eccentricity of the Earth's orbit varies with a period of 90,000 to 100,000 years, the tilt of its rotation axis varies between 22.1° and 24.5° with a 40,000-year period, and the orientation of its rotation axis changes due to precession (see Section 2-6) with a 26,000-year period. These variations can affect climate by altering the amount of solar energy that heats the Earth. They help explain why the Earth periodically undergoes an extended period of low temperatures called an *ice age,* the last of which ended about 11,000 years ago.

As we saw in Sections 8-1 and 8-5, one of the most important factors affecting global temperatures is the abundance of greenhouse gases such as CO_2. Geologic processes can alter this abundance, either by removing CO_2 from the atmosphere (as happens when fresh rock is uplifted and exposed to the air, where it can absorb atmospheric CO_2 to form carbonates) or by supplying new CO_2 (as in volcanic eruptions). But since the 1970s it has become clear that human activity can also affect climate by adding greenhouse gases to the atmosphere.

Our species is having an increasing effect on the biosphere because our population is skyrocketing. Figure 8-26 shows the sharp rise in the human population that began in the late 1700s with the Industrial Revolution and the spread of modern ideas about hygiene. This rise accelerated in the twentieth century thanks to medical and technological advances ranging from antibiotics to high-yield grains. In

figure 8-25

The Earth's Biosphere This image, based on data from the *Nimbus 7* spacecraft, shows the distribution of plant life over the Earth's surface. Land colors designate vegetation: dark green for the rain forests, light green and gold for savannas and farmland, and yellow for the deserts. Ocean colors show where microscopic, free-floating plants called phytoplankton are found. Phytoplankton are most abundant in the red and orange areas and least abundant in the dark blue areas. (NASA/GSFC)

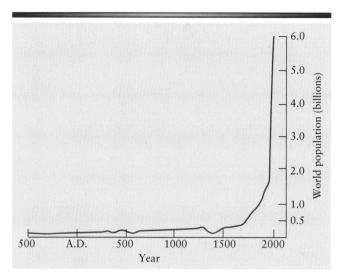

figure 8-26

The Human Population Data and estimates from the U.S. Bureau of the Census, the Population Reference Bureau, and the United Nations Population Fund were combined to produce this graph showing the human population over the past 2500 years. Note how the population began to rise in the eighteenth century and the astonishing population increase in the twentieth century.

figure 8-27 R I **V** U X G

The Deforestation of Amazonia The Amazon, the world's largest rain forest, is being destroyed at a rate of 20,000 square kilometers per year in order to provide land for grazing and farming and as a source for lumber. About 80 percent of the logging is being carried out illegally. (Martin Wendler/Okapia/Photo Researchers)

1960 there were 3 billion people on Earth, and in 1975, 4 billion. The 6-billion mark was passed in 1999, and, according to the United Nations Population Fund, there will be 8.0 billion people on Earth by the year 2025 and 9.4 billion by 2050.

Every human being has basic requirements: food, clothing, and housing. We all need fuel for cooking and heating. To meet these demands, we cut down forests, cultivate grasslands, and build sprawling cities. A striking example of this activity is occurring in the Amazon rain forest of Brazil. Tropical rain forests are vital to our planet's ecology because they absorb significant amounts of CO_2 and release O_2 through photosynthesis. Although rain forests occupy only 7% of the world's land areas, they are home to at least 50% of all plant and animal species on Earth. Nevertheless, to make way for farms and grazing land, people simply cut down the trees and set them on fire—a process called slash-and-burn. Such deforestation, along with extensive lumbering operations, is occurring in Malaysia, Indonesia, and Papua New Guinea. The rain forests that once thrived in Central America, India, and the western coast of Africa are almost gone (Figure 8-27).

All this activity, along with the use of petroleum products in automobiles and of coal in power plants, is adding carbon dioxide to the Earth's atmosphere faster than plants and geological processes can extract it. Each year the amount of atmospheric CO_2 increases by about 6 billion tons! Figure 8-28 shows how the carbon dioxide content of the

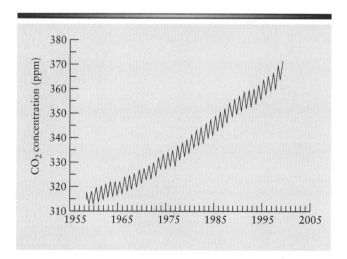

figure 8-28

The Carbon Dioxide Content of the Atmosphere This graph shows measurements of atmospheric carbon dioxide in parts per million (ppm). The sawtooth pattern results from plants absorbing more carbon dioxide during the spring and summer. The CO_2 concentration in the atmosphere has increased by 16% since continuous observations started in 1958. By comparison, measurements of Antarctic ice that froze around 1850 shows that the CO_2 concentration then was 280 to 290 ppm. (NOAA/Scripps Institution of Oceanography)

atmosphere has increased since 1958. What is more, industrial and agricultural activities release other greenhouse gases, principally methane (CH_4) and nitrous oxide (N_2O).

Humans did not create the greenhouse effect, but we are responsible for strengthening it by adding greenhouse gases to the atmosphere. Figure 8-29 shows how the global air temperature has changed since the middle of the nineteenth century. The ten warmest years of the twentieth century all occurred between 1985 and 2000. Simulations predict that temperatures will continue to rise by an additional 0.9 to 3.5°C during the twenty-first century, a phenomenon called **global warming.** They further predict that rainfall will increase dramatically in some parts of the Earth and decrease in others, ice in the Earth's polar regions will melt, the sea level will rise, and low-lying terrain will flood.

Human activity is also having potentially disastrous effects on the upper atmosphere. Certain chemicals released into the air, in particular a class of chemicals called chlorofluorocarbons (CFCs), which have been used in refrigeration and electronics, are destroying the ozone in the stratosphere. We saw in Section 8-5 that ozone molecules absorb ultraviolet radiation from the Sun. Without this high-altitude **ozone layer,** solar ultraviolet radiation would beat down on the Earth's surface with greatly increased intensity. Such ultraviolet radiation breaks apart most of the delicate molecules that form living tissue. Hence, a complete loss of the ozone layer would lead to a catastrophic ecological disaster.

In 1985 scientists discovered an **ozone hole,** a region with an abnormally low concentration of ozone, over Ant-

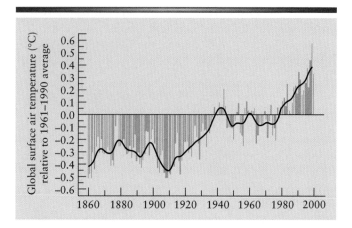

figure 8-29

Changes in Global Temperature Land and sea surface temperatures were combined to produce this graph, which shows a warming trend over the past 140 years. The vertical scale is measured relative to the 1961–1990 average temperature. (Hadley Centre for Climate and Prediction and Research, U.K. Meteorological Office)

arctica. Since then this hole has generally expanded from one year to the next (Figure 8-30). Similar but smaller effects have been observed in the stratosphere above other parts of the Earth. By international agreement, CFCs have been phased out and are being replaced by compounds that do

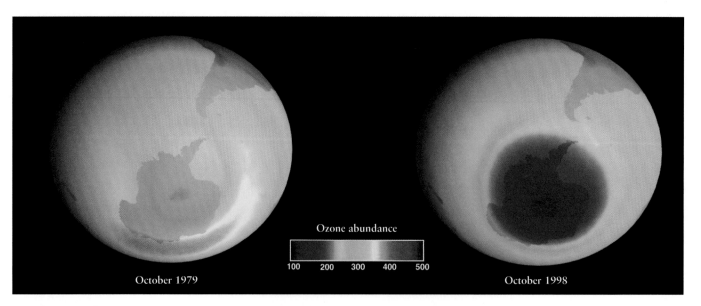

figure 8-30

The Antarctic Ozone Hole These two images compare the distribution of stratospheric ozone over Antarctica in October 1979 and October 1998. False colors in the order of the rainbow show the ozone concentration, with red and orange for high concentration and purple for low. The data show that there has been a net decrease of 50% in ozone over the Antarctic. The amount of ozone at mid-latitudes, where most of the human population lives, decreased by 10 to 20% over the same period. (SVS/GSFC/TOMS/NASA)

not deplete stratospheric ozone, and sunlight naturally produces more ozone in the stratosphere. But even in the best of circumstances, the damage to the ozone layer is not expected to heal for many decades. Much remains to be learned about the behavior of ozone in the atmosphere, and scientists are studying the Antarctic ozone hole with great interest and concern.

As we continue to populate our planet, it is clear that we face many challenges. Age-old modes of behavior that helped ensure the survival of our ancestors may not be the wisest courses of action in the future. Contemplating our behavior and our relationship to the Earth brings to mind a Chinese proverb: If we don't change the direction in which we are going, we are likely to end up where we are headed.

KEY WORDS

albedo, p. 180

asthenosphere, p. 189

atmosphere (atm), p. 196

atmospheric pressure, p. 196

aurora australis (*plural* aurorae), p. 193

aurora borealis, p. 193

biosphere, p. 198

convection, p. 180

convection cell, p. 197

convection current, p. 180

core, p. 183

coronal mass ejection, p. 193

crust (of Earth), p. 183

crystal, p. 186

earthquake, p. 184

epicenter, p. 184

global warming, p. 200

greenhouse effect, p. 182

greenhouse gas, p. 182

igneous rock, p. 190

inner core (of Earth), p. 185

lava, p. 190

lithosphere, p. 189

magma, p. 190

magnetopause, p. 193

magnetosphere, p. 193

mantle, p. 183

melting point, p. 185

mesosphere, p. 197

metamorphic rock, p. 190

mineral, p. 190

northern lights, p. 193

oceanic rift, p. 189

outer core (of Earth), p. 185

outgassing, p. 195

ozone, p. 197

ozone layer, p. 200

ozone hole, p. 200

P wave, p. 184

photosynthesis, p. 195

plastic, p. 186

plate (lithospheric), p. 186

plate tectonics, p. 188

respiration, p. 195

rock, p. 190

S wave, p. 184

seafloor spreading, p. 188

sedimentary rock, p. 190

seismic wave, p. 184

seismograph, p. 184

shock wave, p. 193

southern lights, p. 193

stratosphere, p. 197

subduction zone, p. 189

surface wave, p. 184

thermosphere, p. 197

troposphere, p. 196

Van Allen belts, p. 193

KEY IDEAS

The Earth's Energy Sources: All activity in the Earth's atmosphere, oceans, and surface is powered by three sources of energy.

• Solar energy is the energy source for the atmosphere. In the greenhouse effect, some of this energy is trapped by infrared-absorbing gases in the atmosphere, raising the Earth's surface temperature.

• Tidal forces from the Moon and Sun help to power the motion of the oceans.

• The internal heat of the Earth is the energy source for geologic activity.

The Earth's Interior: Studies of seismic waves (vibrations produced by earthquakes) show that the Earth has a small, solid inner core surrounded by a liquid outer core. The outer core is surrounded by the dense mantle, which in turn is surrounded by the thin low-density crust.

• Seismologists deduce the Earth's interior structure by studying how longitudinal P waves and transverse S waves travel through the Earth's interior.

• The Earth's inner and outer cores are composed of almost pure iron with some nickel mixed in. The mantle is composed of iron-rich minerals.

• Both temperature and pressure steadily increase with depth inside the Earth.

Plate Tectonics: The Earth's crust and a small part of its upper mantle form a rigid layer called the lithosphere.

The lithosphere is divided into huge plates that move about over the plastic layer called the asthenosphere in the upper mantle.

• Plate tectonics, or movement of the plates, is driven by convection within the asthenosphere. Molten material wells up at oceanic rifts, producing seafloor spreading, and is returned to the asthenosphere in subduction zones. As one end of a plate is subducted back into the asthenosphere, it helps to pull the rest of the plate along.

• Plate tectonics is responsible for most of the major features of the Earth's surface, including mountain ranges, volcanoes, and the shapes of the continents and oceans.

• Plate tectonics is involved in the formation of the three major categories of rocks: igneous rocks (cooled from molten material), sedimentary rocks (formed by the action of wind, water, and ice), and metamorphic rocks (altered in the solid state by extreme heat and pressure).

The Earth's Magnetic Field and Magnetosphere: Electric currents in the liquid outer core generate a magnetic field. This magnetic field produces a magnetosphere that surrounds the Earth and blocks the solar wind from hitting the atmosphere.

• A bow-shaped shock wave, where the supersonic solar wind is abruptly slowed to subsonic speeds, marks the outer boundary of the magnetosphere.

• Most of the particles of the solar wind are deflected around the Earth by the magnetosphere.

• Some charged particles from the solar wind are trapped in two huge, doughnut-shaped rings called the Van Allen belts. An excess of these particles can initiate an auroral display.

The Earth's Atmosphere: The Earth's atmosphere differs from those of the other terrestrial planets in its chemical composition, circulation pattern, and temperature profile.

• The Earth's atmosphere evolved from being mostly water vapor to being rich in carbon dioxide. A strong greenhouse effect kept the Earth warm enough for water to remain liquid and to permit the evolution of life.

• The appearance of photosynthetic living organisms led to our present atmospheric composition, about four-fifths nitrogen and one-fifth oxygen.

• The Earth's atmosphere is divided into layers called the troposphere, stratosphere, mesosphere, and thermosphere. Ozone molecules in the stratosphere absorb ultraviolet light.

• Because of the Earth's rapid rotation, the circulation in its atmosphere is complex, with three circulation cells in each hemisphere.

The Biosphere: Human activity is changing the Earth's biosphere, on which all living organisms depend.

• Deforestation and the burning of fossil fuels is increasing the greenhouse effect in our atmosphere and warming the planet.

• Industrial chemicals released into the atmosphere are threatening the ozone layer in the stratosphere.

REVIEW QUESTIONS

1. Describe the various ways in which the Earth is unique among the planets of our solar system.

2. Describe how energy is transferred from the Earth's surface to the atmosphere by both convection and radiation.

3. How does the greenhouse effect influence the temperature of the atmosphere? How does this effect differ from what actually happens in a greenhouse?

4. Describe the interior structure of the Earth.

5. How do we know that the Earth was once entirely molten?

6. The deepest wells and mines go down only a few kilometers. What, then, is the evidence that iron is abundant in the Earth's core? That the Earth's outer core is molten but the inner core is solid?

7. The inner core of the Earth is at a higher temperature than the outer core. Why, then, is the inner core solid and the outer core molten instead of the other way around?

8. Describe the process of plate tectonics. Give specific examples of geographic features created by plate tectonics.

9. Explain how convection in the Earth's interior drives the process of plate tectonics.

10. What is the difference between a rock and a mineral?

11. What are the differences among igneous, sedimentary, and metamorphic rocks? What do these rocks tell us about the sites at which they are found?

12. Why do some geologists suspect that Pangaea was the most recent in a succession of supercontinents?

13. Why do you suppose that active volcanoes, such as Mount St. Helens in Washington State, are usually located in mountain ranges that border on subduction zones?

14. Describe the Earth's magnetosphere. If the Earth did not have a magnetic field, do you think aurorae would be more common or less common than they are today?

15. How do we know that the Earth's magnetic field is not due to magnetized iron in the planet's interior?

16. What are the Van Allen belts?

17. Summarize the history of the Earth's atmosphere. What role has biological activity plated in this evolution?

18. Describe the structure of the Earth's atmosphere. Explain how heat from the Sun and the Earth's rotation affect the circulation of the Earth's atmosphere.

19. Carbon dioxide and ozone each make up only a fraction of a percent of our atmosphere. Why, then, should we be concerned about small increases or decreases in the atmospheric abundance of these gases?

ADVANCED QUESTIONS

Problem-solving tips and tools

You can find most of the Earth data that you need in Table 8-1. You'll need to know that a sphere has surface area $4\pi r^2$ and volume $\frac{4}{3}\pi r^3$, where r is the sphere's radius. You may have to consult an atlas to examine the geography of the South Atlantic. Also, remember that the average density of an object is its mass divided by its volume. Sections 5-3 and 5-4 and Box 5-2 describe how to solve problems involving blackbody radiation, Section 2-6 discusses precession, and Section 4-5 describes the properties of elliptical orbits.

20. The total power in sunlight that reaches the top of our atmosphere is 1.75×10^{17} W. **(a)** How many watts of power are reflected back into space due to the Earth's albedo? **(b)** If the Earth had no atmosphere, all of the solar power that was not reflected would be absorbed by the Earth's surface. In equilibrium, the heated surface would act as a blackbody that radiates as much power as it absorbs from the Sun. How much power would the entire Earth radiate? **(c)** How much power would one square meter of the surface radiate? **(d)** What would be the average temperature of the surface? Give your answer in both the Kelvin and Celsius scales. **(e)** Why is the Earth's actual average temperature higher than the value you calculated in (d)?

21. On average, the temperature beneath the Earth's crust increases at a rate of 20°C per kilometer. At what depth would water boil? (Assume the surface temperature is 20°C and ignore the effect of the pressure of overlying rock on the boiling point of water.)

22. What fractions of Earth's total volume are occupied by the core, the mantle, and the crust?

23. What fraction of the total mass of the Earth lies within the inner core?

24. **(a)** Using data for the mass and size of the Earth listed in Table 8-1, verify that the average density of the Earth is 5500 kg/m³. **(b)** Assuming that the average density of material in the Earth's mantle is about 3500 kg/m³, what must the average density of the core be? Is your answer consistent with the values given in Table 8-2?

25. Africa and South America are separating at a rate of about 3 centimeters per year, as explained in the text. Assuming that this rate has been constant, calculate when these two continents must have been in contact. Today the two continents are 6600 km apart.

26. When the crust under an ocean is pushed toward a continent, the oceanic crust is pushed under the continental crust and undergoes subduction (see the right-hand side of Figure 8-15). What does this tell us about the density of oceanic crust compared to continental crust?

27. The surfaces of Mercury, the Moon, and Mars are riddled with craters formed by the impact of space debris. Many of these are billions of years old. By contrast, there are only a few conspicuous craters on the Earth's surface, and these are generally less than 500 million years old. What do you suppose explains the difference?

28. Fast-moving charged particles can damage living organisms in a number of ways. For example, they can cause mutations if they collide with a DNA molecule within a cell. Explain how the fact that the Earth's interior is molten helps minimize the effect of the solar wind on the biosphere.

29. Most auroral displays have a green color dominated by emission from oxygen atoms at a wavelength of 557.7 nm. **(a)** What minimum energy (in electron volts) must be imparted to an oxygen atom to make it emit this wavelength? **(b)** Why is your answer in (a) a *minimum* value?

30. Describe how the present-day atmosphere and surface temperature of the Earth might be different **(a)** if carbon dioxide had never been released into the atmosphere; **(b)** if carbon dioxide had been released, but life had never evolved on Earth.

31. The Earth's primordial atmosphere probably contained an abundance of methane (CH_4) and ammonia (NH_3), whose molecules were broken apart by ultraviolet light from the Sun and particles in the solar wind. What happened to the atoms of carbon, hydrogen, and nitrogen that were liberated by this dissociation?

32. The photograph below shows the soil at Daspoort Tunnel near Pretoria, South Africa. The whitish layer that extends from lower left to upper right is 2.2 billion years old. Its color is due to a lack of iron oxide. More recent soils typically contain iron oxide and have a darker color. Explain what this tells us about the history of the Earth's atmosphere.

(Courtesy of H. D. Holland)

R I **V** U X G

33. Earth's atmospheric pressure decreases by a factor of one-half for every 5.5-km increase in altitude above sea level. Construct a plot of pressure versus altitude, assuming the pressure at sea level is one atmosphere (1 atm). Discuss the characteristics of your graph. At what altitude is the atmospheric pressure equal to 0.001 atm?

34. The Earth is at perihelion on January 3 and at aphelion on July 4. Because of precession, in 13,000 years the amount of sunlight in summer will be more than at present in the northern hemisphere but less than at present in the southern hemisphere. Explain why.

DISCUSSION QUESTIONS

35. The human population on Earth is currently doubling about every 30 years. Describe the various pressures placed on the Earth by uncontrolled human population growth. Can such growth continue indefinitely? If not, what natural and human controls might arise to curb this growth? It has been suggested that overpopulation problems could be solved by colonizing the Moon or Mars. Do you think this is a reasonable solution? Explain your answer.

36. One scientific study suggests that the continued burning of fossil and organic fuels by humans is releasing enough CO_2 to stimulate the greenhouse effect and eventually melt the polar icecaps. Antarctica has an area of 13 million square kilometers and is covered by an icecap that varies in thickness from 300 meters near the coast to 1800 meters in the interior. Estimate the volume of this icecap. Assuming that water and ice have roughly the same density, estimate the amount by which the water level of the world's oceans would rise if Antarctica's ice were to melt completely. What portions of the Earth's surface would be inundated by such a deluge?

 WEB/CD-ROM QUESTIONS

37. The periodic formation and breakup of supercontinents imply that profound environmental changes occur about every 500 million years. What sort of global changes might accompany the formation and breakup of a supercontinent? How might these cycles affect the evolution of life? Use the World Wide Web to research the life-forms that dominated the Earth when Pangaea existed 250 million years ago and when fragments of the preceding supercontinent were as dispersed as today's continents.

38. Because of plate tectonics, the arrangement of continents in the future will be different from today. Search the World Wide Web for information about "Pangea Ultima," a supercontinent that may form in the distant future. When is it predicted to form? How will it compare to the supercontinent that existed 200 million years ago (see Figure 8-11*a*)?

39. Use the World Wide Web to research how global warming may affect the Earth's future surface temperature. What are some predictions for the surface temperature in

2050? In 2100? What effects may the increased temperature have on human health?

40. Use the World Wide Web to investigate the current status of the Antarctic ozone hole. Is the situation getting better or worse? Is there a comparable hole over the North Pole? Why do most scientists blame the chemicals called CFCs for the existence of the ozone hole?

41. In 1989 representatives of many nations signed a global treaty called the Montreal Protocol to protect the ozone layer. Use the World Wide Web to investigate the current status of this treaty. How many nations are signatories to this treaty? Has the treaty been amended since it was first signed? When are various substances that destroy the ozone layer scheduled to be phased out?

 42. Observing Mountain Range Formation. Access the two animations "The Collision of Two Plates: South America" and "The Collision of Two Plates: The Himalayas" in Chapter 8 of the *Universe* web site or CD-ROM. **(a)** In which case are the mountains formed by tectonic uplift? In which case are the mountains formed by volcanoes from rising lava? **(b)** For each animation, describe which plate is moving in which direction.

OBSERVING PROJECTS

43. When was the last earthquake near your hometown? How far is your hometown from a plate boundary? What kinds of topography (for example, mountains, plains, seashore) dominate the geography of your hometown area? Does the topography and the frequency of earthquakes seem to be consistent with your hometown's proximity to a plate boundary?

 44. Use the *Starry Night* program to view the Earth from space. First select **Viewing Location...** in the **Go** menu and set the viewing location to your city or town. (You use the list of cities provided, or you can use the mouse to click on your approximate position on the world map. Then click the **Set Location** button.) In the Control Panel at the top of the window, set the local time to 12:00:00 P.M. (noon). To see the Earth from space, use the elevation buttons (the ones that look like a rocket landing or taking off) in the Control Panel to raise yourself off the surface until you can see the entire Earth. (You may want to adjust your field of view so that you are looking directly at the Earth.) **(a)** What evidence can you see that the Earth has an atmosphere? **(b)** Using the controls at the right-hand end of the Control Panel, zoom in to show more detail around your city or town. The amount of detail is comparable to the view from a spacecraft a few million kilometers away. Can you see any evidence of plate tectonics? Of the presence of life? What does this tell you about the importance of sending spacecraft to explore planets at close range?

Our Barren Moon

(Eugene Cernan, *Apollo 17*, NASA)

R I **V** U X G

Other than the Earth itself, the most familiar world in the solar system is the Moon. It is so near to us that some of its surface features are visible to the naked eye. But the Moon is also a strange and alien place, with dramatic differences from our own Earth. It has no substantial atmosphere, no global magnetic field, and no liquid water of any kind. On the airless Moon, the sky is black even at midday. And unlike the Earth's surface, the surface of the Moon has remained essentially unaltered for billions of years. The boulder in the photograph—shown being examined by *Apollo 17* astronaut Harrison Schmitt, one of only a dozen humans to have visited another world—is actually one of the most recent features on the lunar surface: It tumbled down a mountainside to its present location a mere 800,000 years ago.

Today the Moon is a geologically dead world. But its early history was violent and chaotic. In this chapter, we will learn how the Moon may have formed some 4.6 billion years ago as the result of a titanic collision between the young Earth and a rogue protoplanet the size of Mars. We will see evidence that the young Moon was pelted incessantly for hundreds of millions of years by large chunks of interplanetary debris, producing a cratered landscape. We will discover that the largest of these impacts formed immense circular basins that then flooded with molten lava. You can see the solidified lava as dark patches that cover much of the face of the Moon. In later chapters we will find other evidence that catastrophic collisions have played an important role in shaping the solar system.

As you read the sections of this chapter, look for the answers to the following questions.

9-1 Is the Moon completely covered with craters?

9-2 Has there been any exploration of the Moon since the *Apollo* program in the 1970s?

9-3 Does the Moon's interior have a similar structure to the interior of the Earth?

9-4 How do Moon rocks compare to rocks found on the Earth?

9-5 How did the Moon form?

9-1 The Moon's airless, dry surface is covered with plains and craters

Compared with the distances between planets, which are measured in tens or hundreds of millions of kilometers, the average distance from the Earth to the Moon is just 384,400 km (238,900 mi). Despite its being so close, the Moon has an angular diameter of just ½°. This tells us that it is a rather small world. The diameter of the Moon is 3476 km (2160 mi), just 27% of the Earth's diameter (Figure 9-1). Table 9-1 lists some basic data about the Moon.

Even without binoculars or a telescope, you can easily see that dark gray and light gray areas cover vast expanses of the Moon's rocky surface. If you observe the Moon over many nights, you will see the Moon go through its phases as the dividing line between the illuminated and darkened portion of the Moon (the **terminator**) shifts. But you will always see the same pattern of dark and light areas, because the Moon keeps the same side turned toward the Earth; you never see the **far side** (or back side) of the Moon. We saw in Section 3-2 that this is the result of the Moon's **synchronous rotation**: It takes precisely the same amount of time for the Moon to rotate on its axis (one lunar "day") as it does to complete one orbit around the Earth.

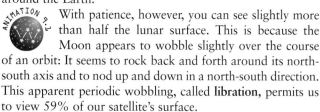 With patience, however, you can see slightly more than half the lunar surface. This is because the Moon appears to wobble slightly over the course of an orbit: It seems to rock back and forth around its north-south axis and to nod up and down in a north-south direction. This apparent periodic wobbling, called **libration**, permits us to view 59% of our satellite's surface.

We stress that the Moon only *appears* to be wobbling. In fact, its rotation axis keeps nearly the same orientation in space as the Moon goes around its orbit. The reason why the Moon seems to rock back and forth is that its orbit around the Earth is slightly elliptical. As a result, the Moon's orbital motion does not keep pace with its rotation at all points around the orbit. Furthermore, because the Moon's rotation axis is not exactly perpendicular to the plane of its orbit, it appears to us to nod up and down.

Clouds always cover some portion of the Earth's surface, but clouds are never seen on the Moon (Figure 9-1). This is because the Moon is too small a world to have an atmosphere.

Its surface gravity is low, only about one-sixth as great as that of the Earth, and thus any gas molecules can easily escape into space (see Box 7-2). Because there is no atmosphere to scatter sunlight, the daytime sky on the Moon appears as black as the nighttime sky on the Earth (see the photograph that opens this chapter).

The absence of an atmosphere means that the Moon can have no liquid water on its surface. Here on Earth, water molecules are kept in the liquid state by the pressure of the atmosphere pushing down on them. To see what happens to liquid water on the airless Moon, consider what happens when the pressure of the atmosphere is reduced. The air pressure in

Figure 9-1 R I **V** U X G

Earth and Moon to Scale The *Galileo* spacecraft recorded this image of the Earth (at lower right) and the Moon (at upper left) on December 16, 1992, while en route to Jupiter. Both the Earth and the Moon were about 6.2 million km (3.9 million mi) from the spacecraft, so this image shows them to scale. The Earth has blue oceans, an atmosphere streaked with white clouds, and continents continually being reshaped by plate tectonics. By contrast, the Moon has no oceans, no atmosphere, and no evidence of plate tectonics. (NASA)

table 9-1	Moon Data	
Distance from Earth (center to center):	Average: 384,400 km = 238,900 mi Maximum (apogee): 405,500 km Minimum (perigee): 363,300 km	
Eccentricity of orbit:	0.0549	
Average orbital speed:	3680 km/h	
Sidereal period (relative to fixed stars):	27.322 days	
Synodic period (new moon to new moon):	29.531 days	
Inclination of lunar equator to orbit:	6.68°	
Inclination of orbit to ecliptic:	5.15°	
Diameter (equatorial):	3476 km = 2160 mi = 0.272 Earth diameter	
Mass:	7.349×10^{22} kg = 0.0123 Earth mass	
Average density:	3344 kg/m^3	
Escape speed:	2.4 km/s	
Surface gravity (Earth = 1):	0.17	
Albedo:	0.11	
Average surface temperatures:	Day: 130°C = 266°F = 403 K Night: −180°C = −292°F = 93 K	

(Lick Observatory, University of California) R I **V** U X G

Denver, Colorado (elevation 1650 m, or 5400 ft), is only 83% as great as at sea level. Therefore, in Denver, less energy has to be added to the molecules in a pot of water to make them evaporate, and water boils at only 95°C (= 368 K = 203°F), as compared with 100°C (= 373 K = 212°F) at sea level. (This is why some foods require longer cooking times at high elevations, where the boiling water used for cooking is not as hot as at sea level.) On the Moon, the atmospheric pressure is essentially zero and the boiling temperature is lower than the temperature of the lunar surface. Under these conditions, water can exist as a solid (ice) or a vapor but not as a liquid.

 If astronauts were to venture out onto the lunar surface without wearing spacesuits, the lack of atmosphere would *not* cause their blood to boil. That's because our skin is strong enough to contain bodily fluids. In the same way, if we place an orange in a sealed container and remove all the air, the orange retains all of its fluid and does not explode. (It even tastes the same afterward.) The reasons why astronauts wore pressurized spacesuits on the Moon (see the photograph that opens this chapter) were to provide oxygen for breathing, to regulate body temperature, and to protect against ultraviolet radiation from the Sun. Since the Moon has no atmosphere, and thus no ozone layer like that of the Earth (Section 8-6), an unprotected astronaut would suffer a sunburn after just 10 seconds of exposure!

Since the Moon has no atmosphere to obscure its surface, you can get a clear view of several different types of lunar features with a small telescope or binoculars. These include *craters*, the dark-colored *maria*, and the light-colored lunar

Figure 9-2 R I **V** U X G

The Moon as Seen from Earth The Moon's diameter (3476 km = 2160 mi) is slightly less than the distance from New York to San Francisco. This image is a composite of photographs made at first quarter and last quarter, when long shadows enhance the surface features. (Lick Observatory, University of California)

highlands or *terrae* (Figure 9-2). Taken together, these different features show the striking differences between the surfaces of the Earth and the Moon.

With an Earth-based telescope, some 30,000 **craters** are visible, with diameters ranging from 1 km to several hundred km. Close-up photographs from lunar orbit have revealed millions of craters too small to be seen from Earth. Following a tradition established in the seventeenth century, the most prominent craters are named after famous philosophers, mathematicians, and scientists, such as Plato, Aristotle, Pythagoras, Copernicus, and Kepler.

Virtually all craters, both large and small, are the result of the Moon's having been bombarded by meteoritic material. For this reason, they are also called **impact craters.** When this theory of the origin of craters was proposed by German astronomer Franz Gruithuisen in 1824, a major sticking point was the observation that nearly all craters are circular. If craters were merely gouged out by high-speed rocks, rocks striking the Moon in any direction except straight downward would have created noncircular craters. A century after Gruithuisen, it was realized that a meteoroid colliding with the Moon generates a shock wave in the lunar surface at the point of impact. Such a shock wave produces a circular crater no matter what direction the meteoroid was moving. (In a similar way, the craters made by artillery shells are almost always circular.)

The shock wave from an impact can spread over an area much larger than the size of the meteoroid that generated the wave. Very large craters more than 100 km across, such as the crater Clavius shown in Figure 9-3, were probably created by the impact of a fast-moving piece of rock only a few kilometers in radius. Many such large craters have a pronounced "central peak," a feature that is characteristic of a high-speed impact (Figure 9-4).

The large dark areas on the lunar surface (clearly visible in Figure 9-2) are called **maria** (pronounced "MAR-ee-uh"). They form a pattern that looks crudely like a human face, sometimes called "the man in the Moon." The singular form of maria is **mare** (pronounced "MAR-ay"), which means "sea" in Latin. These terms date from the seventeenth century, when observers using early telescopes thought these dark features were large bodies of water. They gave the maria fanciful and romantic names such as Mare Tranquillitatis (Sea of Tranquillity), Mare Nubium (Sea of Clouds), Mare Nectaris (Sea of Nectar), Mare Serenitatis (Sea of Serenity), and Mare Imbrium (Sea of Showers) (Figure 9-5). While modern astronomers still use these names, they now understand that the maria are not bodies of water (which, as we have seen, could not exist on the airless Moon) but rather the remains of huge lava flows. The maria get their distinctive appearance from the dark color of the solidified lava.

Figure 9-3 R I **V** U X G

The Crater Clavius This photograph from the 5-meter (200-in.) Palomar telescope shows one of the largest craters on the Moon. Clavius has a diameter of 232 km (144 mi) and a depth of 4.9 km (16,000 ft), measured from the crater's floor to the top of the surrounding rim. The entire New York–Philadelphia metropolitan area would fit nicely within Clavius. (Palomar Observatory)

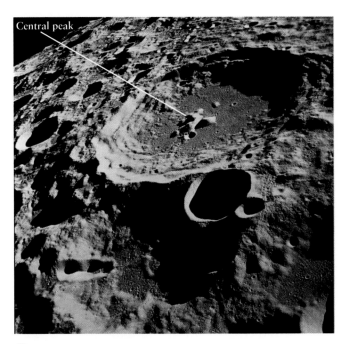

Central peak

figure 9-4 R I **V** U X G

Details of a Lunar Crater This photograph, taken from lunar orbit in 1969, shows a typical view of the Moon's heavily cratered far side. The large crater near the middle of the picture is approximately 80 km (50 mi) in diameter, equal to the length of San Francisco Bay. Note the crater's central peak and the numerous tiny craters that pockmark the lunar surface. (NASA)

figure 9-5 R I **V** U X G

Mare Imbrium from Earth Mare Imbrium is the largest of 14 dark plains that dominate the Earth-facing side of the Moon. Its diameter of 1100 km (700 mi) is about the same as the distance from Denver to Dallas or from Cleveland to Miami. The maria formed after the surrounding light-colored terrain, so they have not been exposed to meteoritic bombardment for as long and hence have fewer craters. (Carnegie Observatories)

Close-up photographs from lunar orbit reveal small craters in the maria, as well as occasional cracks that appear to be ancient lava rivers (Figure 9-6). On the whole, however, the maria have far fewer craters than the surrounding terrain. Since craters are caused by meteoritic bombardment, this means that the maria have not been exposed to that bombardment for as long as the surrounding terrain. Hence, the maria must be relatively young, and the lava flows that formed them must have occurred at a later stage in the Moon's geologic history.

Although they are much larger than craters, maria also tend to be circular in shape (see Figure 9-5). This suggests that, like craters, these depressions in the lunar surface were also created by impacts. It is thought that very large meteoroids or asteroids some tens of kilometers across struck the Moon, forming basins. These depressions subsequently flooded with lava that flowed out from the Moon's interior through cracks in the lunar crust. The solidified lava formed the maria that we see today.

An impact large enough to create a mare basin must have thrust upward tremendous amounts of material around the impact site. This explains the mountain ranges that curve in circular arcs around mare basins. (You can see one such mountain range around the edges of Mare Imbrium in Figure 9-5.) Analy-

sis of lunar rocks as well as observations made from lunar orbit have validated this picture of the origin of lunar mountains.

The older, light-colored terrain that surrounds the maria is called **terrae**, from the Latin for "lands." (Seventeenth-century astronomers thought that this terrain was the dry land that surrounded lunar oceans.) Detailed measurements by astronauts in lunar orbit demonstrated that the maria on the Moon's Earth-facing side are 2 to 5 km below the average lunar elevation, while the terrae extend to several kilometers above the average elevation. Terrae are therefore often called **lunar highlands.** The flat, low-lying, dark gray maria cover only 15% of the lunar surface. The remaining 85% is light gray, heavily cratered lunar highlands.

Remarkably, when the Moon's far side was photographed for the first time by the Soviet *Luna 3* spacecraft in 1959, scientists were surprised to find almost no maria there. Thus, lunar highlands predominate on the side of the Moon that faces away from the Earth (Figure 9-7). Presumably, the crust is thicker on the far side, so even the most massive impacts there were unable to crack through the crust and let lava flood onto the surface.

Something is entirely missing from our description of the Moon's surface—any mention of plate tectonics. Indeed, no evidence for plate tectonics can be found on the Moon. Recall

figure 9-6 R I **V** U X G

Details of Mare Tranquillitatis This photograph from lunar orbit reveals numerous tiny craters and cracks in the surface of a typical mare. What appears to be a dry riverbed running from lower left to middle right was once filled with flowing lava rather than flowing water. Some time after this "river" cooled and solidified, a meteoroid landed on top of it and produced the crater shown by the label. This color photograph also shows that the Moon's surface is not a uniform gray color. (Figures 9-2, 9-3, and 9-5 are black-and-white photographs.) (*Apollo 10*, NASA)

figure 9-7 R I **V** U X G

The Near and Far Sides of the Moon The right side of this composite image from the *Galileo* spacecraft shows the Earth-facing side of the Moon with its large, dark, flat maria. On the left is the Moon's far side, which is almost entirely composed of light-colored lunar highlands. Mare Orientale lies on the boundary between the near and far sides. The broad dark region at lower left is the South Pole–Aitken Basin, a scar some 2100 km (1300 mi) across and 12 km (7 mi) deep. The impact that made this basin is thought to have excavated all the way through the Moon's crust, exposing part of the mantle. (NASA)

from Section 8-3 that the Earth's major mountain ranges were created by collisions between lithospheric plates. In contrast, it appears that the Moon is a one-plate world: Its entire lithosphere is a single, solid plate. There are no rifts where plates are moving apart, no mountain ranges of the sort formed by collisions between plates, and no subduction zones. Thus, while lunar mountain ranges bear the names of terrestrial ranges, such as the Alps, Carpathians, and Apennines, they are entirely different in character. Geologically, the Moon is a dead world.

Without an atmosphere to cause erosion and without plate tectonics, the surface of the Moon changes only very slowly. That is why the Moon's craters, many of which are now known to be billions of years old, remain visible today. On the Earth, by contrast, the continents and seafloor are continually being reshaped by seafloor spreading, mountain building, and subduction. Thus, any craters that were formed by impacts on the ancient Earth's surface have long since been erased. There are about 150 impact craters on the Earth today, but these were formed within the last few hundred million years or less and have not yet been erased.

We have uncovered a general rule:

A planet or satellite with abundant craters has an old surface with little or no geological activity.

We will make use of this rule many times as we explore other worlds in the solar system.

9-2 Manned exploration of the lunar surface was one of the greatest adventures in human history

For thousands of years, storytellers have invented fabulous tales of voyages to the Moon. The seventeenth-century French satirist Cyrano de Bergerac wrote of a spacecraft whose motive power was provided by spheres of morning dew. (After all, reasoned Cyrano, dew rises with the morning sun, so a spacecraft would rise along with it.) In his imaginative tale *Somnium*, Cyrano's contemporary Johannes Kepler (the same

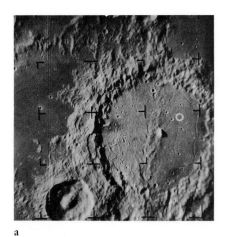
a

b

figure 9-8 R I **V** U X G
The Moon from *Ranger 9* and from Earth (a) Seconds before it crashed onto the Moon on March 24, 1965, *Ranger 9* sent back this picture of the crater Alphonsus. (The white circle shows where the spacecraft hit.) The resolution is far better than in the Earth-based telescopic view shown in **(b)**. Alphonsus is about 129 km (80 mi) in diameter. (NASA)

Kepler who deduced the laws of planetary motion) used the astronomical knowledge of his time to envision what it would be like to walk on the Moon's surface. In 1865 French author Jules Verne published *From the Earth to the Moon*, a story of a spacecraft sent to the Moon from a launching site in Florida.

Almost exactly a hundred years later, Florida was indeed the starting point of the first voyages made by humans to other worlds. Between 1969 and 1972, 12 astronauts walked on the Moon, and fantasy became magnificent reality.

As in Verne's story, politics had much to do with the motivation for going to the Moon. But in fact there were excellent scientific reasons to do so. Because the Earth is a geologically active planet, typical terrestrial surface rocks are only a few hundred million years old, which is only a small fraction of the Earth's age. Thus, all traces of the Earth's origins have been erased. By contrast, rocks on the geologically inactive surface of the Moon have been essentially undisturbed for billions of years. Samples of lunar rocks have helped answer many questions about the Moon and have shed light on the origin of the Earth and the birth of the entire solar system.

Lunar missions began in 1959, when the Soviet Union sent three small spacecraft—*Luna*s 1, 2, and 3—toward the Moon. American attempts to reach the Moon began in the early 1960s with Project Ranger. Equipped with six television cameras, the *Ranger* spacecraft transmitted close-up views of the lunar surface taken in the final few minutes before it crashed onto the Moon. Figure 9-8 shows a view from *Ranger 9* just before it hit the floor of the crater Alphonsus. The *Ranger* images showed far more detail than even the best Earth-based telescope.

A complete high-resolution photographic survey of the lunar surface was the mission of the *Lunar Orbiter* program, under which five spacecraft were launched between August 1966 and August 1967. In all, the five *Lunar Orbiter*s sent back a total of 1950 high-resolution pictures covering 99½% of the Moon. Some of these photographs show features as small as 1 meter across. Because these pictures revealed the lunar surface in close detail, they were essential in helping NASA scientists choose landing sites for the astronauts.

Even after the *Ranger* missions, it was unclear whether the lunar surface was a safe place to land. One leading theory was that billions of years of impacts by meteoroids had ground the Moon's surface to a fine powder, much like that used by dental hygienists to polish teeth. Perhaps a spacecraft attempting to land would simply sink into a quicksand-like sea of dust! Thus, before a manned landing on the Moon could be attempted, it was necessary to try a soft landing of an unmanned spacecraft. This was the purpose of the *Surveyor* program. Between June 1966 and January 1968, five *Surveyor*s successfully touched down on the Moon, sending back pictures and data directly from the lunar surface. These missions demonstrated convincingly that the Moon's surface was as solid as that of the Earth. Figure 9-9 shows an astronaut visiting *Surveyor 3*, two and a half years after it had landed on the Moon.

The first of six manned lunar landings took place on July 20, 1969, when the *Apollo 11* lunar module *Eagle* set down in Mare Tranquillitatis (see Figure 9-6). Astronauts Neil Armstrong and Edwin "Buzz" Aldrin were the first humans to set foot on the Moon. *Apollo 12* also landed in a mare, Oceanus Procellarum (Ocean of Storms). The *Apollo 13* mission experienced a nearly catastrophic inflight explosion that made it impossible for it to land on the Moon. Fortunately, titanic efforts by the astronauts and ground personnel brought it safely home. The remaining four missions, *Apollo 14* through *Apollo 17*, were made in progressively more challenging terrain, chosen to permit the exploration of a wide variety of geologically interesting features. The first era of human exploration of the Moon culminated in December 1972, when *Apollo 17* landed in rugged mountains east of Mare Serenitatis (see the photo that opens this chapter). Figure 9-10 shows one of the *Apollo* bases.

Between 1966 and 1976, a series of unmanned Soviet spacecraft also orbited the Moon and soft-landed on its surface. The first soft landing, by *Luna 9*, came four months before the U.S. *Surveyor 1* arrived. The first lunar satellite, *Luna 10*, orbited the Moon four months before *Lunar Orbiter 1*. In the

figure 9-9 R I **V** U X G

An Astronaut Visits Surveyor 3 *Surveyor 3* successfully soft-landed on the Moon in April 1967. It was visited two and a half years later by *Apollo 12* astronaut Alan Bean, who took pieces of the spacecraft back to the Earth for analysis. (The *Apollo 12* lunar module is visible in the distance.) The retrieved pieces showed evidence of damage by micrometeoroid impacts, demonstrating that small bits of space debris are still colliding with the Moon. (Pete Conrad, *Apollo 12*, NASA)

figure 9-10 R I **V** U X G

The Apollo 15 Base This photograph shows astronaut James Irwin and the *Apollo 15* landing site at the foot of the Apennine Mountains near the eastern edge of Mare Imbrium. The hill in the background is about 5 km away and rises about 3 km high. At the right is a Lunar Rover, an electric-powered vehicle used by astronauts to explore a greater radius around the landing site. The last three *Apollo* missions carried Lunar Rovers. (Dave Scott, *Apollo 15*, NASA)

1970s, robotic spacecraft named *Luna 17* and *Luna 21* landed vehicles that roamed the lunar surface, and *Luna*s *16, 20,* and *24* brought samples of rock back to the Earth.

Unmanned spacecraft continued the exploration of the Moon in the 1990s. The *Galileo* spacecraft was launched in 1989 on a mission to Jupiter. On the way, it followed a convoluted trajectory that took it past the Earth-Moon system in December 1990 and December 1992. During these two passes, *Galileo* recorded color images of the Moon with cameras far more advanced than those of the *Lunar Orbiter*s (see Figure 9-7). In 1994 a small spacecraft named *Clementine* spent more than two months observing the Moon from lunar orbit (see the photograph that opens Chapter 7). Although originally intended to test lightweight imaging systems for future defense satellites, *Clementine* proved to be a remarkable tool for lunar exploration. It carried an array of very high-resolution cameras sensitive to ultraviolet and infrared light as well as the visible spectrum. Different types of materials on the lunar surface reflect different wavelengths, so analysis of *Clementine* images made in different parts of the electromagnetic spectrum has revealed the composition of large areas of the Moon's surface (Figure 9-11).

One remarkable result of the *Clementine* mission was evidence for a patch of ice tens or hundreds of kilometers across at the Moon's south pole. Deep within the South Pole–Aitken Basin (see Figure 9-7) are places where sunlight never reaches and where temperatures are low enough to prevent ice from evaporating. Radio waves beamed from *Clementine* toward these areas were detected by radio tele-

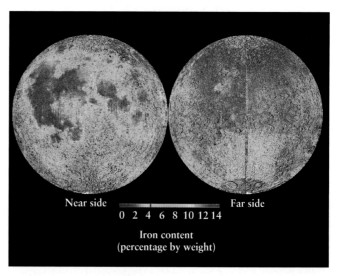

figure 9-11 R I **V** U X G

***Clementine* Maps the Moon's Surface Composition** Images at different wavelengths from the *Clementine* spacecraft were used to make this map of the concentration of iron on the lunar surface. The areas of highest concentration, shown in red, coincide closely with the maria on the near side (compare Figure 9-2). The lowest iron concentration, shown in blue, is found in the lunar highlands. (No data were collected in the gray areas.) The data confirm that the maria were formed from iron-rich lavas. The large green region of intermediate iron concentration on the far side is the South Pole–Aitken Basin (see Figure 9-7). (JSC/NASA)

scopes on the Earth and showed the characteristic signature of reflection from ice.

 Data from the *Lunar Prospector* spacecraft, which went into orbit around the Moon in January 1998, reinforced the *Clementine* findings and indicated that even more ice lies in deep craters at the Moon's north pole. Altogether there may be as much as 6 billion tons of ice at the lunar poles, enough to meet the water requirements of 40,000 future lunar colonists for a century. The ice may have been deposited by comets (see Section 7-5) that crashed into the lunar surface.

One complication is that the *Lunar Prospector* observations confirm only that there is an excess of hydrogen atoms at the lunar poles. These could be trapped within water molecules (H_2O) or within other, more exotic minerals. At the end of the *Lunar Prospector* mission in July 1999, the spacecraft was crashed into a crater at the lunar south pole in the hope that a plume of water vapor might rise from the impact site. Although more than a dozen large Earth-based telescopes watched the Moon during the impact, no such plume was observed, nor was any debris of any kind. Thus, the tantalizing notion that ice may exist at the lunar poles remains to be confirmed. A new generation of European and Japanese unmanned spacecraft, scheduled to be launched toward the Moon starting in 2002, may help provide a definitive answer.

9-3 The Moon has no global magnetic field but has a small core beneath a thick mantle

The Moon landings gave scientists an unprecedented opportunity to explore an alien world. The *Apollo* spacecraft were therefore packed with an assortment of apparatus that the astronauts deployed around the landing sites (Figure 9-12). For example, all the missions carried magnetometers, which confirmed that the present-day Moon has no global magnetic field. However, careful magnetic measurements of lunar rocks returned by the *Apollo* astronauts indicated that the Moon did have a weak magnetic field when the rocks solidified billions of years ago. Hence, the Moon must originally have had a small, molten, iron-rich core. (We saw in Section 8-4 that the Earth's magnetic field is generated by fluid motions in its interior.) The core presumably solidified at least partially as the Moon cooled, so that the lunar magnetic field disappeared.

More recently, scientists used *Lunar Prospector* to probe the character of the Moon's core. The spacecraft made extensive measurements of lunar gravity and of how the Moon responds when it passes through the Earth's magnetosphere (Section 8-4). When combined with detailed observations of how the Earth's tidal forces affect the motions of the Moon, the data indicate that the present-day Moon has a partially liquid, iron-rich core about 700 km (435 mi) in diameter. While 32% of the Earth's mass is in its core, the core of the Moon contains only 2 to 3% of the lunar mass.

Scientists have used seismic waves to probe the Moon's interior, just as is done for the Earth (Section 8-2). The *Apollo* astronauts set up seismometers on the Moon's surface. It was found that the Moon exhibits far less seismic activity than the Earth. Roughly 3000 **moonquakes** were detected per year, whereas a similar seismometer on the Earth would record hundreds of thousands of earthquakes per year. Furthermore, typical moonquakes are very weak. A major terrestrial earthquake is in the range from 6 to 8 on the Richter scale, while a major moonquake measures 0.5 to 1.5 on that scale and would go unnoticed by a person standing near the epicenter.

Analysis of the feeble moonquakes reveals that most originate 600 to 800 km below the surface, deeper than most earthquakes. On Earth, the deepest earthquakes mark the boundary between the solid lithosphere and the plastic asthenosphere. This is because the lithosphere is brittle enough to fracture and produce seismic waves, whereas the deeper rock

ƒigure 9-12 R I **V** U X G

Seismic Equipment on the Moon This view of the *Apollo 11* base shows astronaut Buzz Aldrin standing alongside a seismometer that detected moonquakes and transmitted data back to the Earth. By combining data from seismometers left at different locations on the Moon, scientists have been able to deduce the structure of the Moon's interior. The *Apollo 11* lunar module and some additional equipment are in the background. (Neil Armstrong, *Apollo 11*, NASA)

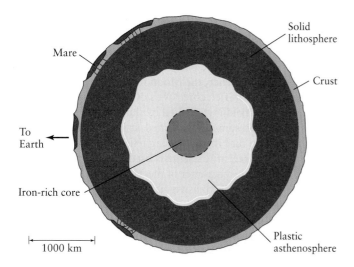

figure 9-13

The Internal Structure of the Moon Like the Earth, the Moon has a crust, a mantle, and a core. The lunar crust has an average thickness of about 60 km on the Earth-facing side but about 100 km on the far side. The crust and solid upper mantle form a lithosphere about 800 km thick. The plastic (nonrigid) asthenosphere extends all the way to the base of the mantle. The iron-rich core is roughly 700 km in diameter.

of the asthenosphere oozes and flows rather than cracks. We may conclude that the Moon's lithosphere is about 800 km thick (Figure 9-13). By contrast, the Earth's lithosphere is only 50 to 100 km thick.

The lunar asthenosphere extends down to the Moon's relatively small iron-rich core, which is more than 1400 km below the lunar surface. The asthenosphere and the lower part of the lithosphere are presumably composed of relatively dense iron-rich rocks. The uppermost part of the lithosphere is a lower-density crust, with an average thickness of about 60 km on the Earth-facing side and up to 100 km on the far side. For comparison, the thickness of the Earth's crust ranges from 5 km under the oceans to about 30 km under major mountain ranges on the continents (see Table 8-2). We saw in Section 4-8 that this slightly lopsided distribution of mass between the near and far sides of the Moon helps to keep the Moon in synchronous rotation, so that we always see the same side of the Moon.

On the whole, the Moon's interior is much more rigid and less fluid than the interior of the Earth. Plate tectonics requires that there be fluid motion just below a planet's surface, so it is not surprising that there is no evidence for plate tectonics on the lunar surface.

Earthquakes tend to occur at the boundaries between tectonic plates, where plates move past each other, collide, or undergo subduction (see Figure 8-13). But there are no plate motions on the Moon. What, then, causes moonquakes? The answer turns out to be the Earth's tidal forces. Just as the tidal forces of the Sun and Moon deform the Earth's oceans and give rise to ocean tides (see Figure 4-25), the Earth's tidal

forces deform the solid body of the Moon. The amount of the deformation changes as the Moon moves around its elliptical orbit and the Earth-Moon distance varies. As a result, the body of the Moon flexes slightly, triggering moonquakes. The tidal stresses are greatest when the Moon's is nearest the Earth—that is, at perigee (see Section 3-5)—which is just when the *Apollo* seismometers reported the most moonquakes. Box 9-1 has more information about how the tidal stresses on the Moon depend on its distance from Earth.

Without tectonics, and without the erosion caused by an atmosphere or oceans, the only changes occurring today on the lunar surface are those due to meteoroid impacts. These impacts were also monitored by the seismometers left on the Moon by the *Apollo* astronauts. These devices, which remained in operation for several years, were sensitive enough to detect a hit by a grapefruit-sized meteoroid anywhere on the lunar surface. The data show that every year the Moon is struck by 80 to 150 meteoroids having masses between 0.1 and 1000 kg (roughly ¼ lb to 1 ton).

Why is the Moon's interior so much more solid than that of the Earth? One central reason is simply that the Moon is much smaller (Figure 9-1). A large turkey or roast taken from the oven will stay warm inside for hours, but a single meatball will cool off much more rapidly. In a similar way, any planet will tend to cool down as it emits electromagnetic radiation into space, but a larger planet (which has less surface area relative to its volume) will cool down more slowly. Because the Moon is so much smaller than the Earth, it has lost much more of its internal heat.

The differences between the Earth and the Moon suggest another general rule for solid worlds:

The smaller the world, the less internal heat it is likely to have retained, and, thus, the less geologic activity it will display on its surface.

In later chapters we will find that this rule explains many key features of the other terrestrial planets, as well as of most of the moons of the solar system. We will also find some remarkable exceptions to this rule, especially among the Moon-sized but geologically active satellites of Jupiter.

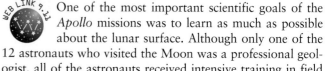

9-4 Lunar rocks reveal a geologic history quite unlike that of Earth

One of the most important scientific goals of the *Apollo* missions was to learn as much as possible about the lunar surface. Although only one of the 12 astronauts who visited the Moon was a professional geologist, all of the astronauts received intensive training in field geology, and they were in constant communication with planetary scientists on Earth as they explored the lunar surface.

The Moon has no atmosphere or oceans to cause "weathering" of the surface. But the *Apollo* astronauts found that the lunar surface has undergone a kind of erosion due to billions of years of relentless meteoritic bombardment. This

Calculating Tidal Forces

The gravitational force between any two objects decreases with increasing distance between the objects. This principle leads to a simple formula for estimating the *tidal* force that the Earth exerts on parts of the Moon.

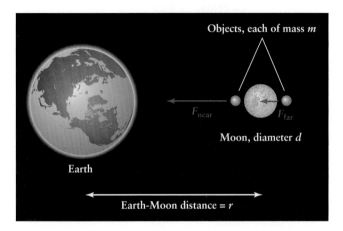

Consider two small objects, each of mass m, on opposite sides of the Moon. Because the two objects are at different distances from the Earth's center, the Earth exerts different forces F_{near} and F_{far} on them (see the accompanying illustration). The consequence of this difference is that the two objects tend to pull away from each other and away from the center of the Moon (see Section 4-8, especially Figure 4-23). The *tidal force* on these two objects, which is the force that tends to pull them apart, is the difference between the forces on the individual objects:

$$F_{tidal} = F_{near} - F_{far}$$

Both F_{near} and F_{far} are both inversely proportional to the square of the distance from the Earth (see Section 4-7). If this distance were doubled, both forces would decrease to $\frac{1}{2}^2 = \frac{1}{4}$ of their original values. But the tidal force F_{tidal}, which is the *difference* between these two forces, decreases even more rapidly with distance: It is inversely proportional to the *cube* of the distance from the Earth to the Moon. That is, doubling the Earth-Moon distance decreases the tidal force to $\frac{1}{2}^3 = \frac{1}{8}$ of its original value. If the Earth's mass is $M_{Earth} = 5.974 \times 10^{24}$ kg, the Moon's diameter is d, and the center-to-center distance from the Earth to the Moon is r, the tidal force is given approximately by the formula

$$F_{tidal} = \frac{2GM_{Earth}md}{r^3}$$

where $G = 6.67 \times 10^{-11}$ newton m²/kg² is the universal constant of gravitation.

EXAMPLE: Let's calculate the tidal force for two 1-kg lunar rocks at the locations shown in the figure. The Moon's diameter d and the average Earth-Moon distance r are given in Table 9-1, but we must convert these values to meters:

$$d = 3476 \text{ km} \times \frac{10^3 \text{ m}}{1 \text{ km}} = 3.476 \times 10^6 \text{ m}$$

$$r = 384,400 \text{ km} \times \frac{10^3 \text{ m}}{1 \text{ km}} = 3.844 \times 10^8 \text{ m}$$

The tidal force is then

$$F_{tidal} = \frac{2(6.67 \times 10^{-11})(5.974 \times 10^{24})(1)(3.476 \times 10^6)}{(3.844 \times 10^8)^3}$$

$$= 4.88 \times 10^{-5} \text{ newton}$$

We can compare this to the weight of a 1-kg rock on the Moon, which is equal to its mass multiplied by the acceleration due to gravity (see Section 4-6). This acceleration is equal to 9.8 m/s² on Earth, but is only 0.17 as great on the Moon. So, the weight of a 1-kg rock on the Moon is

$$0.17 \times 1 \text{ kg} \times 9.8 \text{ m/s}^2 = 1.7 \text{ newtons}$$

Thus, the tidal force on lunar rocks is far less than their weight. This is a good thing: If the tidal force (which tries to pull a rock away from the Moon's center) were greater than the weight (which pulls it toward the Moon's center), loose objects would levitate off the Moon's surface and fly into space!

The tidal force of the Earth on the Moon deforms the solid body of the Moon, so that it acquires *tidal bulges* on its near and far sides. The mass of each bulge must be proportional to the tidal force that lifts lunar material away from the Moon's center. So, the above formula tells us that the mass of each bulge is inversely proportional to the cube of the Earth-Moon distance r:

$$m_{bulge} = \frac{A}{r^3}$$

Determining the value of the constant A requires knowing how easily the Moon deforms, which is a complicated problem beyond our scope.

(Continued on following page)

box 9-1 (*Continued from previous page*)

Our formula for the tidal force F_{tidal} tells us the *net* tidal force on the Moon if we think of the objects of mass m on opposite sides of the Moon as being the tidal bulges. So, we substitute m_{bulge} for m in the tidal force formula:

$$F_{\text{tidal-net}} = \frac{2GM_{Earth}m_{bulge}d}{r^3} = \frac{2GM_{Earth}}{r^3}\frac{A}{r^3}d$$

$$= \frac{2GM_{Earth}Ad}{r^6}$$

Thus, the net tidal force on the Moon is inversely proportional to the *sixth* power of the Earth-Moon distance. Even a small decrease in this distance can cause a substantial increase in the net tidal force.

EXAMPLE: We can compare the net tidal force on the Moon at perigee (when it is closest to the Earth) and at apogee (when it is farthest away). The distances are given in Table 9-1. If we take the ratio of these two forces, the constants (including the unknown quantity A) cancel out:

$$\frac{F_{\text{tidal-net-perigee}}}{F_{\text{tidal-net-apogee}}} = \frac{2GM_{Earth}Ad/r^6_{perigee}}{2GM_{Earth}Ad/r^6_{apogee}} = \left(\frac{r_{apogee}}{r_{perigee}}\right)^6$$

$$= \left(\frac{405{,}500 \text{ km}}{363{,}300 \text{ km}}\right)^6 = (1.116)^6 = 1.93$$

The net tidal force at perigee is almost twice as great as at apogee. This force helps create the stresses that generate moonquakes, so it is not surprising that these quakes are more frequent at perigee than at apogee.

---

bombardment has pulverized the surface rocks into a layer of fine powder and rock fragments called the **regolith,** from the Greek for "blanket of stone" (Figure 9-14). This layer varies in depth from about 2 to 20 m. Just as the rough-surfaced acoustic tile used in ceilings absorbs sound waves, the regolith absorbs most of the sunlight that falls on it. This helps to account for the Moon's very low albedo (see Table 9-1).

In addition to their observations of six different landing sites, the *Apollo* astronauts brought back 2415 individual samples of lunar material totaling 382 kg (842 lb). In addition, the unmanned Soviet spacecraft *Lunas 16, 20,* and *24* returned 300 g (⅔ lb) from three other sites on the Moon. This collection of lunar material has provided important information about the early history of the Moon that could have been obtained in no other way.

All of the lunar samples are igneous rocks; there are no metamorphic or sedimentary rocks. This suggests that most or all of the Moon's surface was once molten. Indeed, Moon rocks are composed mostly of the same minerals that are found in terrestrial volcanic rocks.

Astronauts who visited the maria discovered that these dark regions of the Moon are covered with basaltic rock quite similar to the dark-colored rocks formed by lava from volcanoes on Hawaii and Iceland. The rock of these low-lying lunar plains is called **mare basalt** (Figure 9-15).

In contrast to the dark maria, the lunar highlands are composed of a light-colored rock called **anorthosite** (Figure 9-16). On the Earth, anorthositic rock is found only in such very old mountain ranges as the Adirondacks in the eastern United States. In comparison to the mare basalts, which have more of the heavier elements like iron, manganese, and titanium,

Figure 9-14   R I **V** U X G

The Regolith Billions of years of bombardment by space debris have pulverized the uppermost layer of the Moon's surface into powdered rock. This layer, called the regolith, is utterly bone-dry. It nonetheless sticks together like wet sand, as shown by the sharp outline of an astronaut's bootprint. (*Apollo 11*, NASA)

figure 9-15 R I **V** U X G

Mare Basalt This 1.531-kg (3.376-lb) specimen of mare basalt was brought back by the crew of *Apollo 15*. It is a vesicular basalt, so-called because of the holes, or vesicles, covering 30% of its surface. Gas must have dissolved under pressure in the lava from which this rock solidified. When the lava reached the airless lunar surface, the pressure on the lava dropped and the gas formed bubbles. Some of the bubbles were frozen in place as the rock cooled. (NASA)

figure 9-16 R I **V** U X G

Anorthosite The light-colored lunar terrae (highlands) are composed of this ancient type of rock, which is thought to be the material of the original lunar crust. Lunar anorthosites vary in color from dark gray to white; this sample from the *Apollo 16* mission is a medium gray. It measures 18 × 16 × 7 cm. (NASA)

anorthosite is rich in silicon, calcium, and aluminum. Anorthosite is therefore less dense than basalt. During the period when the Moon's surface was molten, the less dense anorthosite rose to the top, forming the terrae that make up the majority of the present-day lunar surface.

Although anorthosite is the dominant rock type in the lunar highlands, most of the rock samples collected there were not pure anorthosite but rather **impact breccias.** These are composites of different rock types that were broken apart, mixed, and fused together by a series of meteoritic impacts (Figure 9-17).

Every rock found on Earth, even those from the driest desert, has some water locked into its mineral structure. But lunar rocks are totally dry. With the exception of ice at the lunar poles (see Section 9-2), there is no evidence that water has ever existed on the Moon. Because the Moon lacks both an atmosphere and liquid water, it is not surprising that no traces of life have ever been found there.

By carefully measuring the abundances of trace amounts of radioactive elements in lunar samples, geologists have determined that lunar rocks formed more than 3 billion years ago. Of these ancient rocks, however, anorthosite is much older than the mare basalts. Typical anorthositic specimens from the highlands are between 4.0 and 4.3 billion years old. All these extremely ancient specimens are probably samples of the Moon's original crust. In sharp contrast, all the mare basalts are between 3.1 and 3.8 billion years old. With no plate tectonics to bring fresh material to the surface, no new anorthosites or basalts have formed on the Moon for more than 3 billion years.

By correlating the ages of Moon rocks with the density of craters at the sites where the rocks were collected, geologists

have found that the rate of impacts on the Moon has changed over the ages. The ancient, heavily cratered lunar highlands are evidence of an intense bombardment that dominated the Moon's early history. For nearly a billion years, rocky debris left over from the formation of the planets rained down upon

figure 9-17 R I **V** U X G

Impact Breccias These rocks are evidence of the Moon's long history of bombardment from space. Parts of the Moon's original crust were shattered and strewn across the surface by meteoritic impacts. Later impacts compressed and heated this debris, welding it into the type of rock shown here. Such impact breccias are rare on Earth but abundant on the Moon. (NASA)

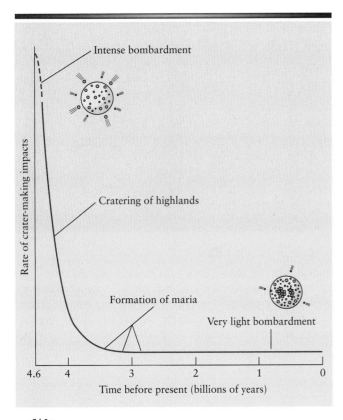

figure 9-18

The Rate of Crater Formation on the Moon This graph shows the rate at which impact craters have been formed on the Moon during different epochs in lunar history. The cratering rate was very high during the first 800 million years after the Moon formed. (The very early history of cratering is not well understood, because the most ancient craters have been covered up or erased by subsequent craters. To indicate this uncertainty, the leftmost part of the curve is shown as a dashed line.) Near the end of this intense bombardment, several large impacts gouged out the mare basins. The impact rate for the past 3.5 billion years has been only about 10^{-4} as great as during the most intense bombardment.

the Moon's young surface. As Figure 9-18 shows, this barrage extended from 4.6 billion years ago, when the Moon's surface solidified, until about 3.8 billion years ago. At its peak, this bombardment from space would have produced a new crater of about 1 km radius somewhere on the Moon about once per century. (If this seems like a long time interval, remember that we are talking about a bombardment that lasted hundreds of millions of years and produced millions of craters over that time.)

The frequency of impacts gradually tapered off, however, as meteoroids and planetesimals were swept up by the newly formed planets. This crater-making era ended about the time when the impact of several large planetesimals gouged out the mare basins. A series of lava flows that occurred between

3.8 and 3.1 billion years ago flooded the mare basins with iron-rich magma that oozed up from the Moon's still-molten mantle. The relative absence of craters in the maria (see Figure 9-5) tells us that the impact rate over the past 3 billion years has been quite low.

9-5 The Moon probably formed from debris cast into space when a huge planetesimal struck the proto-Earth

The lunar rocks collected by the *Apollo* and *Luna* missions have provided important clues about the history of the Moon. Additional information has come from a simple device that the *Apollo 11, 14,* and *15* missions left on the Moon: a set of reflectors, similar to the orange and red ones found in automobile taillights. Since 1969, scientists on Earth have been firing pulses of laser light at these reflectors and measuring very accurately the length of time it takes each pulse to return to Earth. Knowing the speed of light, they can use this data to calculate the distance to the Moon to an accuracy of just 3 centimeters. Remarkably, they have found that the Moon is moving away from the Earth at a very gradual rate of 3.8 cm per year.

This recession was predicted 70 years before *Apollo* by the British astronomer George Darwin, second son of the famous evolutionist Charles Darwin. Darwin began with the notion that the Moon's tidal forces elongate the oceans into a football shape (see Section 4-8, especially Figure 4-25). However, the long axis of this "football" does not point precisely at the Moon. The Earth spins on its axis more rapidly than the Moon revolves around the Earth, and this rapid rotation carries the tidal bulge about 10° ahead of a line connecting the Earth and Moon (Figure 9-19). This misaligned bulge produces a small but steady gravitational force that tugs the Moon forward and lifts it into an ever larger orbit around the Earth. In other words, Darwin concluded that the Moon must be spiraling away from the Earth—which is just what the laser experiments confirmed.

As the Moon moves away from the Earth, its sidereal period is getting longer in accordance with Kepler's third law (see Section 4-7). At the same time, friction between the oceans and the body of the Earth is gradually slowing the Earth's rotation. The length of the Earth's day is therefore increasing, by approximately 0.002 second per century.

These observations mean that in the distant past, the Earth was spinning faster than it is now and the Moon was much closer. Darwin theorized that the early Earth may have been spinning so fast that a large fraction of its mass tore away, and that this fraction coalesced to form the Moon. This is called the **fission theory** of the Moon's origin. Prior to the *Apollo* program, two other theories were in competition with the fission theory. The **capture theory** posits that the Moon was formed elsewhere in the solar system and then drawn into orbit about the Earth by gravitational forces. By contrast, the **co-creation theory** proposes that the Earth and the Moon were formed at the same time but separately.

Forward pull on Moon makes
it spiral into a larger orbit

Moon

10°

Friction between Earth and its
oceans makes Earth rotate more slowly

figure 9-19

The Moon's Tidal Recession The Earth's rapid rotation drags the tidal bulge of the oceans about 10° ahead of a direct alignment with the Moon. The bulge on the side nearest the Moon exerts a small forward force on the Moon that makes it spiral slowly away from the Earth. At the same time, friction between the oceans and the solid Earth slows down the Earth's rotation rate.

One fact used to support the fission theory was that the Moon's average density (3344 kg/m³) is comparable to that of the Earth's outer layers, as would be expected if the Moon had been flung out of a rapidly rotating proto-Earth. However, the fission theory predicts that lunar and terrestrial rocks should be similar in composition. In fact, samples returned from the Moon show that they are significantly different.

Volatile elements (such as potassium and sodium) boil at relatively low temperatures under 900°C, whereas **refractory elements** (such as titanium, calcium, and aluminum) boil at 1400°C or higher. Compared with terrestrial rocks, the lunar rocks have slightly greater proportions of the refractory elements and slightly lesser proportions of the volatile elements. This suggests that the Moon formed from material that was heated to a higher temperature than that from which the Earth formed. Some of the volatile elements boiled away, leaving the young moon relatively enriched in the refractory elements.

Proponents of the capture theory used these differences to argue that the Moon formed elsewhere and was later captured by the Earth. They also noted that the plane of the Moon's orbit is close to the plane of the ecliptic, as would be expected if the Moon had originally been in orbit about the Sun but was later captured by the Earth.

There are difficulties with the capture theory, however. If the Earth did capture the Moon intact, then a host of conditions must have been met entirely by chance. The Moon must have coasted to within 50,000 km of the Earth at exactly the right speed to leave a solar orbit, and must gone into orbit around the Earth without actually hitting our planet.

Proponents of the co-creation theory argued that it is easier for a planet to capture swarms of tiny rocks. In this theory the Moon formed from just such rocky debris. Great numbers of rock fragments orbited the newborn Sun in the plane of the ecliptic along with the protoplanets. Heat generated in collisions could have baked the volatile elements out of these smaller rock fragments. Then, just as planetesimals accreted to form the proto-Earth in orbit about the Sun, the fragments in orbit about the Earth accreted to form the Moon.

The present scientific consensus is that *none* of these three theories—fission, capture, and co-creation—is correct. Instead, the evidence points toward an idea proposed in 1975 by William Hartmann and Donald Davis and independently by Alastair Cameron and William Ward. In this **collisional ejection theory,** the proto-Earth was struck off-center by a Mars-sized object, and this collision ejected debris from which the Moon formed.

The collisional ejection theory agrees with what we know about the early history of the solar system. We saw in Section 7-8 that smaller objects collided and fused together to form the inner planets. Some of these collisions must have been quite spectacular, especially near the end of planet formation, when most of the mass of the inner solar system was contained in a few dozen large protoplanets. The collisional ejection theory proposes that one such protoplanet struck the proto-Earth 4.6 billion years ago. Figure 9-20 shows a supercomputer simulation of this cataclysm. In the simulation, energy released during the collision produces a huge plume of vaporized rock that squirts out from the point of impact. As this ejected material cools, it coalesces to form the Moon (Figure 9-20f).

The collisional ejection theory also agrees with many properties of lunar rocks and of the Moon as a whole. For example, rock vaporized by the impact would have been depleted of volatile elements and water, leaving the moon rocks we now know. If the off-center collision took place after chemical differentiation had occurred on the Earth, so that our planet's iron had sunk to its center, then relatively little iron would have been ejected to become part of the Moon (see Figure 9-20). This explains the Moon's low density and the small size of its iron core. It also explains a curious property of the South Pole–Aitken Basin (see Figure 9-7), where an ancient impact excavated the surface down to a depth of 12 km. Data from the *Clementine* spacecraft show that iron concentration in this basin is only 10% (see Figure 9-11), far less than the 20–30% concentration at a corresponding depth below the Earth's surface.

We can now summarize the entire geologic history of the Moon. The surface of the newborn Moon was probably molten for many years, both from heat released during the impact of rock fragments falling into the young satellite and from the decay of short-lived radioactive isotopes. As the

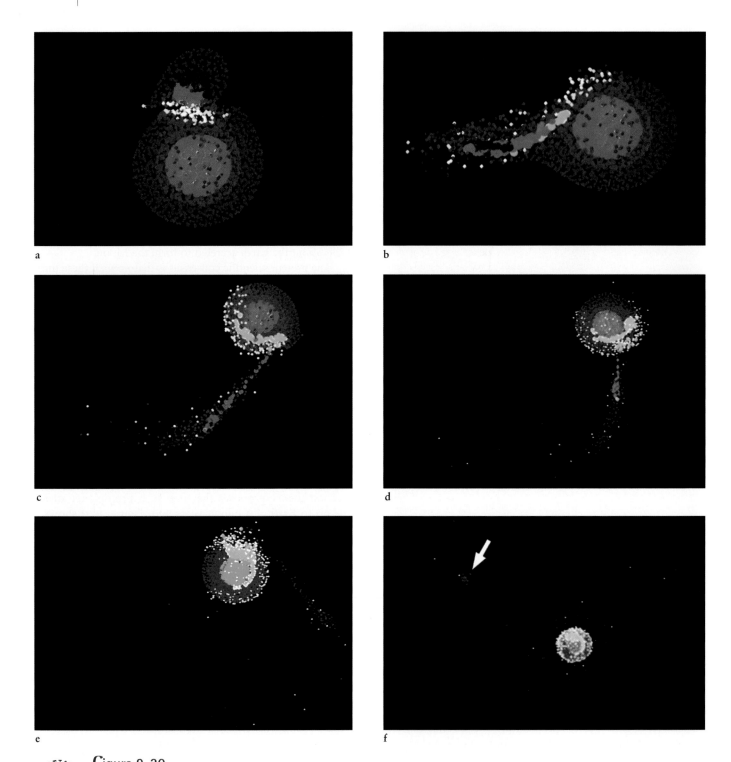

a

b

c

d

e

f

figure 9-20

The Formation of the Moon This supercomputer simulation shows how the Moon could have formed in the aftermath of a collision between a Mars-sized protoplanet and the proto-Earth. To follow the stages of this simulation, successive views zoom out to show increasingly larger volumes of space. Blue and green indicate iron from the cores of the Earth and the protoplanet; red, orange, and yellow indicate rocky mantle material. In this simulation, the impact ejects both mantle and core material, but most of the iron falls back onto the Earth. Some rocky matter does not fall back but coalesces to form the Moon (indicated by the arrow) during its first orbit of the Earth. (Courtesy of W. Benz)

Moon gradually cooled, low-density lava floating on the Moon's surface began to solidify into the anorthositic crust that exists today. The barrage of rock fragments that ended about 3.8 billion years ago produced the craters that cover the lunar highlands.

At the end of this crater-making era, more than a dozen asteroid-size objects, each measuring at least 100 km across, struck the Moon and blasted out the vast mare basins. Meanwhile, heat from the decay of long-lived radioactive elements like uranium and thorium began to melt the inside of the Moon. From 3.8 to 3.1 billion years ago, great floods of molten rock gushed up from the lunar interior, filling the impact basins and creating the basaltic maria we see today.

Very little has happened on the Moon since those ancient times. A few relatively fresh craters have been formed, but the astronauts visited a world that has remained largely unchanged for more than 3 billion years. During that same period on the Earth, by contrast, the continents have been totally transformed time and time again (see <u>Section 8-3</u>).

Many questions and mysteries still remain. The six American and three Soviet lunar landings have brought back samples from only nine locations, barely scratching the lunar surface. We still know very little about the Moon's far side and poles. Are there really vast ice deposits at the poles? How much of the Moon's interior is molten? Just how large is its iron core? How old are the youngest lunar rocks? Did lava flows occur over western Oceanus Procellarum only 2 billion years ago, as crater densities there suggest? Could there still be active volcanoes on the Moon? Are there mineral deposits on the lunar surface from which oxygen for life support and rocket fuel could

Figure 9-21

The Moon—Site of Future Industry? Volcanic material on the lunar surface contains trapped oxygen, which could be released by heating lunar material along with hydrogen gas. Oxygen and water could also be extracted from ice deposits at the lunar poles. These essentials could support a permanent lunar colony and long-term exploration of the Moon. The Moon's weaker gravity and lack of an atmosphere also make it a useful launching pad for sending payloads farther into the solar system. The rocket fuel for this enterprise would use oxygen "mined" from the Moon. (NASA)

be extracted, making the Moon a springboard for human exploration of the solar system (Figure 9-21)? Such questions can be answered only by returning to the Moon.

KEY WORDS

anorthosite, p. 216

capture theory (of Moon's formation), p. 218

co-creation theory (of Moon's formation), p. 218

collisional ejection theory (of Moon's formation), p. 219

crater, p. 208

far side (of the Moon), p. 206

fission theory (of Moon's formation), p. 218

impact breccia, p. 217

impact crater, p. 208

libration, p. 206

lunar highlands, p. 209

mare (plural maria), p. 208

mare basalt, p. 216

moonquake, p. 213

refractory element, p. 219

regolith, p. 216

synchronous rotation, p. 206

terminator, p. 206

terrae, p. 209

volatile element, p. 219

KEY IDEAS

Appearance of the Moon: The Earth-facing side of the Moon displays light-colored, heavily cratered highlands and dark-colored, smooth-surfaced maria. The Moon's far side has almost no maria.

• Virtually all lunar craters were caused by space debris striking the surface. There is no evidence of plate tectonic activity on the Moon.

Internal Structure of the Moon: Much of our knowledge about the Moon has come from human exploration in the 1960s and early 1970s and from more recent observations by unmanned spacecraft.

• Analysis of seismic waves and other data indicates that the Moon has a crust thicker than that of the Earth (and thickest on the far side of the Moon), a mantle with a thickness equal to about 80% of the Moon's radius, and a small iron core.

• The Moon's lithosphere is far thicker than that of the Earth. The lunar asthenosphere probably extends from the base of the lithosphere to the core.

• The Moon has no global magnetic field today, although it had a weak magnetic field billions of years ago.

Geologic History of the Moon: The anorthositic crust exposed in the highlands was formed between 4.0 and 4.3 billion years ago, whereas the mare basalts solidified between 3.1 and 3.8 billion years ago.

• The Moon's surface has undergone very little change over the past 3 billion years.

• Meteoroid impacts have been the only significant "weathering" agent on the Moon. The Moon's regolith, or surface layer of powdered and fractured rock, was formed by meteoritic action.

• All of the lunar rock samples are igneous rocks formed largely of minerals found in terrestrial rocks. The lunar rocks contain no water and also differ from terrestrial rocks in being relatively enriched in the refractory elements and depleted in the volatile elements.

Origin of the Moon: The collisional ejection theory of the Moon's origin holds that the proto-Earth was struck by a Mars-sized protoplanet and that debris from this collision coalesced to form the Moon. This theory successfully explains most properties of the Moon.

• The Moon was molten in its early stages, and the anorthositic crust solidified from low-density magma that floated to the lunar surface. The mare basins were created later by the impact of planetesimals and filled with lava from the lunar interior.

• Tidal interactions between the Earth and Moon are slowing the Earth's rotation and pushing the Moon away from the Earth.

REVIEW QUESTIONS

1. Explain why liquid water cannot exist on the surface of the Moon.

2. What kind of features can you see on the Moon with a small telescope?

3. Describe the differences between the maria and the lunar highlands. Which kind of terrain is more heavily cratered? Which kind of terrain was formed later in the Moon's history? How do we know?

4. What does it mean to say the Moon is a "one-plate world"? What is the evidence for this statement?

5. Why was it necessary to send unmanned spacecraft to land on the Moon before sending humans there?

6. What is the evidence that ice exists at the lunar poles? Is this evidence definitive?

7. Could you use a magnetic compass to navigate on the Moon? Why or why not?

8. Describe the evidence that (**a**) the Moon has a more solid interior than the Earth and (**b**) the Moon's interior is not completely solid.

9. Explain why moonquakes occur more frequently when the Moon is at perigee than at other locations along its orbit.

10. Why is the Earth geologically active while the Moon is not?

11. On the basis of moon rocks brought back by the astronauts, explain why the maria are dark colored but the lunar highlands are light colored.

12. Briefly describe the main differences and similarities between Moon rocks and Earth rocks.

13. Rocks found on the Moon are between 3.1 and 4.6 billion years old. By contrast, the majority of the Earth's surface is made of oceanic crust that is less than 200 million years old, and the very oldest Earth rocks are about 3 billion years old. If the Earth and Moon are essentially the same age, why is there such a disparity in the ages of rocks on the two worlds?

14. Why do most scientists favor the collisional ejection theory of the Moon's formation?

15. Some people who supported the fission theory proposed that the Pacific Ocean basin is the scar left when the Moon pulled away from the Earth. Explain why this idea is probably wrong.

ADVANCED QUESTIONS

Questions preceded by an asterisk (*) involve topics discussed in Box 9-1.

Problem-solving tips and tools

Recall that the average density of an object is its mass divided by its volume. The volume of a sphere is $\frac{4}{3}\pi r^3$, where r is the sphere's radius. The surface area of a sphere of radius r is $4\pi r^2$, while the surface area of a circle of radius r is πr^2. Recall also that the acceleration of gravity on the Earth's surface is 9.8 m/s^2. You may find it useful to know that a 1-pound (1-lb) weight presses down on the Earth's surface with a force of 4.448 newtons. You might want to review Newton's universal law of gravitation in Section 4-7. Consult Table 9-1 and the Appendices for any additional data.

16. In a whimsical moment during the *Apollo 14* mission, astronaut Alan Shepard hit two golf balls over the lunar surface. Give two reasons why they traveled much farther than golf balls do on Earth.

17. If you view the Moon through a telescope, you will find that details of its craters and mountains are more visible when the Moon is near first quarter phase or last quarter phase than when it is at full phase. Explain why.

18. Temperature variations between day and night are much more severe on the Moon than on Earth. Explain why.

19. Using the diameter and mass of the Moon given in Table 9-1, verify that the Moon's average density is about 3344 kg/m³. Explain why this average density implies that the Moon's interior contains much less iron than the interior of the Earth.

20. How much would an 80-kg person weigh on the Moon? How much does that person weigh on the Earth?

*21. In Box 9-1 we calculated the tidal force that the Earth exerts on two 1-kg rocks located on the near and far sides of the Moon. We assumed that the Earth-Moon distance was equal to its average value. Repeat this calculation (a) for the Moon at perigee and (b) for the Moon at apogee. (c) What is the ratio of the tidal force at perigee to the tidal force at apogee?

22. The last manned spacecraft to land on the Moon was the *Apollo 17* lunar module *Challenger* in December 1972. After lifting off from the lunar surface and returning to the command module *America*, in which they would return to the Earth, the *Apollo 17* astronauts sent the unoccupied *Challenger* to crash into the lunar surface. The seismic waves from this impact were detected by seismometers left by the crews of *Apollos 12, 14, 15, 16,* and *17*. Why was it useful to do this? Why was it not enough to detect seismic waves from naturally occurring moonquakes?

23. The youngest lunar anorthosites are 4.0 billion years old, and the youngest mare basalts are 3.1 billion years old. Would you expect to find any impact breccias on the Moon that formed less than 3.1 billion years ago? Explain your answer.

24. In the maria, the lunar regolith is about 2 to 8 meters deep. In the lunar highlands, by contrast, it may be more than 15 meters deep. Explain how the different ages of the maria and highlands can account for these differences.

25. The mare basalts are volcanic rock. Is it likely that active volcanoes exist anywhere on the Moon today? Explain.

26. During the period of most intense meteoritic bombardment, a new 1-km-radius crater formed somewhere on the Moon about once per century. During this same period, what was the probability that such a crater would be created within 1 km of a certain location on the Moon during a 100-year period? During a 10⁶-year period? (*Hint:* If you drop a coin onto a checkerboard, the probability that the coin will land on any particular one of the board's 64 squares is ¹⁄₆₄.)

27. Calculate the round-trip travel time for a pulse of laser light that is fired from a point on the Earth nearest the

Moon, hits a reflector at the point on the Moon nearest the Earth, and returns to its point of origin. Assume that the Earth and Moon are at their average separation from each other.

28. If the tidal recession of the Moon were to continue indefinitely, eventually the length of the day and the lunar month would both be equal to 47 of our present days. At that point the Earth would always keep the same face toward the Moon. Determine the average distance between the Earth and the Moon in this situation. (In fact, the Sun will have come to the end of its lifetime long before this happens.)

29. Before the *Apollo* missions to the Moon, there were two diametrically opposite schools of thought about the history of lunar geology. The "cold moon" theory held that all lunar surface features were the result of impacts. The most violent impacts melted the surface rock, which then solidified to form the maria. The opposite "hot moon" theory held that all lunar features, including maria, mountains, and craters, were the result of volcanic activity. Explain how lunar rock samples show that neither of these theories is entirely correct.

*30. When the Moon originally coalesced, it may have been only one-tenth as far from the Earth as it is now. (a) When the Moon first coalesced, was the Earth's tidal force strong enough to lift rocks off the lunar surface? Explain. (b) Compared with the net tidal force that the Earth exerts on the Moon today, how many times larger was the net tidal force on the newly coalesced Moon? (This strong tidal force kept the one axis of the Moon oriented toward the Earth, and the Moon kept that orientation after it solidified.)

31. If the Earth and the Moon had formed independently in the same region of the solar system, we would expect that iron would be present in each world in about the same percentage. Explain how the different relative sizes of the Earth's core and the Moon's core suggest that the Moon formed instead by collisional ejection.

DISCUSSION QUESTIONS

32. Comment on the idea that without the presence of the Moon in our sky, astronomy would have developed far more slowly.

33. No *Apollo* mission landed on the far side of the Moon. Why do you suppose this was? What would have been the scientific benefits of a mission to the far side?

34. Compare the advantages and disadvantages of exploring the Moon with astronauts as opposed to using mobile, unmanned instrument packages.

35. Describe how you would empirically test the idea that human behavior is related to the phases of the Moon. What problems are inherent in such testing?

36. How would our theories of the Moon's history have been affected if astronauts had discovered sedimentary rock on the Moon?

37. Imagine that you are planning a lunar landing mission. What type of landing site would you select? Where might you land to search for evidence of recent volcanic activity?

 WEB/CD-ROM QUESTIONS

38. Several unmanned missions to the Moon were under development as of this writing. These include *LUNAR-A* and *SELENE* (Institute of Space and Astronautical Science, Japan) and *SMART-1* (European Space Agency). Search the World Wide Web for current information about these missions. When is each mission scheduled to be launched? What investigations will each mission make that have not been made before? What scientific issues may these missions resolve?

 39. Determining the Size of the Planetesimal that Formed the Moon. Access the animation "The Formation of the Moon" in Chapter 9 of the *Universe* web site or CD-ROM. Determine how many times larger the proto-Earth is than the impacting planetesimal. If Mars is about 50% the size of Earth, how does the planetesimal compare in size with present-day Mars?

OBSERVING PROJECTS

Observing tips and tools

 If you do not have access to a telescope, you can learn a lot by observing the Moon through binoculars. Note that the Moon will appear right-side-up through binoculars but inverted through a telescope; if you are using a map of the Moon to aid your observations, you will need to take this into account. Inexpensive maps of the Moon can be purchased from most good bookstores or educational supply stores. You can determine the phase of the Moon either by looking at a calendar (most of which tell you the dates of new moon, first quarter, full moon, and last quarter), by checking the weather page of your newspaper, by consulting the current issue of *Sky & Telescope* or *Astronomy* magazine, by using the *Starry Night* program, or by using the World Wide Web.

40. Use a telescope or binoculars to observe the Moon. Compare the texture of the lunar surface you see on the maria with that of the lunar highlands. How does the visibility of details vary with distance from the terminator (the boundary between day and night on the Moon)?

41. Observe the Moon through a telescope every few nights over a period of two weeks between new moon and full moon. Make sketches of various surface features, such as craters, mountain ranges, and maria. How does the appearance of these features change with the Moon's phase? Which features are most easily seen at a low angle of illumination? Which features show up best with the Sun nearly overhead?

 42. Use the *Starry Night* program to observe the changing appearance of the Moon. First turn off daylight (select **Daylight** in the **Sky** menu) and show the entire celestial sphere (select **Atlas** in the **Go** menu). Center on the Moon by using the **Find...** command in the **Edit** menu. Using the controls at the right-hand end of the Control Panel, set the time step to 3 hours and click on the Forward button (a triangle that points to the right). (**a**) Describe how the phase of the Moon changes. (**b**) Look carefully at features near the left-hand and right-hand limbs (edges) of the Moon. Are these features always at the same position relative to the limb? Explain in terms of libration. (**c**) Does the apparent size of the Moon always stay the same, or does it vary? Explain what this tells you about the shape of the Moon's orbit.

Sun-Scorched Mercury

(NASA)

R I V U X G

These images show two heavily cratered worlds: Mercury (on the left) and the Moon (on the right). While lunar craters have been known since Galileo first turned his telescope on the Moon in 1610, the surface of Mercury was not seen clearly until the 1970s.

Mercury is difficult to observe from Earth: It is too small and too close to the Sun. Earth-based visual observations reveal very few surface details, although radio and radar observations provide information about the planet's temperature and rotation. It was not until 1974 that the first (and so far only) spacecraft to visit Mercury gave us close-up images of the planet. The craters of Mercury, like those of the Moon, are the scars of countless impacts that occurred soon after the planets were formed.

Mercury is by no means a mere clone of the Moon, however. The lava flows on Mercury's surface lack the characteristic dark color of the lunar maria. The surface of Mercury has unusual wrinkles quite unlike anything found on the Moon or, for that matter, the Earth. And unlike the Moon, Mercury has a large iron core and a magnetic field. Mercury thus has a Moon-like surface but an Earthlike interior.

Mercury is also quite unlike our Moon in the way it spins on its axis. While the Moon makes one rotation during one orbit around the Earth, Mercury makes exactly $1\frac{1}{2}$ rotations during one orbit around the Sun. No other planet or satellite rotates in this way.

Despite its small size, Mercury is an important planet to astronomers. If we can understand how Mercury formed and acquired its unique personality, we may be able to understand how the other terrestrial planets—including our own Earth—came to be.

As you read the sections of this chapter, look for the answers to the following questions.

10-1   What makes Mercury such a difficult planet to see?

10-2   What is unique about Mercury's rotation?

10-3   How do the surface features on Mercury differ from those on the Moon?

10-4   Is Mercury's internal structure more like that of the Earth or the Moon?

10-1   Earth-based optical observations of Mercury are difficult

Mercury is often one of the brightest objects in the sky. At its greatest brilliance, it appears brighter than any star, which is why Mercury has been known since ancient times. Its motions played a role in the religious beliefs of the ancient Mayans, Egyptians, Greeks, and Romans.

Although it may be bright, Mercury is so close to the Sun that it is quite difficult to observe. (It is said that Copernicus never saw it.) Mercury moves around the Sun at an average distance of only 0.387 AU (57.9 million kilometers, or 36 million miles) along an orbit that is more eccentric than that of any other planet except Pluto. Figure 10-1 shows the orbits of Mercury and the Earth to scale. Table 10-1 summarizes data about the planet.

Like all the planets, Mercury shines by reflected sunlight. Because it orbits so close to the Sun, it is exposed to sunlight that on average is about 7 times as intense as on Earth. Furthermore, Mercury is always less than 1.5 AU from Earth. These factors explain why Mercury appears bright, despite having a rather low albedo of about 0.12 (that is, it reflects only about 12% of the sunlight that falls on its surface, roughly comparable to weathered asphalt).

Mercury can best be seen with the naked eye when it is as far from the Sun in the sky as it can be, at its greatest eastern or western elongation (Section 4-2). For a few days near the time of **greatest eastern elongation,** Mercury appears low over the western horizon as an "evening star" for a short time after sunset. Near the time of **greatest western elongation,** Mercury can be glimpsed as a "morning star" that rises before the Sun in the brightening eastern sky.

CAUTION! Remember that "greatest *western* elongation" and "greatest *eastern* elongation" refer to Mercury's position in the sky relative to the Sun. When Mercury is at greatest western elongation, it is west of the Sun in the sky; at dawn, when the Sun is rising in the east, Mercury has already risen. When Mercury is at greatest eastern elongation, it is east of the Sun in the sky and is still above the horizon when the Sun sets in the west.

Because its orbit is so close to the Sun, Mercury's maximum elongation is only 28°. The celestial sphere appears to rotate at 15° per hour (360° divided by 24 hours), so Mercury never rises more than 2 hours before sunrise or sets more than 2 hours after sunset. Unfortunately, Mercury's elliptical orbit and its inclination to the ecliptic often place Mercury much less than 28° from the horizon at sunset or sunrise. Hence, some elongations are more favorable for viewing Mercury than others, as Figure 10-2 shows. A total of six or seven greatest elongations (both eastern and western) occur each year (Table 10-2), but usually only two of these will be favorable for viewing the planet.

Naked-eye observations of Mercury are best made at dusk or dawn, but the best telescopic views are obtained in the daytime when the planet is high in the sky, far above the degrading atmospheric effects near the horizon. A yellow filter can eliminate much of the blue light from our sky. The

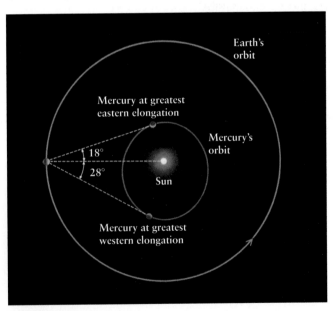

figure 10-1

Mercury's Orbit Mercury moves around the Sun every 88 days in a rather eccentric orbit. The distance between Mercury and the Sun varies from 70 million kilometers at aphelion to only 46 million kilometers at perihelion. As seen from the Earth, the angle between Mercury and the Sun at greatest eastern or western elongation can be as large as 28° (when Mercury is near aphelion) or as small as 18° (near perihelion).

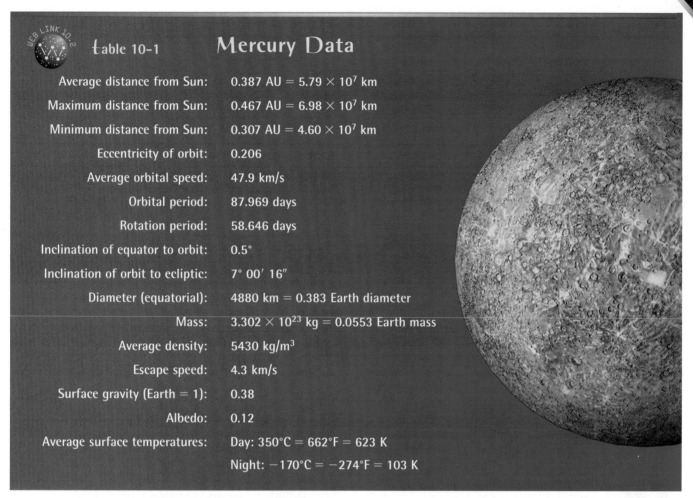

table 10-1	Mercury Data	
Average distance from Sun:	0.387 AU = 5.79×10^7 km	
Maximum distance from Sun:	0.467 AU = 6.98×10^7 km	
Minimum distance from Sun:	0.307 AU = 4.60×10^7 km	
Eccentricity of orbit:	0.206	
Average orbital speed:	47.9 km/s	
Orbital period:	87.969 days	
Rotation period:	58.646 days	
Inclination of equator to orbit:	0.5°	
Inclination of orbit to ecliptic:	7° 00′ 16″	
Diameter (equatorial):	4880 km = 0.383 Earth diameter	
Mass:	3.302×10^{23} kg = 0.0553 Earth mass	
Average density:	5430 kg/m³	
Escape speed:	4.3 km/s	
Surface gravity (Earth = 1):	0.38	
Albedo:	0.12	
Average surface temperatures:	Day: 350°C = 662°F = 623 K	
	Night: −170°C = −274°F = 103 K	

(Astrogeology Team, U.S. Geological Survey) R I **V** U X G

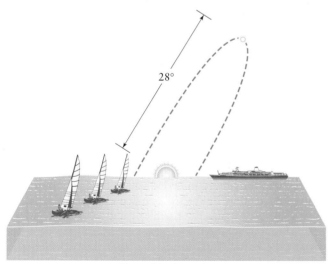

Favorable elongation

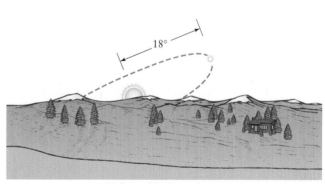

Unfavorable elongation

figure 10-2

Favorable Versus Unfavorable Elongations The tilt of the Earth's axis, the inclination and eccentricity of Mercury's orbit, and the latitude of the observer combine to make an elongation either favorable or unfavorable for viewing Mercury. The elongation at left is favorable; Mercury is at its greatest possible elongation, and its orbital plane is inclined at a large angle to the observer's horizon. The elongation at right is quite unfavorable; Mercury is at its minimum possible elongation, and its orbital plane lies at only a shallow angle above the horizon.

table 10-2	Greatest Elongations of Mercury,* 2001–2005	
Year	Greatest western elongations	Greatest eastern elongations
2001	March 11, July 9, October 29	January 28, May 22, September 18
2002	February 22, June 21, October 13	January 11, May 4, September 1, December 26
2003	February 4, June 3, September 26	April 16, August 14, December 9
2004	January 17, May 14, September 9, December 29	March 29, July 27, November 21
2005	April 26, August 23, December 12	March 12, July 9, November 3

*Mercury can be seen in the eastern sky at dawn for a few days around greatest western elongation.
It can be seen in the western sky at dusk for a few days around greatest eastern elongation.

photographs in Figure 10-3, which are among the finest Earth-based views of Mercury, were taken during daytime.

Mercury travels around the Sun faster than any other object in the solar system. Its sidereal period, or time to complete one full orbit, is only 88 days. Its synodic period, which is the time to go through a complete cycle of configurations as seen from the Earth (see Section 4-2), is 116 days or about one-third of a year. Thus, Mercury passes through inferior conjunction (when it is between the Earth and the Sun) at least three times a year. Therefore, you might expect to see Mercury occasionally silhouetted against the Sun. Such a passage in front of the Sun is called a **solar transit.** In fact, transits of Mercury across the Sun are not very common, because Mercury's orbit is tilted to the plane of the ecliptic by 7°. As a result, Mercury usually lies well above or below the solar disk at the moment of inferior conjunction.

In order for a solar transit to take place, the Sun, Mercury, and the Earth must be in nearly perfect alignment. This arrangement can occur only in May and November, when the Earth is located near the line along which the planes of the Earth's orbit and Mercury's orbit intersect. Even then a solar transit will occur only if Mercury happens to be in inferior conjunction at that time. The first two transits of the twenty-first century are on May 7, 2003, and November 8, 2006. The longest transits occur in May, when Mercury is near aphelion and therefore is traveling comparatively slowly along its orbit. The maximum duration of a solar transit is 9 hours. The photograph of a solar transit in Figure 10-4 is dramatic proof of how tiny Mercury is in comparison with the Sun!

10-2 Mercury rotates slowly and has an unusual spin-orbiting coupling

During the 1880s, the Italian astronomer Giovanni Schiaparelli attempted to make the first map of Mercury. Unfortunately, even today's best telescopes reveal only a few faint, hazy markings. Schiaparelli's views of Mercury were so vague and indistinct that he made an understandable but major error, which went uncorrected for more than half a century. He erroneously concluded that Mercury always kept the same side facing the Sun.

Many objects in our solar system are in synchronous rotation, so that their rotation period equals their period of revolution—a situation also called **1-to-1 spin-orbit coupling.** We saw in Section 4-8 how the Earth's tidal forces keep the Moon in synchronous rotation, so that it always keeps the same side toward the Earth. Tidal forces also keep the two moons of Mars with the same side facing their parent planet, and likewise many of the satellites of Jupiter and Saturn.

It had been suggested as early as 1865 that the Sun's tidal forces would keep Mercury in synchronous rotation. But this is not how Mercury rotates. The first clue about Mercury's true rotation period came in 1962, when astronomers detected radio radiation coming from the innermost planet.

As we learned in Sections 5-3 and 5-4, every dense object (such as a planet) emits electromagnetic radiation with a con-

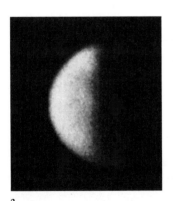
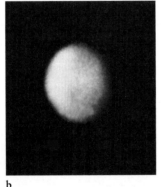

a   b

ƒigure 10-3 R I **V** U X G

Earth-Based Views of Mercury These two views are among the best photographs of Mercury ever produced with an Earth-based telescope. Hazy markings are faintly visible on the tiny planet. Note that Mercury goes through phases like the Moon. It appears larger when it is at first or last quarter, as shown in **(a)**, than when it is nearly full, as shown in **(b)**. Galileo discovered this same relationship for the planet Venus (see Figure 4-8) and used this discovery as compelling evidence that the planets orbit the Sun and not the Earth. (New Mexico State University Observatory)

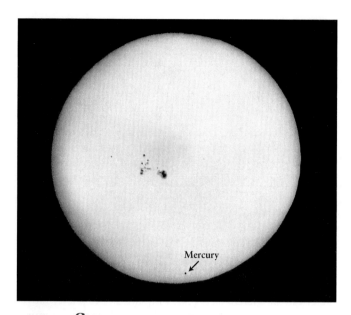

figure 10-4 R I **V** U X G

A Solar Transit of Mercury Roughly a dozen solar transits of Mercury occur in each century. This photograph shows the tiny planet (shown by the arrow) silhouetted against the Sun during the transit of November 14, 1907. The dark sunspots visible near the middle of the Sun's disk are actually much larger than Mercury. (Yerkes Observatory)

tinuous spectrum. The dominant wavelength of this radiation depends on the object's temperature. Only at absolute zero does a dense object emit no radiation at all.

If Mercury exhibited synchronous rotation, one side of the planet would remain in perpetual, frigid darkness, quite near absolute zero. To test this idea, in 1962 the radio astronomer W. E. Howard and his colleagues at the University of Michigan began monitoring radiation from Mercury. (If the planet

had one side that was very cold, its emission should be principally at quite long wavelengths. Hence, a radio telescope was used for these observations.) They were surprised to discover that the temperature on the planet's nighttime side is about 100 K (= −173°C = −280°F), not nearly as cold as expected. These observations were in direct contradiction to the idea of synchronous rotation.

One idea offered to explain Howard's observations was that Mercury rotates synchronously, but that winds in the planet's atmosphere carried warmth from the daytime side of the planet around to the nighttime side. But, in fact, there are no winds on Mercury. Its daytime temperature is too high and its gravity too weak to retain any substantial atmosphere (see Box 7-2).

The definitive observation was made in 1965, when Rolf B. Dyce and Gordon H. Pettengill used the giant 1000-ft radio telescope at the Arecibo Observatory in Puerto Rico (Figure 10-5) to bounce powerful radar pulses off Mercury. The outgoing radiation consisted of microwaves of a very specific wavelength. In the reflected signal echoed back from the planet, the wavelength had shifted as a result of the Doppler effect (see Section 5-9, especially Figure 5-24). As Mercury rotates, one side of the planet approaches the Earth, while the other side recedes from the Earth. Microwaves reflected from the planet's approaching side were shortened in wavelength, whereas those from its receding side were lengthened (Figure 10-6). Thus, the radar pulse went out at one specific wavelength, but it came back spread over a small wavelength range. From the width of this wavelength range, Dyce and Pettengill deduced that Mercury's rotation period is approximately 59 days.

Giuseppe Colombo, an Italian physicist with a long-standing interest in Mercury, found this number intriguing. Colombo noted that Dyce and Pettengill's result for the rotation period is very close to two-thirds of Mercury's sidereal period of 87.969 days:

$$\frac{2}{3}(87.969 \text{ days}) = 58.646 \text{ days}$$

figure 10-5 R I **V** U X G

The Arecibo Radio Telescope With a diameter of 305 m (1000 ft), this is the largest single-dish radio telescope on Earth. The dish is mounted permanently in a bowl-shaped valley in Puerto Rico and is not movable, so it can be used only to examine objects near the zenith. The primary use of the Arecibo telescope is to detect radio waves emitted by objects in space. To measure the rotation period of Mercury, radio waves were emitted by the telescope, and the waves that were reflected back from Mercury were detected. (Arecibo Observatory)

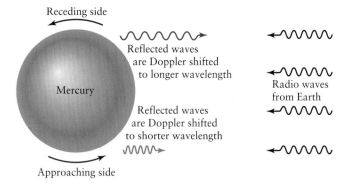

figure 10-6

Measuring Mercury's Rotation Period As Mercury rotates, one side of the planet moves away from the Earth and the other side toward the Earth. If radio waves of a single wavelength are beamed toward Mercury, waves reflected from these two sides will be Doppler shifted to longer and shorter wavelengths, respectively. By measuring the wavelength shift of the reflected radiation, astronomers deduced how fast Mercury rotates and determined its rotation period.

Colombo therefore boldly speculated that Mercury's true rotation period is exactly 58.646 days, or 58 days and 15½ hours. He realized that this figure would mean that Mercury is locked into a **3-to-2 spin-orbit coupling:** The planet

makes *three* complete rotations on its axis for every *two* complete orbits around the Sun. No other planet or satellite in the solar system has this curious relationship between its rotation and its orbital motion.

Figure 10-7 shows how gravitational forces from the Sun cause Mercury's 3-to-2 spin-orbit coupling. The force of gravity decreases with increasing distance, which explains how the Moon is able to raise a tidal bulge in the Earth's oceans (see [Section 4-8](#)). Mercury has no oceans, but it has a natural bulge of its own; thanks to the Sun's tidal forces, the planet is slightly elongated along one axis. The stronger gravitational force that the Sun exerts on the near side of the planet tends to twist the long axis to point toward the Sun, as Figure 10-7a shows. Indeed, Mercury's long axis would always point toward the Sun if its orbit were circular or nearly so. In this case the same side of Mercury would always face the Sun, and there would be synchronous rotation (Figure 10-7b). But Mercury's orbit has a rather high eccentricity, and the twisting effect on Mercury decreases rapidly as the planet moves away from the Sun. As a result, Mercury's long axis points toward the Sun only at perihelion (Figure 10-7c). From one perihelion to the next, alternate sides of Mercury face the Sun.

Colombo's enlightened guess was confirmed by further radar observations and, later, by the *Mariner 10* spacecraft. After making the first-ever close flyby of Mercury in the spring of 1974, *Mariner 10* was placed in a 176-day orbit around the Sun. This orbital period brings the spacecraft back

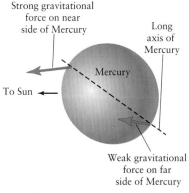

a   Tidal forces on Mercury

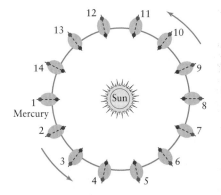

b   1-to-1 spin-orbit coupling

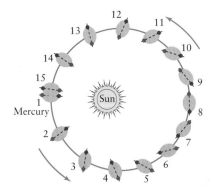

c   3-to-2 spin-orbit coupling

figure 10-7

Spin-Orbit Coupling **(a)** Mercury is not a sphere but is slightly elongated. The gravitational force of the Sun on the near side of Mercury's long axis is greater than the force on the far side, and this tends to twist the long axis to point toward the Sun. **(b)** If Mercury were in a circular orbit, the twisting effect shown in (a) would keep the planet's long axis (shown by a dashed line) always pointed toward the Sun. Hence, Mercury would always keep the same side (shown by a red dot) facing the Sun. **(c)** Because Mercury has a rather eccentric orbit, it exhibits a 3-to-2 spin-orbit coupling; the planet spins 1½ times on its axis during each complete orbit around the Sun, or three times during two complete orbits. The end of Mercury's long axis marked by a red dot faces the Sun at one perihelion (point 1), but at the next perihelion (point 15) the opposite end—marked by a blue dot—faces the Sun.

to the planet every two Mercurian years, just enough time for the planet to complete three rotations. During the second and third flybys, the spacecraft's cameras showed that the same side of Mercury was facing the Sun as on the first flyby. Mercury does indeed rotate three times about its axis for every two orbits it makes of the Sun.

Because of the 3-to-2 spin-orbit coupling, the average time from sunrise to sunset on Mercury is just equal to its orbital period, or just under 88 days. This helps explain the tremendous difference between daytime and nighttime temperatures on Mercury. Not only is sunlight about 7 times more intense on Mercury than on the Earth (because of the smaller size of Mercury's orbit), but that sunlight has almost three Earth months to heat up the surface. As a result, daytime temperatures at the equator are high enough—about 430°C (roughly 700 K, or 800°F)—to melt lead. (In comparison, a typical kitchen oven reaches only about 300°C or 550°F.)

The time from sunset to the next sunrise is also 88 days, and the surface thus has almost three Earth months of darkness during which to cool down. Therefore, nighttime temperatures reach to below −170°C (about 100 K, or −270°F). This is cold enough to freeze carbon dioxide and methane. The surface of Mercury is truly inhospitable!

While the Sun moves slowly across Mercury's sky, it does not always move from east to west as it does as seen from Earth. The reason is that Mercury's speed along its elliptical orbit varies substantially, in accordance with Kepler's second law (see Section 4-5, especially Figure 4-17). It moves fastest (59 km/s) at perihelion and slowest (39 km/s) at aphelion.

As seen from Mercury's surface, the Sun rises in the east and sets in the west, just as it does on the Earth. When Mercury is near perihelion, however, the planet's rapid motion along its orbit outpaces its leisurely rotation about its axis (one rotation in 58.646 days). Hence, the usual east-west movement of the Sun across Mercury's sky is interrupted, and the Sun actually stops and moves backward (from west to east) for a few Earth days. If you were standing on Mercury watching a sunset occurring at perihelion, the Sun would not simply set. It would dip below the western horizon and then come back up, only to set a second time a day or two later.

10-3 Images from *Mariner 10* reveal Mercury's heavily cratered surface

Scientists acquired their first detailed knowledge about Mercury's surface during 1974 and 1975, when the unmanned *Mariner 10* spacecraft made three close flybys of the planet (Figure 10-8a). At each closest approach *Mariner 10* passed over Mercury's darkened, nighttime side, and so was not able to obtain images. The spacecraft's photography was therefore divided into two parts—incoming views, taken prior to closest approach, and outgoing

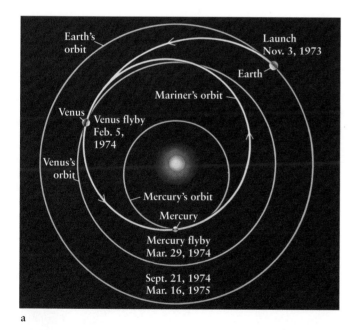

a

b

R I **V** U X G

Figure 10-8

Mariner 10 Views Mercury (a) The unmanned *Mariner 10* spacecraft was placed in a solar orbit that took it past Venus once and past Mercury three times during 1974 and 1975. (b) To make this mosaic of the side of Mercury viewed by *Mariner 10*, scientists combined dozens of images recorded by the spacecraft as it flew toward and away from the planet. The blank swatch running down from Mercury's north pole is a region that was not imaged. (Astrogeology Team, U.S. Geological Survey)

views, taken as *Mariner 10* receded from the planet—which were synthesized to make the mosaic shown in Figure 10-8*b*.

The *Mariner 10* images show only one hemisphere of Mercury's surface, since the time interval between flybys was equal to twice the planet's orbital period and three times its rotation period. Hence, the same side of Mercury was illuminated by the Sun during each flyby.

As *Mariner 10* closed in on Mercury, scientists were struck by the Moonlike pictures appearing on their television monitors. It was obvious that Mercury, like the Moon, is a barren, heavily cratered world, with no evidence for plate tectonics. But craters are not the only features of Mercury's surface; there are also gently rolling plains, long, meandering cliffs, and an unusual sort of jumbled terrain.

Figure 10-9 shows a typical close-up view sent back from *Mariner 10*. The consensus among astronomers is that most of the craters on both Mercury and the Moon were produced during the first 700 million years after the planets formed. Debris remaining after planet formation rained down on these young worlds, gouging out most of the craters we see today. We saw in Section 9-4 that the strongest evidence for this chronology comes from analysis and dating of Moon rocks. *Mariner 10* did not land on Mercury, so we are not able to make the same kind of direct analysis of rocks from the planet's surface.

The differences between Mercury and the Moon are as striking as the similarities. As an example, Figure 10-10 shows the Moon's southern hemisphere. Notice how the lunar craters are densely packed, with one overlapping the next. In sharp contrast, Mercury's surface has extensive low-lying plains (examine the upper half of Figure 10-9). These large, smooth areas are about 2 km lower than the cratered terrain.

We saw in Section 9-4 that the lunar maria were produced by extensive lava flows between 3.1 and 3.8 billion years ago.

Primordial lava flows probably also explain the plains of Mercury. As large meteoroids punctured the planet's thin, newly formed crust, lava welled up from the molten interior to flood low-lying areas.

From the number of craters that pit them, Mercury's plains appear to have been formed near the end of the era of heavy bombardment, a little more than 3.8 billion years ago. Mercury's plains are therefore older than most of the lunar maria. Analysis of *Mariner 10* images made at different wavelengths also shows that the material in Mercury's plains has a lower iron content than the lunar maria. This is presumably why the plains of Mercury do not have the dark color of the Moon's maria. (Contrast the photographs of Mercury and the Moon in the figure that opens this chapter).

Mariner 10 also revealed numerous long cliffs, called **scarps**, meandering across Mercury's surface (Figure 10-11). Some scarps rise as much as 3 km (2 mi) above the surrounding plains and are 20 to 500 km long. These cliffs probably formed as the planet cooled and contracted.

ANALOGY The shrinkage of Mercury's crust is probably a result of a general property of most materials: They shrink when cooled and expand when heated. The lid on a jar of jam makes a useful analogy. When the jar is kept cold in the refrigerator, the metal lid contracts more than the glass jar. This makes a tight seal, but also makes the lid hard to unscrew. You can loosen the lid by running hot water over it; this makes the metal lid expand more than the glass jar, helping to break the seal.

If there had still been molten material beneath Mercury's surface when this contraction took place, lava would leaked onto the surface around the scarps. This does not appear to

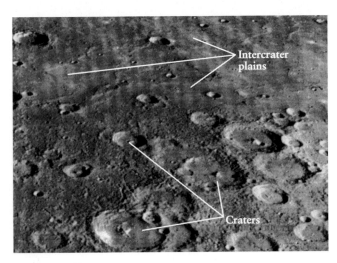

Figure 10-9 R I **V** U X G

Mercurian Craters and Plains This view of Mercury's northern hemisphere was taken by *Mariner 10* at a range of 55,000 km (34,000 mi) from the planet's surface. Numerous craters and extensive intercrater plains appear in this photograph, which covers an area 480 km (300 mi) wide. (NASA)

Figure 10-10 R I **V** U X G

Craters on the Moon This Earth-based photograph shows a 600-km (370-mi) wide portion of the Moon's southern hemisphere during last quarter moon. Note how the lunar craters are more densely packed than those on the surface of Mercury, shown in Figure 10-9. (Carnegie Observatories)

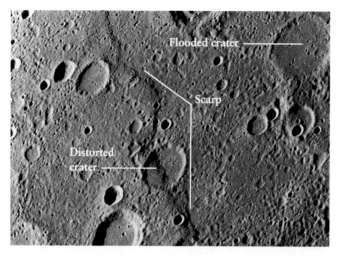

figure 10-11 R I **V** U X G

ANIMATION 10.3

A Scarp A long cliff called Santa Maria Rupes runs across this *Mariner 10* image of a region near Mercury's equator. This cliff, or scarp, is more than a kilometer high and extends for several hundred kilometers. The old, large crater at the center of the photograph was distorted vertically when the scarp formed. At the upper right of the image a crater 70 km (45 mi) in diameter has apparently been flooded by lava that welled up from Mercury's interior. The area shown is approximately 200 km (120 mi) across. (NASA)

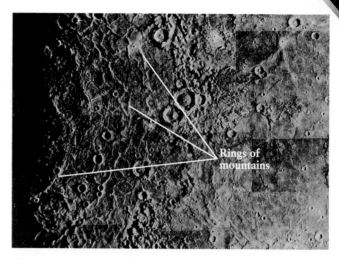

figure 10-12 R I **V** U X G

The Caloris Basin An ancient impact created this 1300-km (810-mi) diameter basin. Its center is hidden in the shadows just beyond the left side of the image. The impact fractured the surface extensively, forming several concentric chains of mountains. The mountains in the outermost ring are up to 2 km (6500 ft) high. (NASA)

have happened. Hence, the scarps must have formed relatively late in Mercury's history, after the lava flows had ended and after the planet had solidified to a substantial depth beneath the surface. With the exception of the scarps, there are no features on Mercury that resemble the boundaries of tectonic plates. Thus, we can regard Mercury's crust, like that of the Moon, as a single plate.

The most impressive feature discovered by *Mariner 10* was a huge impact scar called the Caloris Basin (from the Latin word for "hot"). The Sun is directly over the Caloris Basin during alternating perihelion passages, and thus it is the hottest place on the planet once every 176 days. During each of the *Mariner 10* flybys, the Caloris Basin happened to lie on the terminator (the line dividing day from night; see Section 9-1). Hence, in Figure 10-12 slightly more than half of the basin is hidden on the night side of the planet.

The Caloris Basin, which measures 1300 km (810 mi) in diameter, is both filled with and surrounded by smooth lava plains. The basin was probably gouged out by the impact of a large meteoroid that penetrated the planet's crust, allowing lava to flood out onto the surface and fill the basin. Because relatively few craters pockmark these lava flows, the Caloris impact must be relatively young. It must have occurred toward the end of the crater-making period that dominated the first 700 million years of our solar system's history.

The impact that created the huge Caloris Basin must have been so violent that it shook the entire planet. On the side of Mercury opposite the Caloris Basin, *Mariner 10* discovered a consequence of this impact—a jumbled, hilly region covering about 500,000 square kilometers, about twice the size of the

state of Wyoming (Figure 10-13). Geologists theorize that seismic waves from the Caloris impact became focused as they passed through Mercury. As this concentrated seismic energy

figure 10-13 R I **V** U X G

Unusual, Hilly Terrain The tiny, fine-grained wrinkles in this image are actually closely spaced hills on the side of Mercury opposite the Caloris Basin. The hills are about 5 to 10 km wide and have elevations between 100 and 1800 m (300 and 5900 ft). The large, smooth-floored crater near the center of this photograph is 170 km (106 mi) in diameter. (NASA)

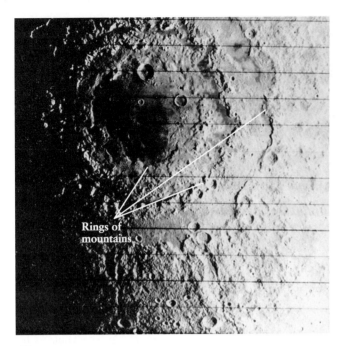

ƒigure 10-14   R I **V** U X G

Mare Orientale on the Moon Mare Orientale is one of several large impact basins on the Moon that resemble the Caloris Basin on Mercury. The outermost ring of peaks forms a circle 900 km (560 mi) in diameter. This image was sent back in 1967 from the *Lunar Orbiter 4* spacecraft. Black lines interrupt the picture because the image was transmitted to the Earth in thin strips. (NASA)

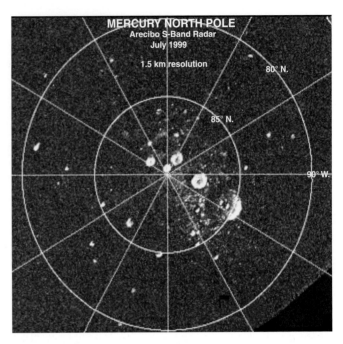

ƒigure 10-15   **R** I V U X G

Ice at Mercury's North Pole? This radar image is centered on Mercury's north pole. White indicates strong reflection of the radar signal, and darker shades indicate progressively weaker reflection. The bright dot at the center of the image, a region of very strong reflection, is thought to be a patch of frozen water (ice). Ice could survive on the floors of craters at Mercury's north pole, where it can remain permanently shaded from the Sun and reach temperatures as low as 150°C (about 125 K, or –235°F). (Courtesy of J. K. Harmon and Martin Slade, JPL/NASA)

reached the far surface of the planet, it deformed the surface and created the hills shown in Figure 10-13.

Features similar to those of the Caloris Basin exist on our Moon. The best example is Mare Orientale, shown in Figure 10-14 (it can also be seen in Figure 9-7). Its diameter is 900 km (560 mi). Just as for the Caloris Basin, the impact that created Mare Orientale fractured the crust and produced encircling rings of mountains. The theory about the seismic upthrusting of jumbled hills on Mercury is supported by the finding of similar (although less extensive) chaotic hills on the side of the Moon opposite Mare Orientale.

In the decades since the *Mariner 10* mission, astronomers have continued to study Mercury with both visible-light and radio telescopes. Radar observations have detected patches at Mercury's north and south poles that are unusually effective at reflecting radio waves (Figure 10-15). It has been suggested that these patches may be regions of water ice deep within craters where the Sun's rays never reach. (We saw in Section 9-2 that ice may also exist in shadowed regions at the Moon's poles.) If this is correct, then Mercury is truly a world of extremes, with perpetual ice caps at one or both poles but with midday temperatures at the equator that are high enough to melt lead.

10-4 Like the Earth, Mercury has an iron core and a magnetic field

Mercury's average density of 5430 kg/m³ is quite similar to that of the Earth, which is 5515 kg/m³. We saw in Section 8-2 that typical rocks from the Earth's surface have a density of only about 3000 kg/m³ because they are composed primarily of lightweight, mineral-forming elements. The higher average density of our planet is caused by the Earth's iron core. By studying how the Earth vibrates during earthquakes, geologists have deduced that our planet's iron core occupies about 17% of the Earth's volume.

Pressures inside a planet increase the density of rock by squeezing its atoms into smaller volumes. Because the Earth is 18 times more massive than Mercury, the weight of this larger mass compresses the Earth's core much more than Mercury's core. If the Earth were uncompressed, its density would be 4400 kg/m³, whereas Mercury's uncompressed density would still be a hefty 5300 kg/m³. This higher uncompressed density means that Mercury has a larger proportion of iron than the Earth: An iron core must occupy about 42% of Mercury's volume. Mercury is therefore the most iron-rich planet in the solar system. Figure 10-16 is a scale drawing of the interior structures of Mercury and the Earth.

Several theories have been proposed to account for Mercury's high iron content. According to one theory, the inner regions of the primordial solar nebula were so warm that only those substances with high condensation temperatures—like iron-rich minerals—could have condensed into solids. Another theory suggests that a brief episode of very powerful solar wind could have stripped Mercury of its low-density mantle shortly after the Sun formed. A third possibility is that during the final stages of planet formation, Mercury was struck by a large planetesimal. Supercomputer simulations show that this cataclysmic collision would have ejected much of the lighter mantle, leaving a disproportionate amount of iron to reaccumulate to form the planet we see today (Figure 10-17).

An important clue to the structure of Mercury's iron core came from the *Mariner 10* magnetometers, which discovered that Mercury has a magnetic field similar to that of the Earth but only about 1% as strong. The Earth's magnetic field (Section 8-4) is produced by electric currents flowing in the liquid portions of the Earth's iron core. (Similarly, the magnetism that surrounds an electromagnet is produced by electricity flowing in a coil of wire.)

For this same effect to take place in Mercury's interior, two things must be true: (1) at least part of Mercury's core must be in a liquid state, and (2) there must be a source of energy to make material flow within this liquid region in order to pro-

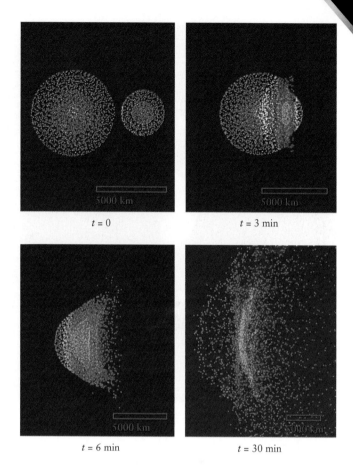

$t = 0$ $t = 3$ min

$t = 6$ min $t = 30$ min

ƒigure 10-17

Stripping Mercury's Mantle by a Collision To account for Mercury's high iron content, one theory proposes that a collision with a planet-sized object stripped Mercury of most of its rocky mantle. These four images show a supercomputer simulation of a head-on collision between proto-Mercury (on the left in the top left image) and a planetesimal one-sixth its mass. Both worlds are shattered by the impact, which vaporizes much of their rocky mantles. Mercury eventually re-forms from the iron-rich debris left behind. (Courtesy of W. Benz, A. G. W. Cameron, and W. Slattery)

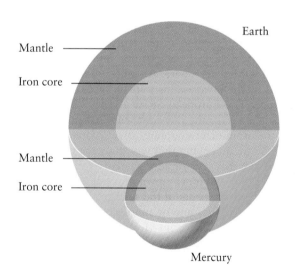

Earth

Mantle

Iron core

Mantle

Iron core

Mercury

ƒigure 10-16

The Internal Structures of Mercury and the Earth
Mercury is the most iron-rich planet in the solar system. The diameter of its iron core is about 75% of the planet's diameter, corresponding to 42% of Mercury's volume. Surrounding the core is a 600-km-thick rocky mantle. For comparison, the diameter of the Earth's iron core is only 55% of the diameter of the planet as a whole, or about 17% of its volume.

duce electric currents. The second requirement is evidence that Mercury, like the Earth, also has a *solid* part of its core: As material deep in the liquid core cools and solidifies to join the solid portion of the core, it releases the energy needed to stir up the motions of the remaining liquid material. Mercury's magnetic field therefore suggests that its core has the same general structure as the Earth's core (recall Figure 8-9).

Mercury's magnetic field interacts with the solar wind much as does the Earth's field, but on a much smaller scale. The solar wind is a constant flow of charged particles (mostly protons and electrons) away from the outer layers of the Sun's upper atmosphere. A planet's magnetic field can repel and deflect the impinging particles, thereby forming an elongated "cavity" in the solar wind called a magnetosphere. (We discussed the Earth's magnetosphere in Section 8-4.)

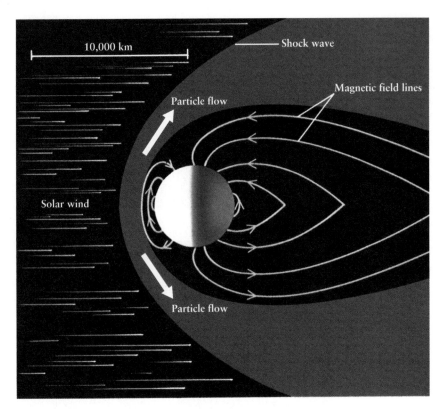

 figure 10-18

Mercury's Magnetosphere
Mercury's weak magnetic field has only 1% the strength of the Earth's field. This is just strong enough to carve out a cavity in the solar wind, preventing the impinging particles from striking the planet's surface directly. Most of the particles of the solar wind are deflected around the planet in a turbulent region colored purple in this scale drawing. (Compare this with the Earth's much larger magnetosphere, shown in Figure 8-20.)

Charged-particle detectors aboard *Mariner 10* mapped the structure of Mercury's magnetosphere (Figure 10-18). When the particles in the solar wind first encounter Mercury's magnetic field, they are abruptly slowed, producing a shock wave that marks the boundary where this sudden decrease in velocity occurs. Most of the particles from the solar wind are deflected around the planet, just as water is deflected to either side of the bow of a ship. Mercury's magnetosphere thus prevents the solar wind from reaching the planet's surface. Because Mercury's magnetic field is much weaker than that of the Earth, it is not able to capture particles from the solar wind. Hence, it has no structures like the Van Allen belts that surround the Earth (recall Figure 8-20).

Much about Mercury still remains unknown. What is the character of the side of the planet that *Mariner 10* did not see? What is the chemical composition of the surface rocks? Was the surface of Mercury once completely molten? What caused Mercury to have such a high iron content? What portion of the planet's core is liquid, and what portion is solid? Is there indeed ice perpetually frozen at the poles? Learning the answers will require extensive observations of Mercury by a spacecraft at close range.

 Three such missions are in the planning stages. *MESSENGER* (*ME*rcury *S*urface, *S*pace *EN*vironment, *GE*ochemistry, and *R*anging) is a U.S. spacecraft whose sophisticated instruments will make detailed measurements of surface composition and may help reveal the nature of Mercury's core. A similar spacecraft is being considered by the Japanese Institute of Space and Astronautical Science. The European Space Agency's *BepiColombo* mission is named for the physicist who first proposed Mercury's 3-to-2 spin-orbit coupling. *BepiColombo* is actually three spacecraft: an orbiter that will study the planet itself, a smaller orbiter that will explore Mercury's magnetosphere, and a lander that will touch down near one of the poles and make direct measurements of the surface and its properties. If all goes as planned, one or more of these missions will begin returning new data from Mercury by 2009. Until then, Mercury will remain one of the most enigmatic worlds of the solar system.

KEY WORDS

1-to-1 spin-orbit coupling, p. 228

3-to-2 spin-orbit coupling, p. 230

greatest eastern elongation, p. 226

greatest western elongation, p. 226

scarp, p. 232

solar transit, p. 228

KEY IDEAS

Motions of Mercury in the Earth's Sky: At its greatest eastern and western elongations, Mercury is never more than 28° from the Sun, so it can be seen with the naked eye only briefly after sunset or before sunrise.

• Solar transits of Mercury occur about a dozen times per century.

Mercury's Rotation: Poor telescopic views of Mercury's surface led to the mistaken impression that the planet always keeps the same side toward the Sun, a configuration called synchronous rotation or 1-to-1 spin-orbit coupling.

• Radio and radar observations in the 1960s revealed that Mercury in fact has 3-to-2 spin-orbit coupling. For an observer on Mercury, the average time from sunset to sunrise (one-half of a solar day) is equal to the time for a complete orbit around the Sun (one Mercurian year).

The Surface of Mercury: The *Mariner 10* spacecraft made several passes near Mercury in the mid-1970s, providing pictures of its surface.

• The Mercurian surface is pocked with craters like those of the Moon, but there are extensive smooth plains between these craters.

• Long cliffs called scarps meander across the surface of Mercury. These scarps probably formed as the planet cooled, solidified, and shrank.

• The impact of a large object long ago formed the huge Caloris Basin and shoved up jumbled hills on the opposite side of the planet.

Mercury's Interior and Magnetic Field: Like the Earth, Mercury has an iron-rich core.

• The iron core of Mercury has a diameter equal to three-fourths of the planet's diameter, whereas the diameter of the Earth's core is only slightly more than one-half of the Earth's diameter.

• Mercury has a weak magnetic field, indicating that at least part of the iron core is liquid. The magnetic field produces a magnetosphere around Mercury that blocks the solar wind from reaching the surface of the planet.

REVIEW QUESTIONS

1. Why can't you see any surface features on Mercury when it is closest to the Earth?

2. Why are naked-eye observations of Mercury best made at dusk or dawn, while telescopic observations are best made during the day?

3. Table 10-2 shows that a greatest western elongation of Mercury is always followed by a greatest eastern elongation, and vice versa. Explain why.

4. What is a solar transit? Why are solar transits of Mercury relatively infrequent?

5. What is 3-to-2 spin-orbit coupling? How is the rotation period of an object exhibiting 3-to-2 spin-orbit coupling related to its orbital period? What aspects of Mercury's orbit cause it to exhibit 3-to-2 spin-orbit coupling? What telescopic observations proved this?

6. Explain why Mercury does not have a substantial atmosphere.

7. Explain why *Mariner 10* was able to photograph only one side of Mercury even though the spacecraft returned to the planet three times.

8. Compare the surfaces of Mercury and our Moon. How are they similar? How are they different?

9. What kind of surface features are found on Mercury? Why are they probably much older than most surface features on the Earth?

10. How could you tell which craters in Figure 10-10 are younger than the others?

11. How do we know that the scarps on Mercury are younger than the lava flows? How can you tell that the scarp in Figure 10-11 is younger than the vertically distorted crater at the center of the figure?

12. Explain why the Sun is directly over the Caloris Basin on Mercury only every other time that the planet is at perihelion.

13. Explain why mountains on Earth can be said to be formed from below, while the mountains that ring the Caloris Basin on Mercury and Mare Orientale on the Moon are best described as having formed from above.

14. Why do astronomers think that Mercury has a very large iron core?

15. Briefly describe at least one possible history that would account for Mercury's large iron core.

16. How is Mercury's magnetosphere similar to that of the Earth? How is it different? Why do you suppose Mercury does not have Van Allen belts?

ADVANCED QUESTIONS

Problem-solving tips and tools

You may need to refresh your memory about the small-angle formula, found in Box 1-1; about Kepler's third law, described in Section 4-5; and about Wien's law and the Doppler effect, discussed in Section 5-4 and Section 5-9, respectively. The circumference of a circle of radius r is $2\pi r$. Box 7-2 discusses the criteria for a planet to be able to retain an atmosphere.

17. Find the largest angular size that Mercury can have as seen from the Earth. In order for Mercury to have this apparent size, at what point in its orbit must it be?

18. Suppose you have a superb telescope that can resolve features as small as 1 arcsec across. What is the size (in kilometers) of the smallest surface features you should be able to see on Mercury? How does your answer compare with the size of the Caloris Basin? (*Hint:* Assume that you choose to observe Mercury when it is at greatest elongation, about 25° from the Sun. As Figure 10-1 shows, at this point in its orbit, the Earth-Mercury distance is about the same as the Earth-Sun distance.)

19. Figure 10-1 shows Mercury with a greatest eastern elongation of 18° and a greatest western elongation of 28°. (a) Make a drawing like Figure 10-1 that shows how the Earth and Mercury would have to be positioned on their orbits to have Mercury at a greatest eastern elongation of 28° and a greatest western elongation of 18°. (b) On June 9, 2000, Mercury was at a greatest eastern elongation of 24°. Was Mercury at perihelion, aphelion, or some other point on its orbit? Explain.

20. (a) Explain why November solar transits of Mercury, which occur near the time of perihelion passage, are more common than May transits. (b) When a transit of Mercury takes place, can it be seen by observers anywhere on Earth? Why or why not?

21. Find the value of λ_{max} for blackbody radiation coming from the sunlit side of Mercury. In what part of the electromagnetic spectrum does this lie?

22. If the albedo of Mercury were increased, would the planet's surface temperature go up or down? Explain your answer.

23. (a) Mercury has a 58.646-day rotation period. What is the speed at which a point on the planet's equator moves due to this rotation? (*Hint:* Remember that speed is distance divided by time. What distance does a point on Mercury's equator travel as the planet makes one rotation?) (b) Use your answer to (a) to answer the following: As a result of rotation, what difference in wavelength is observed for a radio wave of wavelength 12.5 cm (such as is actually used in radar studies of Mercury) emitted from either the approaching or receding edge of the planet?

24. (a) Calculate the minimum molecular weight of a gas that could in theory be retained as an atmosphere by Mercury if the average daytime temperature were 620 K. (b) Are there any abundant gases that meet this minimum criterion? Why doesn't Mercury have an atmosphere of these gases? (*Hint:* The mass of a molecule is equal to its molecular weight, μ, multiplied by the mass of a hydrogen atom, $m_H = 1.67 \times 10^{-27}$ kg. The molecular weight of a molecule equals the sum of the atomic weights of its atoms, which can be looked up in a periodic table of the elements. Thus, for example, the molecular weight of

CO_2 is $12.0 + 16.0 + 16.0 = 44.0$, and so the molecule's mass is 44.0 m_H.)

25. How much would an 80-kg person weigh on Mercury? How does that compare with that person's weight on the Moon? How much does that person weigh on the Earth?

26. When an impact crater is formed, material (called *ejecta*) is sprayed outward from the impact. (Look at the photograph of the Moon on the right-hand side of the figure that opens this chapter. At the lower right of this photograph, you can see the light-colored ejecta surrounding the crater Stevinus.) While ejecta are found surrounding the craters on Mercury, they do not extend as far from the craters as do ejecta on the Moon. Explain why, using the difference in surface gravity between the Moon and Mercury.

27. The orbital period of *Mariner 10* is twice that of Mercury. Use this fact to calculate the length of the semimajor axis of the spacecraft's orbit.

28. Give at least two pieces of evidence that Mercury has undergone chemical differentiation.

29. It is thought that the electric currents that generate a planet's magnetic field are aided by the planet's rotation, which helps to keep liquid material inside the planet in motion. Use this idea to argue why Mercury's magnetic field should be much smaller than the Earth's.

30. Consider the idea that Mercury has a solid iron-bearing mantle that is permanently magnetized like a giant bar magnet. Using the fact that iron demagnetizes at temperatures above 770°C, present an argument against this explanation of Mercury's magnetic field.

DISCUSSION QUESTIONS

31. Before about 350 B.C., the ancient Greeks did not realize that Mercury seen in the morning sky (which they called Apollo) and seen in the evening sky (which they called Hermes) were actually the same planet. Discuss why you think it took some time to realize this.

32. If you were planning a new mission to Mercury, what features and observations would be of particular interest to you?

33. What evidence do we have that the surface features on Mercury were not formed during recent geological history?

WEB/CD-ROM QUESTIONS

34. Search the World Wide Web for the latest information about upcoming missions to Mercury. Which of these missions have been given funding so that they can proceed? When will they be launched, and when will they arrive in orbit around Mercury? What scientific experiments will

they carry? What scientific issues are these instruments intended to resolve?

 35. Elongations of Mercury. Access the animation "Elongations of Mercury" in Chapter 10 of the *Universe* web site or CD-ROM. (**a**) View the animation and notice the dates of the greatest eastern and greatest western elongations. Which time interval is greater: from a greatest eastern elongation to a greatest western elongation, or vice versa? (**b**) Based on what you observe in the animation, draw a diagram to explain your answer to the question in (a).

OBSERVING PROJECTS

Observing tips and tools

 Remember that Mercury is visible in the morning sky when it is at or near greatest western elongation and in the evening sky when at or near greatest eastern elongation. You can consult such magazines as *Sky & Telescope* and *Astronomy*, or the web sites for these magazines, for more detailed information about when and where to look for Mercury during a given month. You can also use the *Starry Night* program on the CD-ROM that accompanies this textbook.

36. Refer to Table 10-2 to determine the dates of the next two or three greatest elongations of Mercury. Consult such magazines as *Sky & Telescope* and *Astronomy*, or the web sites for these magazines, to determine if any of these greatest elongations is going to be a favorable one. If so, make plans to be one of those rare individuals who has actually seen the innermost planet of the solar system. Set aside several evenings (or mornings) around the date of the favorable elongation to reduce the chances of being "clouded out." Select an observing site that has a clear, unobstructed view of the horizon where the Sun sets (or rises). If possible, make arrangements to have a telescope at your disposal. Search for the planet on the dates you have selected, and make a drawing of its appearance through your telescope.

37. This observing project should be performed only under the direct supervision of an astronomer who knows how to point a telescope safely at Mercury. Make arrangements to view Mercury during broad daylight. This is best done by visiting an observatory where the coordinates (right ascension and declination) of Mercury's position can be used to point the telescope. **DO NOT LOOK AT THE SUN! Looking directly at the Sun can cause blindness.**

 38. Use the *Starry Night* program to observe solar transits of Mercury. First turn off daylight (select **Daylight** in the **Sky** menu) and show the entire celestial sphere (select **Atlas** in the **Go** menus). Center on Mercury by using the **Find...** command in the **Edit** menu. Using the controls at the right-hand end of the Control Panel, zoom in until the field of view is 1°. (**a**) In the Control Panel, set the date and time to May 7, 2003, at 12:00:00 A.M., and set the time step to 20 minutes. Step backward or forward through time using the single-step buttons (the leftmost and the rightmost buttons) and record the times at which the solar transit begins and ends. What is the total duration of the solar transit? (**b**) Set the date and time to November 8, 2006, at 12:00:00 A.M. Again step backward or forward through time, record the times when the solar transit begins and ends, and find the total duration of the solar transit. (**c**) The maximum duration of a solar transit of Mercury is 9 hours. Explain why the transits you observed in (a) and (b) last a substantially shorter time.

Cloud-Covered Venus

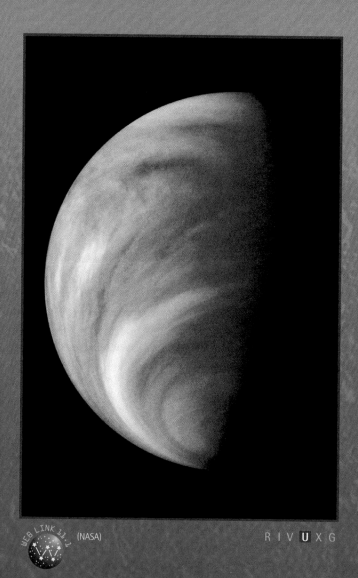

t first glance, Venus looks like the Earth's twin. The two planets have almost the same mass, diameter, average density, and surface gravity. But Venus is closer to the Sun than the Earth, so it is exposed to more intense sunlight. This has helped turn this potentially Earthlike planet into a world that is utterly hostile to living organisms, with a crushingly thick atmosphere and an almost complete lack of water.

As this image from the *Pioneer Venus Orbiter* spacecraft shows, Venus also has a perpetual cloud cover, which for many years kept astronomers from learning much about the planet. The great revolution in the study of Venus has been the use of radio waves and radar, which can penetrate the clouds and reveal what lies beneath. Radar observations not only demonstrated that Venus rotates backward but also gave us pictures of the planet's remarkably flat surface.

Beginning in the 1970s, spacecraft sent to Venus by both the Soviet Union and the United States provided astronomers with their first high-resolution images of the planet's surface. The bizarre topography revealed by these images shows that Venus has followed a very different geological history from the Earth's. Missions to Venus also confirmed that the planet has an extremely hot, dense atmosphere of carbon dioxide with clouds of sulfuric acid droplets. Active volcanoes may be responsible for the high sulfur content of the Venusian clouds.

The greenhouse effect is responsible for Venus's high temperature—the same greenhouse effect that,

WEB LINK 11.1 (NASA)

R I V **U** X G

in a milder form, acts in our own atmosphere and that might someday cause our polar ice caps to melt. Hence, an understanding of our sister planet may give important insights into the future of our own world.

As you read the sections of this chapter, look for the answers to the following questions.

11-1 What makes Venus such a brilliant "morning star" or "evening star"?

11-2 What is strange about the rotation of Venus?

11-3 In what ways does Venus's atmosphere differ radically from our own?

11-4 Why do astronomers suspect that there are active volcanoes on Venus?

11-5 Why is there almost no water on Venus today? Why do astronomers think that water was once very common on Venus?

11-6 Does Venus have the same kind of active surface geology as the Earth?

11-1 The surface of Venus is hidden beneath a thick, highly reflective cloud cover

Venus, familiar for centuries as the "morning star" and "evening star," is easy to identify. Unlike Mercury, it is possible to see Venus without interference from the Sun's glare. Indeed, at times Venus is the brightest object in the night sky. These simple observations tell us quite a bit about Venus's orbit and atmosphere.

Venus can be seen fairly far from the Sun—at its greatest elongation, about 47° away (Figure 11-1)—because its orbit is almost twice as large as that of Mercury. At its greatest eastern elongation, when it is called an "evening star," Venus is

seen high above the western horizon after sunset. At greatest western elongation, Venus is called the "morning star," because it rises nearly 3 hours before the Sun and is positioned high in the eastern sky at dawn. Table 11-1 summarizes some basic facts about Venus and its orbit, and Table 11-2 lists the dates of greatest eastern and western elongation for Venus from 2001 to 2005.

At its greatest brilliance, Venus is 16 times brighter than the brightest star and is outshone only by the Sun and the Moon. There are four reasons why: Venus is relatively large (almost the same size as the Earth), it is close to the Sun, it is close to the Earth, and it reflects 59% of the sunlight that falls on the planet (its albedo is 0.59).

Through Earth-based telescopes, Venus appears almost completely featureless, with neither continents nor mountains (see Figure 4-8). However, astronomers soon realized that

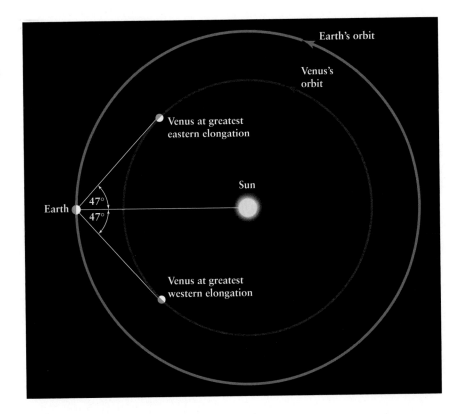

figure 11-1

Venus's Orbit Venus travels around the Sun along a nearly circular orbit with a period of 224.7 days. The average distance between Venus and the Sun is 108 million kilometers (the average distance between the Earth and the Sun is 150 million kilometers). At its greatest eastern elongation, Venus appears as a prominent "evening star"; at its greatest western elongation, it is a prominent "morning star."

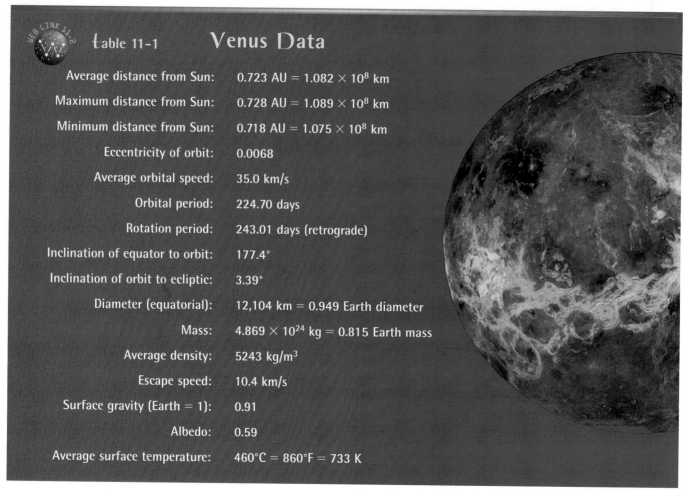

table 11-1 — Venus Data

Average distance from Sun:	0.723 AU = 1.082×10^8 km
Maximum distance from Sun:	0.728 AU = 1.089×10^8 km
Minimum distance from Sun:	0.718 AU = 1.075×10^8 km
Eccentricity of orbit:	0.0068
Average orbital speed:	35.0 km/s
Orbital period:	224.70 days
Rotation period:	243.01 days (retrograde)
Inclination of equator to orbit:	177.4°
Inclination of orbit to ecliptic:	3.39°
Diameter (equatorial):	12,104 km = 0.949 Earth diameter
Mass:	4.869×10^{24} kg = 0.815 Earth mass
Average density:	5243 kg/m³
Escape speed:	10.4 km/s
Surface gravity (Earth = 1):	0.91
Albedo:	0.59
Average surface temperature:	460°C = 860°F = 733 K

(NASA/JPL) **R** I V U X G

table 11-2 — Greatest Elongations of Venus,* 2001–2005

Greatest western elongations	Greatest eastern elongations
	2001 January 17
2001 June 8	2002 August 22
2003 January 11	2004 March 29
2004 August 17	2005 November 3

*Venus can be seen in the eastern sky in the hours before dawn for several months around greatest western elongation. It can be seen in the western sky in the hours after dusk for several months around greatest eastern elongation.

figure 11-2 R I **V** U X G

Venus at Inferior Conjunction This photograph of Venus at inferior conjunction shows an illuminated atmosphere surrounding the planet. If Venus did not have an atmosphere to scatter sunlight, we would see a thin crescent rather than a ring. This photograph was specially processed to reveal the ring, which is too faint to show in the images in Figure 4-8. (Lowell Observatory)

they were not seeing the true surface of the planet. Rather, Venus must be covered by a thick, unbroken layer of clouds. A cloud layer would also explain why Venus reflects such a large fraction of the sunlight that falls on it. Direct evidence that Venus has a thick atmosphere came when nineteenth-century astronomers observed Venus near the time of inferior conjunction. This is when Venus lies most nearly between the Earth and the Sun so that we see the planet lit from behind. As Figure 11-2 shows, sunlight is scattered by Venus's atmosphere, producing a luminescent ring around the planet that would not otherwise be present.

One other aspect of Venus that can be observed in visible light is a solar transit, in which Venus passes directly in front of the Sun at inferior conjunction. This does not happen at every inferior conjunction, because the plane of the orbit of Venus is inclined by 3.39° to the ecliptic (the plane of the Earth's orbit). Indeed, solar transits of Venus are exceedingly rare; the last two were in 1874 and 1882, and not one transit of Venus occurred during the entire twentieth century. Venusian transits usually come in pairs separated by eight years. The next pair will be visible on June 8, 2004, and June 6, 2012. Because Mercury orbits closer to the Sun, it is more likely to pass in front of the Sun's disk as seen from the Earth, and transits of Mercury are more frequent (see Section 10-1).

11-2 Venus's rotation is slow and retrograde

Venus rotates very slowly; the sidereal rotation period of the planet is 243.01 days, even longer than the planet's orbital period. To an imaginary inhabitant of Venus who could somehow see through the cloud cover, stars would move across the sky at a rate of only 1½° per Earth day. As seen from the Earth, by contrast, stars move 1½° across the sky in just 6 minutes.

Not only does Venus rotate slowly, it also rotates in an unusual direction. Most of the planets in the solar system spin on their axes in **prograde rotation**. Prograde means "forward," and planets with prograde rotation spin on their axes in the same direction in which they orbit the Sun (Figure 11-3a). As seen from the surface of a planet with prograde rotation, such as the Earth, the Sun and stars rise in the east and set in the west.

Venus is an exception to the rule: It spins in **retrograde rotation**. That is, the direction in which it spins on its axis is backward compared to the direction in which it orbits the Sun, as Figure 11-3b shows. (Recall that "retrograde" means "backward.") If an observer on the surface of Venus could see through the thick layer of clouds, the Sun and stars would rise in the west and set in the east!

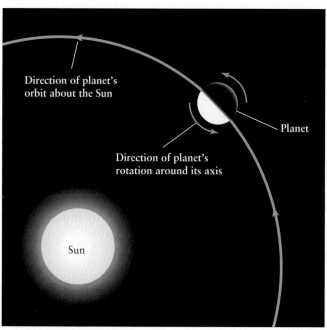

a   Prograde rotation

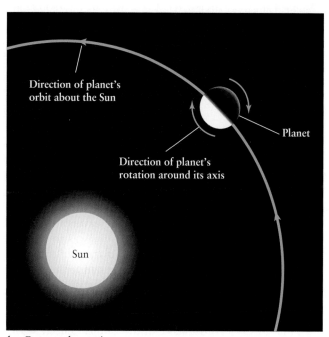

b   Retrograde rotation

ƒigure 11-3

Prograde and Retrograde Rotation   (a) If you could view the solar system from a point several astronomical units above the Earth's North Pole, you would see all of the planets orbiting the Sun in a counterclockwise direction. Most of the planets also rotate counterclockwise on their axes. This is called prograde (forward) rotation. **(b)** Venus is an exception; from your perspective high above the plane of the solar system, you would see Venus rotating clockwise. This is called retrograde (backward) rotation.

CAUTION! Be careful not to confuse retrograde *rotation* with retrograde *motion*. When Venus is near inferior conjunction, so that it is passing the Earth on its smaller, faster orbit around the Sun, it appears to us to move from east to west on the celestial sphere from one night to the next. This apparent motion is called retrograde motion, because it is backward to the apparent west-to-east motion that Venus displays most of the time. (You may want to review the discussion of retrograde motion in Sections 4-1 and 4-2.) By contrast, the retrograde rotation of Venus means that it spins on its axis from east to west, rather than west to east like the Earth.

How do astronomers know how Venus rotates? One approach is to track features in the planet's cloud cover. But winds in the atmosphere cause such features to move relative to Venus's surface, so this approach does not give a direct measurement of how the planet itself rotates. It is necessary to watch the surface of Venus itself. As we have seen, visible light cannot penetrate the planet's perpetual cloud cover. Instead, astronomers must use a different form of electromagnetic radiation.

Clouds may contain gas, dust, haze, water droplets, or other small particles. In general, electromagnetic radiation can pass easily through such a cloud only if the wavelength is large compared to the size of the particles. We then say that the cloud is *transparent* to the radiation. For example, clouds in the Earth's atmosphere are made of water droplets with an average diameter of about 20 μm (2×10^{-5} m, or 20,000 nm). Visible light has wavelengths between 400 to 700 nm, which is less than the droplet size. Hence, visible light cannot easily pass through such clouds, which is why cloudy days are darker than sunny days. But radio waves, with wavelengths of 0.1 m or more, and microwaves, with wavelengths from 10^{-3} m to 0.1 m, can pass through such a cloud with ease. (That is why your cellular phone, radio, or television works just as well on a cloudy or foggy day as on a clear one.)

Like clouds in the Earth's atmosphere, the clouds of Venus are transparent to radio waves and microwaves. Beginning in the early 1960s, advances in radio telescope technology made it possible to send microwave radiation to Venus and detect the waves reflected from its surface. Although they did not record surface features directly, these observations allowed measurements of Venus's rotation. The key was to measure how the wavelengths of the reflected radiation differed from the wavelength of the radiation originally sent to the planet.

If a beam of microwaves is sent from the Earth to Venus at one precise wavelength, the reflected waves come back spread over a small range of wavelengths. We saw in Section 10-2 that this is a consequence of the Doppler effect: Waves reflected from the approaching side of the rotating planet are shortened in wavelength, while waves reflected from the receding side undergo an increase in wavelength (see Figure 10-6). From the amount of the wavelength spread, astronomers can determine how rapidly Venus is rotating; by noticing which side of the planet is approaching and which is receding, they can tell whether the rotation is prograde or retrograde. Such measurements confirm that Venus's rotation is indeed retrograde and reveal the planet's slow rotation.

Why is Venus's retrograde rotation worthy of our attention? Here's why: It adds to the puzzle of how our solar system came to be. If you could view our solar system from a great distance above the Earth's north pole, you would see all the planets orbiting the Sun counterclockwise. Closer examination would reveal that the Sun and most of the planets also *rotate* counterclockwise on their axes; the exceptions are Venus, Uranus, and Pluto. Most of the satellites of the planets also move counterclockwise along their orbits and rotate in the same direction. Thus, the rotation of Venus is not just opposite to its orbital motion; it is opposite to most of the orbital and rotational motions in the solar system!

As we saw in Section 7-7, the Sun and planets formed from a solar nebula that was initially rotating. (Had it not been rotating, all of the material in the nebula would have fallen inward toward the protosun, leaving nothing to form the planets.) Because the planets formed from the material of this rotating cloud, they tend to rotate on their axes in that same direction. It is difficult to imagine how three planets could have bucked this trend. One theory suggests that the impact of a huge planetesimal could have reversed Venus's direction of rotation. There is no direct evidence, however, that such an impact ever took place.

Venus's rotation has other puzzling aspects. The length of an apparent solar day on Venus—that is, from when the Sun is highest in the sky to when it is again highest in the sky (Section 2-7)—is 116.8 Earth days. Remarkably, this is almost exactly one-fifth as long as the planet's 584-day synodic period. (We saw in Section 4-2 that this is the time between successive inferior conjunctions, when Venus and the Earth are closest together.) Is this curious ratio the result of gravitational interactions between Venus and the Earth? Or is it sheer coincidence? No one knows the answer.

The unusual rotation of Venus presumably holds a message about the planet's history and perhaps about the history and formation of the entire solar system. But there is as yet no consensus as to just what that message is.

11-3 Venus has a hot, dense atmosphere and corrosive cloud layers

Venus is bathed in more intense sunlight than Earth because it is closer to the Sun. If Venus had no atmosphere, and if its surface had an albedo like that of Mercury or the Moon, heating by the Sun would bring its surface temperature to around 45°C (113°F)—comparable to that found in the hottest regions on Earth. What was unclear until the mid-twentieth century was how much effect Venus's atmosphere has on the planet's surface temperature.

We saw in Section 8-1 that gases such as water vapor (H_2O) and carbon dioxide (CO_2) in the Earth's atmosphere trap some of the infrared radiation that is emitted from our planet's surface. This elevates the temperatures of both the atmosphere and its surface, a phenomenon called the greenhouse effect. In 1932 Walter S. Adams and Theodore Dunham Jr. at the Mount Wilson Observatory found absorption lines of

CO_2 in the spectrum of sunlight reflected from Venus, indicating that carbon dioxide is also present in the Venusian atmosphere. (Section 7-3 describes how astronomers use spectroscopy to determine the chemical compositions of atmospheres.) Thus, Venus's surface, like the Earth's, should be warmed by the greenhouse effect.

However, Venus's perpetual cloud cover reflects back into space a substantial amount of the solar energy reaching the planet. This effect by itself acts to cool Venus's surface, just as clouds in our atmosphere can make overcast days cooler than sunny days. Some scientists suspected that this cooling kept the greenhouse effect in check, leaving Venus with surface temperatures below the boiling point of water. Venus might then have warm oceans and possibly even life in the form of tropical vegetation. But if the greenhouse effect were strong enough, surface temperatures could be so high that any liquid water would boil away. Then Venus would have a dry, desert-like surface with no oceans, lakes, or rivers.

This scientific controversy was resolved in 1962 when the unmanned U.S. spacecraft *Mariner 2* (Figure 11-4) made the first close flyby of Venus. As we learned in Sections 5-3 and 5-4, a dense object (like a planet's surface) emits radiation whose intensity and spectrum depend on the temperature of the object. *Mariner 2* carried instruments that measured radiation coming from the planet at two microwave wavelengths, 1.35 cm and 1.9 cm. Venus's atmosphere is transparent to both these wavelengths, so *Mariner 2* saw radiation that had been emitted by the planet's surface and then passed through the atmosphere like visible light through glass.

From the amount of radiation that *Mariner 2* detected at these two wavelengths, astronomers concluded that the surface temperature of Venus was more than 400°C—well above the boiling point of water and higher than even daytime temperatures on Mercury. Thus, there cannot be any liquid water on the planet's surface. Water vapor absorbs microwaves at 1.35 cm, so if there were substantial amounts of water vapor in Venus's atmosphere, it would have blocked this wavelength from reaching the detectors on board *Mariner 2*. In fact, the spacecraft did detect strong 1.35-cm emissions from Venus. Thus, the atmosphere, too, must be all but devoid of water.

This picture of Venus as a dry, hellishly hot world was reinforced by subsequent spacecraft launched in the 1960s and 1970s. The United States sent a series of lightweight spacecraft past Venus and studied the planet and its environment with remote sensing devices. The Soviet Union, which had more powerful rockets, sent massive vehicles that plunged directly into the Venusian clouds.

Building spacecraft that could survive a descent into the Venusian atmosphere proved to be more challenging than anyone had expected. Finally, in 1970, the Soviet spacecraft *Venera 7* managed to transmit data for a few seconds directly from the Venusian surface. *Venera 7* and other successful Soviet landers during the early 1970s finally determined the surface temperature to be a nearly constant 460°C (860°F).

The success of the Soviet landers was a major technological achievement, for the pressure at the surface of Venus proved to be 90 atmospheres—that is, 90 times greater than the average air pressure at sea level on Earth. (One atmosphere,

f igure 11-4   R I **V** U X G

WEB LINK 11.4

The *Mariner 2* Spacecraft *Mariner 2* was the first of all spacecraft from Earth to make a successful flyby of another planet. After coming within 34,773 km of Venus on December 14, 1962, it went into a permanent orbit around the Sun. (GSFC/NASA)

or 1 atm, is equal to 1.01×10^5 newtons per square meter, or 14.7 pounds per square inch). This is about the same as the pressure at a depth of 1 kilometer below the surface of the Earth's oceans. The density of the atmosphere at the surface of Venus is likewise high, more than 50 times greater than the sea-level density of our atmosphere. The atmosphere is so massive that once heated by the Sun, it retains its heat throughout the long Venusian night. This explains why the temperatures on the day and night sides of Venus are almost identical.

Atmospheric probes revealed why Venus's surface temperatures are so high: 96.5% of the molecules in Venus's atmosphere are carbon dioxide, with nitrogen (N_2) making up most of the remaining 3.5% (see Table 8-3 for a comparison of the atmospheres of Venus, Earth, and Mars). Because Venus's atmosphere is so thick, and because most of the atmosphere is the greenhouse gas CO_2, the greenhouse effect has run wild. On Earth the greenhouse gases H_2O and CO_2 together make up only about 1% of our relatively sparse atmosphere, and the greenhouse effect has elevated the surface temperature by about 36°C (65°F). But on Venus, the dense shroud of CO_2 traps infrared radiation from the surface so effectively that it has raised the surface temperature by more than 400°C (720°F).

Soviet and American spacecraft also discovered that Venus's clouds are primarily confined to three high-altitude layers. An upper cloud layer lies at altitudes between 68 and 58 km, a denser and more opaque cloud layer from 58 to 52 km, and an

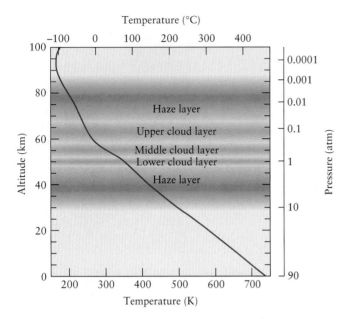

figure 11-5

Temperature and Pressure in the Venusian Atmosphere The scale on the right-hand side of the graph shows how pressure increases as you descend in the Venusian atmosphere. The atmosphere is so dense that the pressure 50 km above the surface is 1 atm, the same as at sea level on Earth. The curve shows how temperature varies with altitude. As you descend, the temperature increases from a minimum of about 170 K (about −100°C, or −150°F) at an altitude of 90 km to a maximum of more than 730 K (about 460°C, or 860°F) at the surface. Venus's cloud layers lie at altitudes between 48 and 68 km (30 to 42 mi) above the surface. By comparison, clouds in the Earth's atmosphere are seldom found at altitudes above 12 km (8 mi) (see Figure 8-23).

even more dense and opaque layer between 52 and 48 km. Above and below the clouds are 20-km-thick layers of haze (Figure 11-5). Below the lowest haze layer, the atmosphere is remarkably clear all the way down to the surface of Venus.

The Soviet and American probes all carried instruments that measured temperature and pressure as they descended to the planet's surface. Figure 11-5 shows how both pressure and temperature decrease smoothly with increasing altitude. As we saw in Section 8-5, the Earth's atmosphere has a much more complicated relationship between temperature and altitude (see Figure 8-23).

Spacecraft measurements show that sulfur, which is not found in any appreciable amount in our atmosphere, plays an important role in the Venusian atmosphere. Sulfur combines with other elements to form gases such as sulfur dioxide (SO_2) and hydrogen sulfide (H_2S), along with sulfuric acid (H_2SO_4), the same acid used in automobile batteries. While the Earth's clouds are composed of water droplets, Venusian clouds contain almost no water. Instead, they are composed of droplets of concentrated sulfuric acid. Thanks to the high

temperatures on Venus, these droplets never rain down on the planet's surface; they simply evaporate at high altitude. (You can see a similar effect on Earth. On a hot day in the desert of the American southwest, streamers of rain called *virga* appear out of the bottoms of clouds but evaporate before reaching the ground.)

The sulfuric acid in Venus's atmosphere makes the clouds a cauldron of chemical reactions hostile to metals and other solid materials. For example, reactions with fluorides and chlorides in surface rocks give rise to hydrofluoric acid (HF) and hydrochloric acid (HCl). Further reactions produce fluorosulfuric acid (HSO_3F), which is one of the most corrosive substances known to chemists, capable of dissolving lead, tin, and most rocks. It would be hard to imagine a less inviting place for humans to visit!

The Venusian atmosphere differs from our own not only in its pressures, temperatures, and chemical composition, but also in the patterns of its global circulation. Astronomers have been able to study these patterns using spacecraft in orbit around Venus. For example, in the early 1980s, the American *Pioneer Venus Orbiter* sent back numerous images of Venus taken at ultraviolet wavelengths, at which its atmospheric markings stand out best (Figure 11-6). By following individual cloud markings, scientists determined that the upper levels of Venus's atmosphere rotate around the planet in just four

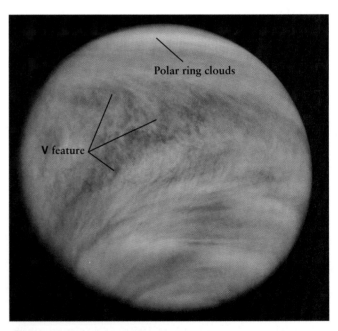

figure 11-6 R I V **U** X G

Venus's Cloud Patterns This false-color ultraviolet image of Venus was recorded by *Pioneer Venus Orbiter*. The dark V feature is produced by the rapid motion of the clouds around the planet's equator. The clouds move in the same retrograde direction as the rotation of the planet itself but at a much greater speed. The polar ring clouds are caused by atmospheric vortices near the pole (see Figure 11-7). (GSFC/NASA)

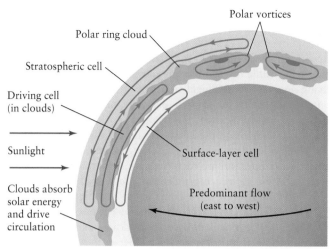

Polar vortices

Polar ring cloud

Stratospheric cell

Driving cell
(in clouds)

Sunlight

Clouds absorb
solar energy
and drive
circulation

Surface-layer cell

Predominant flow
(east to west)

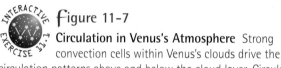

figure 11-7

Circulation in Venus's Atmosphere Strong convection cells within Venus's clouds drive the circulation patterns above and below the cloud layer. Circular wind patterns (vortices) in the polar regions force the cloud layer to bulge upward at polar latitudes, producing a polar ring cloud that is clearly visible in many *Pioneer Venus Orbiter* photographs like that in Figure 11-6. (Adapted from A. Seiff)

days. This confirmed earlier observations made with ground-based telescopes. The rapid motion of Venus's upper atmosphere is in sharp contrast to the slow rotation of the solid planet itself. Like the rotation of the planet, the motion of the atmosphere is in a retrograde direction (from east to west).

The *Pioneer Venus Multiprobe* spacecraft consisted of four probes that were dropped into the Venusian atmosphere in 1978. As they descended, they determined the atmosphere's dominant circulation patterns. Gases in the equatorial regions are warmed by the Sun, then rise upward and travel in the upper cloud layer toward the cooler polar regions. At the polar latitudes, the cooled gases sink to the lower cloud layer, in which they are transported back toward the equator. Recall from Section 8-1 that this process of heat transfer, whereby hot gases rise while cooler gases sink, is called *convection*.

The circulation of Venus's atmosphere is dominated by two huge convection cells, one in the northern hemisphere and another in the southern hemisphere, which circulate gases between the equatorial and polar regions of the planet (Figure 11-7). These convection cells, which are almost entirely contained within the main cloud layers, are called "driving cells" because they propel similar circulation cells above and below the main cloud deck somewhat like a meshed set of gears. The circulation is so effective at transporting heat around Venus's atmosphere that there is almost no temperature difference between the planet's equator and its poles.

As Venus's upper atmosphere makes its four-day rotation around the planet, prevailing winds with speeds of 350 km/h

(220 mi/h) blow from east to west at high altitude. These winds stretch out the driving cells and produce the characteristic V-shaped, chevronlike patterns visible in Figure 11-6. (In a similar way, the Earth's rotation stretches out the convection cells in our atmosphere to produce the complicated circulation patterns shown in Figure 8-24.)

On the Earth, friction between the atmosphere and the ground causes wind speeds at the surface to be much less than at high altitude. The same is true on Venus: The greatest wind speed measured by spacecraft on Venus's surface is only about 5 km/h (3 mi/h). Thus, only slight breezes disturb the crushing pressures and infernal temperatures found on Venus's dry, lifeless surface.

We have seen that the tremendous amount of carbon dioxide in the Venusian atmosphere explains why the surface is so hot. But where did all the CO_2 come from? And why do sulfur compounds play such an important role in Venus's clouds? We will find the answers to these questions by studying the present state and past history of Venus's surface.

11-4 Volcanic eruptions are probably responsible for Venus's clouds

Sulfur compounds make up less than one part per billion of the Earth's atmosphere. By contrast, SO_2 and other sulfur compounds together make up about 0.015% of Venus's atmosphere as a whole and are the dominant component of the Venusian clouds. The apparent reason why sulfur compounds are relatively abundant on Venus is that sulfurous gases are being injected into the atmosphere by volcanic processes.

When *Pioneer Venus Orbiter* arrived at Venus in 1978, its ultraviolet spectrometer recorded unexpectedly high levels of sulfur dioxide and sulfuric acid, which steadily declined over the next several years. A similar, anomalously high abundance of haze particles may have occurred in the late 1950s, as indicated by telescopic observations made at that time from the Earth. In explanation, Larry W. Esposito of the University of Colorado proposed that in both the late 1950s and the late 1970s, energetic volcanic eruptions injected sulfur dioxide into Venus's upper atmosphere. It therefore seems that Venus's sulfur-rich clouds may be regularly replenished by active volcanoes.

This notion is based on the observation that volcanoes on Earth inject sulfurous gases into our atmosphere. For example, during the Mount St. Helens eruption of 1980 (Figure 11-8), geologists monitored the emission of substantial amounts of sulfuric acid and other sulfur compounds. Many of these substances are highly reactive and short-lived, forming sulfate compounds that become part of the planet's surface rocks. For these substances to be relatively abundant on Venus, they must be constantly replenished by new eruptions.

Pioneer Venus Orbiter, the *Venera 11* and *Venera 12* landers, and the Jupiter-bound *Galileo* spacecraft found further evidence of active volcanoes. Antennas on board these spacecraft detected radio bursts thought to be caused by strokes of lightning. Lightning discharges are often seen in the plumes of erupting volcanoes on Earth.

WEB LINK 11.5

ƒigure 11-8 R I **V** U X G

A Volcanic Eruption on Earth The eruption of Mount St. Helens in Washington State on May 18, 1980, had the explosive energy of about 25 million tons of TNT. The eruption released an ash plume 40 km (25 mi) high, and injected into the atmosphere substantial amounts of sulfurous gases such as sulfur dioxide (SO_2), hydrogen sulfide (H_2S), and sulfuric acid (H_2SO_4). Figure 8-14 shows another volcanic eruption. (USGS)

The first views of Venus's volcanoes came from Earth-based radar observations. One of the most powerful tools for this task is the 305-m (1000-ft) dish at the Arecibo Observatory in Puerto Rico (see Figure 10-5). The Arecibo dish is used to transmit toward Venus a powerful, brief burst of microwaves, which easily penetrate the planet's clouds and reflect off its surface. Different types of terrain reflect microwaves more or less efficiently, so astronomers are able to construct a map of the Venusian surface by analyzing the reflected radiation. This method is most successful when Venus is near inferior conjunction. At other times, the Venus-Earth distance is too great and the radar echo too weak to produce good data.

WEB LINK 11.6

Figure 11-9 is a radar view of a large Venusian volcano named Theia Mons. (Most features on Venus are named for women in history and legend. Theia

was one of the Titans of Greek mythology.) It rises to an altitude of 6000 m (20,000 ft) and has gently sloping sides that extend over an area 1000 km (620 mi) in diameter. A volcano having this characteristic shape is called a **shield volcano,** because in profile it resembles an ancient Greek warrior's shield lying on the ground. Shield volcanoes also exist on both Earth and Mars, the best-known example being the Hawaiian Islands. Shield volcanoes on Venus and Mars, as well as the Hawaiian Islands on the Earth, are thought to be the result of **hot-spot volcanism,** whereby a hot region beneath the planet's surface extrudes molten rock over a long period of time.

WEB LINK 11.7

By far the most detailed images of the Venusian volcanoes have come from the highly successful *Magellan* spacecraft, which went into orbit around Venus in 1990. *Magellan* used sophisticated radar techniques to map more than 98% of the planet with a resolution of 100 m. It observed more than 1600 major volcanoes and volcanic features on Venus. Figure 11-10 shows a typical *Magellan* image of a volcano.

ANIMATION 11.2

In addition to mapping the planet with radar, *Magellan* used a radar altimeter that bounced microwaves off the ground directly below the spacecraft. By measuring the time delay of the reflected waves, scientists

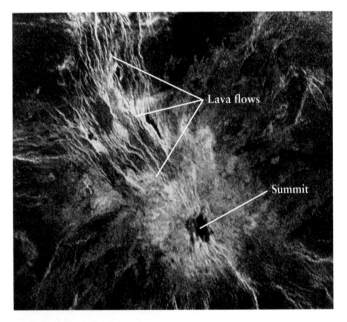

ƒigure 11-9 **R** I V U X G

An Earth-Based Radar Image of a Venusian Volcano The Venusian volcano Theia Mons is centered near the bottom of this image. Dark areas are where the surface is smooth and reflects microwaves only weakly, while bright areas have a rough-textured surface that reflect microwaves more strongly. Bright lava flows extend from the volcano's summit (the dark spot toward the bottom center) toward the upper left of this image. The image covers an area approximately 1700 by 1500 km (1100 by 900 mi), roughly 3 times the size of Texas. (Courtesy of D. B. Campbell, Arecibo Observatory)

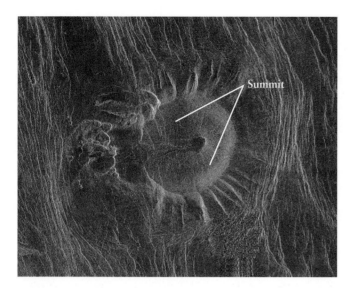

figure 11-10   R I V U X G

A *Magellan* Radar Image of a Venusian Volcano
The orbiting *Magellan* spacecraft provided detailed radar images of the Venusian surface. This image, made from an altitude of 720 km (450 mi), covers an area only 83 by 73 km (52 by 45 mi). The smallest details visible are about 100 m across, about the size of a football stadium. The flat, 35 km- (22 mi-) wide summit of this volcano is surrounded by a number of ridges and valleys, which give the volcano an insect-like appearance. (NASA)

could determine the height and depth of Venus's terrain. By combining all this information, scientists have constructed a detailed three-dimensional map of the Venusian surface. They have used this map and a supercomputer to produce

vivid perspective views of the surface, like the one shown in Figure 11-11. Such views have provided scientists with a tremendous amount of insight into volcanism on Venus.

Most of the volcanoes on Venus are probably inactive at present, just as is the case with most volcanoes on Earth. But *Magellan* found evidence of recent volcanic activity, some of which may be continuing today. The key to estimating the amount of recent volcanic activity is that the radar reflectivity of volcanic materials depends on whether the material is relatively fresh or relatively old. By mapping these reflectivity variations on Venus, *Magellan* found many areas with young lava flows. Some of the youngest material found by *Magellan* caps the volcano Maat Mons, named for an ancient Egyptian goddess (see Figure 11-11). Geologists estimate that the topmost material is no more than 10 million years old and could be much younger. The presence of such young lava flows reinforces the idea that Venus, like the Earth but unlike the Moon or Mercury, has some present-day volcanic activity.

To verify the volcanic nature of the Venusian surface, it is necessary to visit the surface and examine rock samples. Figure 11-12 is a panoramic view taken in 1981 by the Soviet spacecraft *Venera 13*, one of ten unmanned spacecraft that the Soviet Union landed successfully on the surface of Venus. Russian scientists hypothesize that this region was covered with a thin layer of lava that fractured upon cooling to create the rounded, interlocking shapes seen in the photograph. This hypothesis agrees with information obtained from chemical analyses of surface material made by the spacecraft's instruments. These analyses indicate that the surface composition is similar to lava rocks called basalt, which are common on the Earth (see Figure 8-18*a*) and in the maria of the Moon (see Figure 9-15). The results from *Venera 13* and other landers are consistent with the picture of Venus as a world whose surface and atmosphere have been shaped by volcanic activity.

figure 11-11   R I V U X G

Young Lava Flows on Venus The tall peak in this computer-generated perspective view is Maat Mons, the second tallest volcano on Venus. Maat Mons, with a height of 8 km (26,000 ft) but a diameter of 395 km (245 mi), actually has very gentle slopes; the vertical dimension in this image has been exaggerated by a factor of 22.5, equivalent to stretching your house into the shape of the Washington Monument. False color has been added to suggest the actual color of sunlight that penetrates Venus's thick clouds to reach the surface. As in Figure 11-9, dark areas are those that reflect radio waves weakly, while bright areas reflect more strongly. The bright lava flows on the slopes of Maat Mons are estimated to be no more than 10 million years old. (NASA, JPL Multimission Image Processing Laboratory)

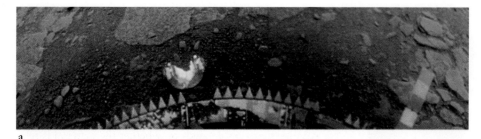

a

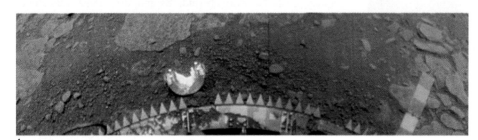

b

A Venusian Landscape (a) This wide-angle color photograph from *Venera 13* shows the rocky surface of Venus. The triangular teeth at the base of the spacecraft are 5 cm (2 inches) apart. The rocks appear orange because the thick, cloudy atmosphere absorbs the blue component of sunlight. The orange tint can also be seen on the spacecraft's striped color calibration bar (see the lower right of the image). The stripes that appear yellow are actually white in color. **(b)** This color-corrected view shows that the rocks are actually gray in color. The rocky plates covering the ground may be fractured segments of a thin layer of lava. (Courtesy of C. M. Pieters and the Russian Academy of Sciences)

11-5 The climate on Venus followed a different evolutionary path from that on Earth

Volcanic activity on Venus has not only shaped the planet's surface features but also influenced the Venusian atmosphere by adding new gases to it. Volcanoes on the Earth do the same thing, but the atmospheres of the two planets have evolved in very different ways. The Earth has abundant water in its oceans but very little carbon dioxide in its relatively thin atmosphere. By contrast, Venus is very dry and its thick atmosphere is mostly carbon dioxide. The challenge is to understand how these two planets, almost identical in size, ended up with atmospheres that are so different in chemical composition and density.

The original atmospheres of both Venus and the Earth were derived at least in part from gases spewed forth, or outgassed, by volcanoes (see Section 8-5). The gases that emanate from present-day volcanoes on Earth, such as Mount St. Helens (see Figure 11-8), are predominantly water vapor (H_2O), carbon dioxide (CO_2), and sulfur dioxide (SO_2). These gases should therefore have been important parts of the original atmospheres of both Venus and the Earth. Much of the water on both planets is also thought to have come from impacts from comets, icy bodies formed in the outer solar system (see Section 7-5).

In fact, water probably once dominated the Venusian atmosphere. Venus and the Earth are similar in size and mass, so Venusian volcanoes may well have outgassed as much water vapor as on the Earth, and both planets would have had about the same number of comets strike their surfaces. Studies of how stars evolve suggest that the early Sun was only about 70% as luminous as it is now, so the temperature in Venus's early atmosphere must have been quite a bit lower.

Water vapor may actually have been able to liquefy and form oceans on Venus.

But if water vapor and carbon dioxide were once so common in the atmospheres of both the Earth and Venus, what became of the Earth's CO_2? And what happened to the water on Venus?

The answer to the first question is that CO_2 is indeed found in abundance on the Earth. But, as we saw in Section 8-5, it is either dissolved in the oceans or chemically bound into carbonate rocks, such as the limestone and marble that formed in the oceans. If the Earth became as hot as Venus, much of its CO_2 would be boiled out of the oceans and baked out of the crust. Our planet would soon develop a thick, oppressive carbon dioxide atmosphere much like that of Venus.

To answer the question about Venus's lack of water, we must return to the early history of the planet. Just as on the present-day Earth, the oceans of Venus kept the amount of atmospheric CO_2 in check by dissolving it in the oceans and binding it up in carbonate rocks. Enough of the liquid water would have vaporized to create a thick cover of water vapor clouds. Since water vapor is a greenhouse gas, this humid atmosphere—perhaps denser than the Earth's present-day atmosphere, but far less dense than the atmosphere that surrounds Venus today—would have efficiently trapped heat from the Sun. At first, this would have had little effect on the oceans of Venus. Although the temperature would have climbed above 100°C, the boiling point of water at sea level on the Earth, the added atmospheric pressure from water vapor would have kept the water in Venus's oceans in the liquid state.

This hot and humid state of affairs may have persisted for several hundred million years. But as the Sun's energy output slowly increased over time, the temperature at the surface would eventually have risen above 374°C (647 K, or 705°F). Above this temperature, no matter what the atmospheric pressure, Venus's oceans would have begun to evaporate, and the added water vapor in the atmosphere would have increased the greenhouse effect. This would have made the temperature even higher and caused the oceans to evaporate faster, producing more water vapor. That in turn would have further intensified the greenhouse effect and made the temperature climb higher still. This is an example of a **runaway greenhouse effect**, in which an increase in temperature causes a further increase in temperature, and so on.

ANALOGY A runaway greenhouse effect is like a house in which the thermostat has accidentally been connected backward. If the temperature in such a house gets above the value set on the thermostat, the heater comes on and makes the house even hotter.

Once Venus's oceans disappeared, so did the mechanism for removing carbon dioxide from the atmosphere. With no oceans to dissolve it, outgassed CO_2 began to accumulate in the atmosphere, making the greenhouse effect "run away" even faster. Temperatures eventually became high enough to "bake out" any CO_2 that was trapped in carbonate rocks. This liberated carbon dioxide formed the thick atmosphere of present-day Venus.

Sulfur dioxide (SO_2) is also a greenhouse gas. Although present in far smaller amounts than carbon dioxide, it would have contributed to Venus's rising temperatures. Like CO_2, it dissolves in water, a process that helps moderate the amount of sulfur dioxide in our atmosphere. But when the oceans evaporated on Venus, the amount of atmospheric SO_2 would have increased. (As we saw in Section 11-4, volcanic eruptions help to sustain the amount of SO_2 in Venus's atmosphere.)

The temperature in Venus's atmosphere would not have increased indefinitely. In time, solar ultraviolet radiation striking molecules of water vapor—the dominant cause of the runaway greenhouse effect—would have broken them into hydrogen and oxygen atoms. The lightweight hydrogen atoms would have then escaped into space (see Box 7-2 for a discussion of why lightweight atoms can escape a planet more easily than heavy atoms). The remaining atoms of oxygen, which is one of the most chemically active elements, would have readily combined with other substances in Venus's atmosphere. Eventually, almost all of the water vapor would have been irretrievably lost from Venus's atmosphere. With all the water vapor gone, and essentially all the carbon dioxide removed from surface rocks, the greenhouse effect would no longer have "run away," and the rising temperature would have leveled off.

Today, the infrared-absorbing properties of CO_2 have stabilized Venus's surface temperature at its present value of 460°C. Only miniscule amounts of water vapor—about 30 parts per million, or 0.003%—remain in the atmosphere. As on Earth, volcanic outgassing still adds small amounts of water vapor to the atmosphere, along with carbon dioxide and sulfur dioxide. But on Venus, these water molecules either combine with sulfur dioxide to form sulfuric acid clouds or break apart due to solar ultraviolet radiation. The great irony is that this state of affairs is the direct result of an earlier Venusian atmosphere that was predominantly water vapor!

Venus provides an object lesson about the importance of water on climate. On Earth, volcanic outgassing releases the greenhouse gases CO_2 and SO_2 into the atmosphere. But because these gases dissolve in rainwater and in the oceans, they are removed from the atmosphere, and their contribution to the greenhouse effect is moderated. On Venus, however, there are no oceans and no rainfall, and nothing moderates the greenhouse effect (Figure 11-13). As we will see in the next section, the absence of water on Venus also has major consequences for the planet's geology.

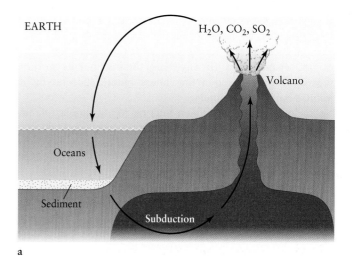

figure 11-13

Chemical Cycles on Earth and Venus On both Earth and Venus, volcanic outgassing injects water vapor (H_2O), carbon dioxide (CO_2), and sulfur dioxide (SO_2) into the atmosphere. **(a)** On Earth, where water is abundant, CO_2 and SO_2 dissolve in the oceans and are incorporated into sedimentary rocks. These are eventually subducted to great depths, where the CO_2 and SO_2 are liberated and escape by volcanic action. **(b)** On Venus there are no oceans. Outgassed CO_2 becomes part of the atmosphere, water vapor either breaks apart by ultraviolet radiation or combines with SO_2 to form sulfuric acid, and SO_2 is either locked up in minerals or adds to Venus's clouds.

11-6 The surface of Venus shows no evidence of plate tectonics

Venus is nearly the same size as Earth. Hence, we might expect it to have many of the same characteristics as the Earth: a molten interior, a magnetic field generated by fluid motions in its interior, and a crust that is reshaped by plate tectonics. But just as spacecraft revealed Venus's atmosphere to be different from what scientists had expected, they also provided many surprises about Venus's surface and interior.

The presence of volcanism on Venus (Section 11-4) strongly suggests that Venus, like Earth, has a molten interior. Unfortunately, we have no seismic data to back up this conclusion. None of the *Venera* spacecraft that landed on Venus carried seismometers. (In any event, the instruments on these spacecraft only lasted a few hours before succumbing to Venus's harsh environment.)

 Despite its molten interior, Venus has no planet-wide magnetic field. One possible explanation is the planet's long rotation period. Venus rotates so slowly that the fluid material within the planet is hardly agitated at all, and so may not move in the fashion that generates a magnetic field. Although it has no magnetic field and hence no magnetosphere, Venus is nonetheless shielded from the solar wind by ions in the planet's upper atmosphere.

Venus's crust also has many surprising characteristics. As we saw in Section 11-4, radar observations by the *Magellan* spacecraft have given us detailed three-dimensional information about the topography of Venus. The map in Figure 11-14 summarizes this information, with a color code to denote elevation. Unlike Earth, where there is a sharp contrast in elevation between the ocean floors and the continents, about 60% of Venus's terrain lies within 500 m of the average elevation. More than 80% of Venus's surface is covered with volcanic plains that are the result of numerous lava flows.

Just a few large highlands rise above the planet's generally level surface. The highlands in the northern hemisphere (at the top of Figure 11-14), named Ishtar Terra, after the Babylonian goddess of love, are approximately the same size as Australia. Ishtar Terra is dominated by a high plateau ringed by high mountains; the highest, Maxwell Montes, rises to an altitude

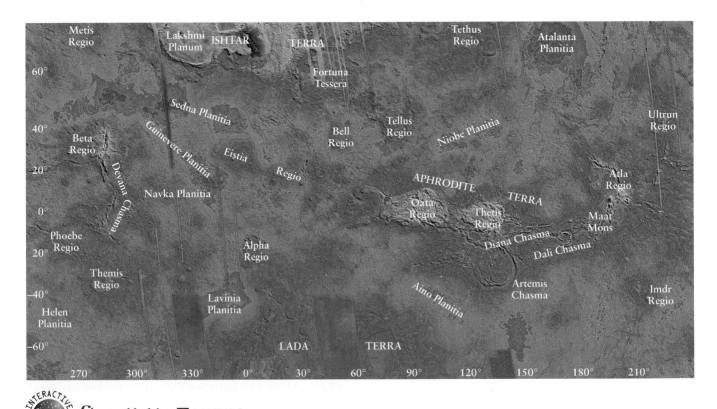

figure 11-14 R I V U X G

A Topographic Map of Venus Radar altimeter measurements by *Magellan* were used to produce this topographic map of Venus, which covers latitudes from 60° north to 60° south. Color indicates elevation—red corresponds to the highest, blue to the lowest. Gray areas were not mapped by *Magellan*. Flat plains of volcanic origin cover about 80% of the planet's surface, with only a few continentlike highlands. One of these highland areas is Aphrodite Terra, the scorpion-shaped feature that extends along the equator between 70° and 210° east longitude. The volcano shown in Figure 11-11, Maat Mons, lies at the far eastern end of Aphrodite Terra. (Peter Ford, MIT; NASA/JPL)

of 12 km (7 mi) above the average surface elevation. For comparison, the summit of Mount Everest on the Earth is 9 km (6 mi) above sea level. The most extensive Venusian highlands are Aphrodite Terra, which extends for 16,000 km around the equator. (Aphrodite was the Greek goddess whom the Romans called Venus.) Aphrodite Terra is 2000 km wide, giving it an area comparable to that of Africa. The global radar image in Figure 11-15 shows that most of Aphrodite Terra is covered by networks of faults and fractures.

Before *Magellan*, geologists wondered if Venus's molten interior had given rise to plate tectonics, like those that have remolded the face of the Earth. If this were the case, then the same tectonic effects might also have shaped the surface of Venus. As we saw in Section 8-3, the Earth's hard outer shell, or lithosphere, is broken into about a dozen large plates that slowly shuffle across the globe. Long mountain ranges, like the Mid-Atlantic Ridge (Figure 8-12), are created where fresh magma wells up from the Earth's interior to push the plates apart.

The images from *Magellan* show *no* evidence of Earthlike tectonics on Venus. On Earth, long chains of volcanic mountains (like the Cascades in North America or the Andes in South America) form along plate boundaries where subduction is taking place. Volcanic features on Venus, by contrast, do not appear in chains. Also absent on Venus is a type of faulting that always occurs with seafloor spreading on the Earth. These faults, which are perpendicular to rifts on the ocean floor, give the Earth's ocean ridges an appearance that distinctly resembles a staircase (examine Figure 8-12). Because Venus lacks any

evidence of subduction or of seafloor spreading, we know that there can have been only limited horizontal displacement of its lithosphere. Thus, like the Moon (Section 9-1) and Mercury (Section 10-3), Venus is a one-plate planet. But there have been local, small-scale deformations and reshaping of the surface. As an example, roughly a fifth of Venus's surface is covered by folded and faulted ridges.

Why is plate tectonics absent on Venus? One key reason may be the absence of water. Crustal motions on Earth are powered by large convection currents in the asthenosphere, the partially molten layer that lies between the lithosphere and the mantle (see Figure 8-15). This layer can flow because it contains water, which lowers the melting-point temperature of the asthenosphere's rock and makes it more plastic. (Water gets into the asthenosphere by being subducted along with descending crustal plates.) But Venus lost all the water in its atmosphere and from its surface during the era of the runaway greenhouse effect. Hence, there may be no asthenosphere at all beneath Venus's crust.

Another explanation may be that Venus's lithosphere is too plastic to sustain plate tectonics. The high surface temperature may soften rock so much that it is unable to sustain the kind of stresses associated with plate tectonics. Any large differences in elevation, such as are found on Earth between the continents and the seafloors, would disappear after a few tens of millions of years. In this model, tall mountains such as Maxwell Montes must be very young, which reinforces the picture of Venus as a volcanically active world (Section 11-4).

Figure 11-16 shows the difference between key geologic processes on Earth and on Venus. On Earth, new crust is formed at midocean ridges where material flows upward from the interior. This crust is eventually reprocessed into the interior at subduction zones. On Venus, by contrast, upwelling magma from the mantle pushes upward on the crust, elevating it and producing volcanic structures like those in Figures 11-9, 11-10, and 11-11. This hot-spot volcanism does not add much new material to the crust. Instead, the crust is simply pushed sideways. The result is that the crust is compressed together over locations where magma is moving downward in the mantle. Such compression folds the surface into ridges and valleys. Features of this kind are found on Venus's broad lava plains (Figure 11-17).

Magellan has produced superb pictures of Venus's impact craters (Figure 11-18). Venus probably has about a thousand craters larger than a few kilometers in diameter, many more than have been found on Earth but only a small fraction of the number on the Moon or Mercury. We saw in Section 9-4 that the number of impact craters is a clue to the age of a planet's surface. Such craters formed at a rapid rate during the early history of the solar system, when considerable interplanetary debris still orbited the Sun. Since that time, impact rates have declined, as shown in Figure 9-18, and old craters have been obliterated as volcanism or tectonics renews a planet's surface (Figure 11-19). Consequently, the more craters a planet has, the older its surface. The age of the Venusian surface seems to be roughly 500 million years. This is about twice the age of the Earth's surface but much younger than the surfaces of the Moon or Mercury, each of which is billions of years old.

VIDEO 11-2

Figure 11-15 ☑ R I V U X G

A Global View of Venus In this mosaic of *Magellan* radar images of Venus, Aphrodite Terra is the elongated, light-colored, wispy feature that wraps one-third of the way around the planet. Aphrodite is roughly parallel to Venus's equator, which extends horizontally across the middle of the picture. (NASA/JPL)

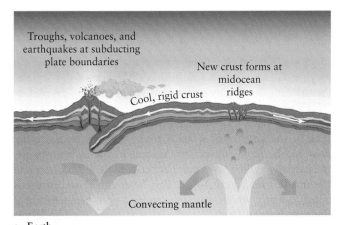

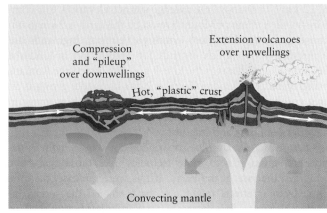

a Earth

b Venus

ƒigure 11-16

Geologic Processes on Earth and Venus **(a)** Plate tectonics is active on Earth, creating new crust at midocean ridges and returning it to the interior at subduction zones. The crust is cool enough, and hence rigid enough, to move in large plates. **(b)** Venus shows no evidence of plate tectonics. The absence of water may limit motions below the crust, and the high surface temperatures may make the crust too hot (and hence too soft) to move as a rigid plate. (Adapted from J. W. Head)

Surprisingly, Venus's craters are uniformly scattered across the planet's surface. We would expect that older regions on the surface—which have been exposed to bombardment for a longer time—would be more heavily cratered, while younger regions would be relatively free of craters. For example, the ancient highlands on the Moon are much more heavily cratered than the younger maria (see Section 9-4). Because such variations are not found on Venus, scientists conclude that the

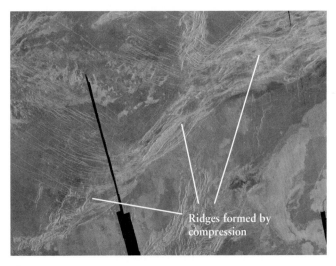

ƒigure 11-17 R I V U X G

Compressional Folding on Venus This *Magellan* radar image shows part of the low-lying Lavinia Planitia region at 50° south latitude, 345° east longitude (see Figure 11-14). Compression of the surface, as suggested by Figure 11-16*b*, has formed a bright belt of ridges that run from lower left to upper right. Lava flows have flooded the surrounding terrain. (Black stripes are where data are missing.) The area shown extends 540 km (338 mi) north to south and 900 km (563 mi) east to west. (NASA/JPL)

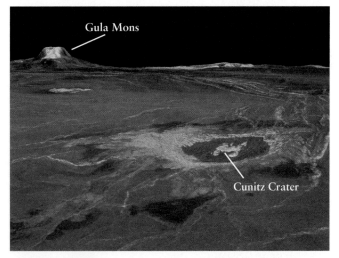

  ƒigure 11-18 R I V U X G

A Venusian Crater This perspective view (constructed from *Magellan* data) shows an impact crater 48 km (30 mi) in diameter, about the same size as St. Louis, Missouri, and its suburbs. This crater, named for the Polish astronomer Maria Cunitz, has a central peak rising out of its dark interior (compare Figure 9-4). The volcano near the horizon, named Gula Mons after a Babylonian earth goddess, is 3000 m (10,000 ft) high. These features lie in western Eistla Regio, just to the left of center in Figure 11-14. (NASA/JPL)

entire surface of the planet has essentially the *same* age. This is very different from the Earth, where geological formations of widely different ages can be found.

Scientists have proposed two competing hypotheses to explain the apparent youth of Venus's surface. According to the **equilibrium resurfacing hypothesis,** proposed by Roger J. Phillips of Washington University, volcanic eruptions are continually covering up craters at about the same rate that craters are formed by impact. Central to this model is the observation that volcanism is widespread on Venus, so that craters are destroyed at roughly equal rates over all parts of the planet.

An alternative model, put forth by Gerald G. Schraber of the U.S. Geological Survey and Robert G. Strom of the University of Arizona, proposes that an intense period of eruptions took place on Venus several hundred million years ago. According to this **global catastrophe hypothesis,** the entire globe was resurfaced over a short period with fresh lava, covering up any older craters. In one version of this hypothesis, upwelling material from the mantle solidifies on the underside of Venus's lithosphere. The increasingly thick lithosphere acts as an insulating blanket enveloping the planet's interior, trapping heat until the mantle becomes hot enough to break the lithosphere apart. The result is widespread volcanism that covers nearly the entire planet with lava. The lithosphere then cools and thickens, and the cycle begins anew. If this picture is correct, then every few hundred million years the entire face of Venus is erased at once, and there can be no very old rocks on the Venusian surface. By contrast, plate tectonics erases the Earth's surface in a piece-

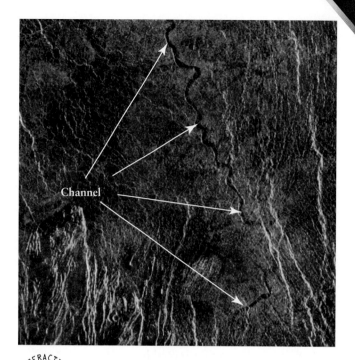

ƒigure 11-20   R I V U X G

The Longest Channel in the Solar System This *Magellan* image shows a portion of a long, meandering, 2 km- (1.2 mi-) wide channel named Baltis Vallis. ("Baltis" is the Syrian word for the planet Venus.) Although it resembles a river on Earth, Baltis Vallis could never have held liquid water, which has not existed on Venus for billions of years. Instead, a moderate increase in Venus's surface temperature could have melted lavas rich in calcium compounds, while the rest of the surface remained solid. The molten lava could have then carved out this and several other riverlike features found on Venus. This image, which depicts an area about 130 by 190 km (81 by 118 mi), shows only a portion of the channel's total length of 6800 km (4200 mi). In comparison, the Nile—the longest river on the Earth—is some 6600 km (4100 mi) long. (NASA/JPL)

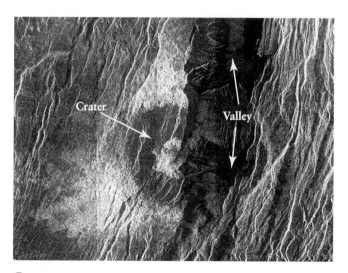

ƒigure 11-19   R I V U X G

A Partially Obliterated Crater In this *Magellan* image, the right half of an old impact crater 37 km (23 mi) in diameter was destroyed when a fault in the crust formed a deep valley 20 km (12 mi) wide. The vertical streaks through the surviving left half of the crater are other fractures or faults that have occurred since the crater was formed. This crater, named for the American economist Emily Balch, lies in Beta Regio (see the left edge of Figure 11-14). (NASA/JPL)

meal fashion, which is why some rocks on the Earth are billions of years old.

The intense eruptions proposed in the global catastrophe hypothesis would also have had tremendous effects on Venus's climate. Such eruptions would have outgassed tremendous amounts of carbon dioxide, water vapor, and sulfur dioxide. Simulations by Mark A. Bullock and David J. Grinspoon of the University of Colorado suggest that this fresh supply of greenhouse gases would eventually have raised Venus's surface temperature to 100°C higher than its present-day value. Such temperatures would have melted some of the solid materials that make up Venus's surface, which could explain certain riverlike features seen on the planet today (Figure 11-20). Atmospheric temperatures would also have increased, causing Venus's sulfuric acid clouds to evaporate. The atmosphere and surface would have returned to their present "mild" temperatures only after hundreds of millions of years.

Although Venus is nearly the Earth's twin in size, its surface and atmosphere have evolved in radically different ways than the Earth. In the long run, however, both planets may meet the same fate. Our Sun's luminosity will continue to increase as it ages, and in about a billion years sunlight will be bright enough to begin evaporating the Earth's oceans. The atmosphere will experience the same sort of runaway greenhouse effect that occurred on Venus long ago. As temperatures drive toward 1000 K, vast amounts of carbon dioxide will be baked out of the Earth's rocks, creating a Venus-like atmosphere. By studying Venus, we may be looking into the distant future of our own world.

KEY WORDS

equilibrium resurfacing hypothesis, p. 255

global catastrophe hypothesis, p. 255

hot-spot volcanism, p. 248

prograde rotation, p. 243

retrograde rotation, p. 243

runaway greenhouse effect, p. 251

shield volcano, p. 248

KEY IDEAS

Motions of Venus in the Earth's Sky: At its greatest eastern and western elongations, Venus is about 47° from the Sun, and it can be seen for several hours after sunset or before sunrise.

Comparison of the Earth and Venus: Venus is similar to the Earth in its size, mass, average density, and surface gravity, but it is covered by unbroken, highly reflective clouds that conceal its other features from Earth-based observers.

Venus's Rotation: Venus rotates slowly in a retrograde direction with a solar day of 117 Earth days and a rotation period of 243 Earth days. There are approximately two Venusian solar days in a Venusian year.

Venus's Atmosphere and Clouds: Spacecraft measurements reveal that 96.5% of the Venusian atmosphere is carbon dioxide. Most of the balance of the atmosphere is nitrogen.

• Venus's clouds consist of droplets of concentrated sulfuric acid. Active volcanoes on Venus may be a continual source of this sulfurous material.

• Venus's clouds are confined to altitudes between 48 and 68 km above the planet's surface. A haze layer extends down to an elevation of 30 km, beneath which the atmosphere is clear.

• The surface pressure on Venus is 90 atm, and the surface temperature is 460°C. Both temperature and pressure decrease as altitude increases.

• The circulation of the Venusian atmosphere is dominated by two huge convection currents in the cloud layers, one in the northern hemisphere and one in the southern hemisphere.

• The upper cloud layers of the Venusian atmosphere move rapidly around the planet in a retrograde direction, with a period of only about four Earth days.

The Greenhouse Effect: Venus's high temperature is caused by the greenhouse effect, as the dense carbon dioxide atmosphere traps and retains energy from sunlight.

• The early atmosphere of Venus contained substantial amounts of water vapor. This caused a runaway greenhouse effect that evaporated Venus's oceans and drove carbon dioxide out of the rocks and into the atmosphere. Almost all of the water vapor was eventually lost by the action of ultraviolet radiation on the upper atmosphere.

• The Earth has roughly as much carbon dioxide as Venus, but it has been dissolved in the Earth's oceans and chemically bound into its rocks.

The Surface of Venus: The surface of Venus is surprisingly flat, mostly covered with gently rolling hills. There are a few major highlands and several large volcanoes.

• The surface of Venus shows no evidence of the motion of large crustal plates, which plays a major role in shaping the Earth's surface.

• The density of craters suggests that the entire surface of Venus is no more than a few hundred million years old. According to the equilibrium resurfacing hypothesis, this happens because old craters are erased by ongoing volcanic eruptions. In the alternative global catastrophe hypothesis, all of Venus was resurfaced at essentially the same time.

REVIEW QUESTIONS

1. As seen from the Earth, the brightness of Venus changes as it moves along its orbit. Describe the main factors that determine Venus's variations in brightness as seen from the Earth.

2. If Venus did not have an atmosphere, how would its appearance as seen from Earth be different?

3. Why was it so difficult to determine the rate and direction of Venus's rotation? How were these finally determined?

4. What roles does the greenhouse effect play in the atmospheres of Venus and the Earth?

5. Why was it difficult to determine Venus's surface temperature from Earth? How was this finally determined?

6. The *Mariner 2* spacecraft did not enter Venus's atmosphere, but it was nonetheless able to determine that the atmosphere is very dry. How was this done?

7. Why is it hotter on Venus than on Mercury?

8. What is the evidence for active volcanoes on Venus?

9. How might Venus's cloud cover change if all of Venus's volcanic activity suddenly stopped? How might these changes affect the overall Venusian environment?

10. Why are there no oceans on Venus? Where has Venus's water gone?

11. Why is there so much carbon dioxide in Venus's atmosphere while very little of this gas is present in the Earth's atmosphere?

12. What is the difference between the greenhouse effect as it exists on Venus today and the runaway greenhouse effect that existed in Venus's early atmosphere?

13. Describe the Venusian surface. What kinds of features would you see if you could travel around on the planet?

14. In what ways does the surface topography of Venus differ from that of the Earth?

15. Why do scientists think that Venus's surface was *not* molded by the kind of tectonic activity that shaped the Earth's surface?

16. Describe how Venus's lack of water and high surface temperatures may help explain the absence of plate tectonic activity.

17. Compare and contrast the kinds of geologic activity that occur on Venus with those that occur on Earth.

18. Describe two competing hypotheses that attempt to explain why Venus's surface is only a few hundred million years old.

ADVANCED QUESTIONS

Problem-solving tips and tools

You should recall that Wien's law (Section 5-4) relates the temperature of a blackbody to λ_{max}, its wavelength of maximum emission. Box 5-4 describes some of the physics of light scattering. Section 5-9 and Box 5-6 explain the Doppler effect and how to do calculations using it. The linear speed of a point on a planet's equator is the planet's circumference divided by its rotation period; recall that the circumference of a circle of radius r is $2\pi r$.

19. Venus takes 440 days to move from greatest western elongation to greatest eastern elongation, but it needs only 144 days to go from greatest eastern elongation to greatest western elongation. With the aid of a diagram like Figure 11-1, explain why.

20. Before about 350 B.C. it was not generally understood that Venus seen in the morning sky (that is, at greatest western elongation) and Venus seen in the evening sky (at greatest eastern elongation) were actually the same planet. Construct a geocentric model of the planets (like that shown in Figure 4-10) in which the "morning Venus" and the "evening Venus" are two distinct planets.

21. During what time of the day or night was the photograph in Figure 11-2 made? How can you tell?

22. Venus's sidereal rotation period is 243.01 days and its orbital period is 224.70 days. Use these data to prove that a solar day on Venus lasts 116.8 days. (*Hint:* Develop a formula relating Venus's solar day to its sidereal rotation period and orbital period similar to the first formula in Box 4-1.)

23. In Section 11-2 we described the relationship between the length of Venus's synodic period and the length of an apparent solar day on Venus. Using this and a diagram, explain why at each inferior conjunction the same side of Venus is turned toward the Earth.

24. If you aim microwaves of wavelength 1.9 cm at the entire face of Venus, what will be the spread in wavelength of the reflected waves caused by the planet's rotation? (See Figure 10-6.)

25. At what wavelength does Venus's surface emit the most radiation? Do astronomers have telescopes that can detect this radiation? Why can't we use such telescopes to view the planet's surface?

26. The *Mariner 2* spacecraft detected more microwave radiation when its instruments looked at the center of Venus's disk than when it looked at the edge, or limb, of the planet. (This effect is called *limb darkening*.) Explain how these observations show that the microwaves are emitted by the planet's surface rather than its atmosphere.

27. Explain how you could estimate the size of the droplets that make up Venus's clouds by beaming radio waves of different wavelengths through the clouds to a spacecraft on the planet's surface.

28. When the *Galileo* spacecraft flew past Venus in 1990 while on its way to Jupiter, it used its infrared camera to view lower-level clouds in the Venusian atmosphere. Why was it necessary to use infrared light to see these clouds?

29. In the classic Ray Bradbury science-fiction story "All Summer in a Day," human colonists on Venus are subjected to continuous rainfall except for one day every few years when the clouds part and the Sun comes out for an hour or so. Discuss how our understanding of Venus's atmosphere has evolved since this story was first published in 1954.

30. A hypothetical planet has an atmosphere that is opaque to visible light but transparent to infrared radiation. How

would this affect the planet's surface temperature? Contrast and compare this hypothetical planet's atmosphere with the greenhouse effect in Venus's atmosphere.

31. Suppose that Venus had no atmosphere at all. How would the albedo of Venus then compare with that of Mercury or the Moon? Explain your answer.

32. Water has a density of 1000 kg/m^3, so a column of water n meters tall and 1 meter square at its base has a mass of $n \times 1000$ kg. On either the Earth or Venus, which have nearly the same surface gravity, a mass of 1 kg weighs about 9.8 newtons (2.2 lb). Calculate how deep you would have to descend into the Earth's oceans for the pressure to equal the atmospheric pressure on Venus's surface, 90 atm or 9×10^6 newtons per square meter.

33. Marine organisms produce sulfur-bearing compounds, some of which escape from the oceans into the Earth's atmosphere. (These compounds are largely responsible for the characteristic smell of the sea.) Even more sulfurous gases are injected into our atmosphere by the burning of sulfur-rich fossil fuels, such as coal, in electric power plants. Both of these processes add more sulfur compounds to the atmosphere than do volcanic eruptions. On lifeless Venus, by contrast, volcanoes are the only source for sulfurous atmospheric gases. Why, then, are sulfur compounds so much rarer in our atmosphere than in the Venusian atmosphere?

34. The Hawaiian islands lie along a nearly straight line (see the accompanying figure). A hot spot under the Pacific plate has remained essentially stationary for 70 million years while the plate has moved to the northwest some 6000 km. The upwelling magma has thus produced a long chain of hot-spot volcanoes. (The oldest, at the northwest end of the chain, have eroded so much that they no longer rise above the ocean surface.) Hot-spot volcanoes are also found on Venus, but they do not form long chains. Explain what this tells us about tectonic activity on Venus.

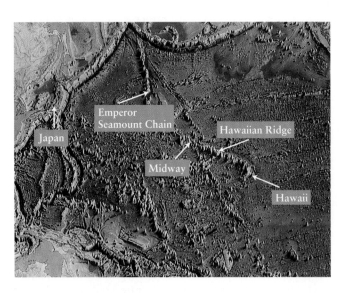

DISCUSSION QUESTIONS

35. Describe the apparent motion of the Sun during a "day" on Venus relative to (a) the horizon and (b) the background stars. (Assume that you can see through the cloud cover.)

36. If you were designing a space vehicle to land on Venus, what special features would you think necessary? In what ways would this mission and landing craft differ from a spacecraft designed for a similar mission to Mercury?

WEB/CD-ROM QUESTIONS

37. A number of European astronomers traveled to Asia and the Pacific islands to observe the transits of Venus in 1761 and 1769. Search the World Wide Web for information about these expeditions. Why were these events of such interest to astronomers? How definitive were the results of these observations?

38. Search the World Wide Web for the latest information about proposed future missions to Venus. Which of these missions have been given funding so that they can proceed? What scientific experiments will they carry? Which scientific issues are these instruments intended to resolve?

 39. Surface Temperature of Venus. Access the Active Integrated Media Module "Wien's Law" in Chapter 5 of the *Universe* web site or CD-ROM. (a) Using the Wien's Law calculator, determine Venus's approximate temperature if it emits blackbody radiation with a peak wavelength of 3866 nm. (b) By trial and error, find the wavelength of maximum emission for a surface temperature of 733 K (for present-day Venus) and a surface temperature of 833 K (as it might be in the event of a global catastrophe that released more greenhouse gases into Venus's atmosphere). In what part of the electromagnetic spectrum do these wavelengths lie?

OBSERVING PROJECTS

Observing tips and tools

 Venus is visible in the morning sky when it is at or near greatest western elongation and in the evening sky when at or near greatest eastern elongation. These are also the times when Venus can be seen at the highest altitude above the horizon during the hours of darkness. Consult such magazines as *Sky & Telescope* and *Astronomy* or their web sites for more detailed information about when and where to look for Venus during a given month. You can also use the *Starry Night* program on the CD-ROM that accompanies this textbook.

40. Refer to Table 11-2 to see if Venus is near a greatest elongation. If so, view the planet through a telescope. Make a sketch of the planet's appearance. From your sketch, can you determine if Venus is closer to us or farther from us than the Sun?

41. Using a small telescope, observe Venus once a week for a month and make a sketch of the planet's appearance on each occasion. From your sketches, can you determine whether Venus is approaching us or moving away from us?

42. This observing project should be performed only under the direct supervision of an astronomer who knows how to point a telescope safely at Venus. Make arrangements to view Venus during broad daylight. This is best done by visiting an observatory where the coordinates (right ascension and declination) of Venus's position can be used to point the telescope. **DO NOT LOOK AT THE SUN! Looking directly at the Sun can cause blindness.**

43. Use the *Starry Night* program to observe solar transits of Venus. First turn off daylight (select **Daylight** in the **Sky** menu) and show the entire celestial sphere (select **Atlas** in the **Go** menu). Center on Venus by using the **Find...** command in the **Edit** menu. Using the controls at the right-hand end of the Control Panel, zoom in until the field of view is 1°. Finally, make sure you can see the green line that represents the ecliptic (if you cannot, select **The Ecliptic** in the **Guides** menu). (**a**) In the Control Panel, set the date and time to June 8, 2004, at 12:00:00 A.M., and set the time step to 20 minutes. Step backward or forward through time using the single-step buttons (the leftmost and rightmost time control buttons) and record the times at which the solar transit begins and ends. What is the total duration of the solar transit? (**b**) During the transit, is Venus precisely on the ecliptic? If not, about how far off is it? (*Hint:* The Sun has an angular diameter of about 30 arcmin.) (**c**) Repeat parts (a) and (b) for the solar transit of Venus on June 6, 2012.

Red Planet Mars

Mars Attacks! Invaders from Mars! Martians, Go Home! The fourth planet from the Sun has inspired many science-fiction films and novels about alien invasions. But why Mars? People have long speculated that life might exist there, because the red planet has many Earthlike characteristics. At the beginning of the twentieth century, some astronomers claimed to have seen networks of linear features on the Martian surface, perhaps "canals" built by an advanced civilization.

The Hubble Space Telescope image of Mars reproduced here shows no sign of canals, however. Indeed, over the past four decades spacecraft have confirmed that Mars has no canals, no signs of life, and no liquid water on its surface. But they have also found evidence that Mars was once a very different place. Spacecraft in Martian orbit have photographed ancient flood channels through which immense amounts of water once coursed, as well as gullies that may have been caused by smaller but still substantial flows. Data from spacecraft that have landed on Mars show that liquid water may have existed there for extended periods—perhaps long enough for Mars to have had large lakes and even an ocean.

Spacecraft have also revealed how the face of Mars has been reshaped. Internal forces have produced gigantic volcanoes and an immense valley that is as long as North America is wide. External forces, in the form of space debris that has collided

(David Crisp and the WFPC-2 Science Team, JPL/Caltech)

R I **V** U X G

with Mars, have strewn much of the planet's surface with craters of all sizes. The challenge to planetary scientists is to understand how Mars has evolved from a warm, wet, geologically active world to the dry, desolate, and nearly airless planet that we see today.

As you read the sections of this chapter, look for the answers to the following questions.

12-1   When is it possible to see Mars in the night sky?

12-2   Why was it once thought that there are canals on Mars?

12-3   How are the northern and southern hemispheres of Mars different from each other?

12-4   What is the evidence that there was once liquid water on Mars?

12-5   Why is the Martian atmosphere so thin?

12-6   What have we learned about the Martian surface from the *Viking* and *Mars Pathfinder* landers?

12-7   What causes the seasonal color changes on Mars?

12-8   As seen from Mars, how do the Martian moons move across the sky?

12-1 The best Earth-based observations of Mars are made during favorable oppositions

Mars is a relatively small planet, with just over half the diameter of the Earth. (Table 12-1 summarizes some basic information about Mars.) Nonetheless, Mars is the easiest of the other terrestrial planets to study with a small telescope. Unlike Mercury, which never appears very far from the Sun (see Section 10-1), Mars can be seen high in the night sky when it is at opposition. And unlike Venus, with its dense atmosphere and perpetual cloud cover (see Section 11-1), Mars has a thin, almost cloudless atmosphere that permits a clear view of the surface. Under the right conditions, a relatively small telescope can reveal substantial detail on the Martian surface (Figure 12-1).

We get the best views of Mars when it is at opposition, so that it is close to Earth and appears high in our night sky. Such oppositions occur at intervals of about 780 days. However, some oppositions provide better views of Mars than others. As Figure 12-2 shows, Mars has a noticeably elongated orbit. As a result, we get the best view of Mars when it is simultaneously at opposition and near the perihelion of its elliptical orbit, a configuration called a **favorable opposition.** The Earth-Mars distance can then be as small as 0.37 AU, or 56 million kilometers (35 million miles), and the angular diameter of Mars can be as large as 26 arcsec—the same as a moderately large (50 km) lunar crater seen from Earth. Under these conditions, Mars appears in the nighttime sky as a brilliant red object 3½ times brighter than Sirius, the brightest star in the sky.

Unfortunately, the time from one favorable opposition to the next is about 15 years. Hence, astronomers take advantage of all oppositions, even the not-so-favorable ones, to observe Mars. When Mars is at opposition but near aphelion, as last happened in 1997, the Earth-Mars distance can be as large as 0.68 AU, or 101 million kilometers (63 million miles). The image in Figure 12-1 shows Mars during the 1997 opposition, when the angular diameter of Mars was only 14 arcsec. This image was made with a moderately large telescope; when

viewed under these conditions through a small amateur telescope, Mars appears as a small, nearly featureless red dot.

 Although Mars appears red to Earth observers, its surface is actually brown. A sunlit brown surface, if surrounded by the blackness of space, is perceived by the human eye as having a red color.

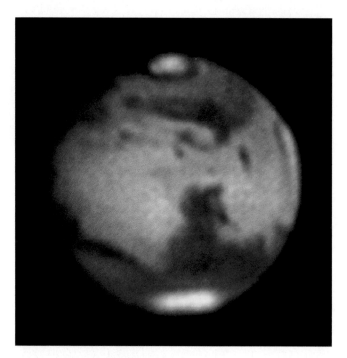

ƒigure 12-1   R I **V** U X G

Mars as Viewed from Earth   This high-quality image of Mars was made by an amateur astronomer using a 41-cm (16-in.) telescope and a CCD detector (see Section 6-4). The image was made during the 1997 opposition, when Mars was 0.669 AU (100 million km, or 62 million mi) away and had an angular diameter of 14 arcsec. Compare this with the Hubble Space Telescope image that opens this chapter, which was taken during the same month and shows the same face of Mars. (Courtesy of Donald Parker)

table 12-1 — Mars Data

Average distance from Sun:	1.524 AU $= 2.279 \times 10^8$ km
Maximum distance from Sun:	1.666 AU $= 2.492 \times 10^8$ km
Minimum distance from Sun:	1.381 AU $= 2.067 \times 10^8$ km
Eccentricity of orbit:	0.093
Average orbital speed:	24.1 km/s
Orbital period:	686.98 days = 1.88 years
Rotation period:	$24^h\ 37^m\ 22^s$
Inclination of equator to orbit:	25.19°
Inclination of orbit to ecliptic:	1.85°
Diameter (equatorial):	6794 km = 0.533 Earth diameter
Mass:	6.418×10^{23} kg = 0.107 Earth mass
Average density:	3934 kg/m³
Escape speed:	5.0 km/s
Surface gravity (Earth = 1):	0.38
Albedo:	0.15
Surface temperatures:	Maximum: 20°C = 70°F = 293 K
	Mean = −53°C = −63°F = 220 K
	Minimum: −140°C = −220°F = 133 K

(NASA, USGS) R I **V** U X G

Three Martian oppositions will take place between the years 2001 and 2005 (Table 12-2). The 2003 opposition is the most favorable, but all three provide good views of the red planet. Figure 4-2 shows the apparent path of Mars across the sky during the 2005 opposition.

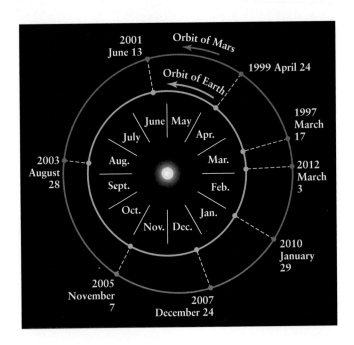

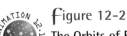

figure 12-2

The Orbits of Earth and Mars The best times to observe Mars are at opposition, when Mars is on the opposite side of our sky from the Sun. The red and blue dots connected by dashed lines show the positions of Mars and Earth at various oppositions. The months shown around the Sun refer to the time of year when the Earth is at each position around its orbit.

table 12-2	Oppositions of Mars, 2001–2005		
	Earth-Mars distance		
Date of opposition	(AU)	(10^6 km)	Angular diameter (arcsec)
2001 June 13	0.450	67.3	20.8
2003 August 28	0.373	55.8	25.1
2005 November 7	0.464	69.4	20.2

12-2 Earth–based observations were once thought to show evidence of intelligent life on Mars

The first reliable record of surface features on Mars was made by the Dutch scientist Christiaan Huygens, who observed the planet during the opposition of 1659. He identified a prominent dark feature that we now call Syrtis Major. (In the image that opens this chapter and in Figure 12-1, Syrtis Major is the dark shark-fin shape that runs from north to south across the center of the planet's disk.) After observing this feature for several weeks, Huygens concluded that the rotation period of Mars is approximately 24 hours, the same as the Earth.

In 1666, the Italian astronomer Gian Domenico Cassini made more refined observations and concluded that a Martian day is about 37½ minutes longer than a solar day on Earth. Cassini was also the first to see the Martian polar caps, which bear a striking superficial resemblance to the Arctic and Antarctic regions on Earth (see the figure that opens this chapter). More than a century later, the German-born English astronomer William Herschel suggested that the Martian polar caps might be made of ice or snow.

Herschel also found that Mars's axis of rotation is not perpendicular to the plane of the planet's orbit, but is tilted by about 25° away from the perpendicular. This is very close to the Earth's 23½° tilt (Figure 2-12). This striking coincidence means that Mars experiences Earthlike seasons, with opposite seasons in the northern and southern Martian hemispheres (see Section 2-5). Because Mars takes nearly two (Earth) years to orbit the Sun, the Martian seasons last nearly twice as long as on Earth.

During spring and summer in a Martian hemisphere, the polar cap shrinks and the dark markings (which sometimes look greenish) become very distinct. Half a Martian year later, with the approach of fall and winter, the dark markings fade and the polar cap grows. These seasonal variations were widely held to mean that there is vegetation on Mars that changes seasonally, just as on the Earth. And if there was plant life, might there not also be intelligent beings?

Observations by the Italian astronomer Giovanni Schiaparelli during the favorable opposition of 1877 fueled speculations about life on Mars. He reported that during moments of "good seeing," when the air is exceptionally calm and clear, he was able to see 40 straight-line features crisscrossing the Martian surface (Figure 12-3). He called these dark linear features *canali*, an Italian word for "channels," which was soon mistranslated into English as "canals." The alleged discovery of canals implied that there were intelligent creatures on Mars capable of substantial engineering feats. This speculation helped motivate the American millionaire Percival Lowell to finance a major new observatory near Flagstaff,

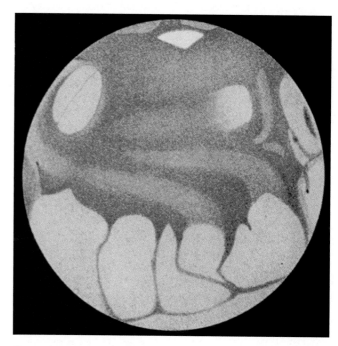

figure 12-3 R I **V** U X G

The Mirage of the Martian Canals Giovanni Schiaparelli studied the Martian surface using a 20-cm (8-in.) telescope, the same size used by many amateur astronomers today. He recorded his observations in drawings like this one, which shows a network of linear features that Percival Lowell and others interpreted as irrigation canals. Later observations with larger telescopes revealed that the "canals" were mere illusions. (Michael Hoskin, ed., *The Cambridge Illustrated History of Astronomy*, Cambridge University Press, 1997, p. 286. Illustration by G. V. Schiaparelli. Courtesy Institute of Astronomy, University of Cambridge, UK)

Arizona, primarily to study Mars. By the end of the nineteenth century, Lowell had reported observations of 160 Martian canals.

Although many astronomers observed Mars, not all saw the canals. In 1894, the American astronomer Edward Barnard, working at Lick Observatory in California, complained that "to save my soul I can't believe in the canals as Schiaparelli draws them." But objections by Barnard and others did little to sway the proponents of the canals.

As the nineteenth century drew to a close, speculation about Mars grew more and more fanciful. Perhaps the reddish-brown color of the planet meant that Mars was a desert world, and perhaps the Martian canals were an enormous planetwide irrigation network. From these ideas, it was a small leap to envision Mars as a dying world with canals carrying scarce water from melting polar caps to farmlands near the equator. The terrible plight of the Martian race formed the basis of inventive science fiction by Edgar Rice Burroughs, Ray Bradbury, and many others. It also led to the notion that the Martians might be a warlike race who schemed to invade the Earth for its abundant resources. (In Roman mythology, Mars was the god of war.) Stories of alien invasions, from H. G. Wells's 1898 novel *The War of the Worlds* down to the present day, all owe their existence to the *canali* of Schiaparelli.

Modern observations of the Martian surface, such as the Hubble Space Telescope image that opens this chapter, show *no* sign whatsoever of linear canals. Why, then, were Schiaparelli, Lowell, and others so certain that the canals were there? One reason is that their observations were made from the Earth's surface, at the bottom of an often turbulent atmosphere that blurs images seen through a telescope. A second reason is that the human eye and brain can readily be deceived. The brain can connect together two dark streaks on the Martian surface to give the perception of a single "canal." In this way did astronomers such as Schiaparelli and Lowell delude themselves.

The episode of the Martian canals was not the first time that a number of scientists came to very wrong conclusions. Like any human activity, science will always involve human error and human frailties. But scientific investigation is fundamentally self-correcting, because new observations and experiments can always call into question the scientific theories of the past. Thus it was during the first decades of the twentieth century, when the Greek astronomer Eugène Antoniadi observed Mars with a state-of-the-art telescope and found that the "canals" were actually unconnected dark spots. And so it was again in the 1960s, when unmanned spacecraft from Earth first observed the Martian surface at close range.

12-3 Unmanned spacecraft found craters, volcanoes, and canyons on Mars

WEB LINK 12.4 Between 1964 and 1969, three American spacecraft—*Mariner 4*, *Mariner 6*, and *Mariner 7*—flew past Mars and sent back the first close-up pictures of the planet's surface. These images showed no trace of the canals of Schiaparelli and Lowell and absolutely no evidence

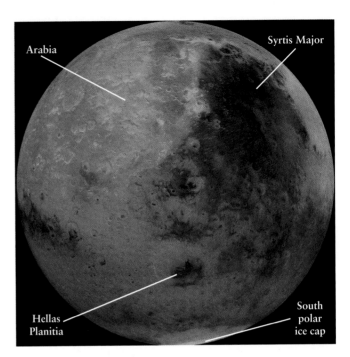

Figure 12-4 R I **V** U X G

Martian Craters This view of one Martian hemisphere is actually a mosaic of about 100 images made by the *Viking Orbiter 1* and *2* spacecraft in 1980. Arabia, the light-colored region at upper left, is dotted with numerous flat-bottomed craters. Hellas Planitia, at lower center, is an impact basin 5 times the size of Texas. Syrtis Major, the dark region at upper right, can also be seen in Figure 12-1 and the image that opens this chapter. (USGS/NASA)

of vegetation. The dark surface markings are just different-colored terrain.

Why did many Earth-based observers report that dark areas on Mars had the greenish hue of vegetation? The probable explanation is a quirk of human color vision. When a neutral gray area is viewed next to a bright red or brown area, the eye perceives the gray to have a blue-green color.

The most remarkable discovery of the first three *Mariner* missions to Mars was that the planet's surface is pockmarked with craters (Figure 12-4). At about the same time as these missions, scientists were learning that craters on the Moon were formed by interplanetary debris striking the lunar surface (see Section 9-4). The number of impacts tapered off about 3.8 billion years ago as the supply of such debris was used up, though mare-gouging impacts continued to about 3 billion years ago. The Martian craters, too, are the result of impacts from space. Since so many craters survive to the present day, at least part of the Martian surface must be extremely ancient.

Although many of the Martian craters are quite large, they had escaped detection from Earth because of the Martian atmosphere. This atmosphere is extremely thin by Earth standards; measurements from spacecraft reveal an average pressure at the Martian surface of only 0.0063 atmosphere, roughly the same as at an altitude of 35 km (115,000 ft) above the Earth's surface. (We saw in Section 8-5 and Section 11-3 that one atmosphere is the average sea-level pressure on the Earth.) But even this thin atmosphere obscures the Martian surface somewhat. That is why even the Hubble Space Telescope cannot see Martian craters distinctly (compare Figure 12-4 with the image that opens this chapter), and why it took close-up images made by spacecraft near Mars to reveal these craters.

Mariners 4, 6, and *7* transformed scientists' picture of Mars from that of an Earthlike world to that of a somewhat larger version of the Moon. But these three spacecraft observed just 10% of the Martian surface during their brief encounters. To get more complete observations required putting a spacecraft into orbit around Mars.

Three such orbiters reached Mars in 1971: NASA's *Mariner 9,* which mapped the planet for a year, and the shorter-lived Soviet *Mars 2* and *Mars 3.* These were followed by NASA's two *Viking Orbiters,* which made images of Mars over a four-year period beginning in 1976; by *Phobos 2,* a Soviet spacecraft that carried out two months of observations in 1989; and by NASA's *Mars Global Surveyor,* which has been studying Mars at close range since 1997. (Several other orbital missions have been attempted, but failed either to achieve Mars orbit or to maintain contact with Earth. Later in this chapter we discuss missions that have landed on Mars.)

Each new mission used increasingly sophisticated technology to study the Martian surface. These missions have revealed Mars to be a remarkably diverse planet, with many geologic features unique in the solar system.

One of the most striking properties of the Martian surface is the difference between the northern and southern hemispheres. As the topographic map in Figure 12-5 shows, the average elevation of the northern hemisphere is about 5 km (3 mi) lower than that of the southern hemisphere. Hence, planetary scientists refer to the **northern lowlands** and **southern highlands.** Furthermore, the northern lowlands are remarkably smooth and free of craters, while craters of many sizes are found across the southern highlands. This implies that the northern lowlands have been resurfaced by some process that eradicated the ancient lowland craters. Thus, the surface of the northern lowlands is relatively young, while that of the southern highlands is relatively old (see Section 9-1). This set of striking

figure 12-5

The Topography of Mars The color coding on this map of Mars shows elevations above (positive numbers) or below (negative numbers) the planet's average radius. To produce this map, an instrument on board *Mars Global Surveyor* fired pulses of laser light at the planet's surface, then measured how long it took each reflected pulse to return to the spacecraft. The labeled features can be seen in Figures 12-4, 12-6, 12-7, and 12-8. The *Viking Lander 1* (VL1), *Viking Lander 2* (VL2), and *Mars Pathfinder* (MP) landing sites are each marked with an X. (MOLA Science Team, NASA/GSFC)

differences between the two Martian hemispheres is called the **crustal dichotomy.**

Figure 12-5 shows that the dividing line between the northern lowlands and southern highlands is *not* precisely along the Martian equator. Thus, "northern hemisphere" is not exactly synonymous with "northern lowlands."

Several explanations for the Martian crustal dichotomy have been proposed. One idea is that a massive object struck the northern hemisphere relatively late in the planet's history. Like the impacts that formed the lunar maria (Section 9-4), such an impact would have excavated the crust to a depth of several kilometers and erased older craters. A smaller but still world-shaking impact of this kind produced Hellas Planitia, an immense basin 2300 km across and 6 km deep in the southern hemisphere that can actually be seen from Earth (see Figure 12-4 and Figure 12-5). One argument against this model is that the boundary between the northern lowlands and southern highlands is too irregular to be the wall of a giant impact crater.

Another idea is that Mars once had plate tectonic activity like the Earth, but that this activity came to an end long ago as Mars cooled. (We saw in Section 9-3 that small bodies tend to lose heat more rapidly than large bodies. Because Mars is substantially smaller than Earth, it should have lost more of its internal heat.) The young terrain in the northern lowlands could have formed by crustal spreading, such as happens along oceanic rifts on Earth like the Mid-Atlantic Ridge (see Figure 8-12).

Oceanic crust on Earth is about 30 km thinner than continental crust (see Table 8-2). Hence, if ancient plate tectonics produced the northern lowlands and southern highlands, we would expect a thinner crust in the northern hemisphere. One way to test this is to carefully monitor the motion of a spacecraft orbiting Mars. If there is a concentration of mass (such as a thicker crust) at one location on the planet, gravitational attraction will make the spacecraft speed up as it approaches the concentration and slow down as it moves away.

By analyzing the orbit of *Mars Global Surveyor* in just this way, a team of scientists headed by David Smith of NASA's Goddard Space Flight Center and Maria Zuber of MIT find that the Martian crust is about 40 km thick under the northern lowlands but about 70 km thick under the southern highlands. However, they find that the boundary between thin and thick crust does *not* line up with the boundary between high and low terrain, which contradicts the plate tectonic model.

It may be that the northern lowlands are the result of a combination of effects, including resurfacing by volcanic eruptions. Such resurfacing has taken place on Venus, where extensive lava flows have erased ancient craters (Section 11-6). The first evidence that Mars, too, has had a volcanic past came when the *Mariner 9* mission discovered several enormous volcanoes. The largest of these, Olympus Mons, is the largest volcano in the solar system. It covers an area as big as the state of Missouri, and rises 24 km (15 mi) above the surrounding plains—nearly 3 times the height of Mount Everest

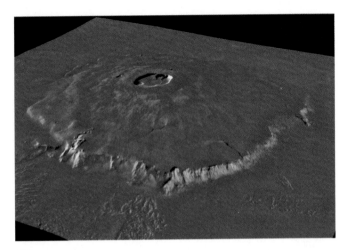

figure 12-6   R I **V** U X G

Olympus Mons This perspective view was created by combining a number of images from the two *Viking Orbiter* spacecraft. Olympus Mons is nearly 3 times as tall as Mount Everest. Its base measures 600 km (370 mi) in diameter, and the scarps (cliffs) that surround the base are 6 km (4 mi) high in spots. The caldera, or volcanic crater, at the summit is approximately 70 km across, large enough to contain the state of Rhode Island. (© Calvin J. Hamilton)

(Figure 12-6). By comparison, the highest volcano on Earth, Mauna Loa in the Hawaiian Islands, has a summit only 8 km (5 mi) above the ocean floor.

Olympus Mons and other Martian volcanoes probably formed by *hot-spot volcanism*. In this process, magma wells upward from a hot spot in a planet's mantle, elevating the overlying surface and producing a volcano (see Section 11-4). On Earth, hot-spot volcanism is the origin of the Hawaiian Islands. These islands are part of long chain of volcanoes that formed as the Pacific tectonic plate moved over a long-lived hot spot. On Mars, by contrast, the huge size of Olympus Mons strongly suggests an absence of plate tectonics. A single hot spot under this volcano probably pumped magma upward through the same vent for millions of years, producing one giant volcano rather than a long chain of smaller ones. The same process presumably gave rise to Maxwell Montes and other large volcanoes on Venus (see Section 11-4).

If Olympus Mons is so gigantic, why didn't astronomers see it through telescopes long before *Mariner 9*? The explanation is that hot spots produce volcanoes with very gently sloping sides. Even when illuminated from the side, as at sunset or sunrise, such volcanoes cast only small shadows. Hence, Olympus Mons does not stand out from its surroundings as seen from Earth.

Hot spots have produced a number of other features on Mars. A 2500-km (1500-mi) wide cluster of large volcanoes lies just east of Olympus Mons in a region called the Tharsis rise. It is actually a dome-shaped bulge that has been lifted 5 to 6 kilometers above the planet's average elevation (see Figure 12-5). Apparently, a massive plume of magma once welled upward from a hot spot under this region, producing the Tharsis rise and its volcanoes. Another hot spot on the opposite side of Mars produced a smaller bulge centered on the volcano Elysium Mons, shown in Figure 12-5. None of the Martian volcanoes are active today.

East of the Tharsis rise, *Mariner 9* discovered a vast chasm that runs roughly parallel to the Martian equator (Figure 12-7). If this canyon were located on Earth, it would stretch from Los Angeles to New York. In honor of the spacecraft that revealed so much of the Martian surface, this canyon has been designated Valles Marineris.

Valles Marineris begins with heavily fractured terrain in the west and ends with ancient cratered terrain in the east. Many geologists suspect that Valles Marineris was caused by the same upwelling of material that formed the Tharsis rise. As the Martian surface bulged upward at Tharsis, there would have been tremendous stresses on the crust, which would have caused extensive fracturing. Thus, Valles Marineris may be a **rift valley**, a feature created when a planet's crust breaks apart along a fault line. Rift valleys are found on Earth; two examples are the Red Sea and the Rhine River valley in Europe. Other, smaller rifts in the Martian crust are found all around the Tharsis rise.

While the motion of material in the mantle has had a large effect on the Martian surface, plate tectonics has not. Mars lacks the global network of ridges and subduction zones that plate tectonics has produced on the Earth. The reason is that the outer layers of the red planet have cooled more extensively than those of the Earth. The Martian crust is tens of kilometers thicker than the Earth's crust, making it impossible for one part of the crust to be subducted beneath another. Hence, the entire crust of Mars makes up a single tectonic plate, as is the case on the Moon (Section 9-1), Mercury (Section 10-3), and Venus (Section 11-6).

Mars Global Surveyor measurements show that present-day Mars has no planetwide magnetic field. However, the spacecraft found magnetized regions in the ancient southern highlands. These portions of the crust presumably formed when Mars did have a planetwide magnetic field. This would have been early in Martian history, when the planet's interior was still hot and currents could flow in its core to produce a magnetic field. As surface material cooled and solidified, it became magnetized by the planetwide field, and this material has retained its magnetization over the eons. Remarkably, different regions of the southern highlands are magnetized in different directions. This suggests that the Martian magnetic field reversed direction from time to time, just as the Earth's magnetic field does (see Section 8-4). *Mars Global Surveyor* has not found magnetized areas in impact basins such as Hellas Planitia. Hence, the Martian magnetic field must have shut off by the time these basins were formed by impact about 3.9 billion years ago.

Today Mars probably has a core with a radius about half that of the planet. Unlike the Earth's core, however, the Martian core is thought to contain substantial amounts of sulfur in addition to iron. (We saw in Section 7-8 that planets can have different chemical compositions, depending on how far from the Sun they formed.) Temperatures in the core should be high enough to melt sulfur and sulfur compounds. However, the electrical properties of these substances are different from those of iron. Hence, while the flow of molten iron in the Earth's outer core causes electric currents that generate our planet's magnetic field, flows within the sulfur-rich Martian core do not.

12-4 Surface features indicate that water once flowed on Mars

Spacecraft orbiting Mars have seen no evidence of rainfall or of liquid water anywhere on the planet's surface. But they have revealed many features that look like dried-up lakes and riverbeds (Figure 12-8). Intricate branched patterns and delicate channels meandering among flat-bottomed craters strongly suggest that water once flowed over the Martian surface. Such dried-up water channels are found throughout the old southern highlands but in relatively few places on the younger northern lowlands. Hence, large amounts of water flowed on the Martian surface in the distant past but not in the recent past.

Spacecraft images have also revealed evidence of powerful flash floods that long ago raged across the surface of Mars (Figure 12-9). In addition to flash floods, there is also evidence that water flowed for sustained periods on the Martian

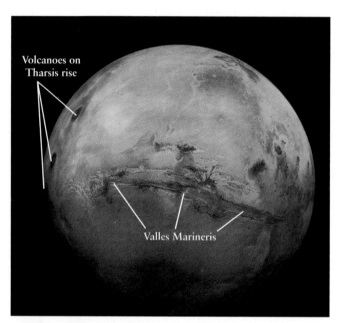

Volcanoes on Tharsis rise

Valles Marineris

  Figure 12-7   R I **V** U X G

Valles Marineris This mosaic of *Viking Orbiter* images shows the side of Mars opposite that shown in Figure 12-4. At the far left is the Tharsis rise, with its three large volcanoes. The huge rift valley of Valles Marineris extends from west to east for more than 4000 km (2500 mi). The valley is 600 km (400 mi) wide at its center, and its deepest part is 8 km (5 mi) beneath the surrounding plateau. Compare this image with Figure 12-5. (USGS/NASA)

Figure 12-8   R I **V** U X G

Ancient River Channels on Mars This *Viking Orbiter* image shows a network of dried riverbeds extending across the cratered southern highlands. The large craters at upper left and upper right are each about 35 km (22 mi) across. Liquid water would evaporate in the sparse, present-day Martian atmosphere, so these channels must date from a time when the Martian atmosphere was thicker and its climate more Earthlike. (Michael Carr, USGS)

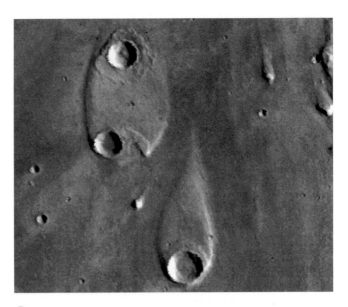

Figure 12-9   R I **V** U X G

Signs of Ancient Floods on Mars These teardrop-shaped islands, each about 40 km (25 mi) long, rise above the floor of an ancient Martian channel called Ares Valles. They were carved out by a torrent of water that once flowed from the bottom of this image toward the top. The size and shape of the islands suggest that the peak flow rate in these floods was more than 1000 times that of the Mississippi River. Similar flood-carved islands are found on Earth in eastern Washington State. (NASA/USGS)

surface. Figure 12-10 shows a canyon that appears to have been shaped gradually by water erosion.

The features shown in Figures 12-8, 12-9, and 12-10 must all be quite old, for liquid water cannot exist anywhere on the Martian surface today. Water is liquid over only a limited temperature range: If the temperature is too low, water becomes ice, and if the temperature is too high, it becomes water vapor. What determines this temperature range is the atmospheric pressure above a body of water. If the pressure is very low, molecules easily escape from the liquid's surface, causing the water to vaporize. Thus, at low pressures, water more easily becomes water vapor. The average surface temperature on Mars is only 220 K ($-53°C$, or $-63°F$), and the average pressure is only 0.0063 atmosphere. With this combination of temperature and pressure, water can exist as a solid (ice) and as a gas (water vapor) but not as a liquid. You can see this same situation inside a freezer, where water vapor swirls around over ice cubes.

In order to keep water on Mars in the liquid state, it would be necessary to increase both the temperature (to keep water from freezing) and the pressure (to keep the liquid water from evaporating). Therefore, during the past, when liquid water existed on Mars, the atmosphere must have been both thicker and warmer than it is today.

If water once flowed on Mars, where might that water be today? It is not in the Martian atmosphere, which is 95.3% carbon dioxide (CO_2) and contains only a trace amount of

water vapor (see Table 8-3). Indeed, it never rains on Mars. If all the water vapor could somehow be squeezed out of the Martian atmosphere, it would not fill one of the five Great Lakes of North America.

Another possibility is that the Martian polar ice caps are reservoirs of frozen water. (You can see these polar ice caps in the image that opens this chapter and in Figure 12-4.) However, the winter temperature at the Martian poles is low enough ($-140°C = -220°F$) to freeze CO_2 as well as water. How much, then, of the ice caps is frozen CO_2 (dry ice) and how much is water ice?

Mariner 9 provided a partial answer to this question in 1972, as it observed spring and summer coming to the northern hemisphere of Mars. During the Martian spring, the north polar cap receded rapidly, consistent with a thin layer of CO_2 frost evaporating in the sunlight. With the arrival of summer, however, the rate of recession slowed abruptly. This indicated that a thicker and less easily evaporated layer of water ice had been exposed. Scientists concluded that the **residual polar caps**—the portions that survive the Martian summers—contain a large quantity of frozen water (Figure 12-11). Calculating the volume of water ice in the residual caps is difficult, however, because we do not know the thickness of the layer of ice. Furthermore, the south polar cap has never been observed to lose its cover of frozen CO_2, so we do not have direct evidence of water ice there. Thus, the exact amount of frozen water stored in the Martian polar caps is still a matter of debate.

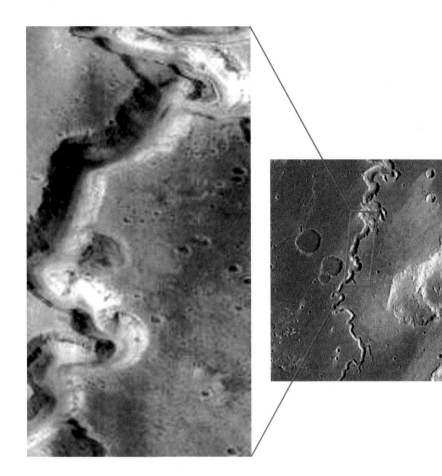

figure 12-10   R I **V** U X G

Signs of Sustained Water Flow on Mars
The *Viking Orbiter 1* image on the right shows a 2.5-km (1.6-mi) wide canyon in the Martian southern highlands. The area outlined in red is shown in more detail in the *Mars Global Surveyor* image on the left. The terraces within the canyon resemble structures found within riverbeds on Earth. In addition to water erosion, a localized collapse of the surface (like a sinkhole on Earth) may have helped shape this canyon. (NASA/JPL/Malin Space Science Systems)

The residual ice caps cannot be the only site of Martian water, however. Close-up views such as Figure 12-9 show that flash-flood features usually emerge from craters or from collapsed, jumbled terrain. It therefore seems likely that frozen water is also stored in **permafrost** under the Martian surface, similar to the layer beneath the tundra in the Earth's far northern regions (Figure 12-12). In the distant past, heat from a meteoroid impact or from volcanic activity occasionally caused a sudden melting of this subsurface ice. The ground then collapsed, and millions of tons of rock pushed the water to the surface.

The total amount of water on Mars is not known. But by examining flood channels at various locations around Mars, Michael Carr of the U.S. Geological Survey has estimated that there is enough water to cover the planet to a depth of 500 meters (1500 ft). (By comparison, Earth has enough water to cover our planet to a depth of 2700 meters, or 8900 ft.) Thus, while the total amount of water on Mars is less than on Earth, there may be enough to have once formed lakes or even oceans. Indeed, some low-lying Martian terrain has layers of sediment that resemble those found in dry lakebeds (Figure 12-13).

Somewhat more controversial is the idea that the flat northern lowlands of Mars are actually the bottom of an ancient ocean. James Head of Brown University and his colleagues have analyzed high-resolution topographical data from *Mars Global Surveyor* (see Figure 12-5), and have found

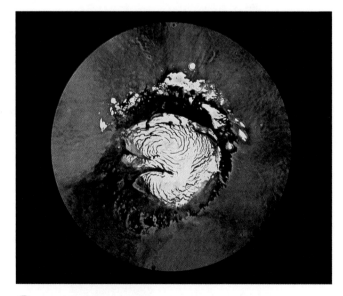

figure 12-11   R I **V** U X G

The North Polar Ice Cap This mosaic of *Viking Orbiter* images shows the Martian northern ice cap streaked with dark dust. During the northern winter, this region is covered by CO_2 frost. This evaporates in the summer, exposing a residual polar cap made primarily of water ice. The CO_2 frost never completely evaporates from the south pole, so it is not known for certain whether a similar layer of water ice exists there. (JPL/NASA)

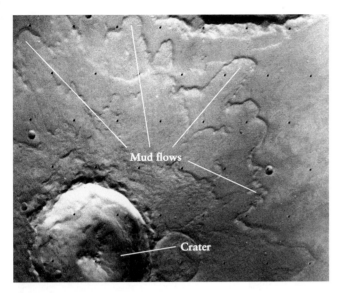

Figure 12-12 R I **V** U X G

A Martian "Mud Splash" Most large, young Martian craters are surrounded by features that resemble mud flows. This is evidence for a subsurface layer of water ice, which would have been melted by a crater-forming impact. No such features appear around craters on the bone-dry Moon (see Figure 9-8). The crater in this *Viking Orbiter 1* image, called Yuty, is about 18 km (11 mi) across. (GSFC/NASA)

evidence of an ancient shoreline that encircles the lowlands. Large channels leading from the highlands to the lowlands could have supplied a northern ocean with runoff from the higher terrain to the south.

It may be that the era of liquid water on Mars is not quite over. Recent *Mars Global Surveyor* images show gullies apparently carved by water flowing down the walls of pits or craters (Figure 12-14). These could have formed when water trapped underground—where the pressure is greater and water can remain a liquid—seeped onto the surface. The gullies appear to be geologically young, so it is possible that even today some liquid water survives below the Martian surface.

Why were conditions on Mars once so different—with a warmer and thicker atmosphere and with liquid water on the surface—from today? To find an answer, we must explore how the Martian atmosphere evolved.

12-5 Earth and Mars began with similar atmospheres that evolved very differently

As nineteenth-century astronomers observed Mars through their telescopes, they were occasionally able to see cloud formations (Figure 12-15). This was the first evidence that Mars has an atmosphere (without an atmosphere, a planet

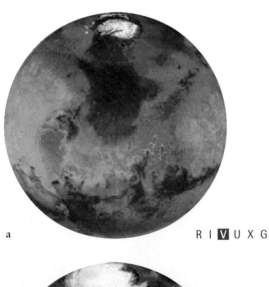

a R I **V** U X G

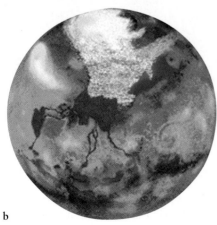

b

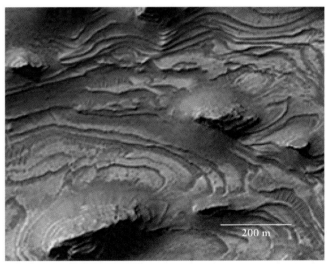

c R I **V** U X G

Figure 12-13

Ancient Oceans and Lakes on Mars? **(a)** This *Viking Orbiter* mosaic shows the dry, frigid surface of present-day Mars. **(b)** If enough water was present when Mars had a more Earthlike climate, there may have been lakes and oceans (shown in blue). This artist's impression depicts where water may have been darkened by suspended sediment (shown in red) or covered by a layer of ice (shown in white). **(c)** *Mars Global Surveyor* images, such as this one of a portion of Valles Marineris, reveal terrain with "stairstep" layers. Such terrain could have been created by sedimentation at the bottom of an ancient body of water. (a, b: Slim Films; c: Malin Space Science Systems/JPL/NASA)

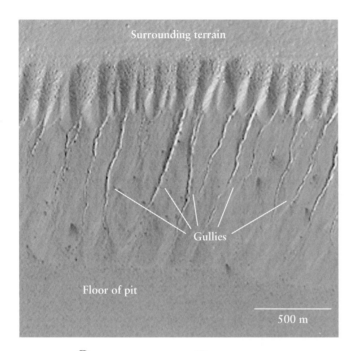

Surrounding terrain

Gullies

Floor of pit

500 m

figure 12-14   R I **V** U X G

Evidence for Recent Liquid Water on Mars When *Mars Global Surveyor* looked straight down into this pit near the Martian south pole, it saw a series of gullies along the pit wall. These appear to have been formed when subsurface water seeped out to the surface. Many such gully systems have been discovered. (Malin Space Science Systems/JPL/NASA)

WEB LINK 12.7

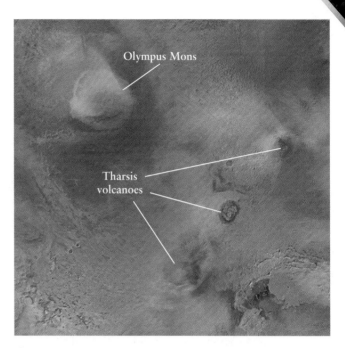

Olympus Mons

Tharsis
volcanoes

figure 12-15   R I **V** U X G

Clouds Top the Volcanoes of Mars On most Martian afternoons, bluish-white clouds form at the summits of Olympus Mons and the volcanoes of the Tharsis rise. You can see other clouds at the center right and bottom of the image that opens this chapter. (Malin Space Science Systems/JPL/NASA)

can have no clouds). It was difficult to measure the Martian atmosphere's spectrum, however, and for many decades no one knew for sure what the atmosphere is made of. When the *Mariner 4* spacecraft flew past Mars in 1965, its measurements indicated that the planet's atmosphere is almost entirely carbon dioxide (CO_2). We now know that the extremely thin atmosphere is about 95.3% carbon dioxide and 2.7% nitrogen, with small amounts of argon, oxygen, carbon monoxide, and water vapor. The clouds themselves are made of tiny water ice crystals (as in wispy cirrus clouds on Earth) as well as CO_2 ice ("dry ice") crystals.

As we have seen, the Martian climate must have been much warmer and more Earthlike in the ancient past. What could have provided this extra warmth? The likeliest mechanism is the greenhouse effect (recall Section 8-1, in particular Figure 8-5).

WEB LINK 12.8

Because the atmospheres of the Earth and Mars are transparent to sunlight, they are not heated directly by the Sun. Instead, they are heated from below by energy reradiated from the ground at longer, infrared wavelengths. Greenhouse gases such as water vapor (H_2O) and carbon dioxide (CO_2) absorb these wavelengths, trapping the heat and warming the atmosphere. Without the greenhouse effect, the average temperature on Earth would be about 36°C lower than it is now, and our entire planet would be frozen over.

In the present-day Martian atmosphere, the greenhouse effect is much less pronounced; it increases the temperature by

only about 5°C. This is because the thin Martian atmosphere permits much of the infrared radiation to escape into space. If the greenhouse effect was once important on Mars, the Martian atmosphere must have been much thicker than today, with much greater amounts of water vapor and carbon dioxide.

Extinct Martian volcanoes suggest how such an ancient atmosphere could have been created. The terrestrial planets obtained their atmospheres primarily through volcanic outgassing (see Section 8-5). Water vapor, carbon dioxide, and nitrogen are among the most common gases in volcanic vapors. These three gases probably constituted the bulk of the atmospheres of both the Earth and Mars 4 billion years ago.

On Earth most of the water is in the oceans. Nitrogen (N_2), which is not very reactive, is still in the atmosphere. Carbon dioxide, however, dissolves in water; thus, rain washed most of the CO_2 out of our planet's atmosphere long ago. The dissolved CO_2 reacted with rocks in rivers and streams, and the residue was ultimately deposited on the ocean floors, where it turned into carbonate rocks, such as limestone. Consequently, most of the Earth's carbon dioxide is tied up in the Earth's crust, and carbon dioxide makes up only a small fraction of the Earth's atmosphere. (About 1 in every 3000 molecules in our atmosphere is a carbon dioxide molecule.)

The amount of carbon dioxide in our atmosphere is sustained in part by plate tectonics. Tectonic activity causes carbonate rocks to cycle through volcanoes, where they are heated and forced to liberate their trapped CO_2. This liberated

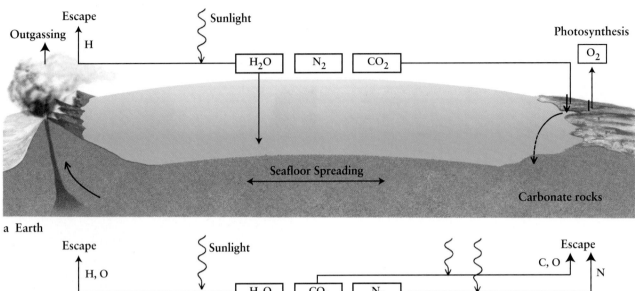

a Earth

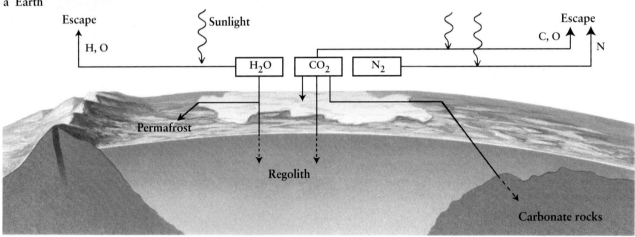

b Mars

Figure 12-16

The Atmospheres of the Earth and Mars (a) Water vapor (H_2O) in the Earth's primitive atmosphere condensed into oceans, while nitrogen (N_2) remained in the air. Carbon dioxide (CO_2) cycles between the atmosphere and being stored in carbonate rocks. (b) On Mars, a thick CO_2 atmosphere initially kept the planet warm enough (through the greenhouse effect) for liquid water to exist. Rainfall eventually washed the CO_2 out of the atmosphere. Today, water ice probably resides in the polar caps and the Martian permafrost. Some water and CO_2 may be chemically bound in the fine-grained Martian regolith. Ultraviolet light from the Sun breaks atmospheric molecules into their constituent atoms, which then escape into space. (Adapted from R. M. Haberle)

gas then rejoins the atmosphere. Without volcanoes, rainfall would remove all the CO_2 from our atmosphere in only a few thousand years. The truly unique feature of our atmosphere—the presence of oxygen—is the result of biological activity, in particular, photosynthesis by plant life (Figure 12-16a).

On Mars, rainfall 4 billion years ago may have washed much of the planet's carbon dioxide from its atmosphere, perhaps creating carbonate rocks in which the CO_2 is today chemically bound. If so, the Martian carbonate rocks are significant reservoirs of the primordial CO_2 outgassed from ancient volcanoes. But any tectonic activity on Mars came to an end as the planet cooled, so its carbonate rocks were not recycled through volcanoes as they are on Earth. The depletion of carbon dioxide from the Martian atmosphere into the rocks was therefore permanent.

As the amount of atmospheric CO_2 declined, the greenhouse effect on Mars weakened and temperatures began to fall. This temperature decrease caused more water vapor to condense into raindrops and fall to the surface, taking even more CO_2 with them and further weakening the greenhouse effect. Thus, a decrease in temperature would have caused a further decrease in temperature—a phenomenon sometimes called a **runaway icehouse effect**. (As we saw in Section 11-5, just the opposite has taken place on Venus, where temperatures have spiraled upward to a hellish level.)

According to calculations by James B. Pollack of NASA's Ames Research Center, a Martian carbon dioxide atmosphere with the same pressure as our atmosphere could have survived for only 10 to 100 million years. However, volcanic activity early in Mars's history may have recycled some carbon dioxide into the atmosphere, thus prolonging the greenhouse effect for perhaps half a billion years. Ultimately, however, rainfall succeeded in removing both water vapor and most of the carbon dioxide from the Martian atmosphere. With only a thin CO_2 atmosphere remaining, surface temperatures on Mars eventually stabilized at their present frigid values.

As both water vapor and carbon dioxide became depleted, ultraviolet light from the Sun could penetrate the thinning Martian atmosphere to strip it of nitrogen. Nitrogen molecules (N_2) normally do not have enough thermal energy to escape from Mars, but they can acquire that energy from ultraviolet photons, which break the molecules in two. Ultraviolet photons can also split carbon dioxide and water molecules, giving their atoms enough energy to escape (Figure 12-16b). Indeed, in 1971 the Soviet *Mars 2* spacecraft found a stream of oxygen and hydrogen atoms (from the breakup of water molecules) escaping into space. Since Mars is being depleted of water in this way, the total amount of water must have been greater in the past, perhaps enough to have inundated the northern lowlands (see Figure 12-13).

Oxygen atoms that did not escape into space could have combined with iron-bearing minerals in the surface, forming rustlike compounds. Such compounds have a characteristic red color and may be responsible for the overall reddish-brown color of the planet.

In the absence of tectonics, these three effects—rainfall, loss of gases into space, and chemical reactions between gases and surface rocks—would have thinned the Martian atmosphere and cooled its surface. Perhaps in this way Mars came to have the sparse atmosphere and frigid conditions we observe today.

12-6 The *Viking* and *Mars Pathfinder* landers explored the barren Martian surface

Spacecraft orbiting Mars have revolutionized our understanding of the planet. But nothing substitutes for data collected from a planet's surface, which is why several spacecraft have been sent to land on Mars. To date three of these spacecraft, all from NASA, have successfully landed on the Martian surface and returned substantial data: the two *Viking Landers* in 1976 and *Mars Pathfinder* in 1997. All three landed in different regions of the northern lowlands.

The two *Viking* spacecraft were launched in 1975 and arrived at Mars almost a year later. Each spacecraft consisted of two modules, an orbiter (introduced in Section 12-4) and a lander. After entering orbit about Mars, each lander separated from its orbiter and descended to the Martian surface with the aid of a heat shield, retrorockets, and a parachute. Each lander is roughly the size of a small automobile (Figure 12-17).

Viking Lander 1 touched down in July 1976 in a flat, moderately cratered plain called Chryse Planitia (see Figure 12-5). From this site it radioed data back to Earth until November 1982. The spacecraft's cameras revealed a landscape littered with jagged rocks, as Figure 12-18 shows. These are probably

figure 12-17   R I **V** U X G

A *Viking Lander* Each *Viking Lander* is about 2 m (6 ft) tall from the base of its footpads to the top of its antenna. (This photograph actually shows a full-size replica under test on Earth.) The extendable arm on the right ends in a scoop for retrieving surface samples. The dish antenna was used to transmit data to one of the *Viking* orbiters overhead, which in turn relayed the information back to Earth. (NASA)

the wreckage of an ancient crater-forming impact, which ejected rocky debris in all directions.

Figure 12-18 also shows that the landscape around *Viking Lander 1* is covered with a fine-grained dust. These particles have piled up in places to form drifts that resemble sand dunes on Earth. The dust particles are much smaller than ordinary household dust on Earth: They are only about 1 μm (10^{-6} m) across, about the size of the particles in cigarette smoke.

figure 12-18   R I **V** U X G

The View from *Viking Lander 1* The rocks of Chryse Planitia resemble volcanic rocks on Earth (see Figure 8-18a), and thus are thought to be pieces of an ancient lava flow that was broken apart by crater-forming impacts. Fine-grained debris covers the rocks and the spaces between them, and has piled up to form dunes. The large pile of rocks to the left of center is about 2 m across and 8 m away from the camera. (Dr. Edwin Bell, II/ NSSDC/GSFC/NASA)

Viking Lander 2 set down in Utopia Planitia in August 1976, on nearly the opposite side of the planet from *Viking Lander 1* but about 1500 kilometers (930 miles) closer to the north pole. (Figure 12-5 shows both landing sites.) It continued to transmit data to Earth until April 1980. The terrain around *Viking Lander 2* is remarkably crater-free, but is nonetheless strewn with rocks like those in Figure 12-18. It is thought that many of these rocks were ejected from a large crater about 200 kilometers east of the landing site. As at the *Viking Lander 1* site, the rocks in Utopia Planitia are covered with dust, though the dust there does not pile up in drifts.

Even before *Viking*, astronomers knew that dust was an important part of the Martian environment. Telescopic observers found that surface features on Mars are sometimes obscured by a yellowish haze. This haze was interpreted to be caused by windblown storms of dust (Figure 12-19). Because the Martian air is so tenuous, a 100-km/h (60-mi/h) wind on Mars is only about as strong as a 10-km/h (6-mi/h) breeze on Earth. Hence, only very small dust particles are carried aloft and spread over the surface by the Martian winds. Airborne dust also affects the color of the Martian sky, which is pink when the air is dusty but blue when the air is nearly clear.

In addition to cameras, each *Viking Lander* had a scoop at the end of a mechanical arm to dig into the Martian regolith and retrieve samples for analysis (see Figure 12-17). Bits of the regolith clung to a magnet mounted on the scoop, indicating that the regolith contains iron. A device called an X-ray spec-

trometer measured the chemical composition of samples at both lander sites and showed them to be rich in iron, silicon, and sulfur. Thus, the Martian regolith can best be described as an iron-rich clay.

The *Viking Landers* each carried a miniature chemistry laboratory for analyzing the regolith. Curiously, when a small amount of water was added to a sample of regolith, the sample began to fizz and release oxygen. This experiment indicated that the regolith contains unstable chemicals called peroxides and superoxides, which break down in the presence of water to release oxygen gas.

The unstable chemistry of the Martian regolith probably comes from solar ultraviolet radiation that beats down on the planet's surface. (Unlike Earth, Mars has no ozone layer to screen out ultraviolet light.) Ultraviolet photons easily break apart molecules of carbon dioxide (CO_2) and water vapor (H_2O) by knocking off oxygen atoms, which then become loosely attached to chemicals in the regolith. Ultraviolet photons also produce highly reactive ozone (O_3) and hydrogen peroxide (H_2O_2), which also become incorporated in the regolith.

Among the most important scientific goals of the *Viking Landers* was to search for evidence of life on Mars. Although it was clear that Mars has neither civilizations nor vast fields of plants, it still seemed possible that simple microorganisms could have evolved on Mars during its early brief epoch of Earthlike climate. If so, perhaps some species had been able to survive to the present day. To test this idea, the *Viking Landers* looked for biologically significant chemical reactions in samples of the Martian regolith. But they failed to detect any reactions of this type. Perhaps life never existed on Mars, or perhaps it existed once but was destroyed by the presence of peroxides and superoxides in the regolith.

On Earth, hydrogen peroxide is commonly used as an antiseptic. When poured on a wound, it fizzes and froths as it reacts with organic material and destroys germs. Thus, the *Viking Landers* may have failed to detect any biochemistry on Mars because the regolith has literally been sterilized! Hence, we cannot yet tell whether life ever existed on Mars, or whether it might yet exist elsewhere on or beneath the planet's surface. (In Chapter 30 we discuss the *Viking Lander* experiments in detail, as well as the controversial proposal that fossil Martian microorganisms may have traveled to Earth aboard meteorites.)

One of the limitations of the *Viking Landers* was that they could collect and examine samples only within the reach of the spacecraft's mechanical arm. A more versatile unmanned explorer, *Mars Pathfinder*, landed on Mars on July 4, 1997. Unlike the *Viking Landers*, *Mars Pathfinder* did not have a heavy and expensive rocket engine to lower it gently to the Martian surface. Instead, the spacecraft was surrounded by airbags like those used in automobiles. Had there been anyone to watch *Mars Pathfinder* land, they would have seen an oversized beach ball hit the surface at 50 km/h (30 mi/h), then bounce for a kilometer before finally rolling to a stop. Unharmed by its wild ride, the spacecraft then deflated its airbags and began to study the geology of Mars.

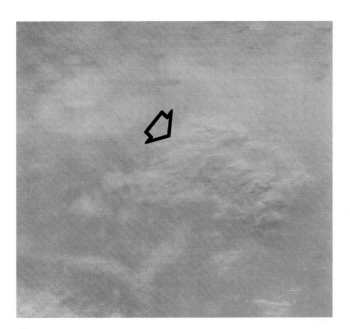

Figure 12-19 R I **V** U X G

A Martian Dust Storm This *Viking Orbiter* image shows a 300-km-wide dust storm (shown by the arrow) moving across the floor of the Argyre Planitia impact basin during early 1977. Over the ages, deposits of dust from these storms have filled in the bottoms of most Martian craters. (NASA/JPL)

a

figure 12-20 R I V̄ U X G

Sojourner and ***Mars Pathfinder*** (a) The technicians in this photo are working on the *Sojourner* rover in a "clean room," which is kept clear of any small particles that might contaminate the rover and cause mechanical problems on Mars. The upper surface of the rover is covered with solar panels that provided power for its electric motors. **(b)** This view from the stationary part of *Mars Pathfinder* shows the surrounding field of dust-covered rocks. Scientists gave the rocks whimsical names such as Jiminy Cricket, Lunchbox, and Pooh Bear. In this image, *Sojourner* is deploying its X-ray spectrometer against a rock nicknamed Moe. The hill on the horizon is about 1 km (0.6 mi) away. (NASA/JPL)

b

On board *Mars Pathfinder* was a small, wheeled robot called *Sojourner* (Figure 12-20). Following directions from Earth, *Sojourner* spent three months exploring the rocks around its landing site. Over those three months, the mission radioed to Earth more than 10 times as many images as had the two *Viking Lander*s combined.

Sojourner carried a camera of its own as well as an X-ray spectrometer for measuring the chemical composition of individual rocks. Each *Viking Lander* also carried an X-ray spectrometer, but it was located within the spacecraft body and could analyze only samples retrieved by the mechanical arm. *Sojourner*, however, carried its spectrometer to rocks several meters from *Mars Pathfinder*, as Figure 12-20b shows.

Mars Pathfinder landed about 800 km (500 mi) from *Viking Lander 1*, in an ancient flood channel in the northern lowlands called Ares Valles. (Figure 12-5 shows the landing site, and Figure 12-9 shows a portion of this plain.) Flood waters can carry rocks of all kinds great distances from their original locations, so scientists expected that this landing site would show great geologic diversity. They were not disappointed. Some rocks resemble the impact breccias found in the ancient highlands of the Moon (see Figure 9-17). Still others have a layered structure like sedimentary rocks on Earth (see Figure 8-18b). Such rocks form gradually at the bottom of bodies of water, which suggests that liquid water

was stable on Mars for a substantial period of time. To explain this, it may be necessary to imagine an ancient Mars that was even warmer and had an even thicker atmosphere than described in Section 12-4.

Many of the rocks at the *Mars Pathfinder* landing site appear to be igneous rock produced by volcanic action. We have seen that Mars has volcanoes that were once active, so this was not a surprise. What was surprising was the chemical composition of these volcanic rocks. Unlike basalts, the most common type of volcanic rock found on Earth (see Figure 8-18a), the rocks that *Sojourner* analyzed with its X-ray spectrometer have a relatively high concentration of silicon. Silicon-rich volcanic rocks, called **andesites**, are less dense than basalts. They are formed when crustal material gets so hot that it becomes molten, allowing less dense materials to float to the top.

Observations from *Mars Global Surveyor* indicate that andesites are found in broad regions of the young northern lowlands, including the *Mars Pathfinder* landing site. By contrast, basalts dominate in the older southern highlands. Thus, at some time in the past, something changed within Mars to alter the kind of lava that flowed from volcanoes. Why this happened is as yet unknown. Whatever the explanation, Mars has apparently had a more dynamic geologic past than heretofore suspected.

12-7 The Martian atmosphere changes dramatically with the seasons

The *Viking Lander* and *Mars Pathfinder* spacecraft all carried instruments to measure meteorological conditions on Mars. The thin, nearly cloudless Martian atmosphere provides little thermal insulation, and there are large swings in air temperature over the course of a Martian day. At the *Mars Pathfinder* site, temperatures ranged from 197 K (−76°C = −105°F) just before dawn to 263 K (−10°C = 14°F) in the "heat" of the afternoon.

Each afternoon, parcels of warm air rise from the heated surface and form whirlwinds called **dust devils.** (A similar phenomenon with the same name occurs in dry or desert terrain on Earth.) Martian dust devils can reach an altitude of 6 km (20,000 ft) and help spread dust particles across the planet's surface. All three landers measured changes in air pressure as dust devils swept past. Often several of these would pass in a single afternoon. Dust devils are large enough to seen by spacecraft orbiting Mars (Figure 12-21). As the sun sets and afternoon turns to evening, the dust devils subside until the following day.

figure 12-22 R I **V** U X G

Winter in Mars's Northern Hemisphere This picture, taken during midwinter, shows a thin frost layer that lasted for about a hundred days at the *Viking Lander 2* site. Freezing carbon dioxide adheres to water-ice crystals and dust grains in the atmosphere, causing them to fall to the ground. (NASA/JPL)

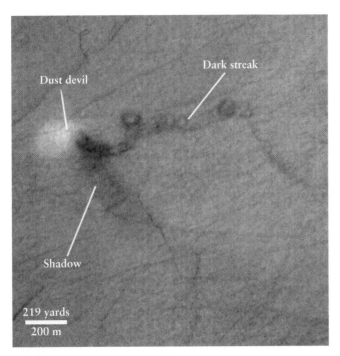

figure 12-21 R I **V** U X G

A Martian Dust Devil This *Mars Global Surveyor* image shows a dust devil as seen from almost directly above. This tower of swirling air and dust casts a long shadow in the afternoon sun. The dust devil had been moving from right to left before the picture was taken, leaving a dark, curlicue-shaped trail in its wake. The area shown is about 1.5 by 1.7 km (about 1 mile on a side). (NASA/JPL/Malin Space Science Systems)

In addition to recording conditions over the course of a Martian day, the two long-lived *Viking Landers* measured how the weather changes over the course of a Martian year. (*Mars Pathfinder*, with its lifetime of just three months, could not do this.) *Viking Landers 1* and 2 both touched down in the northern hemisphere of Mars when it was spring in that hemisphere and autumn in the southern hemisphere. Initially, both spacecraft measured atmospheric pressures around 0.008 atmosphere. After only a few weeks on the Martian surface, however, atmospheric pressure at both landing sites was dropping steadily. Mars seemed to be rapidly losing its atmosphere, and some scientists joked that soon all the air would be gone! The actual explanation proved to be quite straightforward—winter was coming to the southern hemisphere.

During the southern hemisphere's winter, the Martian south pole becomes so cold that large amounts of carbon dioxide condense out of the atmosphere there, covering the ground with flakes of dry ice. The formation of this "snow" removes gas from the Martian atmosphere, thereby lowering the atmospheric pressure across the planet. Several months later, when spring comes to the southern hemisphere, the dry-ice snow evaporates rapidly and the atmospheric pressure returns to its prewinter levels. The pressure drops again in the southern hemisphere's summer, because it is then winter in the northern hemisphere, and dry-ice snow condenses at northern latitudes (Figure 12-22). Therefore, both temperature and atmospheric pressure change significantly with the Martian seasons.

As on the Earth, the seasons on Mars are caused by the tilt of the planet's rotation axis. When the north pole is tilted toward the Sun, it is summer in the northern hemisphere and winter in the southern hemisphere; when the south pole is tilted toward the Sun, the seasons are reversed (see Figures 2-12 and 2-13). But because Mars has a rather elliptical orbit, the seasonal variations in the planet's two hemispheres are somewhat different. Summer comes to the southern hemisphere of Mars at the same time that the planet makes its closest approach to the Sun. In the northern hemisphere, by contrast, summer comes when Mars is farthest from the Sun. Hence, the temperature at the south pole during its summer should be higher than at the north pole when it is summer there. Paradoxically, the *Viking Orbiter*s measured just the opposite! The north pole's temperature climbs to 200 K during the northern summer, while the midsummer temperature at the south pole never gets above 150 K.

Why is summer warmer in the northern hemisphere than in the southern hemisphere? This seemingly backward state of affairs is caused by dust in the atmosphere. When it is summer in the southern hemisphere, the Sun warms the soil surrounding the south polar ice cap, while the ice cap remains quite cold. This temperature contrast gives rise to strong winds that often produce swirling dust storms. A dusty atmosphere admits less sunlight than a clear one, so these storms help to keep the south polar ice cap cold even in midsummer. At the north pole, however, the solar heating during summer is weaker because the planet is far from the Sun. Hence, the temperature difference between the cold north polar ice cap and the surrounding sun-warmed soil is not as great, and the strong winds needed to produce dust storms do not appear. As a result, the north polar ice cap remains exposed to sunlight, the temperature there rises to levels never found at the south pole, and carbon dioxide frost is able to evaporate completely.

Seasonal winds can also explain why Mars changes in appearance over the course of a Martian year. As Figure 12-18 shows, small sand drifts were seen at the *Viking Lander 1* site. Careful scrutiny of pictures taken over several months reveals that these drifts are pushed around slightly by the seasonal winds. New rocks are occasionally exposed and others covered up. Therefore, as seen from Earth, large areas of the planet vary from dark to light with the passage of the seasons. These variations mimic the color changes that would be expected from vegetation, which is why early observers thought they saw signs of plant life on Mars (see Section 12-2).

12-8 The two Martian moons resemble asteroids

 Two moons move around Mars in orbits close to the planet's surface. These satellites are so tiny that they were not discovered until the favorable opposition of 1877. While Schiaparelli was seeing canals, the American astronomer Asaph Hall spotted the two moons, which orbit almost directly above its equator. He named them Phobos ("fear") and Deimos ("panic"), after the mythical horses that drew the chariot of the Greek god of war.

Phobos is the inner and larger of the Martian moons. It circles Mars in just 7 hours and 39 minutes at an average distance of only 9378 km (5828 mi) from the center of Mars. (Recall that our Moon orbits at an average distance of 384,400 km from the Earth's center.) This orbital period is much shorter than the Martian day; hence, Phobos rises in the *west* and gallops across the sky in only 5½ hours, as viewed by an observer near the Martian equator. During this time, Phobos appears several times brighter in the Martian sky than Venus does from the Earth.

Deimos, which is farther from Mars and somewhat smaller than Phobos, appears about as bright from Mars as Venus does from the Earth. Deimos orbits at an average distance of 23,460 km (14,580 mi) from the center of Mars, slightly beyond the right distance for a synchronous orbit—an orbit in which a satellite seems to hover above a single location on the planet's equator. As seen from the Martian surface, Deimos rises in the east and takes about three full days to creep slowly from one horizon to the other.

The *Mariner 9* mission treated astronomers to their first close-up views of the Martian satellites. Numerous additional images from the *Viking Orbiter*s revealed Phobos and Deimos to be jagged and heavily cratered (Figure 12-23). Both moons

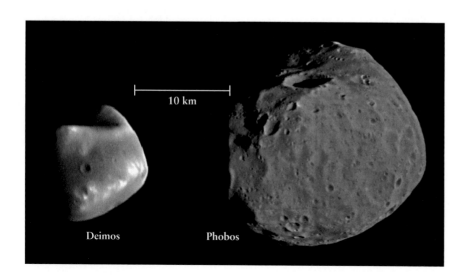

ƒigure 12-23 R I **V** U X G

The Moons of Mars These *Viking Orbiter* images show Deimos and Phobos to the same scale. What appears to be a fold at the top of the image of Deimos is actually a side view of an impact crater 10 km across, comparable to the moon's 22-km average diameter. An only slightly more powerful impact would have broken Deimos apart. (JPL/NASA)

are somewhat elongated from a spherical shape: Phobos is approximately 28 by 23 by 20 km in size, while Deimos is slightly smaller, at roughly 16 by 12 by 10 km.

The *Viking* observations also showed that Phobos and Deimos are both in synchronous rotation. The tidal forces that Mars exerts on these elongated moons cause them to always keep the same sides toward Mars, just as the Earth's tidal forces ensure that the same side of the Moon always faces us (see Section 4-8).

We saw in Section 9-5 that the Moon raises a tidal bulge on the Earth (see Figure 9-19). Because the Earth rotates on its axis more rapidly than the Moon orbits the Earth, this bulge is dragged ahead of a line connecting the Earth and Moon (see Figure 9-19). The gravitational force of this bulge pulls the Moon into an ever larger orbit. In the same way, and for the same reason, Deimos is slowly moving away from Mars. But for fast-moving Phobos, the situation is reversed: Mars rotates on its axis more slowly than Phobos moves around Mars. This slow rotation holds back the tidal bulge that Phobos raises in the solid body of Mars, and this bulge thus is *behind* a line connecting the centers of Mars and Phobos. (Imagine that the bulge in Figure 9-19 is tilted clockwise rather than counter-clockwise.) This bulge exerts a gravitational force on Phobos that pulls it back, slowing the moon's motion and making it slowly spiral inward toward Mars. This inward spiral will cease in about 40 million years when Phobos smashes into the Martian surface.

The origin of the Martian moons is unknown. One idea is that they are asteroids that Mars captured from the nearby asteroid belt (see Section 7-5). Like Phobos and Deimos, many asteroids reflect very little sunlight (typically less than 10%) because of a high carbon content. Phobos and Deimos also have very low densities (1900 kg/m^3 and 1760 kg/m^3 respec-

tively), which are also characteristic of carbon-rich asteroids. Perhaps two of these asteroids wandered close enough to Mars to become permanently trapped by the planet's gravitational field. Alternatively, Phobos and Deimos may have formed in orbit around Mars out of debris left over from the formation of the planets. Some of this debris might have come from the asteroid belt, giving these two moons an asteroidlike character. Future measurements of the composition of the Martian moons—perhaps by landing spacecraft on their surfaces—may shed light on their origin.

The exploration of Mars and its two moons has really just begun. Recent discoveries from *Mars Global Surveyor* and *Mars Pathfinder* have only whetted our appetite to know more about Mars. Fortunately, every 25 months the Earth and Mars are ideally placed for sending a spacecraft from one planet to another with the minimum thrust. For the next several years, one or more spacecraft will be launched toward Mars at each of these opportunities. By 2010, Mars will have been visited by a new generation of orbiters and landers from NASA, the European Space Agency (ESA), and the Japanese Institute of Space and Astronautical Science (ISAS). One of these spacecraft may land on Mars, collect samples with an advanced rover, and return those samples to Earth. Detailed analysis of Martian samples in laboratories on Earth could answer many open questions about the red planet, its formation, and its evolution.

It is ironic that at the end of the nineteenth century, the English novelist H. G. Wells envisioned a fleet of Martian spaceships coming to Earth on a mission of conquest. Today, at the beginning of the twenty-first century, it is we on Earth who are sending robotic spacecraft on a peaceful scientific "invasion" of Mars.

KEY WORDS

andesite, p. 275

crustal dichotomy, p. 266

dust devil, p. 276

favorable opposition, p. 261

northern lowlands, p. 265

permafrost, p. 269

residual polar cap, p. 268

rift valley, p. 267

runaway icehouse effect, p. 272

southern highlands, p. 265

KEY IDEAS

Earth-Based Observations of Mars: The best Earth-based views of Mars are obtained at favorable oppositions, when Mars is simultaneously at opposition and near perihelion.

• Earth-based observers found that a solar day on Mars is nearly the same length as on Earth, that Mars has polar caps that expand and shrink with the seasons, and that the Martian surface undergoes seasonal color changes.

• A few observers reported a network of linear features called canals. These observations, which proved to be illusions, led to many speculations about Martian life.

 Spacecraft Observation of Mars: Spacecraft have returned close-up views of the Martian surface. These show that the Martian surface has numerous craters, several huge volcanoes, a vast rift valley, and dried-up riverbeds—but no canals.

- The heavily cratered southern highlands are older and about 5 km higher in elevation than the smooth northern lowlands. The origin of this crustal dichotomy is not completely understood.

- Martian volcanoes and the Valles Marineris rift valley were formed by upwelling plumes of magma in the mantle. For reasons that are not understood, the chemical composition of ancient Martian lava is different from that of more recent lava.

- Mars has no planetwide magnetic field at present but may have had one in the ancient past.

Water on Mars: Flash-flood features and dried riverbeds on the Martian surface indicate that water once flowed on Mars.

- No liquid water can exist on the Martian surface today. But Mars's polar caps contain frozen water, a layer of permafrost may exist below the Martian regolith, and there may be liquid water beneath the surface.

- The Martian polar caps expand in winter as a thin layer of frozen carbon dioxide (dry ice) is deposited from the atmosphere.

- Mars's primordial atmosphere was thicker and warmer than the present-day atmosphere. It probably contained enough carbon dioxide and water vapor to support a greenhouse effect that permitted liquid water to exist on the planet's surface.

- If enough liquid water once existed on the Martian surface, the northern lowlands could have been flooded to form an ocean.

The Martian Atmosphere: The Martian atmosphere is composed mostly of carbon dioxide. The atmospheric pressure on the surface is less than 1% that of the Earth and shows seasonal variations as carbon dioxide freezes onto and evaporates from the poles.

- Great dust storms sometimes blanket Mars. Fine-grained dust in its atmosphere gives the Martian sky a pinkish-orange tint.

- Seasonal winds blow dust across the face of Mars, covering and uncovering the underlying surface material and causing seasonal color changes. Afternoon dust devils help to transport dust from place to place.

Life on Mars: Chemical reactions in the Martian regolith, together with ultraviolet radiation from the Sun, appear to sterilize the Martian surface. Hence, the *Viking Lander* spacecraft found no evidence for present-day life on Mars.

The Moons of Mars: Mars has two small, football-shaped satellites that move in orbits close to the surface of the planet. They may be captured asteroids or may have formed in orbit around Mars out of solar system debris.

REVIEW QUESTIONS

1. Explain why Mars has the longest synodic period of all the planets, although its sidereal period is only 687 days.

2. Why is it best to view Mars near opposition? Why are some oppositions better than others?

3. Why did Earth observers report that they had seen straight-line features ("canals") on Mars?

4. What is the Martian crustal dichotomy? What theories have been proposed to explain its origin? What are the arguments in favor of or against each theory?

5. Compare the volcanoes of Venus, the Earth, and Mars. Cite evidence that hot-spot volcanism is or was active on all three worlds.

6. Olympus Mons and Valles Marineris are respectively the largest volcano and the largest valley known in the solar system. How, then, could the three *Mariner* spacecraft that flew past Mars in the 1960s have missed seeing these features?

7. Giant impact basins like Hellas Planitia are found on Mars but not on Earth. Why not?

8. What geologic features (or lack thereof) on Mars have convinced scientists that plate tectonics did not significantly shape the Martian surface?

9. If you were an astronaut on Mars, would it be useful to carry a compass for navigation? Why or why not?

10. Why is it impossible for liquid water to exist on Mars today? What was different in the past that allowed liquid water to exist then?

11. Substantial quantities of liquid water have not been present on Mars for a very long time. Explain how we know this.

12. Why is it reasonable to assume that the primordial atmospheres of the Earth and Mars were roughly the same?

13. Why is Mars red?

14. Carbon dioxide accounts for about 95% of the present-day atmospheres of both Mars and Venus. Why, then, is there a strong greenhouse effect on Venus but only a weak greenhouse effect on Mars?

15. Explain why the Martian sky sometimes appears pink.

16. What is the current state of knowledge concerning life on Mars? Is it known definitely whether life currently exists on Mars or whether it once existed there?

17. What discoveries did *Mars Pathfinder* make about Martian rocks? Why weren't the *Viking Landers* able to make these same discoveries 20 years earlier?

18. What is a dust devil? Why would you feel much less breeze from a Martian dust devil than from a dust devil on Earth?

19. Discuss the statement "On Mars, the winters are so cold that the atmosphere freezes." How accurate is this statement?

20. In which Martian hemisphere are the summers warmer? Why is there a difference between the hemispheres? As part

of your explanation, make a drawing like Figure 12-2 that shows Mars in its orbit around the Sun.

21. A full moon on Earth is bright enough to cast shadows. As seen from the Martian surface, would you expect a full Phobos or full Deimos to cast shadows? Why or why not?

ADVANCED QUESTIONS

Problem-solving tips and tools

You will need to remember the small angle formula (Box 1-1) and the formula for the angular resolution of a telescope (Section 6-3). For the relationship between a planet's sidereal and synodic periods, see Box 4-1. You may also need to review the form of Kepler's third law that explicitly includes mass (see Section 4-7 and Box 4-4). The volume of a sphere of radius r is $\frac{4}{3}\pi r^3$.

22. Using Figure 12-2, explain why oppositions of Mars are most favorable when they occur in August.

23. When Mars is at opposition, what is its phase as seen from the Earth? What is the Earth's phase as seen from Mars? Explain with a diagram.

24. According to Table 4-1, the synodic period of Mars—that is, the time between successive oppositions—is 780 days. But the dates given in Table 12-2 show that the intervals between successive oppositions vary. Explain this apparent contradiction.

25. From one opposition of Mars to the next, how many orbits around the Sun does the Earth complete? How many orbits does Mars complete? (Your answers will *not* be whole numbers.)

26. Explain why the image that opens this chapter shows much finer detail than Figure 12-1.

27. For a planet to appear to the naked eye as a disk rather than as a point of light, its angular size would have to be 1 arcmin, or 60 arcsec. (This is the same as the angular separation between lines in the bottom row of an optometrist's eye chart.) (a) How close would you have to be to Mars in order to see it as a disk with the naked eye? Does Mars ever get this close to Earth? (b) Would the Earth ever be visible as a disk to an astronaut on Mars? Would she be able to see the Earth and the Moon separately, or would they always appear as a single object? Explain.

28. (a) Suppose you have a telescope with an angular resolution of 1 arcsec. What is the size (in kilometers) of the smallest feature you could see on the Martian surface during the opposition of 2003? (See Table 12-2.) (b) Suppose you had access to the Hubble Space Telescope, which has an angular resolution of 0.1 arcsec. What is the size (in kilometers) of the smallest feature you could see on Mars with the HST during the 2003 opposition?

29. An amateur astronomer wants to purchase a telescope that will enable her to resolve Hellas Planitia (diameter 2300 km) during the 2005 opposition of Mars. What must be the minimum diameter of the telescope objective? Assume that she observes Mars using a yellow filter, so that the wavelength used is 550 nm. Ignore the blurring of the image by the Earth's atmosphere.

30. The Earth's northern hemisphere is 39% land and 61% water, while its southern hemisphere is only 19% land and 81% water. Thus, the southern hemisphere could also be called the "water hemisphere." The Moon also has two distinct hemispheres, the near side (which has a number of maria) and the far side (which has almost none). How are these hemispheric differences on Earth and on the Moon similar to the Martian crustal dichotomy? How are they different?

31. For a group of properly attired astronauts equipped with oxygen tanks, a climb to the summit of Olympus Mons would actually be a relatively easy (albeit long) hike rather than a true mountain climb. Give two reasons why.

32. *Mars Global Surveyor* (MGS) is in a nearly circular orbit with an orbital period of 117 minutes. (a) Using the data in Table 12-1, find the radius of the orbit. (b) What is the average altitude of MGS above the Martian surface? (c) The orbit of MGS passes over the north and south poles of Mars. Explain how this makes it possible for the spacecraft to observe the entire surface of the planet.

33. Suppose *Mars Global Surveyor* had discovered magnetized regions within the Hellas Planitia impact basin. How would this discovery have affected our understanding of Martian history?

34. On Mars, the difference in elevation between the highest point (the summit of Olympus Mons) and the lowest point (the bottom of the Hellas Panitia basin) is 30 km. On Earth, the corresponding elevation difference (from the peak of Mount Everest to the bottom of the deepest ocean) is only 20 km. Discuss why the maximum elevation difference is so much greater on Mars.

35. (a) The Grand Canyon in Arizona was formed over 15 to 20 million years by the flowing waters of the Colorado River, as well as by rain and wind. Contrast this formation scenario to that of Valles Marineris. (b) Valles Marineris is sometimes called "the Grand Canyon of Mars." Is this an appropriate description? Why or why not?

36. The classic 1950 science-fiction movie *Rocketship X-M* shows astronauts on the Martian surface with oxygen masks for breathing but wearing ordinary clothing. Would this be a sensible choice of apparel for a walk on Mars? Why or why not?

37. Compare the evolution of the atmospheres of Venus, the Earth, and Mars. Explain how the atmospheres of these three planets turned out so differently from one another, even though they may all have started with roughly the same chemical composition.

38. Suppose Mars were moved into the Earth's orbit and the Earth were moved into Mars's orbit. What effect would this have on the atmospheres of the two planets?

39. Suppose that the only information you had about Mars was the image of the surface in Figure 12-18. Describe at least two ways that you could tell from this image that Mars has an atmosphere.

40. Although the *Viking Lander 1* and *Viking Lander 2* landing sites are 1500 km apart and have different geologic histories, the chemical compositions of the dust at both sites are nearly identical. (**a**) What does this suggest about the ability of the Martian winds to transport dust particles? (**b**) Would you expect that larger particles such as pebbles would also have identical chemical compositions at the two *Viking Lander* sites? Why or why not?

41. The landing site for *Viking Lander 1* is in an ancient flood plain, while the site for *Viking Lander 2* is near the southernmost extent of the north polar ice cap. The *Viking* mission planners selected these sites because they suspected water might lie beneath the surface there. Why were these good places to search for potential Martian microorganisms?

42. Is it reasonable to suppose that the polar regions of Mars might harbor life forms, even though the Martian regolith is sterile at the *Viking Lander* sites?

43. The orbit of Phobos has a semimajor axis of 9378 km. Use this information and the orbital period given in the text to calculate the mass of Mars. How does your answer compare with the mass of Mars given in Table 12-1?

44. Why do you suppose that Phobos and Deimos are not round like our Moon?

45. Calculate the angular sizes of Phobos and Deimos as they pass overhead, as seen by an observer standing on the Martian equator. How do these sizes compare with that of the Moon seen from the Earth's surface? Would Phobos and Deimos appear as impressive in the Martian sky as the Moon does in our sky?

46. You are to put a spacecraft into a synchronous circular orbit around the Martian equator, so that its orbital period is equal to the planet's rotation period. Such a spacecraft would always be over the same part of the Martian surface. (**a**) Find the radius of the orbit and the altitude of the spacecraft above the Martian surface. (**b**) Suppose Mars had a third moon that was in a synchronous orbit. Would tidal forces make this moon tend to move toward Mars, away from Mars, or neither? Explain.

DISCUSSION QUESTIONS

47. One explanation that has been proposed for the origin of Valles Marineris is that it is a place where plate tectonics started on Mars and where two parts of the crust began to move apart like tectonic plates. In this picture,

tectonic activity died out on Mars because the planet's interior cooled too rapidly, so that convection in the mantle ceased. What sorts of observations of Mars would you make in an attempt to prove or disprove this explanation?

48. The total cost of the *Mars Global Surveyor* mission was about $154 million. (To put this number into perspective, in 2000 the U.S. Mint spent about $40 million to advertise its new $1 coin. Several recent Hollywood movies have had larger budgets than *Mars Global Surveyor*.) Does this expenditure seem reasonable to you? Why or why not?

49. Compare the scientific opportunities for long-term exploration offered by the Moon and Mars. What difficulties would be inherent in establishing a permanent base or colony on each of these two worlds?

50. Imagine you are an astronaut living at a base on Mars. Describe your day's activities, what you see, the weather, the spacesuit you are wearing, and so on.

 WEB/CD-ROM QUESTIONS

51. Search the World Wide Web for information about ongoing or future Mars missions, including the European Space Agency's *Mars Express* and the Japanese orbiter *Nozomi*. What is the current status of these missions? What are their scientific goals?

52. In 1999 two NASA spacecraft—*Mars Climate Orbiter* and *Mars Polar Lander*—failed to reach their destinations. Search the World Wide Web for information about these missions. What were their scientific goals? How and why did the missions fail? Which future missions, if any, are intended to pick up where these missions left off?

53. Search the World Wide Web for information about possible manned missions to Mars. How long might such a mission take? How expensive would such a project be? What would be the advantages of a manned mission compared to an unmanned one?

 54. **Conjunctions of Mars.** Access and view the animation "The Orbits of Earth and Mars" in Chapter 12 of the *Universe* web site or CD-ROM. (**a**) The animation highlights three dates when Mars is in opposition, so that the Earth lies directly between Mars and the Sun. By using the "Stop" and "Play" buttons in the animation, find two times during the animation when Mars is in *conjunction*, so that the Sun lies directly between Mars and the Earth (see Figure 4-6). For each conjunction, make a drawing showing the positions of the Sun, the Earth, and Mars, and record the month and year when the conjunction occurs. (*Hint:* See Figure 12-2.) (**b**) When Mars is in conjunction, at approximately what time of day does it rise as seen from Earth? At what time of day does it set? Is Mars suitably placed for telescopic observation when in conjunction?

OBSERVING PROJECTS

Observing tips and tools

Mars is most easily seen around an opposition (see Table 12-2). At other times Mars may be visible only in the early morning hours before sunrise or in the early evening just after sunset. Consult such magazines as *Sky & Telescope* and *Astronomy* or their web sites for more detailed information about when and where to look for Mars during a given month. You can also use the *Starry Night* program on the CD-ROM that accompanies this textbook.

55. If Mars is suitably placed for observation, arrange to view the planet through a telescope. Draw a picture of what you see. What magnifying power seems to give you the best image? Can you distinguish any surface features? Can you see a polar cap or dark markings? If not, can you offer an explanation for Mars's bland appearance?

56. Use the *Starry Night* program to observe the appearance of Mars. First turn off daylight (select **Daylight** in the **Sky** menu) and show the entire celestial sphere (select **Atlas** in the **Go** menu). Center on Mars by using the **Find...** command in the **Edit** menu. Using the controls at the right-hand end of the Control Panel, zoom in until the field of view is roughly 30 arcseconds (30″). (a) In the Control Panel, set the time step to 30 minutes and click on the "Forward" button (a triangle that points to the right). Describe what you see. (b) Click on the "stop" button (a black square) in the Control Panel. Change the time step to 30 days and click again on the Forward button. Describe what you see. Using a diagram like Figure 4-6, explain the changes in the apparent size of the planet.

57. Use the *Starry Night* program to observe the moons of Mars. First turn off daylight (select **Daylight** in the **Sky** menu) and show the entire celestial sphere (select **Atlas** in the **Go** menu). Center on Mars by using the **Find...** command in the **Edit** menu. Using the controls at the right-hand end of the Control Panel, zoom in until the field of view is roughly 1 arcminute (1′). In the Control Panel set the time step to 30 minutes and click on the "Forward" button (a triangle that points to the right). You will see the two moons, Phobos and Deimos, orbiting Mars. (a) Click on the "stop" button (a black square) in the Control Panel. Record the date and time in the display, and note the position of Phobos (the inner moon). Step through time using the single-step button (the rightmost time control button) until Phobos returns to approximately the same position, then record the date and time in the display. From your observations, what is the orbital period of Phobos? How does your result compare with the orbital period given in Appendix 3? (b) Repeat part (a) for Deimos (the outer moon).

Jupiter: Lord of the Planets

Jupiter is the largest and most massive planet in the solar system. Our Earth is tiny by comparison: Jupiter is more than 11 times larger in diameter and has over 318 times more mass. Unlike the Earth, Jupiter is composed primarily of the light-weight elements hydrogen and helium, in abundances very similar to those in the Sun.

As this image from the *Cassini* spacecraft shows, Jupiter's rapid rotation stretches its weather systems into colorful bands that extend completely around the planet. Some weather features endure year after year, while others are only short-lived. Jupiter also has an intense magnetic field that is generated in the planet's interior, where hydrogen is so highly compressed that it becomes a metal.

Jupiter formed in the outer reaches of the solar nebula, far from the protosun. Temperatures there were low enough that dust grains acquired thick, frosty coatings of frozen water, methane, and ammonia. Enough of these ice-coated dust grains accreted to make a protoplanet several times more massive than Earth. The protoplanet's strong gravitational pull attracted and retained tremendous quantities of hydrogen and helium, so that Jupiter grew to its present immense size. Today, billions of years later, Jupiter's bulk of hydrogen and helium is apparently still contracting slowly. This may explain why Jupiter radiates substantially more heat than it receives from the Sun.

WEB LINK 13.1 (NASA/JPL/University of Arizona)

R I **V** U X G

Jupiter is also surrounded by an extensive retinue of moons. You can see one of these, Ganymede, in the accompanying image. (To appreciate Jupiter's immense size, note that Ganymede is about 50% larger in diameter than our Moon.) Worlds in their own right, Jupiter's satellites will be discussed separately in Chapter 14.

As you read the sections of this chapter, look for the answers to the following questions.

13-1 Can you stand on the surface of Jupiter, as you can on the Earth or the Moon?

13-2 What is going on in Jupiter's Great Red Spot?

13-3 What is the nature of Jupiter's multicolored clouds?

13-4 What does the chemical composition of Jupiter's atmosphere imply about the planet's origin?

13-5 How do astronomers know about Jupiter's deep interior?

13-6 What is the origin of Jupiter's strong magnetic field?

13-7 Where do you find the highest temperatures in the solar system?

13-1 Huge, massive Jupiter is composed largely of lightweight gases

Jupiter is the largest of the Jovian planets, and the largest object in the solar system other than the Sun itself. Astronomers have known about Jupiter's huge diameter and immense mass for centuries. Given the distance to the planet and its angular size, they used the small-angle formula (Box 1-1) to calculate that Jupiter's diameter is about 11 times bigger than that of the Earth. By observing the orbits of Jupiter's four large moons and applying Newton's form of Kepler's third law (see Section 4-7 and Box 4-4), astronomers also determined that Jupiter is 318 times more massive than the Earth. In fact, Jupiter has 2½ times the combined mass of all the other planets, satellites, asteroids, meteoroids, and comets in the solar system. A visitor from interstellar space might well describe our solar system as

the Sun, Jupiter, and some debris! Table 13-1 lists some basic data about Jupiter.

As for any planet whose orbit lies outside the Earth's orbit, the best time to observe Jupiter is when it is at opposition. At such times, Jupiter can appear nearly 3 times brighter than Sirius, the brightest star in the sky. Only the Moon and Venus can outshine Jupiter at opposition. Through a telescope, Jupiter presents a disk nearly 50 arcsec in diameter, approximately twice the angular diameter of Mars under the most favorable conditions. Because Jupiter takes almost a dozen Earth years to orbit the Sun, it appears to meander slowly across the 12 constellations of the zodiac at the rate of approximately one constellation per year. Successive oppositions occur at intervals of about 13 months (Table 13-2).

As seen through an Earth-based telescope, Jupiter's atmosphere displays colorful bands that extend around the planet parallel to its equator (Figure 13-1). Figure 13-2, a more detailed close-up view from a spacecraft, shows alternating

ƒigure 13-1 R I **V** U X G

Jupiter Seen from Earth Various dark-colored belts and light-colored zones are easily identified in this Earth-based view of the largest planet in our solar system. The Great Red Spot, which is about the same diameter as the Earth, was exceptionally prominent when this photograph was taken. (Courtesy of S. Larson)

ƒigure 13-2 R I **V** U X G

Jupiter Seen from a Spacecraft The *Voyager 1* spacecraft recorded this image at a distance of only 30 million kilometers from Jupiter. Features as small as 600 km across can be seen in the turbulent cloudtops. Complex cloud motions surround the Great Red Spot. (NASA/JPL)

table 13-1	Jupiter Data	
Average distance from Sun:	5.203 AU = 7.783 × 10⁸ km	
Maximum distance from Sun:	5.455 AU = 8.160 × 10⁸ km	
Minimum distance from Sun:	4.950 AU = 7.406 × 10⁸ km	
Eccentricity of orbit:	0.048	
Average orbital speed:	13.1 km/s	
Orbital period:	11.86 years	
Rotation period:	$9^h 50^m 28^s$ (equatorial)	
	$9^h 55^m 29^s$ (internal)	
Inclination of equator to orbit:	3.12°	
Inclination of orbit to ecliptic:	1.30°	
Diameter:	142,984 km = 11.209 Earth diameters (equatorial)	
	133,708 km = 10.482 Earth diameters (polar)	
Mass:	1.899×10^{27} kg = 317.8 Earth masses	
Average density:	1326 kg/m³	
Escape speed:	60.2 km/s	
Surface gravity (Earth = 1):	2.36	
Albedo:	0.44	
Average temperature at cloudtops:	−108°C = −162°F = 165 K	

(NASA) R I **V** U X G

table 13-2	Oppositions of Jupiter, 2001–2005		
	Earth-Jupiter distance		Angular diameter
Date of opposition	(AU)	(10⁶ km)	(arcsec)
2002 January 1	4.19	626	47.0
2003 February 2	4.33	647	45.5
2004 March 4	4.43	662	44.5
2005 April 3	4.46	667	44.2

dark and light bands parallel to Jupiter's equator in subtle tones of red, orange, brown, and yellow. The dark, reddish bands are called **belts,** and the light-colored bands are called **zones.**

In addition to these conspicuous stripes, a huge, red-orange oval called the **Great Red Spot** is often visible in Jupiter's southern hemisphere. This remarkable feature was first seen by the English scientist Robert Hooke in 1664 but may be much older. It appears to be an extraordinarily long-lived storm in the planet's dynamic atmosphere. Many careful observers have also reported smaller spots and blemishes that last for only a few weeks or months in Jupiter's turbulent clouds.

Observations of Jupiter allow astronomers to determine how fast Jupiter rotates. At its equator, Jupiter completes a full

a   b

figure 13-3

Solid Rotation versus Differential Rotation **(a)** When a solid object like a billiard ball rotates, every part of the object takes exactly the same time to complete one rotation. **(b)** A rotating fluid displays differential rotation. To see this, put some grains of sand, bread crumbs, or other small particles in a pot of water, stir the water with a spoon to start it rotating, then take out the spoon. The particles near the center of the pot take less time to make a complete rotation than those away from the center.

rotation in only 9 hours, 50 minutes, and 28 seconds, making it not only the largest and most massive planet in the solar system but also the one with the fastest rotation. However, Jupiter rotates in a strikingly different way from the Earth, the Moon, Mercury, Venus, or Mars. If Jupiter were a solid body like a terrestrial planet (or, for that matter, a billiard ball), all parts of Jupiter's surface would rotate through one complete circle in this same amount of time (Figure 13-3a). But by watching features in Jupiter's cloud cover, Gian Domenico Cassini discovered in 1690 that the polar regions of the planet rotate a little more slowly than do the equatorial regions. (You may recall this Italian astronomer from Section 12-2 as the gifted observer who first determined Mars's rate of rotation.) Near the poles, the rotation period of Jupiter's atmosphere is about 9 hours, 55 minutes, and 41 seconds.

ANALOGY You can see this kind of rotation, called **differential rotation,** in the kitchen. As you stir the water in a pot, different parts of the liquid take different amounts of time to make one "rotation" around the center of the pot (Figure 13-3b). Differential rotation shows that Jupiter cannot be solid throughout its volume but is at least partially fluid, like water in a pot.

If Jupiter has a partially fluid interior, it seems unlikely that it could be made of the same sort of rocky materials that constitute the terrestrial planets. An important clue to Jupiter's composition is its average density, which is only 1326 kg/m^3 (about one-fourth of the Earth's average density). To explain this low average density, Rupert Wildt of the University of Göttingen in Germany suggested in the 1930s that Jupiter is

composed mostly of hydrogen and helium atoms—the two lightest elements in the universe—held together by their mutual gravitational attraction to form a planet. Wildt was motivated in part by his observations of prominent absorption lines of methane and ammonia in Jupiter's spectrum. (We saw in Section 7-3 how spectroscopy plays an important role in understanding the planets.) A molecule of methane (CH$_4$) contains four hydrogen atoms, and a molecule of ammonia (NH$_3$) contains three. The presence of these hydrogen-rich molecules was strong, but indirect, evidence of abundant hydrogen in Jupiter's atmosphere.

Direct evidence for hydrogen and helium in Jupiter's atmosphere, however, was slow in coming. The problem was that neither gas produces prominent spectral lines in the visible sunlight reflected from the planet. To show the presence of these elements conclusively, astronomers had to look for spectral lines in the ultraviolet part of the spectrum. These lines are very difficult to measure from the Earth, because almost no ultraviolet light penetrates our atmosphere (see Figure 6-27). Astronomers first detected the weak spectral lines of hydrogen molecules in Jupiter's spectrum in 1960. The presence of helium was finally confirmed in the 1970s, when spacecraft first flew past Jupiter and measured its hydrogen spectrum in detail. (Collisions between helium and hydrogen atoms cause small but measurable changes in the hydrogen spectrum, which is what the spacecraft instruments detected.)

Today we know that the chemical composition of Jupiter's atmosphere is 86.2% hydrogen molecules (H$_2$), 13.6% helium atoms, and 0.2% methane, ammonia, water vapor, and other gases. The percentages in terms of *mass* are somewhat different because a helium atom is twice as massive as a hydrogen molecule. Hence, by mass, Jupiter's atmosphere is approximately 75% hydrogen, 24% helium, and 1% other substances. We will see evidence in Section 13-5 that Jupiter has a large rocky core made of heavier elements. It is estimated that the breakdown by mass of the planet as a whole (atmosphere plus interior) is approximately 71% hydrogen, 24% helium, and 5% all heavier elements.

CAUTION! Because Jupiter is almost entirely hydrogen and helium, it would be impossible to land a spacecraft there. An astronaut foolish enough to try would notice the hydrogen and helium around the spacecraft becoming denser, the temperature rising, and the pressure increasing as the spacecraft descended. But the hydrogen and helium would never solidify into a surface on which the spacecraft could touch down. Long before reaching Jupiter's rocky core, the pressure of the hydrogen and helium would reach such unimaginably high levels that any spacecraft, even one made of the strongest known materials, would be crushed.

13-2 Images from spacecraft show many details in Jupiter's clouds

Most of our detailed understanding of Jupiter comes from a series of robotic spacecraft that have examined this extraordi-

a *Pioneer 10*, December 1973

b *Pioneer 11*, December 1974

c *Voyager 1*, March 1979

d *Voyager 2*, July 1979

e HST, February 1995

figure 13-4 R I **V** U X G

Five Views of Jupiter These images show major changes in the planet's upper atmosphere over more than 21 years. Differences are apparent even between the two *Voyager* views (**c** and **d**), made only four months apart. Image (**e**) was taken by the 2.4-m Hubble Space Telescope from Earth orbit, at a distance of 960 million (9.6×10^8) km from Jupiter; it is comparable in quality to the *Voyager* images, taken at only 30 million (3×10^7) km from Jupiter but with much smaller cameras. All five pictures show the Great Red Spot in Jupiter's southern hemisphere. The black dot on the *Pioneer 10* view (**a**) is the shadow of Jupiter's moon Io. (a, b, c, and d: NASA/JPL; e: Reta Beebe and Amy Simon, New Mexico State University, and NASA)

nary planet at close range. They found striking evidence of stable, large-scale weather patterns in Jupiter's atmosphere, as well as evidence of dynamic changes on smaller scales.

The first four spacecraft to visit Jupiter each made a single flyby of the planet. *Pioneer 10* flew past Jupiter in December 1973, followed a year later by the nearly identical *Pioneer 11*. In 1979 another pair of spacecraft, *Voyager 1* and *Voyager 2*, sailed past Jupiter. These spacecraft sent back spectacular close-up color pictures of Jupiter's dynamic atmosphere. In 1995, the *Galileo* spacecraft went into orbit around Jupiter and began a multiyear program of observations. From its orbital vantage point, *Galileo* has been able to monitor how Jupiter's atmosphere changes over a period of weeks or months. Most recently, during

2000–2001 the *Cassini* spacecraft viewed Jupiter at close range while en route to its final destination of Saturn (see the image that opens this chapter).

While the general pattern of Jupiter's atmosphere stayed the same during the four years between the *Pioneer* and *Voyager* flybys, there were some remarkable changes in the area surrounding the centuries-old Great Red Spot. During the *Pioneer* flybys, the Great Red Spot was embedded in a broad white zone that dominated the planet's southern hemisphere (Figures 13-4*a* and 13-4*b*). By the time of the *Voyager* missions, a dark belt had broadened and encroached on the Great Red Spot from the north (Figures 13-4*c* and 13-4*d*). In the same way that colliding weather systems in our atmosphere can produce strong winds and turbulent air, the interaction in

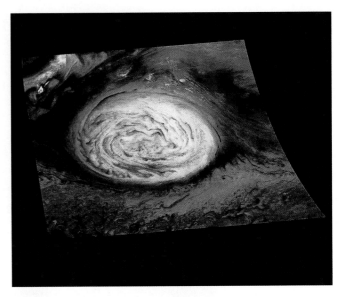

figure 13-5 R I V U X G
Turbulence Around the Great Red Spot
Atmospheric turbulence around the Great Red Spot is clearly seen in this *Voyager 2* image. When this picture was taken in 1979, the Great Red Spot was about 20,000 km long and about 10,000 km wide. For comparison, the Earth's diameter is 12,756 km. The prominent white oval south of (below) the Great Red Spot has been observed since 1938. (NASA/JPL)

figure 13-6 R I V U X G
***Galileo* Views the Great Red Spot** This June 1996 view from the *Galileo* spacecraft is actually a mosaic of images made at different infrared wavelengths, which are reflected preferentially by clouds at different altitudes. The highest clouds, shown in red and white, are found within the Great Red Spot. A green color indicates medium-level clouds, and the blue and black areas around the Great Red Spot denote the lowest clouds. (These are false colors, not the actual colors of the clouds.) (NASA/JPL)

Jupiter's atmosphere between the belt and the Great Red Spot embroiled the entire region in turbulence (Figure 13-5). By 1995, when Jupiter was viewed by the Hubble Space Telescope, the Great Red Spot was once again centered within a white zone (Figure 13-4e).

Over the past three centuries, Earth-based observers have reported many long-term variations in the Great Red Spot's size and color. At its largest, it measured 40,000 by 14,000 km —so large that three Earths could fit side by side across it. At other times (as in 1976 and 1977), the spot almost faded from view. During the *Voyager* flybys of 1979, the Great Red Spot was comparable in size to the Earth (see Figure 13-5).

The *Galileo* spacecraft has helped clarify the vertical structure of the Great Red Spot. This structure can be seen in Figure 13-6, which is a mosaic of 18 infrared images made by *Galileo* using three different filters. This technique reveals clouds at different levels in Jupiter's atmosphere, which reflect different wavelengths of infrared light. Figure 13-6 shows that most of the Great Red Spot is made of clouds at relatively high altitudes. It is surrounded by a collar of very low-level clouds at an altitude perhaps 50 km (30 miles, or 160,000 feet) below the high clouds at the center of the spot. This same kind of structure is seen in high-pressure areas in the Earth's atmosphere, although on a very much smaller scale.

Cloud motions in and around the Great Red Spot reveal that the spot rotates counterclockwise with a period of about six days. Furthermore, winds to the north of the spot blow to

the west, and winds south of the spot move toward the east. The circulation around the Great Red Spot is thus like a wheel spinning between two oppositely moving surfaces (Figure 13-7). This surprisingly stable wind pattern has survived for at least three centuries. Weather patterns on the Earth tend to change character and eventually dissipate when they move between plains and mountains or between land and sea. But because no solid surface or ocean underlies Jupiter's clouds, no such changes can occur for the Great Red Spot—which may explain its persistence.

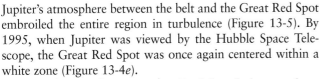
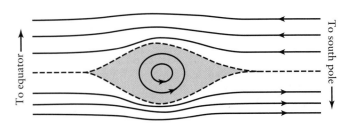

figure 13-7
Circulation Around the Great Red Spot The winds to the north and south of the Great Red Spot blow in opposite directions. Winds within the Great Red Spot itself spin counterclockwise, completing a full revolution in about six days. (Adapted from A. P. Ingersoll)

Other persistent features in Jupiter's atmosphere are the **white ovals**, like the one seen near the Great Red Spot in Figure 13-5. Several white ovals are also visible in Figure 13-4. As in the Great Red Spot, wind flow in white ovals is counterclockwise. White ovals are also apparently long-lived; Earth-based observers have reported seeing them in the same location since 1938.

Most of the white ovals are observed in Jupiter's southern hemisphere, whereas **brown ovals** are more common in Jupiter's northern hemisphere (Figure 13-8). Brown ovals appear dark in a visible-light image like Figure 13-8, but they appear bright in an infrared image. For this reason, brown ovals are understood to be holes in Jupiter's cloud cover. They permit us to see into the depths of the Jovian atmosphere, where the temperature is higher and the atmosphere emits infrared light more strongly. White ovals, by contrast, have relatively low temperatures. They are areas with cold, high-altitude clouds that block our view of the lower levels of the atmosphere.

The *Voyager, Galileo,* and *Cassini* images might suggest a state of incomprehensible turmoil, but, surprisingly, there is also great regularity in the Jovian atmosphere. The computer-generated view in Figure 13-9 shows how Jupiter would look to someone located directly over either the planet's north pole or its south pole. Note the regular spacing of such cloud

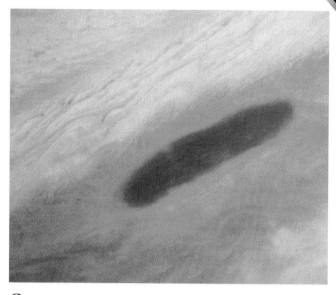

ƒigure 13-8 R I **V** U X G

A Brown Oval Oval-shaped openings in Jupiter's main cloud layer reveal warm, brownish gases below. The length of this oval is roughly equal to the Earth's diameter. *Voyager 1* recorded this image in 1977 at a distance of 4 million kilometers from Jupiter. (NASA/JPL)

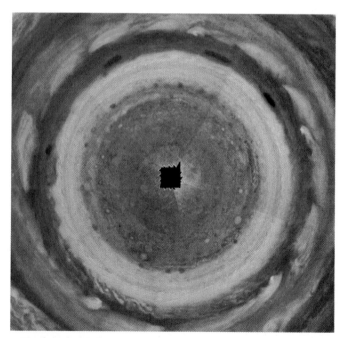

a Northern hemisphere

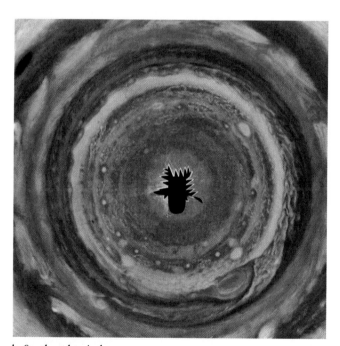

b Southern hemisphere

ƒigure 13-9 R I **V** U X G

Jupiter's Northern and Southern Hemispheres *Voyager* images were combined and computer processed to construct these views that look straight down onto Jupiter's north and south poles. **(a)** In this northern view, light-colored plumes are evenly spaced around the equatorial regions. Several brown ovals are visible. **(b)** In this southern view, the three biggest white ovals are separated by almost exactly 90° of longitude. In both the northern and southern views, the banded belt-zone structure is absent near the poles. The black spots at the poles themselves are areas not photographed by the spacecraft. (NASA/JPL)

features as ripples, plumes, and light-colored wisps. The regular spacing of white ovals in the southern hemisphere is quite apparent. These regularities are probably the result of stable, large-scale weather patterns that encircle Jupiter.

13-3 The motions of Jupiter's atmosphere are affected by solar energy, the planet's internal heat, and differential rotation

Weather patterns on Earth are the result of the motions of air masses. Winter storms in the midwestern United States occur when cold, moist air moves southward from Canada, and tropical storms in the southwest United States are caused by the northeastward motion of hot, moist air from Hawaii. All such motions, as well as the overall global circulation of our atmosphere, are powered by the Sun. Sunlight is absorbed by the Earth's surface, and the heated surface in turn warms the atmosphere and stirs it into motion (see Section 8-1 and Section 8-5). The energy of sunlight also powers the motions of the atmospheres of Venus (Section 11-3) and Mars (Section 12-5). In Jupiter's atmosphere, however, motions are powered both by solar energy and by the planet's own internal energy.

In the late 1960s, astronomers using Earth-based telescopes made the remarkable discovery that Jupiter emits more energy in the form of infrared radiation than it absorbs from sunlight—in fact, about twice as much. (By comparison, the internal heat that the Earth radiates into space is only about 0.005% of what it absorbs from the Sun.) The excess heat presently escaping from Jupiter is thought to be energy left over from the formation of the planet 4.6 billion years ago. As gases from the solar nebula fell into the protoplanet, vast amounts of gravitational energy were converted into thermal energy that heated the planet's interior. Jupiter has been slowly radiating this energy into space ever since. Because Jupiter is so large, it has retained substantial thermal energy even after billions of years, and so still has plenty of energy to emit as infrared radiation.

Heat naturally flows from a hot place to a colder place, never the other way around. Hence, in order for heat to flow upward through the Jovian atmosphere and radiate out into space, the temperature must be warmer deep inside the atmosphere than it is at the cloudtops. Infrared measurements have confirmed that the temperature within Jupiter's atmosphere does indeed increase with increasing depth (Figure 13-10).

ANALOGY You can understand the statement that "heat naturally flows from a hot place to a colder place" by visualizing an ice cube placed on a hot frying pan. Experience tells you that the hot frying pan cools down a little while the ice warms up and melts. If heat were to flow the other way, the ice would get colder and the frying pan would get hotter—which never happens in the real world.

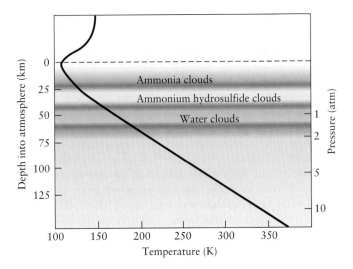

figure 13-10

Temperature and Pressure in Jupiter's Upper Atmosphere
The curve in this graph shows how temperature varies as you descend in the Jovian atmosphere. The scale on the right-hand side of the graph shows how pressure increases with depth. The graph also shows Jupiter's three major cloud layers, along with the colors that predominate at various depths. These cloud layers are not found at all locations on Jupiter; there are some relatively clear, cloud-free areas.

When a fluid is warm at the bottom and cool at the top, like water being warmed in a pot, one effective way for energy to be distributed through the fluid is by the up-and-down motion called *convection*. (We first discussed convection in Section 8-1; see, in particular, Figure 8-4.) In a planet's atmosphere, convection causes warm gases from low levels to ascend and cool, while cooler gases at high altitude descend and heat up. In our atmosphere, the Earth's rotation twists the rising and descending air into several convection cells that encircle the planet (see Figure 8-24). The same thing happens in Jupiter's atmosphere because of the planet's rapid, differential rotation. The light-colored zones, where we see high-elevation clouds, are regions where gas is rising and cooling; the dark-colored belts, where we see deeper into Jupiter's cloud cover, are regions where gases are descending and being heated (Figure 13-11).

Planetary scientists have confirmed this description of the zones and belts by comparing infrared and visible-light images of Jupiter (Figure 13-12). The cold Jovian clouds emit radiation primarily at infrared wavelengths, with an intensity that depends on the cloud temperature (see Section 5-4) and hence on the cloud's depth within the atmosphere (see Figure 13-10). The infrared image in Figure 13-12a shows stronger emission from the belts, where we are looking at lower-lying, warmer levels of Jupiter's atmosphere.

VIDEO 13.4 In addition to the vertical motion of gases within the belts and zones, Jupiter's very rapid rotation creates a global pattern of eastward and westward **zonal winds** with speeds that can exceed 500 km/h (300 mi/h).

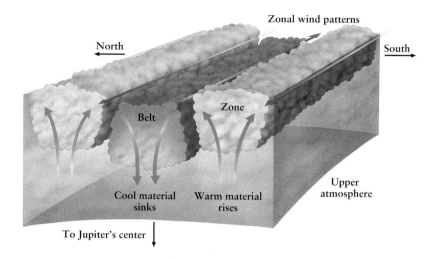

Zonal wind patterns

North

South

Belt

Zone

Cool material sinks

Warm material rises

Upper atmosphere

To Jupiter's center

figure 13-11

The Belt-Zone Structure of Jupiter's Clouds In Jupiter's light-colored zones, gases warmed by heat from the planet's interior rise upward and cool, forming high-altitude clouds. In the dark-colored belts, gases descend and undergo an increase in temperature; the cloud layers seen there are at lower altitudes than in the zones. Jupiter's rapid differential rotation shapes the belts and zones into bands parallel to the planet's equator. The rotation also causes the wind velocities at the boundaries between belts and zones to be predominantly to the east or west.

The atmospheric circulation pattern on the Earth has a similar pattern of eastward and westward flow, as shown in Figure 8-24, but with slower wind speeds. The faster, more energetic winds on Jupiter are presumably a result of the planet's faster rotation, as well as the substantial flow of heat from the planet's interior.

Jupiter's zonal winds are generally strongest at the boundaries between belts and zones. The wind reverses direction between the northern boundary of a belt or zone and the southern boundary. As Figure 13-7 shows, such reversals in wind direction are associated with circulating storms, such as the Great Red Spot.

From spectroscopic observations and calculations of atmospheric temperature and pressure, scientists conclude that Jupiter has three main cloud layers of differing chemical composition. The uppermost cloud layer is composed of

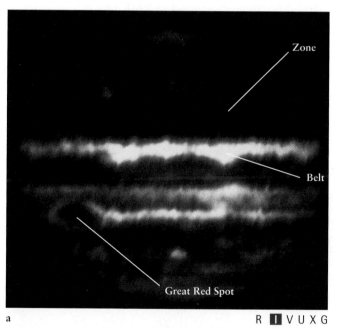

Zone

Belt

Great Red Spot

a   R **I** V U X G

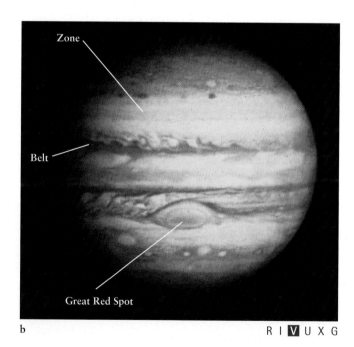

Zone

Belt

Great Red Spot

b   R I **V** U X G

figure 13-12

Jupiter's Warm Belts and Cool Zones **(a)** This infrared image was made with the Hale 5-meter telescope in California. Bright and dark areas in the image correspond to high and low temperatures, respectively. **(b)** This visible-light image from the *Voyager 1* spacecraft was recorded at the same time as the image in (a). Comparing the two images shows that the zones are regions where we see cold, light-colored, high-altitude clouds, while the belts are areas in which we see warm, dark-colored gases at lower levels within Jupiter's atmosphere. The Great Red Spot appears dark in the infrared image, showing that it is a region of cool, high-altitude clouds. (Compare with Figure 13-6.) (NASA/JPL)

crystals of frozen ammonia. About 25 km deeper in the atmosphere, ammonia (NH_3) and hydrogen sulfide (H_2S)—a compound of hydrogen and sulfur—combine to produce ammonium hydrosulfide (NH_4SH) crystals. At even greater depths, the clouds are composed of crystals of frozen water.

The colors of Jupiter's clouds depend on the temperatures of the clouds and, therefore, on the depth of the clouds within the atmosphere. Brown clouds are the warmest and are thus the deepest layers that we can see in the Jovian atmosphere. Whitish clouds form the next layer up, followed by red clouds in the highest layer. Jupiter's whitish zones are therefore somewhat higher than the brownish belts, while the red clouds in the Great Red Spot are among the highest found anywhere on the planet.

Despite all the data from various spacecraft, we still do not know what gives Jupiter's clouds their colors. The crystals of NH_3, NH_4SH, and water ice in Jupiter's three main cloud layers are all white. Thus, other chemicals must cause the browns, reds, and oranges. Some scientists suspect that certain molecules closely related to ammonium hydrosulfide, which can form into long chains that have a yellow-brown color, play an important role in Jupiter's clouds. Others think that compounds of sulfur or phosphorus, which can assume many different colors depending on its temperature, might be involved. The Sun's ultraviolet radiation, which can induce chemical reactions, may help produce these compounds and thus give the Jovian clouds their beautiful colors.

13-4   Jupiter's deep atmosphere has been explored by a comet and by a space probe

From observations of Jupiter's cloud layers, we have been able to learn a great deal about the atmosphere beneath the clouds. But there is no substitute for direct observation, and for many years scientists planned to send a spacecraft to explore deep into Jupiter's atmosphere. The first such probe was released by the *Galileo* spacecraft as it approached Jupiter in 1995. Even before this probe, however, nature provided an exploratory probe of its own—in the form of a comet that plunged into Jupiter.

In March 1993 the professional astronomers Eugene and Carolyn Shoemaker and their amateur colleague David Levy discovered a remarkable comet. As Figure 13-13 shows, it was actually a chain of more than 20 small pieces, strung out like pearls on a necklace. (We described comets briefly in Section 7-5.) It was officially named Comet Shoemaker-Levy 9, because it was the ninth comet that these three astronomers had discovered together. By tracing its orbit backward, astronomers deduced that this comet had originally been a single object until it passed close to Jupiter in July 1992. The planet's gravitational tidal forces then tore the comet into the fragments shown in Figure 13-13. (You may want to review the discussion of tidal forces in Section 4-8.) What is more, Jupiter's gravity had altered the trajectories of the comet fragments, so that they were doomed to collide with Jupiter.

As astronomers around the world watched, 23 fragments of Comet Shoemaker-Levy 9 crashed into Jupiter's atmosphere between July 16 and July 22, 1994. The fragments have been estimated to be no larger than 1 kilometer across, but were accelerated by Jupiter's gravitational pull to a speed on impact of 60 km/s (220,000 km/h = 130,000 mi/h). The largest fragment collided with Jupiter with an energy equivalent to 600 million megatons of TNT, tens of thousands of times greater than the total destructive energy of all of the nuclear weapons on Earth.

The impacts all occurred on the side of Jupiter that faced away from Earth and, therefore, were not immediately visible to astronomers. Thanks to Jupiter's rapid rotation, however, each impact site came into view within a few minutes after the impact itself (Figure 13-14). As each fragment entered Jupiter's atmosphere, it generated a shock wave that vaporized the fragment itself and turned the fragment's debris, along

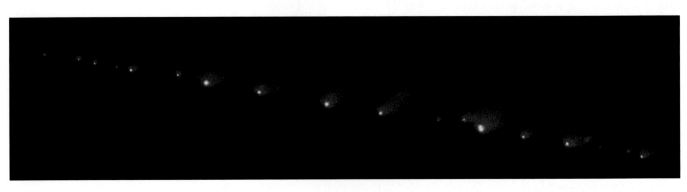

Figure 13-13   R I **V** U X G

Comet Shoemaker-Levy 9 This Hubble Space Telescope image shows icy fragments of Comet Shoemaker-Levy 9. At the time that this image was made in May 1994, the fragments were spread over a distance of 1.1 million kilometers (1.1×10^6 km = 710,000 miles), or approximately 3 times the distance from the Earth to the Moon. These fragments struck Jupiter two months later, producing a sequence of titanic fireballs. (H. A. Weaver and T. E. Smith, Space Telescope Science Institute; NASA)

WEB LINK 13.6

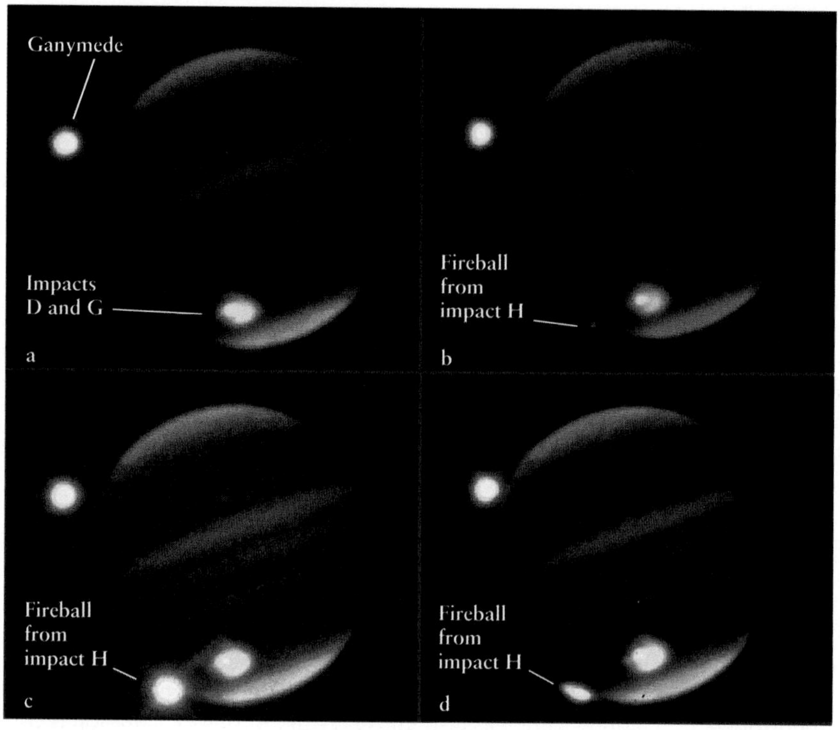
Ganymede
Impacts
D and G
a
Fireball
from
impact H
b
Fireball
from
impact H
c
Fireball
from
impact H
d

WEB LINK 13.7

Figure 13-14 R **I** V U X G

A Jupiter Impact This sequence of infrared images shows the impact of fragment H, the eighth piece of Comet Shoemaker-Levy 9 to strike Jupiter, on July 18, 1994. **(a)** Just before the impact, Jupiter's moon Ganymede can be seen as a bright spot to the left of and above the planet. The bright blob in Jupiter's southern hemisphere shows where fragments D and G struck the planet several hours before. **(b)** Just after the impact of fragment H on the far side of Jupiter, the impact fireball can be seen peeking around the lower left corner of the planet. **(c)** Seven minutes after the image in (b), the fireball is near its maximum brightness. **(d)** The fourth image, recorded 16 minutes after the image in (c), shows the fireball flattening out and fading. These images were made using the 3.5-meter telescope at Spain's Calar Alto Observatory. (Tom Herbst, Max-Planck-Institut für Astronomie, Heidelberg)

with gases from Jupiter's atmosphere, into a huge, 10,000°C fireball. Each impact's fireball, which contained millions of tons of material, then erupted upward to as much as 3000 km (1900 mi) above Jupiter's clouds. As this material fell back downward, it left dark-colored blotches the size of the Earth in Jupiter's upper atmosphere. These "scars" from the different impacts persisted for several weeks after the impacts ended (Figure 13-15).

Astronomers hoped to learn more about the chemical composition of Jupiter's atmosphere by examining the spectra of light emitted from the fireballs. But analyzing the fireball spectra is complicated because of uncertainties about the composition of the comet as well as about how deep into Jupiter's atmosphere the comet fragments penetrated. As an example, spectral lines of sulfur were seen in the light from one of the larger impacts. This could mean that the comet fragment penetrated to the level of the ammonium hydrosulfide (NH_4SH) clouds shown in Figure 13-10, where the energy of the impact would have broken the NH_4SH molecules apart and liberated sulfur atoms to be ejected upward

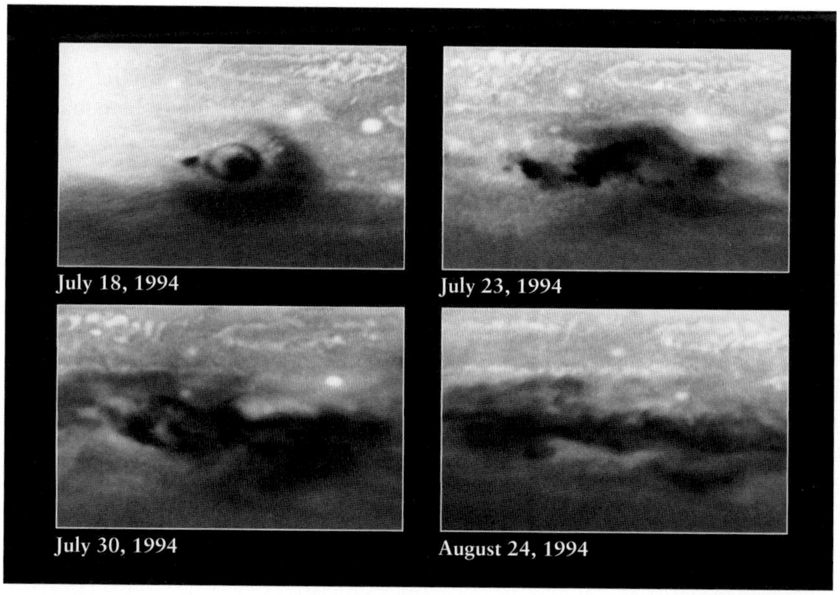
July 18, 1994
July 23, 1994
July 30, 1994
August 24, 1994

Figure 13-15 R I **V** U X G

Debris from an Impact on Jupiter This sequence of Hubble Space Telescope images shows the evolution of one of the Comet Shoemaker-Levy impact sites. The July 18, 1994, image, taken about 90 minutes after the impact of fragment G, shows a well-defined dark crescent. Over the next several weeks Jupiter's high-speed winds gradually erased this and other impact scars. (Heidi Hammel, MIT; NASA)

with the fireball. Alternatively, the fragment might have penetrated to only a shallow depth above the NH_4SH clouds, and the sulfur atoms whose spectrum was seen in the fireball might have been part of the fragment itself. Scientists are continuing to analyze the Comet Shoemaker-Levy 9 impacts with the hope of realizing their full scientific potential.

CAUTION, The impact of the comet fragments on Jupiter led many people to worry about a comet or asteroid striking the Earth, with disastrous consequences. But, as we will see in Chapter 17, such impacts are extraordinarily rare. According to reliable estimates, an object the size of Comet Shoemaker-Levy 9 can be expected to collide with the Earth only about once in 300,000 years. Among the many worries of everyday life, comets crashing into the Earth are very far down the list!

WEB LINK 13.8 Just one year after the comet struck Jupiter, a probe of human design began its own one-way trip into the giant planet's atmosphere. While still 81.52 million kilometers (50.66 million miles) from Jupiter, the *Galileo* spacecraft released the *Galileo Probe*, a cone-shaped body about the size of an office desk. While *Galileo* itself later fired its rockets to place itself into an orbit around Jupiter, the *Galileo Probe* continued on a course that led it on December 7, 1995, to a point in Jupiter's clouds just north of the planet's equator.

Unlike the comet fragments that preceded it, the *Galileo Probe* did not burn up in a fireball. Instead, a heat shield protected the spacecraft as air friction slowed its descent speed from 171,000 km/h (106,000 mi/h) to 40 km/h (25 mi/h) in just 3 minutes. The spacecraft then deployed a parachute and floated down through the atmosphere. For the next hour, the probe observed its surroundings and radioed its findings back to the main *Galileo* spacecraft, which in turn radioed them to Earth. The mission ceased at a point some 200 km (120 mi) below Jupiter's upper cloud layer, where the tremendous pressure (24 atmospheres) and high temperature (152°C = 305°F) finally overwhelmed the probe's electronics.

While the *Galileo Probe* did not have a camera, it did carry a variety of instruments that made several new discoveries. A radio-emissions detector found evidence for lightning discharges that, while less frequent than on Earth, are individually much stronger than lightning bolts in our atmosphere. Other measurements showed that Jupiter's winds, which are brisk in the atmosphere above the clouds, are even stronger beneath the clouds. In fact, the *Galileo Probe* measured a nearly constant wind speed of 650 km/h (400 mi/h) throughout its descent. This shows convincingly that the energy source for the winds is Jupiter's interior heat. If the winds were driven primarily by solar heating, as is the case on Earth, the wind speed would have decreased with increasing depth.

A central purpose of the *Galileo Probe* mission was to test the three-layer cloud model depicted in Figure 13-10 by making direct measurements of Jupiter's clouds. But the probe saw only traces of the NH_3 and NH_4SH cloud layers and found no sign at all of the low-lying water clouds. One explanation of this surprising result is that the probe by sheer chance entered

a **hot spot,** an unusually warm and cloud-free part of Jupiter's atmosphere. Indeed, observations from the Earth showed strong infrared emission from the probe entry site, as would be expected where a break in the cloud cover allows a view of Jupiter's warm interior (see Figure 13-12).

The probe also made measurements of the chemical composition of Jupiter's atmosphere. The relative proportions of hydrogen and helium proved to be almost exactly the same as in the Sun, in line with the accepted picture of how Jupiter formed from the solar nebula (Section 7-8). A number of heavy elements, however, including carbon, nitrogen, and sulfur, proved to be significantly more abundant on Jupiter than in the Sun. This was an expected result. Over the past several billion years, Jupiter's strong gravity should have pulled in many pieces of interplanetary debris. (Comet Shoemaker-Levy 9 is only the most recent.) Thanks to this debris, Jupiter built up a small but appreciable abundance of heavy elements.

Water is a major constituent of small bodies in the outer solar system and thus should likewise be present inside Jupiter in noticeable quantities. Surprisingly, the amount of water actually detected by the *Galileo Probe* was less than half the expected amount. One proposed explanation for this disparity is that the hot spot where the probe entered the atmosphere was also unusually dry. If this explanation is correct, Jupiter's hot spots are analogous to hot, dry deserts on Earth. An alternative idea is that more water than expected has rained out of Jupiter's upper atmosphere and coalesced deep within the planet. For now, the enigma of Jupiter's missing water remains unresolved.

Another surprising but potentially important result from the *Galileo Probe* concerns three elements, argon (Ar), krypton (Kr), and xenon (Xe). These elements are three of the **noble gases,** so called because their atoms are chemically aloof: They do not combine with other atoms to form molecules. (The noble gases, which also include helium, appear in the rightmost column of the periodic table in Box 5-5.) All three of these elements appear in tiny amounts in the Earth's atmosphere and in the Sun, and so must also have present in the solar nebula. But the *Galileo Probe* found that argon, krypton, and xenon are about 3 times as abundant in Jupiter's atmosphere as in the Sun.

If these elements were incorporated into Jupiter directly from the gases of the solar nebula, they should be equally as abundant in Jupiter as in the Sun (which formed from the gas at the center of the solar nebula; see Section 7-7). So, the excess amounts of argon, krypton, and xenon must have entered Jupiter in the form of solid planetesimals, just as did carbon, nitrogen, and sulfur. But at Jupiter's distance from the Sun, the presumed temperature of the solar nebula was too high to permit argon, krypton, or xenon to solidify. How, then, did the excess amounts of these noble gases become part of Jupiter?

Several hypotheses have been proposed to explain these results. One idea is that icy bodies, including solid argon, krypton, and xenon, formed in an interstellar cloud even before the cloud collapsed to form the solar nebula. A second notion is that the solar nebula may actually have been colder that previously thought. If either of these ideas proves to be correct, it would mean that Jovian planets can form closer to their stars than had heretofore been thought. This would help to explain why astronomers find massive, presumably Jovian planets orbiting close to stars other than the Sun (see Section 7-9).

A third hypothesis proposes that our models of the solar nebula are correct, but that Jupiter originally formed much farther from the Sun. Out in the cold, remote reaches of the solar nebula, argon, krypton, and xenon could have solidified into icy particles and fallen into the young Jupiter, thus enriching it in those elements. Interactions between Jupiter and the material of the solar nebula could then have forced the planet to spiral inward, eventually reaching its present distance from the Sun. As we saw in Section 7-9, a model of this kind has been proposed to explain why extrasolar planets orbit so close to their stars. More research will be needed to decide which of these hypotheses is closest to the truth.

The story of the *Galileo Probe* exemplifies two of the most remarkable aspects of scientific research. First, when new scientific observations are made, the results often pose many new questions at the same time that they provide answers to some old ones. And second, it sometimes happens that investigating what might seem to be trivial details—such as the abundances of trace elements in Jupiter's atmosphere—can reveal clues about truly grand issues, such as how the planets formed.

13-5 Jupiter's oblateness divulges the planet's rocky core

Neither Comet Shoemaker-Levy 9 nor the *Galileo Probe* penetrated more than a few hundred kilometers into Jupiter's atmosphere. To learn about the planet's structure at greater depths, astronomers must use indirect clues. One such clue is the shape of Jupiter, which indicates the size of the rocky core buried deep beneath Jupiter's atmosphere.

Even a casual glance through a small telescope shows that Jupiter is not spherical but is slightly flattened or **oblate**. (You can see this in Figure 13-1 and Figure 13-2.) The diameter across Jupiter's equator (142,980 km) is 6.5% larger than its diameter from pole to pole (133,700 km). Thus, Jupiter is said to have an **oblateness** of 6.5%, or 0.065. By comparison, the oblateness of the Earth is just 0.34%, or 0.0034.

If Jupiter did not rotate, it would be a perfect sphere. A massive, nonrotating object naturally settles into a spherical shape so that every atom on its surface experiences the same gravitational pull aimed directly at the object's center. Because Jupiter is rotating, however, the body of the planet tends to fly outward and away from the axis of rotation.

ANALOGY You can demonstrate this effect for yourself. Stand with your arms hanging limp at your sides, then spin yourself around. Your arms will naturally tend to fly outward, away from the vertical axis of rotation of your body. Likewise, if you drive your car around a circular road, you feel thrown toward the outside of the circle; the car as a whole is rotating around the center of the circle, and you tend to move

away from the rotation axis. (We discussed the physical principles behind this in Box 4-3.) For Jupiter, this same effect deforms the planet into its nonspherical, oblate shape.

The oblateness of a planet also depends on how the planet's mass is distributed over its volume. The challenge to planetary scientists is to create a model of a planet's mass distribution that gives the right value for the oblateness. One such model suggests that 2.6% of Jupiter's mass is concentrated in a dense, rocky core. Although this core has about 8 times the mass of the Earth, the crushing weight of the remaining bulk of Jupiter compresses it to a diameter of just 11,000 km, slightly smaller than the Earth's diameter of 12,800 km. The pressure at the center of the core is estimated to be about 70 million atmospheres, and the central temperature is probably about 22,000 K. By contrast, the temperature at Jupiter's cloudtops is only 165 K.

Much of the rocky core was probably the original "seed" around which the proto-Jupiter accreted. However, additional amounts of rocky material were presumably added later by meteoritic material that fell into the planet and sank to the center. Material from icy planetesimals, too, would have sunk deep within Jupiter. In the model we have been describing, these "ices"—principally water (H_2O), methane (CH_4), ammonia (NH_3), and other molecules made by chemical reactions among these substances—form a layer some 3000 km thick that surrounds the rocky core. (Because these substances are less dense than rock, they "float" on top of the rocky core.) Despite the high pressures in this layer, the temperature is so high that the "ices" are probably in a liquid state.

13-6 Metallic hydrogen in Jupiter's interior endows the planet with a strong magnetic field

The oblateness of Jupiter gives astronomers information about the size of Jupiter's rocky core. It does not, however, provide much insight into the hydrogen-rich bulk of Jupiter between its rocky core and its colorful cloudtops. Most of our knowledge of this part of the planet's interior comes from observations of Jupiter with radio telescopes. By using these telescopes to detect radio waves from Jupiter, astronomers have found evidence of electric currents flowing within the planet's hydrogen-rich interior.

Astronomers began discovering radio emissions coming from Jupiter in the 1950s. A small portion of these emissions is **thermal radiation,** with the sort of spectrum that would be expected from a dense object due to its temperature. (We discussed the electromagnetic radiation from heated objects in Section 5-3 and Section 5-4.) However, most of the radio energy emitted from Jupiter is **nonthermal radiation.** This has a very different sort of spectrum from radiation from a heated object and is found in two distinct wavelength ranges.

At wavelengths of a few meters, Jupiter emits sporadic bursts of radio waves. This **decametric** (ten-meter) **radiation** is probably caused by electrical discharges associated with powerful electric currents in Jupiter's ionosphere. These discharges result from electromagnetic interactions between Jupiter and its large satellite Io. At shorter wavelengths of a few tenths of a meter, Jupiter emits a nearly constant stream of electromagnetic radiation. The properties of this **decimetric (tenth-meter) radiation** show that it is emitted by electrons moving through a strong magnetic field at speeds comparable to the speed of light. The magnetic field deflects the electrons into spiral-shaped trajectories, and the spiraling electrons radiate energy like miniature radio antennas. Energy produced in this fashion is called **synchrotron radiation.** Astronomers now realize that synchrotron radiation is very important in a wide range of situations throughout the universe, from pulsars and quasars to the radio emissions of entire galaxies.

If some of Jupiter's radiation is indeed synchrotron radiation, then Jupiter must have a magnetic field. Measurements by the *Pioneer* and *Voyager* spacecraft verified this and indicated that the field at Jupiter's equator is 14 times stronger than at the Earth's equator. As for the Earth, Jupiter's field is thought to be generated by motions of an electrically conducting fluid in the planet's interior (see Section 8-4). But wholly unlike Earth, the moving fluid within Jupiter is an exotic form of hydrogen that acts like a liquid metal.

A hydrogen atom consists of a single proton orbited by a single electron. Deep inside Jupiter, pressures are so great and the atoms are so close that electrons can hop from one atom to another. Their motion, directed by Jupiter's rotation and by convection within the planet, creates an electric current, just as the ordered movement of electrons in a copper wire constitutes an electric current. In other words, the highly compressed hydrogen deep inside Jupiter behaves like a metal; thus, it is called **liquid metallic hydrogen.**

Laboratory experiments show that hydrogen becomes a liquid metal when the pressure exceeds about 1.4 million atmospheres. Recent calculations suggest that this transition occurs at a depth of about 7000 km below Jupiter's cloudtops. Thus, the internal structure of Jupiter consists of four distinct regions (Figure 13-16): a rocky core, a roughly 3000-km thick layer of liquid "ices," a layer of helium and liquid metallic hydrogen about 56,000 km thick, and a layer of helium and ordinary hydrogen about 7000 km thick. The colorful cloud patterns that we can see through telescopes are located in the outermost 100 km of the outer layer.

CAUTION! Jupiter is sometimes misleadingly called a "gas giant" planet. It is true that Jupiter's primary constituents—hydrogen and helium—are gases under normal conditions on Earth. But Figure 13-16 shows that most of Jupiter's volume, as well as most of its mass, is actually in the form of *liquid* hydrogen and helium. Jupiter might be better described as a "liquid giant"!

As Figure 13-16 shows, much of Jupiter's enormous bulk is electrically conductive liquid metal. Because of Jupiter's rapid rotation, electric currents in this metallic hydrogen generate a powerful magnetic field, in much the same way that liquid metals in the Earth's core produce the Earth's magnetic field.

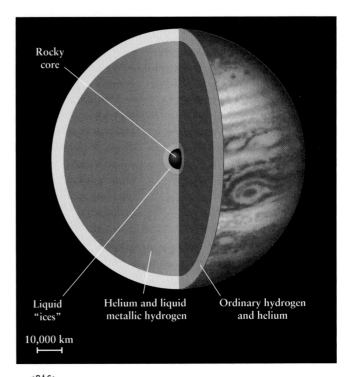

figure 13-16

Jupiter's Internal Structure Jupiter probably has a four-layer internal structure. The rocky core (shown in black) is surrounded by a hot liquid layer (shown in blue) of water, methane, ammonia, and related substances, somewhat misleadingly called "ices." Surrounding this is a thick layer of helium and liquid metallic hydrogen (shown in orange) and a relatively shallow outer layer of ordinary hydrogen and helium (shown in yellow).

13-7 Jupiter has an extensive magnetosphere

Jupiter's strong magnetic field surrounds the planet with a magnetosphere so large that it envelops the orbits of many of its moons. The *Pioneer* and *Voyager* spacecraft, which carried instruments to detect charged particles and magnetic fields, revealed the awesome dimensions of the Jovian magnetosphere. The shock wave that surrounds Jupiter's magnetosphere is nearly 30 million kilometers across. If you could see Jupiter's magnetosphere from the Earth, it would cover an area in the sky 16 times larger than the full Moon. Figure 13-17 shows the structure of Jupiter's magnetosphere.

The *Pioneer* and *Voyager* spacecraft reported the outer boundary of Jupiter's magnetosphere (on the left in Figure 13-17) at distances ranging from 3 to 7 million kilometers (2 to 4 million miles) above the planet. The "downstream" side of the magnetosphere (on the right in Figure 13-17) extends for more than a billion kilometers (650 million miles), even beyond the orbit of Saturn. The size of Jupiter's magnetosphere is sensitive to fluctuations in the solar wind: It expands and

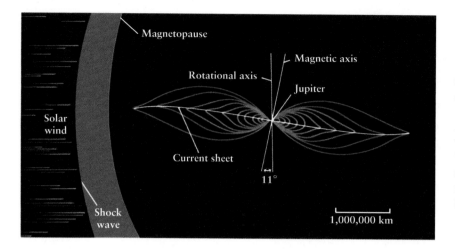

ƒigure 13-17

Jupiter's Magnetosphere Jupiter's magnetosphere is enveloped by a shock wave, where the supersonic solar wind is abruptly slowed to subsonic speeds. Most of the particles of the solar wind are deflected around Jupiter in the turbulent region between the shock wave and magnetopause, shown here in solid purple. Particles trapped inside Jupiter's magnetosphere are spread out into a vast current sheet by the planet's rapid rotation. Jupiter's axis of rotation is inclined to its magnetic axis by about 11°.

contracts by a factor of 2, rapidly and frequently. By contrast, dramatic changes in the size of the Earth's magnetosphere are extremely rare despite occasional "gusts" in the solar wind.

 The inner regions of our magnetosphere are dominated by two huge Van Allen belts (recall Figure 8-20) that are filled with charged particles. In a similar manner, Jupiter entraps immense quantities of charged particles in belts of its own. An astronaut venturing into these belts would quickly receive a lethal dose of radiation. In addition, Jupiter's rapid rotation spews the charged particles out into a huge **current sheet,** which lies close to the plane of Jupiter's magnetic equator (see Figure 13-17). Jupiter's magnetic axis is inclined 11° from the planet's axis of rotation, and the orientation of Jupiter's magnetic field is the reverse of the Earth's magnetic field—a compass would point toward the south on Jupiter.

Particles trapped in the Earth's magnetic field stream onto the north and south magnetic poles, creating the aurora (see Section 8-4). The same effect takes place on Jupiter. Images from the *Galileo* spacecraft show glowing auroral rings centered on each of the Jovian magnetic poles. As on Earth, the emission comes from a region high in Jupiter's atmosphere, about 300 to 600 km (185 to 370 mi) above the cloudtops.

Figure 13-18 shows radio emissions from charged particles in the densest regions of Jupiter's magnetosphere. These emissions vary slightly with a period of 9 hours, 55 minutes, and 30 seconds, as Jupiter's rotation changes the angle at which we view its magnetic field. Because the magnetic field is anchored deep within the planet, this variation reveals Jupiter's **internal rotation period.** This period, indicative of the rotation of the bulk of the planet's mass, is slightly slower than the atmospheric rotation at the equator (period 9 hours, 50 minutes, 28 seconds) but about the same as the atmospheric rotation at the poles (period 9 hours, 55 minutes, 41 seconds). The faster motion of the atmosphere at the equator relative to the interior is driven by Jupiter's internal heat.

The *Voyager* spacecraft discovered that the inner regions of Jupiter's magnetosphere contain a hot, gaslike mixture of charged particles called a **plasma.** A plasma is formed when a gas is heated to such extremely high temperatures that electrons

are torn off the atoms of the gas. As a result, a plasma is a mixture of positively charged ions and negatively charged electrons. The plasma that envelops Jupiter consists primarily of electrons and protons, with some ions of helium, sulfur, and oxygen. These charged particles are caught up in Jupiter's rapidly rotating magnetic field and are accelerated to high speeds.

The fast-moving particles in Jupiter's plasma exert a substantial pressure that holds off the solar wind. The *Voyager* data suggest that the pressure balance between the solar wind and the hot plasma inside the Jovian magnetosphere is precarious. A gust in the solar wind can blow away some of the plasma, at which point the magnetosphere deflates rapidly to as little as one-half its original size. Charged-particle detectors carried by the *Voyager* spacecraft recorded several bursts of hot plasma that may have been associated with such deflations of the magnetosphere. However, additional electrons and ions accelerated by Jupiter's rotating magnetic field soon replenish the plasma, and the magnetosphere expands again.

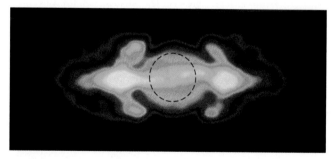

ƒigure 13-18   **R** I V U X G

A Radio View of Jupiter The Very Large Array (see Figure 6-25) was used to produce this false-color map of synchrotron emission from Jupiter at a wavelength of 21 cm. The emission comes from a region about five Jupiter diameters wide and elongated parallel to the planet's magnetic equator. Jupiter is at the center of the image, indicated by the dashed circle. The strongest emission, shown in white, comes from electrons trapped in the densest regions of Jupiter's current sheet. (NRAO)

KEY WORDS

belts, p. 285

brown ovals, p. 289

current sheet, p. 297

decametric radiation, p. 295

decimetric radiation, p. 296

differential rotation, p. 286

Great Red Spot, p. 285

hot spot, p. 294

internal rotation period, p. 297

liquid metallic hydrogen, p. 296

noble gas, p. 294

nonthermal radiation, p. 295

oblate, oblateness, p. 295

plasma, p. 297

synchrotron radiation, p. 296

thermal radiation, p. 295

white ovals, p. 289

zonal winds, p. 290

zones, p. 285

KEY IDEAS

Jupiter's Composition and Structure: Jupiter, whose mass is equivalent to the mass of 318 Earths, is composed of 71% hydrogen, 24% helium, and 5% all other elements. Jupiter has a higher percentage of heavy elements than does the Sun.

• Jupiter probably has a rocky core with a mass of about 8 Earth masses. This core is surrounded by a layer of liquid "ices" (water, ammonia, methane, and associated compounds). On top of this is a layer of helium and liquid metallic hydrogen that makes up most of the planet's mass. Jupiter's outermost layer is composed primarily of ordinary hydrogen and helium. All of Jupiter's visible features are near the top of this outermost layer.

• Jupiter has the shortest rotation period of any planet. The rotation is so rapid that the planet is flattened into a noticeably oblate shape. The rotation of Jupiter's interior is revealed by variations in Jupiter's radio emission.

• Jupiter emits about twice as much heat as it receives from the Sun. Presumably the planet's interior is still cooling.

Jupiter's Atmosphere: The visible "surface" of Jupiter is actually the tops of its clouds. The planet's rapid rotation twists the clouds into dark belts (where gas is descending and warming) and light zones (where gas is rising and cooling) that run parallel to the equator. Strong zonal winds run along the belts and zones.

• The outer layers of the Jovian atmosphere show differential rotation: The equatorial regions rotate slightly faster than the polar regions. The polar rotation rate is nearly the same as the internal rotation rate.

• The colored ovals visible in the Jovian atmosphere represent gigantic storms. Some, such as the Great Red Spot, are quite stable and persist for many years.

• There are presumed to be three cloud layers in Jupiter's atmosphere. The reasons for the distinctive colors of these different layers are not yet known. The *Galileo Probe*, so far the only spacecraft to directly explore Jupiter's atmosphere, detected only two of these cloud layers. The

reason may be that the probe fell into an unusually warm region, or hot spot.

Jupiter's Magnetic Field and Magnetosphere: Jupiter has a strong magnetic field created by currents in the metallic hydrogen layer. Its huge magnetosphere contains a vast current sheet of electrically charged particles.

• Charged particles in the densest portions of Jupiter's magnetosphere emit synchrotron radiation at radio wavelengths.

• The Jovian magnetosphere encloses a low-density plasma of charged particles whose temperature is higher than at the center of the Sun. The magnetosphere exists in a delicate balance between pressures from the plasma and from the solar wind. When this balance is disturbed, the size of the magnetosphere fluctuates drastically.

REVIEW QUESTIONS

1. Jupiter was at opposition on November 28, 2000. On that date Jupiter appeared to be in the constellation Taurus. Approximately when will Jupiter next be at opposition in this same region of the celestial sphere? Explain your answer.

2. Which planet, Mars or Jupiter, passes closer to the Earth? On which planet is it easier to see details with an Earth-based telescope? Explain your answers.

3. In what ways are the motions of Jupiter's atmosphere like the motion of water stirred in a pot (see Figure 13-3*b*)? In what ways are they different?

4. Compare Jupiter's chemical composition with that of the Sun. What does this tell us about Jupiter's formation?

5. Astronomers can detect the presence of hydrogen in stars by looking for the characteristic absorption lines of hydrogen in the star's visible spectrum (Figure 5-19). They can also detect hydrogen in glowing gas clouds by looking for hydrogen's characteristic emission lines (Figure 5-16). Explain why neither of these techniques helped

Earth-based astronomers to detect hydrogen in Jupiter's atmosphere.

6. Is the chemical composition of Jupiter as a whole the same as that of its atmosphere? Explain any differences in terms of chemical differentiation.

7. What are the belts and zones in Jupiter's atmosphere? Is the Great Red Spot more like a belt or a zone? Explain your answer.

8. Give one possible explanation why weather systems on Jupiter are longer-lived than weather systems on Earth.

9. What are white ovals and brown ovals? What can we infer about them from infrared observations?

10. Compare and contrast the source of energy for motions in our atmosphere with the energy source for motions in Jupiter's atmosphere.

11. What is thought to be the source of Jupiter's excess internal heat?

12. Did the Comet Shoemaker-Levy 9 impacts on Jupiter provide unambiguous information about the composition of Jupiter's atmosphere? Why or why not?

13. Comets like Comet Shoemaker-Levy 9 are much more likely to hit Jupiter than they are to hit Earth. Give two justifications for this statement, one in terms of Jupiter's diameter, the other in terms of Jupiter's mass.

14. Which data from the *Galileo Probe* were in agreement with astronomers' predictions? Which data were surprising?

15. Fewer than one in every 10^5 atoms in Jupiter's atmosphere is an argon atom, and fewer than one in 10^8 is an atom of krypton or xenon. If these atoms are so rare, why are scientists concerned about them? How do the abundances of these elements in Jupiter's atmosphere compare to the abundances in the Sun? What hypotheses have been offered to explain these observations?

16. Why is Jupiter oblate? What do astronomers learn from the value of Jupiter's oblateness?

17. What is liquid metallic hydrogen? What is its significance for Jupiter?

18. Describe Jupiter's internal structure, and compare it with the internal structure of the Earth.

19. What data and what techniques have been used to determine the internal structure of Jupiter?

20. Compare and contrast Jupiter's magnetosphere with the magnetosphere of a terrestrial planet like Earth. Why is the size of the Jovian magnetosphere highly variable, while that of the Earth's magnetosphere is not?

21. If Jupiter does not have any observable solid surface and its atmosphere rotates differentially, how are astronomers able to determine the planet's internal rotation rate?

22. What is a plasma? Where are plasmas found in the vicinity of Jupiter?

ADVANCED QUESTIONS

Problem-solving tips and tools

Box 4-4 describes how to use a very useful form of Kepler's third law. Newton's universal law of gravitation, discussed in Section 4-7, is the basic equation from which you can calculate a planet's surface gravity. Box 7-2 discusses escape speed, and Sections 5-3 and 5-4 discuss the properties of thermal radiation (including the Stefan-Boltzmann law, which relates the temperature of a body to the amount of thermal radiation that it emits).

23. The angular diameter of Jupiter at opposition varies little from one opposition to the next (see Table 13-2). By contrast, the angular diameter of Mars at opposition is quite variable (see Table 12-2). Explain why is there a difference between these two planets.

24. Jupiter's equatorial diameter and the rotation period at Jupiter's equator are both given in Table 13-1. Use these data to calculate the speed at which an object at the cloudtops along Jupiter's equator moves around the center of the planet.

25. Using orbital data for a Jovian satellite of your choice (see Appendix 3), calculate the mass of Jupiter. How does your answer compare with the mass quoted in Table 13-1?

26. Roughly speaking, Jupiter's composition (by mass) is three-quarters hydrogen and one-quarter helium. The mass of a single hydrogen atom is given in Appendix 7; the mass of a single helium atom is about 4 times greater. Use these numbers to calculate how many hydrogen atoms and how many helium atoms there are in Jupiter.

27. Estimate the wind velocities in the Great Red Spot, which rotates with a period of about six days.

28. If Jupiter emitted just as much energy (as infrared radiation) as it receives from the Sun, the average temperature of the planet's cloudtops would be about 107 K. Given that Jupiter actually emits twice this much energy, find what the average temperature must actually be.

29. (a) From Figure 13-10, what is the temperature in Jupiter's atmosphere at the depth where the pressure is 1 atmosphere, the same as at the Earth's surface? Give your answer in the Kelvin, Celsius, and Fahrenheit scales. (b) At the depth where the temperature in Jupiter's atmosphere is the same as room temperature (20°C = 68°F), what is the atmospheric pressure? (c) The pressure at the surface of the Earth's oceans is 1 atmosphere, and increases by 1 atmosphere for every 10 meters that you descend below the surface. At what depth in the Earth's oceans is the pressure the same as the value you found in (b)?

30. Consider a hypothetical future spacecraft that would float, suspended from a balloon, for extended periods in Jupiter's upper atmosphere. If it is desired to have this

spacecraft return to the Earth after completing its mission, to what speed would the spacecraft's rocket motor have to accelerate it in order to escape Jupiter's gravitational pull? Compare with the escape speed from the Earth, equal to 11.2 km/s.

31. The *Galileo Probe* had a mass of 339 kg. On Earth, its weight (the gravitational force exerted on it by Earth) was 3320 newtons, or 747 lb. What was the gravitational force that Jupiter exerted on the *Galileo Probe* when it entered Jupiter's clouds?

32. From the information given in Section 13-5, calculate the average density of Jupiter's rocky core. How does this compare with the average density of the Earth? With the average density of the Earth's solid inner core? (See Table 8-1 and Table 8-2 for data about the Earth.)

33. In the outermost part of Jupiter's outer layer (shown in yellow in Figure 13-16), hydrogen is principally in the form of molecules (H_2). Deep within the liquid metallic hydrogen layer (shown in orange in Figure 13-16), hydrogen is in the form of single atoms. Recent laboratory experiments suggest that there is a gradual transition between these two states, and that the transition layer overlaps the boundary between the ordinary hydrogen and liquid metallic hydrogen layers. Use this information to redraw Figure 13-16 and to label the following regions in Jupiter's interior: (i) ordinary (nonmetallic) hydrogen molecules; (ii) nonmetallic hydrogen with a mixture of atoms and molecules; (iii) liquid metallic hydrogen with a mixture of atoms and molecules; (iv) liquid metallic hydrogen atoms.

DISCUSSION QUESTIONS

34. Describe some of the semipermanent features in Jupiter's atmosphere. Compare and contrast these long-lived features with some of the transient phenomena seen in Jupiter's clouds.

35. Suppose you were asked to design a mission to Jupiter involving an unmanned airplanelike vehicle that would spend many days (months?) flying through the Jovian clouds. What observations, measurements, and analyses should this aircraft be prepared to make? What dangers might the aircraft encounter, and what design problems would you have to overcome?

36. What sort of experiment or space mission would you design in order to establish definitively whether Jupiter has a rocky core?

37. The classic science-fiction films *2001: A Space Odyssey* and *2010: The Year We Make Contact* both involve manned spacecraft in orbit around Jupiter. What kinds of observations could humans make on such a mission that cannot be made by robotic spacecraft? What would be the risks associated with such a mission? Do you think that a manned Jupiter mission would be as worthwhile as a manned mission to Mars? Explain your answers.

 ## WEB/CD-ROM QUESTIONS

38. Search the World Wide Web for recent information about the orbiting *Galileo* spacecraft. What new observations of Jupiter has it conducted? What discoveries has it made?

39. On Jupiter, the noble gases argon, krypton, and xenon provide important clues about Jupiter's past. On Earth, xenon is used in electronic strobe lamps because it emits a very white light when excited by an electric current. Argon, by contrast, is one of the gases used to fill ordinary incandescent lightbulbs. Search the World Wide Web for information about why argon is used in this way. Why do premium, long-life lightbulbs use krypton rather than argon?

 40. Moving Weather Systems on Jupiter. Access and view the video "The Great Red Spot" in Chapter 13 of the *Universe* web site or CD-ROM. **(a)** Near the bottom of the video window you will see a white oval moving from left to right. By stepping through the video one frame at a time, estimate how long it takes this oval to move a distance equal to its horizontal dimension. (*Hint:* You can keep track of time by noticing how many frames it takes a feature in the Great Red Spot, at the center of the video window, to move in a complete circle around the center of the spot. The actual time to move in such a circle is about six days.) **(b)** The horizontal dimension of the white oval is about 4000 km. At what approximate speed (in kilometers per hour) does the white oval move?

OBSERVING PROJECTS

Observing tips and tools

 Like Mars, Jupiter is most easily seen around opposition. The dates of Jupiter's oppositions are listed in Table 13-2. There is an opposition of Jupiter every 13 months, and the planet is visible in the night sky for several months before or after opposition. At other times, Jupiter may be visible in either the predawn morning sky or the early evening sky. Consult such magazines as *Sky & Telescope* and *Astronomy* or their web sites for more detailed information about when and where to look for Jupiter during a given month. You can also use the *Starry Night* program on the CD-ROM that accompanies this textbook. A relatively small telescope with an objective diameter of 15 to 20 cm (6 to 8 inches), used with a medium-power eyepiece to give a magnification of 25× or so, should enable you to see some of the dark belts.

41. If Jupiter is visible in the night sky, make arrangements to view the planet through a telescope. What magnifying power seems to give you the best view? Draw a picture of what you see. Can you see any belts and zones? How many? Can you see the Great Red Spot?

42. Make arrangements to view Jupiter's Great Red Spot through a telescope. Consult the *Sky & Telescope* web site, which lists the times when the center of the Great Red Spot passes across Jupiter's central meridian. The Great Red Spot is well placed for viewing for 50 minutes before and after this happens. You will need a refractor with an objective of at least 15 cm (6 in.) diameter or a reflector with an objective of at least 20 cm (8 in.) diameter. Using a pale blue or green filter can increase the color contrast and make the spot more visible. For other useful hints, see the article "Tracking Jupiter's Great Red Spot" by Alan MacRobert (*Sky & Telescope*, September 1997).

43. Use the *Starry Night* program to observe the appearance of Jupiter. First turn off daylight (select **Daylight** in the **Sky** menu) and show the entire celestial sphere (select **Atlas** in the **Go** menu). Center on Jupiter by using the **Find...** command in the **Edit** menu. Using the controls at the right-hand end of the Control Panel, zoom in until Jupiter nearly fills the field of view. In the Control Panel, set the time step to 10 minutes. Use the single-step buttons to observe Jupiter's rotation. Step through enough time to determine the rotation period of the planet. (*Hint:* You may want to track the motion of an easily recognizable feature, such as the Great Red Spot.) How does your answer compare with the rotation period given in Table 13-1?

The Galilean Satellites of Jupiter

In 1610, Galileo discovered four satellites circling Jupiter. He called them the Medicean Stars, in order to curry the favor of a wealthy Florentine patron of the arts and sciences. We now call them the **Galilean satellites.** Individually they are named Io, Europa, Ganymede, and Callisto, after four mythical lovers and companions of the god whom the Greeks called Zeus and the Romans called Jupiter.

Almost four centuries after Galileo, astronomers discovered these four satellites anew, thanks to the *Voyager* flybys in 1979 and the orbiting *Galileo* spacecraft in the late 1990s. They found that while Io, Europa, Ganymede, and Callisto are comparable in size to our Moon or the planet Mercury, they are entirely different from any world previously seen or even imagined. (The montage of spacecraft images shows the Galilean satellites along with part of Jupiter.)

Io, the innermost of the four, is the most geologically active world in the solar system. Its surface is covered with colorful sulfur compounds deposited by dozens of active volcanoes. Next outward is Europa, which is enveloped by a layer of ice. Crisscrossing the ice are fractures that suggest a dynamic geologic history, and beneath the ice may lie a worldwide ocean a hundred kilometers deep. Ganymede and Callisto are each covered with a thick layer of ice; the difference is that Callisto has a cratered, geologically dead surface, while Ganymede shows signs of tectonic activity and has a magnetic field of its own.

The four Galilean satellites, along with several other, smaller moons of Jupiter, form a solar system in miniature. By exploring these fascinating satellites, we advance our understanding of the physical processes that shape the worlds that orbit the Sun, including our own Earth.

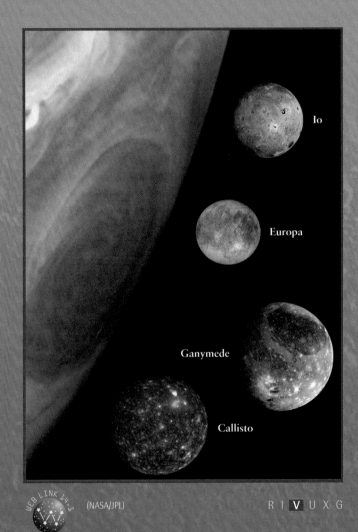

WEB LINK 14.1 (NASA/JPL)   R I **V** U X G

Io

Europa

Ganymede

Callisto

As you read the sections of this chapter, look for the answers to the following questions.

14-1 What is special about the orbits of the Galilean satellites?

14-2 Are all the Galilean satellites made of rocky material, like the Earth's moon?

14-3 How did the Galilean satellites form?

14-4 Why does Io have active volcanoes? How does volcanic activity on Io differ from that on Earth?

14-5 How does Io act like an electric generator?

14-6 What is the evidence that Europa has an ocean beneath its surface?

14-7 Do the icy satellites Ganymede and Callisto show any sign of geologic activity?

14-8 Why were Jupiter's rings discovered only recently?

14-1 The Galilean satellites are easily seen with Earth-based telescopes

When viewed through an Earth-based telescope, the **Galilean satellites** look like pinpoints of light (Figure 14-1). All four satellites orbit Jupiter in nearly the same plane as the planet's equator. From Earth, we always see that plane nearly edge-on, so the satellites appear to move back and forth relative to Jupiter (see Figure 4-12). The orbital periods are fairly short, ranging from 1.8 (Earth) days for Io to 16.7 days for Callisto, so we can easily see the satellites move from one night to the next and even during a single night. These motions led Galileo to realize that he was seeing objects orbiting around Jupiter, in much the same way that Copernicus said that planets move around the Sun (see Section 4-2).

Each of the four Galilean satellites is bright enough to be visible to the naked eye. Why, then, did Galileo need a telescope to discover them? The reason is that as seen from the Earth, the angular separation between Jupiter and the satellites is quite small—never more than 10 arcminutes for Callisto and even less for the other three Galilean satellites. To the naked eye, these satellites are lost in the overwhelming glare of Jupiter. But a small telescope or even binoculars increases the apparent angular separation by enough to make the Galilean satellites visible.

The brightness of each satellite varies slightly as it orbits Jupiter. This is because the satellites also spin on their axes as they orbit Jupiter, and dark and light areas on their surfaces are alternately exposed to and hidden from our view. Remarkably, each Galilean satellite goes through one complete cycle of brightness during one orbital period. This tells us that each satellite is in *synchronous rotation*, so that it rotates exactly once on its axis during each trip around its orbit (see Figure 3-4b). As we saw in Section 4-8, our Moon's synchronous rotation is the result of gravitational forces exerted on the Moon by Earth. Likewise, Jupiter's gravitational forces keep the Galilean satellites in synchronous rotation.

Synchronous rotation means that the rotation period and orbital period are in a 1-to-1 ratio for each Galilean satellite. Remarkably, there is also a simple ratio of the different orbital periods of the three inner Galilean satellites, Io, Europa, and Ganymede. During the 7.155 days that Ganymede takes to complete one orbit around Jupiter, Europa makes two orbits and Io makes four orbits. Thus, the orbital periods of Io, Europa, and Ganymede are in the ratio 1:2:4, which you can verify from the data in Table 14-1.

This special relationship among the satellites' orbits is maintained by the gravitational forces that they exert on one another. Those forces can be quite strong, because the three inner Galilean satellites pass relatively close to one another; at their closest approach, Io and Europa are separated by only two-thirds the distance from the Earth to the Moon. Such a close approach occurs once for every two of Io's orbits, so Europa's gravitational pull acts on Io in a rhythmic way. Indeed, Io, Europa, and Ganymede all exert rhythmic gravitational tugs on one another. Just as a drummer's rhythm keeps

figure 14-1   R I **V** U X G

The Galilean Satellites Through a Small Telescope This photograph, taken by an amateur astronomer with a small telescope, shows the four Galilean satellites alongside an overexposed image of Jupiter. One satellite appears to the left of Jupiter, while the other three appear to the planet's right. (Courtesy of J. Jenkins)

Table 14-1 The Galilean Satellites Compared with the Moon, Mercury, and Mars

	Average distance from Jupiter (km)	Orbital period (days)	Diameter (km)	Mass (kg)	Mass (Moon = 1)	Average density (kg/m³)	Albedo
Io	421,600	1.769	3642	8.932×10^{22}	1.22	3529	0.63
Europa	670,900	3.551	3120	4.791×10^{22}	0.65	3018	0.64
Ganymede	1,070,000	7.155	5268	1.482×10^{23}	2.02	1936	0.43
Callisto	1,883,000	16.689	4800	1.077×10^{23}	1.47	1851	0.17
Moon	—	—	3476	7.349×10^{22}	1.00	3344	0.11
Mercury	—	—	4880	3.302×10^{23}	4.49	5430	0.12
Mars	—	—	6794	6.419×10^{23}	8.73	3934	0.15

5000 km

Io | Europa | Ganymede | Callisto

Jupiter | 421,600 km | 670,900 km | 1,070,000 km | 1,883,800 km

Average distance from Jupiter

Note: Jupiter is shown to the same scale as the distances of the satellites from Jupiter. Compared to this scale, the images of the satellites themselves have been enlarged 74×.

(NASA/JPL) R I **V** U X G

musicians on the same beat, this gravitational rhythm maintains the simple ratio of orbital periods among the satellites.

By contrast, Callisto orbits at a relatively large distance from the other three large satellites. Hence, the gravitational forces on Callisto from the other satellites are relatively weak, and there is no simple relationship between Callisto's orbital period and the period of Io, Europa, or Ganymede.

Because the orbital planes of the Galilean satellites are nearly edge-on to our line of sight, we see these satellites undergoing transits, eclipses, and occultations. In a transit, a satellite passes between us and Jupiter, and we see the satellite's shadow as a black dot against the planet's colorful cloud-tops. (Figure 13-4a shows Io in transit across the face of Jupiter.) In an eclipse, one of the satellites disappears and then reappears as it passes into and out of Jupiter's enormous shadow. In an **occultation,** a satellite passes completely behind Jupiter, and the satellite is temporarily blocked from our Earth-based view. (The word *occultation* comes from a Latin verb meaning "to cover.")

Before spacecraft first ventured to Jupiter, astronomers used eclipses to estimate the diameters of the Galilean satellites. When a satellite emerges from Jupiter's shadow, it does not blink on instantly. Instead, there is a brief interval during which the satellite gets progressively brighter as more of its surface is exposed to sunlight. The planet's diameter is calculated from the duration of this interval and the satellite's orbital speed (which is known from Kepler's laws).

Occultations can also be used to determine the diameter of a satellite by measuring the time it takes to disappear behind Jupiter or to reappear from behind the planet. In addition, the Galilean satellites occasionally occult one another, because all four orbit Jupiter in the plane of the planet's equator. Every six years, when the Earth passes through this equatorial plane, there is a period of a few days during which Earth observers can see mutual occultations of the satellites. Once again, timing the occultations enables astronomers to calculate the diameters of the satellites.

Table 14-1 lists up-to-date values of these diameters. As the table shows, the smallest Galilean satellite (Europa) is slightly smaller than our Moon; the largest satellite (Ganymede) is larger than Mercury and more than three-quarters the size of Mars. Thus, the Galilean satellites truly are worlds in their own right.

14-2 Data from spacecraft reveal the unique properties of the Galilean satellites

Even the finest images made with Earth-based telescopes have revealed relatively few details about the Galilean satellites. Almost everything we know about these satellites has come from observations made at close range by spacecraft.

The first close-range observations of the Galilean satellites were made by the *Pioneer 10* and *Pioneer 11* spacecraft as they flew past Jupiter in 1973 and 1974. The images from these missions were of relatively low resolution, however, and much more information has come from three subsequent missions. *Voyager 1* and *Voyager 2* flew past Jupiter and its satellites in 1979 and recorded tens of thousands of images, one of which is shown in Figure 14-2. An even more extensive, multi-year investigation of Jupiter and its satellites is being carried out by the *Galileo* spacecraft, which entered orbit around Jupiter in 1995.

One key goal of these missions was to determine the densities of the Galilean satellites to very high accuracy. Given the density of a planet or satellite, astronomers can draw conclusions about its chemical composition and internal structure. To find the densities, scientists first determined the satellite masses. They did this by measuring how the gravity of Io, Europa, Ganymede, and Callisto deflected the trajectories of the *Voyager* and *Galileo* spacecraft. Given the masses and diameters of the satellites, they then calculated each satellite's average density (mass divided by volume). Table 14-1 lists our current knowledge about the sizes and masses of the Galilean satellites.

Of the four Galilean satellites, Europa proves to be the least massive: Its mass is only two-thirds that of our Moon. Ganymede is by far the most massive of the four, with more than double the mass of our Moon. In fact, Ganymede is the most massive satellite anywhere in the solar system. In second place is Callisto, with about 1½ times the Moon's mass.

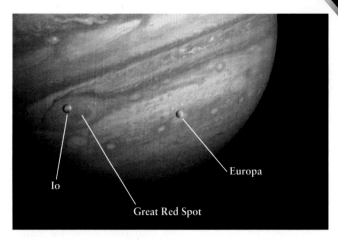

Figure 14-2   R I **V** U X G

Io and Europa from *Voyager 1* *Voyager 1* recorded this image of Jupiter and the Galilean satellites Io and Europa in February 1979, when it was 20 million kilometers (12.4 million miles) from the giant planet. The satellites are actually smaller than some of Jupiter's cloud features, such as the Great Red Spot (see Section 13-2). The *Voyager*s also recorded many images of the Galilean satellites at closer range. (NASA/JPL)

It would be quite incorrect, however, to think of Ganymede and Callisto as merely larger versions of our Moon. The reason is that both Ganymede and Callisto have very low average densities of less than 2000 kg/m^3. By comparison, typical rocks in the Earth's crust have densities around 3000 kg/m^3, and the average density of our Moon is 3344 kg/m^3. The low densities of Ganymede and Callisto mean that these satellites cannot be composed primarily of rock. Instead, they are thought to be made of roughly equal parts of rock and water ice. Water ice is substantially less dense than rock, but it becomes as rigid as rock under high pressure (such as is found in the interiors of the Galilean satellites). Furthermore, water molecules are relatively common in the solar system. Enough water would have been available in the early solar system to make up a substantial portion of a large satellite such as Ganymede or Callisto.

Water ice cannot be a major constituent of the two inner Galilean satellites, however. The innermost satellite, Io, has the highest average density, 3529 kg/m^3, slightly greater than the density of our Moon. The next satellite out, Europa, has an average density of 3018 kg/m^3. Both of these values are close to the densities of typical rocks in the Earth's crust. Hence, it is reasonable to suppose that both Io and Europa are made primarily of rocky material.

The most definitive evidence for water ice on Jupiter's satellites has come from spectroscopic observations. The spectra of sunlight reflected from Europa, Ganymede, and Callisto show absorption at the infrared wavelengths characteristic of water ice molecules. You can see this in Figure 7-4, which compares the spectrum of Europa to the spectrum of ice. Because Europa's density shows that it is composed mostly of rock, its ice must be limited to the satellite's outer regions.

Among the Galilean satellites, only Io shows no trace of water ice on its surface or in its interior.

14-3 The Galilean satellites formed like a solar system in miniature

As Table 14-1 shows, the more distant a Galilean satellite is from Jupiter, the greater the proportion of ice within the satellite and the lower its average density. But why should this be? An important clue is that the average densities of the planets show a similar trend: Moving outward from the Sun, the average density of the planets steadily decreases, from more than 5000 kg/m^3 for Mercury to less than 1000 kg/m^3 for Saturn (recall Table 7-1). The best explanation for this similarity is that the Galilean satellites formed around Jupiter in much the same way that the planets formed around the Sun, although on a much smaller scale.

We saw in Section 7-8 that planets formed from dust grains that coalesced in the solar nebula. In the inner parts of the nebula, close to the glowing protosun, only dense, rocky grains were able to survive. These accumulated over time into the dense, rocky inner planets. But in the cold outer reaches of the nebula, dust grains were able to retain icy coatings of low-density material like water and ammonia. These low-density substances are much more abundant than rocky materials, so they formed larger protoplanets than in the inner parts of the nebula. These large protoplanets in turn attracted large amounts of hydrogen and helium gas from the nebula, producing the immense Jovian planets.

The gas accumulating around Jupiter's protoplanetary core formed a rotating "Jovian nebula." The central part of this nebula became the huge envelope of hydrogen and helium that makes up most of Jupiter's bulk. But in the outer parts of this nebula, dust grains could have accreted to form small solid bodies. These grew to become the Galilean satellites.

Although the "Jovian nebula" was far from the protosun, not all of it was cold. Jupiter, like the protosun, must have emitted substantial amounts of radiation due to Kelvin-Helmholtz contraction, which we discussed in Section 7-7. (We saw in Section 13-3 that even today, Jupiter emits more energy due to its internal heat than it absorbs from sunlight.) Hence, temperatures very close to Jupiter must have been substantially higher than at locations farther from the planet. Calculations by NASA scientists James Pollack, Fraser Fanale, and collaborators show that only rocky material would condense at the orbital distances of Io and Europa, but frozen water could be retained and incorporated into satellites at the distances of Ganymede and Callisto. In this way Jupiter ended up with two distinct classes of Galilean satellites (Figure 14-3).

The analogy between the solar nebula and the "Jovian nebula" is not exact. Unlike the Sun, Jupiter is a "failed star." Its internal temperatures and pressures never became high enough to ignite nuclear reactions that convert hydrogen into helium. (Jupiter's mass would have had to be about 80 times larger for these reactions to have begun.) Furthermore, the

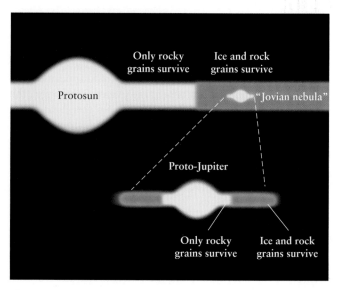

Figure 14-3

Formation of the Galilean Satellites Heat from the protosun made it impossible for icy grains to survive within the innermost 2.5 to 4 AU of the solar nebula. Within that distance, only rocky grains remained to form the terrestrial planets. In the same way, Jupiter's heat evaporated any icy grains that were too close to the center of the "Jovian nebula." Hence, the two inner Galilean satellites were formed primarily from rock, while the outer two incorporated both rock and ice.

icy worlds that formed in the outer region of the "Jovian nebula"—namely, Ganymede and Callisto—were of relatively small mass. Hence, they were unable to attract gas from the nebula and become Jovian planets in their own right. Nonetheless, it is remarkable how the main processes of "planet" formation seem to have occurred twice in our solar system: once in the solar nebula around the Sun, and once in microcosm in the gas cloud around Jupiter.

The diverse densities of the Galilean satellites suggest that these four worlds may be equally diverse in their geology. This turns out to be entirely correct. Because each of the Galilean satellites is unique, we devote the next several sections to each satellite in turn.

14-4 Io is covered with colorful sulfur compounds ejected from active volcanoes

Before the two *Voyager* spacecraft flew past Jupiter, the nature of Io was largely unknown. On the basis of Io's size and density, however, it was expected that Io would be a world much like our own Moon—geologically dead, with little internal heat available to power tectonic or volcanic activity. Hence, it was thought that the surface of Io would be extensively cratered, because there would have been little geologic activity to erase those craters over the satellite's history.

The true nature of Io proved to be utterly different from these naive predictions. On March 5, 1979, *Voyager 1* came within 21,000 kilometers (13,000 miles) of Io and began sending back a series of bizarre and unexpected pictures of the satellite (Figure 14-4). These images showed that Io has *no* impact craters at all! Instead, the surface is pockmarked by irregularly shaped pits and is blotched with color, giving Io an appearance totally unlike that of any other world in the solar system. Baffled by what they were seeing, scientists at the Jet Propulsion Laboratory (JPL) in Pasadena, California (where the pictures were being received and enhanced by computer processing), jokingly compared Io to a pizza or a rotten orange. In fact, the scientists were seeing confirmation of a brilliant prediction published only a few days earlier: Io's surface is the result of intense volcanic activity.

Three days before *Voyager 1* flew past Io, the journal *Science* published an article by Stanton Peale of the University of California, Santa Barbara, and Patrick Cassen and Ray Reynolds of NASA's Ames Research Center. In this article, they reported their conclusion that Io's interior must be kept hot by Jupiter's tidal forces.

As we saw in Section 4-8, tidal forces are differences in the gravitational pull on different parts of a planet or satellite. These forces tend to deform the shape of the planet or satellite (see Figure 4-24). Io, for example, is somewhat distorted from a spherical shape by tidal forces from Jupiter. This deformation is not constant, however. As Io moves around its orbit, Europa and Ganymede exert gravitational tugs on it in a regular, rhythmic fashion, thanks to the 1:2:4 ratio among the orbital periods of these three satellites. The result is that Io's orbit is distorted into an ellipse, and, therefore, the distance between Io and Jupiter varies as Io goes around its orbit. As the distance changes, so does the strength of the tidal forces that Jupiter exerts on Io. The varying tidal stresses on Io alternately squeeze and flex it. Just as a ball of clay or bread dough gets warm as you knead it between your fingers, this continually varying tidal stress causes **tidal heating** of Io's interior.

Tidal heating adds energy to Io at a rate of about 10^{14} watts, equivalent to 24 tons of TNT exploding every second. As this energy makes its way to the satellite's surface, it provides about 2.5 watts of power to each square meter of Io's surface. By comparison, the average global heat flow through the Earth's crust is 0.06 watts per square meter. Only in volcanically active areas on Earth do we find heat flows that even come close to Io's average. Thus, Peale, Cassen, and Reynolds predicted "widespread and recurrent surface volcanism" on Io.

No one expected that *Voyager 1* would obtain images of erupting volcanoes on Io. After all, a spacecraft making a single trip past the Earth would be highly unlikely to catch a large volcano actually erupting. But that is exactly what happened at Io. Several days after the Jupiter flyby, Linda Morabito, a navigation engineer at JPL, noticed a large umbrella-shaped cloud protruding from Io in one image (Figure 14-5a). She had discovered a plume of gas from a volcanic eruption. Careful reexamination of the close-up images revealed seven more giant volcanic plumes like those shown in Figure 14-5b. The Peale-Cassen-Reynolds prediction was confirmed resoundingly: Io is by far the most volcanic world in the solar system.

Io's volcanic plumes rise to astonishing heights of 70 to 280 km above Io's surface. To reach such altitudes, the material must emerge from volcanic vents with speeds between 300 and 1000 m/s (1100 to 3600 km/h, or 700 to 2200 mi/h). Even the most violent terrestrial volcanoes, like Vesuvius, Krakatoa, and Mount St. Helens (see Figure 11-8), have eruption speeds of only around 100 m/s (360 km/h, or 220 mi/h). Scientists thus began to suspect that the processes that produce Io's volcanic plumes must be fundamentally different from volcanic processes here on Earth.

An important clue about Io's volcanism came from the infrared spectrometers aboard *Voyager 1*, which detected abundant sulfur and sulfur dioxide in Io's volcanic plumes. This led to the idea that the plumes are actually more like geysers than volcanic eruptions. In a geyser on Earth, water seeps

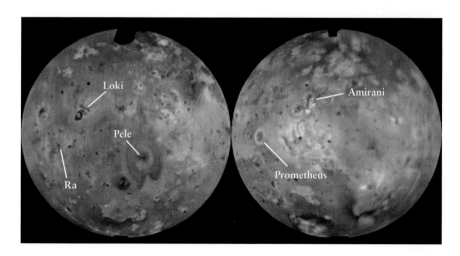

 figure 14-4 R I **V** U X G

Io These mosaics of Io's two hemispheres were built up from individual *Voyager* images. The extraordinary range of colors is probably caused by deposits of sulfur and sulfur compounds ejected from Io's numerous volcanoes. The labels show several of these volcanoes, which are named for sun gods and fire gods of different cultures. The black areas at the very top of each hemisphere are regions that the *Voyagers* did not observe. (NASA/JPL)

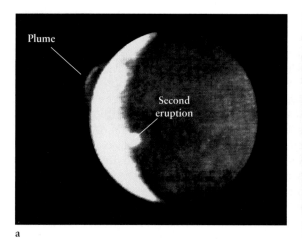

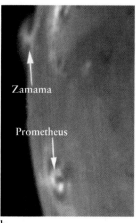

ꬵigure 14-5 R I **V** U X G

Volcanic Plumes on Io (a) Io's volcanic eruptions were first discovered on this *Voyager 1* image. The volcanic plume at upper left rises to 260 km (150 mi) above Io's surface. The volcano producing this eruption is Pele, which is visible in Figure 14-4. A second plume can be seen on the boundary between the day and night hemispheres of Io. **(b)** This close-up from the *Galileo* spacecraft shows the plumes from the volcanoes Prometheus (labeled in Figure 14-4) and Zamama. The plumes rise to about 100 km (60 mi) above Io's surface and are about 250 km (150 mi) wide. Particles in the plumes scatter sunlight, giving the plumes a bluish color (see Box 5-4). (a: NASA; b: Planetary Image Research Laboratory/University of Arizona/JPL/NASA)

down to volcanically heated rocks, where it changes to steam and erupts explosively through a vent. The planetary geologists Susan Kieffer, Eugene Shoemaker, and Bradford Smith have suggested that sulfur dioxide rather than water could be the principal propulsive agent driving volcanic plumes on Io. Sulfur dioxide is a solid at the frigid temperatures found on most of Io's surface, but it should be molten at depths of only a few kilometers. Just as the explosive conversion of water into steam produces a geyser on the Earth, the conversion of liquid sulfur dioxide into a high-pressure gas could result in eruption velocities of up to 1000 m/s.

 Io's dramatic coloration (see Figure 14-4) is probably due to sulfur and sulfur dioxide, which is ejected in volcanic plumes and later falls back to the surface. Sulfur is normally bright yellow, which explains the dominant color of Io's surface. But if sulfur is heated and suddenly

cooled, as would happen if it were ejected from a volcanic vent and allowed to fall to the surface, it can assume a range of colors from orange and red to black. Indeed, these colors are commonly found around active volcanic vents (Figure 14-6). Whitish surface deposits, by contrast, are probably due to sulfur dioxide (SO_2). Volcanic vents on Earth commonly discharge SO_2 in the form of an acrid gas. But on Io, when hot SO_2 gas is released by an eruption into the cold vacuum of space, it crystallizes into white snowflakes. This sulfur dioxide "snow" then rains back onto Io's surface, presumably forming the whitish deposits seen on the satellite's surface (examine Figure 14-4 and Figure 14-6a).

The plumes are not the whole story of volcanism on Io, however. The sources of the plumes appear in *Voyager* images as black spots 10 to 50 km in diameter, which are actually volcanic vents (see Figure 14-4). Many of these black spots,

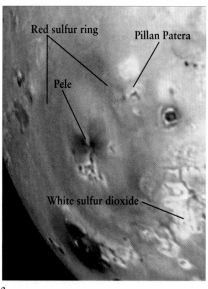

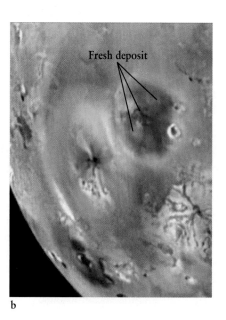

ꬵigure 14-6 R I **V** U X G

Colors on Io's Changing Surface (a) This image from the *Galileo* spacecraft shows a portion of Io's southern hemisphere as it appeared in April 1997. The active volcanic vent Pele is surrounded by a red ring with diffuse edges that resembles paint sprayed from a can. These are probably due to sulfur ejected from Pele. The whitish deposits at the lower right are probably sulfur dioxide snow. **(b)** A few months later, a volcanic plume erupted from a vent called Pillan Patera. When the plume material fell back to the surface, it left a dark circular stain some 400 km (250 mi) in diameter—roughly the size of the state of Arizona—that partially covered the red ring from Pele. Other than Earth, Io is the only world in the solar system that shows such noticeable changes over such short time spans. (NASA/JPL)

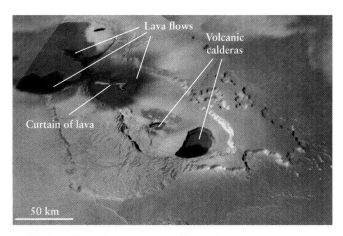

Lava flows

Volcanic calderas

Curtain of lava

50 km

 figure 14-7 R I **V** U X G

Io's Lava Flows and a Curtain of Fire The craters in this mosaic of *Galileo* spacecraft images are actually a nested chain of volcanic calderas (pits) with black lava flows. The red streak at upper left is a fissure in the surface about 40 km (25 mi) long, through which lava is erupting upward like a fountain. The erupting lava reaches the amazing height of 1500 m (5000 ft). The glow from this lava fountain was so intense that it was visible with Earth-based telescopes. (University of Arizona/JPL/NASA)

that lies underneath the Earth's crust is a plastic solid rather than a true liquid.) The global extent of this "magma ocean" would explain why volcanic activity is spread uniformly over Io's surface, rather than being concentrated in pockets as on Earth (see Section 8-3). It could also explain the origin of Io's mountains, which reach heights up to 10 km (30,000 ft). In this model, these mountains are blocks of Io's crust that are tilting before sinking into the depths of the magma ocean.

Io's worldwide volcanic activity is remarkably persistent. When *Voyager 2* flew through the Jovian system in July 1979, four months after *Voyager 1*, almost all of the volcanic plumes seen by *Voyager 1* were still active. And when *Galileo* first viewed Io in 1996, about half of the volcanoes seen by the *Voyager* spacecraft were still ejecting material, while other, new volcanoes had become active.

Io may have as many as 300 active volcanoes, each of which ejects an estimated 10,000 tons of material per second. Altogether, volcanism on Io may eject as much as 10^{13} tons of matter each year. This is sufficient to cover the satellite's entire surface to a depth of 1 meter in a century, or to cover an area of 1000 square kilometers in a few weeks (see Figure 14-6). Thanks to this continual "repaving" of the surface, there are probably no long-lived features on Io, and any impact craters are quickly obliterated.

which cover 5% of Io's surface, are surrounded by dark lava flows. Studying these volcanic areas has been a key project for the *Galileo* spacecraft since it began monitoring the Galilean satellites in 1996. Figure 14-7 shows one of Io's most active volcanic regions, where in 1999 lava was found spouting to altitudes of thousands of meters along a fissure 40 km (25 mi) in length. Thus, Io has two distinct styles of volcanic activity: unique geyserlike plumes and lava flows that are more dramatic versions of volcanic flows on Earth.

While sulfur and sulfur compounds are the key constituents of Io's volcanic plumes, the lava flows must be made of something else. Infrared measurements from the *Galileo* spacecraft (Figure 14-8) show that lava flows on Io are at temperatures of 1700 to 2000 K (about 1450 to 1750°C, or 2600 to 3150°F). At these temperatures sulfur could not remain molten but would evaporate almost instantly. Io's lavas are also unlikely to have the same chemical composition as typical lavas on Earth, which have temperatures of only 1300 to 1450 K. Instead, lavas on Io are probably **ultramafic lavas.** These are enriched in magnesium and iron, which gives the lava a higher melting temperature.

Solidified ultramafic lavas are found on Earth, but primarily in lava beds that formed billions of years ago when the Earth's interior was much hotter than today. The presence of molten ultramafic lava on Io suggests that its interior, too, is substantially hotter than that of the present-day Earth. A recent hypothesis proposes that Io has a 100-km (60-mi) thick crust that floats atop a worldwide ocean of liquid magma 800 km (500 mi) deep. (As we saw in Section 8-2, the mantle

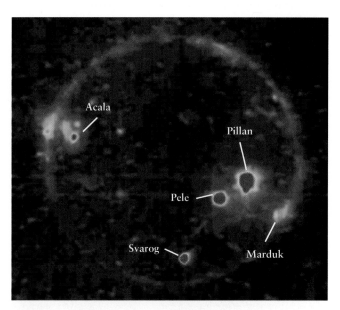

Acala

Pillan

Pele

Svarog

Marduk

 figure 14-8 R **I** **V** U X G

Io's Glowing Volcanoes When Io passes into the darkness of Jupiter's shadow, the satellite's volcanoes can be seen glowing at both infrared and visible wavelengths. This image from the *Galileo* spacecraft shows the infrared and visible glow from several volcanic regions, including Pele and Pillan (see Figure 14-6). These regions are substantially hotter than typical lava flows on Earth. (Planetary Image Research Laboratory/University of Arizona/JPL/NASA)

14-5 Jupiter's magnetic field makes electric currents flow in Io's molten interior

Most of the material ejected from Io's volcanoes falls back onto the satellite's surface. But some does not, by virtue of Io's location deep within Jupiter's magnetosphere. The magnetosphere contains charged particles, some of which collide with Io and its volcanic plumes. The impact of these collisions knocks ions out of the plumes and off the surface, and these ions become part of Jupiter's magnetosphere. The result is the **Io torus,** a huge doughnut-shaped ring of ionized gas, or plasma (see Section 13-7), that circles Jupiter at the distance of Io's orbit. The plasma's glow can be detected from Earth (Figure 14-9a).

Jupiter's magnetic field has other remarkable effects on Io. As Jupiter rotates, its magnetic field rotates with it, and this field sweeps over Io at high speed. This generates a voltage of 400,000 volts across the satellite, which causes 5 million amperes of electric current to flow through Io.

ANALOGY Whenever you go shopping with a credit card, you use the same physical principle that generates an electric current within Io. The credit card's number is imprinted as a magnetic code in the brown stripe on the back of the card. The salesperson who swipes your card through the card reader is actually moving the card's magnetic field past a coil of wire within the reader. This generates within the coil an electric current that carries the same coded information about your credit card number. That information is transmitted to the credit card company, which (it is hoped) approves your purchase. An electric generator at a power plant works in the same way. A coil of wire is moved through a strong magnetic field, which makes current flow in the coil. This current is delivered to transmission lines and eventually to your home.

In fact, a current flows not only through Io but also through the sea of charged particles in Jupiter's magnetosphere and through Jupiter's atmosphere. This forms a gigantic, oval-shaped electric circuit that connects Io and Jupiter in the same way that wires connect the battery and the lightbulb in a flashlight (Figure 14-9b). Jupiter's aurora, which we discussed in Section 13-7, is particularly strong at the locations where this current strikes the upper atmosphere of Jupiter.

Part of the electric current that flows between Io and Jupiter is made up of electrons that spiral around Jupiter's magnetic field lines. As they spiral, the electrons act like miniature antennas and emit radio waves. This is the source of Jupiter's bursts of decametric radio radiation, which we discussed in Section 13-6.

The electric current within and near Io also creates a weak magnetic field, which was first detected by *Voyager 1. Galileo* made a more sensitive measurement when it flew to within 900 km (560 mi) of Io's surface on December 7, 1995. Remarkably, the *Galileo* measurements suggest that the magnetic field near Io may be too strong—comparable to that of Mercury, although weaker than that of Earth—to be generated by Io's electric current alone. This has led scientists to speculate that Io generates its own magnetic field through the motions of molten material in its interior. (We saw in Section 8-4 that such motions within our own planet produce the Earth's magnetic field.) If these speculations are correct, Io is the smallest world in the solar system to generate its own magnetic field.

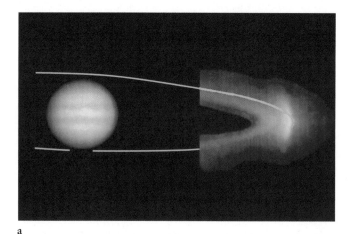

a

Figure 14-9 R ■ V U X G

The Io Torus (a) An Earth-based telescope was used to record this false-color infrared image of Io's plasma torus. Green shows emission from "cold" ions at about 10,000 K, while purple shows emission from "warm" ions at 600,000 K. Because of Jupiter's glare, only the outer edge of the torus could be photographed; an artist added the visible-light picture of Jupiter and the yellow line indicating the rest of the torus. **(b)** Material

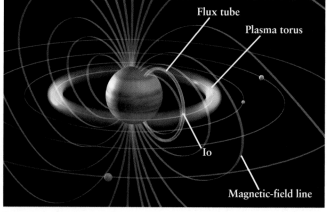

b

from Io is the main source of plasma for the torus and for Jupiter's magnetosphere as a whole. A flux tube (a region of concentrated magnetic field) carries plasma between Jupiter and Io, forming an immense electric circuit. (a: Courtesy of J. Trauger; b: Alfred T. Kamajian/Torrence V. Johnson, "The Galileo Mission to Jupiter and Its Moons," *Scientific American*, February 2000, p.44)

If Io has undergone enough tidal heating to melt its interior, chemical differentiation must have taken place and the satellite should have a dense core. (We described chemical differentiation in Section 7-8 and Box 7-1.) To test for this, scientists measured how Io's gravity deflected the trajectory of *Galileo* as the spacecraft flew past. From these measurements, they could determine not only Io's mass but also the satellite's oblateness (how much it deviates from a spherical shape because of its rotation). The amount of oblateness indicates the size of the core; the greater the fraction of the satellite's mass contained in its core, the less oblate the satellite will be. The *Galileo* observations suggest that Io has a dense core composed of iron and iron sulfide (FeS), with a radius of about 900 km (about half the satellite's overall radius). Surrounding the core is a mantle of partially molten rock, on top of which is Io's visible crust.

14-6 Europa is covered with a smooth layer of ice that may cover a worldwide ocean

WEB LINK 14-8 Europa, the second of the Galilean satellites, is the smoothest body in the solar system. There are no mountains and no surface features greater than a few hundred meters high (Figure 14-10). There are almost no craters, indicating a young surface that has been reprocessed by geologic activity. The dominant surface feature is a worldwide network of stripes and cracks (Figure 14-11). Like Io, Europa is an exception to the general rule that a small world should be cratered and geologically dead (see Section 9-3).

figure 14-10 R **I** **V** U X G

Europa Dark lines crisscross Europa's smooth, icy surface in this composite of visible and infrared images from the *Galileo* spacecraft. These are fractures in Europa's crust that can be as much as 20 to 40 km (12 to 25 mi) wide. Only one young impact crater, below and to the right of center, can be seen in this image. This indicates that Europa has a very young surface on which all older craters have been erased. (NASA/JPL)

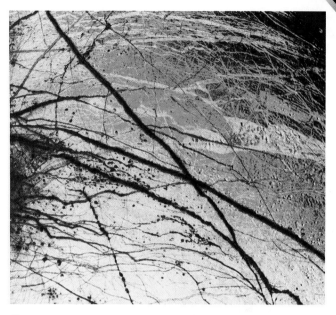

figure 14-11 R **I** **V** U X G

Fractures in Europa's Crust This false-color view (a composite of visible and infrared images from *Galileo*) shows a network of fractures and dark, linear features on part of Europa's icy surface. Smooth ice plains, which are the basic terrain found on Io, are shown in blue. The red and brown colors in the fractures indicate the presence of minerals in the ice. The area shown is about 1260 km (780 mi) across. (NASA/JPL)

Neither *Voyager 1* nor *Voyager 2* flew very near Europa, so most of our knowledge of this satellite comes from the close passes made by *Galileo*. (Both Figure 14-10 and Figure 14-11 are *Galileo* images.) But even before these spacecraft visited Jupiter, spectroscopic observations from Earth indicated that Europa's surface is almost pure frozen water (see Figure 7-4). This was confirmed by instruments on board *Galileo*, which showed that Europa's infrared spectrum is a close match to that of a thin layer of fine-grained water ice frost on top of a surface of pure water ice. (The brown areas in Figure 14-10 show where the icy surface contains deposits of rocky material from meteoritic impacts, from Europa's interior, or from a combination of these sources.)

The purity of Europa's ice suggests that water is somehow brought upward from the moon's interior to the surface, where it solidifies to make a fresh, smooth layer of ice. Indeed, some *Galileo* images show what appear to be lava flows on Europa's surface, although the "lava" in this case is mostly ice. This idea helps to explain why Europa has very few craters (any old ones have simply been covered up) and why its surface is so smooth. Europa's surface may thus represent a water-and-ice version of plate tectonics.

CAUTION! Although Europa's surface is almost pure water ice, keep in mind that Europa is not merely a giant ice ball. The satellite's density shows that rocky material makes up about 85 to 90% of Europa's mass. Hence, only a small fraction of the mass, about 10 to 15%, is water ice. Because the surface is icy, we can conclude that the rocky material is found within Europa's interior.

Europa is too small to have retained much of the internal heat that it had when it first formed. But something is powering the geologic processes that erase craters and bring fresh water to Europa's surface. If not internal heat, what is the source of this energy? The most likely answer, just as for Io, is Jupiter's tidal forces. The rhythmic gravitational tugs exerted by Io and Ganymede on Europa deform its orbit into an ellipse; the varying Europa-Jupiter distance causes tidal stresses that vary correspondingly, which makes Europa flex. But because it is farther from Jupiter, tidal effects on Europa are only about one-fourth as strong as those on Io, which may explain why no ongoing volcanic activity has yet been seen on Europa.

Some features on Europa's surface, such as the fracture patterns shown in Figure 14-11, may be the direct result of the crust's being stretched and compressed by tidal flexing. Figure 14-12 shows other features, such as networks of ridges and a young, very smooth circular area, that were probably caused instead by the internal heat that tidal flexing generates. The rich variety of terrain depicted in Figure 14-12, with stress ridges going in every direction, shows that Europa has a complex geologic history.

Among the unique structures found on Europa's surface are **ice rafts**. The area shown in Figure 14-13 was apparently subjected to folding, producing the same kind of linear features as those in Figure 14-12. But a later tectonic disturbance broke the surface into small chunks of crust a few kilometers across, which then "rafted" into new positions. A similar sort of raft-ing happens in the Earth's Arctic Ocean every spring, when the winter's accumulation of surface ice breaks up into drifting ice floes. The existence of such structures on Europa strongly suggests that there is a subsurface layer of liquid water or soft ice over which the ice rafts can slide with little resistance.

The ridges, faults, ice rafts, and other features strongly suggest that Europa has substantial amounts of internal heat. This heat could prevent water from freezing beneath Europa's surface, creating a worldwide ocean beneath the crust. If this picture is correct, geologic processes on Europa may be an exotic version of those on Earth, with the roles of solid rock and molten magma being played by solid ice and liquid water.

A future spacecraft may use radar to penetrate through Europa's icy crust to search for definitive proof of liquid water beneath the surface. But magnetic field measurements made by the *Galileo* spacecraft have already provided some key evidence favoring this picture. Unlike the Earth or Jupiter, Europa does not seem to create a steady magnetic field of its own. But as Europa moves through Jupiter's intense magnetic field, electric currents are induced within the satellite's interior, just as they are within Io (see Section 14-5), and these currents generate a weak but measurable field. (The strength and direction of this induced field varies as Europa moves through different parts of the Jovian magnetosphere, which would not be the case if Europa generated its field by itself.)

To explain these observations, there must be an electrically conducting fluid beneath Europa's crust—a perfect description of an underground ocean of water with dissolved minerals. (Pure water is a very poor conductor of electricity, so other substances must be present to make the water conducting.) If some of this water should penetrate upward to Europa's surface through cracks in the crust, it would vaporize and spread

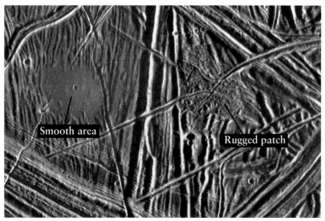

a

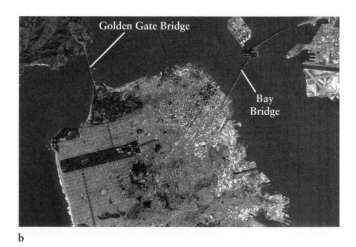

b

Figure 14-12 R I **V** U X G

Europa and San Francisco in Close-up The *Galileo* image of part of Europa's surface **(a)** is shown to the same scale as the picture of San Francisco **(b)**. Both images show an area 13 by 18 km (8 by 11 mi). Folding and faulting on Europa's surface has formed a network of overlapping ridges. To the left of center in (a), a fluid has erupted onto the surface, covering the ridges and forming a flat, smooth area about 3 km (2 mi) in diameter. The rugged patch 4 km (2.5 mi) across to the right of center was created by some subsurface disturbance. Europa's interior must have enough energy to power the formation of these complex geologic structures. (NASA/JPL)

WEB LINK 14.9

figure 14-13   R I **V** U X G

Ice Rafts on Europa This composite *Galileo* image shows portions of Europa's icy crust that have broken into "rafts" and been moved around by an underlying liquid or plastic layer. The linear features on each raft indicate how the individual pieces of crust must once have fit together, like a pattern on the shards of a broken vase. The colors are probably due to mineral contaminants that were released from beneath the surface when the crust broke apart. The image shows an area 70 by 30 km (44 by 19 mi); the largest crust fragment is about 13 km (8 mi) across. (Planetary Image Research Laboratory/University of Arizona/JPL/NASA)

the dissolved minerals across the terrain. This could explain the reddish-brown colors in Figure 14-13.

By combining measurements of Europa's induced magnetic field, gravitational pull, and oblateness, scientists conclude that Europa's outermost 100 to 200 km are ice and water. (It is not clear how much of this is liquid and how much solid.) Within this outer shell is a rocky mantle surrounding a metallic core some 600 km (400 mi) in radius.

Remarkably, Europa also has an extremely thin atmosphere of oxygen. (By Earth standards, this atmosphere would qualify as a near-vacuum.) Oxygen in the Earth's atmosphere is produced by plants through photosynthesis (see Section 8-5). But Europa's oxygen atmosphere is probably the result of ions from Jupiter's magnetosphere striking the satellite's icy surface. These collisions break apart water molecules, liberating atoms of hydrogen (which escape into space) and oxygen.

The existence of a warm, subsurface ocean on Europa, if proved, would make Europa the only world in the solar system other than the Earth on which there is liquid water. This would have dramatic implications. On Earth, water and warmth are essentials for the existence of life. Perhaps single-celled organisms have evolved in the water beneath Europa's

crust, where they would use dissolved minerals and organic compounds as food sources. Future missions to Europa may search for evidence of life within this exotic satellite.

14-7 Ganymede and Callisto have heavily cratered, icy surfaces

Unlike Io and Europa, the two outer Galilean satellites have cratered surfaces. In this respect, Ganymede and Callisto bear a superficial resemblance to our own Moon (see the illustration with Table 14-1). But unlike the Moon's craters, the craters on both Ganymede and Callisto are made primarily of ice rather than rock, and Ganymede has a number of surface features that indicate a geologically active past.

Ganymede is the largest satellite in our solar system and is even larger than the planet Mercury. Like our Moon, Ganymede has two distinct kinds of terrain, called **dark terrain** and **bright terrain** (Figure 14-14). On the Moon, the dark maria are younger than the light-colored lunar highlands (see Section 9-1). On Ganymede, by contrast, the dark terrain is older, as indicated by its higher density of craters. The bright terrain is much less cratered and therefore younger. Because ice is more reflective than rock, even

figure 14-14   R I **V** U X G

Ganymede This view from the *Galileo* spacecraft shows the hemisphere of Ganymede that always faces away from Jupiter. This hemisphere is dominated by a huge, dark, circular area called Galileo Regio. The lighter-colored terrain is younger, as shown by the smaller number of craters found there. Note that craters in general appear bright, suggesting that the impacts that formed the craters excavated the surface to reveal ice underneath. (NASA/JPL)

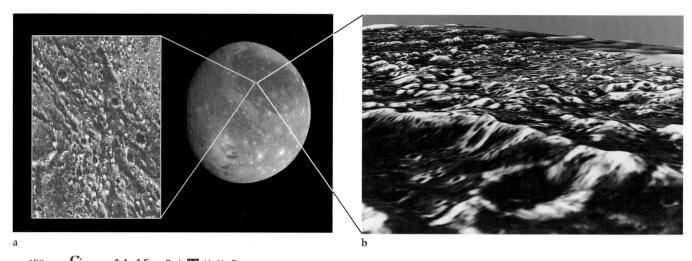

a

b

figure 14-15 R I **V** U X G

WEB LINK 14-12

Ancient Dark Terrain on Ganymede **(a)** The inset is a close-up from the *Galileo* spacecraft of Ganymede's dark-colored Galileo Regio. The area shown is 46 by 64 km (29 by 38 mi), about the same size as the state of Rhode Island. Smaller craters dot the landscape, and long, deep furrows extend from the upper left to the lower right of the image. **(b)** This perspective view of part of Galileo Regio was created from two *Galileo* images by computer image processing. (NASA/JPL)

Ganymede's dark terrain is substantially brighter than the lunar surface.

Voyager and *Galileo* images show noticeable differences between young and old impact craters on Ganymede. The youngest craters are surrounded by bright rays of freshly exposed ice that has been excavated by the impact (see the lower left of Figure 14-14). Older craters are darker, perhaps because of chemical processes triggered by exposure to sunlight.

The largest single feature on Ganymede is the circular island of dark terrain called Galileo Regio, seen at the upper right of Figure 14-14. It covers nearly one-third of the hemisphere of Ganymede that faces away from Jupiter. Figure 14-15 shows that Galileo Regio is strewn with impact craters, reinforcing the idea that the dark terrain is billions of years old. Galileo Regio is also marked by long, deep furrows. These are deformations of Ganymede's crust that have partially erased some of the oldest craters. Since other craters lie on top of the furrows, the stresses that created the furrows must have occurred long ago.

Linear features are also seen in the bright terrain. The *Voyager* missions discovered that the bright terrain is marked by long grooves, some of which extend for hundreds of kilometers and are as much as a kilometer deep. The higher-resolution *Galileo* images found many even finer grooves, as Figure 14-16 shows. These show that the bright terrain has been subjected to a variety of tectonic stresses. *Galileo* also found a number of small craters overlying the grooves, some of which are visible in Figure 14-16. The density of craters shows that the bright terrain is also rather old, although not so old as dark terrain such as Galileo Regio.

After the *Voyager* missions, it was thought that the bright terrain represented fresh ice that had flooded through cracks

in Ganymede's crust, analogous to the maria on our Moon. However, the high-resolution *Galileo* images of this terrain showed no evidence for the kinds of surface features that would be expected if this terrain was volcanic in origin. A more recent idea is that the bright terrain has been badly frac-

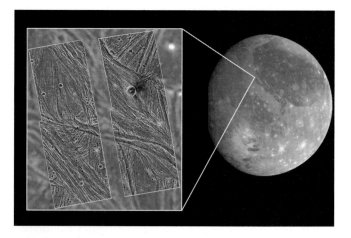

figure 14-16 R I **V** U X G

WEB LINK 14-13

Bright Terrain on Ganymede The inset shows an area of Ganymede's bright terrain adjacent to Galileo Regio. The two tilted strips are *Galileo* images; these are shown on top of the best *Voyager* image of the same area, emphasizing the tremendous improvement in resolution afforded by *Galileo*. The bright terrain is marked by long grooves going in different directions. The grooves must be quite ancient, since numerous impact craters have formed on top of them. The area shown in the inset is 120 by 100 km (75 by 68 mi). (NASA/JPL)

tured by the stresses that produced the grooves. Fracturing ice makes it more reflective; to see this, put a few cubes of ice in a glass, then fill the glass with warm water to make the ice cubes crack. The fracture lines in each cube are clearly visible because they reflect light. In the same way, the fracturing of Ganymede's bright terrain can explain why it is about 25% more reflective than the dark terrain.

The furrows in the dark terrain and the grooves in its bright terrain suggest that Ganymede was once geologically active. But no one expected to find geologic activity on Ganymede today. It is too small to retain much of the heat left over from its formation or from radioactive elements, and it orbits too far from Jupiter to have any substantial tidal heating. So, it came as quite a shock when *Galileo* discovered that Ganymede has its own magnetic field, and that this field is twice as strong as that of Mercury. This field is sufficiently strong to trap charged particles, giving Ganymede its own "mini-magnetosphere" within Jupiter's much larger magnetosphere.

The presence of a magnetic field shows that electrically conducting material must be in motion within Ganymede, which means that the satellite must still have substantial internal heat. A warm interior causes differentiation, and *Galileo* measurements indeed showed that Ganymede is highly differentiated. It has a metallic core some 500 km in radius, surrounded by a rocky mantle and by an outer shell of ice some 800 km thick.

One explanation for Ganymede's internal heat is that gravitational forces from the other Galilean satellites may have affected its orbit. While Ganymede's present-day orbit around Jupiter is quite circular, calculations show that the orbit could have been more eccentric in the past. The resulting large variations in the distance between Jupiter and Ganymede would have caused substantial tidal heating of Ganymede's interior. It may have retained a substantial amount of that heat down to the present day.

In contrast to Io, Europa, and Ganymede is Callisto, Jupiter's outermost Galilean satellite. Images from *Voyager* showed that Callisto has numerous impact craters scattered over an icy crust (Figure 14-17). Callisto's ice is not as reflective as that on Europa or Ganymede, however; the surface appears to be covered with some sort of dark mineral deposit.

CAUTION! Figure 14-17, as well as the images that accompany Table 14-1, shows that Callisto is the darkest of any of the Galilean satellites. Nonetheless, its surface is actually more than twice as reflective as that of our own Moon. Callisto's surface may be made of "dirty" ice, but even dirty ice reflects more light than the gray rock found on the Moon. If you could somehow replace our own Moon with Callisto, a full moon as seen from Earth would be 4 times brighter than it is now (thanks also in part to Callisto's larger size).

The *Voyager* images led scientists to conclude that Callisto is a dead world, with an ancient surface that has never been reshaped by geologic processes. There is no simple relationship between Callisto's orbital period and those of the other Galilean satellites, so it would seem that Callisto never experienced any tidal heating and hence was unable to power any geologic processes at its surface. But once the *Galileo* space-

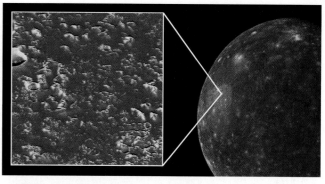

figure 14-17   R I **V** U X G

Callisto Numerous craters pockmark Callisto's icy surface, as seen in this mosaic of views from *Voyager 1*. The inset, a high-resolution *Galileo* image, shows a portion of a huge impact basin called Valhalla. Most of the very smallest craters in this close-up view have been completely obliterated, and the terrain between the craters is blanketed by dark, dusty material. The area shown in the inset is about 11 km (7 mi) across. (Left: Arizona State University/JPL/NASA; right: JPL/NASA)

craft went into orbit around Jupiter and began returning high-resolution images, scientists realized that Callisto's nature is not so simple. In fact, Callisto proves to be the most perplexing of the Galilean satellites.

One curious discovery from *Galileo* is that while Callisto has numerous large craters, there are very few with diameters less than 1 km. Such small craters are abundant on Ganymede, which has probably had the same impact history as Callisto. This implies that most of Callisto's small craters have been eroded away. How this could have happened is not known. Another surprising result is that Callisto's surface is covered by a blanket of dark, dusty material (see the inset in Figure 14-17). Where this material came from, and how it came to be distributed across the satellite's surface, is not understood.

Callisto's most unexpected feature is that, like Europa, it appears to have an induced magnetic field that varies as the satellite moves through different parts of Jupiter's magnetosphere. Callisto must have a layer of electrically conducting material in order to generate such a field. One model suggests that this material is in the form of a subsurface ocean like Europa's but only about 10 km (6 mi) deep. However, without tidal heating, Callisto's interior is probably so cold that a layer of liquid water would quickly freeze. One proposed explanation is that Callisto's ocean is a mixture of water and a form of "antifreeze." In cold climates on Earth, antifreeze is added to the coolant in automobile radiators to lower the liquid's freezing temperature and prevent it from solidifying. Ammonia in Callisto's proposed subsurface ocean could play the same role.

The presence of a liquid layer beneath Callisto's surface is difficult to understand. Even with ammonia acting as an antifreeze, this layer cannot be too cold or it would solidify. This argues for a relatively *warm* interior. But *Galileo* measurements suggest that Callisto's interior is not highly differentiated. This implies that the interior is *cold*, and may never

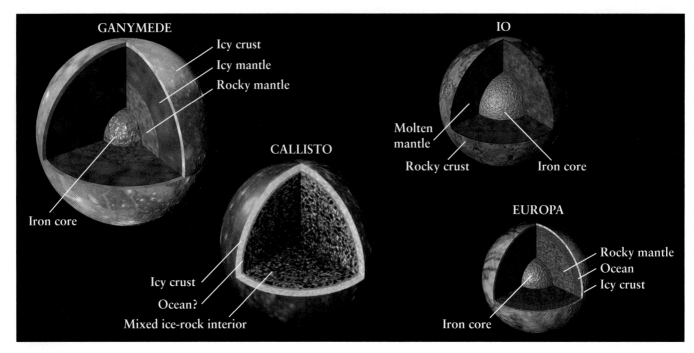

figure 14-18

Interiors of the Galilean Satellites These cross-sectional diagrams show the probable internal structures of the four Galilean satellites, based on information from the *Galileo* mission. (NASA/JPL)

have been heated enough to undergo complete chemical differentiation. The full story of Callisto's interior, it would seem, has yet to be told.

Although Ganymede and Callisto have many differences, one feature they share is an extremely thin and tenuous atmosphere. The molecules that make up Ganymede's atmosphere are oxygen (O_2) and ozone (O_3). These are thought to be released from gas bubbles trapped in Ganymede's surface ice. On Callisto, by contrast, the atmosphere is composed of carbon dioxide (CO_2). This gas may have evaporated from Callisto's poles, where temperatures are low enough to form CO_2 ice.

The diagrams in Figure 14-18 summarize the probable interior structures of the Galilean satellites. They demonstrate the remarkable variety of these terrestrial worlds.

14-8 The *Voyager* spacecraft discovered several small satellites and a ring around Jupiter

Besides the Galilean satellites, Jupiter has 24 other small satellites. (Eleven of these were discovered in 2000. These discoveries had not been confirmed as of early 2001.) Each is named for a mythological character associated with Jupiter. Table 14-2 gives some details of the satellites' orbits.

The four innermost of these satellites lie within Io's orbit. As Figure 14-19 shows, these satellites have irregular shapes

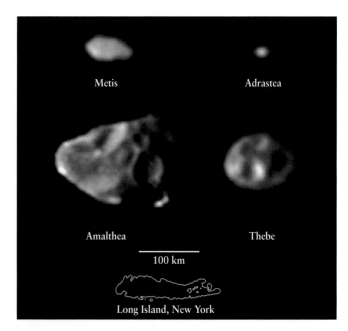

figure 14-19 R I **V** U X G

Jupiter's Four Inner Satellites First observed in 1892, Amalthea was the last satellite of any planet to be discovered without the aid of photography. Metis, Adrastea, and Thebe, by contrast, are so small that they were not discovered until 1979. These images from the *Galileo* spacecraft are the first to show these three satellites as anything more than points of light. (NASA/JPL; Cornell University)

table 14-2	Jupiter's Family of Satellites			

	Average radius of orbit		Orbital period (days)	Year of discovery
	(km)	(Jupiter radii)		
Inner satellites:				
Metis	127,960	1.7922	0.2948	1979
Adrastea	128,980	1.8065	0.2983	1979
Amalthea	181,300	2.539	0.4981	1892
Thebe	221,900	3.108	0.6745	1979
Galilean satellites:				
Io	421,600	5.905	1.769	1610
Europa	670,900	9.397	3.551	1610
Ganymede	1,070,000	14.99	7.155	1610
Callisto	1,883,000	26.37	16.689	1610
Outer satellites:				
Leda	11,094,000	155.4	238.72	1974
Himalia	11,480,000	160.8	250.57	1904
Lysithea	11,720,000	164.2	259.22	1938
Elara	11,737,000	164.4	259.65	1905
Ananke	21,200,000	296.9	631[R]	1951
Carme	22,600,000	316.5	692[R]	1938
Pasiphae	23,500,000	329.1	735[R]	1908
Sinope	23,700,000	331.9	758[R]	1914
S/1999 J1	24,200,000	338.5	768[R]	2000

Note: The superscript R on the orbital period of a satellite means that it orbits Jupiter in a retrograde direction. Eleven additional satellites (S/2000 J1 through S/2000 J11) were discovered in 2000. Their average distances from Jupiter range from 7 to 24 million km, and their sizes range from 3 to 8 km. Nine are in retrograde orbits. The discoveries of these satellites had not been confirmed as of this writing (early 2001), so they do not appear in the above table.

much like a potato (or an asteroid). The largest of the inner satellites, Amalthea, measures about 270 by 150 km (170 by 95 mi)—about 10 times larger than the moons of Mars, but only about one-tenth the size of the Galilean satellites. Amalthea has a distinct reddish color due to sulfur that Jupiter's magnetosphere removed from Io's volcanic plumes, then deposited onto Amalthea's surface.

Jupiter's three other inner satellites are even smaller than Amalthea. They were first discovered in images from *Voyager 1* and *Voyager 2*. The *Voyager*s also made the quite unexpected discovery that Jupiter is encircled by a very faint system of rings (Figure 14-20). These rings are quite unlike the rings of Saturn, which can easily be seen with even a small telescope. The difference is that Saturn's rings are made up of a great many icy, reflective objects a centimeter or larger in size, as we will discuss in Chapter 15. By contrast, Jupiter's rings are composed of particles of rock that have an average size of only about 1 μm (= 0.001 mm = 10^{-6} m) and that reflect very little light. The dimness of these rings makes them challenging to

observe from Earth, which explains why their presence was first revealed by a spacecraft. The ring particles originate from meteorite impacts on the four small, inner satellites.

The rings, the four inner satellites, and the Galilean satellites all orbit Jupiter in the plane of the planet's equator. Furthermore, these are all **prograde orbits,** which means that these objects orbit Jupiter in the same direction as Jupiter's rotation. This is what we would expect if the ring and inner satellites formed from the same primordial, rotating cloud as Jupiter itself. (In a similar way, the planets all orbit the Sun in nearly the same plane and in the same direction as the Sun's rotation. As we saw in Section 7-7, this reinforces the idea that the Sun and planets formed from the same solar nebula.)

In contrast, the 20 outer satellites all circle Jupiter along orbits that are inclined at steep angles to the planet's equatorial plane. All are quite small, with estimated diameters from 3 km for the tiniest to 185 km for the largest. These outer satellites are thought not to have formed along with Jupiter; rather, they are probably wayward

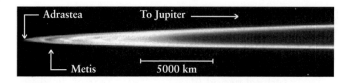

Adrastea   To Jupiter →

Metis   5000 km

figure 14-20 R I **V** U X G

Jupiter's Main Ring This view from the *Galileo* spacecraft shows Jupiter's main ring almost edge-on. The ring's outer radius is close to the orbit of Adrastea, the second closest of Jupiter's moons. Not shown in this image is an even larger and even more tenuous pair of "gossamer rings," one of which extends out to the orbit of Amalthea and the other out to Thebe's orbit. (NASA/JPL; Cornell University)

asteroids captured by Jupiter's powerful gravitational field. One piece of evidence for this is that 14 of the outermost satellites are in **retrograde orbits:** They orbit Jupiter in a direction opposite the planet's rotation. Calculations show that it is easier for Jupiter to capture a satellite into a retrograde orbit than a prograde one.

Saturn, Uranus, and Neptune also have rings and families of satellites whose orbits lie in the equatorial planes of their planets. All of these other planets' rings and most of their satellites are in prograde orbits, suggesting that they, too, formed at the same time as their planets. We will see in Chapters 15 and 16 that these other systems of rings and satellites are no less remarkable than the diverse collection of objects in orbit around Jupiter.

KEY WORDS

bright terrain (Ganymede), p. 313

dark terrain (Ganymede), p. 313

Galilean satellites, p. 303

Io torus, p. 310

ice rafts (Europa), p. 312

occultation, p. 304

prograde orbit, p. 317

retrograde orbit, p. 318

tidal heating, p. 307

ultramafic lava, p. 309

KEY IDEAS

Nature of the Galilean Satellites: The four Galilean satellites orbit Jupiter in the plane of its equator. All are in synchronous rotation.

• The orbital periods of the three innermost Galilean satellites, Io, Europa, and Ganymede, are in the ratio 1:2:4.

• The two innermost Galilean satellites, Io and Europa, have roughly the same size and density as our Moon. They are composed principally of rocky material. The two outermost Galilean satellites, Ganymede and Callisto, are roughly the size of Mercury. Lower in density than either the Moon or Mercury, they are made of roughly equal parts ice and rock.

• The Galilean satellites probably formed in a similar fashion to our solar system but on a smaller scale.

Io: Io is covered with a colorful layer of sulfur compounds deposited by frequent explosive eruptions from volcanic vents. These eruptions resemble terrestrial geysers.

• The energy to heat Io's interior and produce the satellite's volcanic activity comes from tidal forces that flex the satellite. This tidal flexing is aided by the 1:2:4 ratio of orbital periods among the inner three Galilean satellites.

• The Io torus is a ring of electrically charged particles circling Jupiter at the distance of Io's orbit. Interactions between this ring and Jupiter's magnetic field produce strong radio emissions.

• Io may have a magnetic field of its own.

Europa: While composed primarily of rock, Europa is covered with a smooth layer of water ice.

• The surface has hardly any craters, indicating a geologically active history. Other indications are a worldwide network of long cracks and ice rafts that indicate a subsurface layer of liquid water or soft ice.

• As for Io, tidal heating is responsible for Europa's internal heat.

• An ocean may lie beneath Europa's frozen surface. Minerals dissolved in this ocean may explain Europa's induced magnetic field.

Ganymede: The heavily cratered, icy surface of Ganymede is composed of islands of dark, ancient surface separated by regions of heavily grooved, lighter-colored, younger terrain.

• Ganymede is highly differentiated, and probably has a metallic core. It has a surprisingly strong magnetic field and a magnetosphere of its own.

• While there is at present little tidal heating of Ganymede, it may have been heated in this fashion in the past.

Callisto: Callisto has a heavily cratered crust of water ice. The surface shows little sign of geologic activity, because there was never any significant tidal heating of Callisto.

However, some unknown processes have erased the smallest craters and blanketed the surface with a dark, dusty substance.

• Magnetic field data seems to suggest that Callisto has a shallow subsurface ocean.

Other Satellites and Jupiter's Rings: Jupiter has a total of 16 satellites, including the four Galilean satellites.

• Inside Io's orbit are four small satellites and a faint system of dust rings. Like the Galilean satellites, these orbit in the plane of Jupiter's equator.

• Nine small satellites move in much larger orbits that are noticeably inclined to the plane of Jupiter's equator. Five of these orbit in the direction opposite to Jupiter's rotation.

REVIEW QUESTIONS

1. Why can't the Galilean satellites be seen with the naked eye?

2. In what ways does the system of Galilean satellites resemble our solar system? In what ways is it different?

3. No spacecraft from Earth has ever landed on any of the Galilean satellites. How, then, can we know anything about the chemical compositions of these satellites?

4. In what ways did the formation of the Galilean satellites mimic the formation of the planets? In what ways were the two formation processes different?

5. All of the Galilean satellites orbit Jupiter in the same direction. Furthermore, the planes of their orbits all lie within 0.5° of Jupiter's equatorial plane. Explain why these observations are consistent with the idea that the Galilean satellites formed from a "Jovian nebula."

6. What is the source of energy that powers Io's volcanoes? How is it related to the orbits of Io and the other Galilean satellites?

7. Io has no impact craters on its surface, while our Moon is covered with craters. What is the explanation for this difference?

8. Long before the *Voyager* flybys, Earth-based astronomers reported that Io appeared brighter than usual for the few hours after it emerged from Jupiter's shadow. From what we know about the material ejected from Io's volcanoes, suggest an explanation for this brief brightening of Io.

9. Despite all the gases released from its interior by volcanic activity, Io does not possess a thick atmosphere. Explain why not.

10. How do lavas on Io differ from typical lavas found on Earth? What does this difference tell us about Io's interior?

11. What is the Io torus? What is its source?

12. What is the origin of the electric current that flows through Io?

13. How was the *Galileo* spacecraft used to determine the internal structure of Io and the other Galilean satellites?

14. What surface features on Europa provide evidence for geologic activity?

15. What is the evidence for an ocean of liquid water beneath Europa's icy surface? What is the evidence that substances other than water are dissolved in this ocean?

16. What aspects of Europa lead scientists to speculate that life may exist there?

17. Why is ice an important constituent of Ganymede and Callisto, but not of the Earth's Moon?

18. How do scientists know that the dark terrain on Ganymede is younger than the bright terrain?

19. In what ways is Ganymede like our own Moon? In what ways is it different? What are the reasons for the differences?

20. Why were scientists surprised to learn that Ganymede has a magnetic field? What does this field tell us about Ganymede's history?

21. Why are numerous impact craters found on Ganymede and Callisto but not on Io or Europa?

22. Describe the surprising aspects of Callisto's surface and interior that were revealed by the *Galileo* spacecraft. Why did these come as a surprise?

23. Compare and contrast the surface features of the four Galilean satellites. Discuss their relative geological activity and the evolution of these four satellites.

24. The larger the orbit of a Galilean satellite, the less geologic activity that satellite has. Explain why.

25. Explain how the 1:2:4 ratio of the orbital periods of Io, Europa, and Ganymede is related to the geologic activity on these satellites.

26. How would you account for the existence of the satellites of Jupiter other than the Galilean ones? How would you account for the existence of the ring?

ADVANCED QUESTIONS

Problem-solving tips and tools

Newton's form of Kepler's third law (Box 4-4) relates the masses of two objects in orbit around each other to the period and size of the orbit. The small-angle formula is discussed in Box 1-1. The best seeing conditions on the Earth give a limiting angular resolution of $\frac{1}{4}$ arcsec. Because the orbits of the Galilean satellites are almost perfect circles, you can easily calculate the orbital speeds of these satellites from the data listed in Table 14-1 or Table 14-2. Data about Jupiter itself are given in Table 13-1.

27. Using the orbital data in Table 14-1, demonstrate that the Galilean satellites obey Kepler's third law.

28. What is the size of the smallest feature you should be able to see on a Galilean satellite through a large telescope under conditions of excellent seeing when Jupiter is near opposition? How does this compare with the best *Galileo* images, which have resolutions around 25 meters?

29. Astronomers who observed Io with the Hubble Space Telescope in 1992 claimed that they could see features as small as 150 km across. When the observations were made, Io was 4.45 AU from the Earth. What was the angular resolution of the Hubble Space Telescope at this time? (Subsequent repair missions to the Hubble Space Telescope have further improved its resolution.)

30. Using the diameter of Io (3642 km) as a scale, estimate the height to which the plume of Pele rises above the surface of Io in Figure 14-5a. (You will need to make measurements on this figure using a ruler.) Compare your answer to the value given in the figure caption.

31. Jupiter, its magnetic field, and the charged particles that are trapped in the magnetosphere all rotate together once every 10 hours. Io takes 1.77 days to complete one orbit. Using a diagram, explain why particles from Jupiter's magnetosphere hit Io primarily from behind (that is, on the side of Io that trails as it orbits the planet).

32. Assuming material is ejected from Io into Jupiter's magnetosphere at the rate of 1 ton per second (1000 kg/s), how long will it be before Io loses 10% of its mass? How does your answer compare with the age of the solar system?

33. How long does it take for Ganymede to enter or leave Jupiter's shadow? Assume that the shadow has a sharp edge.

34. (a) To an observer floating in Jupiter's Great Red Spot, Io would rise in the east and set in the west, but Metis would rise in the *west* and set in the *east*. Explain how this is possible. (b) At what distance from the center of Jupiter would a satellite have to orbit so that it would neither rise nor set as seen by the observer in (a)?

DISCUSSION QUESTIONS

35. If you could replace our Moon with Io, and if Io could maintain its present amount of volcanic activity, what changes would this cause in our nighttime sky? Do you think that Io could in fact remain volcanically active in this case? Why or why not?

36. In the classic science-fiction film *2010: The Year We Make Contact*, an alien intelligence causes Jupiter to contract so much that nuclear reactions begin at its center. As a result, Jupiter becomes a star like the Sun. Is this possible in principle? Explain your answer.

37. Speculate on the possibility that Europa, Ganymede, or Callisto might harbor some sort of life. Explain your reasoning.

38. Suppose you were planning four missions that would land a spacecraft on each of the Galilean satellites. What kinds of questions would you want these missions to answer? What kinds of data would you want your spacecraft to send back? In view of the different environments on the four satellites, how would the designs of the four spacecraft differ? Be specific about the possible hazards and problems each spacecraft might encounter in landing on the four satellites. If only one of these missions could be funded, which one would you choose? Why?

 WEB/CD-ROM QUESTIONS

39. Search the World Wide Web for recent information about the orbiting *Galileo* spacecraft. What new discoveries has it made about Jupiter's satellites and rings?

40. Various spacecraft missions have been proposed to explore Europa in greater detail. Search the World Wide Web for information about these. How would these missions test for the presence of an ocean beneath Europa's surface?

41. The seventeenth satellite of Jupiter, S/1999 J1, was discovered in 2000 by analyzing observations made in 1999. Search the World Wide Web for information about how this satellite was discovered. How was it determined that S/1999 J1 is actually in orbit around Jupiter?

 42. **The Surface of Ganymede.** Access and view the video "Jupiter's Moon Ganymede" in Chapter 14 of the *Universe* web site or CD-ROM. Describe the different surface features that you see, and explain how each type of feature was formed.

OBSERVING PROJECTS

Observing tips and tools

You can easily find the apparent positions of the Galilean satellites for any date and time using the *Starry Night* software on the CD-ROM that accompanies this textbook. For even more detailed information about satellite positions, consult the "Satellites of Jupiter" section in the *Astronomical Almanac* for the current year.

43. Observe Jupiter through a pair of binoculars. Can you see all four Galilean satellites? Make a drawing of what you observe. If you look again after an hour or two, can you see any changes?

44. Observe Jupiter through a small telescope on three or four consecutive nights. Make a drawing each night showing the positions of the Galilean satellites relative to Jupiter. Record the time and date of each observation.

Consult the sources listed above in the "Observing tips and tools" to see if you can identify the satellites by name.

45. Make arrangements to observe an eclipse, a transit, or an occultation of one of the Galilean satellites. Consult a listing of such phenomena in the "Satellites of Jupiter" section in the *Astronomical Almanac* for the current year. Choose the phenomenon you would like to see and calculate its scheduled time by converting the universal time given in the *Astronomical Almanac* to your local time.

46. If you are fortunate enough to have access to a large telescope with a primary mirror 1 meter or more in diameter, make arrangements to view Jupiter through that telescope. Describe what you see. Under conditions of excellent seeing when Jupiter is near opposition, the Galilean satellites should look like tiny discs rather than starlike pinpoints of light.

 47. Use the *Starry Night* program to observe the Galilean satellites of Jupiter. First turn off daylight (select **Daylight** in the **Sky** menu) and show the entire celestial sphere (select **Atlas** in the **Go** menu). Center on Jupiter by using the **Find...** command in the **Edit** menu. Using the controls at the right-hand end of the Control Panel, zoom in or out until the field of view is roughly 30 arcminutes (30′). In the Control Panel, set the time step to 1 hour and click on the "Forward" button (a triangle that points to the right). You will see the four Galilean satellites orbiting Jupiter. (**a**) Are all four satellites ever on the same side of Jupiter? (**b**) Observe the satellites passing in front of and behind Jupiter. (Zoom in as needed.) Explain how your observations tell you that all four satellites orbit Jupiter in the same direction.

The Spectacular Saturnian System

S aturn is arguably the most beautiful world in the solar system. To the unaided eye, it appears as a featureless point of light. But through even a small telescope, Saturn stands out because of the thin, flat rings that encircle it. And when observed at close range, as was done by the *Voyager* spacecraft that flew past Saturn in the early 1980s, the ringed planet is revealed to be as complex and diverse as it is beautiful.

The *Voyager* explorations of Saturn's hydrogen-rich atmosphere and cloud patterns helped scientists appreciate the differences in how Saturn and neighboring Jupiter evolved. Like Jupiter, Saturn has a substantial magnetic field that guides charged particles toward the planet's magnetic poles. These particles cause the glowing auroral rings shown in the accompanying Hubble Space Telescope image. But other aspects of Saturn's atmosphere and interior have taken a very different course than on Jupiter.

The *Voyager* spacecraft also glimpsed surface features on many of Saturn's satellites. Perhaps the most intriguing satellite is Titan, on which the *Voyagers* saw no surface features at all. This is because Titan is enveloped by a hazy, nitrogen-rich atmosphere substantially thicker than our own. There may be rainfall in Titan's atmosphere; if so, it rains liquid hydrocarbons, not water.

Some of Saturn's other, smaller satellites exert gravitational forces that profoundly affect the struc-

WEB LINK 15.1   (J. Trauger, JPL; NASA)   R I V **U** X G

ture and appearance of the rings. Still other satellites show evidence of volcanic activity, but with frigid "lavas" made of water and ammonia. This retinue of moons, along with the elegant system of rings and the distinctive properties of the planet itself, makes the Saturnian system an essential destination on our tour of the solar system.

As you read the sections of this chapter, look for the answers to the following questions.

15-1   Who discovered that Saturn has rings?

15-2   Are Saturn's rings actually solid bands that encircle the planet?

15-3   How uniform and smooth are Saturn's rings?

15-4   How do Saturn's satellites affect the shape of its rings?

15-5   Why are the color variations in Saturn's atmosphere less dramatic than those on Jupiter?

15-6   Why is Saturn more oblate than Jupiter?

15-7   Why does Saturn, like Jupiter, emit more radiation than it receives from the Sun?

15-8   How is it possible for Saturn's moon Titan to have an atmosphere?

15-9   What kinds of geologic activity are seen on Saturn's other satellites?

15-10   What plans are there for future exploration of the Saturnian system?

15-1   Earth-based observations reveal three broad rings encircling Saturn

Saturn is the second largest of the Jovian planets, with a mass greater than that of all seven smaller planets combined. (Table 15-1 lists some basic data about Saturn.) Although it moves around the Sun in an orbit nearly twice the size of Jupiter's, Saturn is so large and so reflective that it appears to the naked eye as a very bright star. But to appreciate Saturn's true majesty, it must be viewed with a telescope (Figure 15-1).

In 1610, Galileo became the first person to see Saturn through a telescope. He saw few details, but he did notice two puzzling lumps protruding from opposite edges of the planet's disk. Curiously, these lumps disappeared in 1612, only to reappear in 1613. Other observers saw similar appearances and disappearances over the next several decades.

In 1655, the Dutch astronomer Christiaan Huygens began to observe Saturn with a better telescope than was available to any of his predecessors. (We discussed in Section 12-2 how Huygens used this same telescope to observe Mars.) On the basis of his observations, Huygens suggested that Saturn was surrounded by a thin, flattened ring. At times this ring was edge-on as viewed from the Earth, making it almost impossible to see. At other times Earth observers viewed Saturn from an angle either above or below the plane of the ring, and the ring was visible, as in Figure 15-1. (The lumps that Galileo saw were the parts of the ring to either side of Saturn, blurred by the poor resolution of his rather small telescope.) Astronomers confirmed this brilliant deduction over the next several years as they watched the ring's appearance change just as Huygens had predicted.

As the quality of telescopes improved, astronomers realized that Saturn's "ring" is actually a *system* of rings, as Figure 15-2 shows. In 1675, the Italian astronomer Gian Domenico Cassini (whom we also met in Section 12-2) discovered a dark division in the ring. The **Cassini division** is an apparent gap about 4500 kilometers wide that separates the outer **A ring** from the brighter **B ring** closer to the planet. In the mid-1800s, astronomers managed to identify the faint **C ring,** or *crepe* ring, that lies just inside the B ring.

Saturn and its rings are best seen when the planet is at or near opposition (Table 15-2). A modest telescope gives a good view of the A and B rings, but a large telescope and excellent observing conditions are needed to see the C ring.

Earth-based views of the Saturnian ring system change as Saturn slowly orbits the Sun. Huygens was the first to understand that this change occurs because the rings lie in the plane of Saturn's equator, and this plane is tilted 27° from the plane of Saturn's orbit. As Saturn orbits the Sun, its rotation axis and the plane of its equator keep the same orientation in space around the orbit. (The same is true for the Earth; see Figure 2-12.) Hence, over the course of a Saturnian year, the rings are viewed from various angles by an Earth-based

Figure 15-1   R I **V** U X G

Saturn Seen from Earth   The astronomer Stephen Larson took 16 photographs of Saturn on the same night in 1974 with the 1.5-m telescope at the Catalina Observatory in Arizona. He then combined them to make this superb image. Note the faint stripes in Saturn's cloudtops as well as the prominent gap in the rings, called the Cassini division. (NASA)

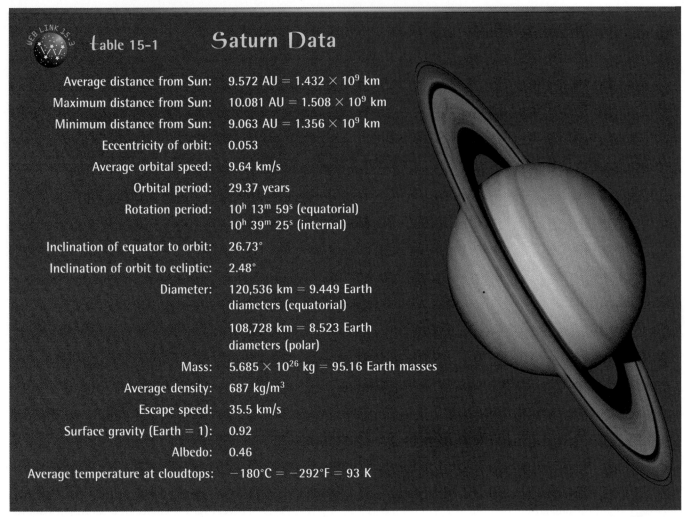

table 15-1 Saturn Data

Average distance from Sun:	9.572 AU = 1.432×10^9 km
Maximum distance from Sun:	10.081 AU = 1.508×10^9 km
Minimum distance from Sun:	9.063 AU = 1.356×10^9 km
Eccentricity of orbit:	0.053
Average orbital speed:	9.64 km/s
Orbital period:	29.37 years
Rotation period:	10^h 13^m 59^s (equatorial)
	10^h 39^m 25^s (internal)
Inclination of equator to orbit:	26.73°
Inclination of orbit to ecliptic:	2.48°
Diameter:	120,536 km = 9.449 Earth diameters (equatorial)
	108,728 km = 8.523 Earth diameters (polar)
Mass:	5.685×10^{26} kg = 95.16 Earth masses
Average density:	687 kg/m^3
Escape speed:	35.5 km/s
Surface gravity (Earth = 1):	0.92
Albedo:	0.46
Average temperature at cloudtops:	−180°C = −292°F = 93 K

(USGS/JPL/NASA) R I **V** U X G

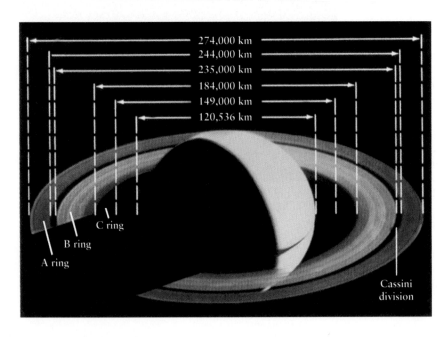

figure 15-2 R I **V** U X G

Saturn's System of Rings This *Voyager 1* image shows many details of Saturn's rings. Saturn's equatorial diameter is labeled, as are the diameters of the inner and outer edges of the rings. Closest to the planet is the C ring, so faint that it is almost invisible in this view. Next outward from Saturn is the bright B ring, whose width is about twice the Earth's diameter. The A ring is the outermost ring that can be seen from Earth. The 4500-km-wide Cassini division lies between the B ring and the A ring. Saturn is visible through the rings (look near the bottom of the image), which shows that the rings are not solid. (NASA/JPL)

table 15-2   Oppositions of Saturn, 2001–2005

Date of opposition	Earth-Saturn distance		Angular diameter (arcsec)
	(AU)	(10^6 km)	
2001 December 3	8.08	1209	20.6
2002 December 17	8.05	1205	20.7
2003 December 31	8.05	1204	20.7
2005 January 13	8.08	1208	20.6

observer (Figure 15-3). At certain times Saturn's north pole is tilted toward the Earth and the observer looks "down" on the "top side" of the rings. Half a Saturnian year later, Saturn's south pole is tilted toward us and the "underside" of the rings is exposed to our Earth-based view.

When our line of sight to Saturn is in the plane of the rings, the rings are viewed edge-on and seem to disappear entirely (see the image at the lower right corner of Figure 15-3). This disappearance indicates that the rings are very thin. In fact, they are thought to be only a few tens of meters thick. In proportion to their diameter, Saturn's rings are thousands of times thinner than the sheets of paper used to print this book.

The last edge-on presentation of Saturn's rings was in 1995–1996, and the next will occur in 2008–2009. Until 2008, Earth observers will see the "underside" of the rings (as in the image at the lower left corner of Figure 15-3).

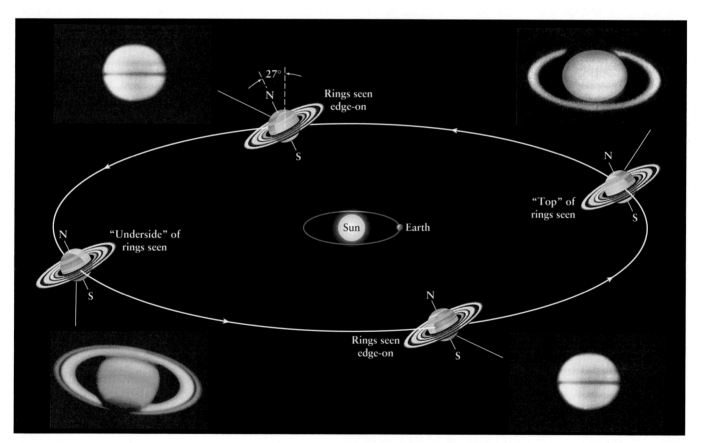

figure 15-3   R I **V** U X G

The Changing Appearance of Saturn's Rings Saturn's rings are aligned with its equator, which is tilted 27° from Saturn's orbital plane. As Saturn moves around its orbit, the rings maintain the same orientation in space, so Earth-based observers see the rings at various angles. The accompanying Earth-based photographs show Saturn at various points in its orbit. Note that the rings seem to disappear entirely when viewed edge-on, which occurs about every 15 years. (Lowell Observatory)

The Spectacular Saturnian System | 325

15-2 Saturn's rings are composed of numerous icy fragments

Astronomers have long known that Saturn's rings could not possibly be solid sheets of matter. In 1857, the Scottish physicist James Clerk Maxwell proved theoretically that if the rings were solid, differences in Saturn's gravitational pull on different parts of the rings would cause the rings to tear apart. He concluded that Saturn's rings are made of "an indefinite number of unconnected particles."

In 1895, James Keeler at the Allegheny Observatory in Pittsburgh became the first to confirm by observation that the rings are not rigid. He did this by photographing the spectrum of sunlight reflected from Saturn's rings. As the rings orbit Saturn, the spectral lines from the side approaching us are blueshifted by the Doppler effect (recall Section 5-9, especially Figure 5-24). At the same time, spectral lines from the receding side of the rings are redshifted. Keeler noted that the size of the wavelength shift increased inward across the rings—the closer to the planet, the greater the shift. This variation in Doppler shift proved that the inner portions of Saturn's rings are moving around the planet more rapidly than the outer portions. Indeed, the orbital speeds across the rings are in complete agreement with Kepler's third law: The square of the orbital period about Saturn at any place in the rings is proportional to the cube of the distance from Saturn's center. (See Section 4-7 and Box 4-4.) This is exactly what would be expected if the rings consisted of numerous tiny "moonlets," or **ring particles,** each individually circling Saturn.

Saturn's rings are quite bright; they reflect 80% of the sunlight that falls on them. (By comparison, Saturn itself reflects 46% of incoming sunlight.) Astronomers therefore long suspected that the ring particles are made of ice and ice-coated rock. This hunch was confirmed in the 1970s, when the American astronomers Gerard P. Kuiper and Carl Pilcher identified absorption features of frozen water in the rings' near-infrared spectrum. The *Voyager 1* and *Voyager 2* spacecraft made even more detailed infrared measurements as they flew past Saturn in 1980 and 1981, respectively. These indicate that the temperature of the rings ranges from −180°C (−290°F) in the sunshine to less than −200°C (−300°F) in Saturn's shadow. Water ice is in no danger of melting or evaporating at these temperatures.

To determine the sizes of the particles that make up Saturn's rings, astronomers analyzed the radio signals received from the *Voyager* spacecraft as the spacecraft passed behind the rings. How effectively radio waves can travel through the rings depends on the relationship between the wavelength and the particle size. The results showed that most of the particles range in size from pebble-sized fragments about 1 cm in diameter to chunks about 5 m across, the size of large boulders. Most abundant are snowball-sized particles about 10 cm in diameter.

It seems reasonable to suppose that all this material is ancient debris that failed to accrete (fall together) into satellites. The total amount of material in the rings is quite small. If Saturn's entire ring system were compressed together to make a satellite, it would be no more than 100 km (60 mi) in diameter. But, in fact, the ring particles are so close to Saturn that they will never be able to form moons.

To see why, imagine a collection of small particles orbiting a planet. Gravitational attraction between neighboring particles tends to pull the particles together. However, because the various particles are at differing distances from the parent planet, they also experience different amounts of gravitational pull from the planet. This difference in gravitational pull is a **tidal force** that tends to keep the particles separated. (We discussed tidal forces in detail in Section 4-8. You may want to review that section, and in particular Figure 4-23.)

The closer a pair of particles is to the planet, the greater the tidal force that tries to pull the pair apart. At a certain distance from the planet's center, called the **Roche limit,** the disruptive tidal force is just as strong as the gravitational force between the particles. (The concept of this limit was developed in the mid-1800s by the French mathematician Edouard Roche.) Inside the Roche limit, the tidal force overwhelms the gravitational pull between neighboring particles, and these particles cannot accrete to form a larger body. Instead, they tend to spread out into a ring around the planet. Indeed, most of Saturn's system of rings visible in Figure 15-2 lies within the planet's Roche limit. Likewise, Jupiter's main ring (see Figure 14-20) lies within the Roche limit of Jupiter.

All large planetary satellites are found outside their planet's Roche limit. If any large satellite were to come inside its planet's Roche limit, the planet's tidal forces would cause the satellite to break up into fragments. (We will see in Chapter 16 that such a catastrophic tidal disruption may be the eventual fate of Neptune's large satellite, Triton.)

It may seem that it would be impossible for any object to hold together inside a planet's Roche limit. But the ring particles inside Saturn's Roche limit survive and do not break apart. The reason is that the Roche limit applies only to objects held together by the gravitational attraction of each part of the object for the other parts. By contrast, the forces that hold a rock or a ball of ice together are chemical bonds between the object's atoms and molecules. These chemical forces are much stronger than the disruptive tidal force of a nearby planet, so the rock or ball of ice does not break apart. In the same way, people walking around on the Earth's surface (which is inside the Earth's Roche limit) are in no danger of coming apart, because we are held together by comparatively strong chemical forces rather than gravity.

We will see in Chapter 16 that the rings of Uranus and Neptune are also made of many individual particles orbiting inside each planet's Roche limit. Unlike the rings of Saturn, however, these rings are quite dim and difficult to see from the Earth.

15-3 Saturn's rings consist of thousands of narrow, closely spaced ringlets

The photograph in Figure 15-1, which was made in 1974, indicates what astronomers understood about the structure of Saturn's rings in the mid-1970s. Each of the A, B, and C rings appeared to be rather uniform, with little or no evidence of any internal structure. Within a few years, however, close-up observations from spacecraft revealed the true complexity of the rings, as well as their chemical composition.

 Three spacecraft—*Pioneer 11, Voyager 1,* and *Voyager 2*—flew past Saturn in 1979, 1980, and 1981, respectively, after each had first made a close flyby of Jupiter. (*Pioneer 10* also flew past Jupiter but did not visit Saturn. Instead, it was aimed so that Jupiter's gravity deflected its path out of the plane of the solar system and into interstellar space.) *Pioneer 11* had only a relatively limited capability to make images, but cameras on board the two *Voyager* spacecraft sent back a number of pictures showing the detailed structure of Saturn's rings.

 Figure 15-4 shows one such *Voyager* image. Some of the features seen in this image were expected, including the **Encke gap,** a 270-km-wide division in the outer A ring observed by the German astronomer Johann Franz Encke in 1838. But to the amazement of scientists, images like the one in Figure 15-4 revealed that the broad A, B, and C rings are not uniform at all but instead consist of hundreds upon hundreds of closely spaced bands or **ringlets.** These ringlets are arrangements of ring particles that have evolved from the combined gravitational forces of neighboring particles, of Saturn's moons, and of the planet itself.

The *Voyager* cameras also sent back high-quality pictures of the narrow **F ring,** which was first detected by *Pioneer 11.* The F ring, which you can see in Figure 15-4, is only about 100 km wide and lies 4000 km beyond the outer edge of the A ring. Close-up views showed that the F ring often consists of several intertwined strands (Figure 15-5). One *Voyager 1* image displayed a total of five strands, each about 10 km across. Planetary scientists suspect that gravitational forces from tiny, unseen moons within the F ring produce this curious and complex structure.

Spacecraft have done more than show Saturn's rings in greater detail; they have also viewed the rings from perspectives not possible from Earth. Through Earth-based telescopes, we can see only the sunlit side of the rings. From this perspective, the B ring appears very bright, the A ring moderately bright, the C ring dim, and the Cassini division dark (see Figure 15-1 and Figure 15-2). The fraction of sunlight reflected back toward the Sun (the albedo) is directly related to the concentration and size of the particles in the ring. The B ring is bright because it has a high concentration of relatively large, icy, reflective particles, whereas the darker Cassini division has a lower concentration of such particles.

The *Voyagers* expanded our knowledge of the ring particles by imaging the shaded side of the rings, which cannot be seen from Earth. Figure 15-6 shows such an image. Because the spacecraft was looking back toward the Sun, this picture shows sunlight that has passed through the rings. The B ring looks darkest in this image because little sunlight gets through its dense collection of particles. If the Cassini division were completely empty, it would look black, because we would then see through it to the blackness of empty space. But the Cassini division looks *bright* in Figure 15-6. This shows that the Cassini division is not empty but contains a relatively

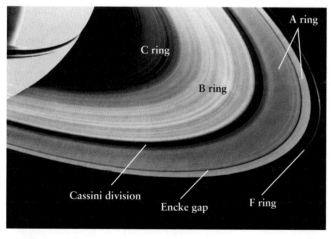

 Figure 15-4 R I **V** U X G

Details of Saturn's Rings The colors in this *Voyager 1* image have been enhanced to emphasize small differences between different portions of the rings. The broad Cassini division is clearly visible, as is the narrow Encke gap in the outer A ring. The very thin F ring lies just beyond the outer edge of the A ring. (NASA/JPL)

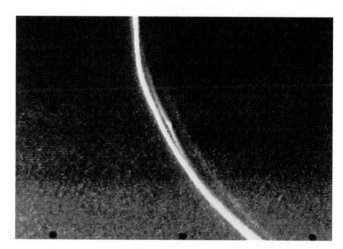

Figure 15-5 R I **V** U X G

Details of the F Ring This *Voyager 1* image shows several strands that together constitute the F ring. The total width of the F ring is about 100 km. The narrowness of the F ring is the result of gravitational forces exerted by two small satellites (see Section 15-4). (NASA/JPL)

ƒigure 15-6 R I **V** U X G

The View from the Far Side of the Rings This false-color view of the side of Saturn's rings away from the Sun was taken by *Voyager 1*. The Cassini division appears white, not black like the empty gap between the A and F rings. Hence, the Cassini division cannot be empty, but must contain a number of relatively small particles that scatter sunlight like dust motes. (NASA/JPL)

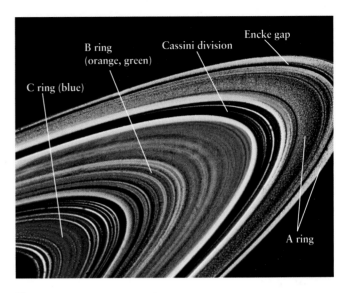

ƒigure 15-7 R I **V** **U** X G

Color Variations in the Rings Computer processing has severely exaggerated the subtle color variations in this *Voyager 2* view of the sunlit side of Saturn's rings (actually a composite of visible-light and ultraviolet images). Note the distinct color variations across both the A and B rings. These color variations are indicative of slight differences in chemical composition among particles in different parts of the rings. (NASA/JPL)

small number of particles. (In the same way, dust particles in the air in your room become visible when a shaft of sunlight passes through them.)

The process by which light bounces off particles in its path is called **light scattering.** The proportion of light scattered in various directions depends upon both the size of the particles and the wavelength of the light. (See Box 5-4 for examples of light scattering on Earth.) Because visible light has a much smaller wavelength than radio waves, light scattering allows scientists to measure the sizes of particles too small to detect using the radio technique described in Section 15-2. As an example, by measuring the amount of light scattered from the rings at various wavelengths and from different angles as the two *Voyager* spacecraft sped past Saturn, scientists showed that the F ring contains a substantial number of tiny particles about 1 μm (= 10^{-6} m) in diameter. This is about the same size as the particles found in smoke.

There are also subtle differences in color from one ring to the next, as the computer-enhanced image in Figure 15-7 shows. Although the main chemical constituent of the ring particles is frozen water, trace amounts of other chemicals—perhaps coating the surfaces of the ice particles—are probably responsible for the different colors. These trace chemicals have not yet been identified.

The color variations in Figure 15-7 suggest that the icy particles do not migrate substantially from one ringlet to another. Had such migration taken place, the color differences would have been smeared out over time. The color differences may also indicate that different sorts of material were added to the rings at different times. In this scenario, the rings did

not all form at the same time as Saturn but were added to over an extended period. New ring material could have come from small satellites that shattered after being hit by a stray asteroid or comet, like the comet that ran into Jupiter in 1994 (see Section 13-4).

In addition to revealing new details about the A, B, C, and F rings, the *Voyager* cameras also discovered three new ring systems: the D, E, and G rings. The drawing in Figure 15-8 shows the layout of all of Saturn's known rings along with the orbits of some of Saturn's satellites. The **D ring** is Saturn's innermost ring system. It consists of a series of extremely faint ringlets located between the inner edge of the C ring and the Saturnian cloudtops. The **E ring** and the **G ring** both lie far from the planet, well beyond the narrow F ring. Both of these outer ring systems are extremely faint, fuzzy, and tenuous. Each lacks the ringlet structure so prominent in the main ring systems. The E ring encloses the orbit of Enceladus, one of Saturn's icy satellites. Some scientists suspect that water geysers on Enceladus are the source of ice particles in the E ring, much as Io's volcanoes produce a torus of material along its orbit around Jupiter (see Section 14-5).

15-4 Saturn's inner satellites affect the appearance and structure of its rings

Why do Saturn's rings have such a complex structure? The answer is that, like any object in the universe, the particles that make up the rings are affected by the force of gravity. Saturn's

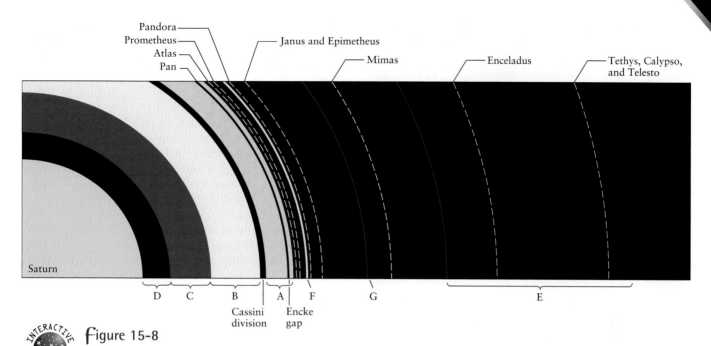

Pandora · Prometheus · Atlas · Pan · Janus and Epimetheus · Mimas · Enceladus · Tethys, Calypso, and Telesto

Saturn

D · C · B · A · F · G · E

Cassini division · Encke gap

*f*igure 15-8

The Arrangement of Saturn's Rings This scale drawing shows the location of Saturn's rings along with the orbits of the 11 inner satellites. Only the A, B, and C rings can readily be seen from the Earth. The remaining rings are very faint and were discovered during spacecraft flybys (although the D and E rings may have been glimpsed beforehand by Earth-based observers). Note that the satellites Janus and Epimetheus share the same orbit, as do the three satellites Tethys, Calypso, and Telesto.

gravitational pull keeps the ring particles in orbit around the planet. But the satellites of Saturn also exert gravitational forces on the rings. These forces can shape the orbits of the ring particles and help give rise to the rings' structure.

Astronomers have long known that one of Saturn's moons, Mimas, has a gravitational effect on its ring system. Mimas is a moderate-sized satellite that orbits Saturn every 22.6 hours. According to Kepler's third law, ring particles in the Cassini division should orbit Saturn approximately every 11.3 hours. Consequently, a given set of ring particles in the Cassini division will find itself between Saturn and Mimas on every second orbit (that is, every 22.6 hours). In other words, particles in the Cassini division are in a 2-to-1 resonance with Mimas. Because of these repeated alignments, the combined gravitational forces of Saturn and Mimas cause small particles to deviate from their original orbits, sweeping them out of the Cassini division. This helps explain why the Cassini division is relatively empty (although not completely so, as Figure 15-6 shows). This effect is somewhat analogous to the rhythmic tidal effects that act among the Jovian satellites Io, Europa, and Ganymede (recall Section 14-1 and Section 14-4).

While Mimas tends to spread ring particles apart, two other satellites exert gravitational forces that herd particles together. The *Voyager* spacecraft discovered these satellites orbiting on either side of the narrow F ring (Figure 15-9). Pandora, the outer of the two satellites, moves around Saturn at a slightly slower speed than the particles in the ring. As ring particles pass Pandora, they experience a tiny gravitational tug that slows them down slightly. Hence, these parti-

cles lose a little energy and fall into orbits a bit closer to Saturn. At the same time, Prometheus, the inner satellite, orbits the planet somewhat faster than the F ring particles. The gravitational pull of Prometheus makes the ring particles

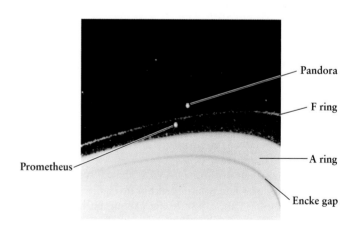

*f*igure 15-9   R I **V** U X G

The F Ring and Its Two Shepherds The tiny satellites Prometheus and Pandora—both less than 150 km across—orbit Saturn on either side of its F ring. The gravitational effects of these two shepherd satellites focus and confine the particles in the F ring to a band about 100 km wide. At the instant shown in this *Voyager 2* image, Prometheus and Pandora were less than 1800 km (1100 mi) apart; Prometheus passed Pandora about 2 hours later. The satellites lap each other every 25 days. (NASA/JPL)

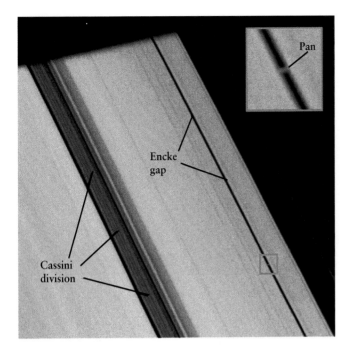

ƒigure 15-10   R I **V** U X G

Pan and the Encke Gap This *Voyager 2* image shows Saturn's tiny moon Pan orbiting within the Encke gap, a division in the A ring (see also Figure 15-4). By exerting gravitational forces on the ring particles to either side, Pan helps maintain the Encke gap's 270-km width. Pan's existence was predicted theoretically from the properties of the Encke gap; it was discovered in 1990 by Mark Showalter of NASA after a computer search of 30,000 *Voyager* images made a decade earlier. (Calvin J. Hamilton and NASA/JPL)

15-5 Saturn's atmosphere extends to a greater depth and has higher wind speeds than Jupiter's atmosphere

If we ignore its distinctive rings, we find that Saturn has many similarities to its larger cousin Jupiter. Saturn is only slightly smaller in equatorial diameter than Jupiter (120,536 km versus 142,984 km), although it has only 30% as much mass. (See Table 7-1 for more comparisons between Jupiter and Saturn.) Both planets exhibit differential rotation, taking less time to complete one rotation at the equator than near the poles. The chemical compositions of their atmospheres are also similar. Earth-based spectroscopic observations along with data from spacecraft confirm that, like Jupiter, Saturn has a hydrogen-rich atmosphere with trace amounts of methane (CH_4), ammonia (NH_3), and water vapor (H_2O). These compounds are the simplest combinations of hydrogen with carbon, nitrogen, and oxygen. And like Jupiter, Saturn's atmosphere probably has three distinct cloud layers: an upper layer of frozen ammonia crystals, a middle layer of crystals of ammonium hydrosulfide (NH_4SH), and a lower layer of water ice crystals.

Although their atmospheres have similar structures and compositions, Saturn and Jupiter are by no means identical in appearance. Saturn's clouds lack the colorful contrast of Jupiter's clouds. Nevertheless, many Earth-based photographs and most spacecraft views show faint belts and zones on Saturn (Figure 15-11).

Saturn has no long-lived storm systems analogous to Jupiter's Great Red Spot (see Section 13-1). But on rare occasions, Earth-based observers have reported seeing storms in Saturn's clouds that last for several days or even months.

speed up a little, giving them some extra energy and nudging them into slightly higher orbits. The competition between the pulls exerted by these two satellites focuses the F ring particles into a well-defined, narrow band. Because of their confining influence, Prometheus and Pandora are called **shepherd satellites**. We will see in Chapter 16 that Uranus is encircled by narrow rings that are likewise kept confined by shepherd satellites.

The Encke gap in Saturn's A ring, shown in Figure 15-4, actually has a satellite orbiting within it (Figure 15-10). This miniature moon, Pan, is just 20 km (12 mi) in diameter. In a fashion similar to the F ring's shepherds, Pan's gravitational forces cause ring particles outside its orbit to move farther out and cause those inside its orbit to move farther inward toward Saturn. In this way, Pan helps to maintain a gap in the rings that is 270 km (170 mi) wide, many times wider than Pan itself. Gravitational forces from Pan also cause wavelike disturbances in the ring material on either side of the gap. Other small satellites presumably orbit within Saturn's rings and cause small gaps of their own, but their existence has yet to be confirmed.

ƒigure 15-11   R I **V** U X G

Saturn's Belts and Zones in True Color Astronomers combined Hubble Space Telescope images made through red, green, and blue filters to produce this view of Saturn's true colors. While belts and zones are clearly visible, there is much less color contrast than on Jupiter (see Figure 13-2). (Space Telescope Science Institute/Hubble Heritage Team)

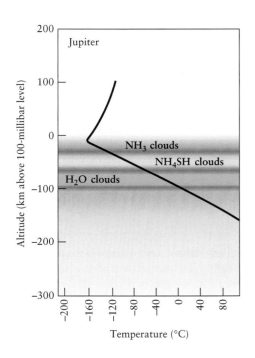

f igure 15-12   R I **V** U X G

A New Storm on Saturn The white arrowhead-shaped feature near Saturn's equator is a giant storm the size of the Earth that appeared in August 1994. The storm apparently formed when warm gases rose upward through Saturn's atmosphere. The lifted gases cooled as they rose, causing gaseous ammonia to solidify into crystals that appear as white clouds. On Earth, similar rapid lifting of air occurs in thunderstorms and can cause droplets of water to solidify into hailstones. The storm lasted several months; it had changed little when this Hubble Space Telescope image was made in December 1994. (Reta Beebe, New Mexico State University; D. Gilmore and L. Bergeron, Space Telescope Science Institute; and NASA)

Figure 15-12 shows one such "white spot" that appeared near Saturn's equator in 1994. About 20 temporary white spots have been seen on Saturn over the past two centuries.

The different appearances of Saturn and Jupiter are related to the different masses of the two planets. Jupiter's strong surface gravity compresses its three cloud layers into a region just 75 km deep in the planet's upper atmosphere. Saturn has a smaller mass and hence weaker surface gravity, so its atmosphere is subjected to less compression. As a result, the same three cloud layers are spread out over a range of nearly 300 km (Figure 15-13). The colors of Saturn's clouds are less dramatic than Jupiter's because deeper cloud layers are partly obscured by the hazy atmosphere above them.

By following features in the Saturnian clouds, scientists have determined the wind speeds in the planet's upper atmosphere. As in Jupiter's atmosphere, there are counterflowing easterly and westerly currents in Saturn's upper atmosphere. However, Saturn's equatorial jet is much broader—and much faster—than that of Jupiter. In fact, wind speeds near Saturn's equator approach 500 m/s (1800 km/h, or 1100 mi/h). This is approximately two-thirds the speed of sound in Saturn's atmosphere, and faster than winds on any other planet. Why Saturn should have such fast-moving winds is not yet completely understood.

15-6 Saturn's internal structure is similar to that of Jupiter, but contains less liquid metallic hydrogen

If we look beneath the atmosphere of Saturn and consider the planet as a whole, we find it to be both similar to and different from Jupiter. Saturn's average density, 687 kg/m³, is only about half that of Jupiter (1326 kg/m³) and is, in fact, the lowest of any planet in the solar system. Saturn's low density means that it is composed mostly of hydrogen and helium. Saturn's smaller mass, less than a third that of Jupiter, means

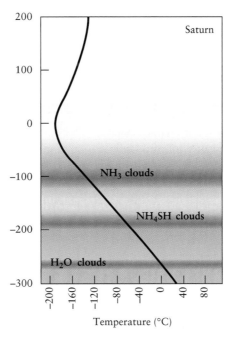

f igure 15-13

The Upper Atmospheres of Jupiter and Saturn These graphs show temperature versus depth in the atmospheres of Jupiter and Saturn, as well as the probable arrangements of the cloud layers. (Zero on the vertical axis is chosen to be the point in each atmosphere where the pressure is 100 millibars, or one-tenth of Earth's atmospheric pressure.) Because Saturn has weaker surface gravity than Jupiter, its atmospheric layers are more spread out. Beneath the cloud layers of both planets, the atmosphere is composed almost entirely of hydrogen and helium. (Adapted from A. P. Ingersoll)

that there is less gravitational force tending to compress its hydrogen and helium. This explains why the density is so low.

Saturn is also the most *oblate*—that is, the most flattened from pole to pole by its rotation—of all the planets (examine Table 7-1 and Figure 15-11). Saturn's oblateness is about 9.8%, or 0.098, which means the planet's diameter at its equator is about 9.8% larger than its diameter from pole to pole. We saw in Section 13-5 that the oblateness of a planet is directly related to its speed of rotation and to how the planet's mass is distributed over its volume. It also depends on the planet's total mass. Saturn rotates at about the same rate as Jupiter but has less mass, and therefore less gravity to pull its material inward. Hence, it follows that Saturn's rotation should cause material at its equator to bulge outward more than is the case on Jupiter.

If Saturn and Jupiter had the same internal structure, we would actually expect Saturn to be even more oblate than it really is. We can account for this discrepancy if Saturn has a different mass distribution than Jupiter. Detailed calculations suggest that about 10% of Saturn's mass is contained in its rocky core, whereas Jupiter's rocky core contains only about 2.6% of the planet's entire mass.

Saturn's massive core and low average density mean that its density must decrease rapidly outward from the core toward the surface. This suggests that Saturn generally resembles Jupiter in its internal structure. It probably has a solid, rocky inner core and an outer core of liquid "ices." These are substances like water, methane, ammonia, and their compounds that are normally solid at the low temperatures found in the outer solar system but that are liquid inside Saturn's warm interior. The core is enveloped by a mantle of liquid helium and liquid metallic hydrogen, which in turn is surrounded by a layer of helium and ordinary hydrogen in both liquid and gaseous states (Figure 15-14).

Compared with Jupiter, Saturn has less mass, less gravity, and less internal pressure. The lower pressure means that less of the hydrogen within Saturn can be compressed to form liquid metallic hydrogen, which is why its mantle is relatively small (see Figure 15-14). As for Jupiter (see Section 13-6), Saturn's liquid metallic hydrogen is thought to be the source of the planet's magnetic field; the smaller volume of the mantle may explain why Saturn's field is substantially weaker than that of Jupiter.

Saturn's magnetosphere contains far fewer charged particles than Jupiter's. There are two reasons for this deficiency. First, Saturn lacks a moon like Io, which ejects a ton of material into Jupiter's magnetosphere each second from its surface and volcanic plumes. Second, and more important, many charged particles are absorbed by the icy particles that make up Saturn's rings. This depletes the concentration of particles in the inner magnetosphere. The charged particles that do exist in Saturn's magnetosphere are concentrated in radiation belts similar to the Van Allen belts in the Earth's magnetosphere. There are enough of these particles, however, to produce aurorae around Saturn's north and south poles (see the image that opens this chapter; compare Figure 8-21a).

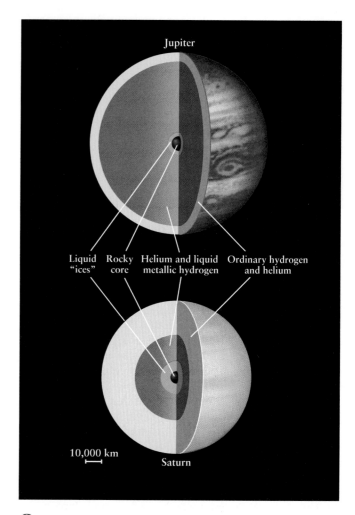

ƒigure 15-14

The Internal Structures of Jupiter and Saturn These diagrams of the interiors of Jupiter and Saturn are drawn to the same scale. The interiors of both planets have four distinct layers. Each planet's rocky core is surrounded by an outer core of liquid "ices," a layer of liquid metallic hydrogen, and a layer of liquid and gaseous molecular hydrogen. Note that Saturn's rocky core is larger than Jupiter's. Saturn also has a smaller volume of liquid metallic hydrogen, which helps explain why Saturn's magnetic field is weaker than that of Jupiter.

15-7 Like Jupiter, Saturn emits more radiation than it receives from the Sun

Both Jupiter and Saturn radiate more energy into space than they receive as sunlight. This means that both planets have internal sources of heat. As we saw in Section 13-3, a tremendous amount of thermal energy was trapped inside Jupiter as the planet accreted from the solar nebula 4.6 billion years ago. Since then, Jupiter has been slowly cooling off as this energy escapes in the form of infrared radiation.

Because Saturn is smaller than Jupiter, it should have begun with less internal heat trapped inside, and it should have radiated that heat away more rapidly. (We saw in Section 9-3 why a small world cools down faster than a large one.) Hence, we would expect Saturn to be radiating very little energy today. In fact, when we take Saturn's smaller mass into account, we find that Saturn releases about 25% *more* energy from its interior on a per-kilogram basis than does Jupiter. How is Saturn able to emit so much energy?

A second puzzle is Saturn's apparent deficiency of helium. Before the *Voyager* flybys in the 1980s, astronomers suspected both Jupiter and Saturn to have compositions similar to that of the original solar nebula, which is assumed to be the same as the Sun's composition. Each of these giant planets is both massive enough and cool enough to have retained all the gases that originally accreted from the solar nebula. Indeed, measurements from the *Galileo* spacecraft show that Jupiter's atmospheric composition is quite similar to that of the Sun: 86.2% hydrogen, 13.6% helium, and 0.2% all other elements. But Saturn is surprisingly different: The *Voyager* spacecraft reported that its atmosphere has a serious helium deficiency. The chemical composition of Saturn's atmosphere is 96.3% hydrogen, 3.3% helium, and 0.4% all other elements. If Saturn formed from the same material as Jupiter, its helium abundance should be about the same. So where did Saturn's helium go?

The explanation of both Saturn's excess heat output and its helium deficiency may stem directly from its being smaller than Jupiter. As a result, Saturn probably did cool more rapidly, triggering a process analogous to the way rain develops here on Earth. When the air is cool enough, humidity in the Earth's atmosphere condenses into raindrops that fall to the ground. On Saturn, however, it is droplets of liquid helium that condense within the planet's cold, hydrogen-rich outer layers. In this scenario, helium is deficient in Saturn's upper atmosphere simply because it has fallen farther down into the planet.

ANALOGY An analogy to helium "rainfall" within Saturn is what happens when you try to sweeten tea by adding sugar. If the tea is cold, the sugar does not dissolve well and tends to sink to the bottom of the glass even if you stir the tea with a spoon. But if the tea is hot, the sugar dis-solves with only a little stirring. In the same way, it is thought that the descending helium droplets once again dissolve in hydrogen once they reach the warmer depths of Saturn's interior. Jupiter's entire interior is warm enough to permit helium to dissolve in hydrogen, which is why helium "rainfall" seems not to take place there.

As the helium droplets descend, their gravitational energy is converted into thermal energy (heat) that eventually escapes from the planet's surface. This "raining out" of helium from Saturn's upper layers is calculated to have begun 2 billion years ago. The amount of thermal energy released by this process adequately accounts for the extra heat radiated by Saturn since that time.

15-8 Titan has a thick, opaque atmosphere rich in methane, nitrogen, and hydrocarbons

WEB LINK 15-9 Unlike Jupiter, Saturn has only one large satellite that is comparable in size to our own Moon. This satellite, Titan, has a diameter of 5150 km. Of all the satellites, only Jupiter's satellite Ganymede (see Section 14-7) is larger. Like Ganymede, Titan has a low density that suggests it is made of a mixture and ice and rock. (See Table 7-2 for a comparison between Titan and the six other large satellites of our solar system.)

Titan (Figure 15-15) was discovered by Christiaan Huygens in 1665, but its unique features were not revealed until the twentieth century. By the early 1900s, scientists had begun to suspect that Titan might have an atmosphere, because it is cool enough and massive enough to retain heavy gases. (Box 7-2 explains the criteria for a planet or satellite to retain an atmosphere.) These suspicions were confirmed in 1944, when Gerard Kuiper discovered the spectral lines of methane in sunlight reflected from Titan. (We discussed Titan's spectrum and its interpretation in Section 7-3.) Titan is now known to be the only satellite in the solar system with an appreciable atmosphere.

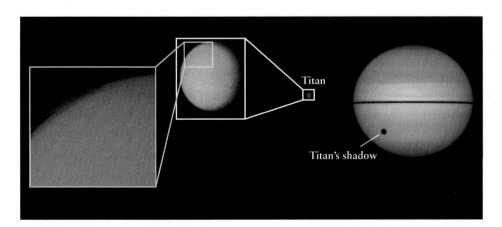

Figure 15-15 R I **V** U X G

Titan The Hubble Space Telescope image at far right shows Titan and the shadow it casts on Saturn's clouds. (This image was recorded in 1995, when the rings appeared nearly edge-on.) The *Voyager* image in the middle shows the thick, unbroken, nearly featureless haze that surrounds Titan. At far left is a color-enhanced *Voyager* image that shows a bluish haze extending well above the satellite's visible edge, or limb. (Left: JPL/NASA; center: NASA; right: Erich Karkoschka, LPL/STScI/NASA)

Because of its atmosphere, Titan was a primary target for the *Voyager* missions. To the chagrin of mission scientists and engineers, however, *Voyager 1* spent hour after precious hour of its limited observation time sending back featureless images (see Figure 15-15). Titan's atmosphere is so thick—about 200 km (120 mi) deep, or about 10 times the depth of the Earth's troposphere—that it blocks any view of the surface. The haze surrounding Titan is so dense that little sunlight penetrates to the ground; noon on Titan is probably no brighter than a moonlit midnight on Earth.

By examining how the radio signal from *Voyager 1* was affected by passing through Titan's atmosphere, scientists inferred that the atmospheric pressure at Titan's surface is 50% greater than the atmospheric pressure at sea level on the Earth. Titan's surface gravity is weaker than that of Earth, so to produce this pressure Titan must have considerably more gas in its atmosphere than Earth. Indeed, calculations show that about 10 times more gas (by mass) lies above a square meter of Titan's surface than above a square meter of Earth.

Voyager data suggest that roughly 90% of Titan's atmosphere is nitrogen. Most of this nitrogen probably came from ammonia (NH_3), a compound of nitrogen and hydrogen that is quite common in the outer solar system. Ammonia is easily broken into hydrogen and nitrogen atoms by the Sun's ultraviolet radiation. Titan's gravity is too weak to retain hydrogen atoms (see Box 7-2), so these atoms escape into space and leave the more massive nitrogen atoms behind.

The second most abundant gas on Titan is methane, which is the principal component of the "natural gas" used on Earth as fuel. The interaction of methane with ultraviolet light from the Sun produces a variety of other carbon-hydrogen compounds, or **hydrocarbons.** For example, the *Voyager* infrared spectrometers detected small amounts of ethane (C_2H_6), acetylene (C_2H_2), ethylene (C_2H_4), and propane (C_3H_8) in Titan's atmosphere. Ethane is the most abundant of these byproducts. It condenses into droplets as it is produced and falls to Titan's surface. Over the history of the solar system, enough ethane may have been produced to create rivers and lakes. To test this idea, astronomers have made observations at infrared wavelengths (which can penetrate all the way through Titan's atmosphere) in an effort to map the satellite's surface. Images produced in this way suggest that there may indeed be bodies of liquid hydrocarbons on Titan (Figure 15-16). These images also confirm that Titan, like most satellites in the solar system, is in synchronous rotation.

Nitrogen in Titan's atmosphere can combine with airborne hydrocarbons to produce other compounds, such as hydrogen cyanide (HCN). Hydrogen cyanide, along with other molecules, can join together in long, repeating molecular chains to form substances called **polymers.** Droplets of some polymers remain suspended in the atmosphere, forming an **aerosol.** (Common aerosols on the Earth include fog, mist, and paint sprayed from a can.) The polymers in Titan's airborne aerosol are thought to be responsible for the reddish-brown color seen in Figure 15-15. The heavier polymer particles do not remain airborne but settle down onto Titan's solid surface, probably covering it with a thick layer of sticky, tarlike goo.

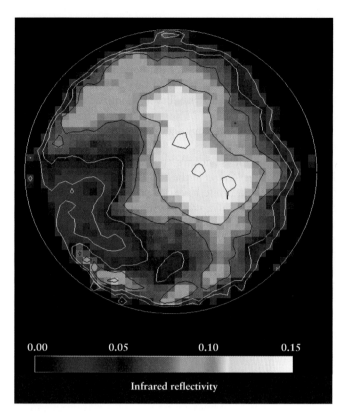

0.00   0.05   0.10   0.15

Infrared reflectivity

ƒigure 15-16 R ■ V U X G

Hydrocarbon Seas on Titan? The false colors in this infrared view of Titan indicate what fraction of infrared light from the Sun is reflected by different parts of the surface. The dark blue patches, with a reflectivity of less than 0.03, or 3%, could be either seas of liquid hydrocarbons or large expanses of solid organic material. The more reflective yellow area may be a "continent" of rock and water ice. Titan's diameter is shown by the white circle. (Seran G. Gibbard, Bruce Macintosh, and Claire Max, IGPP/LLNL)

Some of the compounds of hydrogen, nitrogen, and carbon found in Titan's atmosphere and on its surface are the building blocks of the organic molecules on which life is based. There is little reason to suspect that life exists on Titan, however; its surface temperature of 95 K ($-178°C = -288°F$) is prohibitively cold. Nevertheless, a more detailed study of Titan's chemistry may shed light on the origins of life on the Earth.

15-9 **The icy surfaces of Saturn's six moderate-sized moons provide clues to their histories**

Twenty-eight confirmed moons have been found orbiting Saturn (Table 15-3). All are named for mythological characters associated with Saturn, the Roman god of agriculture. It is helpful to sort these satellites into three groups according to size.

Table 15-3 | Saturn's Satellites

	Distance from center of Saturn		Orbital period (days)	Size (km)	Density (kg/m³)
	(km)	(Saturn radii)			
Pan	133,570	2.22	0.573	20	—
Atlas	137,640	2.28	0.602	20 × 20 × 20	—
Prometheus	139,350	2.31	0.613	140 × 100 × 80	—
Pandora	141,700	2.35	0.629	110 × 90 × 70	—
Epimetheus	151,422	2.51	0.694	140 × 120 × 100	—
Janus	151,472	2.51	0.695	220 × 200 × 160	—
Mimas	185,520	3.08	0.942	392	1400
Enceladus	238,020	3.95	1.370	500	1200
Tethys	294,660	4.89	1.888	1060	1200
Calypso	294,660	4.89	1.888	34 × 28 × 26	—
Telesto	294,660	4.89	1.888	24 × 22 × 22	—
Dione	377,400	6.26	2.737	1120	1400
Helene	377,400	6.26	2.737	36 × 32 × 30	—
Rhea	527,040	8.74	4.518	1530	1300
Titan	1,221,850	20.25	15.945	5150	1880
Hyperion	1,481,000	24.55	21.277	410 × 260 × 220	—
Iapetus	3,561,300	59.02	79.331	1460	1200
Phoebe	12,952,000	214.7	550.48[R]	220	—

This table lists some facts about 18 of Saturn's satellites. In 2000 ten additional small satellites (S/2000 S1 through S/2000 S10) were discovered and confirmed; their orbits are still tentative as of this writing (early 2001) and do not appear in this table.

All the larger satellites are spherical, and their diameters are listed in the Size column. For the smaller satellites, which are not spherical, three dimensions—width, length, and height—are given. The masses, and hence the densities, of the smaller satellites are not yet known.

Of the moons for which rotation rates are known, almost all rotate synchronously (that is, the rotation period is the same as the orbital period, so the moon always keeps the same face toward Saturn.) The exceptions are Phoebe and Hyperion.

The F ring shepherds are Prometheus and Pandora. Janus and Epimetheus are called co-orbital satellites, because they move in almost the same orbit. Tethys, Calypso, and Telesto are also co-orbital satellites, as are Dione and Helene. In the latter two cases, the tiny satellites occupy specific locations along the orbits of the larger moons, where a balance exists between the gravitational pulls of Saturn and the larger moon. We will discuss these locations, called the Lagrangian points, in Chapter 17.

The superscript R on the orbital period of Phoebe means that it orbits Saturn in a retrograde direction.

One planet-sized satellite, Titan, is actually one of the terrestrial worlds of our solar system. With its diameter of 5150 km, Titan is intermediate in size between Mercury and Mars.

Six moderate-sized satellites were all discovered before 1800. These six satellites have properties and diameters that lead us to group them in pairs: Mimas and Enceladus (400 to 500 km in diameter), Tethys and Dione (roughly 1000 km in diameter), and Rhea and Iapetus (about 1500 km in diameter).

Twenty-eight tiny satellites may be either captured asteroids (as in the case of Phoebe, Saturn's outermost satellite) or jagged fragments of ice and rock produced by impacts and collisions. The shepherd satellites, for example, may be such fragments.

The most curious and mysterious of these worlds may be Saturn's six moderate-sized satellites (Figure 15-17). These six satellites have very low average densities of less than 1400 kg/m³. This implies that unlike the satellites of the inner solar system or the Galilean satellites of Jupiter, they are composed primarily of ices (mostly frozen water and ammonia) with very little rock. The motions of these mid-sized satellites are not surprising: All six orbit the planet in the plane of Saturn's equator and in the direction of the planet's rotation, and all six exhibit synchronous rotation. What is surprising is the curious variety of surface features that the *Voyager* spacecraft discovered on these satellites.

Mimas, the innermost of the six middle-sized moons, has the heavily cratered surface we would expect on such a small world. Its most remarkable feature is a crater so large (Figure 15-17a) that the impact that formed it must have come close to shattering Mimas into fragments.

a Mimas   b Enceladus   c Tethys

d Dione   e Rhea   f Iapetus

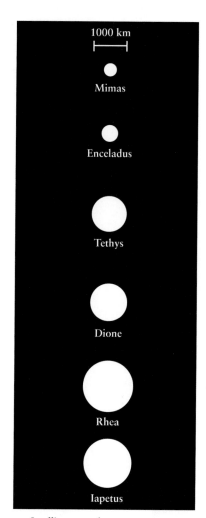

1000 km

Mimas

Enceladus

Tethys

Dione

Rhea

Iapetus

g Satellites to scale

ƒigure 15-17 R I **V** U X G

Six Mid-Sized Satellites The images from the *Voyager 1* and *Voyager 2* spacecraft above show the variety of surface features found on Saturn's six moderate-sized satellites. The satellites are *not* shown to scale (JPL/NASA). **(a)** Mimas, the smallest and innermost of the six, has a huge impact crater 130 km (80 mi) in diameter. Mimas itself is only 400 km (250 mi) in diameter. **(b)** Ice flows and cracks on the surface of Enceladus are strong evidence for recent geologic activity. The youngest, crater-free ice flows are estimated to be less than 100,000 years old. **(c)** A portion of the surface of Tethys (toward the lower right in this image) is more lightly cratered than the rest. Some sort of volcanism involving "lavas" of water and ammonia may have resurfaced this region. **(d)** Bright, wispy streaks crisscross Dione's trailing hemisphere (on the right in this image). The leading hemisphere has a more ordinary cratered surface. **(e)** Rhea has the same sort of difference between hemispheres as Dione. The cratered leading hemisphere is shown here. **(f)** Mysterious Iapetus also has differences between its hemispheres, but of a unique kind: The leading hemisphere (at upper left in this image) is extremely dark, while the trailing hemisphere is highly reflective. **(g)** These outlines show the relative sizes of Saturn's moderate-sized satellites.

VIDEO 15.4

In contrast to Mimas's heavily cratered surface, its neighbor and near twin, Enceladus, has extensive crater-free regions (Figure 15-17b). Apparently some form of geologic activity has resurfaced these areas within the past 100 million years, and may still be under way. Geologic activity may also explain why Enceladus has an amazingly high albedo of 0.95, making it the most reflective large object in our solar system. Some process has freed the satellite's icy surface of rock and dust, whose presence on the other icy satellites is the likeliest explanation for their lower albedos. Evidence for even more dramatic geologic activity comes from studies of Saturn's E ring, which suggest that some kind of eruptions on Enceladus supply particles to this ring (see Section 15-3). However, the *Voyager* spacecraft did not see any such eruptions.

What could be powering the geologic activity on Enceladus? One guess is tidal heating, which provides the internal heat for the inner Galilean satellites of Jupiter (see Section 14-4). The satellite Dione orbits Saturn in 65.7 hours, almost exactly

twice Enceladus's orbital period of 32.9 hours, and the 1:2 ratio of orbital periods should set Enceladus up to be caught in a tidal tug-of-war between Saturn and Dione. Calculations suggest that the amount of tidal heating produced in this way is modest, but should be enough to melt the ices of Enceladus at least part of the time. Curiously, there is also a 1:2 ratio between the orbital periods of Mimas and Saturn's moon Tethys, but the cratered surface of Mimas shows no sign that this moon was ever tidally heated.

Tethys and Dione are the next-largest pair of Saturn's six moderate-sized satellites. Most of Tethys, like Mimas, is heavily cratered. But Tethys also has a curious, roughly circular plain where craters are more sparse than on the rest of the satellite (Figure 15-17c). Like the lunar maria (see Section 9-1), this region has apparently been resurfaced by lava flows. The difference is that the "lava" on Tethys is probably a mixture of water and ammonia, which flows like a viscous fluid at the ultracold temperatures found on Saturn's icy satellites. Dione, slightly larger than Tethys, also has a strange surface dichotomy: The surfaces of its leading hemisphere (that is, the hemisphere facing toward its direction of orbital motion) and trailing hemisphere are quite different. Dione's leading hemisphere is heavily cratered, but its trailing hemisphere is covered by a network of strange, wispy markings (Figure 15-17*d*). These wisps may be troughs and valleys in the icy surface, or they may be fresh deposits of frozen water formed by volcanic outgassing from Dione's interior.

Because Tethys and Dione have about double the diameter of Enceladus, the energy to power their geologic activity probably came from within. These larger satellites retain heat more effectively, including heat from the decay of radioactive materials in each satellite's interior. (Such materials are scarce within the icy satellites of Saturn, but only relatively little heat is needed to power ice volcanism.)

Rhea and Iapetus are the largest and most distant of Saturn's moderate-sized moons. Rhea resembles Dione in that its leading hemisphere is heavily cratered (Figure 15-17*e*); its trailing hemisphere is covered by a network of odd wispy markings. Whatever process produced the wisps on one side of Dione may also have been active on the corresponding side of Rhea.

Iapetus has been known to be unusual ever since its discovery in 1671: Unlike the other satellites, its brightness varies greatly as it orbits Saturn. Images from the *Voyager* spacecraft show extreme differences between the leading and trailing hemispheres of Iapetus (Figure 15-17*f*). The leading hemisphere is as black as asphalt (albedo = 0.05), but the trailing hemisphere is highly reflective (albedo = 0.50), like the surfaces of the other moderate-sized moons.

The dark material covering Iapetus's leading hemisphere may have come from Phoebe, the most distant of Saturn's known satellites. Phoebe moves in a retrograde direction along an orbit tilted well away from the plane of Saturn's equator. This motion leads scientists to suspect that Phoebe is a captured asteroid. Its dark surface does indeed resemble the surfaces of a particular class of carbon-rich asteroids. Bits and pieces of this dark charcoal-like material drifting toward Saturn may have been swept up onto the leading hemisphere of Iapetus. However, a sticking point with this scenario is that the dark material on Iapetus is reddish in color, while Phoebe's surface is not. For now, the two-sided nature of Iapetus remains a mystery.

15-10 A Saturn-orbiting spacecraft and a Titan lander will provide a wealth of new information

Images and data from the *Voyager* spacecraft revolutionized our view of Saturn's complex system of rings and satellites. But there is much still to be learned about the Saturnian system. Of particular interest are the atmosphere and surface of Titan, which may harbor important clues about the origin of life on Earth. The next stage of Saturn exploration is about to begin, thanks to the *Cassini* spacecraft (Figure 15-18).

ƒigure 15-18

Cassini and Huygens The *Cassini* spacecraft will go into orbit around Saturn in 2004. Approximately two stories tall and weighing (on Earth) more than 6 tons, *Cassini* will use its suite of instruments to examine Saturn, its rings, and satellites over at least a four-year period. It will also deploy a probe called *Huygens* (seen at the bottom of this artist's conception), which will land on Titan's surface. (NASA)

Launched in October 1997, *Cassini* is a joint project involving NASA, the European Space Agency, the Italian Space Agency, and scientists from 17 nations. When it arrives at Saturn in 2004, *Cassini* will go into orbit around the giant planet and begin an observation program that will last at least four years. Long-term monitoring of Saturn's atmosphere and rings will allow scientists to better understand their complicated dynamics, and high-resolution imaging of the satellites will allow us to explore their exotic geology. Titan will come in for special attention: *Cassini* will make more than 30 flybys of Titan and will use radar to map the satellite's surface to a resolution of 500 meters.

Before going into orbit around Saturn, *Cassini* will deploy a probe that will enter Titan's atmosphere. This probe, named *Huygens*, will spend 2½ hours descending through the atmosphere on a parachute and making measurements of its surroundings. As it descends, the probe will radio its findings to *Cassini* to be relayed to Earth. If it survives the impact of a 25-km/h (15-mi/h) landing, *Huygens* will give an on-site report on the nature of Titan's surface—be it water ice, solid ammonia, liquid ethane, or hydrocarbon "goo." In this way Titan, one of the most curious worlds in the solar system, may give up some of its secrets.

KEY WORDS

A ring, p. 323

aerosol, p. 334

B ring, p. 323

C ring, p. 323

Cassini division, p. 323

D ring, p. 328

E ring, p. 328

Encke gap, p. 327

F ring, p. 327

G ring, p. 328

hydrocarbon, p. 334

light scattering, p. 328

polymer, p. 334

ring particles, p. 326

ringlets, p. 327

Roche limit, p. 326

shepherd satellite, p. 330

tidal force, p. 326

KEY IDEAS

Appearance of Saturn's Rings: Saturn is circled by a system of thin, broad rings lying in the plane of the planet's equator. This system is tilted away from the plane of Saturn's orbit, which causes the rings to be seen at various angles by an Earth-based observer over the course of a Saturnian year.

Structure of the Rings: Three major, broad rings can be seen from the Earth. The faint C ring lies nearest Saturn. Just outside it is the much brighter B ring, then a dark gap called the Cassini division, and then the moderately bright A ring. Other, fainter rings were observed by the *Voyager* spacecraft.

• The principal rings of Saturn are composed of numerous particles of ice and ice-coated rock ranging in size from a few micrometers to about 10 m.

• Most of the rings exist inside the Roche limit of Saturn, where disruptive tidal forces are stronger than the gravitational forces attracting the ring particles to each other.

• Each of the major rings is composed of a great many narrow ringlets.

• The faint, narrow F ring, which is just outside the A ring, is kept narrow by the gravitational pull of shepherd satellites.

Atmosphere and Internal Structure: Saturn's internal structure and atmosphere are similar to those of Jupiter. However, Saturn's core makes up a larger fraction of its volume, and its liquid metallic hydrogen mantle is shallower than that of Jupiter.

• Saturn's atmosphere contains less helium than Jupiter's atmosphere. This lower abundance may be the result of helium raining downward into the planet.

• If helium rain does fall, the resulting conversion of gravitational energy into thermal energy would account for Saturn's surprisingly strong heat output.

• Saturn has belts and zones like those of Jupiter. The cloud layers in Saturn's atmosphere are spread out over a greater range of altitude than those of Jupiter.

Titan: The largest Saturnian satellite, Titan, is a terrestrial world with a dense nitrogen atmosphere. A variety of hydrocarbons are produced there by the interaction of sunlight with methane. These compounds form an aerosol layer in Titan's atmosphere and possibly cover some of its surface with lakes of ethane.

 Other Satellites: Six moderate-sized moons circle Saturn in regular orbits: Mimas, Enceladus, Tethys, Dione, Rhea, and Iapetus. They are probably composed largely of ice, but their surface features and histories vary significantly. Saturn also has more than twenty much smaller satellites, some of which may be captured asteroids.

REVIEW QUESTIONS

1. When Saturn is at different points in its orbit, we see different aspects of its rings because the planet has a 27° tilt. If the tilt angle were different, would it be possible to see the upper and lower sides of the rings at all points in Saturn's orbit? If so, what would the tilt angle have to be? Explain your answers.

2. Saturn is the most distant of the planets visible without a telescope. Is there any way we could infer this from naked-eye observations? Explain. (*Hint:* Think about how Saturn's position on the celestial sphere must change over the course of weeks or months.)

3. Describe the structure of Saturn's rings. What evidence is there that ring particles do not migrate significantly between ringlets?

4. If Saturn's rings are not solid, why do they look solid when viewed through a telescope?

5. During the planning stages for the *Pioneer 11* mission, when relatively little was known about Saturn's rings, it was proposed to have the spacecraft fly through the Cassini division. Why would this have been a bad idea?

6. The space shuttle and other manned spacecraft orbit the Earth well within the Earth's Roche limit. Explain why these spacecraft are not torn apart by tidal forces.

7. Although the *Voyager* spacecraft did not collect any samples of Saturn's ring particles, measurements from these spacecraft allowed scientists to determine the sizes of the particles. Explain how this was done.

8. Why is the term "shepherd satellite" appropriate for the objects so named? Explain how a shepherd satellite operates.

9. Compare the atmospheres of Jupiter and Saturn. Why does Saturn's atmosphere look "washed out" in comparison to that of Jupiter?

10. Compare the interiors of Jupiter and Saturn. What are the similarities? What are the differences?

11. It has been claimed that Saturn would float if one had a large enough bathtub. Using the mass and size of Saturn given in Table 15-1, confirm that the planet's average density is about 690 kg/m³, and comment on this somewhat fanciful claim.

12. Explain why Saturn is more oblate than Jupiter, even though Saturn rotates more slowly.

13. Both Jupiter and Saturn emit more energy than they receive from the Sun in the form of sunlight. Compare the internal energy sources of the two planets that produce this emission.

14. On a warm, humid day, water vapor remains in the atmosphere. But if the temperature drops suddenly, the water vapor forms droplets, clouds appear, and it begins to rain. Relate this observation to the explanation why there is relatively little helium in Saturn's atmosphere compared to the atmosphere of Jupiter.

15. Describe Titan's atmosphere. What effect has the Sun's ultraviolet radiation had on Titan's atmosphere?

16. Why do scientists suspect that there may be liquid hydrocarbons on Titan?

17. Which of Saturn's moderate-sized satellites show evidence of geologic activity? What might be the energy source for this activity?

18. Explain why debris from Phoebe would be expected to pile up only on the leading hemisphere of Iapetus. (*Hint:* How do the orbits of these two satellites compare? How does the orbital motion of debris falling slowly inward toward Saturn compare with the orbital motion of Iapetus?)

19. Saturn's equator is tilted by 27° from the ecliptic, while Jupiter's equator is tilted by only 3°. Use these data to explain why we see fewer transits, eclipses, and occultations of Saturn's satellites than of the Galilean satellites.

ADVANCED QUESTIONS

Problem-solving tips and tools

The small-angle formula is given in Box 1-1. The Doppler effect is the subject of Section 5-9, while Section 6-3 discusses factors affecting the angular resolution of a telescope. Saturn's satellites, whose orbital parameters are given in Table 15-3, obey Newton's form of Kepler's third law (see Section 4-7 and Box 4-4). For a discussion of escape speed and how planets retain their atmospheres, see Box 7-2.

20. As seen from Earth, the intervals between successive edge-on presentations of Saturn's rings alternate between about 13 years, 9 months, and about 15 years, 9 months. Why do you think these two intervals are not equal?

21. **(a)** Use Newton's form of Kepler's third law to calculate the orbital periods of particles at the outer edge of the A ring and at the inner edge of the B ring.
(b) Saturn's rings orbit in the same direction as Saturn's rotation. If you were floating along with the cloudtops at Saturn's equator, would the outer edge of the A ring and the inner edge of the B ring appear to move in the same or opposite directions? Explain.

 22. This *Voyager 2* close-up image of Saturn's rings shows a number of dark, straight features called *spokes*. As these features orbit around Saturn, they tend to retain their shape like the rigid spokes on a rotating bicycle wheel. (The black dots were added by the *Voyager* camera system to help scientists calibrate the electronic image.) The spokes rotate at the same rate as Saturn's magnetic field and are thought to be clouds of tiny, electrically charged particles that are kept in orbit by magnetic forces. Explain why the spokes could *not* maintain their shape if they were kept in orbit by gravitational forces alone.

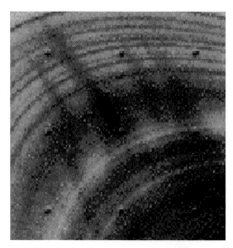

(NASA/JPL) R I **V** U X G

23. It is well known that the Cassini division involves a 2-to-1 resonance with Mimas. Does the location of the Encke gap—133,500 km from Saturn's center—correspond to a resonance with one of the other satellites? If so, which one?

24. (a) An astronaut floating above Saturn's cloudtops would see a blue sky, even though Saturn's atmosphere has a very different chemical composition than Earth's. Explain why. **(b)** The image of Saturn in Figure 15-15 appears bluish around the edges of the planet. Explain why this should be. (*Hint:* For both (a) and (b), see Box 5-4.)

25. Find the escape speed on Titan. What is the limiting molecular weight of gases that could be retained by Titan's gravity? (*Hint:* Use the ideas presented in Box 7-2 and assume an average atmospheric temperature of 95 K.)

26. Many of the gases in the atmosphere of Titan, such as methane, ethane, and acetylene, are highly flammable. Why, then, doesn't Titan's atmosphere catch fire? (*Hint:* What gas in our atmosphere is needed to make wood, coal, or gasoline burn?)

27. Although Jupiter's satellite Ganymede is about the same size and mass as Titan, Ganymede does not have an appreciable atmosphere. Suggest a reason for this difference.

28. At infrared wavelengths, the Hubble Space Telescope can see details on Titan's surface as small as 580 km (360 mi) across. Determine the angular resolution of the Hubble Space Telescope using infrared light. If visible light is used, is the angular resolution better, worse, or the same? Explain your answer.

29. (a) To an observer on Enceladus, what is the time interval between successive oppositions of Dione? Explain your answer. **(b)** As seen from Enceladus, what is the angular diameter of Dione at opposition? How does this compare to the angular diameter of the Moon as seen from Earth (about ½°)?

DISCUSSION QUESTIONS

30. Suppose that Saturn were somehow moved to an orbit around the Sun with semimajor axis 1 AU, the same as the Earth's. Discuss what long-term effects this would have on the planet and its rings.

31. Comment on the suggestion that Titan may harbor life-forms.

32. Jupiter's satellite Io and Saturn's satellite Enceladus are both geologically active, and both are in 2-to-1 resonances with other satellites. However, the amount of geologic activity of Enceladus is far less than on Io. Discuss some possible reasons for this difference.

33. Imagine that you are in charge of planning the *Cassini* orbital tour of the Saturnian system. In your opinion, what objects in the system should be examined, what data should be collected, and what kinds of questions should the mission attempt to answer?

WEB/CD-ROM QUESTIONS

34. Search the World Wide Web, especially the web sites for NASA's Jet Propulsion Laboratory and the European Space Agency, for information about the current status of the *Cassini* mission. When will *Cassini* arrive at Saturn? What are the current plans for its tour of Saturn's satellites? What ideas are being considered for the *Cassini* extended mission, to begin in 2008?

35. The two *Voyager* spacecraft were launched from Earth along a trajectory that took them directly to Jupiter. The force of Jupiter's gravity then gave the two spacecraft a "kick" that helped push them onward to Saturn. The much larger *Cassini* spacecraft, by contrast, was first launched on a trajectory that took it past Venus. Search the World Wide Web, especially the web sites for NASA's Jet Propulsion Laboratory and the European Space Agency, for information about the trajectory that *Cassini* is taking through the solar system. Explain why this trajectory is so different than that of the *Voyager*s.

36. In 2000 astronomers reported the discovery of several new moons of Saturn. Search the World Wide Web for information about these. How did astronomers discover them? Have the observations been confirmed? How large are these moons? What sorts of orbits do they appear to follow?

37. The Rotation Rate of Saturn. Access and view the video "Saturn from the Hubble Space Telescope" in Chapter 15 of the *Universe* web site or CD-ROM. The total amount of time that actually elapses in this video is 42.6 hours. Using this information, identify and follow an atmospheric feature and determine the rotation period of Saturn. How does your answer compare with the value given in Table 15-1?

OBSERVING PROJECTS

Observing tips and tools

To determine the best time of night to view Saturn, consult the current issue of *Sky & Telescope* or *Astronomy* magazine or their web sites or use the *Starry Night* software on the CD-ROM that accompanies this textbook. If your goal is to view Saturn's satellites, consult the section entitled "Satellites of Saturn" in the *Astronomical Almanac* for the current year. This includes a diagram showing the orbits of Mimas, Enceladus, Tethys, Dione, Rhea, Titan, and Hyperion. Plan your observing session by looking up the dates and times of the most recent greatest eastern elongations of the various satellites. You will have to convert from universal time (UT), which is the same as Greenwich Mean Time, to your local time zone. Then, using the tick marks along the orbits in the diagram, estimate the positions of the satellites relative to Saturn at the time you will be at the telescope. Another useful resource is the "Celestial Calendar" section of *Sky & Telescope*. During months when Saturn is visible in the night sky, this section of the magazine includes a chart of Saturn's satellites.

38. View Saturn through a small telescope. Make a sketch of what you see. Estimate the angle at which the rings are tilted to your line of sight. Can you see the Cassini division? Can you see any belts or zones in Saturn's clouds? Is there a faint, starlike object near Saturn that might be Titan? What observations could you perform to test whether the starlike object is a Saturnian satellite?

39. If you have access to a moderately large telescope, make arrangements to observe several of Saturn's satellites. At the telescope, you should have no trouble identifying Titan. Tethys, Dione, and Rhea are about one-sixth as bright as Titan and should be the next easiest satellites to find. Can you confidently identify any of the other satellites?

 40. Use the *Starry Night* program to observe the changing appearance of Saturn. First turn off daylight (select **Daylight** in the **Sky** menu) and show the entire celestial sphere (select **Atlas** in the **Go** menu). Center on Saturn by using the **Find...** command in the **Edit** menu. Using the controls at the right-hand end of the Control Panel, zoom in until Saturn nearly fills the field of view. In the Control Panel, set the time step to 1 year. Use the single-step time control buttons to observe the changing aspect of the rings. During which of the next 30 years will we see the rings edge-on?

The Outer Worlds

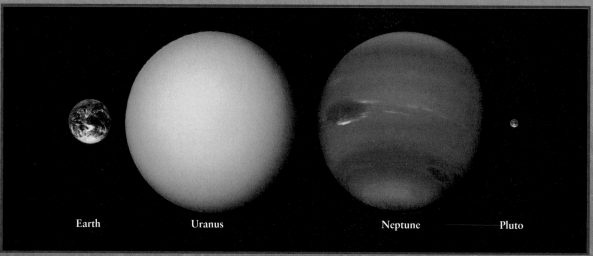

Earth   Uranus   Neptune   Pluto

(Alan Stein, Southwest Research Institute; Marc Buie, Lowell Observatory; NASA; and ESA)   R I **V** U X G

Beyond Saturn, in the cold, dark recesses of the solar system, orbit three planets that have long been shrouded in mystery. These planets—Uranus, Neptune, and Pluto (all shown here to the same scale as the Earth)—are so distant, so dimly lit by the Sun, and so slow in their motion against the stars that they were unknown to ancient astronomers and were discovered only after the invention of the telescope. Even then, little was known about Uranus and Neptune until *Voyager 2* flew past these planets during the 1980s.

Surrounded by a system of small moons and thin, dark rings, Uranus is tipped on its side so that its axis of rotation lies nearly in its orbital plane. This remarkable orientation suggests that Uranus may have been the victim of a staggering impact by a massive planetesimal. Neptune is, at first glance, a denser, more massive version of Uranus, but it is a far more active world. It has an internal energy source that Uranus seemingly lacks, as well as atmospheric bands and storm activity resembling those on Jupiter.

Neptune also has dark rings, small, icy moons, and an intriguing large satellite, Triton, which is nearly devoid of impact craters and has geysers that squirt nitrogen-rich vapors. Triton's retrograde orbit suggests that this strange world may have been gravitationally captured by Neptune.

A close cousin of Triton is Pluto, the smallest and most remote of the planets. Pluto still harbors many mysteries, because it has not yet been visited by a spacecraft. It is just one of thousands of small, icy worlds that orbit the Sun at the outskirts of the solar system.

As you read the sections of this chapter, look for the answers to the following questions.

16-1   How did Uranus and Neptune come to be discovered?

16-2   What gives Uranus its distinctive greenish-blue color?

16-3   Why are the clouds on Neptune so much more visible than those on Uranus?

16-4   Are Uranus and Neptune merely smaller versions of Jupiter and Saturn?

16-5   What is so unusual about the magnetic fields of Uranus and Neptune?

16-6   Why are the rings of Uranus and Neptune so difficult to see?

16-7   Do the moons of Uranus show any signs of geologic activity?

16-8   What makes Neptune's moon Triton unique in the solar system?

16-9   What plans are there to send a spacecraft to Pluto?

16-10   Are there other planets beyond Pluto?

16-1   Uranus was discovered by chance, but Neptune's existence was predicted by applying Newtonian mechanics

Uranus was discovered quite by accident by William Herschel, a German-born musician who immigrated to England in 1757 and became fascinated by astronomy. Using a telescope that he built himself, Herschel was systematically surveying the sky on March 13, 1781, when he noticed a faint, fuzzy object that he first thought to be a distant comet. By the end of 1781, however, his observations had revealed that the object's orbit was relatively circular and was larger than Saturn's orbit. Comets, by contrast, can normally be seen only when they follow elliptical orbits that bring them much closer to the Sun.

Herschel had discovered the seventh planet from the Sun. In doing so, he had doubled the radius of the known solar system from 9.5 AU (the semimajor axis of Saturn's orbit) to 19.2 AU (the distance from the Sun to Uranus). Herschel originally named his discovery Georgium Sidus (Latin for "Georgian star") in honor of the reigning monarch, George III. The name Uranus—in Greek mythology, the personification of Heaven—came into currency only some decades later.

Although Herschel received the credit for discovering Uranus, he was by no means the first person to have seen it. At opposition, Uranus is just barely bright enough to be seen with the naked eye under good observing conditions, so it was probably seen by the ancients. Many other astronomers with telescopes had sighted this planet before Herschel; it is plotted on at least 20 star charts drawn between 1690 and 1781. But all these other observers mistook Uranus for a dim star. Herschel was the first to track its motion relative to the stars and recognize it as a planet. This was no small task, because Uranus moves very slowly on the celestial sphere, just over 4° in the space of a year (compared to about 12° for Saturn and about 35° for Jupiter).

It was by carefully tracking Uranus's slow motions that astronomers were led to discover Neptune. By the beginning of the nineteenth century, it had become painfully clear to astronomers that they could not accurately predict the orbit of Uranus using Newtonian mechanics. By 1830 the discrepancy between the planet's predicted and observed positions had become large enough (2 arcmin) that some scientists suspected that Newton's law of gravitation might not be accurate at great distances from the Sun.

In 1843, John Couch Adams, a 24-year-old recent graduate of Cambridge University in England, began to explore an earlier and sounder suggestion. Perhaps the gravitational pull of an as yet undiscovered planet was causing Uranus to deviate slightly from its predicted orbit. After spending two years attempting to calculate the orbit of this unknown object, Adams concluded that Uranus had indeed caught up with and had passed a more distant planet. Uranus had accelerated slightly as it approached the unknown planet, then decelerated slightly as it receded from the planet. Adams predicted that the unknown planet would be found at a certain position in the constellation of Aquarius (the Water Carrier).

In October 1845, Adams submitted his calculations to George Airy, the Astronomer Royal of Great Britain. Airy was unconvinced, and the matter was dropped. A few months later, however, the French astronomer Urbain Jean Joseph Le Verrier independently made the same calculations and came to the same conclusion as Adams. Their predicted positions for the unknown planet differed by less than 1°.

Shocked into action, Airy promptly persuaded James Challis, the director of the observatory at Cambridge, to begin a thorough search. The search was hampered by Challis's lack of enthusiasm, as well as by his lack of an up-to-date star map for the part of the sky where Adams had predicted the new planet would be found. Many uncharted stars had to be observed, then followed over many nights to see whether any of them showed a planetlike motion with respect to other stars.

Meanwhile, Le Verrier wrote to Johann Gottfried Galle at the Berlin Observatory. Galle received the letter on

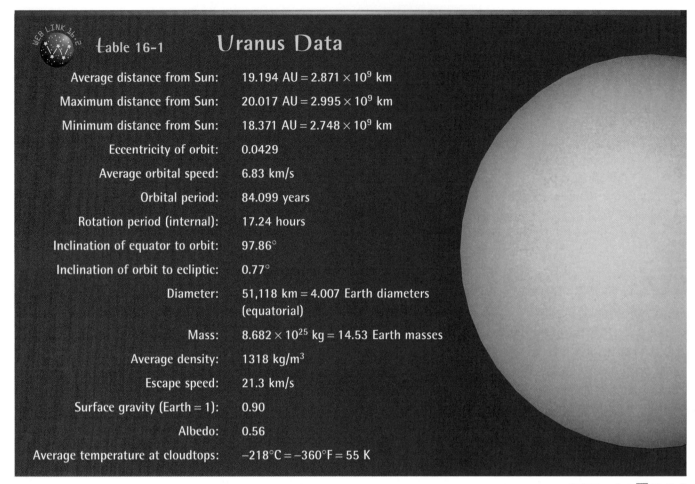

table 16-1	Uranus Data
Average distance from Sun:	19.194 AU = 2.871×10^9 km
Maximum distance from Sun:	20.017 AU = 2.995×10^9 km
Minimum distance from Sun:	18.371 AU = 2.748×10^9 km
Eccentricity of orbit:	0.0429
Average orbital speed:	6.83 km/s
Orbital period:	84.099 years
Rotation period (internal):	17.24 hours
Inclination of equator to orbit:	97.86°
Inclination of orbit to ecliptic:	0.77°
Diameter:	51,118 km = 4.007 Earth diameters (equatorial)
Mass:	8.682×10^{25} kg = 14.53 Earth masses
Average density:	1318 kg/m^3
Escape speed:	21.3 km/s
Surface gravity (Earth = 1):	0.90
Albedo:	0.56
Average temperature at cloudtops:	−218°C = −360°F = 55 K

(NASA) R I **V** U X G

September 23, 1846, and he sighted the planet that very night. Because he had the finest star charts available, Galle easily located an uncharted star with the expected brightness in the predicted location.

Le Verrier proposed that the planet be called Neptune. After years of debate between English and French astronomers, the credit for its discovery came to be divided equally between Adams and Le Verrier.

At opposition, Neptune can be bright enough to be visible through small telescopes. The first person thus to have seen it was probably Galileo. His drawings from January 1613, when he was using his telescope to observe Jupiter and its four large satellites, show a "star" less than 1 arcmin from Neptune's location. Galileo even noted in his observation log that on one night this star seemed to have moved in relation to the other stars. But Galileo would have been hard pressed to identify Neptune as a planet, because its motion against the background stars is so slow (just over 2° per year).

Through a large, modern telescope, both Uranus and Neptune are dim, uninspiring sights. Each planet appears as a hazy, featureless disk with a faint greenish-blue tinge. Although Uranus and Neptune are both about 4 times larger in diameter than Earth, they are so distant that their angular diameters as seen from the Earth are tiny—no more than

4 arcsec for Uranus and just over 2 arcsec for Neptune. To an Earth-based observer, Uranus is roughly the size of a golf ball seen at a distance of 1 kilometer.

From 2001 through 2005, Uranus and Neptune are together in the same part of the sky, moving slowly from Capricornus (the Sea Goat) into Aquarius. During these years, both planets are at opposition in either July or August, which are thus the best months to view them with a telescope. Tables 16-1 and 16-2 give basic data about Uranus and Neptune.

16-2 Uranus is nearly featureless and has an unusually tilted axis of rotation

Scientists had hoped that *Voyager 2* would reveal cloud patterns in Uranus's atmosphere when it flew past the planet in January 1986. But even images recorded at close range showed Uranus to be remarkably featureless (Figure 16-1). Faint cloud markings became visible in these images only after extreme computer enhancement.

Voyager 2 data confirmed that the Uranian atmosphere is dominated by hydrogen (82.5%) and helium (15.2%), similar to the atmospheres of Jupiter and Saturn. Uranus differs,

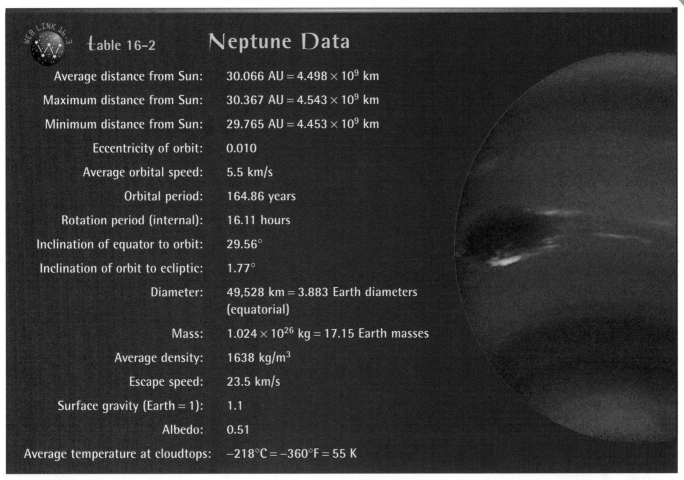

Table 16-2   Neptune Data

Average distance from Sun:	30.066 AU = 4.498×10^9 km
Maximum distance from Sun:	30.367 AU = 4.543×10^9 km
Minimum distance from Sun:	29.765 AU = 4.453×10^9 km
Eccentricity of orbit:	0.010
Average orbital speed:	5.5 km/s
Orbital period:	164.86 years
Rotation period (internal):	16.11 hours
Inclination of equator to orbit:	29.56°
Inclination of orbit to ecliptic:	1.77°
Diameter:	49,528 km = 3.883 Earth diameters (equatorial)
Mass:	1.024×10^{26} kg = 17.15 Earth masses
Average density:	1638 kg/m³
Escape speed:	23.5 km/s
Surface gravity (Earth = 1):	1.1
Albedo:	0.51
Average temperature at cloudtops:	−218°C = −360°F = 55 K

(NASA)   R I **V** U X G

however, in that 2.3% of its atmosphere is methane (CH_4), 10 times the percentage found on Jupiter and Saturn. In fact, Uranus has a higher percentage of all heavy elements—including carbon atoms, which are found in molecules of methane—than Jupiter and Saturn. (In Section 16-4 we will see why this is so.)

Methane preferentially absorbs the longer wavelengths of visible light, so sunlight reflected from Uranus's upper atmosphere is depleted of its reds and yellows. This gives the planet its distinct greenish-blue appearance. As on Saturn's moon Titan (see Section 15-8), ultraviolet light from the Sun turns some of the methane gas into a hydrocarbon haze, making it difficult to see the lower levels of the atmosphere.

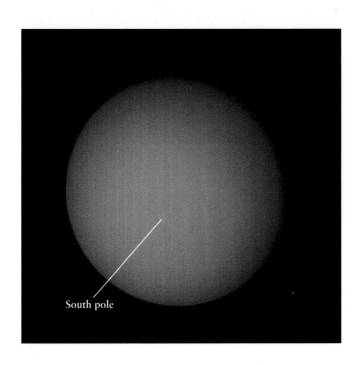

South pole

figure 16-1   R I **V** U X G

Uranus from *Voyager 2* This image looks nearly straight down onto Uranus's south pole, which was pointed almost directly at the Sun when *Voyager 2* flew past in 1986. None of the *Voyager 2* images of Uranus shows any pronounced cloud patterns. The color is due to methane in the planet's atmosphere, which absorbs red light but reflects green and blue. (JPL/NASA)

Ammonia (NH_3), which constitutes 0.01 to 0.03% of the atmospheres of Jupiter and Saturn, is almost completely absent from the Uranian atmosphere. The reason is that Uranus is colder than Jupiter or Saturn: The temperature in its upper atmosphere is only 55 K (–218°C = –360°F). Ammonia freezes at these very low temperatures, so any ammonia has long since precipitated out of the atmosphere and into the planet's interior. For the same reason, Uranus's atmosphere is also lacking in water. Hence, the substances that make up the clouds on Jupiter and Saturn—ammonia, ammonium hydrosulfide (NH_4SH), and water—are not available on Uranus. This helps to explain the bland, uniform appearance of the planet shown in Figure 16-1.

The few clouds found on Uranus are made primarily of methane, which condenses into droplets only if the pressure is sufficiently high. Therefore, methane clouds form only at lower levels within the atmosphere, where they are difficult to see.

Voyager 2 also confirmed that Uranus's rotation axis is tilted in a unique and bizarre way. Herschel found the first evidence of this in 1787, when he discovered two moons orbiting Uranus in a plane that is almost perpendicular to the plane of the planet's orbit around the Sun. Because the large moons of Jupiter and Saturn were known to orbit in the same plane as their planet's equator and in the same direction as

their planet's rotation, it was thought that the same must be true for the moons of Uranus. Thus, Uranus's equator must be almost perpendicular to the plane of its orbit, and its rotation axis must lie very nearly in that plane (Figure 16-2).

Careful measurement shows that Uranus's axis of rotation is tilted by 98°, as compared to 23½° for Earth (compare Figure 16-2 with Figure 2-12). A tilt angle greater than 90° means that Uranus exhibits retrograde (backward) rotation, shown in Figure 11-3. Along with Venus and Pluto, it is one of only three planets in the solar system to do so. Modern astronomers suspect that Uranus acquired its large tilt angle billions of years ago, when another massive body collided with Uranus while the planet was still forming.

The radical tilt of its axis means that as Uranus moves along its 84-year orbit, its north and south poles alternately point toward or away from the Sun. This produces highly exaggerated seasonal changes. For example, when *Voyager 2* flew by in 1986, Uranus's south pole was pointed toward the Sun. Most of the planet's southern hemisphere was bathed in continuous sunlight, while most of the northern hemisphere was subjected to a continuous, frigid winter night. But over the following quarter of a Uranian year later (21 of our years), sunlight has gradually been returning to the northern hemisphere, triggering immense storms there and giving the planet a much less bland appearance. Figure 16-3 shows some of these storms.

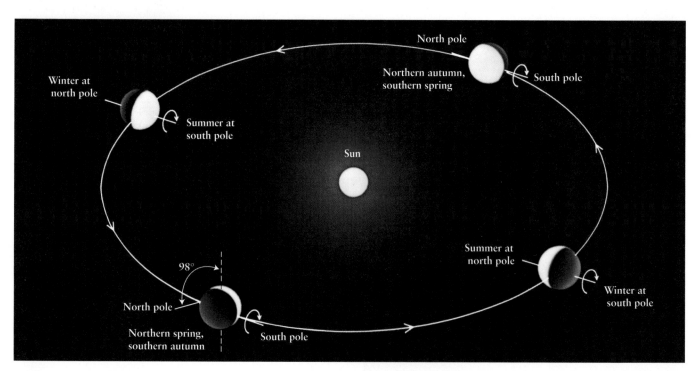

figure 16-2

Exaggerated Seasons on Uranus For most planets, the rotation axis is roughly perpendicular to the plane of the planet's orbit around the Sun. But for Uranus, the rotation axis is tilted by 98° from the perpendicular, which causes severely exaggerated seasons. For example, during midsummer at Uranus's south pole, the Sun appears nearly overhead for many Earth years, while the planet's northern regions are subjected to a long, continuous winter night. Half an orbit later, the seasons are reversed.

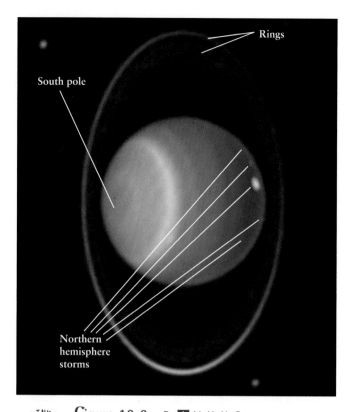

South pole

Rings

Northern hemisphere storms

figure 16-3 R **I** V U X G

Sunlight Produces Uranian Storms When the Hubble Space Telescope took this false-color infrared image of Uranus in 1998, sunlight was returning to the planet's northern hemisphere. (Compare with Figure 16-1, which shows Uranus 12 years earlier.) The resulting heating of the northern hemisphere's atmosphere triggered a series of storms, each of which was more than 1000 km (550 mi) across. These are much more visible at infrared wavelengths than with visible light. This image also shows Uranus's system of rings (see Section 16-6) and a few of its moons. (Erich Karkoschka/STScI/NASA)

By following the motions of clouds and storm systems on Uranus, scientists find that the planet's winds flow to the east—that is, in the same direction as the planet's rotation—at northern and southern latitudes, but to the west near the equator. This is quite unlike the situation on Jupiter and Saturn, where the zonal winds alternate direction many times between the north and south poles (see Section 13-3 and Section 15-5). The fastest Uranian winds (about 700 km/h, or 440 mi/h) are found at the equator.

Although Uranus's equatorial region was receiving little sunlight at the time of the *Voyager 2* flyby, the atmospheric temperature there (about 55 K = −218°C = −359°F) was not too different from that at the sunlit pole. Heat must therefore be efficiently transported from the poles to the equator. This north-south heat transport, which is perpendicular to the wind flow, may have mixed and homogenized the atmosphere to make Uranus nearly featureless.

The rotation period of Uranus's atmosphere is about 16 hours. Like Jupiter and Saturn, Uranus rotates differentially, so this period depends on the latitude. This can be measured by tracking the motions of clouds. To determine the rotation period for the underlying body of the planet, scientists looked to Uranus's magnetic field, which is presumably anchored in the planet's interior, or at least in the deeper and denser layers of its atmosphere. Data from *Voyager 2* indicate that Uranus's internal period of rotation is 17.24 hours.

16-3 Neptune is a cold, bluish world with Jupiterlike atmospheric features

At first glance, Neptune appears to be the twin of Uranus, as can be seen from the image that opens this chapter and by comparing Tables 16-1 and 16-2. But these two planets are by no means identical. While Neptune and Uranus have almost the same diameter, Neptune is 18% more massive. Neptune's axis of rotation also has a more modest 29½° tilt. When *Voyager 2* flew past Neptune in August 1989, it revealed that the planet has a more active and dynamic atmosphere than Uranus. This activity suggests a powerful source of energy within Neptune.

The *Voyager 2* data showed that Neptune has essentially the same atmospheric composition as Uranus: 80% hydrogen, 19% helium, 2% methane, and almost no ammonia or water vapor. As for Uranus, the presence of methane gives Neptune a characteristic bluish-green color. The temperature in the upper atmosphere is also the same as on Uranus, about 55 K. That this should be so, even though Neptune is much farther from the Sun, is further evidence that Neptune has a strong internal heat source.

Unlike Uranus, however, Neptune has clearly visible cloud patterns in its atmosphere. At the time that *Voyager 2* flew past Neptune, the most prominent feature in the planet's atmosphere was a giant storm called the **Great Dark Spot** (Figure 16-4). The Great Dark Spot had a number of similarities to Jupiter's Great Red Spot (see Section 13-1 and Section 13-2): The storms on both planets were comparable in size to the Earth's diameter, both appeared at about the same latitude in the southern hemisphere, and the winds in both storms circulated in a counterclockwise direction. But Neptune's Great Dark Spot appears not to have been as long-lived as the Great Red Spot on Jupiter. When the Hubble Space Telescope first viewed Neptune in 1994, the Great Dark Spot had disappeared. Another dark storm appeared in 1995 in Neptune's northern hemisphere.

Voyager 2 also saw a few conspicuous whitish clouds on Neptune. These clouds are thought to be produced when winds carry methane gas into the cool, upper atmosphere, where it condenses into crystals of methane ice. *Voyager 2* images show these high-altitude clouds casting shadows onto lower levels of Neptune's atmosphere (Figure 16-5).

Thanks to its greater distance from the Sun, Neptune receives less than half as much energy from the Sun as Uranus.

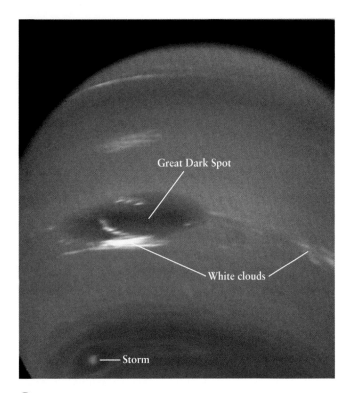

Figure 16-4 R I **V** U X G

Neptune from *Voyager 2* This image looks down onto Neptune's southern hemisphere. At the time the picture was taken (1989), the Great Dark Spot measured about 12,000 by 8000 km—comparable in size to the Earth. A smaller storm appears toward the lower left. The white clouds are thought to be composed of crystals of methane ice. The color contrast in this image has been exaggerated to emphasize the differences between dark and light areas in Neptune's atmosphere. (NASA/JPL)

Figure 16-5 R I **V** U X G

Cirrus Clouds over Neptune *Voyager 2* recorded this image of clouds near Neptune's terminator (the border between day and night on the planet.) Note the shadows cast on the main cloud deck by the white streaks of high-altitude clouds. Like thin, wispy, high-altitude cirrus clouds in the Earth's atmosphere, these clouds are thought to be made of ice crystals. The difference is that Neptune's cirrus clouds are probably made of methane ice crystals, not water ice crystals as on Earth. (NASA/JPL)

But with less solar energy available to power atmospheric motions, why are there high-altitude clouds and huge, dark storms on Neptune but not on Uranus? At least part of the answer is probably that Neptune is still slowly contracting, thus converting gravitational energy into thermal energy that heats the planet's core. The evidence for this is that Neptune, like Jupiter, emits more energy than it receives from the Sun. The combination of a warm interior and a cold outer atmosphere can cause convection in Neptune's atmosphere, producing the up-and-down motion of gases that generates clouds and storms.

Uranus, by contrast, appears to have little or no internal source of thermal energy; measurements to date show that it radiates just as much energy into space as it receives from the Sun. Therefore, Uranus lacks the dynamic atmospheric activity found on Neptune and has a much blander appearance (compare Figure 16-1 and Figure 16-4). It is not known why Uranus should be different from the other Jovian planets.

Neptune also resembles Jupiter in having faint belts and zones parallel to the planet's equator (Figure 16-6). As on Jupiter, the light-colored zones are regions where clouds have been lifted to relatively high altitudes, while the dark belts are where air is descending.

As those on Uranus, most of Neptune's clouds are probably made of droplets of liquid methane. Because these form fairly deep within the atmosphere, they are more difficult to see than the ones on Jupiter. Hence, Neptune's belts and zones are less pronounced than those on Jupiter, although more so than those on Uranus (thanks to the extra cloud-building energy from Neptune's interior).

Although Neptune displays more evidence of up-and-down motion in its atmosphere than Uranus, the global pattern of east and west winds is almost identical on the two planets. This is rather strange. The two planets are heated very differently by the Sun, thanks to their different distances from the Sun and the different tilts of their axes of rotation, so we might have expected the wind patterns on Uranus and Neptune also to be different. Perhaps the explanation of these wind patterns will involve understanding how heat is transported not only within the atmospheres of Uranus and Neptune but within their interiors as well.

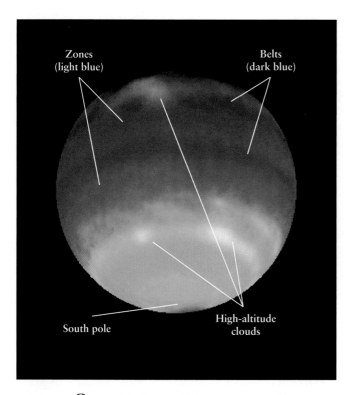

Zones
(light blue)

Belts
(dark blue)

South pole

High-altitude
clouds

Figure 16-6   R I **V** U X G

Neptune's Banded Structure This enhanced-color
Hubble Space Telescope image shows Neptune's belts in
dark blue and zones in light blue. White areas denote high-altitude
clouds, presumably of methane ice; the very highest clouds (near the
top of the image) are shown in yellow-red. The green belt near the
south pole is a region where the atmosphere absorbs blue light,
perhaps indicating some differences in chemical composition.
(Lawrence Sromovsky, University of Wisconsin-Madison; and STScI/NASA)

16-4   Uranus and Neptune contain a higher proportion of heavy elements than Jupiter and Saturn

At first glance, Uranus and Neptune might seem to be simply smaller and less massive versions of Jupiter or Saturn (see Table 7-1). But like many first impressions, this one is misleading because it fails to take into account what lies beneath the surface. We saw in Section 13-6 that the bulk of Jupiter is composed primarily of hydrogen and helium, in nearly the same abundance as the Sun. The same is true of Saturn (see Section 15-7). But Uranus and Neptune must have a different composition, because their average densities are too high.

If Uranus and Neptune had solar abundances of the elements, the smaller masses of these planets would produce less gravitational compression, and Uranus and Neptune would both have lower average densities than Jupiter or Saturn. In fact, however, Uranus and Neptune have average densities

(1320 kg/m^3 and 1640 kg/m^3, respectively) that are comparable to or greater than those of Jupiter (1330 kg/m^3) or Saturn (690 kg/m^3). Therefore, we can conclude that Uranus and Neptune contain greater proportions of the heavier elements than do Jupiter or Saturn.

This picture is not what we might expect. According to our discussion in Section 7-8 of how the solar system formed, hydrogen and helium should be relatively more abundant as distance from the vaporizing heat of the Sun increases. But Uranus and Neptune contain a greater percentage of heavy elements, and, therefore, a *smaller* percentage of hydrogen and helium, than Jupiter and Saturn.

Recall from Section 7-8 that the Jovian planets probably formed in a two-step process: First, planetesimals accreted to form each planet's core; second, hydrogen and helium were gravitationally attracted onto the core to form the planet's envelope. Apparently, the cores of Uranus and Neptune were able to attract relatively little hydrogen or helium, leaving these planets deficient in these two elements.

One possible explanation is that Uranus and Neptune may have taken longer to form their cores than Jupiter and Saturn because the outer regions of the solar nebula were thinner and planetesimals were more widely dispersed. By the time these cores had formed, much of the hydrogen and helium gas in the solar nebula would already have dissipated into interstellar space, leaving relatively little to be incorporated into Uranus and Neptune. One observation that reinforces this idea is that Uranus and Neptune are comparable in mass to the cores of Jupiter and Saturn. All four Jovian planets may have begun with cores of similar size, but Uranus and Neptune failed to accumulate tremendous envelopes of hydrogen and helium, as did Jupiter and Saturn.

One difficulty with this picture is explaining why Uranus and Neptune are even as massive as they are. At the locations of Uranus and Neptune, 19 and 30 AU from the Sun, respectively, the solar nebula was probably so sparse that these planets would have needed longer than the age of the solar system to accumulate their mass. In other words, Uranus and Neptune should not exist at all! One possible solution is that Uranus and Neptune actually formed in denser regions between 4 and 10 AU from the Sun, where they could have grown rapidly to their present sizes. Had they remained at these locations, they could eventually have accumulated enough hydrogen and helium to become as large as Jupiter or Saturn. But before that could happen, gravitational interactions with Jupiter and Saturn would have pushed Uranus and Neptune outward into their present large orbits. Because the solar nebula was very sparse at those greater distances, Uranus and Neptune stopped growing and remained at the sizes we see today.

Figure 16-7 shows one model for the internal structures of Uranus and Neptune. (Compare this to the internal structures of Jupiter and Saturn, shown in Figure 15-14.) In this model each planet has a rocky core roughly the size of Earth, although more massive. Each planet's core is surrounded by a mantle of liquid "ices," including both water and the ammonia that froze out of the planet's upper layers and descended

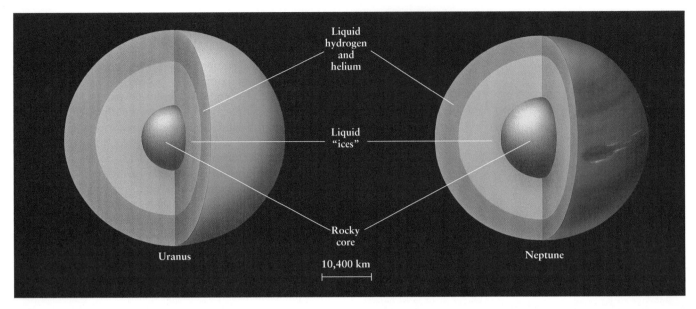

Liquid
hydrogen
and
helium

Liquid
"ices"

Rocky
core

Uranus

Neptune

10,400 km

figure 16-7

The Internal Structures of Uranus and Neptune In the model shown here, both Uranus and Neptune have three regions in their interiors. At the center is a rocky core, resembling a terrestrial planet. Surrounding this is a layer of liquid water with ammonia dissolved in it, topped by an outer layer of liquid molecular hydrogen and liquid helium. The atmospheres are very thin gaseous shells on top of the hydrogen-helium layer. The two planets have nearly the same diameter, but because Neptune is more massive than Uranus, it may have a somewhat larger rocky core.

into the interior. (This means that the mantle is chemically similar to household window cleaning fluid.) Around the mantle is a layer of liquid molecular hydrogen and liquid helium, with a small percentage of liquid methane. This layer is relatively shallow compared to those on Jupiter and Saturn, and the pressure is not high enough to convert the liquid hydrogen into liquid metallic hydrogen.

16-5 The magnetic fields of both Uranus and Neptune are oriented at unusual angles

Astronomers were quite surprised by the data sent back from *Voyager 2*'s magnetometer as the spacecraft sped past Uranus and Neptune. These data, as well as radio emissions from charged particles in their magnetospheres, showed that the magnetic fields of both Uranus and Neptune are tilted at steep angles to their axes of rotation. Uranus's **magnetic axis,** the line connecting its north and south magnetic poles, is inclined by 59° from its axis of rotation; Neptune's magnetic axis is tilted by 47°. By contrast, the magnetic axes of Earth, Jupiter, and Saturn are all nearly aligned with their rotation axes; the angle between their magnetic and rotation axes is 12° or less (Figure 16-8). Scientists were also surprised to find that the magnetic fields of Uranus and Neptune are offset from the centers of the planets.

CAUTION! The drawing of the Earth's magnetic field at the far left of Figure 16-8 may seem to be mislabeled, because it shows the *south* pole of a magnet at the Earth's *North* Pole. But, in fact, this is correct, as you can understand by thinking about how magnets work. On a magnet that is free to swivel, like the magnetized needle in a compass, the "north pole" is called that because it tends to point north on Earth. Likewise, a compass needle's south pole points toward the south on Earth. Furthermore, opposite magnetic poles attract. If you take two magnets and put them next to each other, they try to align themselves so that one magnet's north pole is next to the other magnet's south pole. Now, if you think of a compass needle as one magnet and the entire Earth as the other magnet, it makes sense that the compass needle's north pole is being drawn toward a magnetic *south* pole—which happens to be located near the Earth's geographic North Pole. The Earth's magnetic pole nearest its geographic North Pole is called the "magnetic north pole." Note that the magnets drawn inside Jupiter, Saturn, and Neptune are oriented opposite to the Earth's. On any of these planets, the north pole of a compass needle would point south, not north!

Why are the magnetic axes and axes of rotation of Uranus and Neptune so badly misaligned? And why are the magnetic fields offset from the centers of the planets? One possibility is that their magnetic fields might be undergoing a reversal; geological data show that the Earth's magnetic field has switched

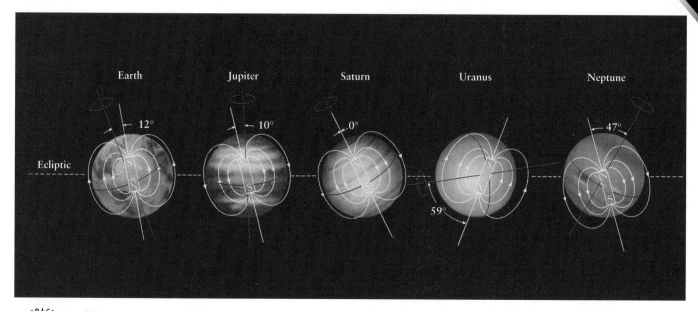

figure 16-8

The Magnetic Fields of Five Planets This drawing shows how the magnetic fields of the Earth, Jupiter, Saturn, Uranus, and Neptune are tilted relative to their rotation axes (shown in red). For both Uranus and Neptune, the magnetic fields are offset from the planet's center and steeply inclined to the rotation axis.

north to south and back again many times in the past. Another possibility is that the misalignments resulted from catastrophic collisions with planet-sized bodies. As we discussed in Section 16-2, the tilt of Uranus's rotation axis and its system of moons suggest that just such collisions occurred long ago. As we will see in Section 16-7, Neptune may have gravitationally captured its largest moon, Triton, but no one knows if that incident was responsible for offsetting Neptune's magnetic axis.

Because neither Uranus nor Neptune is massive enough to compress hydrogen to a metallic state, their magnetic fields cannot be generated in the same way as those of Jupiter and Saturn. Instead, under the high pressures found in the watery mantles of Uranus and Neptune, dissolved molecules such as ammonia lose one or more electrons and become electrically charged (that is, they become ionized; see Section 5-8). Water is a good conductor of electricity when it has such electrically charged molecules dissolved in it, and electric currents in this fluid are probably the source of the magnetic fields of Uranus and Neptune.

16-6 Uranus and Neptune each have a system of thin, dark rings

Like the planet itself, Uranus's rings were discovered by accident. On March 10, 1977, Uranus was scheduled to move in front of a faint star, as seen from the Indian Ocean. A team of astronomers headed by James L. Elliot of Cornell University observed this event, called an **occultation**, from a NASA airplane equipped with a telescope. They hoped that by measuring how long the star was hidden, they could accurately measure Uranus's size. In addition, by carefully measuring how the starlight faded when Uranus passed in front of the star, they planned to deduce important properties of Uranus's upper atmosphere.

To everyone's surprise, the background star briefly blinked on and off several times just before the star passed behind Uranus and again immediately after (Figure 16-9). The astronomers concluded that Uranus must be surrounded by a series of nine narrow rings. During the *Voyager 2* flyby, two more rings were discovered.

Unlike Saturn's rings, the rings of Uranus are dark and narrow: Most are less than 10 km wide. Typical particles in Saturn's rings are chunks of reflective ice the size of snowballs (a few centimeters across), but typical particles in Uranus's rings are 0.1 to 10 meters in size and are no more reflective than lumps of coal. As a result, the Uranian rings are very dark; they reflect only a few percent of the sunlight that falls on them. It is not surprising that these narrow, dark rings escaped detection for so long. Figure 16-10 is a *Voyager 2* image of the rings from the side of the planet away from the Sun, where light scattering from exceptionally small ring particles makes them more visible. (We discussed light scattering in Section 5-6 and Box 5-4.) Figure 16-3 shows the Sun-facing side of the rings.

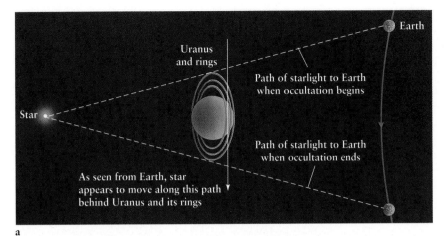

figure 16-9

How the Rings of Uranus Were Discovered (a) As seen from Earth, Uranus occasionally appears to move in front of a distant star. Such an event is called an occultation. (b) This graph suggests how the intensity of a star's light varies as Uranus passes in front of it. The star's light is completely blocked when it is behind the planet. But before and after the occultation by Uranus, the starlight dims temporarily as the rings pass in front of the star. It was in this way that the rings were discovered during an occultation in 1977.

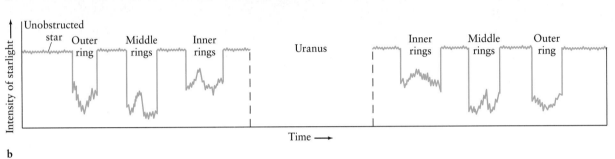

b

All of Uranus's rings are located less than 2 Uranian radii from the planet's center, well within the planet's Roche limit (see Section 15-2). Some sort of mechanism, possibly one involving shepherd satellites, efficiently confines particles to their narrow orbits. (In Section 15-4 we discussed how shepherd satellites help keep planetary rings narrow.) *Voyager 2* searched for shepherd satellites but found only two. The others may be so small and dark that they have simply escaped detection. Or perhaps they aren't there, and our ideas need revision.

 Like Uranus, Neptune is surrounded by a system of thin, dark rings that were first detected in stellar occultations. Figure 16-11 is a *Voyager 2* image of Neptune's rings.

It is so cold at Uranus and Neptune that the ring particles can retain methane ice. Scientists speculate that eons of impacts by electrons trapped in the magnetospheres of the planets have converted this methane ice into dark carbon compounds. This **radiation darkening** can account for the low reflectivity of the rings around both Uranus and Neptune.

16-7 Some of Uranus's satellites show evidence of past tidal heating

Before the *Voyager 2* flyby of Uranus, five moderate-sized satellites—Titania, Oberon, Ariel, Umbriel, and Miranda—were known to orbit the planet (Figure 16-12). Most are named after sprites and spirits in Shakespeare's plays. They range in diameter from about 1600 km (1000 mi) for Titania and Oberon to less than 500 km (300 mi) for Miranda. All these moons have average densities around 1500 kg/m³, which is consistent with a mixture of ice and rock.

Voyager 2 discovered ten other small Uranian satellites, most of which have diameters of less than 100 km (60 mi); six

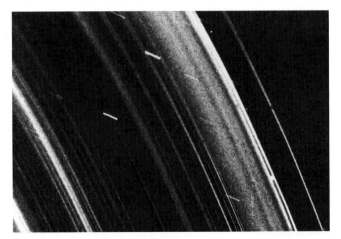

figure 16-10 R I **V** U X G

The Shaded Side of Uranus's Rings This view, taken when *Voyager 2* was in Uranus's shadow, looks back toward the Sun. Numerous fine-grained dust particles in the spaces between the main rings gleam in the sunlight. This dust is probably debris from collisions between larger particles in the main rings. Thanks to these collisions, the Uranian rings will probably be gone in a few hundred million years. The short horizontal streaks are star images, blurred because of the spacecraft's motion during the exposure. (NASA/JPL)

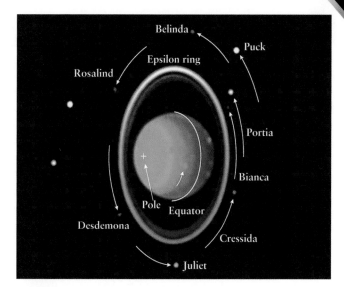

figure 16-11 R I **V** U X G

Neptune's Rings The two main rings of Neptune and a faint, inner ring can easily be seen in this composite of two *Voyager 2* images. The ring particles vary in size from a few micrometers (1 μm = 10⁻⁶ m) to about 10 meters. There is also a faint sheet of particles whose outer edge is located between the two main rings and extends inward toward the planet. The overexposed image of Neptune itself has been blocked out. (NASA)

figure 16-13 R ■ V U X G

Uranus's Rings and Small Satellites This false-color infrared image from the Hubble Space Telescope shows eight of Uranus's small satellites. All the satellites shown here were discovered by the *Voyager 2* spacecraft when it flew past Uranus in 1986. They all lie within 86,000 km of the planet's center, only about one-fifth of the distance from the Earth to our Moon. None is more than 160 km (100 mi) in diameter. The arcs show how far each satellite moves around its orbit in 90 minutes. (Erich Karkoschka, University of Arizona; and NASA)

more were found using ground-based telescopes between 1997 and 2000. Only Jupiter and Saturn have more known satellites than Uranus. Figure 16-13 shows several of the smaller satellites. Table 16-3 summarizes information about all the moons of Uranus.

The Uranian moons are all quite dark. This may be due to radiation darkening of methane ice on the surfaces of the moons, the same mechanism proposed to explain the darkness

of Uranus's rings. Umbriel is one of the darkest moons in our solar system, but some of the craters on Oberon appear to have exposed fresh, undarkened ice. Perhaps Umbriel has simply had the luck to escape recent impacts.

Umbriel and Oberon both appear to be geologically dead worlds, with surfaces dominated by impact craters. By contrast, Ariel's surface appears to have been cracked at some time in the past, allowing some sort of ice lava to flood low-

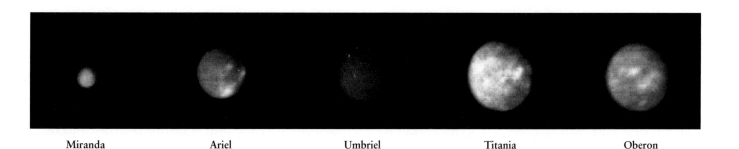

Miranda	Ariel	Umbriel	Titania	Oberon

figure 16-12 R I **V** U X G

Uranus's Principal Satellites This "family portrait"—actually a montage of five *Voyager 2* images—shows Uranus's five largest moons to the same scale and correctly displays their respective reflectivities. (The darkest satellite, Umbriel, actually has a higher albedo than the Earth's Moon.) All five satellites have grayish surfaces, with only slight variations in color. (NASA/JPL)

table 16-3	Uranus's Satellites			
	Average distance from center of Uranus (km)	Orbital period (days)	Diameter (km)	Density (kg/m³)
Cordelia	49,752	0.335	26	—
Ophelia	53,763	0.376	32	—
Bianca	59,166	0.435	44	—
Cressida	61,767	0.464	66	—
Desdemona	62,658	0.474	58	—
Juliet	64,358	0.493	84	—
Portia	66,097	0.513	110	—
Rosalind	69,927	0.558	58	—
1986 U10	75,000	0.62	40	—
Belinda	75,256	0.624	67	—
Puck	86,004	0.762	154	—
Miranda	129,800	1.413	473	1200
Ariel	191,240	2.520	1160	1670
Umbriel	266,000	4.144	1170	1400
Titania	435,840	8.706	1580	1710
Oberon	582,600	13.463	1520	1630
Caliban	7,168,900	579^R	60	—
Stephano	7,942,000	676^R	30	—
Sycorax	12,213,600	1289^R	120	—
Prospero	16,114,000	1953^R	50	—
Setebos	18,205,000	2345^R	40	—

This table lists some facts about Uranus's 21 confirmed satellites. The diameters given for the 16 smallest satellites are estimates. The masses of these satellites are unknown, so no densities are given.

The superscript R on the orbital periods of the five outermost satellites means that they orbit Uranus in a retrograde direction (opposite to the planet's rotation).

lying areas. A similar process appears to have taken place on Titania. This geologic activity may be due to a combination of the satellites' internal heat and tidal heating (see Section 14-4). Although there are no present-day resonances among the orbits of Uranus's satellites, they may have existed in the past and given rise to such tidal heating.

Unique among Uranus's satellites is Miranda, which has a landscape unlike that of any other world in the solar system. Much of the surface is heavily cratered, as we would expect for a satellite only 470 km in diameter, but several regions have unusual and dramatic topography (Figure 16-14).

It has been suggested that one or more long-ago impacts shattered this satellite, only to have the fragments reassemble themselves. In this theory, part of Miranda's present-day surface is material that was in the moon's interior prior to the shattering impact. But detailed analysis of Miranda's geology suggests a different scenario, in which Miranda's orbit was once in resonance with that of more massive Umbriel or Ariel. The resulting tidal heating melted Miranda's interior, causing

dense rocks in some locations on the surface to settle toward the satellite's center as blocks of less dense ice were forced upward toward the surface, thus creating Miranda's resurfaced terrain. The orbital resonance ceased before this process could run its full course, which explains why some ancient, heavily cratered regions remain on Miranda's surface.

16-8 Triton is a frigid, icy world with a young surface and a tenuous atmosphere

Neptune has eight known satellites, six of which were discovered by *Voyager 2* during its 1989 flyby. All are named for mythological beings related to bodies of water. (Neptune itself is named for the Roman god of the sea.) Table 16-4 summarizes our knowledge of Neptune's satellites.

Most of these worlds are small, icy bodies about which little is known. They are probably similar to the smaller satel-

figure 16-14   R I **V** U X G

Miranda This composite of *Voyager 2* images shows the tortured face of Miranda's southern hemisphere. Part of the surface is ancient and heavily cratered, while other parts are dominated by parallel networks of valleys and ridges. At the very bottom of the image—where a "bite" seems to have been taken out of Miranda—is a range of enormous cliffs that jut upward to an elevation of 20 km, twice as high as Mount Everest. (NASA/JPL)

lites of Uranus. The one striking exception is Triton, Neptune's largest satellite, which is in many ways quite unlike any other world in the solar system.

Triton is in a retrograde orbit: It goes around Neptune oppositely to the direction in which the planet rotates. Furthermore, the plane of Triton's orbit is inclined by 23° from the plane of Neptune's equator. It is difficult to imagine how a satellite might form out of the same cloud of material as a planet and end up orbiting in a direction opposite the planet's rotation and in such a tilted plane. Hence, Triton probably formed elsewhere in the solar system, collided long ago with a now-vanished satellite of Neptune, and was captured by Neptune's gravity.

Triton is not the only satellite in the solar system that is thought to have been captured into a retrograde orbit; a few of the small outer satellites of Jupiter and Saturn have retrograde orbits and are probably captured asteroids (see Section 14-8 and Section 15-9). But with a diameter of 2706 km, a bit smaller than our Moon, Triton is certainly the largest such captured satellite.

Figure 16-15 shows the icy, reflective surface of Triton as imaged by *Voyager 2*. There is a conspicuous absence of large craters, which immediately tells us that Triton has a young surface on which the scars of ancient impacts have largely been erased by tectonic activity. There are areas that resemble frozen lakes and may be the calderas of extinct ice volcanoes. Some of Triton's surface features resemble the long cracks seen on Europa (Section 14-6) and Ganymede (Section 14-7). Still other features are unique to Triton. For example, in the upper portion of Figure 16-15 you can see dimpled, wrinkled terrain that resembles the skin of a cantaloupe.

Triton's tectonically active history is probably related to its having been captured into orbit around Neptune. After its capture, Triton most likely started off in a highly elliptical orbit, but today the satellite's orbit is quite circular. The satellite's original elliptical orbit would have been made circular by tidal forces exerted on Triton by Neptune's gravity. These tidal forces would also have stretched and flexed Triton, causing enough tidal heating to melt much of the interior. The resulting volcanic activity (with lavas made of ice rather than molten rock) would have obliterated Triton's original surface features, including craters.

table 16-4	Neptune's Satellites			
	Average distance from center of Neptune (km)	Orbital period (days)	Diameter (km)	Density (kg/m^3)
Naiad	48,227	0.294	58	—
Thalassa	50,075	0.312	80	—
Despina	52,526	0.335	150	—
Galatea	61,953	0.429	160	—
Larissa	73,548	0.555	190	—
Proteus	117,647	1.122	420	—
Triton	354,760	5.877^R	2706	2054
Nereid	5,513,400	360.14	340	—

This table lists some facts about Neptune's eight confirmed satellites. The diameters given for the four innermost satellites are estimates. The masses of all satellites except Triton are unknown, so no densities are given.

The superscript R on the orbital period of Triton means that it orbits Uranus in a retrograde direction (opposite to the planet's rotation).

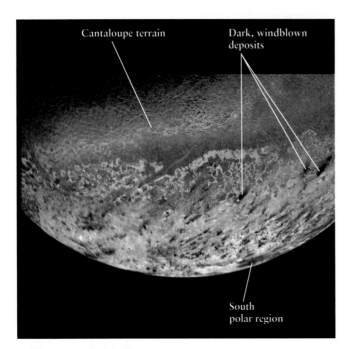

Cantaloupe terrain

Dark, windblown deposits

South polar region

figure 16-15 R I **V** U X G

Triton Several high-resolution *Voyager 2* images were assembled to create this mosaic of Triton, Neptune's largest satellite. The pinkish material surrounding the south polar region is probably nitrogen frost. Some of this presumably evaporates when summer comes to the south pole; the northward flow of the evaporated gas may cause the dark surface markings seen here. Farther north is a brown area of "cantaloupe terrain" that resembles the skin of a melon. (NASA/JPL)

There may still be warmth in Triton's interior today. *Voyager 2* observed plumes of dark material being ejected from the surface to a height of 8 km (5 mi). These plumes may have been generated from a hot spot far below Triton's surface, similar to geysers on the Earth. Alternatively, the energy source for the plumes may be sunlight that warms the surface, producing subsurface pockets of gas and creating fissures in the icy surface through which the gas can escape.

Triton's surface temperature is only 38 K (−235°C = −391°F), the lowest of any world yet visited by spacecraft. This temperature is low enough to solidify nitrogen, and indeed the spectrum of sunlight reflected from Triton's surface shows absorption lines due to nitrogen ice as well as methane ice. But Triton is also warm enough to allow some nitrogen to evaporate from the surface, like the steam that rises from ice cubes when you first take them out of the freezer. *Voyager 2* confirmed that Triton has a very thin nitrogen atmosphere with a surface pressure of only 1.6×10^{-5} atmosphere, about the same as at an altitude of 100 km above the Earth's surface. Despite its thinness, Triton's atmosphere has noticeable effects. *Voyager 2* saw areas on Triton's surface where dark material has been blown downwind by a steady breeze (see Figure 16-15). It also observed dark material ejected from

the geyserlike plumes being carried as far as 150 km by high-altitude winds.

Just as tidal forces presumably played a large role in Triton's history, they also determine its future. Triton raises a tidal bulge on Neptune, just as our Moon distorts the Earth (recall Figure 9-19). In the case of the Earth-Moon system, the gravitational pull of the Earth's tidal bulge causes the Moon to spiral away from the Earth. But because Triton's orbit is retrograde, the tidal bulge on Neptune exerts a force on Triton that makes the satellite slow down rather than speed up. (In Figure 9-19, imagine that the moon is orbiting to the right rather than to the left.) This is causing Triton to spiral gradually in toward Neptune. In approximately 100 million years, Triton will be inside Neptune's Roche limit, and the satellite will eventually be torn to pieces by tidal forces. When this happens, the planet will develop a spectacular ring system—overshadowing by far its present-day set of narrow rings—as rock fragments gradually spread out along Triton's former orbit.

Prior to *Voyager 2*, only one other satellite was known to orbit Neptune. Nereid, which was first sighted in 1949, is in a prograde orbit. Hence, it orbits Neptune in the direction opposite to Triton. Nereid also has the most eccentric orbit of any satellite in the solar system; its distance from Neptune varies from 1.4 million to 9.7 million kilometers. One possible explanation is that when Triton was captured by Neptune's gravity, the interplay of gravitational forces exerted on Nereid by both Neptune and Triton moved Nereid from a relatively circular orbit (like those of Neptune's other, smaller moons) into its present elliptical one.

16-9 Pluto was discovered after a laborious search of the heavens

Speculations about a ninth planet date back to the late 1800s, when a few astronomers suggested that Neptune's orbit was being perturbed by an unknown object. Encouraged by the fame of Adams and Le Verrier, several people set out to discover "Planet X." Two prosperous Boston gentlemen, William Pickering and Percival Lowell, were prominent in this effort. (Lowell also enthusiastically promoted the idea of canals of Mars; see Section 12-2.) Modern calculations show that there are, in fact, no unaccounted perturbations of Neptune's orbit. It is thus not surprising that no planet was found at the positions predicted by Pickering, Lowell, and others. Yet the search continued.

Before he died in 1916, Lowell urged that a special wide-field camera be constructed to help search for Planet X. After many delays, the camera was finished in 1929 and installed at the Lowell Observatory in Flagstaff, Arizona, where a young astronomer, Clyde W. Tombaugh, had joined the staff to carry on the project. On February 18, 1930, Tombaugh finally discovered the long-sought planet. It was disappointingly faint—a thousand times dimmer than the dimmest object visible with the naked eye and 250 times dimmer than Neptune at opposition—and presented no discernible disk. The planet was named for Pluto, the

figure 16-16   R I **V** U X G

Pluto's Motion Across the Sky Pluto was discovered in 1930 by searching for a dim, starlike object that moves slowly in relation to the background stars. These two photographs were taken one day apart. Even when its apparent motion on the celestial sphere is fastest, Pluto moves only about 1 arcmin per day relative to the stars. (Lick Observatory)

mythological god of the underworld, whose name has Percival Lowell's initials as its first two letters. The discovery was publicly announced on March 13, 1930, the 149th anniversary of the discovery of Uranus. The two photographs in Figure 16-16 show one day's motion of Pluto.

Pluto's orbit about the Sun is more elliptical and more steeply inclined to the plane of the ecliptic than the orbit of any other planet (see Figure 7-1). In fact, Pluto's orbit is so eccen-tric that this planet is sometimes closer to the Sun than Neptune. This was the case from 1979 until 1999; indeed, when Pluto was at perihelion in 1989, it was more than 10^8 km closer to the Sun than was Neptune. Table 16-5 lists additional data about Pluto.

Pluto is so far away that it subtends only a very small angle of 0.15 arcsec. Hence, it is extraordinarily difficult to resolve any surface features on the planet. But by observing

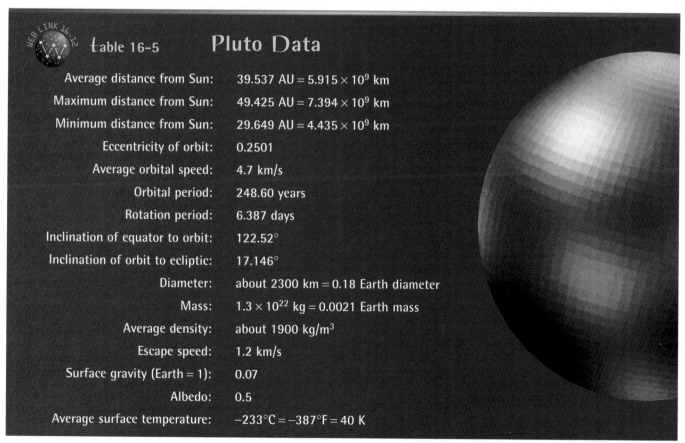

table 16-5 **Pluto Data**

Average distance from Sun:	39.537 AU = 5.915×10^9 km
Maximum distance from Sun:	49.425 AU = 7.394×10^9 km
Minimum distance from Sun:	29.649 AU = 4.435×10^9 km
Eccentricity of orbit:	0.2501
Average orbital speed:	4.7 km/s
Orbital period:	248.60 years
Rotation period:	6.387 days
Inclination of equator to orbit:	122.52°
Inclination of orbit to ecliptic:	17.146°
Diameter:	about 2300 km = 0.18 Earth diameter
Mass:	1.3×10^{22} kg = 0.0021 Earth mass
Average density:	about 1900 kg/m³
Escape speed:	1.2 km/s
Surface gravity (Earth = 1):	0.07
Albedo:	0.5
Average surface temperature:	−233°C = −387°F = 40 K

(Alan Stern, Southwest Research Institute; Marc Buie, Lowell Observatory; NASA; and ESA)   R I **V** **U** X G

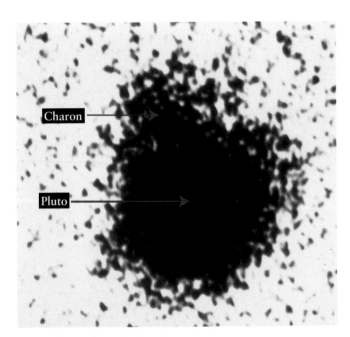

figure 16-17   R I **V** **U** X G

Pluto's Surface The small upper images of Pluto show opposite hemispheres of the planet as viewed by the Hubble Space Telescope. The large images show dark and bright areas on these two hemispheres as determined by computer processing of the Hubble images. The bright regions at the top and bottom of each hemisphere may be polar ice caps; the bright regions nearer Pluto's equator may be impact basins where more reflective subsurface ice has been exposed. (Alan Stern, Southwest Research Institute; Marc Buie, Lowell Observatory; NASA; and ESA)

Pluto with the Hubble Space Telescope over the course of a solar day on Pluto (6.3872 Earth days) and using computer image processing, Alan Stern and Marc Buie generated the maps of Pluto's surface shown in Figure 16-17. These maps show bright polar ice caps as well as regions of different reflectivity near the planet's equator. Observations of Pluto's rotation confirm that the planet's rotation axis is tipped by more than 90°, so that Pluto has retrograde rotation like Uranus.

To determine the nature of Pluto's surface, higher resolution images are needed of the sort that can only be obtained with a spacecraft flyby. At present, however, no missions of this sort have been funded.

16-10 Pluto and its moon, Charon, may be typical of a thousand icy objects that orbit far from the Sun

In 1978, while examining some photographs of Pluto, James W. Christy of the U.S. Naval Observatory noticed that the image of the planet on a photographic plate appeared slightly elongated. Pluto seemed to have a lump on one side (Figure 16-18). He promptly examined a number of other photographs of Pluto and found a series of images that showed the lump moving clockwise around Pluto with a period of about 6 days.

Christy concluded that the lump was actually a satellite of Pluto. He proposed that the newly discovered moon be christened Charon (pronounced KAR-en), after the mythical boatman who ferried souls across the River Styx to Hades, the domain ruled by Pluto. (Christy also chose the name because of its similarity to Charlene, his wife's name.) The average distance between Charon and Pluto is a scant 19,640 km, less than 5% of the distance between the Earth and our Moon. The best pictures of Pluto and Charon have been obtained using the Hubble Space Telescope (Figure 16-19).

Observations show that Charon's orbital period of 6.3872 days is the same as the rotational period of Pluto *and* the rotational period of Charon. In other words, both Pluto and Charon rotate synchronously with their orbital motion,

figure 16-18   R I **V** U X G

The Discovery of Pluto's Moon, Charon Charon was discovered in 1978 when James Christy noticed a "lump" protruding from the top of this greatly magnified image of Pluto. (This picture is a photographic negative, so black and white are reversed.) The small irregular blobs surrounding Pluto and Charon are part of the photographic emulsion. This image was made with a telescope just a few kilometers from the Lowell Observatory, where Pluto itself was discovered. (U.S. Naval Observatory)

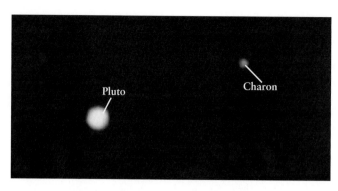

 figure 16-19 R I **V** U X G

Pluto and Charon—A "Double Planet"
Pluto and Charon resemble each other in mass and size more than any other planet-satellite pair in the solar system. They are also the closest such pair: the distance between them is only 19,640 km (12,200 mi), or about 1½ times the Earth's diameter. When this Hubble Space Telescope image was made, the angular separation between Pluto and Charon as seen from Earth was only 0.9 arcsec. (R. Albrecht, ESA/ESO Space Telescope Coordinating Facility; and NASA)

and so both always keep the same face toward each other. As seen from the Charon-facing side of Pluto, Charon neither rises nor sets but remains perpetually suspended above the horizon. Likewise, Pluto neither rises nor sets as seen from Charon.

Soon after the discovery of Charon, astronomers witnessed an alignment that occurs only once every 124 years. From 1985 through 1990, Charon's orbital plane appeared nearly edge-on as seen from the Earth, allowing astronomers to view mutual eclipses of Pluto and its moon. As the bodies passed in front of each other, their combined brightness diminished in ways that revealed their sizes and surface characteristics. Data from the eclipses gave Pluto's diameter as about 2300 km and Charon's as about 1200 km. For comparison, our Moon's diameter (3476 km) nearly equals the diameters of Pluto and Charon added together. The brightness variations during eclipses are consistent with the surface map in Figure 16-17 and suggest that Charon has a bright southern polar cap.

The average densities of Pluto and Charon, at about 2000 kg/m³, are essentially the same as that of Triton (2070 kg/m³). All three worlds are therefore probably composed of a mixture of rock and ice, as we might expect for small bodies that formed in the cold outer reaches of the solar nebula.

Pluto's spectrum shows absorption lines of various solid ices that cover the planet's surface, including nitrogen (N_2), methane (CH_4), and carbon monoxide (CO). Stellar occultation measurements have shown that Pluto, like Triton, has a very thin atmosphere. Exposed to a daytime temperature of around 40 K, N_2 and CO ices turn to gas more easily than frozen CH_4. For this reason, most of Pluto's tenuous atmosphere probably consists of N_2 and CO. Charon's weaker gravity has apparently allowed most of the N_2, CH_4, and CO to escape into space; only water ice is found on its surface.

Pluto and Charon are remarkably like each other in mass, size, and density. Throughout the rest of the solar system, planets are many times larger and more massive than any of their satellites. The exceptional similarities between Pluto and Charon suggest to some astronomers that this binary system may have formed when Pluto collided with a similar body. Perhaps chunks of matter were stripped from the second body, leaving behind a mass, now called Charon, that was captured into orbit by Pluto's gravity. Alternatively, Charon may have been captured by Pluto during a close encounter between the two worlds.

For either of these scenarios to be feasible, there must have been many Plutolike objects in the outer regions of the solar system. Some astronomers estimate that there must have been at least a thousand Plutos in order for a collision or close encounter between two of them to have occurred at least once since the solar system formed 4.6 billion years ago.

An overview of the solar system reveals that small objects are vastly more common than large objects. In other words, the number of objects in the solar system increases dramatically with decreasing mass. This trend suggests that for a thousand Plutolike worlds, there also must have been 10 to 50 icy, Earth-sized worlds in the vicinity of Uranus and Neptune. A direct hit by an object with about the mass of the Earth could have knocked Uranus on its side. None of these objects remains near the orbits of Neptune or Uranus today, because gravitational deflections by these two giant planets long ago pushed all the remaining Plutolike worlds far from the Sun.

In 1992 astronomers David Jewitt and Jane Luu discovered a small object 42 AU from the Sun. Named 1992 QB_1, its diameter is estimated to be only 240 km. Like Pluto, 1992 QB_1 has a reddish color, possibly because frozen methane has been degraded by eons of radiation exposure. As of this writing, more than 300 such objects have been discovered orbiting beyond Neptune, and it has been estimated that there are of thousands of other tiny worlds yet to be discovered in the outer reaches of the solar system.

Observers may soon have the chance to test this theory by looking for these Plutolike worlds. Attempts to find distant Plutos with ordinary telescopes is frustrating because these tiny, remote bodies do not reflect enough sunlight to be seen. They would, however, be weak sources of infrared radiation. Even though these icy objects probably have surface temperatures of only about 60 K, they are nevertheless warmer than the blank background sky. Highly sensitive, wide-angle infrared telescopes may therefore be able to detect a population of Plutos at distances of about 100 AU from the Sun.

Some excellent tools for this search will be the pair of 8-m telescopes Gemini North and Gemini South (one on Mauna Kea in Hawaii, the other on Cerro Pachón, Chile) and the 8-m Subaru (the Japan National Large Telescope on Mauna Kea). All three of these are becoming fully operational as this book is going to press (2001), and will be able to operate in the infrared as well as in the visible. If the search is successful, astronomers will have to account for a new class of objects in their theories about the formation of the solar system.

KEY WORDS

Great Dark Spot, p. 347

magnetic axis, p. 350

occultation, p. 351

radiation darkening, p. 352

KEY IDEAS

Discovery of the Outer Planets: Uranus was discovered by chance, while Neptune was discovered at a location predicted by applying Newtonian mechanics. Pluto was discovered after a long search. If another planet exists beyond Pluto, it is either very small or very far away.

Atmospheres of Uranus and Neptune: Both Uranus and Neptune have atmospheres composed primarily of hydrogen and helium, with about 2% methane.

• Methane absorbs red light, giving Uranus and Neptune their greenish-blue color.

• No white ammonia clouds are seen on Uranus or Neptune. Presumably the low temperatures have caused almost all the ammonia to precipitate into the interiors of the planets. All of these planets' clouds are composed of methane.

• Much more cloud activity is seen on Neptune than on Uranus. This is because Uranus lacks a substantial internal heat source.

Interiors and Magnetic Fields of Uranus and Neptune: Both Uranus and Neptune may have a rocky core surrounded by a mantle of water and ammonia. Electric currents in these mantles may generate the magnetic fields of the planets.

• Uranus's magnetic axis is inclined by 59° from its axis of rotation, while Neptune's is inclined by 47°. The magnetic and rotational axes of all the other planets are more nearly parallel. The magnetic fields of Uranus and Neptune are also offset from the centers of the planets.

Uranus's Unusual Rotation: Uranus's axis of rotation lies nearly in the plane of its orbit, producing greatly exaggerated seasonal changes on the planet.

• The unusual orientation of Uranus's rotational and magnetic axes may be the result of a collision with a planetlike object early in the history of our solar system. Such a collision could have knocked Uranus on its side.

Ring Systems of Uranus and Neptune: Uranus and Neptune are both surrounded by systems of thin, dark rings. The low reflectivity of the ring particles may be due to radiation-darkened methane ice.

Satellites of Uranus and Neptune: Uranus has five satellites similar to the moderate-sized moons of Saturn, plus at least 12 more small satellites. Neptune has eight satellites, one of which (Triton) is comparable in size to our Moon or the Galilean satellites of Jupiter.

• Triton has a young, icy surface indicative of tectonic activity. The energy for this activity may have been provided by tidal heating that occurred when Triton was captured by Neptune's gravity into a retrograde orbit.

• Triton has a tenuous nitrogen atmosphere.

Pluto and Charon: Pluto and its moon, Charon, move together in a highly elliptical orbit steeply inclined to the plane of the ecliptic. They are the only worlds in the solar system not yet visited by spacecraft.

• Several hundred small, icy worlds have been discovered beyond Neptune. Pluto, Charon, and Triton may be part of this population.

REVIEW QUESTIONS

1. Why do you suppose that the discovery of Neptune is rated as one of the great triumphs of science, whereas the discoveries of Uranus and Pluto are not?

2. Could astronomers in antiquity have seen Uranus? If so, why was it not recognized as a planet?

3. (a) Draw a figure like Figure 16-2, and indicate on it where Uranus was in 1986 and 1998. Explain your reasoning. (*Hint:* See Figures 16-1 and 16-2.) (b) In approximately what year will the Sun next be highest in the sky as seen from Uranus's south pole? Explain your reasoning.

4. Why do you suppose the tilt of Uranus's rotation axis was deduced from the orbits of its satellites and not by observing the rotation of the planet itself?

5. Describe the seasons on Uranus. Why are the Uranian seasons different from those on any other planet?

6. A number of storms in the Uranian atmosphere can be seen in Figure 16-3, but none are visible in Figure 16-1. How can you account for the difference?

7. Why are Uranus and Neptune distinctly blue-green in color, while Jupiter or Saturn are not?

8. Why are fewer white clouds seen on Uranus and Neptune than on Jupiter and Saturn?

9. Discuss some competing explanations of why Uranus and Neptune are substantially smaller than Jupiter and Saturn.

10. How do the orientations of Uranus's and Neptune's magnetic axes differ from those of other planets?

11. Briefly describe the evidence supporting the idea that Uranus was struck by a large planetlike object several billion years ago.

12. Compare the rings that surround Jupiter, Saturn, Uranus, and Neptune. Briefly discuss their similarities and differences.

13. The 1977 occultation that led to the discovery of Uranus's rings was visible from the Indian Ocean. Explain why it could not be seen from other parts of the Earth's night side.

14. As *Voyager 2* flew past Uranus, it produced images only of the southern hemispheres of the planet's satellites. Why do you suppose this was?

15. Why do astronomers think that the energy needed to resurface parts of Miranda came from tidal heating rather than the satellite's own internal heat?

16. Using the data in Table 16-3, explain why Uranus's satellites Caliban and Sycorax (both discovered in 1997) were probably captured from space rather than having formed at the same time as the planet itself.

17. If you were floating in a balloon in Neptune's upper atmosphere, in what part of the sky would you see Triton rise? Explain your reasoning.

18. Briefly describe the evidence supporting the idea that Triton was captured by Neptune.

19. Why is it reasonable to suppose that Neptune will someday be surrounded by a broad system of rings, perhaps similar to those that surround Saturn?

20. How can astronomers distinguish a faint solar system object like Pluto from background stars within the same field of view?

21. Describe the circumstantial evidence supporting the idea that Pluto is one of thousands of similar icy worlds that once occupied the outer regions of the solar system.

ADVANCED QUESTIONS

Problem-solving tips and tools

See Box 1-1 for the small-angle formula. For Question 24, you will need to recall that the volume of a sphere of radius r is $\frac{4}{3}\pi r^3$. You can find Newton's formula for the gravitational force between two objects in Section 4-7, a discussion of tidal forces in Section 4-8, Wien's law for blackbody radiation in Section 5-4, and a discussion of the transparency of the Earth's atmosphere to various wavelengths of light in Section 6-7.

22. At certain points in its orbit, a stellar occultation by Uranus would *not* reveal the existence of the rings. What points are those? How often does this circumstance arise? Explain using a diagram.

23. At what planetary configuration is the gravitational force of Neptune on Uranus at a maximum? For this configuration, calculate the gravitational force exerted by the Sun on Uranus and by Neptune on Uranus. Then calculate the fraction by which the sunward gravitational pull on Uranus is reduced by Neptune at that configuration. Based on your calculations, do you expect that Neptune has a relatively large or relatively small effect on Uranus's orbit?

24. According to one model for the internal structure of Uranus, the rocky core and the surrounding shell of water and methane ices together make up 80% of the planet's mass. This interior region extends from the center of Uranus to about 70% of the planet's radius. (a) Find the average density of this interior region. (b) How does your answer to (a) compare with the average density of Uranus as a whole? Is this what you would expect? Why?

25. Show that the ratio of the orbital periods of Neptune and Pluto is very close to 2:3. (This ratio is thought to result from gravitational interactions between Neptune and Pluto. These interactions prevent Neptune and Pluto from ever getting very close to each other.)

26. Suppose you were standing on Pluto. Describe the motions of Charon relative to the Sun, the stars, and your own horizon. Would you ever be able to see a total eclipse of the Sun? (*Hint:* You will need to calculate the angles subtended by Charon and by the Sun as seen by an observer on Pluto.) In what circumstances would you *never* see Charon?

27. It is thought that Pluto's tenuous atmosphere may become even thinner as the planet moves toward aphelion (which it will reach in 2113), then regain its present density as it again moves toward perihelion. Why should this be?

28. The brightness of sunlight is inversely proportional to the square of the distance from the Sun. For example, at a distance of 4 AU from the Sun, sunlight is only $(\frac{1}{4})^2 = \frac{1}{16} = 0.0625$ as bright as at 1 AU. Compared with the brightness of sunlight on the Earth, what is its brightness (a) on Pluto at perihelion and (b) on Pluto at aphelion? (c) How much brighter is it on Pluto at perihelion compared with aphelion? (Even this brightness is quite low. Noon on Pluto is about as dim as it is on the Earth a half hour after sunset on a moonless night.)

29. If Earth-based telescopes can resolve angles down to 0.25 arcsec, how large could an object be at Pluto's average distance from the Sun and still not present a resolvable disk?

30. The observations of Pluto shown in Figure 16-18 were made using blue and ultraviolet light. What

advantages does this have over observations made with red or infrared light?

31. Calculate the maximum angular separation between Pluto and Charon as seen from Earth. (Assume that Pluto is at its minimum distance from the Sun and that Pluto is at opposition as seen from Earth.) Compare your answer with the angular separation given in the caption to Figure 16-19.

32. Presumably Pluto and Charon raise tidal bulges on each other. Explain why the average distance between Pluto and Charon is probably constant, rather than increasing like the Earth-Moon distance or decreasing like the Neptune-Triton distance. Include a diagram like Figure 9-19 as part of your answer.

33. Suppose you wanted to search for planets beyond Pluto. Why might it be advantageous to do your observations at infrared rather than visible wavelengths? (Use Wien's law to calculate the wavelength range best suited for your search.) Could such observations be done at an observatory on the Earth's surface? Explain.

DISCUSSION QUESTIONS

34. Discuss the evidence presented by the outer planets that suggests that catastrophic impacts of planetlike objects occurred during the early history of our solar system.

35. Some scientists are discussing the possibility of placing spacecraft in orbit about Uranus and Neptune. What kinds of data should be collected, and what questions would you like to see answered by these missions?

36. If Triton had been formed along with Neptune rather than having been captured, would you expect it to be in a prograde or retrograde orbit? Would you expect the satellite to show signs of tectonic activity? Explain your answers.

37. Would you expect the surfaces of Pluto and Charon to be heavily cratered? Explain why or why not.

WEB/CD-ROM QUESTIONS

38. **Miranda.** Access and view the video "Uranus's Moon Miranda" in Chapter 16 of the *Universe* web site or CD-ROM. Discuss some of the challenges that would be involved in launching a spacecraft from Earth to land on the surface of Miranda.

39. The discovery image of Charon (Figure 16-18) was made by an astronomer at the U.S. Naval Observatory. Why do you suppose the U.S. Navy carries out work in astronomy? Search the World Wide Web for the answer.

40. Search the World Wide Web for a list of small objects that orbit beyond Neptune. (These are called "trans-Neptunian objects.") Use the list to learn about the current status of 1992 QB$_1$ and similar objects. What kinds of

orbits do these objects have? How do these orbits compare with that of Pluto? What are the largest and smallest objects of this sort that have so far been found, and how large are they?

41. **Separation of Pluto and Charon.** Pluto is located about 4.5 billion km from Earth and has a maximum observable separation from Charon of about 0.9 arcseconds. Access the AIMM (Active Integrated Media Module) called "Small-Angle Toolbox" in Chapter 1 of the *Universe* web site or CD-ROM. Use this AIMM and the above data to determine the distance between Pluto and Charon. How does your answer compare with the value given in the text?

OBSERVING PROJECTS

Observing tips and tools

During the period 2001–2005, Uranus will be at opposition in August, Neptune in late July and early August, and Pluto in June. You can find Uranus and Neptune with binoculars if you know where to look (a good star chart is essential), but Pluto is so dim that it can be a challenge to spot even with a 25-cm (10-inch) telescope. Each year, star charts that enable you to find these planets are printed in the issue of *Sky & Telescope* for the month in which each planet is first visible in the nighttime sky. You can also locate the outer planets using the *Starry Night* program on the CD-ROM that accompanies this textbook.

42. Make arrangements to view Uranus through a telescope. The planet is best seen at or near opposition. Use a star chart at the telescope to find the planet. Are you certain that you have found Uranus? Can you see a disk? What is its color?

43. If you have access to a large telescope, make arrangements to view Neptune. Like Uranus, Neptune is best seen at or near opposition and can most easily be found using a star chart. Can you see a disk? What is its color?

44. If you have access to a large telescope (at least 25 cm in diameter), make arrangements to view Pluto. Using the star chart from *Sky & Telescope* referred to above, view the part of the sky where Pluto is expected to be seen and make a careful sketch of all of the stars that you see. Repeat this process on a later night. Can you identify the "star" that has moved?

45. Use the *Starry Night* program to observe the five large satellites of Uranus. First turn off daylight (select **Daylight** in the **Sky** menu) and show the entire celestial sphere (select **Atlas** in the **Go** menu). Center on Uranus by using the **Find...** command in

the **Edit** menu. Using the controls at the right-hand end of the Control Panel, zoom in or out until the field of view is roughly 1 arcminute (1′). Select **Planet List** in the **Window** menu, and then click on the triangle to the left of the name **Uranus**. This will reveal a list of the planet's five large satellites. Then, in the "Orbit" column on the right-hand side of the Planet List, click to the right of each satellite's name. As you click, marks will appear in this column and the satellites' orbits will appear in the main window. Set the time step in the Control Panel to 2 hours. (**a**) In the Control Panel, click on the "Forward" button (a triangle that points to the right). Describe how the satellites move, and relate your observations to Kepler's third law (see Sections 4-5 and 4-7). (**b**) From time to time, Miranda lies directly between Ariel and Uranus. Using the single-step time control buttons, determine how much time elapses between successive instances of this configuration. How does this compare to the orbital periods of Miranda and Ariel? Relate your observations to the ideas of sidereal period and synodic period (see Section 4-2).

46. Use the *Starry Night* program to observe Neptune and its two outer satellites. First turn off daylight (select **Daylight** in the **Sky** menu) and show the entire celestial sphere (select **Atlas** in the **Go** menu). Center on Neptune by using the **Find...** command in the **Edit** menu. (**a**) Using the controls at the right-hand end of the Control Panel, zoom in until you can clearly see surface features on the planet. Set the time step in the Control Panel to 2 hours, then click on the "Forward" button (a triangle that points to the right). In what direction does Neptune appear to rotate? Illustrate your answer with a diagram. (**b**) Select **Planet List** in the **Window** menu and click on the triangle to the left of the name **Neptune**. This will reveal a list of the planet's two outer satellites. Then, in the "Orbit" column on the right-hand side of the Planet List, click to the right of each satellite's name. As you click, check marks will appear in this column and the satellites' orbits will appear in the main window. Zoom out until you can see the orbit of Triton, the inner of the two satellites, and click on the "Forward" button in the Control Panel. Make a diagram of the orbit that you see. How does the direction of Triton's orbital motion compare to the direction of Neptune's rotation? Does the orbit appear to lie in the same plane as Neptune's equator? How can you tell? (**c**) Zoom farther out until you can see all of Nereid's elongated orbit. To follow the motion of this rather dim satellite, double-click on the name **Nereid** in the Planet List palette, then click on the **Centre and Lock** button. Then click on the "Forward" button in the Control Panel. Can you see Nereid move? Now change the time step to 10 days and watch Nereid go through several orbits. Make a diagram of the orbit that you see. How does the direction of Nereid's orbital motion compare to the direction of Neptune's rotation? Does the orbit appear to lie in the same plane as Nereid's equator? How can you tell?

47. Use the *Starry Night* program to observe Pluto and Charon. First turn off daylight (select **Daylight** in the **Sky** menu) and show the entire celestial sphere (select **Atlas** in the **Go** menu). Center on Pluto by using the **Find...** command in the **Edit** menu. Using the controls at the right-hand end of the Control Panel, zoom in as far as possible. Select **Planet List** in the **Window** menu and click on the triangle to the left of the name **Pluto**. Then, in the "Orbit" column on the right-hand side of the Planet List, click to the right of the name **Charon**. A mark will appear in this column and Charon's orbit will appear in the main window. In the Control Panel, set the time step to 3 hours. (**a**) Use the single-step time control buttons to step through enough time to determine the period of Charon's orbit, How does your answer compare to the value given in Section 16-10? (**b**) What is the apparent shape of Charon's orbit around Pluto? How can you reconcile this with the fact that Charon's actual orbit is nearly a perfect circle?

Vagabonds of the Solar System

(Courtesy of Johnny Horne)

R I **V** U X G

On the Tuesday night after Easter in the year 1066, a strange new star appeared in the European sky. Seemingly trailing fire, this "star" hung in the night sky for weeks. Some, including the English king Harold, may have taken it as an ill omen; others, such as Harold's rival William, Duke of Normandy, may have regarded it as a sign of good fortune. Perhaps they were both right, because by the end of the year Harold was dead and William was installed on the English throne.

Neither man ever knew that he was actually seeing the trail of a city-sized ball of ice and dust, part of which evaporated as it rounded the Sun to produce a long tail. They were seeing a comet—remarkably, the same Comet Halley that last appeared in 1986 and will next be seen in 2061. Although we now understand that comets are natural phenom-

ena rather than supernatural omens, they still have the power to awe and inspire us, as did Comet Hale-Bopp (shown here) in 1997.

Comets, along with asteroids and meteoroids, are relics left over from the formation of the solar system. As such, they are as important to astronomy as fossils are to paleontology. While comets are made of ice and dust, asteroids are rocky bits of the solar nebula that never formed into a full-sized planet. Some asteroids break into fragments, as do some comets, and some of these fragments fall to the Earth as meteorites. On extraordinarily rare occasions, an entire asteroid strikes Earth. Such an asteroid impact may have led to the extinction of the dinosaurs some 65 million years ago. Thus, these minor members of our solar system can have major consequences for our planet.

As you read the sections of this chapter, look for the answers to the following questions.

17-1　How and why were the asteroids first discovered?

17-2　Why didn't the asteroids coalesce to form a single planet?

17-3　What do asteroids look like?

17-4　How might an asteroid have caused the extinction of the dinosaurs?

17-5　What are the differences among meteoroids, meteors, and meteorites?

17-6　What do meteorites tell us about the way in which the solar system formed?

17-7　Why do comets have tails?

17-8　Where do comets come from?

17-9　What is the connection between comets and meteor showers?

17-1　A search for a planet between Mars and Jupiter led to the discovery of asteroids

After Herschel's discovery of Uranus in 1781 (which we described in Section 16-1), many astronomers began to wonder if there were other, as yet undiscovered, planets. If these planets were too dim to be seen by the naked eye, they might still be visible through telescopes. These planets would presumably be found close to the plane of the ecliptic, because all the other planets orbit the Sun in or near that plane (see Section 7-1). But how far from the Sun might these additional planets be found?

Astronomers in the late eighteenth century had a simple rule of thumb, called the *Titius-Bode Law*, relating the sizes of planetary orbits: From one planet to the next, the semimajor axis of the orbit increases by a factor between approximately 1.4 and 2. For example, the semimajor axis of Mercury's orbit is 0.39 AU. Venus's orbit has a semimajor axis of 0.72 AU, which is larger by a factor of $(0.72)/(0.39) = 1.85$. (Unlike Newton's laws, the Titius-Bode "Law" is probably not a fundamental law of nature, but just a reflection of how our solar system happened to form.)

The glaring exception is Jupiter (semimajor axis 5.20 AU), which is more than *3 times* farther from the Sun than Mars (semimajor axis 1.52 AU). Table 7-1 depicts this large space between the orbits of Mars and Jupiter. If the rule of thumb is correct, astronomers reasoned, there should be a "missing planet" with an orbit about 1.4 to 2 times larger than that of Mars. Its semimajor axis should therefore fall somewhere in the range between 2 to 3 AU.

Six German astronomers, who jokingly called themselves the "Celestial Police," organized an international group to begin a careful search for this missing planet. Before their search got under way, however, surprising news reached them from Giuseppe Piazzi, a Sicilian astronomer. Piazzi had been carefully mapping faint stars in the constellation of Taurus when on January 1, 1801, the first night of the nineteenth century, he noticed a dim, previously uncharted star. This star's position shifted slightly over the next several nights. Suspecting that he might have found the "missing planet," Piazzi excitedly wrote to Johann Bode, the director of the Berlin Observatory and a member of the Celestial Police.

Unfortunately, Piazzi's letter did not reach Bode until late March. By that time, Piazzi's object had moved around its orbit and was too near the Sun to be visible in the night sky. With no way of knowing where to look after it emerged from the Sun's glare, astronomers feared that Piazzi's object might have been lost.

Upon hearing of this dilemma, the brilliant young German mathematician Karl Friedrich Gauss took up the challenge. He developed a general method of computing an object's orbit from only three separate measurements of its position on the celestial sphere. (With slight modifications, this same method is used by astronomers today.) In November 1801, Gauss predicted that Piazzi's object would be found in the predawn sky in the constellation of Virgo. And indeed Piazzi's object was sighted again on December 31, 1801, only a short distance from the position Gauss had calculated. Piazzi named the object Ceres (pronounced SEE-reez), after the patron goddess of Sicily in Roman mythology.

Ceres orbits the Sun once every 4.6 years at an average distance of 2.77 AU, just where astronomers had expected to find the missing planet. But Ceres is very small; its diameter is estimated to be a scant 918 km (Figure 17-1). Hence, Ceres reflects only a little sunlight, which is why it cannot be seen with the naked eye even at opposition. (It can be seen with binoculars, but it looks like just another faint star.) Because of its small size, it is known today as a **minor planet**, or **asteroid**.

On March 28, 1802, Heinrich Olbers discovered another faint, starlike object that moved against the background stars. He called it Pallas, after the Greek goddess of wisdom. Like Ceres, Pallas orbits the Sun every 4.6 years at an average distance of 2.77 AU, but its orbit is more steeply inclined from the plane of the ecliptic and is somewhat more eccentric. Pallas is even smaller and dimmer than Ceres, with an estimated diameter of only 522 km. Obviously, Pallas was also not the missing planet.

Did the missing planet even exist? Some astronomers speculated that perhaps there had once been such a full-size planet, but it had somehow broken apart or exploded to produce

ƒigure 17-1   R I **V** U X G

Ceres Compared with the Earth and Moon A drawing of Ceres, the largest of the asteroids and the first one discovered, is shown here to the same scale as the Earth and the Moon. Ceres is too small to be considered a planet, and it cannot be regarded as a moon because it does not orbit any other body. To denote their status, asteroids are also called minor planets. (NASA)

ƒigure 17-2   R I **V** U X G

The Trail of an Asteroid Telescopes used for photographing the sky are motorized so that they follow the apparent motion of the stars across the sky. Because asteroids orbit the Sun, their positions change with respect to the stars. Thus, asteroids can be detected by their blurred trails on time exposure photographs of the stars. This asteroid trail was recorded quite by accident by the Hubble Space Telescope. (R. Evans and K. Stapelfeldt, Jet Propulsion Laboratory; and NASA)

a population of asteroids orbiting between Mars and Jupiter. The search was on to discover this population. Two more such asteroids were discovered in the next few years, Juno in 1804 and Vesta in 1807. Several hundred more were discovered beginning in the mid-1800s, by which time telescopic equipment and techniques had improved.

The next major breakthrough came in 1891, when the German astronomer Max Wolf began using photographic techniques to search for asteroids. Before this, asteroids had to be painstakingly discovered by scrutinizing the skies for faint, uncharted, starlike objects whose positions move slightly from one night to the next. With photography, an astronomer simply aims a camera-equipped telescope at the stars and takes a long exposure. If an asteroid happens to be in the field of view, its orbital motion during the long exposure leaves a distinctive trail on the photographic plate (Figure 17-2). Using this technique, Wolf alone discovered 228 asteroids.

 Today, dozens of new asteroids are discovered every month, mostly by amateur astronomers. After the discoverer reports a new find to the Minor Planet Center of the Smithsonian Astrophysical Observatory, the new asteroid is first given a provisional designation. For example, asteroid 1980 JE was the fifth ("E") to be discovered during the second half of May (the tenth half-month of the year, "J") 1980. If the asteroid is located again on at least four succeeding oppositions (a process than can take decades), the asteroid is assigned an official sequential number (Ceres is 1, Pallas is 2, and so forth), and the discoverer is given the privilege of suggesting a name for the asteroid. The suggested name must then be approved by the International Astronomical Union.

For example, the asteroid 1980 JE was officially named 3834 Zappafrank (after the American musician Frank Zappa) in 1994.

Ceres—or, in modern nomenclature, 1 Ceres—is unquestionably the largest asteroid. With a diameter of 934 km, 1 Ceres accounts for about 30% of the mass of all the asteroids combined. Only three asteroids—1 Ceres, 2 Pallas, and 4 Vesta—have diameters greater than 300 km. Thirty other asteroids have diameters between 200 and 300 km, and 200 more are bigger than 100 km across.

Astronomers estimate that approximately 100,000 asteroids are bright enough to be detected by Earth-based telescopes. The vast majority are smaller than 1 km across. Like Ceres, Pallas, Vesta, and Juno, most asteroids orbit the Sun at distances between 2 and 3.5 AU. This region of our solar system between the orbits of Mars and Jupiter is called the **asteroid belt** (Figure 17-3).

You may wonder how spacecraft such as *Voyager 1*, *Voyager 2*, and *Galileo* were able to cross the asteroid belt to reach Jupiter. Aren't asteroids hazards to space navigation, as they are sometimes depicted in science-fiction movies? Wouldn't these spacecraft have been likely to run into an asteroid? Happily, the answer to both questions is no! While it is true that there are more than 10^5 asteroids, they are spread over a belt with a total area (as seen from above the plane of the ecliptic) of about 10^{17} square kilometers—about 10^8 times greater than the Earth's entire surface area. Hence, the average distance between asteroids in the ecliptic plane is about 10^6 kilometers, about twice the distance between the

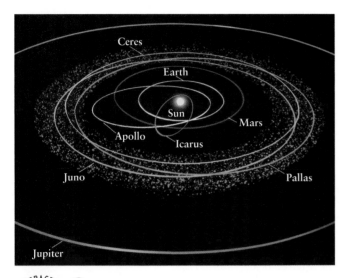

Figure 17-3

The Asteroid Belt Most asteroids orbit the Sun in a belt about 1.5 AU wide between the orbits of Mars and Jupiter. The large asteroids 1 Ceres, 2 Pallas, and 3 Juno all have roughly circular orbits that lie within this belt. By contrast, some asteroids, such as 1862 Apollo and 1566 Icarus, have eccentric orbits that carry them inside the orbit of Earth.

Earth and Moon. Furthermore, many asteroids have orbits that are tilted out of the ecliptic. The *Galileo* spacecraft did pass within a few thousand kilometers of two asteroids, but these passes were intentional and required that the spacecraft be carefully aimed.

What of the nineteenth-century notion of a missing planet? This idea has long since been discarded, because the combined matter of all the asteroids (including an estimate for those not yet officially known) would produce an object barely 1500 km in diameter. This is considerably smaller than anything that could be considered a missing planet. It seems more reasonable to suppose that asteroids are debris left over from the formation of the solar system.

17-2 Jupiter's gravity helped shape the asteroid belt

Where did the asteroid belt come from? Why did asteroids, rather than a planet, form between Mars and Jupiter? Important insights into these questions come from supercomputer simulations of the formation of the terrestrial planets, like the one depicted in Figure 7-17 but more elaborate. A typical simulation starts off with a billion (10^9) or so planetesimals, each with a mass of 10^{15} kg or more, so that their combined mass equals that of the terrestrial planets. The computer then follows these planetesimals as they collide and coalesce to form the planets. By selecting different speeds and positions

for the planetesimals at the start of the simulation, a scientist can study various ways in which the inner solar system might have turned out.

If the effects of Jupiter's gravity are not included in a simulation, an Earth-sized planet usually forms in the asteroid belt, giving us five terrestrial planets instead of four. But when Jupiter's gravity is added, this fifth terrestrial planet is less likely to form. Jupiter's strong pull "clears out" the asteroid belt by disrupting the orbits of planetesimals in the belt, ejecting most of them from the solar system altogether. As a result, the asteroid belt becomes depleted of planetesimals before a planet has a chance to form. The few planetesimals that remain in the simulation—only a small fraction of those that originally orbited between Mars and Jupiter—become the asteroids that we see today.

Jupiter's gravitational influence, however, probably cannot explain why asteroid orbits have the wide variety of semimajor axes, eccentricities, and inclinations to the ecliptic shown in Figure 17-3. Even with the influence of Jupiter, collisions between planetesimals in the solar nebula would have tended to leave the asteroids in roughly circular orbits close to the ecliptic plane, like the orbits of the planets. Therefore, something else must have "stirred up" the asteroids to put them into their current state. Recent analyses suggest that one or more Mars-sized objects—that is, terrestrial planets intermediate in size between the Earth and the Moon—did in fact form within the asteroid belt. Planetesimals that passed within close range of these Mars-sized objects would have experienced strong gravitational forces, and these forces could have deflected the planetesimals into the eccentric or inclined orbits that many asteroids follow today.

Mars-sized objects would also have helped clear out the asteroid belt by deflecting planetesimals into orbits that crossed Jupiter's orbit. Once a planetesimal wandered close to Jupiter, it could be ejected from the solar system by the giant planet's gravitational force.

In this scheme, the Mars-sized objects disappeared when gravitational forces from Jupiter either accelerated them out of the solar system entirely or caused them to slow down and fall into the Sun. A Mars-sized object colliding with the Earth is thought to have produced the Moon (see Figure 9-20). Perhaps this object was an "escapee" from the asteroid belt.

Jupiter's gravity continues to influence the asteroid belt down to the present day. The American astronomer Daniel Kirkwood found the first evidence for this in 1867, when he discovered gaps in the asteroid belt. These features, today called **Kirkwood gaps,** can best be seen in a histogram like Figure 17-4, which shows asteroid orbital periods. The gaps in this histogram show regions where there are relatively few asteroids. Curiously, these gaps occur for asteroid orbits whose periods are simple fractions (such as ⅓, ⅖, 3⁄7, and ½) of Jupiter's orbital period.

To understand why the Kirkwood gaps exist, imagine an asteroid within the belt that circles the Sun once every 5.93 years, exactly half of Jupiter's orbital period. On every second trip around the Sun, the asteroid finds itself lined up between Jupiter and the Sun, always at the same location and with the same orientation. Because of these repeated alignments, called a *2-to-1 resonance,* Jupiter's gravity deflects the asteroid from

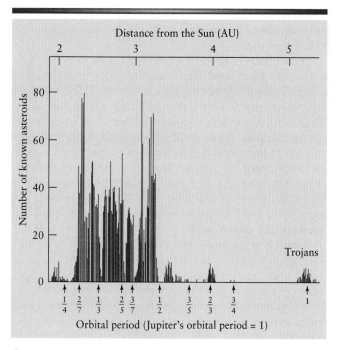

Figure 17-4

The Kirkwood Gaps This graph displays the numbers of asteroids at various distances from the Sun. Notice that very few asteroids have orbits whose orbital periods are equal to the simple fractions ⅓, ⅖, 3/7, and ½ of Jupiter's 11.86-year orbital period. Repeated alignments with Jupiter and its strong gravitational pull deflect asteroids away from these orbits. The Trojan asteroids (at the far right in the graph) follow the same orbit as Jupiter and thus have the same orbital period. We describe these in Section 17-4.

sect each other. A collision between asteroids must be an awe-inspiring event. Typical collision speeds are estimated to be from 1 to 5 km/s (3600 to 18,000 km/h, or 2000 to 11,000 mi/h), which is more than sufficient to shatter rock. Recent observations show that these titanic impacts play an important role in determining the nature of asteroids.

State-of-the-art optical telescopes, like the Hubble Space Telescope, have good enough resolution to reveal surface details on large asteroids such as 1 Ceres, 2 Pallas, and 4 Vesta (Figure 17-5). These asteroids resemble miniature terrestrial planets. They have enough mass, and hence enough gravity, to pull themselves into roughly spherical shapes. Vesta has a surface of basaltic rock resulting from ancient lava flows, suggesting that it underwent chemical differentiation (see Section 7-8 and Box 7-1). These large asteroids are likely to have extensively cratered surfaces due to collisions with other asteroids in the asteroid belt.

Optical telescopes reveal much less detail about the smaller asteroids. But by studying how the brightness of an asteroid changes as it rotates, astronomers can infer the asteroid's shape. An elongated asteroid appears dimmest when its long axis points toward the Sun so that it presents less of its surface to reflect sunlight. By contrast, a rotating spherical asteroid will have a more constant brightness. Such observations show that most smaller asteroids are not spherical at all. Instead, they retain the odd shapes produced by previous collisions with other asteroids.

An important recent innovation in studying asteroids has been the use of radar. An intense radio beam is sent toward an asteroid, and the reflected signal is detected and analyzed to determine the asteroid's shape (Figure 17-6). The amount of

its original 5.93-year orbit, ultimately ejecting it from the asteroid belt. Similar resonances clear out the Cassini division in Saturn's rings (see Section 15-4) and drive the tidal heating of Jupiter's moon Io (see Section 14-4). According to Kepler's third law, a period of 5.93 years corresponds to a semimajor axis of 3.28 AU. Because of Jupiter's gravity, almost no asteroids orbit the Sun at this average distance.

Another Kirkwood gap corresponds to an orbital period of one-third Jupiter's period, or 3.95 years (a *3-to-1 resonance*). Additional gaps exist for other simple ratios between the periods of asteroids and Jupiter. Any Mars-sized bodies that existed in the asteroid belt were probably ejected after wandering into one of the Kirkwood gaps.

17-3 Asteroids occasionally collide with one another

Although the asteroid belt is mostly empty space, collisions between asteroids should occur from time to time. Asteroids move in orbits with a variety of different eccentricities and inclinations (see Figure 17-3), and some of these orbits inter-

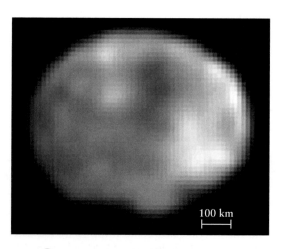

Figure 17-5 R I **V** U X G

A Telescope View of a Large Asteroid With a diameter of 530 km (330 mi), 4 Vesta is the third largest of all the asteroids. This Hubble Space Telescope image shows dark areas that are presumably lava flows, similar to the Moon's maria. The asteroid's nonspherical shape may be the result of a massive impact that tore away a substantial portion of Vesta. (B. Zellner, Georgia Southern University; P. Thomas, Cornell University; and STScI/NASA)

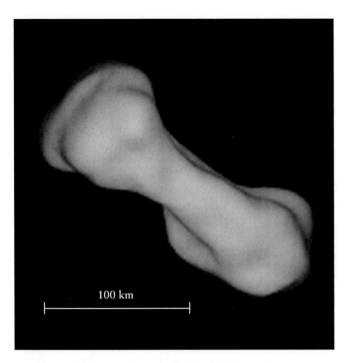

figure 17-6 R I V U X G

A Radar View of a Medium-Sized Asteroid This radar image reveals the curious dog-bone shape of the asteroid 216 Kleopatra. The Arecibo radio telescope (see Figure 10-5) was used both to send radio waves toward Kleopatra and to detect the waves that were reflected back. Scientists then created this image by computer analysis of the reflected waves. (Arecibo Observatory, JPL/NASA)

the radio signal that is reflected also gives clues about the texture of the surface.

The most powerful technique for studying asteroids is to send a spacecraft to make observations at close range. The Jupiter-bound *Galileo* spacecraft gave us our first close-up view of an asteroid in 1991, when it flew past 951 Gaspra. Two years later *Galileo* made a close pass by 243 Ida.

More recently, the *NEAR Shoemaker* spacecraft made a close approach to 253 Mathilde in 1997 (Figure 17-7), and *Deep Space 1* came within 26 km (16 mi) of 9969 Braille in 1999. *NEAR Shoemaker* went into orbit around 433 Eros in 2000 (see Figure 7-7) and touched down on Eros the following year. It was the first spacecraft to orbit and land on an asteroid.

What have spacecraft and Earth-based radar told us about the asteroids? To appreciate the answer, we should first consider what scientists expected to find with these new tools. A few years ago, the consensus among planetary scientists was that the smaller asteroids were solid, rocky objects. The surfaces were presumed to be cratered by impacts, but only relatively small craters were expected. The reason is that an impact forceful enough to make a very large crater would probably fracture a rocky asteroid into two or more smaller asteroids.

The truth proves to be rather different. It is now suspected that most asteroids larger than about a kilometer are not entirely solid, but are actually composed of several pieces. An extreme example is Mathilde (see Figure 17-7). This asteroid has such a low density (1300 kg/m³) that it cannot be made of solid rock, which typically has densities of 2500 kg/m³ or higher. Instead, it is probably a "rubble pile" of small fragments that fit together loosely. Billions of years of impacts have totally shattered this asteroid. But the impacts were sufficiently gentle that the resulting fragments drifted away relatively slowly, only to be pulled back onto the asteroid by gravitational attraction.

Another surprising discovery is that some asteroids have extremely large craters, comparable in size to the asteroid itself. An example is the immense crater on Mathilde, shown in Figure 17-7. The object that formed this crater probably collided with Mathilde at a speed of 1 to 5 km/s (3600 to 18,000 km/h, or 2000 to 11,000 mi/h), which is more than sufficient to shatter rock. Why didn't this collision break the asteroid completely apart, sending the pieces flying in all directions? The explanation may be that the asteroid had already been broken into a loose collection of fragments before this major impact took place. Such a fragmented asteroid can more easily absorb the energy of a collision than can a solid, rocky one.

 If you fire a rifle bullet at a wine glass, the glass will shatter into tiny pieces. But if you fire the same bullet into a bag full of sand (which has much the same

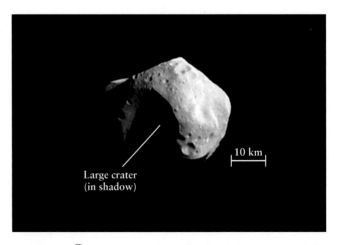

figure 17-7 R I V U X G

A Spacecraft View of a Small Asteroid The asteroid 253 Mathilde has an albedo of only 0.04, making it half as reflective as a charcoal briquette. It has an irregular shape, with dimensions 66 × 48 × 46 km. The partially shadowed crater indicated by the label is about 20 km across and 10 km deep. The *NEAR Shoemaker* spacecraft imaged Mathilde at close range during its close flyby in June 1997. (Johns Hopkins University, Applied Physics Laboratory)

chemical composition as glass), the main damage will be a hole in the bag. In a similar way, an asteroid that is a collection of small pieces can survive a major impact, while a solid asteroid cannot.

Relatively low-speed collisions between asteroids probably produce rubble piles like Mathilde. But higher-speed collisions may shatter an asteroid permanently. Evidence for this was pointed out in 1918 by the Japanese astronomer Kiyotsugu Hirayama, who drew attention to families of asteroids that share nearly identical orbits. Each of these groupings, now called **Hirayama families,** presumably resulted from a parent asteroid that was broken into fragments by a high-speed, high-energy collision with another asteroid.

17-4 Asteroids are found outside the asteroid belt—and have struck the Earth

While Jupiter's gravitational pull depletes certain orbits in the asteroid belt, it actually captures asteroids at two locations much farther from the Sun. The gravitational forces of the Sun and Jupiter work together to hold asteroids at two locations called the **stable Lagrange points.** One of these points is one-sixth of the way around Jupiter's orbit ahead of the planet, while the other point is the same distance behind the planet (Figure 17-8). The French mathematician Joseph Louis Lagrange predicted the existence of these points in 1772; asteroids were first discovered there in 1906.

The asteroids trapped at Jupiter's Lagrange points are called **Trojan asteroids,** named individually after heroes of the Trojan War. Approximately 460 Trojan asteroids have been catalogued so far. An asteroid has also been discovered at one of the Lagrange points of Mars.

Some asteroids have highly elliptical orbits that bring them into the inner regions of the solar system. Asteroids that cross Mars's orbit, or whose orbits lie completely within that of Mars, are called **near-Earth objects** or **NEOs.**

Occasionally a near-Earth object may pass relatively close to the Earth. On October 30, 1937, Hermes passed the Earth at a distance of 900,000 km (560,000 mi), only a little more than twice the distance to the Moon. The closest call to date occurred on December 9, 1994, when an asteroid called 1994 XM_1 passed within a mere 105,000 km (60,000 mi) of Earth (Figure 17-9). Because 1994 XM_1 is only about 10 meters across (about the size of a house), even if it had been aimed directly at the Earth it would probably have broken up in the atmosphere before reaching the surface.

Almost 300 asteroids are known with orbits that cross that of the Earth, and astronomers estimate that there are hundreds of thousands more that we have not yet detected. Fortunately, space is a big place and the Earth is a rather small target, so asteroids strike the Earth only very rarely.

However, collisions between asteroids produce numerous smaller chunks of rock, and many of these do eventually rain down on the terrestrial planets. Fortunately for us, the major-

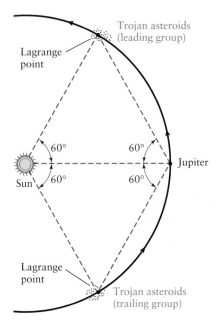

figure 17-8

The Trojan Asteroids The combined gravitational forces of Jupiter and the Sun tend to trap asteroids at the two stable Lagrange points along Jupiter's orbit. Note that the Sun, Jupiter, and either one of these points lie at the vertices of an equilateral triangle. Asteroids at these points are named after Homeric heroes of the Trojan War.

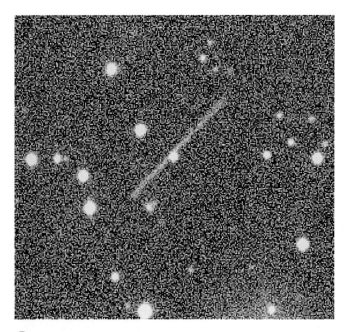

figure 17-9   R I **V** U X G

The Near-Earth Object 1994 XM₁ This image was made on December 9, 1994, 12 hours before asteroid 1994 XM_1 made its closest approach to the Earth. The asteroid was moving so rapidly during the exposure that it appears as a streak rather than as a point of light. (J. V. Scotti, The Spacewatch Project, University of Arizona)

figure 17-10 R I **V** U X G

The Barringer Crater This 1.2-km (¾-mi) wide crater was formed by a meteoroid impact about 50,000 years ago. About 100 other impact craters between 3 and 150 km across have been identified on the Earth. None of these is more than 500 million years old, because the reshaping of the Earth's surface by erosion and by tectonic processes tends to erase craters over time. (Meteor Crater Enterprises)

world. In every case, geological dating reveals that this apparently worldwide layer of iridium-rich clay was deposited about 65 million years ago.

Paleontologists were quick to realize the significance of this date, because it was 65 million years ago that all the dinosaurs rather suddenly became extinct. In fact, two-thirds of all the species on the Earth disappeared within a relatively brief span of time.

The Alvarez discovery suggests a startling explanation for this dramatic mass extinction: Perhaps an asteroid hit the Earth at that time. An asteroid 10 km in diameter slamming into the Earth could have thrown enough dust into the atmosphere to block out sunlight for several years. As the temperature dropped drastically and plants died for lack of sunshine, the dinosaurs would have perished, along with many other creatures in the food chain. The dust eventually settled, depositing an iridium-rich layer around the world. Tiny, rodentlike creatures capable of ferreting out seeds and nuts were among the animals that managed to survive this catastrophe, setting the stage for the rise of mammals in the eons that followed. As the dominant mammals on the planet, we may owe our very existence to an ancient asteroid.

In 1992 a team of geologists suggested that this asteroid crashed into a site in Mexico. They based this conclusion on glassy debris and violently shocked grains of rock ejected from the 180-km-diameter Chicxulub Crater on the Yucatán Peninsula. From the known rate at which radioactive potassium decays, the scientists

ity of these fragments, usually called **meteoroids,** are quite small. But on rare occasions a large fragment collides with our planet. When such a collision takes place, the result is an impact crater whose diameter depends on the mass and speed of the impinging object.

A relatively young, pristine terrestrial impact crater is the Barringer Crater near Winslow, Arizona (Figure 17-10). This crater was formed 50,000 years ago when an iron-rich meteoroid, measuring approximately 50 m (160 ft) across, struck the ground with a speed of more than 11 km/s (40,000 km/h, or 25,000 mi/h). The resulting blast was equivalent to the detonation of a 20-megaton hydrogen bomb and left a crater 1.2 km (¾ mi) wide and 200 m (650 ft) deep at its center.

Iron is an important constituent of asteroids and meteoroids. Another element, iridium, is common in iron-rich minerals but rare in other types of rocks. Measurements of iridium in the Earth's crust can thus tell us the rate at which meteoritic material has been deposited on the Earth over the ages. The geologist Walter Alvarez and his physicist father, Luis Alvarez, from the University of California, Berkeley, made such measurements in the late 1970s.

Working at a site of exposed marine limestone in the Apennine Mountains in Italy, the Alvarez team discovered an exceptionally high abundance of iridium in a dark-colored layer of clay between limestone strata (Figure 17-11). Since this discovery in 1979, a comparable layer of iridium-rich material has been uncovered at numerous sites around the

figure 17-11 R I **V** U X G

Iridium-Rich Clay This photograph of strata in the Apennine Mountains of Italy shows a dark-colored layer of iridium-rich clay. This is sandwiched between older layers of white limestone (lower right) and younger layers of grayish limestone (upper left). The iridium-rich layer may be the result of an asteroid whose impact 65 million years ago caused the extinction of the dinosaurs. The coin is the size of a U.S. quarter. (Courtesy of W. Alvarez)

claim to have pinpointed the date when the asteroid struck—64.98 million years ago.

Some geologists and paleontologists are not convinced that an asteroid impact led to the extinction of the dinosaurs. But many scientists agree that this hypothesis fits the available evidence better than any other explanation that has been offered so far.

17-5 Meteorites are classified as stones, stony irons, or irons, depending on their composition

A meteoroid, like an asteroid, is a chunk of rock in space. There is no official dividing line between meteoroids and asteroids, but the term *asteroid* is generally applied only to objects larger than a few hundred meters across.

A **meteor** is the brief flash of light (sometimes called a *shooting star*) that is visible at night when a meteoroid enters the Earth's atmosphere (Figure 17-12). Frictional heat is generated as the meteoroid plunges through the atmosphere, leaving a fiery trail across the night sky.

WEB LINK 17.10 If a piece of rock survives its fiery descent through the atmosphere and reaches the ground, it is called a **meteorite**. As incredible as it may seem, an estimated total of 300 tons of extraterrestrial matter falls on the Earth each day. Most of this is dust rather than meteorites, however, and is hardly noticeable. It is fortunate that large meteorite falls are rather rare events, because these fast-moving rock fragments can cause substantial damage when they strike the ground (Figure 17-13). Because they are so rare, meteorites are prized by scientists and collectors alike. Almost all meteorites are asteroid fragments and thus provide information about the chemical composition of asteroids. A very few meteorites have been identified as pieces of the Moon or Mars that were blasted off the surfaces of those worlds by asteroid impacts and eventually landed on the Earth.

People have been finding specimens of meteorites for thousands of years, and descriptions of them appear in ancient Chinese, Greek, and Roman literature. Many civilizations have regarded meteorites as objects of veneration. The sacred Black Stone, enshrined at the Ka'aba in the Great Mosque at Mecca, may be a relic of a meteorite impact.

The extraterrestrial origin of meteorites was hotly debated until as late as the eighteenth century. Upon hearing a lecture by two Yale professors, President Thomas Jefferson is said to have remarked, "I could more easily believe that two Yankee professors could lie than that stones could fall from Heaven." Although several falling meteorites had been widely witnessed and specimens from them had been collected, many scientists were reluctant to accept the idea that rocks could fall to the Earth from outer space. Conclusive evidence for the extraterrestrial origin of meteorites came on April 26, 1803, when fragments pelted the French town of L'Aigle. This event was analyzed by the noted physicist Jean-Baptiste Biot, and his conclusions helped finally to convince scientists that meteorites were indeed extraterrestrial.

figure 17-12 R I **V** U X G

A Meteor A meteor is produced when a meteoroid—a piece of interplanetary rock or dust—strikes the Earth's atmosphere at high speed, usually between 5 and 30 km/s. Air friction heats the meteoroid and causes it to glow as it disintegrates. Because the meteoroids are moving so fast, most of them burn up completely while still at high altitude, about 100 km (60 mi) above Earth. This long exposure (notice the star trails) shows the trail of an exceptionally bright meteor called a fireball. (Courtesy of R. A. Oriti)

figure 17-13 R I **V** U X G

A Meteorite "Fender-Bender" On the evening of October 9, 1992, Michelle Knapp of Peekskill, New York, heard a noise from outside that sounded like a car crash. She discovered that the trunk of her 1980 Chevrolet Malibu had been smashed. The damage was done by a 12-kilogram (27-pound) meteorite, which was lying beside the car and was still warm to the touch. Meteorite damage is not covered by automobile insurance, but Ms. Knapp was offered several tens of thousands of dollars for the meteorite (and the car). (R. A. Lanheinrich, Fossils and Meteorites, Ilion, N.Y.)

figure 17-14   R I **V** U X G

A Stony Meteorite Of all meteorites that fall on the Earth, 95% are stones. Many freshly discovered specimens, like the one shown here, are coated with dark fusion crusts. This particular stone fell in Texas. (From the collection of R. A. Oriti)

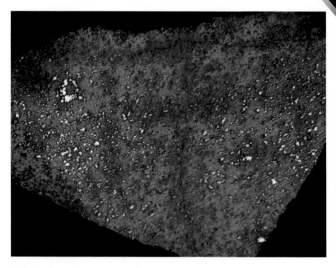

figure 17-15   R I **V** U X G

A Cut and Polished Stone When stony meteorites are cut and polished, some are found to contain tiny specks of iron mixed in the rock. This specimen was discovered in California. (From the collection of R. A. Oriti)

Meteorites are classified into three broad categories: stones, stony irons, and irons. As their name suggests, **stony meteorites,** or **stones,** look like ordinary rocks at first glance, but they are sometimes covered with a **fusion crust** (Figure 17-14). This crust is produced by the melting of the meteorite's outer layers during its fiery descent through the atmosphere. When a stony meteorite is cut in two and polished, tiny flecks of iron can sometimes be found in the rock (Figure 17-15).

Although stony meteorites account for about 95% of all the meteoritic material that falls to the Earth, they are the most difficult meteorite specimens to find. If they lie undiscovered and become exposed to the weather for a few years, they look almost indistinguishable from common terrestrial rocks. Meteorites with a high iron content can be found much more easily, because they can be located with a metal detector. Consequently, iron and stony iron meteorites dominate most museum collections.

Stony iron meteorites consist of roughly equal amounts of rock and iron. Figure 17-16, for example, shows the mineral olivine suspended in a matrix of iron. Only about 1% of the meteorites that fall to the Earth are stony irons.

Iron meteorites (Figure 17-17), or **irons,** account for about 4% of the material that falls on the Earth. Iron meteorites usually contain no stone, but many contain from 10% to 20% nickel. Before humans began extracting iron from ores around 2000 B.C., the only source of the metal was from iron meteorites. Because these are so rare, iron was regarded as a precious metal like gold or silver.

In 1808 Count Alois von Widmanstätten, director of the Imperial Porcelain Works in Vienna, discovered a conclusive test for the most common type of iron meteorite. About 75% of all iron meteorites have a unique crystalline structure. These **Widmanstätten patterns** (pronounced VIT-mahn-shtetten)

become visible when the meteorite is cut, polished, and briefly dipped into a dilute solution of acid (Figure 17-18). Nickel-iron crystals of this type can form only if the molten metal cools slowly over many millions of years. Therefore, Widmanstätten patterns are never found in counterfeit meteorites, or "meteorwrongs."

Widmanstätten patterns suggest that some asteroids were originally at least partly molten inside and remained partly molten long after they were formed. As soon as an asteroid

figure 17-16   R I **V** U X G

A Stony Iron Meteorite Stony irons account for about 1% of all the meteorites that fall to the Earth. This particular specimen, a variety of stony iron called a pallasite, fell in Chile. (Chip Clark)

figure 17-17 R I **V** U X G

An Iron Meteorite Irons are composed almost entirely of nickel-iron minerals. The surface of a typical iron is covered with small depressions caused by ablation (removal by melting) during its high-speed descent through the atmosphere. This specimen was found in Australia. Note the one-cent coin for scale. (From the collection of R. A. Oriti)

figure 17-18 R I **V** U X G

Widmanstätten Patterns When cut, polished, and etched with a weak acid solution, most iron meteorites exhibit interlocking crystals in designs called Widmanstätten patterns. This iron meteorite was found in Australia. (From the collection of R. A. Oriti)

accreted from planetesimals 4.6 billion years ago, rapid decay of short-lived radioactive isotopes could have heated the asteroid's interior to temperatures above the melting point of rock. If the asteroid was large enough—at least 200 to 400 km in diameter—to insulate the interior and prevent cooling, the interior would have remained molten over the next few million years and chemical differentiation would have occurred (see Section 7-8 and Box 7-1). Iron and nickel would have sunk toward the asteroid's center, forcing less dense rock upward toward the asteroid's surface. Like a terrestrial planet, such a **differentiated asteroid** would have a distinct core and crust. After a differentiated asteroid cooled and its core solidified, collisions with other asteroids would have fragmented the parent body into meteoroids. Iron meteorites are therefore specimens from a differentiated asteroid's core, while stony irons come from regions between a differentiated asteroid's core and its crust.

Unlike irons and stony irons, stony meteorites may or may not come from differentiated asteroids. If an asteroid was too small, it would not have retained its internal heat long enough for chemical differentiation to occur. Some stony meteorites are fragments of such **undifferentiated asteroids** (also called *primitive* asteroids), while others are relics of the crust of differentiated asteroids.

17-6 Some meteorites retain traces of the early solar system

A class of rare stony meteorites called **carbonaceous chondrites** shows no evidence of ever having been melted as part of asteroids. As suggested by their name, carbonaceous chon-

drites contain substantial amounts of carbon and carbon compounds, including complex organic molecules and as much as 20% water bound into the minerals. These compounds would have been broken down and the water driven out if these meteorites had been subjected to heating and melting. Carbonaceous chondrites may therefore be samples of the original material from which our solar system was created. The asteroid 253 Mathilde, shown in Figure 17-7, has a very dark gray color and the same sort of spectrum as a carbonaceous chondrite. It, too, is likely composed of material that predates the formation of the solar system.

Amino acids, the building blocks of proteins upon which terrestrial life is based, are among the organic compounds occasionally found inside carbonaceous chondrites. Interstellar organic material has thus probably been falling on our planet since it formed 4.6 billion years ago. Some scientists suspect that carbonaceous chondrites may have played a role in the origin of life on Earth.

Shortly after midnight on February 8, 1969, the night sky around Chihuahua, Mexico, was illuminated by a brilliant blue-white light moving across the heavens. As the light crossed the sky, it disintegrated in a spectacular, noisy explosion that dropped thousands of rocks and pebbles over terrified onlookers. Within hours, teams of scientists were on their way to collect specimens of a carbonaceous chondrite, collectively named the Allende meteorite after the locality (Figure 17-19).

Specimens were scattered in an elongated ellipse approximately 50 km long by 10 km wide. Most fragments were coated with a fusion crust, but minerals immediately beneath the crust showed no signs of damage. Surface material is peeled away as it becomes heated during flight through the atmosphere; this process forms the fusion crust. Because the

ƒigure 17-19   R I **V** U X G

A Piece of the Allende Meteorite This carbonaceous chondrite fell near Chihuahua, Mexico, in February 1969. The meteorite's dark color is due to its high abundance of carbon. Radioactive age-dating indicates that this meteorite is 4.57 billion years old. Geologists therefore conclude that this meteorite is a specimen of primitive planetary material and that its age may even predate the formation of the planets. The ruler is 15 cm (6 in.) long. (Courtesy of J. A. Wood)

heat has little time to penetrate the meteorite's interior, compounds there are left intact.

One of the most striking discoveries to come from the Allende meteorite was evidence suggesting that a nearby star exploded into a **supernova** 4.6 billion years ago. In a supernova, a massive star reaches the end of its life cycle and blows itself apart in a cataclysm that hurls matter outward at tremendous speeds. During this detonation, violent collisions between nuclei produce a host of radioactive elements, including ^{26}Al, a radioactive isotope of aluminum.

Researchers found clear evidence for the former presence of ^{26}Al in the Allende meteorite. Chemical analyses revealed a high abundance of a stable isotope of magnesium (^{26}Mg), which is produced by the radioactive decay of ^{26}Al. Some astronomers interpret this as evidence for a supernova in our vicinity at about the time the Sun was born. Indeed, by compressing interstellar gas and dust, the supernova's shock wave may have triggered the birth of our solar system.

17-7   A comet is a dusty chunk of ice that partially vaporizes as it passes near the Sun

Just as heat from the protosun produced two classes of planets, the terrestrial and the Jovian, two main types of small bodies formed in the solar system. Near the Sun, interplanetary debris consists of the rocky objects called asteroids or meteoroids. Far from the Sun, where temperatures in the early solar system were low enough to permit ices of water, methane, ammonia, and carbon dioxide to form, interplanetary debris took the form of loose collections of ices and small rocky particles. These objects are like icy versions of "rubble pile" asteroids such as 253 Mathilde (see Section 17-3). They could be referred to as "dirty snowballs," a name coined by the Harvard astronomer Fred Whipple. More commonly, however, they are called **comets.**

Asteroids travel around the Sun along roughly circular orbits that are largely confined to the asteroid belt and that lie close to the plane of the ecliptic. In sharp contrast, many comets travel around the Sun along highly elliptical orbits inclined at random angles to the ecliptic. Comets are just a few kilometers across, so when they are far from the Sun they are difficult to spot with even a large telescope. As a comet approaches the Sun, however, solar heat begins to vaporize the comet's ices, liberating gases as well as dust particles. The liberated gases begin to glow, producing a fuzzy, luminous ball called a **coma** that is typically 1 million km in diameter. Some of these luminous gases stream outward into a long, flowing **tail.** This tail, which can be more than 100 million kilometers in length—comparable to the distance from the Earth to the Sun—is one of the most awesome sights that can be seen in the night sky (see Figure 17-20 and the image that opens this chapter).

ƒigure 17-20   R I **V** U X G

Comet Hyakutake Using binoculars, the Japanese amateur astronomer Yuji Hyakutake first noticed this comet on the morning of January 30, 1996. Two months later, Comet Hyakutake came within 0.1 AU (15 million km, or 9 million mi) of the Earth and became one of the brightest comets of the twentieth century. This photograph was taken in April 1996, when the comet's tail extended more than 30° across the sky (more than 60 times the diameter of the full Moon). The stars appear blurred because the camera followed the comet's motion during the exposure. (Courtesy of Gary Goodman)

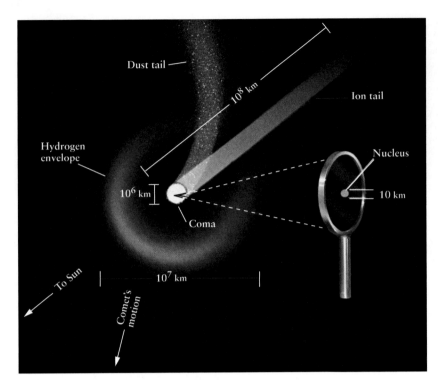

figure 17-21

The Structure of a Comet The solid part of a comet is the nucleus. It is typically about 10 km in diameter. The coma typically measures 1 million kilometers in diameter, and the hydrogen envelope usually spans 10 million kilometers. A comet's tail can be as long as 1 AU.

 Many comets are discovered when they are still far from the Sun, long before their tails grow to full size. Most of these discoveries are made by amateur astronomers, many of whom use only binoculars or small telescopes to aid in their search.

Figure 17-21 depicts the structure of a comet. The solid part of the comet, from which the coma and tail emanate, is called the **nucleus**. It is a mixture of ice and dust that typically measures a few kilometers across. Before 1986, no one had seen the nucleus of a comet. It is small, dim, and buried in the glare of the coma. In that year, the first close-up pictures of a comet's nucleus were obtained by a spacecraft that flew past Comet Halley (Figure 17-22). Comet Halley's potato-shaped nucleus is darker than coal, reflecting only about 4% of the light that strikes it. This dark color is probably caused by a layer of carbon-rich compounds and dust that is left behind as the comet's ice evaporates.

Several 15-km-long jets of dust particles were seen emanating from bright areas on Halley's nucleus. These seem to be active only when exposed to the Sun. The bright areas, which probably cover about 10% of the surface of the nucleus, are presumably places where the dark layer covering the nucleus is particularly thin. When exposed to the Sun, these areas evaporate rapidly, producing the jets. When the nucleus's rotation brings the bright areas into darkness, away from the Sun, the jets shut off.

A part of a comet that is not visible to the human eye is the **hydrogen envelope**, a huge sphere of tenuous gas surrounding the nucleus. This hydrogen comes from water molecules (H_2O) that escape from the comet's evaporating ice and then break apart when they absorb ultraviolet photons from the Sun. The hydrogen atoms also absorb solar ultraviolet photons, which excite the atoms from the ground state into an excited state (see Section 5-8). When the atoms return to the ground state, they emit ultraviolet photons.

Unfortunately for astronomers, these emissions from a comet's hydrogen envelope cannot be detected by Earth-based

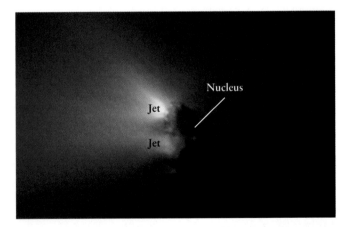

figure 17-22 R I **V** U X G

The Nucleus of Comet Halley This close-up picture, taken in March 1986 by a camera on board the European Space Agency spacecraft *Giotto*, shows the dark, potato-shaped nucleus of Comet Halley. Sunlight from the left illuminates half of the nucleus, which is about 15 km long and 8 km wide. Two bright jets of dust extend from the nucleus toward the Sun. This suggests that most of the vaporization takes place on the sunlit side. (Max-Planck-Institut für Aeronomie)

R I **V** U X G

R I V **U** X G

ƒigure 17-23

Comet Kohoutek and Its Hydrogen Envelope
These two photographs of Comet Kohoutek, which
made an appearance in 1974, are reproduced to the
same scale. The picture on the left shows the comet
in visible light. The ultraviolet view on the right
reveals a huge hydrogen cloud surrounding the
comet's head. (Johns Hopkins University, Naval Research
Laboratory)

telescopes because our atmosphere is largely opaque to ultraviolet light (see Section 6-7). Instead, cameras above the
Earth's atmosphere must be used. Figure 17-23 shows two
views of Comet Kohoutek, one as it appeared to the Earth-
based observers and one as photographed by an ultraviolet
camera aboard a rocket. The ultraviolet image shows the
enormous extent of the hydrogen envelope, which can span
10 million kilometers.

As the diagram in Figure 17-21 suggests, comet tails
always point away from the Sun. This is true regardless of
the direction of the comet's motion (Figure 17-24). The
implication that something from the Sun was "blowing" the
comet's gases radially outward led Ludwig Biermann to
predict the existence of the solar wind a full decade before it
was actually discovered in 1962 by instruments on the
Mariner 2 spacecraft.

In fact, the Sun usually produces two comet tails—an **ion
tail** and a **dust tail** (Figure 17-25). Ionized atoms and molecules—that is, atoms and molecules missing one or more
electrons—are swept directly away from the Sun by the solar
wind to form the relatively straight ion tail. The distinct blue
color of the ion tail is caused by emissions from carbon-
bearing molecules such as CN and C_2. The dust tail is formed
when photons of light strike dust particles freed from the
evaporating nucleus. (Figure 7-14 shows a dust particle of
this type.) Light exerts a pressure on any object that absorbs
or reflects it. This pressure, called **radiation pressure,** is quite
weak but is strong enough to make fine-grained dust particles in a comet's coma drift away from the comet, thus producing a dust tail. The solar wind has less of an effect on
dust particles than on (much smaller) ions, so the dust tail
ends up being curved rather than straight. On rare occasions
a comet is oriented in such a way that its dust tail appears to
stick out in front of the comet (Figure 17-26).

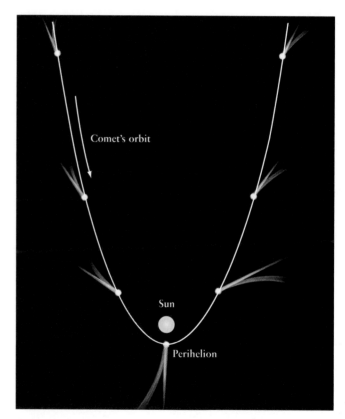

ƒigure 17-24

The Orbit and Tail of a Comet The solar wind and the pressure
of sunlight blow a comet's dust particles and ionized atoms away
from the Sun. Consequently, a comet's tail points generally away
from the Sun. In particular, the tail does *not* always stream behind
the nucleus. At the upper right of this figure, the comet is moving
upward along its orbit and is literally chasing its own tail.

ƒigure 17-25 R I **V** U X G

The Two Tails of Comet Hale-Bopp This image shows Comet Hale-Bopp on March 8, 1997, when it was 1.39 AU from the Earth and 1.00 AU from the Sun. The white dust shines because it reflects sunlight, while the molecules in the ion tail emit their own light with a characteristic blue color. (Compare this with Figure 4-22.) When this picture was taken, the ion tail extended more than 10° across the sky. The red object to the right is the North America Nebula, a star-forming region some 1500 light-years beyond the solar system. (Courtesy of Tony and Daphne Hallas Astrophotos)

After the comet passes perihelion, it recedes from the Sun back into the cold regions of the outer solar system. The ices stop vaporizing, the coma and tail disappear, and the comet goes back into an inert state—until the next time its orbit takes it toward the inner solar system. An object in an elliptical orbit spends most of its time far from the Sun, so it is only during a relatively brief period before and after perihelion that a comet can have a prominent tail. An example is Comet Halley, which orbits the Sun along a highly elliptical path that stretches from just inside the Earth's orbit to slightly beyond the orbit of Neptune (Figure 17-27). The comet's tail is visible to the naked eye or with binoculars only during a few months around perihelion, which last occurred in 1986 and will happen again in 2061.

Other comets have orbits that are larger and more elongated, with even longer periods. The orbit of Comet Hyakutake (see Figure 17-20) takes it to a distance of about 2000 AU from the Sun, 50 times the size of Pluto's orbit, with an orbital period of around 30,000 years!

Over the next several years, a small fleet of spacecraft will investigate comets at close range and actually return samples of comet material to the

Earth. NASA's *Deep Space 1*, launched in 1998, may make a close flyby of Comet Borrelly in 2001. A second NASA spacecraft, *Stardust*, was launched in 1999 and will encounter Comet Wild 2 after five years of interplanetary travel. It will collect a sample of material from the comet's coma and return this sample to Earth in 2006. NASA's *CONTOUR* spacecraft (for Comet Nucleus *Tour*) is scheduled for a 2002 launch. Its flight plan calls for it to rendezvous with three different comets and to analyze the composition of dust released from their nuclei. Another NASA mission, *Deep Impact,* is scheduled for a 2004 launch and a 2005 rendezvous with Comet Tempel 1. It will launch a 500-kg metal cylinder at the comet's

ƒigure 17-26 R I **V** U X G

The Antitail of Comet Hale-Bopp In January 1998, nine months after passing perihelion, Comet Hale-Bopp exhibited an "antitail." Actually, this antitail was merely the end of the dust tail. The Earth and the comet were oriented in such a way that the end of the arched dust tail looked like a spike sticking out of the comet's head. (European Southern Observatory)

2012. If all goes well, *Rosetta* will go into orbit around the comet nucleus and release a small probe that will actually land on the nucleus.

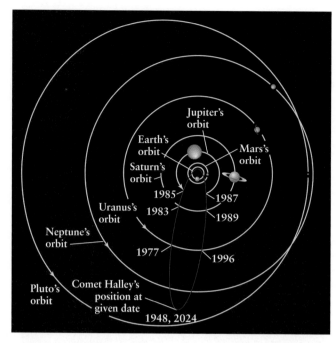

ƒigure 17-27

Comet Halley's Eccentric Orbit Like most comets, Comet Halley has an elongated orbit and spends most of its time far from the Sun. A tail (see Figure 4-22) appears only when the comet is close to perihelion. Comet Halley has been observed at intervals of about 76 years (the period of its orbit) since 88 B.C. We saw in Section 4-7 how the English scientist Edmund Halley became the first to realize that all of these observations were of the same comet.

nucleus, excavating a large crater and revealing its internal composition. Most ambitious of all is the European Space Agency's *Rosetta* mission, to be launched in 2003. It will fly past two asteroids on its way to intercept Comet Wirtanen in

17-8 Comets originate either from a belt beyond Pluto or from a vast cloud in near interstellar space

Why do comets have such elliptical orbits, while asteroids have relatively circular ones? The reason cannot have anything to do with comets being icy and asteroids being rocky, because the gravitational forces that shape an object's orbit do not depend on its chemical composition, only on its mass. Instead, the answer has to do with how the gravitational pulls of the planets helped to shape the orbits of asteroids and comets in different ways.

As we saw in Section 17-2, the asteroids formed in the space between Mars and Jupiter. The relatively small terrestrial planets, with their correspondingly small masses, exert only weak gravitational forces on asteroids. Hence, most asteroids, like the planets, remained in fairly circular orbits. (The exceptions are those few asteroids that wandered too close to Mars, whose gravitational force deflected these asteroids into becoming the near-Earth objects described in Section 17-4.) Jupiter's gravity acted primarily to bunch the asteroids away from the Kirkwood gaps.

Comets, by contrast, first formed in the outer solar system just beyond Neptune, where temperatures were low enough to permit ices to condense into chunks several kilometers across. Because Neptune is much more massive than the terrestrial planets, its gravitational forces had a major effect, deflecting the comets in a wide variety of directions.

There are now two large reservoirs of comets, the Kuiper belt and the Oort cloud. The **Kuiper belt** lies in the plane of the ecliptic and extends from around the orbit of Pluto to about 500 AU from the Sun (Figure 17-28). The existence of

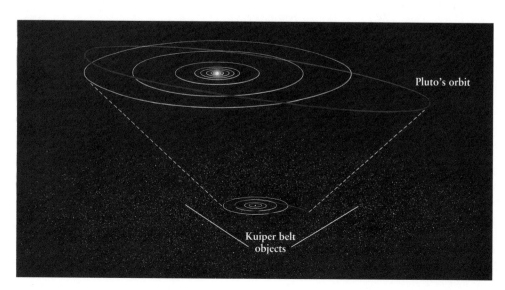

ƒigure 17-28

The Kuiper Belt The Kuiper belt of comets lies in the ecliptic, like the asteroid belt, and extends from around the orbit of Pluto to about 500 AU from the Sun. The upper part of this figure shows the orbits of the planets; the lower part, drawn to a smaller scale, shows the much greater size of the Kuiper belt.

a

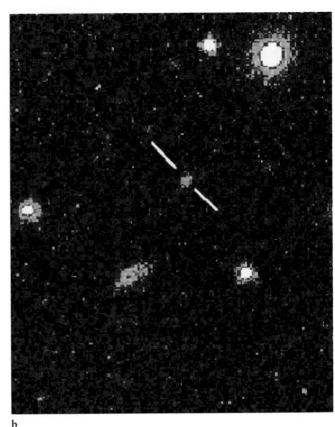

b

figure 17-29 R I **V** U X G

A Kuiper Belt Object These images were the first to show 1993 SC, the sixth Kuiper belt object to be discovered. The images were taken 4.6 hours apart, during which time 1993 SC (shown between the white lines) moved from the position in **(a)** to that in **(b)** against the background stars. The colors in this image are false: Blue represents dark areas, while green, yellow, red, and white denote successively brighter areas. (Alan Fitzsimmons, Queen's University of Belfast)

this belt was first proposed by the American astronomer Gerard Kuiper in 1951; the first Kuiper belt object to be discovered was seen in 1992. (This was the object 1992 QB₁, which we described in Section 16-10.) Nearly 300 of these objects have now been observed (Figure 17-29). It is suspected that there could be as many as 35,000 objects in the Kuiper belt with diameters of 100 km or greater and many more smaller objects. Like other planetesimals, most Kuiper belt objects probably formed at the same distance from the Sun as we find them today.

Jupiter-family comets, which orbit the Sun in fewer than 20 years and return again and again to the inner solar system at predictable intervals, are thought to come from the Kuiper belt. Computer simulations show that gravitational perturbations from Neptune can occasionally launch a Kuiper belt comet into a highly elliptical orbit that takes the comet close to the Sun, where it produces a visible tail. A few of these perturbed Kuiper belt objects are thought to have been captured by the gravitational pulls of Jupiter and Sat-

urn, becoming the small outer satellites of these planets. Some astronomers also regard Pluto and its moon Charon as Kuiper belt objects.

The majority of comets are **intermediate-period comets,** with orbital periods between 20 and 200 years (including Comet Halley), and **long-period comets,** which take roughly 1 to 30 million years to complete one orbit of the Sun. Long-period comets travel along extremely elongated orbits and consequently spend most of their time at distances of roughly 10^4 to 10^5 AU from the Sun, or about one-fifth of the way to the nearest star. These orbits extend far beyond the Kuiper belt. Furthermore, intermediate-period and long-period comets have orbital planes that are often steeply inclined to the plane of the ecliptic. (Comet Halley's orbital plane is inclined by 162° to the ecliptic, which means that it orbits the Sun in the opposite direction to the planets.)

The best explanation for these observations is that intermediate-period and long-period comets come from a reservoir that extends from the Kuiper belt to some 50,000 AU from the Sun. This reservoir, first hypothesized by Dutch astronomer Jan Oort in 1950 and now called the **Oort cloud,** does not lie in the ecliptic; rather, it is a spherical distribution centered on the Sun. This explains why many intermediate-period and long-period comets have steeply inclined orbits.

Because astronomers discover long-period comets at the rate of about one per month, it is reasonable to suppose that there is an enormous population of comets in the Oort cloud. Estimates of the number of "dirty snowballs" in the Oort cloud range as high as 6 trillion (6×10^{12}). Only such a large reservoir of comet nuclei would explain why we see so many long-period comets, even though each one takes several million years to travel once around its orbit. Because the Oort cloud is so distant, it has not yet been possible to detect objects in the Oort cloud directly.

The Oort cloud was probably created 4.6 billion years ago from numerous icy planetesimals that orbited the Sun in the vicinity of the newly formed Jovian planets. During near-collisions with the giant planets, many of these chunks of ice and dust were catapulted by gravity into highly elliptical orbits, in much the same way that *Pioneer* and *Voyager* spacecraft were flung far from the Sun during their flybys of the Jovian planets. Gravitational perturbations from nearby stars tilted the planes of the orbits in all directions, giving the Oort cloud its spherical shape.

The distinction between long-period, intermediate-period, and Jupiter-family comets can be blurred by the effects of gravitational perturbations. During a return trip toward the Sun, an encounter with a Jovian planet may force a comet into a much larger orbit. Alternatively, such an encounter can move a long-period comet into a smaller orbit (Figure 17-30). Several comets have been perturbed in this way into orbits that always remain within the inner solar system. None of these has a prominent tail, however.

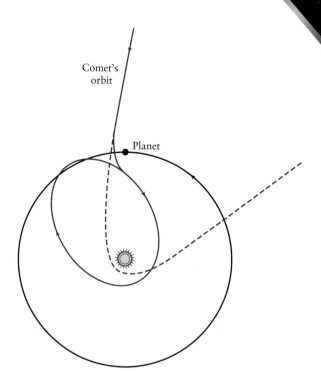

figure 17-30

Transforming a Comet's Orbit The gravitational force of a planet can cause major changes in a planet's orbit. Initially on a highly elliptical orbit (dashed line), comets are sometimes deflected into less elliptical paths. In some cases these comets end up with orbits that keep them within the inner solar system.

17-9 Comets eventually break apart, and their fragments give rise to meteor showers

Because they evaporate away part of their mass each time they pass near the Sun, comets cannot last forever. A typical comet may lose about 0.5% to 1% of its ice each time it passes near the Sun. Hence, the ice completely vaporizes after about 100 or 200 perihelion passages, leaving only a swarm of dust and pebbles. Astronomers have observed some comet nuclei in the process of fragmenting (Figure 17-31).

Comet Shoemaker-Levy 9, which we described in Section 13-4, broke apart for a different reason. As the comet swung by Jupiter in July 1992, the giant planet's tidal forces tore the nucleus into more than 20 fragments (see Figure 13-13). Jupiter's gravity then deflected the trajectories of these fragments so that they plummeted into the planet's atmosphere (see Figure 13-14 and Figure 13-15).

As a comet's nucleus evaporates, residual dust and rock fragments form a **meteoritic swarm,** a loose collection of

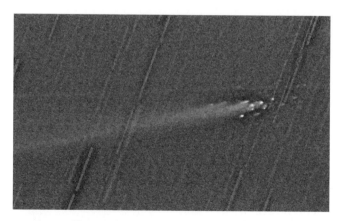

figure 17-31 R I **V** U X G

The Fragmentation of Comet LINEAR The Lincoln Near Earth Asteroid Research (LINEAR) project discovered this comet in 1999. As it passed the perihelion of its orbit in July 2000, Comet LINEAR broke into more than a dozen small fragments. These continue to orbit the Sun along the same trajectory that the comet followed prior to its breakup. This image was made using the Very Large Telescope (see Figure 6-17). (European Southern Observatory)

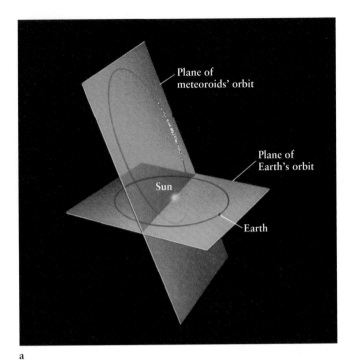

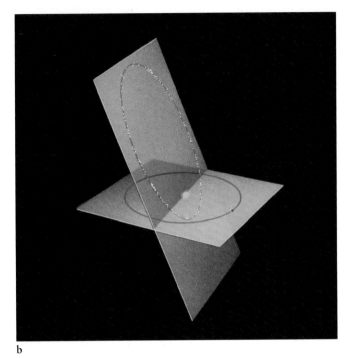

a

b

figure 17-32

Meteoritic Swarms Rock fragments and dust from "burned out" comets continue to circle the Sun.
(a) If the comet is only recently extinct, the particles will still be tightly concentrated in a compact swarm.
The most spectacular meteor showers occur when the Earth passes through such a swarm. **(b)** Over the
ages, the particles spread out along the old comet's elliptical orbit. This produces the most predictable
meteor showers, because the Earth passes through the evenly distributed swarm on each trip around the
Sun. If the plane of the original comet's orbit was steeply inclined to the plane of the Earth's orbit, as
shown here, the meteor shower will strike the Earth from a direction well out of the ecliptic.

debris that continues to circle the Sun along the comet's orbit (Figure 17-32). If the Earth's orbit happens to pass through this swarm, a **meteor shower** is seen as the dust particles strike the Earth's upper atmosphere.

Nearly a dozen meteor showers can be seen each year (Table 17-1). Note that the **radiants** for these showers—that is, the places among the stars from which the meteors appear to come—are not confined to the constellations of the zodiac. This means that the swarms to do not all orbit in the plane of the ecliptic. The reason is that the orbits of their parent comets were inclined at random angles to the plane of the ecliptic (see Figure 17-32).

On June 30, 1908, a spectacular explosion occurred over the Tunguska region of Siberia that released about 10^{15} joules of energy, equivalent to a nuclear detonation of several hundred kilotons. The blast knocked a man off his porch some 60 km away and could be heard more than 1000 km away. Millions of tons of dust were injected into the atmosphere, darkening the sky as far away as California.

Preoccupied with political upheaval and World War I, Russia did not send a scientific expedition to the site until 1927. Researchers found that trees had been seared and felled radially outward in an area about 50 km in diameter (Figure 17-33). There was no clear evidence of a crater. In fact, the trees at "ground zero" were left standing upright, although they were completely stripped of branches and leaves. Because no significant meteorite samples were found, for many years it was assumed that a small comet had struck Earth.

Recently, however, several teams of astronomers have argued that the Tunguska explosion was actually caused by a stony asteroid traveling at supersonic speed. They arrived at this conclusion after assessing the effects of various impactor sizes, speeds, and compositions. Even data from above-ground nuclear detonations of the 1940s and 1950s were worked into the calculations. The Tunguska event is well-matched by a stony asteroid about 80 m (260 ft) in diameter entering the Earth's atmosphere at 22 km/s (80,000 km/h = 50,000 mph). A small comet would have broken up too high in the atmosphere to cause significant damage on the ground.

Large objects, such as that which formed the Chicxulub Crater and may have led to the demise of the dinosaurs (Section 17-3), occasionally strike the Earth with the destructive force of as much as a million megatons of TNT. An asteroid 1 or 2 kilometers in diameter, striking the Earth at

	Date of maximum intensity	Typical hourly rate	Average speed (km/s)	Radiant constellation
Shower				
Quadrantids	January 3	40	40	Boötes
Lyrids	April 22	15	50	Lyra
Eta Aquarids	May 4	20	64	Aquarius
Delta Aquarids	July 30	20	40	Aquarius
Perseids	August 12	50	60	Perseus
Orionids	October 21	20	66	Orion
Taurids	November 4	15	30	Taurus
Leonids	November 16	15	70	Leo
Geminids	December 13	50	35	Gemini
Ursids	December 22	15	35	Ursa Minor

table 17-1   Prominent Yearly Meteor Showers

The date of maximum intensity is the best time to observe a particular shower, although good displays can often be seen a day or two before or after the maximum. The typical hourly rate is given for an observer under optimum viewing conditions. The average speed refers to how fast the meteoroids are moving when they strike the atmosphere.

30 km/s, would destroy an area the size of California. Such an impact would throw more than 10^{13} kg of microscopic dust particles into the atmosphere, which would decrease the amount of sunlight reaching the Earth's surface by enough to threaten the health of the world's agriculture. The loss of a year's crops could lead to the demise of a quarter of the world's population and would place our civilization—although perhaps not the survival of the species—in grave jeopardy.

The good news is that such large objects strike the Earth *very* infrequently. An asteroid 1 km in diameter might be expected to hit the Earth only once every 300,000 years. Statistically, the likelihood that an average North American would by killed by such an event is only 0.2% of the probability of dying in an automobile accident. While the threat of asteroid or comet impact exists, a catastrophic impact is very unlikely to occur in our lifetimes.

figure 17-33

Aftermath of the Tunguska Event In 1908 a stony asteroid with a mass of about 10^8 kilograms entered the Earth's atmosphere over the Tunguska region of Siberia. Its passage through the atmosphere made a blazing trail in the sky some 800 km long. The asteroid apparently exploded before reaching the surface, blowing down trees for hundreds of kilometers around "ground zero." (Courtesy of Sovfoto)

KEY WORDS

amino acids, p. 374

asteroid, p. 365

asteroid belt, p. 366

carbonaceous chondrite, p. 374

coma (of a comet), p. 375

comet, p. 375

differentiated asteroid, p. 374

dust tail, p. 377

fusion crust, p. 373

Hirayama family, p. 370

hydrogen envelope, p. 376

intermediate-period comet, p. 380

ion tail, p. 377

iron meteorite (iron), p. 373

Jupiter-family comet, p. 380

Kirkwood gaps, p. 367

Kuiper belt, p. 379

long-period comet, p. 380

meteor, p. 372

meteor shower, p. 382

meteorite, p. 372

meteoritic swarm, p. 381

meteoroid, p. 371

minor planet, p. 365

near-Earth object (NEO), p. 370

nucleus (of a comet), p. 376

Oort cloud, p. 381

radiant (of a meteor shower), p. 382

radiation pressure, p. 377

stable Lagrange points, p. 370

stony iron meteorite, p. 373

stony meteorite (stone), p. 373

supernova, p. 375

tail (of a comet), p. 375

Trojan asteroid, p. 370

undifferentiated asteroid, p. 374

Widmanstätten patterns, p. 373

KEY IDEAS

Discovery of the Asteroids: Astronomers first discovered the asteroids while searching for a "missing planet."

• Thousands of asteroids with diameters ranging from a few kilometers up to 1000 kilometers orbit within the asteroid belt between the orbits of Mars and Jupiter.

Origin of the Asteroids: The asteroids are the relics of planetesimals that failed to accrete into a full-sized planet, thanks to the effects of Jupiter and other Mars-sized objects.

• Even today, gravitational perturbations by Jupiter deplete certain orbits within the asteroid belt. The resulting gaps, called Kirkwood gaps, occur at simple fractions of Jupiter's orbital period.

• Jupiter's gravity also captures asteroids in two locations, called Lagrangian points, along Jupiter's orbit.

Asteroid Collisions: Asteroids undergo collisions with each other, causing them to break up into smaller fragments.

• Some asteroids, called near-Earth objects, move in elliptical orbits that cross the orbits of Mars and Earth. If such an asteroid strikes Earth, it forms an impact crater whose diameter depends on both the mass and the speed of the asteroid.

• An asteroid may have struck the Earth 65 million years ago, possibly causing the extinction of the dinosaurs and many other species.

Meteoroids, Meteors, and Meteorites: Small rocks in space are called meteoroids. If a meteoroid enters the Earth's atmosphere, it produces a fiery trail called a meteor. If part of the object survives the fall, the fragment that reaches the Earth's surface is called a meteorite.

• Meteorites are grouped into three major classes, according to composition: iron, stony iron, and stony meteorites. Irons and stony irons are fragments of the core of an asteroid that was large enough and hot enough to have undergone chemical differentiation, just like a terrestrial planet. Some stony meteorites come from the crust of such differentiated meteorites, while others are fragments of small asteroids that never underwent differentiation.

• Rare stony meteorites called carbonaceous chondrites may be relatively unmodified material from the solar nebula. These meteorites often contain organic material and may have played a role in the origin of life on Earth.

• Analysis of isotopes in certain meteorites suggests that a nearby supernova may have triggered the formation of the solar system 4.6 billion years ago.

 Comets: A comet is a chunk of ice with imbedded rock fragments that generally moves in a highly elliptical orbit about the Sun.

• As a comet approaches the Sun, its icy nucleus develops a luminous coma, surrounded by a vast hydrogen envelope. An ion tail and a dust tail extend from the comet, pushed away from the Sun by the solar wind and radiation pressure, respectively.

• Fragments of "burned out" comets produce meteoritic swarms. A meteor shower is seen when the Earth passes through a meteoritic swarm.

Origin of Comets: Comets are thought to originate from two regions, the Kuiper belt and the Oort cloud.

• The Kuiper belt lies in the plane of the ecliptic at distances between 40 and 500 AU from the Sun. It is thought to contain many tens of thousands of comet nuclei. Many Jupiter-family comets probably come from the Kuiper belt, and a number of objects have been observed in the Kuiper belt.

• The Oort cloud contains billions of comet nuclei in a spherical distribution that extends out to 50,000 AU from the Sun. Intermediate-period and long-period comets are thought to originate in the Oort cloud. As yet no objects in the Oort cloud have been detected directly.

REVIEW QUESTIONS

1. How did the first asteroids come to be discovered? How did this discovery differ from what astronomers had expected to find?

2. How do modern astronomers discover new asteroids?

3. Describe the asteroid belt. Does it lie completely within the plane of the ecliptic? What are its inner and outer radii?

4. What are Kirkwood gaps? What causes them?

5. Compare the explanation of the Kirkwood gaps in the asteroid belt to the way in which Saturn's moons help produce divisions in that planet's rings (see Section 15-4).

6. What is the evidence that some asteroids are made of a loose conglomeration of smaller pieces?

7. The asteroid 243 Ida, which was viewed by the *Galileo* spacecraft, is a member of a Hirayama family. Discuss what this tells us about the history of this asteroid. Where might you look to find other members of the same family?

8. What are the Trojan asteroids, and where are they located?

9. What is the difference between a meteoroid, a meteor, and an meteorite?

10. Is there anywhere on the Earth where you might find large numbers of stony meteorites that are not significantly weathered? If so, where? If not, why not?

11. Scientists can tell that certain meteorites came from the interior of an asteroid rather than from its outer layers. Explain how this is done.

12. Why are some asteroids differentiated while others are not?

13. Suppose you found a rock you suspect might be a meteorite. Describe some of the things you could do to determine whether it was a meteorite or a "meteorwrong."

14. What is the evidence that carbonaceous chondrites are essentially unaltered relics of the early solar system? What do they suggest about how the solar system may have formed?

15. With the aid of a drawing, describe the structure of a comet.

16. Why is the phrase "dirty snowball" an appropriate characterization of a comet's nucleus?

17. Why do the ion tail and dust tail of a comet point in different directions?

18. Why do comets have prominent tails for only a short time during each orbit?

19. What is the Kuiper belt? How does it compare with the asteroid belt? What are the similarities? What are the differences?

20. What is the Oort cloud? How might it be related to planetesimals left over from the formation of the solar system?

21. Why are comets more likely to break apart at perihelion than at aphelion?

22. Why do astronomers think that meteorites come from asteroids, while meteor showers are related to comets?

23. Why are asteroids, meteorites, and comets all of special interest to astronomers who want to understand the early history of the solar system?

ADVANCED QUESTIONS

Problem-solving tips and tools

We discussed retrograde motion in Section 4-1 and described its causes in Section 4-2. You will need to use Kepler's third law, described in Section 4-5 and Box 4-2, in some of the problems below. Box 7-2 discusses the concept of escape speed. A spherical object of radius r intercepts an amount of sunlight proportional to its cross-sectional area, equal to πr^2. The volume of a sphere of radius r is $\frac{4}{3}\pi r^3$.

24. When Olbers discovered Pallas in March 1802, the asteroid was moving from east to west relative to the stars. At what time of night was Pallas highest in the sky over Olbers's observatory? Explain your reasoning.

25. Consider the Kirkwood gap whose orbital period is two-fifths of Jupiter's period. Calculate the distance from the Sun to this gap. Does your answer agree with Figure 17-4?

26. Suppose that a binary asteroid (two asteroids orbiting each other) is observed in which one member is 16 times brighter than the other. Suppose that both members have the same albedo and that the larger of the two is 120 km in diameter. What is the diameter of the other member?

27. The accompanying image from the *Galileo* spacecraft shows the asteroid 243 Ida, which has dimensions $56 \times 24 \times 21$ km. *Galileo* discovered a tiny moon called Dactyl, just $1.6 \times 1.4 \times 1.2$ km in size, which orbits Ida at a distance of about 100 km. (In Greek mythology, the Dactyli were beings who lived on the slopes of Mount Ida.) Describe a scenario that could explain how Ida came to have a moon.

Ida

Dactyl

(JPL/NASA)   R I **V** U X G

28. Assume that Ida's tiny moon Dactyl (see Question 27) has a density of 2500 kg/m³. **(a)** Calculate the mass of Dactyl in kilograms. For simplicity, assume that Dactyl is a sphere 1.4 km in diameter. **(b)** Calculate the escape speed from the surface of Dactyl. If you were an astronaut standing on Dactyl's surface, could you throw a baseball straight up so that it would never come down? Professional baseball pitchers can throw at speeds around 40 m/s (140 km/h, or 90 mi/h); your throwing speed is probably a bit less.

29. Imagine that you are an astronaut standing on the surface of a Trojan asteroid. How will you see the phase of Jupiter change with the passage of time? How will you see Jupiter move relative to the distant stars? Explain your answers.

30. Use the percentages of stones, irons, and stony iron meteorites that fall to Earth to estimate what fraction of their parent asteroids' interior volume consisted of an iron core. Assume that the percentages of stones and irons that fall to the Earth indicate the fractions of a parent asteroid's interior volume occupied by rock and iron, respectively. How valid do you think this assumption is?

31. *Sun-grazing comets* come so close to the Sun that their perihelion distances are essentially zero. Find the orbital periods of Sun-grazing comets whose aphelion distances are **(a)** 100 AU, **(b)** 1000 AU, **(c)** 10,000 AU, and **(d)** 100,000 AU. Assuming that these comets can survive only a hundred perihelion passages, calculate their lifetimes. (*Hint:* Remember that the semimajor axis of an orbit is one-half the length of the orbit's long axis.)

32. Comets are generally brighter a few weeks after passing perihelion than a few weeks before passing perihelion. Explain why might this be. (*Hint:* Water, including water ice, does an excellent job of retaining heat.)

33. The hydrogen clouds of comets are especially bright at a wavelength of 122 nm. Use Figure 5-22 to explain why.

34. A *very* crude model of a typical comet nucleus is a cube of ice (density 1000 kg/m³) 10 km on a side. **(a)** What is the mass of this nucleus? **(b)** Suppose 1% of the mass of the nucleus evaporates away to form the comet's tail. Suppose further that the tail is 100 million (10^8) km long and 1 million (10^6) km wide. Estimate the average density of the tail (in kg/m³). For comparison, the density of the air you breathe is about 1.2 kg/m³. **(c)** In 1910 the Earth actually passed through the tail of Comet Halley. At the time there was some concern among the general public that this could have deleterious effects on human health. Was this concern justified? Why or why not?

DISCUSSION QUESTIONS

35. From the abundance of craters on the Moon and Mercury, we know that numerous asteroids and meteoroids struck the inner planets early in the history of our solar system. Is it reasonable to suppose that numerous comets also pelted the planets 3.5 to 4.5 billion years ago? Speculate about the effects of such a cometary bombardment, especially with regard to the evolution of the primordial atmospheres on the terrestrial planets.

36. In the 1998 movie *Armageddon*, an asteroid "the size of Texas" is on a collision course with Earth. The asteroid is first discovered by astronomers just 18 days prior to impact. To avert disaster, a team of astronauts blasts the asteroid into two pieces just 4 hours before impact. Discuss the plausibility of this scenario. (*Hint:* On average, the state of Texas extends for about 750 km from north to south and from east to west. How does this compare with the size of the largest known asteroids?)

37. Suppose astronomers discovered that the near-Earth object 1994 XM_1 had been disturbed in such a way as to put it on a collision course with Earth. Describe what humanity could do within the framework of present technology to counter such a catastrophe.

WEB/CD-ROM QUESTIONS

38. Search the World Wide Web to find out why some scientists disagree with the idea that a tremendous impact led to the demise of the dinosaurs. (They do not dispute that the impact took place, only what its consequences were.) What are their arguments? From what you learn, what is your opinion?

39. Several scientific research programs are dedicated to the search for near-Earth objects (NEOs), especially those that might someday strike our planet. Search the World Wide Web for information about at least one of these programs. How does this program search for NEOs? How many NEOs has this program discovered? Will any of these pose a threat in the near future?

40. Estimating the Speed of a Comet. Access and view the video "Two Comets and an Active Sun" in Chapter 17 of the *Universe* web site or CD-ROM. **(a)** Why don't the comets reappear after passing the Sun? **(b)** The white circle shows the size of the Sun, which has diameter 1.39×10^6 km. Using this to set the scale, step through the video and measure how the position of one of the comets changes. Use your measurements and the times displayed in the video to estimate the comet's speed in km/h and km/s. (Assume that the comet moved in the same plane as that shown in the video.) As part of your answer, explain the technique and calculations you used. **(c)** How does your answer in (b) compare with the orbital speed of Mercury, the innermost and fastest-moving planet (see Table 10-1)? Why is there a difference? **(d)** If the comet's motion was not in the same plane as that shown in the video, was its actual speed more or less than your estimate in (b)? Explain.

see Table 10-1

OBSERVING PROJECTS

Observing tips and tools

Meteors: Informative details concerning upcoming meteor showers appear on the web sites for *Sky & Telescope* and *Astronomy* magazines.

Comets: Because astronomers discover dozens of comets each year, there is usually a comet visible somewhere in the sky. Unfortunately, they are often quite dim, and you will need to have access to a moderately large telescope (at least 35 cm, or 14 in.). You can get up-to-date information from the Minor Planet Center web site. Also, if there is an especially bright comet in the sky, useful information about it might be found at the web sites for *Sky & Telescope* and *Astronomy* magazines.

Asteroids: At opposition, some of the largest asteroids are bright enough to be seen through a modest telescope. Check the Minor Planet Center web site to see if any bright asteroids are near opposition. If so, check the current issue as well as the most recent January issue of *Sky & Telescope* magazine for a star chart showing the asteroid's path among the constellations. You will need such a chart to distinguish the asteroid from background stars. Also, you can locate Ceres, Pallas, Juno, and Vesta using the *Starry Night* program on the CD-ROM that accompanies this textbook.

41. Make arrangements to view a meteor shower. Table 17-1 lists the dates of major meteor showers. Choose a shower that will occur near the time of a new moon. Set your alarm clock for the early morning hours (1 to 3 A.M.). Get comfortable in a reclining chair or lie on your back so that you can view a large portion of the sky. Record how long you observe, how many meteors you see, and what location in the sky they seem to come from. How well does your observed hourly rate agree with published estimates, such as those in Table 17-1? Is the radiant of the meteor shower apparent from your observations?

42. If a comet is visible with a telescope at your disposal, make arrangements to view it. Can you distinguish the comet from background stars? Can you see its coma? Can you see a tail?

43. Make arrangements to view an asteroid. Observe the asteroid on at least two occasions, separated by a few days. On each night, draw a star chart of the objects in your telescope's field of view. Has the position of one starlike object shifted between observing sessions? Does the position of the moving object agree with the path plotted on published star charts? Do you feel confident that you have in fact seen the asteroid?

44. Use the *Starry Night* program to observe several comets. (**a**) First turn off daylight (select **Daylight** in the **Sky** menu), show the entire celestial sphere (select **Atlas** in the **Go** menu), and turn off the planets and Sun (select **Planets/Sun** in the **Sky** menu). In the Control Panel, set the date to March 22, 1997. Select **Planet List** in the **Window** menu and click on the triangle to the left of the name **Comets.** This will reveal a list of comets. Double-click on the name **Hale-Bopp** to center on that comet. Zoom in until you can see the comet and its tail. Predict in what direction the Sun is located relative to the comet, and explain how you made your prediction. To check your prediction, zoom out as far as possible, turn the planets and Sun back on (again select **Planets/Sun** in the **Sky** menu) and use the hand cursor to scroll the display in the direction of your prediction. Were you correct? (**b**) Using the Planet List, double-click on the name of another comet to center on it, but do not zoom in. Then move the sky with the hand cursor and locate the Sun. From your observations, predict the direction of the comet's tail on the sky, and explain how you made your prediction. Center again on the comet and zoom in until you can see its tail. Was your prediction correct?

Our Star, the Sun

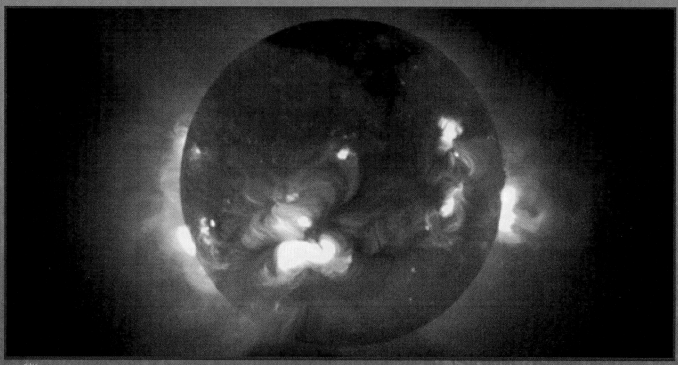

(Institute of Space and Astronomical Science, Yohkoh Project, SXT Group/NASA)

R I V U X G

The Sun is by far the brightest object in the sky. By earthly standards, the temperature of its glowing surface is remarkably high, about 5800 K. Yet astronomers have found regions of the Sun that reach far higher temperatures of several million kelvins. Gases at such temperatures emit X rays and can be seen in an image like this one from an X-ray telescope. Bright, X-ray-emitting regions, like those in this image, can spawn solar flares—powerful eruptions that propel solar material across space to reach the Earth and other planets.

In recent decades, astronomers have used tools such as X-ray telescopes to make a wealth of discoveries about the Sun. We have learned that the Sun shines because at its core hundreds of millions of tons of hydrogen are converted to helium every second. We have found the by-products of this transmutation—strange, ethereal particles called neutrinos—streaming outward from the Sun into space. We have discovered that the Sun has a surprisingly violent atmosphere, with a host of features such as sunspots whose numbers rise and fall on a predictable 11-year cycle. By studying the Sun's vibrations, we have begun to probe beneath its surface into hitherto unexplored realms. And we have just begun to investigate how changes in the Sun's activity can affect the Earth's environment as well as our technological society.

As you read the sections of this chapter, look for the answers to the following questions.

18-1   What is the source of the Sun's energy?

18-2   What is the internal structure of the Sun?

18-3   How can astronomers measure the properties of the Sun's interior?

18-4   How can we be sure that thermonuclear reactions are happening in the Sun's core?

18-5   Does the Sun have a solid surface?

18-6   Since the Sun is so bright, how is it possible to see its dim outer atmosphere?

18-7   Where does the solar wind come from?

18-8   What are sunspots? Why do they appear dark?

18-9   What is the connection between sunspots and the Sun's magnetic field?

18-10   What causes eruptions in the Sun's atmosphere?

18-1   The Sun's energy is generated by thermonuclear reactions in its core

The Sun is the largest member of the solar system, with almost a thousand times more mass than all the planets, moons, asteroids, comets, and meteoroids put together. But the Sun is also a star. In fact, it is a remarkably typical star, with a mass, size, surface temperature, and chemical composition that are approximately midway between the extremes exhibited by the myriad other stars in the heavens. Table 18-1 lists essential data about the Sun.

For most people, what matters most about the Sun is the energy that it radiates into space. Without the Sun's warming rays, our atmosphere and oceans would freeze into an icy layer coating a desperately cold planet, and life on Earth would be impossible. To understand why we are here, we must understand the nature of the Sun.

What makes the Sun such an important source of energy? An important part of the answer is that the Sun has a far higher surface temperature than any of the planets or moons. The Sun's spectrum is close to that of an idealized blackbody with a temperature of 5800 K (see Section 5-3 and Section 5-4, especially Figure 5-11). Thanks to this high temperature, each square meter of the Sun's surface emits a tremendous amount of radiation, principally at visible wavelengths. Indeed, the Sun is the only object in the solar system that emits substantial amounts of visible light. The light that we see from the Moon and planets is actually sunlight that struck those worlds and was reflected toward Earth.

The Sun's size also helps us explain its tremendous energy output. Because the Sun is so large, the total number of square meters of radiating surface—that is, its surface area—is immense. Hence, the total amount of energy emitted by the Sun each second, called its **luminosity**, is very large indeed: about 3.9×10^{26} watts, or 3.9×10^{26} joules of energy emitted per second. (We discussed the relation among the Sun's surface temperature, radius, and luminosity in Box 5-2.) Astronomers denote the Sun's luminosity by the symbol $L_{\odot}$. A circle with a dot in the center is the astronomical symbol for the Sun and was also used by ancient astrologers.

These ideas lead us to a more fundamental question: What keeps the Sun's visible surface so hot? Or, put another way, what is the fundamental source of the tremendous energies that the Sun radiates into space? For centuries, this was one of the greatest mysteries in science. The mystery deepened in the nineteenth century, when geologists and biologists found convincing evidence that life had existed on Earth for at least several hundred million years. (We now know that the Earth is 4.6 billion years old and that life has existed on it for most of its history.) Since life as we know it depends crucially on sunlight, the Sun must be as old. This posed a severe problem for physicists. What source of energy could have kept the Sun shining for so long (Figure 18-1)?

One attempt to explain solar energy was made in the mid-1800s by the English physicist Lord Kelvin (for whom the temperature scale is named) and the German scientist Hermann von Helmholtz. They argued that the tremendous weight of the Sun's outer layers should cause the Sun to contract gradually, compressing its interior gases. Whenever a gas is compressed, its temperature rises. (You can demonstrate this with a bicycle pump: As you pump air into a tire, the temperature of the air increases and the pump becomes warm to the touch.) Kelvin and Helmholtz thus suggested that gravitational contraction could cause the Sun's gases to become hot enough to radiate energy out into space.

This process, called *Kelvin-Helmholtz contraction*, actually does occur during the earliest stages of the birth of a star like the Sun (see Section 7-7). But Kelvin-Helmholtz contraction cannot be the major source of the Sun's energy today. If it were, the Sun would have had to have been much larger in the relatively recent past. Helmholtz's own calculations showed that the Sun could have started its initial collapse from the solar nebula no more than about 25 million years ago. But the geological and fossil record shows that the Earth is far older than that, and so the Sun must be as well. Hence, this model of a Sun that shines because it shrinks cannot be correct.

On Earth, a common way to produce heat and light is by burning fuel, such as a log in a fireplace or coal in a power plant. Is it possible that a similar process explains the energy released by the Sun? The answer is no, because this process could not continue for a long enough time to explain the age

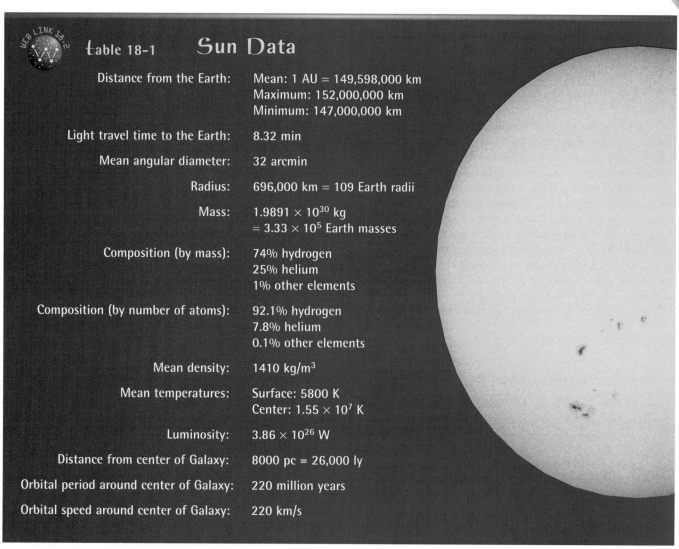

table 18-1	Sun Data	
Distance from the Earth:	Mean: 1 AU = 149,598,000 km Maximum: 152,000,000 km Minimum: 147,000,000 km
Light travel time to the Earth:	8.32 min
Mean angular diameter:	32 arcmin
Radius:	696,000 km = 109 Earth radii
Mass:	1.9891×10^{30} kg $= 3.33 \times 10^{5}$ Earth masses
Composition (by mass):	74% hydrogen 25% helium 1% other elements
Composition (by number of atoms):	92.1% hydrogen 7.8% helium 0.1% other elements
Mean density:	1410 kg/m³
Mean temperatures:	Surface: 5800 K Center: 1.55×10^{7} K
Luminosity:	3.86×10^{26} W
Distance from center of Galaxy:	8000 pc = 26,000 ly
Orbital period around center of Galaxy:	220 million years
Orbital speed around center of Galaxy:	220 km/s

(NOAO)   R I **V** U X G

of the Earth. The chemical reactions involved in burning release roughly 10^{-19} joule of energy per atom. Therefore, the number of atoms that would have to undergo chemical reactions each second to generate the Sun's luminosity of 3.9×10^{26} joules per second is approximately

$$\frac{3.9 \times 10^{26} \text{ joules per second}}{10^{-19} \text{ joule per atom}} = 3.9 \times 10^{45} \text{ atoms per second}$$

figure 18-1   R I **V** U X G

The Sun The Sun's visible surface has a temperature of about 5800 K. At this temperature, all solids and liquids vaporize to form gases. It was only in the twentieth century that scientists discovered what has kept the Sun so hot for billions of years: the thermonuclear fusion of hydrogen nuclei in the Sun's core. (Jeremy Woodhouse/PhotoDisc)

From its mass and chemical composition, we know that the Sun contains about 10^{57} atoms. Thus, the length of time required to consume the entire Sun by burning would be

$$\frac{10^{57} \text{ atoms}}{3.9 \times 10^{45} \text{ atoms per second}} = 3 \times 10^{11} \text{ seconds}$$

There are about 3×10^{7} seconds in a year. Hence, in this model, the Sun would burn itself out in a mere 10,000 (10^{4}) years! This is far shorter than the known age of the Earth, so chemical reactions also cannot explain how the Sun shines.

The source of the Sun's luminosity could be explained if there were a process like burning but that released much more energy per atom. Then the rate at which atoms would have to be consumed would be far less, and the lifetime of the Sun could be long enough to be consistent with the known age of the Earth. Albert Einstein discovered the key to such a process in 1905. According to his *special theory of relativity,* a quantity m of mass can in principle be converted into an amount of energy E according to a now-famous equation:

Einstein's mass-energy equation

$$E = mc^2$$

m = quantity of mass, in kg

c = speed of light = 3×10^{8} m/s

E = amount of energy into which the mass can be converted, in joules

The speed of light c is a large number, so c^2 is huge. Therefore, a small amount of matter can release an awesome amount of energy.

Inspired by Einstein's ideas, astronomers began to wonder if the Sun's energy output might come from the conversion of matter into energy. The Sun's low density of 1410 kg/m^3 indicates that it must be made of the very lightest atoms, primarily hydrogen and helium. In the 1920s, the British astronomer Arthur Eddington showed that temperatures near the center of the Sun must be so high that atoms become completely ionized. Hence, at the Sun's center we expect to find hydrogen nuclei and electrons flying around independent of each other.

Another British astronomer, Robert Atkinson, suggested that under these conditions hydrogen nuclei could fuse together to produce helium nuclei in a *nuclear reaction* that transforms a tiny amount of mass into a large amount of energy. Experiments in the laboratory using individual nuclei show that such reactions can indeed take place. The process of converting hydrogen into helium is called **hydrogen burning,** even though nothing is actually burned in the conventional sense. (Ordinary burning involves chemical reactions that rearrange the outer electrons of atoms but have no effect on the atoms' nuclei.) Hydrogen burning provides the devastating energy released in a hydrogen bomb (see Figure 1-4).

The fusing together of nuclei is also called **thermonuclear fusion,** because it can take place only at extremely high temperatures. The reason is that all nuclei have a positive electric charge and so tend to repel one another. But in the extreme heat and pressure at the Sun's center, hydrogen nuclei (protons) are moving so fast that they can overcome their electric repulsion and actually touch one another. When that happens, thermonuclear fusion can take place.

ANALOGY You can think of protons as tiny electrically charged spheres that are coated with a very powerful glue. If the spheres are not touching, the repulsion between their charges pushes them apart. But if the spheres are forced into contact, the strength of the glue "fuses" them together.

CAUTION! Be careful not to confuse thermonuclear fusion with the similar-sounding process of *nuclear fission.* In nuclear fusion, energy is released by joining together nuclei of lightweight atoms such as hydrogen. In nuclear fission, by contrast, the nuclei of very massive atoms such as uranium or plutonium release energy by fragmenting into smaller nuclei. Nuclear power plants produce energy using

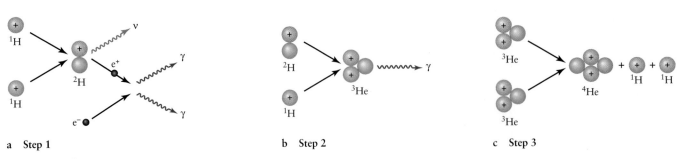

a Step 1   b Step 2   c Step 3

ANIMATION 18.1

figure 18-2

Fusing Hydrogen into Helium Hydrogen burning in the Sun usually takes place in three energy-releasing steps. (See Box 18-1.) **(a)** Two protons (hydrogen nuclei, ^{1}H) fuse to form a hydrogen isotope with one proton and one neutron (^{2}H), a neutral, nearly massless neutrino (ν), and a positively-charged electron, or positron (e$^+$). This positron encounters an ordinary electron (e$^-$), annihilating both particles and converting them into gamma-ray photons (γ). **(b)** The ^{2}H nucleus from the first step combines with a third proton, forming a helium isotope (^{3}He) and releasing another gamma-ray photon. **(c)** Two ^{3}He nuclei collide, forming a different helium isotope with two protons and two neutrons (^{4}He) and releasing two protons.

fission, not fusion. (Generating power using fusion has been a goal of researchers for decades, but no one has yet discovered a commercially viable way to do this.)

We learned in Section 5-8 that the nucleus of a hydrogen atom (H) consists of a single proton. The nucleus of a helium atom (He) consists of two protons and two neutrons. In the nuclear process that Atkinson described, four hydrogen nuclei combine to form one helium nucleus, with a concurrent release of energy:

$$4 \text{ H } \rightarrow \text{He} + \text{energy}$$

In several separate reactions, two of the four protons are changed into neutrons, and eventually combine with the remaining protons to produce a helium nucleus (Figure 18-2). Each time this process takes place, a small fraction (0.7%) of the combined mass of the hydrogen nuclei does not show up in the mass of the helium nucleus. This "lost" mass is converted into energy. Box 18-1 describes how to use Einstein's mass-energy equation to calculate the amount of energy released.

Converting Mass into Energy

figure 18-2 shows the steps involved in the thermonuclear fusion of hydrogen at the Sun's center. In these steps, four protons are converted into a single nucleus of ^{4}He, an isotope of helium with two protons and two neutrons. The reaction depicted in Figure 18-2a also produces a neutral, nearly massless particle called the *neutrino*. Neutrinos respond hardly at all to ordinary matter, so they travel almost unimpeded through the Sun's massive bulk. Hence, the energy that neutrinos carry is quickly lost into space. This loss is not great, however, because the neutrinos carry relatively little energy. (See Section 18-4 for more about these curious particles.)

Most of the energy released by thermonuclear fusion appears in the form of gamma-ray photons. The energy of these photons remains trapped within the Sun for a long time, thus maintaining the Sun's intense internal heat. Some gamma-ray photons are produced by the reaction in Figure 18-2b. Others appear when an electron in the Sun's interior annihilates a positively charged electron, or **positron**, which is a by-product of the reaction in Figure 18-2a. An electron and a positron are respectively matter and antimatter, and they convert entirely into energy when they meet. (You may have thought that "antimatter" was pure science fiction, but it is being created and annihilated in the Sun as you read these words.)

We can summarize the thermonuclear fusion of hydrogen as follows:

$$4 \text{ H } \rightarrow \text{He} + \text{neutrinos} + \text{gamma-ray photons}$$

To calculate how much energy is released in this process, we use Einstein's mass-energy formula: The energy released is equal to the amount of mass consumed multiplied by c^2, where c is the speed of light. To see how much mass is consumed, we compare the combined mass of four hydrogen atoms (the ingredients) to the mass of one helium atom (the product):

$$
\begin{array}{rcl}
4 \text{ hydrogen atoms} & = & 6.693 \times 10^{-27} \text{ kg} \\
-1 \text{ helium atom} & = & -6.645 \times 10^{-27} \text{ kg} \\
\hline
\text{Mass lost} & = & 0.048 \times 10^{-27} \text{ kg}
\end{array}
$$

Thus, a small fraction (0.7%) of the mass of the hydrogen going into the nuclear reaction does not show up in the mass of the helium. This lost mass is converted into an amount of energy $E = mc^2$:

$$E = mc^2 = (0.048 \times 10^{-27} \text{ kg}) (3 \times 10^8 \text{ m/s})^2$$
$$= 4.3 \times 10^{-12} \text{ joule}$$

This is the amount of energy released by the formation of a single helium atom. It would light a 10-watt lightbulb for almost one-half of a trillionth of a second.

EXAMPLE: Let's consider the conversion of 1 kg of hydrogen into helium. Although a kilogram of hydrogen goes into this reaction, 0.7% of the mass is lost, and only 0.993 kg of helium comes out. Using Einstein's equation, we find that the missing 0.007 kg of matter has been transformed into an amount of energy equal to

$$E = mc^2 = (0.007 \text{ kg}) (3 \times 10^8 \text{ m/s})^2 = 6.3 \times 10^{14} \text{ joules}$$

For comparison, this equals the energy released by burning 20,000 metric tons (2×10^7 kg) of coal!

The Sun's luminosity is 3.9×10^{26} joules per second. To generate this much power, hydrogen must be consumed at a rate of

$$\frac{3.9 \times 10^{26} \text{ joules per second}}{6.3 \times 10^{14} \text{ joules per kilogram}}$$
$$= 6 \times 10^{11} \text{ kilograms per second}$$

That is, the Sun converts 600 million metric tons of hydrogen into helium within its core every second.

You may have heard the idea that mass is always conserved (that is, it is neither created nor destroyed), or that energy is always conserved in a reaction. Einstein's ideas show that neither of these statements is quite correct, because mass can be converted into energy and vice versa. A more accurate statement is that the total amount of mass *plus* energy is conserved. Hence, the destruction of mass in the Sun does not violate any laws of nature.

For every four hydrogen nuclei converted into a helium nucleus, 4.3×10^{-12} joule of energy is released. This may seem like only a tiny amount of energy, but it is about 10^7 times larger than the amount of energy released in a typical chemical reaction, such as occurs in ordinary burning. Thus, thermonuclear fusion can explain how the Sun could have been shining for billions of years.

To produce the Sun's luminosity of 3.9×10^{26} joules per second, 6×10^{11} kg (600 million metric tons) of hydrogen must be converted into helium each second. This rate is prodigious, but there is a literally astronomical amount of hydrogen in the Sun. In particular, the Sun's core contains enough hydrogen to have been giving off energy at the present rate for as long as the solar system has existed, about 4.6 billion years, and to continue doing so for another 5 billion years.

 The Sun's energy actually comes from a *series* of nuclear reactions, called the **proton-proton chain,** in which hydrogen nuclei combine to form helium (see Figure 18-2 and Box 18-1). The proton-proton chain is also the energy source for many of the stars in the sky. In stars with central temperatures that are much hotter than that of the Sun, however, hydrogen burning proceeds according to a different set of nuclear reactions, called the **CNO cycle,** in which carbon, nitrogen, and oxygen nuclei absorb protons to produce helium nuclei. Still other thermonuclear reactions, such as helium burning, carbon burning, and oxygen burning, occur late in the lives of many stars.

18-2 A theoretical model of the Sun shows how energy gets from its center to its surface

While thermonuclear fusion is the source of the Sun's energy, this process cannot take place everywhere within the Sun. As we have seen, extremely high temperatures—in excess of 10^7 K—are required for atomic nuclei to fuse together to form larger nuclei. The temperature of the Sun's visible surface, about 5800 K, is too low for these reactions to occur there. Hence, thermonuclear fusion can be taking place only within the Sun's interior. But precisely where does it take place? And how does the energy produced by fusion make its way to the surface, where it is emitted into space in the form of photons?

To answer these questions, we must understand conditions in the Sun's interior. Ideally, we would send an exploratory spacecraft to probe deep into the Sun; in practice, the Sun's intense heat would vaporize even the sturdiest

spacecraft. Instead, astronomers use the laws of physics to construct a theoretical model of the Sun. (We discussed the use of models in science in Section 1-1.) Let's see what ingredients go into building a model of this kind.

Note first that the Sun is not undergoing any dramatic changes. The Sun is not exploding or collapsing, nor is it significantly heating or cooling. The Sun is thus said to be in both *hydrostatic equilibrium* and *thermal equilibrium.*

To understand what is meant by **hydrostatic equilibrium,** imagine a slab of material in the solar interior (Figure 18-3a). In equilibrium, the slab on average will move neither up nor down. (In fact, there are upward and downward motions of material inside the Sun, but these motions average out.) Equilibrium is maintained by a balance among three forces that act on this slab:

1. The downward pressure of the layers of solar material above the slab

2. The upward pressure of the hot gases beneath the slab

3. The slab's weight—that is, the downward gravitational pull it feels from the rest of the Sun

The pressure from below must balance both the slab's weight and the pressure from above. Hence, the pressure below the slab must be greater than that above the slab. In other words, pressure has to increase with increasing depth. For the same reason, pressure increases as you dive deeper into the ocean (Figure 18-3b) or as you move toward lower altitudes in our atmosphere.

Hydrostatic equilibrium also tells us about the density of the slab. If the slab is too dense, its weight will be too great and it will sink; if the density is too low, the slab will rise. To prevent this, the density of solar material must have a certain value at each depth within the solar interior. (The same principle applies to objects that float beneath the surface of the ocean. Scuba divers wear weight belts to increase their average density so that they will neither rise nor sink but will stay submerged at the same level.)

Another consideration is that the Sun's interior is so hot that it is completely gaseous. Gases compress and become more dense when you apply greater pressure to them, so density must increase along with pressure as you go to greater depths within the Sun. Furthermore, when you compress a gas, its temperature tends to rise, so the temperature must also increase as you move toward the Sun's center.

While the temperature in the solar interior is different at different depths, the temperature at each depth remains constant in time. This principle is called **thermal equilibrium.** For the Sun to be in thermal equilibrium, all the energy generated by thermonuclear reactions in the Sun's core must be transported to the Sun's glowing surface, where it can be radiated into space. If too much energy flowed from the core to the surface to be radiated away, the Sun's interior would cool down; alternatively, the Sun would heat up if too little energy flowed to the surface.

But exactly how is energy transported from the Sun's center to its surface? There are three methods of energy transport:

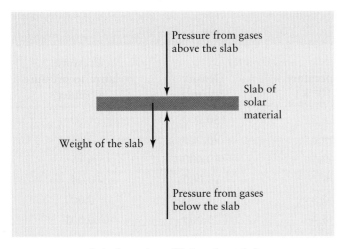

a In hydrostatic equilibrium, forces balance

b In hydrostatic equilibrium, forces balance

figure 18-3

Hydrostatic Equilibrium (a) The Sun's interior is in hydrostatic equilibrium, which means that, on average, material tends to move neither up nor down. For any slab of material in the solar interior, the pressure from material below the slab (which tends to push the slab up) must balance the combined effects of the slab's weight and the pressure from material above the slab (both of which tend to push the slab down). Hence, the pressure must increase with increasing depth. **(b)** The same principle applies to a fish in water. In equilibrium, the forces balance and the fish neither rises nor sinks. (Ken Usami/PhotoDisc)

conduction, convection, and *radiative diffusion.* Only the last two are important inside the Sun.

If you heat one end of a metal bar with a blowtorch, energy flows to the other end of the bar so that it too becomes warm. The efficiency of this method of energy transport, called **conduction,** varies significantly from one substance to another. For example, copper is a good conductor of heat, but wood is not (which is why copper pots often have wooden handles). Conduction is not an efficient means of energy transport inside stars like the Sun, for which the average density is relatively low.

Inside stars like our Sun, energy moves from center to surface by two other means: convection and radiative diffusion. **Convection** is the circulation of fluids—gases or liquids—between hot and cool regions. Hot gases rise toward a star's surface, while cool gases sink back down toward the star's center. This physical movement of gases transports heat energy outward in a star, just as the physical movement of water boiling in a pot transports energy from the bottom of the pot (where the heat is applied) to the cooler water at the surface.

In **radiative diffusion,** photons created in the thermonuclear inferno at a star's center diffuse outward toward the star's surface. Individual photons are absorbed and reemitted by atoms and electrons inside the star. The overall result is an outward migration from the hot core, where photons are constantly created, toward the cooler surface, where they escape into space.

To construct a model of a star like the Sun, astrophysicists express the ideas of hydrostatic equilibrium, thermal equilib-

rium, and energy transport as a set of equations. To ensure that the model applies to the particular star under study, they also make use of astronomical observations of the star's surface. (For example, to construct a model of the Sun, they use the data that the Sun's surface temperature is 5800 K, its luminosity is 3.9×10^{26} W, and the gas pressure and density at the surface are almost zero.) The astrophysicists then use a computer to solve their set of equations and calculate conditions layer by layer in toward the star's center. The result is a model of how temperature, pressure, and density increase with increasing depth below the star's surface.

Table 18-2 and Figure 18-4 show a theoretical model of the Sun that was calculated in just this way. Different models of the Sun use slightly different assumptions, but all models give essentially the same results as those shown here. From such computer models we have learned that at the Sun's center the density is 160,000 kg/m³ (14 times the density of lead!), the temperature is 1.55×10^7 K, and the pressure is 3.4×10^{11} atm. (One atmosphere, or 1 atm, is the average atmospheric pressure at sea level on Earth.)

Table 18-2 and Figure 18-4 show that the solar luminosity rises to 100% at about one-quarter of the way from the Sun's center to its surface. In other words, the Sun's energy production occurs within a volume that extends out only to 0.25 R$_\odot$. (The symbol R$_\odot$ denotes the *solar radius,* or radius of the Sun as a whole, equal to 696,000 km.) Outside 0.25 R$_\odot$, the density and temperature are too low for thermonuclear reactions to take place. Also note that 94% of the total mass of the Sun is found within the inner 0.5 R$_\odot$. Hence, the outer 0.5 R$_\odot$ contains only a relatively small amount of material.

table 18-2   A Theoretical Model of the Sun

Distance from the Sun's center (solar radii)	Fraction of luminosity	Fraction of mass	Temperature ($\times 10^6$ K)	Density (kg/m^3)	Pressure (relative to pressure at center)
0.0	0.00	0.00	15.5	160,000	1.00
0.1	0.42	0.07	13.0	90,000	0.46
0.2	0.94	0.35	9.5	40,000	0.15
0.3	1.00	0.64	6.7	13,000	0.04
0.4	1.00	0.85	4.8	4,000	0.007
0.5	1.00	0.94	3.4	1,000	0.001
0.6	1.00	0.98	2.2	400	0.0003
0.7	1.00	0.99	1.2	80	4×10^{-5}
0.8	1.00	1.00	0.7	20	5×10^{-6}
0.9	1.00	1.00	0.3	2	3×10^{-7}
1.0	1.00	1.00	0.006	0.00030	4×10^{-13}

Note: The distance from the Sun's center is expressed as a fraction of the Sun's radius ($R_\odot$). Thus, 0.0 is at the center of the Sun and 1.0 is at the surface. The fraction of luminosity is that portion of the Sun's total luminosity produced within each distance from the center; this is equal to 1.00 for distances of 0.25 $R_\odot$ or more, which means that all of the Sun's nuclear reactions occur within 0.25 solar radius from the Sun's center. The fraction of mass is that portion of the Sun's total mass lying within each distance from the Sun's center. The pressure is expressed as a fraction of the pressure at the center of the Sun.

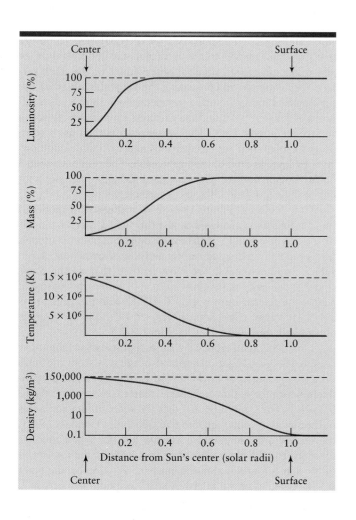

How energy flows from the Sun's center toward its surface depends on how easily photons move through the gas. If the solar gases are comparatively transparent, photons can travel moderate distances before being scattered or absorbed, and energy is thus transported by radiative diffusion. If the gases are comparatively opaque, photons cannot get through the gas easily and heat builds up. Convection then becomes the most efficient means of energy transport. The gases start to churn, with hot gas moving upward and cooler gas sinking downward.

From the center of the Sun out to about 0.71 $R_\odot$, energy is transported by radiative diffusion. Hence, this region is called the **radiative zone**. Beyond about 0.71 $R_\odot$, the temperature is low enough (a mere 2×10^6 K or so) for electrons and hydrogen nuclei to join into hydrogen atoms. These atoms are very effective at absorbing photons, much more so than free electrons or nuclei, and this absorption chokes off the outward flow of photons. Therefore, beyond about

figure 18-4

A Theoretical Model of the Sun's Interior From top to bottom, these graphs depict what percentage of the Sun's total luminosity is produced within each distance from the center, what percentage of the total mass lies within each distance from the center, the temperature at each distance, and the density at each distance. (See Table 18-2 for a numerical version of this model.)

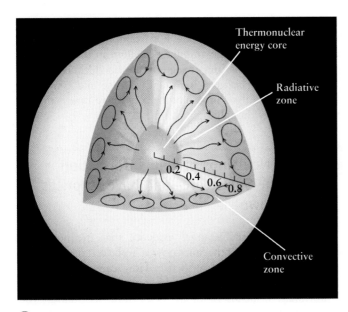

figure 18-5

The Sun's Internal Structure Thermonuclear reactions occur in the Sun's core, which extends out to a distance of 0.25 R$_\odot$ from the center. Energy is transported outward, via radiative diffusion, to a distance of about 0.71 R$_\odot$. In the outer layers between 0.71 R$_\odot$ and 1.00 R$_\odot$, energy flows outward by convection.

0.71 R$_\odot$, radiative diffusion is not an effective way to transport energy. Instead, convection dominates the energy flow in this outer region, which is why it is called the **convective zone.** Figure 18-5 shows these aspects of the Sun's internal structure.

Although energy travels through the radiative zone in the form of photons, the photons have a difficult time of it. Table 18-2 shows that the material in this zone is extremely dense, so photons from the Sun's core take a long time to diffuse through the radiative zone. As a result, it takes approximately 170,000 years for energy created at the Sun's center to travel 696,000 km to the solar surface and finally escape as sunlight. The energy flows outward at an average rate of

50 centimeters per hour, or about 20 times slower than a snail's pace.

Once the energy escapes from the Sun, it travels much faster—at the speed of light. Thus, solar energy that reaches you today took only 8 minutes to travel the 150 million kilometers from the Sun's surface to the Earth. But this energy was actually produced by thermonuclear reactions that took place in the Sun's core hundreds of thousands of years ago.

18-3 Astronomers probe the solar interior using the Sun's own vibrations

 We have described how astrophysicists construct models of the Sun. But since we cannot see into the Sun's opaque interior, how can we check these models to see if they are accurate? What is needed is a technique for probing the Sun's interior. A very powerful technique of just this kind involves measuring vibrations of the Sun as a whole. This field of solar research is called **helioseismology.**

Vibrations are a useful tool for examining the hidden interiors of all kinds of objects. Food shoppers test whether melons are ripe by tapping on them and listening to the vibrations. Geologists can determine the structure of the Earth's interior by using seismographs to record vibrations during earthquakes.

Although there are no true "sunquakes," the Sun does vibrate at a variety of frequencies, somewhat like a ringing bell. These vibrations were first noticed in 1960 by Robert Leighton of the California Institute of Technology, who made high-precision Doppler shift observations of the solar surface. These measurements revealed that parts of the Sun's surface move up and down about 10 meters every 5 minutes (Figure 18-6). Since the mid-1970s, several astronomers have reported slower vibrations, having periods ranging from 20 to 160 minutes. The detection of extremely slow vibrations has inspired astronomers to organize networks of telescopes around and in orbit above the Earth to monitor the Sun's vibrations on a continuous basis.

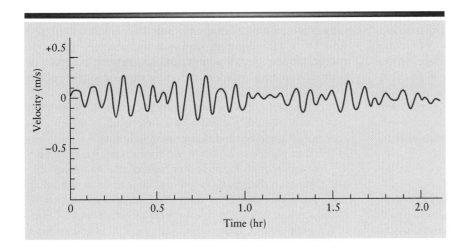

figure 18-6

Vibrations of the Sun's Surface Doppler shift measurements give the speed at which areas of the Sun's surface bob up and down. The data shown here extend over 2 hours and show numerous oscillations with a period of about 5 minutes. The size of the oscillations varies gradually due to interference between a variety of vibrations with different wavelengths. (Adapted from O. R. White)

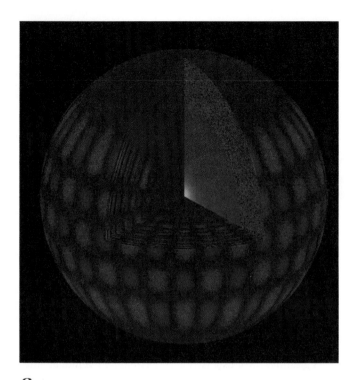

figure 18-7

A Sound Wave Resonating in the Sun This computer-generated image shows one of the millions of ways in which the Sun vibrates because of sound waves resonating in its interior. The regions that are moving outward are colored blue, those moving inward, red. The cutaway shows how deep these oscillations are thought to extend. (National Solar Observatory)

 The vibrations of the Sun's surface can be compared with sound waves. If you could somehow survive within the Sun's outermost layers, you would first notice a deafening roar, somewhat like a jet engine, produced by turbulence in the Sun's gases. But superimposed on this noise would be a variety of nearly pure tones. You would need greatly enhanced hearing to detect these tones, however; the strongest has a frequency of just 0.003 hertz, 13 octaves below the lowest frequency audible to humans. (Recall from Section 5-2 that one hertz is one oscillation per second.)

In 1970, Roger Ulrich, at UCLA, pointed out that sound waves moving upward from the solar interior would be reflected back inward after reaching the solar surface. However, as a reflected sound wave descended back into the Sun, the increasing density and pressure would bend the wave severely, turning it around and aiming it back toward the solar surface. In other words, sound waves bounce back and forth between the solar surface and layers deep within the Sun. These sound waves can reinforce each other if their wavelength is the right size, just as sound waves of a particular wavelength resonate inside an organ pipe.

The Sun oscillates in millions of ways as a result of waves resonating in its interior. Figure 18-7 is a computer-generated illustration of one such mode of vibration. Helioseismologists

can deduce information about the solar interior from measurements of these oscillations. For example, they have been able to set limits on the amount of helium in the Sun's core and convective zone and to determine the thickness of the transition region between the radiative zone and convective zone. They have also found that the convective zone is thicker than previously thought.

18-4 Neutrinos provide information about the Sun's core—and have surprises of their own

We have seen circumstantial evidence that thermonuclear fusion is the source of the Sun's power. To be certain, however, we need more definitive evidence. How can we show that thermonuclear fusion really is taking place in the Sun's core? Unfortunately, helioseismology is of little assistance. The vibrations that astronomers see on the Sun's surface do not penetrate that far into the interior.

Although the light energy that we receive from the Sun originates in the core, it provides few clues about conditions there. The problem is that this energy has changed form repeatedly during its passage from the core: It appeared first as photons diffusing through the radiative zone, then as heat transported through the outer layers by convection, and then again as photons emitted from the Sun's glowing surface. As a result of these transformations, much of the information that the Sun's radiated energy once carried about conditions in the core has been lost.

 If you make a photocopy of a photocopy of a photocopy of an original document, the final result may be so blurred as to be unreadable. In an analogous way, because solar energy is transformed many times while en route to Earth, the story it could tell us about the Sun's core is hopelessly blurred.

But there is a way for scientists to learn about conditions in the Sun's core and to get direct evidence that thermonuclear fusion really does happen there. The trick is to detect the subatomic by-products of thermonuclear fusion reactions.

As part of the process of hydrogen burning, protons change into neutrons and release **neutrinos** (see Figure 18-2a and Box 18-1). Like photons, neutrinos are particles that carry energy, have no electric charge, and have almost no mass. (If neutrinos are truly massless, they must travel through space at the speed of light, just as photons do. Some experiments suggest that neutrinos might have a tiny mass, probably less than one ten-thousandth the mass of an electron, in which case they would travel slightly slower than light.)

Unlike photons, however, neutrinos interact only very weakly with matter. Even the vast bulk of the Sun offers little impediment to their passage, so neutrinos must be streaming out of the core and into space. Indeed, the conversion of hydrogen into helium at the Sun's center produces 10^{38} neutrinos each second. Every second, about 10^{14} neutrinos from

of its neutrons into a proton, creating a radioactive atom of argon (^{37}Ar).

The rate at which argon is produced is related to the *neutrino flux*—that is, the number of neutrinos from the Sun arriving at the Earth per square meter per second. By counting the number of newly created argon atoms, Davis was able to determine the neutrino flux from the Sun. (Other subatomic particles besides neutrinos can also induce reactions that create radioactive atoms. By placing the experiment deep underground, however, the body of the Earth absorbs essentially all such particles—with the exception of neutrinos.)

Davis and his collaborators found that solar neutrinos created one radioactive argon atom in the tank every three days. But this rate corresponds to only *one-third* of the neutrino flux predicted from standard models of the Sun. This troubling discrepancy between theory and observation is called the **solar neutrino problem.** It has motivated scientists around the world to conduct new experiments to measure the flux of solar neutrinos.

 The various nuclear reactions in the proton-proton chain produce neutrinos of differing energies. The vast majority of the neutrinos from the Sun are created during the first reaction, in which two protons combine to form a heavy isotope of hydrogen (see Figure 18-2*a*). But these neutrinos have too little energy to convert chlorine into argon. Davis's experiment responded only to high-energy neutrinos produced by reactions that occur only part of the time near the end of the proton-proton chain. (Figure 18-2 does not show these reactions.)

 To measure the flux of low-energy neutrinos produced by the first reaction of the proton-proton chain, teams of physicists have constructed two neutrino detectors that use several tons of gallium rather than cleaning fluid. Low-energy neutrinos convert gallium (^{71}Ga) into a radioactive isotope of germanium (^{71}Ge). By chemically separating the germanium from the gallium and counting the radioactive atoms, the physicists have been able to measure the low-energy neutrino flux from the Sun. Both detectors—GALLEX in Italy and SAGE (Soviet-American Gallium Experiment) in Russia—have detected only 50% to 60% of the expected neutrino flux.

An experiment in Japan, called Kamiokande, was designed to detect solar neutrinos with energies even higher than those in the Davis experiment. A large underground tank containing 3000 tons of water was surrounded by 1000 light detectors. From time to time, a high-energy solar neutrino struck an electron in one of the water molecules, dislodging it and sending it flying like a pin hit by a bowling ball. The recoiling electron produced a flash of light, which was sensed by the detectors. Kamiokande detected about 45% of the expected flux of neutrinos. By analyzing the flashes, scientists could tell the direction from which the neutrinos were coming and confirmed that they emanated from the Sun. This gave direct evidence that thermonuclear fusion is indeed occurring in the Sun's core.

All these experiments suggest that the Sun is emitting fewer neutrinos than predicted by standard solar models. This discrepancy between theory and observation suggests that

ƒigure 18-8 R I $\boxed{\text{V}}$ U X G

The Davis Solar Neutrino Experiment This tank of cleaning fluid lies 1.5 km below the surface in the Homestake Gold Mine in South Dakota. The Earth shields the tank from stray particles, so that only solar neutrinos can convert chlorine atoms in the fluid into radioactive argon atoms. The number of argon atoms created in the tank is a direct measure of the neutrino flux from the Sun. The detected flux was only about one-third the predicted value, a state of affairs called the "solar neutrino problem." (Courtesy of R. Davis, Brookhaven National Laboratory)

the Sun—that is, **solar neutrinos**—must pass through each square meter of the Earth.

If it were possible to detect these solar neutrinos, we would have direct evidence that thermonuclear reactions really do take place in the Sun's core. Beginning in the 1960s, scientists began to build neutrino detectors for precisely this purpose.

The challenge is that neutrinos are exceedingly difficult to detect. Just as neutrinos pass unimpeded through the Sun, they also pass through the Earth *almost* as if it were not there. We stress the word "almost," because neutrinos can and do interact with matter, albeit infrequently.

On rare occasions a neutrino will strike a neutron and convert it into a proton. This effect was the basis of the original solar neutrino detector, designed and built by Raymond Davis of the Brookhaven National Laboratory in the 1960s. This device used 100,000 gallons of perchloroethylene (C_2Cl_4), a fluid used in dry cleaning, in a huge tank buried deep underground (Figure 18-8). Most of the solar neutrinos that entered Davis's tank passed right through it with no effect whatsoever. But occasionally a neutrino struck the nucleus of one of the chlorine atoms (^{37}Cl) in the cleaning fluid and converted one

either something is wrong with standard solar models or we do not fully understand neutrinos. Some scientists have proposed nonstandard solar models, while others theorize that neutrinos change their character as they travel through space.

One proposed solution to the neutrino problem is that the Sun's core is cooler than predicted by solar models. If the Sun's central temperature is only 10% less than the current estimate, fewer neutrinos would be produced and the neutrino flux would agree with experiments. However, a lower central temperature would cause other obvious features, such as the Sun's size and surface temperature, to be different from what we observe.

Many astronomers and physicists suspect that the solution to the neutrino problem lies not in how neutrinos are produced, but rather in what happens to them between the Sun's core and detectors on the Earth. Physicists have found that there are actually three types of neutrinos. Only one of these types is produced in the Sun, and it is only this type that can be detected by the experiments we have described. But if some of the solar neutrinos change in flight into a different type of neutrino, our detectors would record only a fraction of the total neutrino flux. This effect is called *neutrino oscillation*. In June 1998, scientists at the Super-Kamiokande neutrino observatory in Japan (a larger and more sensitive device than Kamiokande) revealed evidence that neutrino oscillation does indeed take place.

 Super-Kamiokande and the Sudbury Neutrino Observatory in Canada may help solve the solar neutrino problem. While both detectors are sensitive only to high-energy neutrinos, they can detect neutrinos of all three types and can even determine how many of them have undergone neutrino oscillation. The data from these and other solar neutrino experiments may finally reveal the truth about the Sun's core.

18-5 The photosphere is the lowest of three main layers in the Sun's atmosphere

Although the Sun's core is hidden from our direct view, we can easily see sunlight coming from the high-temperature gases that make up the Sun's atmosphere. These outermost layers of the Sun prove to be the sites of truly dramatic activity, much of which has a direct impact on our planet. By studying these layers, we gain further insight into the character of the Sun as a whole.

In a visible-light photograph of the Sun like Figure 18-9, it appears that the Sun has a definite surface. This is actually an illusion; the Sun is gaseous throughout its volume because of its high internal temperature. As you move farther away from the Sun's center, the gases simply become less and less dense.

Why, then, does the Sun appear to have a sharp, well-defined surface? The reason is that essentially all of the Sun's visible light emanates from a single, thin layer of gas called the **photosphere** ("sphere of light"). Just as you can see only a certain distance through the Earth's atmosphere before objects vanish in the haze, we can see only about 400 km into the

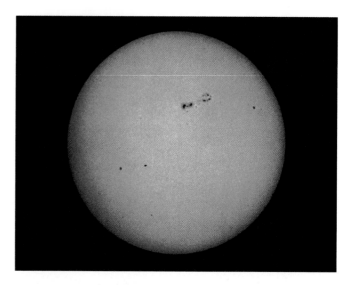

figure 18-9   R I **V** U X G

The Photosphere The photosphere is the layer in the solar atmosphere from which the Sun's visible light is emitted. Note that the Sun appears darker around its limb, or edge; here we are seeing the upper photosphere, which is relatively cool and thus glows less brightly. (The dark sunspots, which we discuss in Section 18-8, are also relatively cool regions.) (Celestron International)

photosphere. This distance is so small compared with the Sun's radius of 696,000 km that the photosphere appears to be a definite surface.

The photosphere is actually the lowest of the three layers that together constitute the solar atmosphere. Above it are the *chromosphere* and the *corona*, both of which are transparent to visible light. We can see them only using special techniques, which we discuss later in this chapter. Everything below the photosphere is called the *solar interior*.

The photosphere is heated from below by energy streaming outward from the solar interior. Hence, temperature should decrease as you go upward in the photosphere, just as in the solar interior (see Table 18-2 and Figure 18-4). We know this is the case because the photosphere appears darker around the edge, or *limb*, of the Sun than it does toward the center of the solar disk, an effect called **limb darkening** (examine Figure 18-9). This happens because when we look near the Sun's limb, we do not see as deeply into the photosphere as we do when we look near the center of the disk (Figure 18-10). The high-altitude gas we observe at the limb is not as hot and thus does not glow as brightly as the deeper, hotter gas seen near the disk center.

The spectrum of the Sun's photosphere confirms how its temperature varies with altitude. As we saw in Section 18-1, the photosphere shines like a nearly perfect blackbody with an average temperature of about 5800 K. However, superimposed on this spectrum are many dark absorption lines (see Figure 5-12). As discussed in Section 5-6, we see an *absorption line spectrum* of this sort whenever we view a hot, glow-

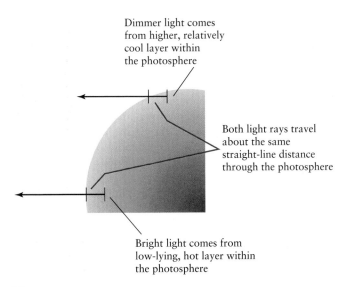

Dimmer light comes from higher, relatively cool layer within the photosphere

Both light rays travel about the same straight-line distance through the photosphere

Bright light comes from low-lying, hot layer within the photosphere

ƒigure 18-10

The Origin of Limb Darkening Light coming from both the Sun's limb and the center of the Sun's disk travels about the same straight-line distance through the photosphere to reach a telescope on Earth. Because of the Sun's curved shape, light from the limb comes from a greater height within the photosphere, where the temperature is lower and the gases glow less brightly. This gives the limb its darker appearance.

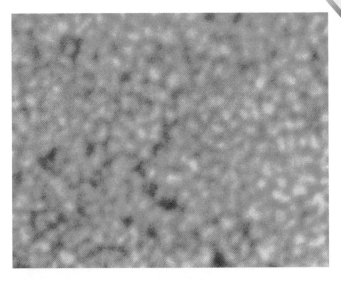

ƒigure 18-11   R I **V** U X G

Solar Granulation High-resolution photographs of the Sun's surface, such as this one, reveal a blotchy pattern called granulation. The individual yellow blobs in this image are granules, each of which measures about 1000 km (600 mi) across. Granules are actually convection cells in the Sun's outer layers (see Figure 18-12). (NOAO)

ing object through a relatively cool gas. In this case, the hot object is the lower part of the photosphere; the cooler gas is in the upper part of the photosphere, where the temperature declines to about 4400 K. All the absorption lines in the Sun's spectrum are produced in this relatively cool layer, as atoms selectively absorb photons of various wavelengths streaming outward from the hotter layers below.

  You may find it hard to think of 4400 K as "cool." But keep in mind that the ratio of 4400 K to 5800 K, the temperature in the lower photosphere, is the same as the ratio of the temperature on a Siberian winter night to that of a typical day in Hawaii.

We can learn still more about the photosphere by examining it with a telescope—but only when using special dark filters to prevent eye damage. *Looking directly at the Sun without the correct filter, whether with the naked eye or with a telescope, can cause permanent blindness!* Under good observing conditions, astronomers using such filter-equipped telescopes can often see a blotchy pattern in the photosphere, called **granulation** (Figure 18-11). Each light-colored **granule** measures about 1000 km (600 mi) across and is surrounded by a darkish boundary. The difference in brightness between the center and the edge of a granule corresponds to a temperature drop of about 300 K.

Granulation is caused by convection of the gas in the photosphere. Gas from lower levels rises upward in granules, cools off, spills over the edges of the granules, and then plunges back

down into the Sun (Figure 18-12). This can occur only if the gas is heated from below, like a pot of water being heated on a stove (see Section 18-2). Along with limb darkening and the Sun's absorption line spectrum, granulation shows that the upper part of the photosphere must be cooler than the lower part.

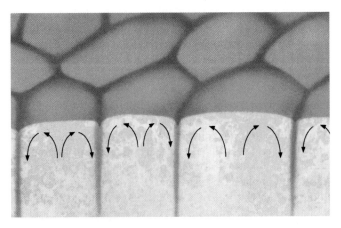

ƒigure 18-12

Convection in the Photosphere Hot gas rising upward produces the bright granules shown in Figure 18-11. Cooler gas sinks downward along the boundaries between granules; this gas glows less brightly, giving the boundaries their dark appearance. The circulating convective motion shown here transports heat from the Sun's interior outward into the solar atmosphere.

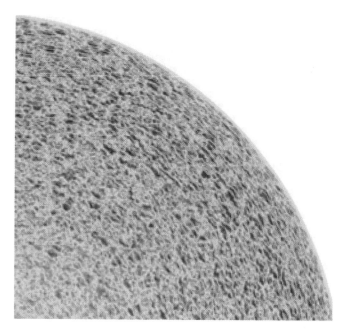

ƒigure 18-13 R I ▮ U X G

Supergranules and Large-Scale Convection Supergranules display relatively little contrast between their center and edges, so they are hard to observe in ordinary images. But they can be seen in a Doppler image like this one. Light from gas that is approaching us (that is, rising) is shifted toward shorter wavelengths, while light from receding gas (that is, descending) is shifted toward longer wavelengths (see Section 5-9). The colors in this image are false: areas of rising and falling gas are colored blue and red, respectively. (David Hathaway, MSFC/NASA)

Time-lapse photography reveals more of the photosphere's dynamic activity. Granules form, disappear, and reform in cycles lasting only a few minutes. At any one time, about 4 million granules cover the solar surface. A typical granule occupies about 10^6 square kilometers, equal to the areas of Texas and Oklahoma combined.

Superimposed on the pattern of granulation are even larger convection cells called **supergranules** (Figure 18-13). As in granules, gases rise upward in the middle of a supergranule, move horizontally outward toward its edge, and descend back into the Sun. The difference is that a typical supergranule is about 35,000 km in diameter, large enough to enclose several hundred granules. This large-scale convection moves at only about 0.4 km/s (1400 km/h, or 900 mi/h), about one-tenth the speed of gases churning in a granule. A given supergranule lasts about a day.

ANALOGY Similar patterns of large-scale and small-scale convection can be found in the Earth's atmosphere. On the large scale, air rises gradually at a low-pressure area, then sinks gradually at a high-pressure area, which might be hundreds of kilometers away. This is analogous to the flow in a supergranule. Thunderstorms in our atmo-

sphere are small but intense convection cells within which air moves rapidly up and down. Like granules, they last only a relatively short time before they dissipate.

Although the photosphere is a very active place, it actually contains relatively little material. Careful examination of the spectrum shows that it has a density of only about 10^{-4} kg/m³, roughly 0.01% the density of the Earth's atmosphere at sea level. The photosphere is made primarily of hydrogen and helium, the most abundant elements in the solar system (see Table 7-3).

Despite being such a thin gas, the photosphere is surprisingly opaque to visible light. If it were not so opaque, we could see into the Sun's interior to a depth of hundreds of thousands of kilometers, instead of the mere 400 km that we can see down into the photosphere. What makes the photosphere so opaque is that its hydrogen atoms sometimes acquire an extra electron, becoming **negative hydrogen ions**. The extra electron is only loosely attached and can be dislodged if it absorbs a photon of any visible wavelength. Hence, negative hydrogen ions are very efficient light absorbers, and there are enough of these light-absorbing ions in the photosphere to make it quite opaque. Because it is so opaque, the photosphere's spectrum is close to that of an ideal blackbody.

18-6 The chromosphere is characterized by spikes of rising gas

An ordinary visible-light photograph such as Figure 18-9 gives the impression that the Sun ends at the top of the photosphere. But during a total solar eclipse, the Moon blocks the photosphere from our view, revealing a glowing, pinkish layer of gas above the photosphere (Figure 18-14). This is the tenuous **chromosphere** ("sphere of color"), the second of the three major levels in the Sun's atmosphere. The chromosphere is only about one ten-thousandth (10^{-4}) as dense as the photosphere, or about 10^{-8} as dense as our own atmosphere. No wonder it is normally invisible!

Unlike the photosphere, which has an absorption line spectrum, the chromosphere has a spectrum dominated by *emission* lines. An emission line spectrum is produced by the atoms of a hot, thin gas (see Section 5-6 and Section 5-8). As their electrons fall from higher to lower energy levels, the atoms emit photons.

One of the strongest emission lines in the chromosphere's spectrum is the H_α line at 656.3 nm, which is emitted by a hydrogen atom when its single electron falls from the $n = 3$ level to the $n = 2$ level (recall Figure 5-21b). This wavelength is in the red part of the spectrum, which gives the chromosphere its characteristic pinkish color. The spectrum also contains emission lines of singly ionized calcium, as well as lines due to ionized helium and ionized metals. In fact, helium was originally discovered in the chromospheric spectrum in 1868, almost thirty years before helium gas was first isolated on Earth.

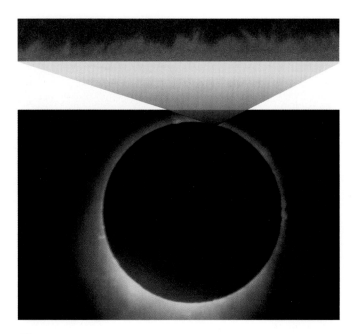

figure 18-14 R I **V** U X G

The Chromosphere This photograph, taken during a total solar eclipse, shows the Sun's chromosphere glowing around the edge of the Moon. It appears pinkish because its hot gases emit light at only certain discrete wavelengths, principally the H$_\alpha$ emission of hydrogen at a red wavelength of 656.3 nm. The expanded area above shows spicules, jets of chromospheric gas that surge upward into the Sun's outer atmosphere. In making this image, a filter was used that is transparent only to H$_\alpha$ emission. (NOAO)

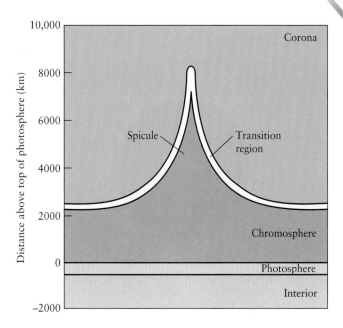

figure 18-15

The Solar Atmosphere This schematic diagram shows the three layers of the solar atmosphere. The lowest, the photosphere, is about 400 km thick. The chromosphere above it extends to an altitude of about 2000 km, with spicules jutting up to nearly 10,000 km above the photosphere. The third, outermost layer is the corona, which we discuss in Section 18-7. It extends many millions of kilometers out into space. (Adapted from J. A. Eddy)

Analysis of the chromospheric spectrum shows that temperature *increases* with increasing height in the chromosphere. This is just the opposite of the situation in the photosphere, where temperature decreases with increasing height. The temperature is about 4400 K at the top of the photosphere; 2000 km higher, at the top of the chromosphere, the temperature is nearly 25,000 K.

The photospheric spectrum is dominated by absorption lines at certain wavelengths, while the spectrum of the chromosphere has emission lines at these same wavelengths. In other words, the photosphere appears dark at the wavelengths at which the chromosphere emits most strongly, such as the H$_\alpha$ wavelength of 656.3 nm. By viewing the Sun through a special filter that is transparent to light only at the wavelength of H$_\alpha$, astronomers can screen out light from the photosphere and make the chromosphere visible. (The same technique can be used with other wavelengths at which the chromosphere emits strongly, including nonvisible wavelengths.) This makes it possible to see the chromosphere at any time, not just during a solar eclipse.

The top photograph in Figure 18-14 is a high-resolution image of the Sun's chromosphere taken through an H$_\alpha$ filter. This image shows numerous vertical spikes, which are actually jets of rising gas called **spicules.** A typical spicule lasts just 15 minutes or so: It rises at the rate of about 20 km/s

(72,000 km/h, or 45,000 mi/h), reaches a height of several thousand kilometers, then collapses and fades away (Figure 18-15). Approximately 300,000 spicules exist at any one time, covering about 1% of the Sun's surface.

Spicules are generally located directly above the edges of supergranules (see Figure 18-13). This is a surprising result, because chromospheric gases are rising in a spicule while photospheric gases are *descending* at the edge of a supergranule. What, then, is pulling gases upward to form spicules? As we will see in Section 18-10, the answer proves to be the Sun's intense magnetic field. But before we delve into how this happens, let us to complete our tour of the solar atmosphere by exploring its outermost, least dense, most dynamic, and most bizarre layer—the region called the corona.

18-7 The corona ejects mass into space to form the solar wind

The **corona,** or outermost region of the Sun's atmosphere, begins at the top of the chromosphere. It extends out to a distance of several million kilometers. Despite its tremendous extent, the corona is only about one-millionth (10^{-6}) as bright as the photosphere—no brighter than the full moon. Hence,

figure 18-16 R I **V** U X G

The Solar Corona This striking photograph of the corona was taken during the total solar eclipse of July 11, 1991. Numerous streamers extend for millions of kilometers above the solar surface. The unearthly light of the corona is one of the most extraordinary aspects of experiencing a solar eclipse. (Courtesy of R. Christen and M. Christen, Astro-Physics, Inc.)

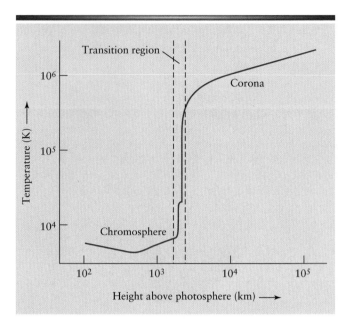

figure 18-17

Temperatures in the Sun's Upper Atmosphere This graph shows how temperature varies with altitude in the Sun's chromosphere and corona. In order to show a large range of values, both the vertical and horizontal scales are nonlinear. In the narrow transition region between the chromosphere and corona, the temperature rises abruptly by a hundredfold, from about 10^4 K to about 10^6 K. (Adapted from A. Gabriel)

the corona can be viewed only when the light from the photosphere is blocked out, either by use of a specially designed telescope or during a total eclipse.

Figure 18-16 is an exceptionally detailed photograph of the Sun's corona taken during a solar eclipse. It shows that the corona is not merely a spherical shell of gas surrounding the Sun. Rather, numerous streamers extend in different directions far above the solar surface. The shapes of these streamers vary on time scales of days or weeks. (For other views of the corona during solar eclipses, see Figure 3-10 and the photograph that opens Chapter 3.)

The corona has an emission line spectrum, characteristic of a hot, thin gas. When this spectrum was first measured in the nineteenth century, astronomers found a number of emission lines at wavelengths that had never been seen in the laboratory. Their explanation was that the corona contained elements that had not yet been detected on the Earth. However, laboratory experiments in the 1930s revealed that these unusual emission lines were in fact caused by the same atoms found elsewhere in the universe—but in highly ionized states. For example, a prominent green line at 530.3 nm is caused by highly ionized iron atoms, each of which has been stripped of 13 of its 26 electrons. In order to strip that many electrons from atoms, temperatures in the corona must reach 2 million kelvins (2×10^6 K) or even higher. Figure 18-17 shows how temperature in both the chromosphere and corona varies with altitude.

CAUTION! The corona is actually not very "hot"—that is, it contains very little thermal energy. The reason is that the corona is nearly a vacuum. In the corona there are only about 10^{11} atoms per cubic meter, compared with about 10^{23} atoms per cubic meter in the Sun's photosphere and about 10^{25} atoms per cubic meter in the air that we breathe. Because of the corona's high temperature, the atoms there are moving at very high speeds. But because there are so few atoms in the corona, the total amount of energy in these moving atoms (a measure of how "hot" the gas is) is rather low. If you flew a spaceship into the corona, you would have to worry about becoming overheated by the intense light coming from the photosphere, but you would notice hardly any heating from the corona's ultrathin gas.

ANALOGY The situation in the corona is similar to that inside a conventional oven that is being used for baking. Both the walls of the oven and the air inside the oven are at the same high temperature, but the air contains very few atoms and thus carries little energy. If you put your hand in the oven momentarily, the lion's share of the heat you feel is radiation from the oven walls.

The low density of the corona explains why it is so dim compared with the photosphere. In general, the higher the temperature of a gas, the brighter it glows. But because there are so few atoms in the corona, the net amount of light that it emits is very feeble compared with the light from the much cooler, but also much denser, photosphere.

The Earth's gravity keeps our atmosphere from escaping into space. In the same way, the Sun's powerful gravitational attraction keeps most of the gases of the photosphere, chromosphere, and corona from escaping. But the corona's high temperature means that its atoms and ions are moving at very high speeds, around a million kilometers per hour. As a result, some of the coronal gas does escape. This outflow of gas, which we first encountered in Section 7-8, is called the **solar wind.**

Each second the Sun ejects about a million tons (10^9 kg) of material into the solar wind. But the Sun is so massive that, even over its entire lifetime, it will eject only a few tenths of a percent of its total mass. The solar wind is composed almost entirely of electrons and nuclei of hydrogen and helium. About 0.1% of the solar wind is made up of ions of more massive atoms, such as silicon, sulfur, calcium, chromium, nickel, iron, and argon. The aurorae seen at far northern or southern latitudes on Earth are produced when electrons and ions from the solar wind enter our upper atmosphere.

Special telescopes enable astronomers to see the origin of the solar wind. To appreciate what sort of telescopes are needed, note that because the temperature of the coronal gas is so high, ions in the corona are moving very fast (see Box 7-2). When ions collide, the energy of the impact is so great that the ion's electrons are boosted to very high energy levels. As the electrons fall back to lower levels, they emit high-energy photons in the ultraviolet and X-ray portions of the spectrum—wavelengths at which the photosphere and chromosphere are relatively dim. Hence, telescopes sensitive to these short wavelengths are ideal for studying the corona and the flow of the solar wind.

The Earth's atmosphere is opaque to ultraviolet light and X rays, so telescopes for these wavelengths must be placed on board spacecraft (see Section 6-7, especially Figure 6-27). Figure 18-18 shows a composite ultraviolet view of the corona from the spacecraft *SOHO* (*Solar and Heliospheric Observatory*), a joint project of the European Space Agency (ESA) and NASA.

Figure 18-18 reveals that the corona is not uniform in temperature or density. The densest, highest-temperature regions appear bright, while the thinner, lower-temperature regions are dark. Note the large dark area, called a **coronal hole** because it is almost devoid of luminous gas. Particles streaming away from the Sun can most easily flow outward through these particularly thin regions. Therefore, it is thought that coronal holes are the main corridors through which particles of the solar wind escape from the Sun.

 Evidence in favor of this picture has come from the *Ulysses* spacecraft, another joint ESA/NASA mission. In 1994 and 1995, *Ulysses* became the first spacecraft to fly over the Sun's north and south poles, where there are apparently permanent coronal holes. The spacecraft indeed measured a stronger solar wind emanating from these holes.

The temperatures in the corona and the chromosphere are not at all what we would expect. Just as you feel warm

figure 18-18   R I V U X G

The Ultraviolet Corona This false-color view of the corona is a composite of two ultraviolet images from *SOHO*. The black ring separates the two images. The outer image shows material from the corona streaming outward to join the solar wind. The inner image shows the origin of these streamers in the lower regions of the corona. The dark feature running across the Sun's disk from the top is a coronal hole, a region where the coronal gases are thinner and at a lower temperature than average. (Inner image: *SOHO* Extreme Ultraviolet Imaging Telescope Consortium. Outer image: Ultraviolet Coronagraph Spectrometer Consortium)

if you stand close to a campfire but cold if you move away, we would expect that the temperature in the corona and chromosphere would *decrease* with increasing altitude and, hence, increasing distance from the warmth of the Sun's photosphere. Why, then, does the temperature in these regions *increase* with increasing altitude? This has been one of the major unsolved mysteries in astronomy for the past half-century. As astronomers have tried to resolve this dilemma, they have found important clues in one of the Sun's most familiar features—sunspots.

18-8   Sunspots are low-temperature regions in the photosphere

Granules, supergranules, spicules, and the solar wind occur continuously. These features are said to be aspects of the *quiet Sun.* But other, more dramatic features appear periodically, including massive eruptions and regions of concentrated magnetic fields. When these are present, astronomers refer to the *active Sun.* One aspect of the active Sun that can easily be seen with even a small telescope (although only with an appropriate filter attached) is the feature known as sunspots.

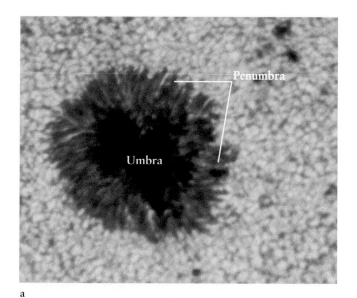

a

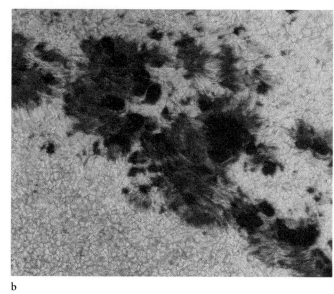

b

figure 18-19   R I **V** U X G

Sunspots **(a)** This high-resolution photograph of the photosphere shows a mature sunspot. The dark center of the spot is called the umbra. It is bordered by the penumbra, which is less dark and has a featherlike appearance. **(b)** In this view of a typical sunspot group, several sunspots are close enough to overlap. In both images you can see granulation in the surrounding, undisturbed photosphere. (NOAO)

Sunspots are irregularly shaped dark regions in the photosphere. Sometimes sunspots appear in isolation (Figure 18-19a), but frequently they are found in sunspot groups (Figure 18-19b; see also Figure 18-9). Although sunspots vary greatly in size, typical ones measure a few tens of thousands of kilometers across—comparable to the diameter of the Earth. Sunspots are not permanent features of the photosphere but last between a few hours and a few months.

Each sunspot has a dark central core, called the *umbra*, and a brighter border called the *penumbra*. We used these same terms in Section 3-4 to refer to different parts of the Earth's or the Moon's shadow. But a sunspot is not a shadow; it is a region in the photosphere where the temperature is relatively low, which makes it appear darker than its surroundings. If the surrounding photosphere is blocked from view, a sunspot's umbra appears red and the penumbra appears orange. As we saw in Section 5-4, Wien's law relates the color of a blackbody (which depends on the wavelength at which it emits the most light) to the blackbody's temperature. The colors of a sunspot indicate that the temperature of the umbra is typically 4300 K and that of the penumbra is typically 5000 K. While high by earthly standards, these temperatures are quite a bit lower than the average photospheric temperature of 5800 K.

The Stefan-Boltzmann law (see Section 5-4) tells us that the energy flux from a blackbody is proportional to the fourth power of its temperature. This law lets us compare the amounts of light energy emitted by a square meter of a sun-

spot's umbra and by a square meter of undisturbed photosphere. The ratio is

$$\frac{\text{flux from umbra}}{\text{flux from photosphere}} = \left(\frac{4300\ \text{K}}{5800\ \text{K}}\right)^4 = 0.30;$$

That is, the umbra emits only 30% as much light as an equally large patch of undisturbed photosphere. This is why sunspots appear so dark.

On rare occasions, a sunspot group is large enough to be seen without a telescope. Chinese astronomers recorded such sightings 2000 years ago, and a huge sunspot group visible to the naked eye (with an appropriate filter) was seen in 1989. But it was not until Galileo introduced the telescope into astronomy (see Section 4-3) that anyone was able to examine sunspots in detail.

Galileo discovered that he could determine the Sun's rotation rate by tracking sunspots as they moved across the solar disk (Figure 18-20). He found that the Sun rotates once in about four weeks. A typical sunspot group lasts about two months, so a specific one can be followed for two solar rotations.

Further observations by the British astronomer Richard Carrington in 1859 demonstrated that the Sun does not rotate as a rigid body. Instead, the equatorial regions rotate more rapidly than the polar regions. This phenomenon is known as **differential rotation**. Thus, while a sunspot near the solar equator takes only 25 days to go once

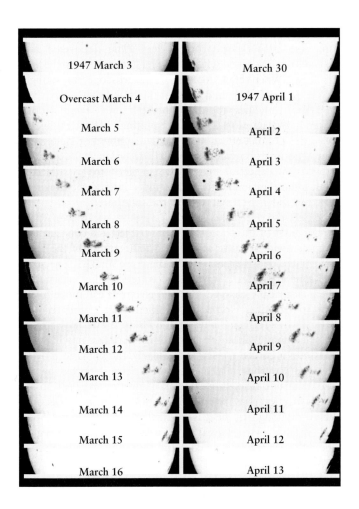

VIDEO 18.4

figure 18-20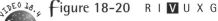

Tracking the Sun's Rotation with Sunspots By observing the same group of sunspots from one day to the next, Galileo found that the Sun rotates once in about four weeks. The equatorial regions of the Sun actually rotate somewhat faster than the polar regions, a phenomenon called differential rotation. This series of photographs shows the same sunspot group over 1½ solar rotations. (The Carnegie Observatories)

around the Sun, a sunspot at 30° north or south of the equator takes 27½ days. The rotation period at 75° north or south is about 33 days, while near the poles it may be as long as 35 days.

The average number of sunspots on the Sun is not constant, but varies in a predictable **sunspot cycle** (Figure 18-21a). This phenomenon was first reported by the German astronomer Heinrich Schwabe in 1843 after many years of observing. As Figure 18-21a shows, the average number of sunspots varies with a period of about 11 years. A period of exceptionally many sunspots is a **sunspot maximum,** as occurred in 1979, 1989, and 2000. Conversely, the Sun is almost devoid of sunspots at a **sunspot minimum,** as occurred in 1976, 1986, and 1996 and is projected to occur in 2007.

The locations of sunspots also vary with the same 11-year sunspot cycle. At the beginning of a cycle, just after a sunspot minimum, sunspots first appear at latitudes around 30°

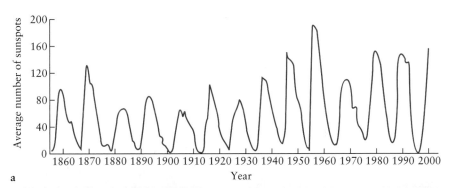

a

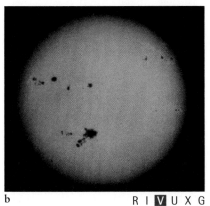

b   R I **V** U X G

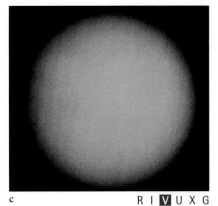

c   R I **V** U X G

figure 18-21

The Sunspot Cycle **(a)** The number of sunspots on the Sun varies with a period of about 11 years. The most recent sunspot maximum occurred in 2000, and the most recent sunspot minimum occurred in 1996. **(b)** This photograph, taken near sunspot maximum in 1989, shows a number of sunspots and large sunspot groups. The sunspot group visible near the bottom of the Sun's disk has about the same diameter as the planet Jupiter. **(c)** Near sunspot minimum, as in this 1986 photograph, essentially no sunspots are visible. (NOAO)

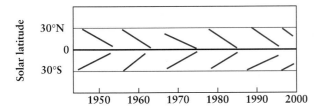

figure 18-22

Variations in the Average Latitude of Sunspots At the beginning of each sunspot cycle, most sunspots are found about 30° north or south of the Sun's equator. Sunspots that appear later in each cycle typically form closer and closer to the equator.

north and south of the solar equator (Figure 18-22). Over the succeeding years, the sunspots occur closer and closer to the equator, until at the end of the cycle virtually all sunspots lie on the solar equator.

18-9 Sunspots are produced by a 22-year cycle in the Sun's magnetic field

Why should the number of sunspots vary with an 11-year cycle? Why should their average latitude vary over the course of a cycle? And why should sunspots exist at all? The first step

toward answering these questions came in 1908, when the American astronomer George Ellery Hale discovered that sunspots are associated with intense magnetic fields on the Sun. When Hale focused a spectroscope on sunlight coming from a sunspot, he found that many spectral lines appear to be split into several closely spaced lines (Figure 18-23). This "splitting" of spectral lines is called the **Zeeman effect**, after the Dutch physicist Pieter Zeeman, who first observed it in his laboratory in 1896. Zeeman showed that a spectral line splits when the atoms are subjected to an intense magnetic field. The more intense the magnetic field, the wider the separation of the split lines.

Hale's discovery showed that sunspots are places where the hot gases of the photosphere are bathed in a concentrated magnetic field. Because of the high temperature of the Sun's atmosphere, many of the atoms there are ionized. The solar atmosphere is thus a gaseous mixture called a **plasma**, in which electrically charged ions and electrons can move freely. Like any moving, electrically charged objects, they can be deflected by magnetic fields. Figure 18-24 shows how a magnetic field in the laboratory bends a beam of fast-moving electrons into a curved trajectory. Similarly, the paths of moving ions and electrons in the photosphere are deflected by the Sun's magnetic field. In particular, magnetic forces act on the hot plasma that rises from the Sun's interior due to convection. Where the magnetic field is particularly strong, these forces push the hot plasma away. The result is a localized region where the gas is relatively cool and thus glows less brightly—in other words, a sunspot.

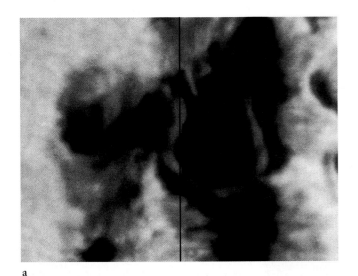

a

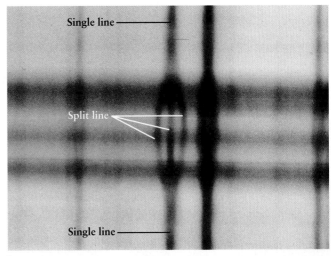

b

figure 18-23 R I **V** U X G

Sunspots Have Strong Magnetic Fields **(a)** Astronomers measure the magnetic field of a sunspot using a spectrograph. A black line in this image of a sunspot shows where the slit of the spectrograph was aimed. **(b)** This is a portion of the resulting spectrum, including a dark absorption line caused by iron atoms in the photosphere. Within the sunspot, this line is split into three components. The separation in wavelength between these components can be used to calculate the strength of the magnetic field in the sunspot. Typical sunspot fields are more than 5000 times stronger than the Earth's magnetic field at its north and south poles. (NOAO)

figure 18-24 R I **V** U X G

Magnetic Fields Deflect Moving, Electrically Charged Objects
In this laboratory experiment, a beam of negatively charged electrons (shown by a blue arc) is aimed straight upward. The entire apparatus is inside a large magnet, and the magnetic field deflects the beam into a curved path. This effect is used to focus the electron beam in a television picture tube and to make electrons oscillate so that they produce microwaves in a microwave oven. (Courtesy of Central Scientific Company)

To get a fuller picture of the Sun's magnetic fields, astronomers take images of the Sun at two wavelengths, one just less than and one just greater than the wavelength of a magnetically split spectral line. From the difference between these two images, they can construct a picture called a **magnetogram,** which displays the magnetic fields in the solar atmosphere. Figure 18-25*a* is an ordinary white-light photograph of the Sun taken at the same time as the magnetogram in Figure 18-25*b*. In the magnetogram, dark blue indicates areas of the photosphere with one magnetic polarity (north), and yellow indicates areas with the opposite (south) magnetic polarity. This image shows that many sunspot groups have roughly comparable areas covered by north and south magnetic polarities (see also Figure 18-25*c*). Thus, a sunspot group resembles a giant bar magnet, with a north magnetic pole at one end and a south magnetic pole at the other.

If different sunspot groups were unrelated to one another, their magnetic poles would be randomly oriented, like a bunch of compass needles all pointing in random directions. As Hale discovered, however, there is a striking regularity in the magnetization of sunspot groups. As a given sunspot group moves with the Sun's rotation, the sunspots in front are called the "preceding members" of the group. The spots that follow behind are referred to as the "following members." Hale compared the sunspot groups in the two solar hemispheres, north or south of the Sun's equator. He found that the preceding members in one solar hemisphere all have the same magnetic polarity, while the preceding members in the other hemisphere have the opposite polarity. Furthermore, in the hemisphere where the Sun has its north magnetic pole, the preceding members of all sunspot groups have north magnetic polarity. In the opposite hemisphere, where the Sun has its south magnetic pole, the preceding members all have south magnetic polarity.

Along with his colleague Seth B. Nicholson, Hale also discovered that the Sun's polarity pattern completely reverses itself every 11 years—the same interval as the time from one

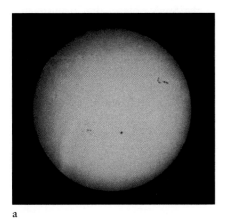
a

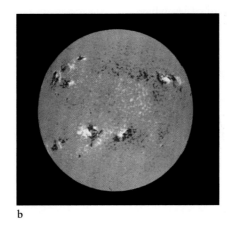
b

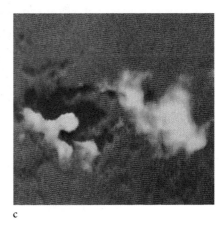
c

figure 18-25 R I **V** U X G

Mapping the Sun's Magnetic Field This visible-light image of the Sun **(a)** was taken at the same time as this magnetogram **(b)**. Dark blue and yellow areas in the magnetogram have north and south magnetic polarity, respectively. Regions with weak magnetic fields, which include much of the Sun's disk, are shown in blue-green. Notice how regions with strong magnetic fields in (b) correlate with the locations of sunspots in (a). **(c)** This close-up magnetogram of a large sunspot group shows that the two ends of the group have opposite magnetic polarities (colored blue and yellow), like the two ends of a giant bar magnet. (NOAO)

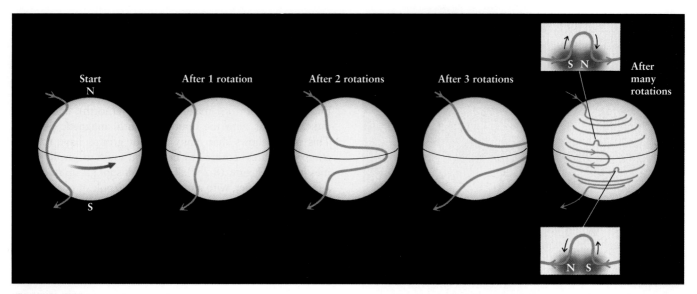

figure 18-26

Babcock's Magnetic Dynamo The sunspot cycle may be due in part to the Sun's differential rotation. Magnetic field lines tend to move along with the plasma in the Sun's outer layers. A field line that starts off running from the Sun's north magnetic pole (N) to its south magnetic pole (S), as in the drawing on the far left, soon becomes distorted because the Sun's equator rotates faster than the poles. After many rotations the magnetic field is wrapped around the Sun like twine wrapped around a ball. The insets on the far right show how sunspot groups appear where the concentrated magnetic field breaks through the solar surface. In each sunspot group, the preceding members—in the insets, the sunspots on the right—have the same polarity as the Sun's magnetic pole in that hemisphere (in this drawing, N in the upper hemisphere and S in the lower hemisphere).

solar maximum to the next. The hemisphere that has preceding north magnetic poles during one 11-year sunspot cycle will have preceding south magnetic poles during the next 11-year cycle, and vice versa. The north and south magnetic poles of the Sun itself also reverse every 11 years. Thus, the Sun's magnetic pattern repeats itself only after two sunspot cycles, which is why astronomers speak of a **22-year solar cycle.**

In 1960, the American astronomer Horace Babcock proposed a description that seems to account for many features of this 22-year solar cycle. Babcock's scenario, called a **magnetic-dynamo model,** makes use of two basic properties of the Sun's photosphere—differential rotation and convection. Differential rotation causes the magnetic field in the photosphere to become wrapped around the Sun (Figure 18-26). As a result, the magnetic field becomes concentrated at certain latitudes on either side of the solar equator. Convection in the photosphere creates tangles in the concentrated magnetic field, and "kinks" erupt through the solar surface. Sunspots appear where the magnetic field protrudes through the photosphere. The theory suggests that sunspots should appear first at northern and southern latitudes and later form nearer to the equator. This is just what is observed (see Figure 18-22). Note also that as shown in Figure 18-26, the preceding member of a sunspot group has the same polarity as the Sun's magnetic pole in that hemisphere, just as Hale observed.

Differential rotation eventually undoes the twisted magnetic field. The preceding members of sunspot groups move toward the Sun's equator, while the following members migrate toward the poles. Because the preceding members from the two hemispheres have opposite magnetic polarities, their magnetic fields cancel each other out when they meet at the equator. The following members in each hemisphere have the opposite polarity to the Sun's pole in that hemisphere; hence, when they converge on the pole, the following members first cancel out and then reverse the Sun's overall magnetic field. The fields are now completely relaxed. Once again, differential rotation begins to twist the Sun's magnetic field, but now with all magnetic polarities reversed. In this way, Babcock's model helps to explain the change in field direction every 11 years.

Recent discoveries in helioseismology (Section 18-3) offer new insights into the Sun's magnetic field. By comparing the speeds of sound waves that travel with and against the Sun's rotation, helioseismologists have been able to determine the Sun's rotation rate at different depths and latitudes. As shown in Figure 18-27, the Sun's surface pattern of differential rotation persists through the convective zone. Farther in, within the radiative zone, the Sun seems to rotate like a rigid object with a period of 27 days at all latitudes. Astronomers suspect that the Sun's magnetic field originates in a relatively thin layer where the radiative and convective zones meet and slide past each other due to their different rotation rates.

Much about sunspots and solar activity remains mysterious. The best calculations predict that a sunspot should break

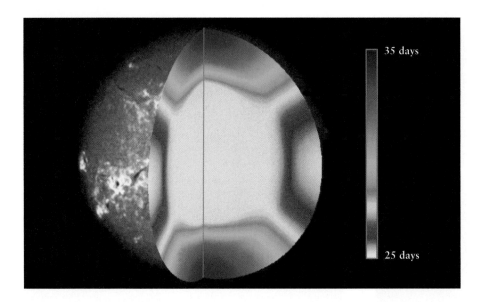

WEB LINK 18.12

figure 18-27

Rotation Rates in the Solar Interior This cutaway picture of the Sun shows how the solar rotation rate varies with depth and latitude. Colors represent rotation periods according to the scale on the right. The surface rotation pattern, which varies from 25 days at the equator to 35 days near the poles, persists throughout the Sun's convective zone. The radiative zone seems to rotate like a rigid sphere. (Courtesy of K. Libbrecht, Big Bear Solar Observatory)

up and disperse fairly quickly, but, in fact, sunspots can persist for many weeks. There are also perplexing irregularities in the solar cycle. For example, the overall reversal of the Sun's magnetic field is often piecemeal and haphazard. One pole may reverse polarity long before the other. For several weeks the Sun's surface may have two north magnetic poles and no south magnetic pole at all.

What is more, there seem to be times when all traces of sunspots and the sunspot cycle vanish for many years. For example, virtually no sunspots were seen from 1645 through 1715. Curiously, during these same years Europe experienced record low temperatures, often referred to as the Little Ice Age, whereas the western United States was subjected to severe drought. By contrast, there was apparently a period of increased sunspot activity during the eleventh and twelfth centuries, during which the Earth was warmer than it is today. Thus, variations in solar activity appear to affect climates on the Earth. The origin of this Sun-Earth connection is a topic of ongoing research.

18-10 The Sun's magnetic field also produces other forms of solar activity

Astronomers now understand that the Sun's magnetic field does more than just explain the presence of sunspots. It is also responsible for the existence of spicules, as well as a host of other dramatic phenomena in the chromosphere and corona.

In a plasma, magnetic field lines and the material of the plasma tend to move together. This means that as convection pushes material toward the edge of a supergranule, it pushes magnetic field lines as well. The result is that vertical magnetic field lines pile up around a supergranule. Plasma that "sticks" to these magnetic field lines thus ends up lifted upward, forming a spicule (see Figure 18-14).

The tendency of plasma to follow the Sun's magnetic field may also explain why the temperature of the corona is so high. Spacecraft observations show magnetic field arches extending tens of thousands of kilometers into the corona, with streamers of electrically charged particles moving along each arch (Figure 18-28). If two arches come near to each other, their flowing charges can interact to form a gigantic

WEB LINK 18.13

figure 18-28 R I V X G

Magnetic Heating of the Corona This false-color ultraviolet image from NASA's *TRACE* spacecraft (*Transition Region and Coronal Explorer*) shows magnetic field loops suspended high above the solar surface. The largest loops are about 200,000 km (120,000 mi, or 15 times the Earth's diameter) across. "Short circuits" among the field lines have raised the gas temperature to as high as 2.7×10^6 K, making it glow at ultraviolet wavelengths. (Stanford-Lockheed Institute for Space Research; *TRACE*; and NASA)

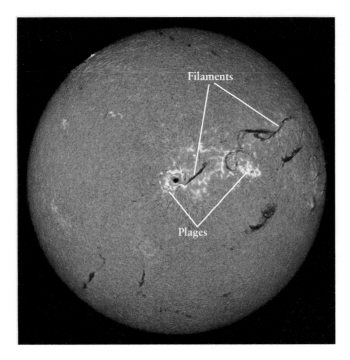

Figure 18-29 R I V U X G

Plages and Filaments in the Chromosphere The bright areas in this H$_\alpha$ image, called plages, are hotter than the surrounding chromosphere. The dark filaments extend far above the normal extent of the chromosphere; when seen against the background of space, they appear as in Figure 18-30. Plages and filaments are associated with strong magnetic fields and are always found near sunspots. (Sacramento Peak Observatory)

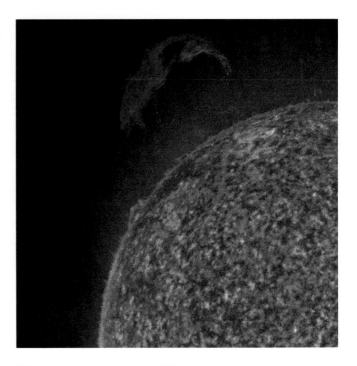

Figure 18-30 R I V U X G

A Solar Prominence A huge prominence arches above the solar surface in this X-ray image from the *SOHO* spacecraft. The image was recorded using light at a wavelength of 30.4 nm, emitted by singly ionized helium atoms at a temperature of about 60,000 K. The gas ejected from this 1997 prominence did not strike the Earth. (Solar and Heliospheric Observatory)

"short circuit." Just as a short circuit on the Earth can make sparks fly, these coronal short circuits release a tremendous amount of energy into the corona. (A single arch contains as much energy as a hydroelectric power plant would generate in a million years.) The amount of energy released in this way appears to be more than enough to maintain the corona's temperature.

Magnetic heating can also explain why the parts of the corona that are most prominent in X-ray images are often those that lie on top of sunspots. (Some examples are the bright regions in the image that opens this chapter.) The intense magnetic field of the sunspots helps give the overlying coronal gas such a high temperature that it emits X rays.

Spicules and coronal heating occur even when the Sun is quiet. But magnetic fields can also explain many aspects of the active Sun in addition to sunspots. Figure 18-29 is an image of the chromosphere made with an H$_\alpha$ filter during a sunspot maximum. The bright areas are called **plages** (from the French word for "beach"). These are bright, hot regions in the chromosphere that tend to form just before the appearance of new sunspots. They are probably created by magnetic fields that push upward from the Sun's interior, compressing and heating a portion of the chromosphere. The dark streaks, called **filaments,** are relatively cool and dense parts of the chromo-

sphere that have been pulled along with magnetic field lines as they arch to high altitudes.

When seen from the side, so that they are viewed against the dark background of space, filaments appear as bright, arching columns of gas called **prominences** (Figure 18-30). They can extend for tens of thousands of kilometers above the photosphere. Some prominences last for only a few hours, while others persist for many months. The prominence in Figure 18-30 was so energetic that it broke free of the magnetic fields that confined it and burst into space.

Violent, eruptive events on the Sun, called **solar flares,** occur in complex sunspot groups. Within only a few minutes, temperatures in a compact region may soar to 5×10^6 K and vast quantities of particles and radiation are blasted out into space. These eruptions can also cause disturbances that spread outward in the solar atmosphere, like the ripples that appear when you drop a rock into a pond.

The most energetic flares carry as much as 10^{30} joules of energy, equivalent to 10^{14} one-megaton nuclear weapons being exploded at once! However, the energy of a solar flare does not come from thermonuclear fusion in the solar atmosphere; instead, it appears to be released from the intense magnetic field around a sunspot group.

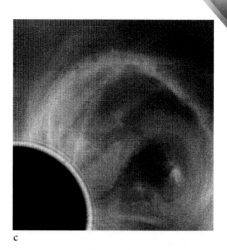

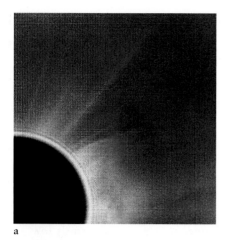

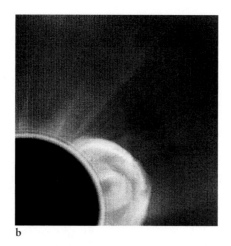

a   b   c

ƒigure 18-31   R I **V** U X G

A Coronal Mass Ejection These visible-light images of a coronal mass ejection are from the *Solar Maximum Mission* spacecraft. The light of the Sun's disk has been blocked to make the corona visible. **(a)** Before the ejection, the corona appears quiet. **(b)** As the ejection begins, it is already hundreds of thousands of kilometers across. **(c)** Just 16 minutes after the image in (b), the leading edge of the ejection has expanded outward by almost a million kilometers. (High Altitude Observatory/*Solar Maximum Mission*)

As energetic as solar flares are, they are dwarfed by **coronal mass ejections** (Figure 18-31). In such an event, more than 10^{12} kilograms (a billion tons) of high-temperature coronal gas is blasted into space at speeds of hundreds of kilometers per second. A typical coronal mass ejection lasts a few hours. These explosive events seem to be related to large-scale alterations in the Sun's magnetic field. Such events occur every few months; smaller eruptions may occur almost daily.

If a solar flare or coronal mass ejection happens to be aimed toward Earth, a stream of high-energy electrons and nuclei reaches us a few days later.

When this plasma arrives, it can interfere with satellites, pose a health hazard to astronauts in orbit, and disrupt electrical and communications equipment on the Earth's surface. Telescopes on Earth and on board spacecraft now monitor the Sun continuously to provide warnings of dangerous levels of solar particles.

The numbers of plages, filaments, solar flares, and coronal mass ejections all vary with the same 11-year cycle as sunspots. But unlike sunspots, coronal mass ejections never completely cease, even when the Sun is at its quietest. Astronomers are devoting substantial effort to understanding these and other aspects of our dynamic Sun.

KEY WORDS

The term preceded by an asterisk (*) is discussed in Box 18-1.

22-year solar cycle, p. 410

chromosphere, p. 402

CNO cycle, p. 394

conduction, p. 395

convection, p. 395

convective zone, p. 397

corona, p. 403

coronal hole, p. 405

coronal mass ejection, p. 413

differential rotation, p. 406

filament, p. 412

granulation, p. 401

granule, p. 401

helioseismology, p. 397

hydrogen burning, p. 392

hydrostatic equilibrium, p. 394

limb darkening, p. 400

luminosity (of the Sun), p. 390

magnetic-dynamo model, p. 410

magnetogram, p. 409

negative hydrogen ion, p. 402

neutrino, p. 398

photosphere, p. 400

plage, p. 412

plasma, p. 408

*positron, p. 393

prominence, p. 412

proton-proton chain, p. 394

radiative diffusion, p. 395

radiative zone, p. 396

solar flare, p. 412

solar neutrino, p. 399

solar neutrino problem, p. 399

solar wind, p. 405

spicule, p. 403

sunspot, p. 406

sunspot cycle, p. 407

sunspot maximum, p. 407

sunspot minimum, p. 407

supergranule, p. 402

thermal equilibrium, p. 394

thermonuclear fusion, p. 392

Zeeman effect, p. 408

KEY IDEAS

Hydrogen Burning in the Sun's Core: The Sun's energy is produced by hydrogen burning, a thermonuclear fusion process in which four hydrogen nuclei combine to produce a single helium nucleus.

• The energy released in a nuclear reaction corresponds to a slight reduction of mass according to Einstein's equation $E = mc^2$.

• Thermonuclear fusion occurs only at very high temperatures; for example, hydrogen burning occurs only at temperatures in excess of about 10^7 K. In the Sun, fusion occurs only in the dense, hot core.

Models of the Sun's Interior: A theoretical description of a star's interior can be calculated using the laws of physics.

• The standard model of the Sun suggests that hydrogen burning takes place in a core extending from the Sun's center to about 0.25 solar radius.

• The core is surrounded by a radiative zone extending to about 0.71 solar radius. In this zone, energy travels outward through radiative diffusion.

• The radiative zone is surrounded by a rather opaque convective zone of gas at relatively low temperature and pressure. In this zone, energy travels outward primarily through convection.

Solar Neutrinos and Helioseismology: Conditions in the solar interior can be inferred from measurements of solar neutrinos and of solar vibrations.

• Neutrinos emitted in thermonuclear reactions in the Sun's core have been detected, but in smaller numbers than expected. Recent experiments may explain why this is so.

• Helioseismology is the study of how the Sun vibrates. These vibrations have been used to infer pressures, densities, chemical compositions, and rotation rates within the Sun.

The Sun's Atmosphere: The Sun's atmosphere has three main layers: the photosphere, the chromosphere, and the corona. Everything below the solar atmosphere is called the solar interior.

• The visible surface of the Sun, the photosphere, is the lowest layer in the solar atmosphere. Its spectrum is similar to that of a blackbody at a temperature of 5800 K. Convection in the photosphere produces granules.

• Above the photosphere is a layer of less dense but higher-temperature gases called the chromosphere. Spicules extend upward from the photosphere into the chromosphere along the boundaries of supergranules.

• The outermost layer of the solar atmosphere, the corona, is made of very high-temperature gases at extremely low density. Activity in the corona includes coronal mass ejections and coronal holes. The solar corona blends into the solar wind at great distances from the Sun.

The Active Sun: The Sun's surface features vary in an 11-year cycle. This is related to a 22-year cycle in which the surface magnetic field increases, decreases, and then increases again with the opposite polarity.

• Sunspots are relatively cool regions produced by local concentrations of the Sun's magnetic field. The average number of sunspots increases and decreases in a regular cycle of approximately 11 years, with reversed magnetic polarities from one 11-year cycle to the next. Two such cycles make up the 22-year solar cycle.

• The magnetic-dynamo model suggests that many features of the solar cycle are due to changes in the Sun's magnetic field. These changes are caused by convection and the Sun's differential rotation.

• A solar flare is a brief eruption of hot, ionized gases from a sunspot group. A coronal mass ejection is a much larger eruption that involves immense amounts of gas from the corona.

REVIEW QUESTIONS

1. What is hydrogen burning? Why is hydrogen burning fundamentally unlike the burning of a log in a fireplace?

2. Why do thermonuclear reactions occur only in the Sun's core?

3. Describe how the net result of the reactions in Figure 18-2 is the conversion of four protons into a single helium nucleus. What other particles are produced in this process? How many of each particle are produced?

4. Give an everyday example of hydrostatic equilibrium. Give an example of thermal equilibrium. Explain how these equilibrium conditions apply to each example.

5. If thermonuclear fusion in the Sun were suddenly to stop, what would eventually happen to the overall radius of the Sun? Justify your answer using the ideas of hydrostatic equilibrium and thermal equilibrium.

6. Give some everyday examples of conduction, convection, and radiative diffusion.

7. Describe the Sun's interior. Include references to the main physical processes that occur at various depths within the Sun.

8. Suppose thermonuclear fusion in the Sun stopped abruptly. Would the intensity of sunlight decrease just as abruptly? Why or why not?

9. Explain how studying the oscillations of the Sun's surface can give important, detailed information about physical conditions deep within the Sun.

10. What is a neutrino? Why is it useful to study neutrinos coming from the Sun? What do they tell us that cannot be learned from other avenues of research?

11. Unlike all other types of telescopes, neutrino detectors are placed deep underground. Why?

12. Describe the dangers in attempting to observe the Sun. How have astronomers learned to circumvent these observational problems?

13. Briefly describe the three layers that make up the Sun's atmosphere. In what ways do they differ from each other?

14. What is solar granulation? Describe how convection gives rise to granules.

15. High-resolution spectroscopy of the photosphere reveals that absorption lines are blueshifted in the spectrum of the central, bright regions of granules but are redshifted in the spectrum of the dark boundaries between granules. Explain how these observations show that granulation is due to convection.

16. What is the difference between granules and supergranules?

17. What are spicules? Where are they found? How can you observe them? What causes them?

18. How do astronomers know that the temperature of the corona is so high?

19. How do astronomers know when the next sunspot maximum and minimum will occur?

20. Why do astronomers say that the solar cycle is really 22 years long, even though the number of sunspots varies over an 11-year period?

21. Explain how the magnetic-dynamo model accounts for the solar cycle.

22. Explain why the surface of the Sun appears black in Figure 18-28.

23. Why should solar flares and coronal mass ejections be a concern for businesses that use telecommunication satellites?

ADVANCED QUESTIONS

The question preceded by an asterisk (*) involves the topic discussed in Box 18-1.

> **Problem-solving tips and tools**
>
> You may have to review Wien's law and the Stefan-Boltzmann law, which are the subject of Section 5-4. Section 5-5 discusses the properties of photons. As we described in Box 5-2, you can simplify calculations by taking ratios, such as the ratio of the flux from a sunspot to the flux from the undisturbed photosphere. When you do this, all the cumbersome constants cancel out. Figure 5-7 shows the various parts of the electromagnetic spectrum. We introduced the Doppler effect in Section 5-9 and Box 5-6. For information about the planets, see Table 7-1.

24. How much energy would be released if each of the following masses were converted *entirely* into their equivalent energy: (a) a carbon atom with a mass of 2×10^{-26} kg, (b) 1 kilogram, and (c) a planet as massive as the Earth (6×10^{24} kg)?

25. Use the luminosity of the Sun (given in Table 18-1) and the answers to the previous question to calculate how long the Sun must shine in order to release an amount of energy equal to that produced by the complete mass-to-energy conversion of (a) a carbon atom, (b) 1 kilogram, and (c) the Earth.

26. Assuming that the current rate of hydrogen burning in the Sun remains constant, what fraction of the Sun's mass will be converted into helium over the next 5 billion years? How will this affect the overall chemical composition of the Sun?

27. (a) Estimate how many kilograms of hydrogen the Sun has consumed over the past 4.6 billion years, and estimate the amount of mass that the Sun has lost as a result. Assume that the Sun's luminosity has remained constant during that time. (b) In fact, the Sun's luminosity when it first formed was only about 70% of its present value. With this in mind, explain whether your answers to part (a) are an overestimate or an underestimate.

*28. (a) A positron has the same mass as an electron (see Appendix 7). Calculate the amount of energy released by the annihilation of an electron and positron. (b) The products of this annihilation are two photons, each of equal energy. Calculate the wavelength of each photon, and confirm from Figure 5-7 that this wavelength is the gamma-ray range.

29. Sirius is the brightest star in the night sky. It has a luminosity of 23.5 $L_\odot$, that is, it is 23.5 times as luminous as the Sun and burns hydrogen at a rate 23.5 times greater than the Sun. How many kilograms of hydrogen does Sirius convert into helium each second?

30. (Refer to the preceding question.) Sirius has 2.3 times the mass of the Sun. Do you expect that the lifetime of Sirius will be longer, shorter, or the same length as that of the Sun? Explain your reasoning.

31. What would happen if the Sun were not in a state of both hydrostatic and thermal equilibrium? Explain your reasoning.

32. Using the mass and size of the Sun given in Table 18-1, verify that the average density of the Sun is 1410 kg/m³. Compare your answer with the average densities of the Jovian planets.

33. Use the data in Table 18-2 to calculate the *average* density of material within 0.1 $R_\odot$ of the center of the Sun. (You will need to use the mass and radius of the Sun as a whole, given in Table 18-1.) Explain why your answer is not the same as the density at 0.1 $R_\odot$ given in Table 18-2.

34. In a typical solar oscillation, the Sun's surface moves up or down at a maximum speed of 0.1 m/s. An astronomer sets out to measure this speed by detecting the Doppler shift of an absorption line of iron with wavelength 557.6099 nm. What is the maximum wavelength shift that she will observe?

35. The amount of energy required to dislodge the extra electron from a negative hydrogen ion is 1.2×10^{-19} J. (a) The extra electron can be dislodged if the ion absorbs a photon of sufficiently short wavelength. (Recall from Section 5-5 that the higher the energy of a photon, the shorter its wavelength.) Find the longest wavelength (in nm) that can accomplish this. (b) In what part of the electromagnetic spectrum does this wavelength lie? (c) Would a photon of visible light be able to dislodge the extra electron? Explain. (d) Explain why the photosphere, which contains negative hydrogen ions, is quite opaque to visible light but is less opaque to light with wavelengths longer than the value you calculated in (a).

36. Astronomers often use an H$_\alpha$ filter to view the chromosphere. Explain why this can also be accomplished with filters that are transparent only to the wavelengths of the H and K lines of ionized calcium.

37. Calculate the wavelengths at which the photosphere, chromosphere, and corona emit the most radiation. Explain how the results of your calculations suggest the best way to observe these regions of the solar atmosphere. (*Hint:* Treat each part of the atmosphere as a perfect blackbody. Assume average temperatures of 50,000 K and 1.5×10^6 K for the chromosphere and corona, respectively.)

38. On November 15, 1999, the planet Mercury passed in front of the Sun as seen from Earth. The *TRACE* spacecraft made these time-lapse images of this event using ultraviolet light (top) and visible light (bottom). (Mercury moved from left to right in these images. The time between successive views of Mercury is 6 to 9 minutes.) Explain

why the Sun appears somewhat larger in the ultraviolet image than in the visible-light image.

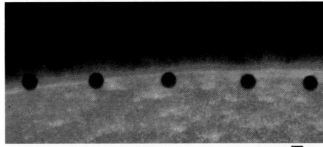

R I V **U** X G

R I **V** U X G

(K. Schrijver, Stanford-Lockheed Institute for Space Research; *TRACE*; and NASA)

39. Find the wavelength of maximum emission of the umbra of a sunspot and the wavelength of maximum emission of a sunspot's penumbra. In what part of the electromagnetic spectrum do these wavelengths lie?

40. (a) Find the ratio of the energy flux from a patch of a sunspot's penumbra to the energy flux from an equally large patch of undisturbed photosphere. Which patch is brighter? (b) Find the ratio of the energy flux from a patch of a sunspot's penumbra to the energy flux from an equally large patch of umbra. Again, which patch is brighter?

41. Suppose that you want to determine the Sun's rotation rate by observing its sunspots. Is it necessary to take the Earth's orbital motion into account? Why or why not?

42. The amount of visible light emitted by the Sun varies only a little over the 11-year sunspot cycle. But the amount of X rays emitted by the Sun can be ten times greater at solar maximum than at solar minimum. Explain why these two types of radiation should be so different in their variability.

DISCUSSION QUESTIONS

43. Discuss the extent to which cultures around the world have worshiped the Sun as a deity throughout history. Why do you suppose there has been such widespread veneration?

44. In the movie *Star Trek IV: The Voyage Home*, the starship *Enterprise* flies on a trajectory that passes close to the Sun's surface. What features should a real spaceship have to survive such a flight? Why?

45. Discuss some of the difficulties in correlating solar activity with changes in the Earth's climate.

46. Describe some of the advantages and disadvantages of observing the Sun (**a**) from space and (**b**) from the Earth's South Pole. What kinds of phenomena and issues might solar astronomers want to explore from these locations?

WEB/CD-ROM QUESTIONS

47. Search the World Wide Web for the latest information from the neutrino detectors at the Super-Kamiokande Observatory and the Sudbury Neutrino Observatory. What are the most recent results from these detectors? What is the current thinking about the solar neutrino problem? What is the status of a new detector called Borexino?

48. Search the World Wide Web for information about features in the solar atmosphere called *sigmoids*. What are they? What causes them? How might they provide a way to predict coronal mass ejections?

49. Determining the Lifetime of a Solar Granule. Access and view the video "Granules on the Sun's Surface" in Chapter 18 of the *Universe* Web site or CD-ROM. Your task is to determine the approximate lifetime of a solar granule on the photosphere. Select an area, then slowly and rhythmically repeat "start, stop, start, stop" until you can consistently predict the appearance and disappearance of granules. While keeping your rhythm, move to a different area of the video and continue monitoring the appearance and disappearance of granules. When you are confident you have the timing right, move your eyes (or use a partner) to the clock shown in the video. Determine the length of time between the appearance and disappearance of the granules and record your answer.

OBSERVING PROJECTS

50. Use a telescope with a solar filter to observe the surface of the Sun. Do you see any sunspots? Sketch their appearance. Can you distinguish between the umbrae and penumbrae of the sunspots? Can you see limb darkening? Can you see any granulation?

51. If you have access to an H_α filter attached to a telescope especially designed for viewing the Sun safely, use this instrument to examine the solar surface. How does the appearance of the Sun differ from that in white light? What do sunspots look like in H_α? Can you see any prominences? Can you see any filaments? Are the filaments in the H_α image near any sunspots seen in white light? (Note that the amount of activity that you see will be much greater at some times during the solar cycle than at others.)

52. Use the *Starry Night* program to measure the Sun's rotation. Set the time in the Control Panel to 12:00:00 P.M., center on the Sun (select **Find...** in the **Edit** menu), and use the controls in the

Control Panel to zoom in until the sunspots are clearly visible. Set the time step to 1 day. Using the single-step time control buttons, step through enough time to determine the rotation rate of the Sun. Which part of the actual Sun's surface rotates at the rate shown in *Starry Night*? (The program does not show the Sun's differential rotation.)

Observing tips and tools

At the risk of repeating ourselves, we remind you to *never look directly at the Sun, because it can easily cause permanent blindness.* You can view the Sun safely without a telescope just by using two pieces of white cardboard. First, use a pin to poke a small hole in one piece of cardboard; this will be your "lens," and the other piece of cardboard will be your "viewing screen." Hold the "lens" piece of cardboard so that it is face-on to the Sun and sunlight can pass through the hole. With your other hand, hold the "viewing screen" so that the sunlight from the "lens" falls on it. Adjust the distance between the two pieces of cardboard so that you see a sharp image of the Sun on the "viewing screen." This image is perfectly safe to view. It is actually possible to see sunspots with this low-tech apparatus.

For a better view, use a telescope with a solar filter that fits on the front of the telescope. A standard solar filter is a piece of glass coated with a thin layer of metal to give it a mirrorlike appearance. This coating reflects almost all the sunlight that falls on it, so that only a tiny, safe amount of sunlight enters the telescope. An H_α filter, which looks like a red piece of glass, keeps the light at a safe level by admitting only a very narrow range of wavelengths. (Filters that fit on the back of the telescope are *not* recommended. The telescope focuses concentrated sunlight on such a filter, heating it and making it susceptible to cracking—and if the filter cracks when you are looking through it, your eye will be ruined instantly and permanently.)

To use a telescope with a solar filter, first aim the telescope away from the Sun, then put on the filter. Keep the lens cap on the telescope's secondary wide-angle "finder scope" (if it has one), because the heat of sunlight can fry the finder scope's optics. Next, aim the telescope toward the Sun, using the telescope's shadow to judge when you are pointed in the right direction. You can then safely look through the telescope's eyepiece. When you are done, make sure you point the telescope away from the Sun before removing the filter and storing the telescope.

Note that the amount of solar activity that you can see (sunspots, filaments, flares, prominences, and so on) will depend on where the Sun is in its 11-year sunspot cycle.

Searching for Neutrinos Beyond the Textbooks

John Bahcall's wide-ranging expertise includes models of the Galaxy, dark matter, atomic and nuclear physics applied to astronomy, stellar evolution, and quasar spectra. After receiving his Ph.D. in physics from Harvard in 1961, he joined the faculty of the California Institute of Technology. Since 1971 he has been a professor of natural sciences at the Institute for Advanced Study in Princeton, New Jersey.

Dr. Bahcall received the National Medal of Science, the nation's highest scientific award, in 1998. His other awards include the Dannie Heinemann Prize of the American

Astronomical Society and the American Institute of Physics for his research on solar neutrinos, and the NASA Distinguished Public Service Medal for his work with the Hubble Space Telescope, He has served as president of the American Astronomical Society and as chair of the Astronomy and Astrophysics Survey Committee of the National Academy of Sciences.

In attempting to understand the Sun, physicists, chemists, and astronomers have been confronted with a mystery—the case of the missing neutrinos. In the early 1960s, Ray Davis and I proposed to test the theory of how the Sun shines. Ray, a chemist at Brookhaven National Laboratory, had developed a neutrino detector that uses a cleaning fluid containing chlorine. Using standard theories of physics and astronomy, I calculated the rate at which neutrinos are produced in the Sun. I could then predict the rate at which neutrinos should be captured in the largest detector Ray could build. If my calculations matched experiment, they would confirm that the Sun shines by nuclear fusion in its interior.

The actual experiment used 100,000 gallons of perchloroethylene, about enough to fill an Olympic-sized swimming pool. Ray and his collaborators put their detector in a deep gold mine, to shield it from other particles that hit the surface of the Earth. To everyone's surprise, Ray's chlorine detector captured many fewer neutrinos than I had predicted. The results were challenged and checked repeatedly over the following three decades, but always with the same result: Many neutrinos appear to be missing! The case of the missing neutrinos has grown stronger with time. Three other experiments, each with a different type of detector, searched for neutrinos from the Sun. They all found fewer than I predicted.

What could be wrong? Where have the neutrinos gone? There are three possibilities. Either the experiments are wrong, the standard model of how the Sun (and other stars) shine is wrong, or something happens to the neutrinos after they are produced. New experiments are now under way in Japan, in Italy, and in Canada to test these hypotheses.

Most people working in the field think that the last of the three explanations is most likely to be correct: Only physics beyond the standard textbooks can describe what has happened to solar neutrinos. Somehow, most physicists think, neutrinos created in the solar interior change into neutrinos that are more difficult to detect as they pass out of the Sun and travel to the Earth. They change their personalities, so to speak! If this is correct, it would be the first experimental

demonstration of a process beyond the standard model of particle physics.

So far, evidence for the new physics is only circumstantial. We know only that the results differ markedly from predictions based on our understanding now of how the Sun shines. Future experiments designed to search for a "smoking gun"—unequivocal evidence of processes not in the physics textbooks—will use the fact that neutrinos come in different types. Most easily detected are the so-called *electron-type* neutrinos; more difficult to detect are *muon-type* and *tau-type* neutrinos.

Suppose some of these neutrinos from the Sun convert into neutrinos that are easier to detect as they pass through the Earth at night on their way to the detector. The change would make the Sun appear brighter (in terms of neutrinos) at night than during the day. If that were seen, it would provide a dramatic demonstration that unconventional physics is occurring.

If such a smoking gun is found, it could offer a clue to new laws of particle physics. Recent experiments in Japan using neutrinos produced in the Earth's atmosphere by energetic particles from outside our solar system (cosmic rays) have already shown that neutrinos have a tiny mass, which allows them to convert from one type to another. Observations of solar neutrinos may lead us to further revise the laws that govern the smallest scales of matter.

I do not know the correct explanation for the missing neutrinos. However, the particle physics ("split personality") explanation has a mathematical beauty and simplicity that are very attractive. If the deity has not chosen this solution to the mystery, then he or she has missed an excellent opportunity to enrich the laws of the universe.

However, the real message of the experiments on solar neutrinos is even more remarkable. It is that working on the frontier of science, you may stumble across something that is beautiful and unexpected. We do not yet know precisely what we have discovered with solar neutrinos, but we do know that future experiments will solve the mystery for us. Their outcome might point the way to a better understanding of fundamental physics and of the stars.

The Search for Extraterrestrial Life

One of the most compelling questions in science is also one of the simplest: Are we alone? That is, is there life beyond the Earth? As yet, we have no definitive answer to this question. None of our spacecraft has found life elsewhere in the solar system, and radio telescopes have yet to detect signals of intelligent origin coming from space. Reports of aliens visiting our planet and abducting humans make compelling science fiction, but none of these reports has ever been verified.

Yet there are reasons to suspect that life might indeed exist beyond the Earth. One is that biologists find living organisms in some of the most "unearthly" environments on our planet. An example (shown here) is at the bottom of the Gulf of Mexico, where the crushing pressure and low temperature make methane—normally a gas—form into solid, yellowish mounds. Amazingly, these mounds teem with colonies of pink, eyeless, alien-looking worms the size of your thumb. If life can flourish here, might it not also flourish in the equally hostile conditions found on other worlds?

In this chapter we will look for places in our solar system where life may once have originated, and where it may exist today. We will see how scientists estimate the chances of finding life beyond our solar system, and how they search for signals from other intelligent species. And we will learn

(Dr. Charles Fisher, Eberly College of Science, Pennsylvania State University)

R I **V** U X G

how a new generation of telescopes may make it possible to detect even single-celled organisms on worlds many light-years away.

As you read the sections of this chapter, look for the answers to the following questions.

30-1 What role could comets and meteorites have played in the origin of life on Earth?

30-2 Have spacecraft found any evidence for life elsewhere in our solar system?

30-3 Do meteorites from Mars give conclusive proof that life originated there?

30-4 How likely is it that other civilizations exist in our Galaxy?

30-5 How do astronomers search for evidence of civilizations on planets orbiting other stars?

30-6 Will it ever be possible to see Earthlike planets orbiting other stars?

30-1 The chemical building blocks of life are found throughout space

Suppose you were the first visitor to a new and alien planet. How would you recognize which of the strange objects around you were living, and which were inanimate? Questions such as these are central to **astrobiology** (also called **exobiology**), the study of life in the universe. Most astrobiologists suspect that if we find living organisms on other worlds, they will be "life as we know it"— that is, their biochemistry will be based on the unique properties of the carbon atom, as is the case for all terrestrial life.

Why carbon? The reason is that carbon has the most versatile chemistry of any element. Carbon atoms can form chemical bonds to create especially long and complex molecules. These carbon-based compounds, called **organic molecules,** include all the molecules of which living organisms are made.

Organic molecules can be linked together to form elaborate structures, such as chains, lattices, and fibers. Some of these structures are capable of complex, self-regulating chemical reactions. Furthermore, the primary constituents of organic molecules—carbon, hydrogen, nitrogen, oxygen, sulfur, and phosphorus—are among the most abundant elements in the universe. The versatility and abundance of carbon suggest that extraterrestrial life is also likely to be based on organic chemistry.

If life is based on organic molecules, then these molecules must initially be present on a planet in order for life to arise from nonliving matter. We now understand that many organic molecules originate from nonbiological processes in interstellar space. One such molecule is carbon monoxide (CO), which is made when a carbon atom and an oxygen atom collide and bond together. Carbon monoxide is found in abundance within giant interstellar clouds that lie along the spiral arms of our Milky Way Galaxy as well as other galaxies (see Figure 1-7). Carbon atoms have also combined with other elements to produce an impressive array of interstellar organic molecules, including ethyl alcohol (CH_3CH_2OH), formaldehyde (H_2CO), methyl cyanoacetylene (CH_3C_3N), and acetaldehyde (CH_3CHO). Radio astronomers have detected these by looking for the telltale microwave emission lines of carbon-based chemicals in interstellar clouds.

The planets of our solar system coalesced from interstellar material (see Section 7-7), and some of the organic molecules in that material must have ended up on the planets' surfaces. Evidence for this comes from meteorites called **carbonaceous chondrites,** like the one in Figure 30-1. These are ancient meteorites that date from the formation of the solar system and that are often found to contain a variety of carbon-based molecules. The spectra of comets (see Section 7-5)—which are also among the oldest objects in the solar system—show that they, too, contain an assortment of organic compounds.

Comets and meteoroids were much more numerous in the early solar system than they are today, and they were correspondingly more likely to collide with a planet. These collisions would have seeded the planets with organic compounds from the very beginning of our solar system's history. Similar processes are thought to take place in other planetary systems,

figure 30-1   R I **V** U X G

A Carbonaceous Chondrite Carbonaceous chondrites are primitive meteorites that date back to the very beginning of the solar system. Chemical analyses of newly fallen specimens disclose that they are rich in organic molecules, many of which are the chemical building blocks of life. This sample is a piece of the Allende meteorite, a large carbonaceous chondrite that fell in Mexico in 1969. (From the collection of Ronald A. Oriti)

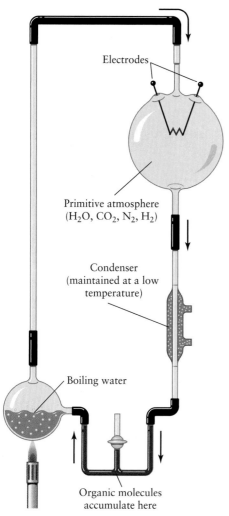

Electrodes

Primitive atmosphere
(H_2O, CO_2, N_2, H_2)

Condenser
(maintained at a low
temperature)

Boiling water

Organic molecules
accumulate here

ƒigure 30-2

The Miller-Urey Experiment (Updated) Modern versions of this classic experiment prove that numerous organic compounds important to life can be synthesized from gases that were present in the Earth's primordial atmosphere. This experiment supports the hypothesis that life on the Earth arose as a result of ordinary chemical reactions.

which are thought to form in basically the same way as did our own (see Figure 7-13).

Comets and meteorites would not have been the only sources of organic material on the young planets of our solar system. In 1952, the American chemists Stanley Miller and Harold Urey demonstrated that under conditions that are thought to have prevailed on the primitive Earth, simple chemicals can combine to form the chemical building blocks of life. In a closed container, they prepared a sample of "atmosphere": a mixture of hydrogen (H_2), ammonia (NH_3), methane (CH_4), and water vapor (H_2O), the most common molecules in the solar system. Miller and Urey then exposed this mixture of gases to an elec-

tric arc (to simulate atmospheric lightning) for a week. At the end of this period, the inside of the container had become coated with a reddish-brown substance rich in amino acids and other compounds essential to life.

Since Miller and Urey's original experiment, most scientists have come to the conclusion that Earth's primordial atmosphere was composed of carbon dioxide (CO_2), nitrogen (N_2), and water vapor outgassed from volcanoes, along with some hydrogen. Modern versions of the Miller-Urey experiment (Figure 30-2) using these common gases have also succeeded in synthesizing a wide variety of organic compounds. The combination of comets and meteorites falling from space and chemical synthesis in the atmosphere could have made the chemical building blocks of life available in substantial quantities on the young Earth.

CAUTION! It is important to emphasize that scientists have *not* created life in a test tube. While organic molecules may have been available on the ancient Earth, biologists have yet to figure out how these molecules gathered themselves into cells and developed systems for self-replication. Nevertheless, because so many chemical components of life are so easily synthesized under conditions that simulate the primordial Earth, it seems reasonable to suppose that life could have originated as the result of chemical processes. Furthermore, because the molecules that combine to form these compounds are rather common, it seems equally reasonable that life could have originated in the same way on other planets.

Organic building blocks are commonplace throughout the universe, but this does not guarantee that life is equally commonplace. If a planet's environment is hostile, life may never get started or may quickly be extinguished. But we now have evidence that Jupiter-sized planets orbit other stars (see Section 7-9 and Geoff Marcy's essay "Alien Planets" at the end of Chapter 7) and that additional planetary systems are forming around young stars (see Section 7-7, especially Figure 7-13). It seems probable that there are Earthlike planets orbiting other stars, and that conditions on some of these worlds may be suitable for life as we know it.

30-2 Europa and Mars are promising places for life to have evolved

If life evolved on Earth from nonliving organic molecules, might the same process have taken place elsewhere in our solar system? Scientists have carefully scrutinized the planets and satellites in an attempt to answer this question, and most of the answers have been disappointing.

One major problem is that liquid water is essential for the survival of life as we know it. The water need not be pleasant by human standards—terrestrial organisms have been found in water that is boiling hot, fiercely acidic, or ice cold (see the image that opens this chapter)—but it must be liquid. In order for water on a planet's surface to remain liquid, the temperature cannot be too hot or too cold.

Furthermore, there must be a relatively thick atmosphere to provide enough pressure to keep liquid water from evaporating. Of all the worlds of the present-day solar system, only Earth has the right conditions for water to remain liquid on its surface. But there is now compelling evidence that Europa, one of the large satellites of Jupiter (see Table 7-2), has an ocean of water *beneath* its icy surface. As it orbits Jupiter, Europa is caught in a tug-of-war between Jupiter's gravitational influence and those of the other large satellites. This flexes the interior of Europa, and this flexing generates enough heat to keep subsurface water from freezing. Chunks of ice on the surface can float around on this underground ocean, rearranging themselves into a pattern that reveals the liquid water beneath.

No one knows whether life exists in Europa's ocean. But interest in this exotic little world is great, and scientists have proposed several missions to explore Europa in more detail.

The next best possibility for the existence of life is Mars. The present-day Martian atmosphere is so thin that water can exist only as ice or as a vapor. However, images made from Martian orbit show dried-up streambeds, flash flood channels, and sediment deposits. These features are evidence that the Martian atmosphere was once thicker and that water once coursed over the planet's surface. Could life have evolved on Mars during its "wet" period? If so, could life—even in the form of microorganisms—have survived as the Martian atmosphere thinned and the surface water either froze or evaporated?

In 1976, two spacecraft landed in different parts of Mars in search of answers to these questions. *Viking Lander 1* and *Viking Lander 2* each carried a scoop at the end of a mechanical arm to retrieve surface samples (Figure 30-3). These samples were deposited into a compact on-board biological laboratory that carried out three different tests for Martian microorganisms.

1. The *gas-exchange experiment* was designed to detect any processes that might be broadly considered as respiration. A surface sample was placed in a sealed container along with a controlled amount of gas and nutrients. The gases in the container were then monitored to see if their chemical composition changed.

2. The *labeled-release experiment* was designed to detect metabolic processes. A sample was moistened with nutrients containing radioactive carbon atoms. If any organisms in the sample consumed the nutrients, their waste products should include gases containing the telltale radioactive carbon.

3. The *pyrolytic-release experiment* was designed to detect photosynthesis, the biological process by which terrestrial plants use solar energy to help synthesize organic compounds from carbon dioxide. In the *Viking* experiments, a surface sample was placed in a container along with radioactive carbon dioxide and exposed to artificial sunlight. If plantlike photosynthesis occurred, microorganisms in the sample would take in some of the radioactive carbon from the gas.

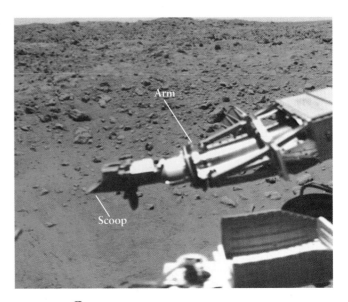

Figure 30-3 R I **V** U X G
Digging in the Martian Surface This view from the *Viking Lander 1* spacecraft shows the mechanical arm with its small scoop against the backdrop of the Martian terrain. The scoop was able to dig about 30 cm (12 in.) beneath the surface. (NASA)

The first data returned from these experiments caused great excitement, for in almost every case, rapid and extensive changes were detected inside the sealed containers. Further analysis of the data, however, led to the conclusion that these changes were due solely to nonbiological chemical processes. It appears that the Martian surface is rich in unstable chemicals that react with water to release oxygen gas. Because the present-day surface of Mars is bone-dry, these chemicals had nothing to react with until they were placed inside the moist interior of the *Viking Lander* laboratory.

At best, the results from the *Viking Lander* biological experiments were inconclusive. Perhaps life never existed on Mars at all. Or perhaps it did originate there, but failed to survive the thinning of the Martian atmosphere, the unstable chemistry of the planet's surface, and exposure to ultraviolet radiation from the Sun. (Unlike Earth, Mars has no ozone layer to block ultraviolet rays.) Another possibility is that Martian microorganisms have survived only in certain locations that the *Viking Lander*s did not sample, such as isolated spots on the surface or deep beneath the ground.

The exploration of Mars, including the search for microorganisms, is just beginning. (See the essay by Matthew Golombek that follows this chapter.) In 2003, a British spacecraft called *Beagle 2* will land on Mars and conduct an entirely different set of biological experiments (Figure 30-4). The spacecraft's name commemorates HMS *Beagle*, the surveying ship from which Charles Darwin made many of the observations that led to the theory of evolution.

Unlike the *Viking Lander*s, *Beagle 2* will test for the presence of either living or dead microorganisms. To do this, it will

f̌igure 30-4

The _Beagle 2_ Lander This lander (shown here in an artist's rendering) will be launched with the European Space Agency's _Mars Express_ spacecraft. On arrival at Mars, _Mars Express_ will go into orbit around the planet while _Beagle 2_ descends to the surface. A parachute and airbags will cushion the landing, after which the spacecraft will unfold to the configuration shown here. Each circular section is about the size of an extra-large pizza. (_Beagle 2_/European Space Agency)

bore into the interiors of rocks to gather pristine, undisturbed samples, then heat these samples in the presence of oxygen gas. All carbon compounds will decompose and form carbon dioxide (CO_2) when treated in this way, but biologically important molecules will signal their presence by decomposing at a lower temperature. As a further test for the chemicals of life, _Beagle 2_ will check to see how many of the CO_2 molecules contain the isotope ^{12}C (which appears preferentially in biological molecules) and how many contain ^{13}C (which does not). (See Box 5-5 for a description of isotopes.) If these experiments give a positive result, it will indicate that Martian rocks contain microorganisms that either survive to the present day or that died out at some point in the past.

Beagle 2 will also carry an experiment to search for traces of methane in the Martian atmosphere. Microorganisms on Earth can gain energy by converting carbon dioxide to methane, and presumably Martian microorganisms could do the same. Left to itself, methane rapidly decomposes in the Martian atmosphere. Hence, if _Beagle 2_ finds any methane, it will necessarily have been freshly formed—which would strongly suggest that life exists on Mars today.

In 1976, while the _Viking Landers_ were carrying out their biological experiments on the Martian surface, the companion _Viking Orbiter_ spacecraft photographed some surface features that could have been crafted by _intelligent_ life on Mars. The _Viking Orbiter 1_ image in Figure 30-5a shows what appears to be a humanlike face, perhaps the product of an advanced and artistic civilization. However, when the more advanced _Mars Global Surveyor_ spacecraft viewed the surface in 1998 using a superior camera (Figure 30-5b), it found no evidence for facial features.

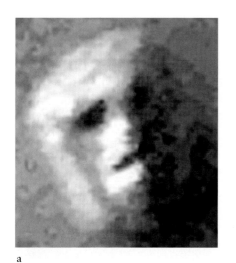
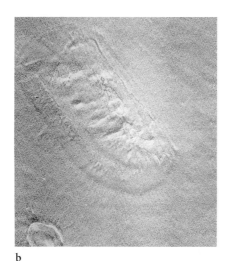
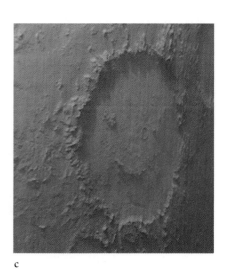

a   b   c

f̌igure 30-5 R I **V** U X G

A "Face" on Mars? **(a)** This 1976 image from _Viking Orbiter 1_ shows a Martian surface feature that resembles a human face. Some suggested that this feature might have been made by intelligent beings. **(b)** This 1998 _Mars Global Surveyor_ (MGS) image, made under different lighting conditions with a far superior camera, reveals the "face" to be just an eroded hill. **(c)** This MGS image shows features of natural origin within a 215-km (134-mi) wide crater. Can you see this "face"? (a: NSSDC/NASA and Dr. Michael H. Carr; b, c: Malin Space Science Systems/NASA)

Scientists are universally convinced that the "face" and other apparent patterns in the *Viking Orbiter* images were created by shadows on wind-blown hills. Microscopic life may once have existed on Mars, and may yet exist today, but there is no evidence that the red planet has ever been the home of intelligent beings.

30-3 Meteorites from Mars have been scrutinized for life-forms

While spacecraft can carry biological experiments to other worlds such as Mars, many astrobiologists look forward to the day when a spacecraft will return Martian samples to laboratories on Earth. (See the essay by Matthew Golombek that follows this chapter.) Until that day arrives, we have the next best thing: A dozen meteorites that appear to have formed on Mars have been found at a variety of locations on the Earth.

These meteorites are called **SNC meteorites** after the names given to the first three examples found (Shergotty, Nakhla, and Chassigny). What identifies SNC meteorites as having come from Mars is the chemical composition of trace amounts of gas trapped within them. This composition is very different from that of the Earth's atmosphere, but is a nearly perfect match to the composition of the Martian atmosphere found by the *Viking Landers*.

How could a rock have got from Mars to Earth? When a large piece of space debris collides with a planet's surface and forms an impact crater, most of the material thrown upward by the impact falls back onto the planet's surface. But some extraordinarily powerful impacts produce large craters—on Mars, roughly 100 km in diameter or larger. These tremendous impacts eject some rocks with such speed that they escape the planet's gravitational attraction and fly off into space.

There are numerous large craters on Mars, so a good number of Martian rocks have probably been blasted into space over the planet's history. These ejected rocks then go into elliptical orbits around the Sun. A few such rocks will have orbits that put them on a collision course with the Earth, and these are the ones that scientists find as SNC meteorites.

Using a technique called radioactive age-dating, scientists find that most SNC meteorites are between 200 million and 1.3 billion years old, much younger than the 4.6-billion-year age of the solar system. But one SNC meteorite, denoted by the serial number ALH 84001 and found in Antarctica in 1984, was discovered in 1993 to be 4.5 billion years old (Figure 30-6a). Thus, ALH 84001 is a truly ancient piece of Mars. Analysis of ALH 84001 suggests that it was fractured by an impact between 3.8 and 4.0 billion years ago, was ejected from Mars by another impact 16 million years ago, and landed in Antarctica a mere 13,000 years ago.

ALH 84001 is the only known specimen of a rock that was on Mars during the era when liquid water existed on the planet's surface. Scientists have therefore investigated its chemical composition carefully, in the hope that this rock may contain clues to the amount of water that once flowed on the Martian surface. One such clue is the presence of rounded grains of minerals called carbonates, which can form only in the presence of water.

In 1996, David McKay and Everett Gibson of the NASA Johnson Space Center, along with several collaborators, reported the results of a two-year study of the carbonate grains in ALH 84001. They made three remarkable findings. First, in and around the carbonate grains were large numbers of elongated, tubelike structures resembling fossilized microorganisms

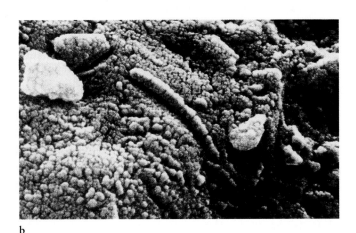

a   R I U X G   b

Figure 30-6

A Meteorite from Mars **(a)** This 1.9-kg meteorite, known as ALH 84001, formed on Mars some 4.5 billion years ago. About 16 million years ago a massive impact blasted it into space, where it drifted in orbit around the Sun until landing in Antarctica 13,000 years ago. The small cube at lower right is 1 cm (0.4 in.) across. **(b)** This electron microscope image, magnified some 100,000 times, shows tubular structures about 100 nanometers (10^{-7} m) in length found within the Martian meteorite ALH 84001. One controversial interpretation is that these are the fossils of microorganisms that lived on Mars billions of years ago. (a: NASA Johnson Space Center; b: *Science*, NASA)

(Figure 30-6*b*). Second, the carbonate grains contain very pure crystals of iron sulfide and magnetite. These two compounds are rarely found together (especially in the presence of carbonates) but can be produced by certain types of bacteria. Indeed, about one-fourth of the magnetite crystals found in ALH 84001 are of a type that on Earth are formed only by bacteria. Third, the carbonates contain organic molecules—just the sort, in fact, that result from the decay of microorganisms.

McKay and Gibson concluded that the structures seen in Figure 30-6*b* are fossilized remains of microorganisms. If so, these organisms lived and died on Mars billions of years ago, during the era when liquid water was abundant.

Are McKay and Gibson's conclusions correct? Their claims of ancient life on Mars are extraordinary, and they require extraordinary proof. With only one rock like ALH 84001 known to science, however, such proof is hard to come by, and many scientists are skeptical. They argue that the structures found in ALH 84001 could have been formed in other ways that do not require the existence of Martian microorganisms. Studies by *Beagle 2* of rocks on Mars may help resolve the controversy.

For now, the existence of microscopic life on Mars in the distant past remains an open question. What is without question, however, is that ALH 84001 has fired the imagination of scientists and the public and generated new excitement about the search for extraterrestrial life.

30-4 The Drake equation helps scientists estimate how many civilizations may inhabit our Galaxy

We have seen that only a few locations in our solar system may have been suitable for the origin of life. But what about other planetary systems? The development of life on the Earth seems to suggest that extraterrestrial life, including intelligent species, might evolve on terrestrial planets around other stars, given sufficient time and hospitable conditions. How can we learn whether such worlds exist, given the tremendous distances that separate us from them? This is the great challenge facing the **search for extraterrestrial intelligence**, or **SETI**.

A tenet of modern folklore is the belief that alien civilizations do exist, and that their spacecraft have visited Earth. Indeed, surveys show that between one-third and one-half of all Americans believe in unidentified flying objects (UFOs). A somewhat smaller percentage believes that aliens have landed on Earth. But, in fact, there is *no* scientifically verifiable evidence of alien visitations. As an example, many UFO proponents believe that the U.S. government is hiding evidence of an alien spacecraft that crashed near Roswell, New Mexico, in 1947. However, the bits of "spacecraft wreckage" found near Roswell turn out to be nothing more than remnants of an unmanned research balloon. To find real evidence of the presence or absence of intelligent civilizations on worlds orbiting other stars, we must look elsewhere.

With our present technology, sending even a small unmanned spacecraft to another star requires a flight time of tens of thousands of years. Speculative design studies have been made for unmanned probes that could reach other stars within a century or less, but these are prohibitively expensive. Instead, many astronomers hope to learn about extraterrestrial civilizations by detecting radio transmissions from them. Radio waves are a logical choice for interstellar communication because they can travel immense distances without being significantly degraded by the interstellar medium, the thin gas and dust found between the stars (see Section 7-6).

Over the past several decades, astronomers have proposed various ways to search for alien radio transmissions, and several searches have been undertaken. In 1960, Frank Drake first used a radio telescope at the National Radio Astronomy Observatory in West Virginia to listen to two Sunlike stars, Tau Ceti and Epsilon Eridani, without success. More than 60 more extensive SETI searches have taken place since then, using radio telescopes around the world. Occasionally, a search has detected an unusual or powerful signal. But none has ever repeated, as a signal of intelligent origin might be expected to do. To date, we have no confirmed evidence of radio transmissions from another world.

 Should we be discouraged by this failure to make contact? What are the chances that a radio astronomer might someday detect radio signals from an extraterrestrial civilization? The first person to tackle this issue was Frank Drake, who proposed that the number of technologically advanced civilizations in the Galaxy could be estimated by a simple equation. This is now called the **Drake equation**:

Drake equation

$$N = R_* \, f_{\mathrm{p}} \, n_{\mathrm{e}} \, f_{\mathrm{l}} \, f_{\mathrm{i}} \, f_{\mathrm{c}} \, L$$

$N =$ number of technologically advanced civilizations in the Galaxy whose messages we might be able to detect

$R_* =$ the rate at which solar-type stars form in the Galaxy

$f_{\mathrm{p}} =$ the fraction of stars that have planets

$n_{\mathrm{e}} =$ the number of planets per solar system that are Earthlike (that is, suitable for life)

$f_{\mathrm{l}} =$ the fraction of those Earthlike planets on which life actually arises

$f_{\mathrm{i}} =$ the fraction of those life-forms that evolve into intelligent species

$f_{\mathrm{c}} =$ the fraction of those species that develop adequate technology and then choose to send messages out into space

$L =$ the lifetime of that technologically advanced civilization

The Drake equation is enlightening because it expresses the number of extraterrestrial civilizations in a simple series of terms. We can estimate some of these terms from what we know about stars and stellar evolution. For example, the first two factors, R_* and f_p, can be determined by observation. In estimating R_*, we should probably exclude stars with masses greater than about 1.5 times that of the Sun. These more massive stars use up the hydrogen in their cores in 3 billion (3×10^9) years or less. On Earth, by contrast, human intelligence developed only within the last million years or so, some 4.6 billion years after the formation of the solar system. If that is typical of the time needed to evolve higher life-forms, then a star of 1.5 solar masses or more probably fades away or explodes into a supernova before creatures as intelligent as we can evolve on any of that star's planets.

Although stars less massive than the Sun have much longer lifetimes, they, too, seem unsuited for life because they are so dim. Only planets very near a low-mass star would be sufficiently warm for life as we know it, and a planet that close is subject to strong tidal forces from its star. We saw in Section 4-8 how the Earth's tidal forces keep the Moon locked in synchronous rotation, with one face continually facing the Earth. In the same way, a planet that orbits too close to its star would have one hemisphere that always faced the star, while the other hemisphere would be in perpetual, frigid darkness.

This leaves us with stars not too different from the Sun. (Like Goldilocks sampling the three bears' porridge, we must have a star that is not too hot and not too cold, but just right.) Based on statistical studies of star formation in the Milky Way, some astronomers estimate that roughly one of these Sunlike stars forms in the Galaxy each year, thus setting R_* at 1 per year.

As we saw in Sections 7-7 and 7-8, the planets in our solar system formed as a natural consequence of the birth of the Sun. We have also seen evidence suggesting that planetary formation may be commonplace around single stars (see Figure 7-13). Many astronomers suspect that most Sunlike stars probably have planets, and so they give f_p a value of 1.

Unfortunately, the rest of the terms in the Drake equation are very uncertain. Let's play with some hypothetical values. The chances that a planetary system has an Earthlike world suitable for life are not known. Were we to consider our own solar system as representative, we could put n_e at 1. Let's be more conservative, however, and suppose that one in ten solar-type stars is orbited by a habitable planet, making $n_e = 0.1$. From what we know about the evolution of life on the Earth, we might assume that, given appropriate conditions, the development of life is a certainty, which would make $f_l = 1$. This is an area of intense interest to astrobiologists.

For the sake of argument, we might also assume that evolution might naturally lead to the development of intelligence (a conjecture that is hotly debated) and also make $f_i = 1$. It's anyone's guess as to whether these intelligent extraterrestrial beings would attempt communication with other civilizations in the Galaxy, but were we to assume they would, f_c would be put at 1 also.

The last variable, L, involving the longevity of a civilization, is the most uncertain of all. Looking at our own

example, we see a planet whose atmosphere and oceans are increasingly polluted by creatures that possess nuclear weapons. If we are typical, perhaps L is as short as 100 years. Putting all these numbers together, we arrive at

$$N = \frac{1}{\text{year}} \times 1 \times 0.1 \times 1 \times 1 \times 1 \times 100 \text{ years} = 10$$

In other words, out of the hundreds of billions of stars in the Galaxy, we would estimate that there are only ten technologically advanced civilizations from which we might receive communications.

A wide range of values has been proposed for the terms in the Drake equation, and these various guesses produce vastly different estimates of N. Some scientists argue that there is exactly one advanced civilization in the Galaxy and that we are it. Others speculate that there may be hundreds or thousands of planets inhabited by intelligent creatures. If we wish to know whether our Galaxy is devoid of other intelligence, teeming with civilizations, or something in between, we must keep searching the skies.

30-5 Radio searches for alien civilizations are under way

Even if only a few alien civilizations are scattered across the Galaxy, we have the technology to detect radio transmissions from them. But if other civilizations are trying to communicate with us using radio waves, what frequency are they using? This is an important question, because if we fail to tune our radio telescopes to the right frequency, we might never know whether the aliens are out there.

A reasonable choice would be a frequency that is fairly free of interference from extraneous sources. SETI pioneer Bernard Oliver was the first to draw attention to a range of relatively noise-free frequencies in the neighborhood of the microwave emission lines of hydrogen (H) and hydroxide (OH) (Figure 30-7). This region of the microwave spectrum is called the **water hole**, because H and OH together make H_2O, or water.

In 1989, NASA began work on the High Resolution Microwave Survey (HRMS), an ambitious project to scan the entire sky at frequencies spanning the water hole from 10^3 to 10^4 MHz. HRMS would have observed more than 800 nearby solar-type stars over a narrower frequency range in the hope of detecting signals that were either pulsed (like Morse code) or continuous (like the carrier wave for a TV or radio broadcast). The sophisticated signal-processing technology of HRMS would have been able to sift through tens of millions of individual frequency channels simultaneously. It would even have been able to detect the minute Doppler shifts in a signal coming from an alien planet as that planet spun on its axis and moved around its star.

Sadly, just one year after HRMS began operation in 1992, the U.S. Congress imposed a mandate requiring that NASA no longer support HRMS or any other radio searches for extra-

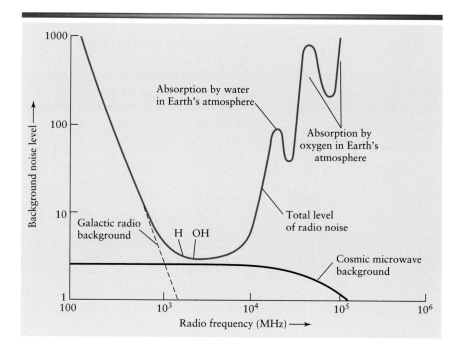

figure 30-7

The Water Hole This graph shows the background noise level from the sky at various radio and microwave frequencies. The so-called water hole is a range of radio frequencies from about 1000 to 10,000 megahertz (Mhz) that happen to have relatively little cosmic noise. Some scientists suggest that this noise-free region would be well suited for interstellar communication. At lower frequencies there is substantial emission from interstellar gas (labeled "Galactic radio background"), and at higher frequencies radio waves tend to be absorbed by the Earth's atmosphere. Within the water hole itself, the principal source of noise is the afterglow of the Big Bang, called the cosmic background radiation. To put this graph in perspective, a frequency of 100 MHz corresponds to "100" on a FM radio, and 10^3 MHz is a frequency used for various types of radar. (Adapted from C. Sagan and F. Drake)

terrestrial intelligence. This decision, which was made on budgetary grounds, saved a few million dollars—an entirely negligible amount compared to the total NASA budget. Ironically, the senator who spearheaded this move was from the state of Nevada, where tax dollars have been spent to signpost a remote desert road as "The Extraterrestrial Highway."

 Even though NASA funding is no longer available, several teams of scientists remain actively involved in SETI programs. Funding for these projects has come from nongovernmental organizations such as the Planetary Society and from private individuals. Since 1995 the SETI Institute in California has been carrying out Project Phoenix, the direct successor to HRMS. When complete, this project will have surveyed a thousand Sunlike stars within 200 light-years at millions of radio frequencies. At Harvard University, BETA (the *Billion-channel ExtraTerrestrial Assay*) is scanning the sky at even more individual frequencies within the water hole. Other multifrequency searches are being carried out under the auspices of the University of Western Sydney in Australia and the University of California.

 A major challenge facing SETI is the tremendous amount of computer time needed to analyze the mountains of data returned by radio searches. To this end, scientists at the University of California, Berkeley, have recruited more than 2 million personal computer users to participate in a project called SETI@home. Each user receives actual data from a detector called SERENDIP IV (*Search for Extraterrestrial Radio Emissions from Nearby, Developed, Intelligent Populations*) and a data analysis program that also acts as a screensaver. When the computer's screensaver is on, the program runs, the data are analyzed, and the results are reported via the Internet to the researchers

at Berkeley. The program then downloads new data to be analyzed. As of this writing, SETI@home users have provided as much computer time as a single computer working full-time for 400,000 years!

All current SETI projects make use of existing radio telescopes and must share telescope time with astronomical researchers. But by 2005, the SETI Institute plans to put into operation a radio telescope that will be dedicated solely to the search for intelligent signals. This telescope, called the Allen Telescope Array, will actually be hundreds of relatively small and inexpensive radio dishes working together. Perhaps this new array will be the first to detect a signal from a distant civilization.

30-6 Infrared telescopes in space will soon begin searching for Earthlike planets

Although no longer involved in SETI, NASA is planning a major effort to search for Earthlike planets suitable for the evolution of an advanced civilization. Such a search poses a major challenge. Planets the size of the Earth are too small to be detected by the indirect methods we described in Section 7-9. They are also too dim to be seen in visible light against the glare of their parent star.

Instead, an orbiting telescope called Terrestrial Planet Finder—currently targeted for a 2011 launch—will search for such planets by detecting their infrared radiation. The rationale is that stars like the Sun emit much less infrared radiation than visible light, while planets are relatively strong emitters

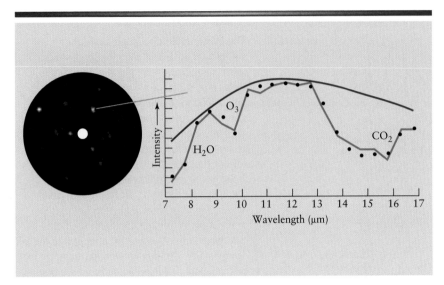

figure 30-8

The Spectrum of a Simulated Planet The image on the left is a simulation of what the Terrestrial Planet Finder infrared telescope might see when it is deployed around 2011. The white dot at the center is a nearby Sunlike star, and the smaller dots around it are planets orbiting the star. On the right is the simulated infrared spectrum of one of the planets, showing broad absorption lines of water vapor (H_2O), ozone (O_3), and carbon dioxide (CO_2). While all these molecules can be created by nonbiological processes, the presence of life will change the relative amounts of each molecule in the planet's atmosphere. Thus, the infrared spectrum of such planets will make it possible to identify worlds on which life may have evolved. (Jet Propulsion Laboratory)

of infrared. Hence, observing in the infrared makes it less difficult (although still technically challenging) to detect planets orbiting a star.

Over its planned six-year lifetime, Terrestrial Planet Finder will search for planets around the brightest 1000 stars within 15 parsecs (50 light-years) of the Sun. It will also analyze the infrared spectra of any planets that it finds, in the hope of seeing the characteristic absorption of atmospheric gases such as ozone, carbon dioxide, and water vapor (Figure 30-8). The relative amounts of these gases, as determined from a planet's spectrum, can reveal whether life is present on that planet.

To achieve enough resolution to detect individual planets, Terrestrial Planet Finder will need to make use of interferometry. We discussed this technique for improving the resolution of telescopes in Section 6-6. By combining the light from four widely spaced 8-m dishes, Terrestrial Planet Finder will make the sharpest infrared images of any telescope in history. The European Space Agency is planning a similar planet-finding interferometry mission called Darwin.

A more speculative project using interferometry is Planet Imager. If funded, this will be an infrared telescope with sufficient resolution that some detail would be visible in the image of an extrasolar planet. One concept for such a mission would consist of five Terrestrial Planet Finder–type telescopes flying in a geometrical formation some 6000 km across (equal to the radius of the Earth). All five telescopes would collect light from the same extrasolar planet, then reflect it onto a single 8-m mirror. The combined light would go to detectors on board a sixth spacecraft (Figure 30-9).

Sometime in the next few decades, missions such as Terrestrial Planet Finder and Planet Imager may answer the question "Are there worlds like Earth orbiting other stars?" If the answer is yes, radio searches for intelligent signals will gain even more impetus.

The potential rewards from such searches are great. Detecting a message from an alien civilization could dramati-

cally change the course of our own civilization, through the sharing of scientific information with another species or an awakening of social or humanistic enlightenment. In only a few years our technology, industry, and social structure might advance the equivalent of centuries into the future. Such changes would touch every person on the Earth. Mindful of these profound implications, scientists push ahead with the search for extraterrestrial intelligence.

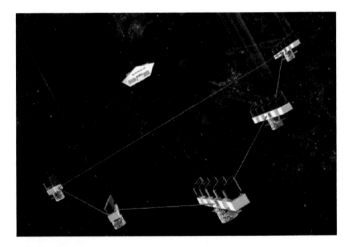

figure 30-9

Planet Imager This illustration shows the concept for Planet Imager, an infrared telescope array designed to create images of Earthlike planets orbiting other stars. Five Terrestrial Planet Finder–type telescopes—each of which has four 8-m mirrors—collect infrared light from a distant planet and reflect it onto a single 8-m mirror, which in turn reflects the combined light onto a spacecraft with infrared-sensitive detectors. To achieve the necessary resolution, the entire Planet Imager array will need to be about 6000 km across, very much larger than shown here. The technology to build Planet Imager does not yet exist. (Jet Propulsion Laboratory)

KEY WORDS

astrobiology, p. 690

carbonaceous chondrite, p. 690

Drake equation, p. 695

exobiology, p. 690

organic molecules, p. 690

search for extraterrestrial intelligence
  (SETI), p. 695

SNC meteorite, p. 694

water hole, p. 696

KEY IDEAS

Organic Molecules in the Universe: All life on Earth, and presumably on other worlds, depends on organic (carbon-based) molecules. These molecules occur naturally throughout interstellar space.

• The organic molecules needed for life to originate were probably brought to the young Earth by comets or meteorites. Another likely source for organic molecules is chemical reactions in the Earth's primitive atmosphere. Similar processes may occur on other worlds.

Life in the Solar System: Besides Earth, only two worlds in our solar system—the planet Mars and Jupiter's satellite Europa—may have had the right conditions for the origin of life.

• Mars once had liquid water on its surface, though it has none today. Life may have originated on Mars during the liquid water era.

• The *Viking Lander* spacecraft searched for microorganisms on the Martian surface, but found no conclusive sign of their presence. The *Beagle 2* mission to Mars will carry out a different set of biological experiments on samples taken from the interiors of rocks.

• An ancient Martian rock that came to Earth as a meteorite shows circumstantial evidence that microorganisms once existed on Mars. Additional rock samples are needed to provide corroboration.

• Europa appears to have extensive liquid water beneath its icy surface. Future missions may search for the presence of life there.

Radio Searches for Extraterrestrial Intelligence: Astronomers have carried out a number of searches for radio signals from other stars. No signs of intelligent life have yet been detected, but searches are continuing and using increasingly sophisticated techniques.

• The Drake equation is a tool for estimating the number of intelligent, communicative civilizations in our Galaxy.

Telescope Searches for Earthlike Planets: A new generation of orbiting infrared telescopes may be able to detect terrestrial planets around nearby stars. If such planets are found, their spectra may reveal the presence or absence of life.

REVIEW QUESTIONS

1. What is meant by "life as we know it"? Why do astrobiologists suspect that extraterrestrial life is likely to be of this form?

2. How have astronomers discovered organic molecules in interstellar space? Does this discovery mean that life of some sort exists in the space between the stars?

3. Mercury, Venus, and the Moon are all considered unlikely places to find life. Suggest why this should be.

4. Summarize the differences in philosophy between the biological experiments on board the *Viking Lander*s and those that will be carried on *Beagle 2*.

5. Suppose someone brought you a rock that he claimed was a Martian meteorite. What scientific tests would you recommend be done to test this claim?

6. Why are most searches for extraterrestrial intelligence made using radio telescopes? Why are most of these carried out at frequencies between 10^3 MHz and 10^4 MHz?

7. Explain why Terrestrial Planet Finder and Planet Imager, both of which are planned to be infrared telescopes, need to be placed in space.

ADVANCED QUESTIONS

Problem-solving tips and tools

The small-angle formula, discussed in Box 1-1, will be useful. Section 5-2 gives the relationship between wavelength and frequency, while Section 5-9 and Box 5-6 discuss the Doppler effect. Section 6-3 gives the relationship between the angular resolution of a telescope, the telescope diameter, and the wavelength used. You will find useful data about the planets in Appendix 1.

8. In 1802, when it seemed likely to many scholars that there was life on Mars, the German mathematician Karl Friedrich Gauss proposed that we signal the Martian inhabitants by drawing huge geometric patterns in the snows of Siberia. His plan was never carried out. **(a)** Suppose patterns had been drawn that were 1000 km across. What minimum diameter would the objective of a Martian telescope need to have to be able to resolve these patterns? Assume that the observations are made at a wavelength of 550 nm, and assume that Earth and Mars are at their minimum separation. **(b)** Ideally, the patterns used would be ones that could not be mistaken for natural formations. They should also indicate that they were created by an advanced civilization. What sort of patterns would you have chosen?

9. Assume that all the terms in the Drake equation have the values given in the text, except for *N* and *L*. **(a)** If there are 1000 civilizations in the Galaxy today, what must be the average lifetime of a technological civilization? **(b)** What if there are a million such civilizations?

10. **(a)** Of the visually brightest stars in the sky listed in Appendix 5, which might be candidates for having Earthlike planets on which intelligent civilizations have evolved? Explain your selection criteria. **(b)** Repeat part (a) for the nearest stars, listed in Appendix 4.

11. It has been suggested that extraterrestrial civilizations would choose to communicate at a wavelength of 21 cm. Hydrogen atoms in interstellar space naturally emit at this wavelength, so astronomers studying the distribution of hydrogen around the Galaxy would already have their radio telescopes tuned to receive extraterrestrial signals. **(a)** Calculate the frequency of this radiation in megahertz. Is this inside or outside the water hole? **(b)** Discuss the merits of this suggestion.

12. Imagine that a civilization in another planetary system is sending a radio signal toward Earth. As our planet moves in its orbit around the Sun, the wavelength of the signal we receive will change due to the Doppler effect. This gives SETI scientists a way to distinguish stray signals of terrestrial origin (which will not show this kind of wavelength change) from interstellar signals. **(a)** Use the data in Appendix 1 to calculate the speed of the Earth in its orbit. For simplicity, assume the orbit is circular. **(b)** If the alien civilization is transmitting at a frequency of 3000 MHz, what wavelength (in meters) would we receive if the Earth were moving neither toward nor away from their planet? **(c)** The maximum Doppler shift occurs if the Earth's orbital motion takes it directly toward or directly away from the alien planet. How large is that maximum wavelength shift? Express your answer both in meters and as a percentage of the unshifted wavelength you found in (b). **(d)** Discuss why it is important that SETI radio receivers be able to measure frequency and wavelength to very high precision.

13. Suppose that Planet Imager has an effective diameter of 6000 km and uses infrared radiation with a wavelength of 10 μm. If it is used to observe an Earthlike planet orbiting the star Epsilon Eridani, 3.22 parsecs (10.5 light-years) from Earth, what is the size of the smallest detail that Planet Imager will be able to resolve on the face of that planet? Give your answer in kilometers.

DISCUSSION QUESTIONS

14. Suppose someone told you that the *Viking Lander*s failed to detect life on Mars simply because the tests were designed to detect terrestrial life-forms, not Martian life-forms. How would you respond?

15. Science-fiction television shows and movies often depict aliens as looking very much like humans. Discuss the likelihood that intelligent creatures from another world would have **(a)** a biochemistry similar to our own, **(b)** two legs and two arms, and **(c)** about the same dimensions as a human.

16. The late, great science-fiction editor John W. Campbell exhorted his authors to write stories about organisms that think as well as humans, but not *like* humans. Discuss the possibility than an intelligent being from another world might be so alien in its thought processes that we could not communicate with it.

17. If a planet always kept the same face toward its star, just as the Moon always keeps the same face toward Earth, most of the planet's surface would be uninhabitable. Discuss why.

18. How do you think our society would respond to the discovery of intelligent messages coming from a civilization on a planet orbiting another star? Explain your reasoning.

19. What do you think will set the limit on the lifetime of our technological civilization? Explain your reasoning.

20. The first of all Earth spacecraft to venture into interstellar space were *Pioneer 10* and *Pioneer 11*, which were launched in 1972 and 1973, respectively. Their missions took them past Jupiter and Saturn and eventually beyond the solar system. Both spacecraft carry a metal plaque with artwork (reproduced here) that shows where the spacecraft is from and what sort of creatures designed it. If an alien civilization were someday to find one of these spacecraft, which of the features on the plaque do you think would be easily understandable to them? Explain.

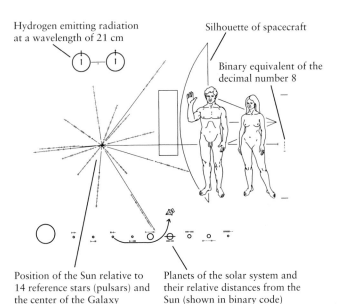

Hydrogen emitting radiation at a wavelength of 21 cm

Silhouette of spacecraft

Binary equivalent of the decimal number 8

Position of the Sun relative to 14 reference stars (pulsars) and the center of the Galaxy

Planets of the solar system and their relative distances from the Sun (shown in binary code)

WEB/CD-ROM QUESTIONS

21. Any living creatures in the subsurface ocean of Europa would have to survive without sunlight. Instead, they might obtain energy from Europa's inner heat. Search the World Wide Web for information about "black smokers," which are associated with high-temperature vents at the bottom of Earth's oceans. What kind of life is found around black smokers? How do these life-forms differ from the more familiar organisms found in the upper levels of the ocean?

22. Like other popular media, the World Wide Web is full of claims of the existence of "extraterrestrial intelligence"—namely, UFO sightings and alien abductions. (**a**) Choose a web site of this kind and analyze its content using the idea of *Occam's razor*, the principle that if there is more than one viable explanation for a phenomenon, one should choose the simplest explanation that fits all the observed facts. (**b**) Read what a skeptical web site has to say about UFO sightings. A good example is the web site of the Committee for the Scientific Investigation of Claims of the Paranormal, or CSICOP. After considering what you have read on both sides of the UFO debate, discuss your opinions about whether aliens really have landed on Earth.

23. The Drake Equation. Access the Active Integrated Media Module "The Drake Equation" in Chapter 30 of the *Universe* web site or CD-ROM. (**a**) For each of the terms in the Drake equation, choose a value that seem reasonable to you. How did you choose these values? Using the module, what do you find for the number of civilizations in our Galaxy? From your calculation, are civilizations common or uncommon in our Galaxy? (**b**) Using the module, choose a set of values that give $N = 10^6$ (a million civilizations). What values did you use? Which of these seem reasonable to you, and why?

OBSERVING PROJECT

24. Use the *Starry Night* program to view the Earth as it might be seen by a visiting spacecraft. First select **Viewing Location...** in the **Go** menu and set the viewing location to your city or town. (You use the list of cities provided, or you can use the mouse to click on your approximate position on the world map. Then click the **Set Location** button.) In the Control Panel at the top of the window, set the local time to 12:00:00 P.M. (noon). To see the Earth from space, use the elevation buttons (the ones that look like a rocket landing or taking off) in the Control Panel to raise yourself off the surface until you can see the entire Earth. (You may want to adjust your field of view so that you are looking directly at the Earth.) (**a**) Describe any features you see that suggest life could exist on Earth. Explain your reasoning. (**b**) Using the controls at the right-hand end of the Control Panel, zoom in to show more detail around your city or town. The amount of detail is comparable to the view from a spacecraft a few million kilometers away. Can you see any evidence that life does exist on Earth? (**c**) From a distance of a few million kilometers, are there any measurements that a spacecraft could carry out to prove that life exists on Earth? Explain your reasoning.

MATTHEW GOLOMBEK

Exploring Mars

Dr. Matt Golombek, chief scientist of the *Mars Pathfinder* mission, is a research geologist at the Jet Propulsion Laboratory, California Institute of Technology, NASA's lead center for planetary exploration. He received his undergraduate degree from Rutgers University and his master's degree and Ph.D. from the University of Massachusetts, Amherst. Dr. Golombek was a postdoctoral fellow at the Lunar and Planetary Institute, Houston, before joining JPL.

Dr. Golombek has conducted research in the structural geology of the Earth, planets, and satellites. He is presently NASA's Mars Exploration Program Landing Site Scientist, leading the effort to select landing sites for Mars missions.

Dr. Golombek cowrote the book *Mars: Uncovering the Secrets of the Red Planet* for the National Geographic Society and has served as editor and associate editor of professional journals. He has received numerous awards, including the NASA Exceptional Scientific Achievement Medal. In his honor, asteroid 6456 was named Golombek.

Mars is the most Earthlike planet in our solar system. It is the first planet humans will visit and the only planet with abundant water that can support life—including people, perhaps—in the future. The geologic record on Mars suggests a warmer early climate in which liquid water (a requirement for life) may have been stable. Further, meteorites from Mars may contain evidence of primitive life. Mars is also a unique terrestrial planet with evidence of major climatic change and a geologic record of rocks on the surface that spans the entire history of the solar system. An exploration program to the red planet will allow not only the investigation of fundamental geologic, climatologic, and exobiologic topics but can also address one of the most compelling scientific questions: "Are we alone in the universe?"

The underlying questions that drive the exploration of Mars are: Did life ever develop on Mars? What is the current climate and how has it changed with time? How have the surface and interior (geology) of the planet evolved through time and what resources are available for future use? Water plays a central role in all these questions.

Robotic spacecraft are among the most complex devices built by humans, and launching and operating them in space are somewhat risky and difficult endeavors. Because of the orbital motions of Mars and Earth, spacecraft can be launched to Mars along a favorable trajectory only once every 26 months. The spacecraft must withstand a nine-month interplanetary journey, observe the planet with carefully designed instruments, and send data back to Earth. These spacecraft typically take three to ten years to design and build, require highly coordinated work by hundreds of scientists and engineers, cost hundreds of millions of dollars to complete, and have a few percent chance of exploding during launch on even the most reliable launch vehicles available.

Future missions are designed to provide an understanding of Mars as a whole. Missions such as NASA's *Mars Surveyor* orbiter being readied for launch in 2001, Europe's *Mars Express* mission to be launched in 2003, Japan's *Nozomi* orbiter departing for Mars in 2003, and other NASA missions under study will obtain imaging data to investigate the geological processes and history of the surface, spectral imaging data directed at understanding the chemistry and mineralogy of surface materials, sounder data to probe the subsurface distribution of water, atmospheric data to study the present state and circulation of the atmosphere, and other data directed at understanding the interaction of the solar wind and other particles with the uppermost atmosphere of Mars.

Remotely sensed data from orbit will always carry an inherent ambiguity or uncertainty, so information collected from a probe on the Martian surface is required to provide "ground truth." NASA's two *Mars Exploration Rover*s being readied for launch in 2003, Britain's *Beagle 2* lander to be launched on *Mars Express* in 2003, and France's four *Netlander*s scheduled for launch in 2005 will provide such "ground truth" and make simultaneous measurements of global atmospheric and interior processes.

An extremely high-priority mission being studied is a "sample-return" mission to bring rock, soil, and atmosphere samples of Mars to the Earth. Much more sophisticated tests can be done with samples in Earth laboratories than by the limited instruments that can be miniaturized and placed on landed spacecraft. Analyses of these samples will allow extensive testing of our knowledge about the evolution of Mars. These studies will also provide information for the design of eventual human exploration missions.

Was Mars once warm and wet like the Earth? Did life begin on Mars? If it did, what happened to it? Could it have come to the Earth in chunks of rocks ejected from Mars? Could we all be Martians? Is life common or rare in the universe? These are the questions we hope to address by exploring our fascinating neighbor.

Appendices

Appendix 1 | The Planets: Orbital Data

Planet	Semimajor axis (AU)	Semimajor axis (10^6 km)	Sidereal period (years)	Sidereal period (days)	Synodic period (days)	Average orbital speed (km/s)	Orbital eccentricity	Inclination of orbit to ecliptic(°)
Mercury	0.3871	57.91	0.2408	87.969	115.88	47.9	0.206	7.00
Venus	0.7233	108.2	0.6152	224.70	583.92	35.0	0.0068	3.39
Earth	1.0000	149.60	1.0000	365.256	—	29.79	0.017	0.00
Mars	1.5236	227.93	1.8808	686.98	779.94	24.1	0.093	1.85
Jupiter	5.2026	778.30	11.856		398.9	13.1	0.048	1.30
Saturn	9.5719	1431.9	29.369		378.1	9.64	0.053	2.48
Uranus	19.194	2877.4	84.099		369.7	6.83	0.043	0.77
Neptune	30.066	4497.8	164.86		367.5	5.5	0.010	1.77
Pluto	39.537	5914.7	248.60		366.7	4.7	0.250	17.12

Appendix 2 | The Planets: Physical Data

Planet	Equatorial diameter (km)	Equatorial diameter (Earth = 1)	Mass (kg)	Mass (Earth = 1)	Average density (kg/m³)	Rotation period (days)*	Inclination of equator to orbit (°)	Surface gravity (Earth = 1)	Albedo	Escape speed (km/s)
Mercury	4880	0.383	3.302×10^{23}	0.0553	5430	58.646	0.5	0.38	0.12	4.3
Venus	12,104	0.949	4.868×10^{24}	0.8150	5243	243.01^R	177.4	0.91	0.59	10.4
Earth	12,756	1.000	5.974×10^{24}	1.000	5515	1.000	23.45	1.000	0.39	11.2
Mars	6794	0.533	6.418×10^{23}	0.1074	3934	1.026	25.19	0.38	0.15	5.0
Jupiter	142,984	11.209	1.899×10^{27}	317.8	1326	0.414	3.12	2.36	0.44	60.2
Saturn	120,536	9.449	5.685×10^{26}	95.16	687	0.444	26.73	1.1	0.47	35.5
Uranus	51,118	4.007	8.682×10^{25}	14.53	1318	0.718^R	97.86	0.92	0.56	21.3
Neptune	49,528	3.883	1.024×10^{26}	17.15	1638	0.671	29.56	1.1	0.51	23.5
Pluto	2300	0.180	1.31×10^{22}	0.002	2000	6.387^R	122.52	0.07	0.5	1.2

* For Jupiter, Saturn, Uranus, and Neptune, the internal rotation period is given. A superscript R means that the rotation is retrograde (opposite the planet's orbital motion).

Appendix 3 | Satellites of the Planets

Planet	Satellite	Discoverer(s)	Average distance from center of planet (km)	Orbital (sidereal) period (days)*	Orbital eccentricity	Diameter of satellite (km)	Mass (kg)
EARTH	Moon	—	384,400	27.322	0.0549	3476	7.349×10^{22}
MARS	Phobos	Hall (1877)	9378	0.319	0.01	$28 \times 23 \times 20$	1.1×10^{16}
	Deimos	Hall (1877)	23,460	1.263	0.00	$16 \times 12 \times 10$	1.8×10^{15}
JUPITER**	Metis	Synott (1979)	127,960	0.2948	0.00	40	1×10^{17}
	Adrastea	Jewitt et al. (1979)	128,980	0.2983	0 (?)	$24 \times 20 \times 16$	1.9×10^{16}
	Amalthea	Barnard (1892)	181,300	0.4981	0.00	$270 \times 200 \times 155$	7.2×10^{18}
	Thebe	Synott (1979)	221,900	0.6745	0.01	100	8×10^{17}
	Io	Galileo (1610)	421,600	1.769	0.00	3642	8.932×10^{22}
	Europa	Galileo (1610)	670,900	3.551	0.01	3120	4.791×10^{22}
	Ganymede	Galileo (1610)	1,070,000	7.155	0.00	5268	1.482×10^{23}
	Callisto	Galileo (1610)	1,883,000	16.689	0.01	4800	1.077×10^{23}
	Leda	Kowal (1974)	11,094,000	238.72	0.15	16	6×10^{15}
	Himalia	Perrine (1904)	11,480,000	250.57	0.16	180	1×10^{19}
	Lysithea	Nicholson (1938)	11,720,000	259.22	0.11	40	8×10^{16}
	Elara	Perrine (1905)	11,737,000	259.65	0.21	80	8×10^{17}
	Ananke	Nicholson (1951)	21,200,000	631[R]	0.17	30	4×10^{16}
	Carme	Nicholson (1938)	22,600,000	692[R]	0.21	44	1×10^{17}
	Pasiphae	Melotte (1908)	23,500,000	735[R]	0.38	70	2×10^{17}
	Sinope	Nicholson (1914)	23,700,000	758[R]	0.28	40	8×10^{16}
	S/1999 J1	Spacewatch (2000)	24,200,000	768[R]	0.12	10	(?)
SATURN***	Pan	Showalter (1990)	133,570	0.573	0.00	20	(?)
	Atlas	Terrile (1980)	137,640	0.602	0 (?)	$20 \times 20 \times 20$	(?)
	Prometheus	Collins et al. (1980)	139,350	0.613	0.00	$140 \times 100 \times 80$	3×10^{17}
	Pandora	Collins et al. (1980)	141,700	0.629	0.00	$110 \times 90 \times 70$	2×10^{17}
	Epimetheus	Walker (1966)	151,422	0.694	0.01	$140 \times 120 \times 100$	6×10^{17}
	Janus	Dolfuss (1966)	151,472	0.695	0.01	$220 \times 200 \times 160$	2×10^{18}
	Mimas	Herschel (1789)	185,520	0.942	0.02	392	3.8×10^{19}
	Enceladus	Herschel (1789)	238,020	1.370	0.00	500	7.3×10^{19}
	Tethys	Cassini (1684)	294,660	1.888	0.00	1060	6.2×10^{20}
	Calypso	Smith et al. (1980)	294,660	1.888	0 (?)	$34 \times 28 \times 26$	(?)
	Telesto	Smith et al. (1980)	294,660	1.888	0 (?)	$24 \times 22 \times 22$	(?)
	Dione	Cassini (1684)	377,400	2.737	0.00	1120	1.1×10^{21}
	Helene	Laques et al. (1980)	377,400	2.737	0.01	$36 \times 32 \times 30$	(?)
	Rhea	Cassini (1672)	527,040	4.518	0.00	1530	2.3×10^{21}
	Titan	Huygens (1655)	1,221,850	15.945	0.03	5150	1.4×10^{23}
	Hyperion	Bond (1848)	1,481,000	21.277	0.10	$410 \times 260 \times 220$	2×10^{19}
	Iapetus	Cassini (1671)	3,561,300	79.331	0.03	1460	1.6×10^{21}
	Phoebe	Pickering (1898)	12,952,000	550.48[R]	0.16	220	4×10^{18}

Planet	Satellite	Discoverer(s)	Average distance from center of planet (km)	Orbital (sidereal) period (days)*	Orbital eccentricity	Diameter of satellite (km)	Mass (kg)
URANUS	Cordelia	*Voyager 2 (1986)*	49,752	0.335	0 (?)	26	(?)
	Ophelia	*Voyager 2 (1986)*	53,763	0.376	0 (?)	32	(?)
	Bianca	*Voyager 2 (1986)*	59,166	0.435	0 (?)	44	(?)
	Cressida	*Voyager 2 (1986)*	61,767	0.464	0 (?)	66	(?)
	Desdemona	*Voyager 2 (1986)*	62,658	0.474	0 (?)	58	(?)
	Juliet	*Voyager 2 (1986)*	64,358	0.493	0 (?)	84	(?)
	Portia	*Voyager 2 (1986)*	66,097	0.513	0 (?)	110	(?)
	Rosalind	*Voyager 2 (1986)*	69,927	0.558	0 (?)	58	(?)
	1986 U10	Karkoschka (1999)	75,000	0.62	0 (?)	40	(?)
	Belinda	*Voyager 2 (1986)*	75,260	0.624	0 (?)	67	(?)
	Puck	*Voyager 2 (1986)*	86,010	0.762	0 (?)	154	(?)
	Miranda	Kuiper (1948)	129,780	1.413	0.00	473	6.6×10^{19}
	Ariel	Lassell (1851)	191,240	2.520	0.00	1160	1.4×10^{21}
	Umbriel	Lassell (1851)	265,790	4.144	0.00	1170	1.2×10^{21}
	Titania	Herschel (1787)	435,840	8.706	0.00	1580	3.5×10^{21}
	Oberon	Herschel (1787)	582,600	13.463	0.00	1520	3.0×10^{21}
	Caliban	Gladman (1997)	7,168,900	579[R]	0 (?)	60	(?)
	Stephano	Gladman (1999)	7,942,000	676[R]	0 (?)	30	(?)
	Sycorax	Nicholson (1997)	12,213,000	1289[R]	0 (?)	120	(?)
	Prospero	Holman (1999)	16,114,000	1953[R]	0 (?)	50	(?)
	Setebos	Kavelaars (1999)	18,205,000	2345[R]	0 (?)	40	(?)
NEPTUNE	Naiad	*Voyager 2 (1989)*	48,227	0.294	0 (?)	58	(?)
	Thalassa	*Voyager 2 (1989)*	50,075	0.312	0 (?)	80	(?)
	Despina	*Voyager 2 (1989)*	52,526	0.335	0 (?)	150	(?)
	Galatea	*Voyager 2 (1989)*	61,953	0.429	0 (?)	160	(?)
	Larissa	*Voyager 2 (1989)*	73,548	0.555	0 (?)	190	(?)
	Proteus	*Voyager 2 (1989)*	117,647	1.122	0 (?)	420	(?)
	Triton	Lassell (1846)	354,760	5.877[R]	0.00	2706	2.15×10^{22}
	Nereid	Kuiper (1949)	5,513,400	360.14	0.75	340	(?)
PLUTO	Charon	Christy (1978)	19,640	6.387	0.00	1190	1.9×10^{21}

* A superscript R means that the satellite orbits in a retrograde direction (opposite to the planet's rotation).

** *Eleven additional small satellites of Jupiter (S/2000 J1 through S/2000 J11)were discovered in 2000. Their average distances from Jupiter range from 7 to 24 million km, and their sizes range from 3 to 8 km. Nine are in retrograde orbits. The discoveries of these satellites had not been confirmed as of this writing (early 2001), so they do not appear in the above table.*

*** *In 2000, ten additional small satellites of Saturn (S/2000 S1 through S/2000 S10) were discovered beyond the orbit of Phoebe. While the existence of these satellites has been confirmed, their orbits are still tentative as of this writing (early 2001) and so do not appear in this table.*

Appendix 4 | The Nearest Stars

Name*	Parallax (arcsec)	Distance (light-years)	Spectral type	Radial velocity** (km/s)	Proper motion (arcsec/year)	Apparent visual magnitude	Absolute visual magnitude	Luminosity (Sun = 1)
Sun			G2 V			−26.7	+4.83	1.00
Proxima Centauri	0.772	4.22	M5.5 V	−22	3.853	+11.01	+15.45	8.2×10^{-4}
Alpha Centauri A	0.742	4.40	G2 V	−25	3.710	−0.01	+4.34	1.77
Alpha Centauri B	0.742	4.40	K0 V	−21	3.724	+1.35	+5.70	0.55
Barnard's Star	0.549	5.94	M4 V	−111	10.358	+9.54	+13.24	3.6×10^{-3}
Wolf 359	0.418	7.80	M6 V	+13	4.689	+13.45	+16.6	3.5×10^{-4}
Lalande 21185	0.392	8.32	M2 V	−84	4.802	+7.49	+10.46	0.023
L 726-8 A	0.381	8.56	M5.5 V	+29	3.366	+12.41	+15.3	9.4×10^{-4}
L 726-8 B	0.381	8.56	M6 V	+32	3.366	+13.2	+16.1	5.6×10^{-4}
Sirius A	0.379	8.61	A1 V	−9	1.339	−1.44	+1.45	26.1
Sirius B	0.379	8.61	white dwarf	−9	1.339	+8.44	+11.3	2.4×10^{-3}
Ross 154	0.336	9.71	M3.5 V	−12	0.666	+10.37	+13.00	4.1×10^{-3}
Ross 248	0.316	10.32	M5.5 V	−78	1.626	+12.29	+14.8	1.5×10^{-3}
Epsilon Eridani	0.311	10.49	K2 V	+17	0.977	+3.72	+6.18	0.40
Lacaille 9352	0.304	10.73	M1.5 V	+10	6.896	+7.35	+9.76	0.051
Ross 128	0.300	10.87	M4 V	−31	1.361	+11.12	+13.50	2.9×10^{-3}
L 789-6	0.294	11.09	M5 V	−60	3.259	+12.33	+14.7	1.3×10^{-3}
61 Cygni A	0.287	11.36	K5 V	−65	5.281	+5.20	+7.49	0.16
61 Cygni B	0.285	11.44	K7 V	−64	5.172	+6.05	+8.33	0.095
Procyon A	0.286	11.40	F5 IV–V	−4	1.259	+0.40	+2.68	7.73
Procyon B	0.286	11.40	white dwarf	−4	1.259	+10.7	+13.0	5.5×10^{-4}
BD +59° 1915 A	0.281	11.61	M3 V	−1	2.238	+8.94	+11.18	0.020
BD +59° 1915 B	0.281	11.61	M3.5 V	+1	2.313	+9.70	+11.97	0.010
Groombridge 34 A	0.280	11.65	M1.5 V	+12	2.918	+8.09	+10.33	0.030
Groombridge 34 B	0.280	11.65	M3.5 V	+11	2.918	+11.07	+13.3	3.1×10^{-3}
Epsilon Indi	0.276	11.82	K5 V	−40	4.704	+4.69	+6.89	0.27
GJ 1111	0.276	11.82	M6.5 V	−5	1.288	+14.79	+17.0	2.7×10^{-4}
Tau Ceti	0.274	11.90	G8 V	−17	1.922	+3.49	+5.68	0.62
GJ 1061	0.270	12.08	M5.5 V	−20	0.836	+13.03	+15.2	1.0×10^{-3}
L 725-32	0.269	12.12	M4.5 V	+28	1.372	+12.10	+14.25	1.7×10^{-3}
BD +05° 1668	0.263	12.40	M3.5 V	+18	3.738	+9.84	+11.94	0.011
Kapteyn's Star	0.255	12.79	M0 V	+246	8.670	+8.86	+10.89	0.013
Lacaille 8760	0.253	12.89	M0 V	+28	3.455	+6.69	+8.71	0.094
Krüger 60 A	0.250	13.05	M3 V	−33	0.990	+9.85	+11.9	0.010
Krüger 60 B	0.250	13.05	M4 V	−32	0.990	+11.3	+13.3	3.4×10^{-3}

This table, compiled from the Hipparcos General Catalogue and from data reported by the Research Consortium on Nearby Stars, lists all known stars within 4.00 parsecs (13.05 light-years).

* Stars that are components of binary systems are labeled A and B.

** A positive radial velocity means the star is receding; a negative radial velocity means the star is approaching.

Name	Designation	Distance (light-years)	Spectral type	Radial velocity (km/s)*	Proper motion (arcsec/year)	Apparent visual magnitude	Apparent visual brightness (Sirius = 1)**	Absolute visual magnitude	Luminosity (Sun = 1)
Sirius A	α CMa A	8.61	A1 V	−9	1.339	−1.44	1.000	+1.45	26.1
Canopus	α Car	313	F0 I	+21	0.031	−0.62	0.470	−5.53	8.2×10^{-4}
Arcturus	α Boo	36.7	K2 III	−5	2.279	−0.05	0.278	−0.31	190
Alpha Centauri A	α Cen A	4.4	G2 V	−25	3.71	−0.01	0.268	+4.34	1.77
Vega	α Lyr	25.3	A0 V	−14	0.035	+0.03	0.258	+0.58	61.9
Capella	α Aur	42.2	G8 III	+30	0.434	+0.08	0.247	−0.48	180
Rigel	α Ori A	773	B8 Ia	+21	0.002	+0.18	0.225	−6.69	7.0×10^{5}
Procyon	α CMi A	11.4	F5 IV–V	−4	1.259	+0.4	0.184	+2.68	7.73
Achernar	α Eri	144	B3 IV	+19	0.097	+0.45	0.175	−2.77	5250
Betelgeuse	α Ori	427	M2 Iab	+21	0.029	+0.45	0.175	−5.14	4.1×10^{4}
Hadar	β Cen	525	B1 II	−12	0.042	+0.61	0.151	−5.42	8.6×10^{4}
Altair	α Aql	16.8	A7 IV–V	−26	0.661	+0.76	0.132	+2.2	11.8
Aldebaran	α Tau A	65.1	K5 III	+54	0.199	+0.87	0.119	−0.63	370
Spica	α Vir	262	B1 V	+1	0.053	+0.98	0.108	−3.55	2.5×10^{4}
Antares	α Sco A	604	M1 Ib	−3	0.025	+1.06	0.100	−5.28	3.7×10^{4}
Pollux	β Gem	33.7	K0 III	+3	0.627	+1.16	0.091	+1.09	46.6
Fomalhaut	α PsA	25.1	A3 V	+7	0.368	+1.17	0.090	+1.74	18.9
Deneb	α Cyg	3230	A2 Ia	−5	0.002	+1.25	0.084	−8.73	3.2×10^{5}
Mimosa	β Cru	353	B0.5 III	+20	0.05	+1.25	0.084	−3.92	3.4×10^{4}
Regulus	α Leo A	77.5	B7 V	+4	0.249	+1.36	0.076	−0.52	331

Data in this table were compiled from the Hipparcos General Catalogue.

* A positive radial velocity means the star is receding; a negative radial velocity means the star is approaching.

** This is the ratio of the star's apparent brightness to that of Sirius, the brightest star in the night sky.

Note: Acrux, or α Cru (the brightest star in Crux, the Southern Cross), appears to the naked eye as a star of apparent magnitude +0.87, the same as Aldebaran, but it does not appear in this table because Acrux is actually a binary star system. The blue-white component stars of this binary system have apparent magnitudes of +1.4 and +1.9, and thus they are dimmer than any of the stars listed here.

Appendix 6 | Some Important Astronomical Quantities

Astronomical unit	$1\ \text{AU} = 1.4960 \times 10^{11}$ m $= 1.4960 \times 10^{8}$ km
Light-year	$1\ \text{ly} = 9.4605 \times 10^{15}$ m $= 63{,}240$ AU
Parsec	$1\ \text{pc} = 3.2616$ ly $= 3.0857 \times 10^{16}$ m $= 206{,}265$ AU
Year	$1\ \text{y} = 365.2564$ days $= 3.156 \times 10^{7}$ s
Solar mass	$1\ \text{M}_\odot = 1.989 \times 10^{30}$ kg
Solar radius	$1\ \text{R}_\odot = 6.9599 \times 10^{8}$ m
Solar luminosity	$1\ \text{L}_\odot = 3.90 \times 10^{26}$ W

Appendix 7 | Some Important Physical Constants

Speed of light	$c = 2.9979 \times 10^{8}$ m/s
Gravitational constant	$G = 6.6726 \times 10^{-11}$ N m^2/kg^2
Planck's constant	$h = 6.6261 \times 10^{-34}$ J s $= 4.1357 \times 10^{-15}$ eV s
Boltzmann constant	$k = 1.3807 \times 10^{-23}$ J/K $= 8.6174 \times 10^{-5}$ eV/K
Stefan-Boltzmann constant	$\sigma = 5.6705 \times 10^{-8}$ W m^{-2} K^{-4}
Mass of electron	$m_e = 9.1094 \times 10^{-31}$ kg
Mass of proton	$m_p = 1.6726 \times 10^{-27}$ kg
Mass of neutron	$m_n = 1.6749 \times 10^{-27}$ kg
Mass of hydrogen atom	$m_H = 1.6735 \times 10^{-27}$ kg
Rydberg constant	$R = 1.0968 \times 10^{7}$ m^{-1}
Electron volt	$1\ \text{eV} = 1.6022 \times 10^{-19}$ J

Appendix 8 | Some Useful Mathematics

Area of a rectangle of sides a and b	$A = ab$
Volume of a rectangular solid of sides a, b, and c	$V = abc$
Hypotenuse of a right triangle whose other sides are a and b	$c = \sqrt{a^2 + b^2}$
Circumference of a circle of radius r	$C = 2\pi r$
Area of a circle of radius r	$A = \pi r^2$
Surface area of a sphere of radius r	$A = 4\pi r^2$
Volume of a sphere of radius r	$V = {}^4\!/_3 \pi r^3$
Value of π	$\pi = 3.1415926536$

Glossary

You can find more information about each term in the indicated chapter or chapters.
See the Key Words section in each chapter for the specific page on which the meaning of a given term is discussed.

A ring   One of three prominent rings encircling Saturn. (Chapter 15)

absolute zero   A temperature of –273°C (or 0 K), at which all molecular motion stops; the lowest possible temperature. (Chapter 5)

absorption line spectrum   Dark lines superimposed on a continuous spectrum. (Chapter 5)

acceleration   The rate at which an object's velocity changes due to a change in speed, a change in direction, or both. (Chapter 4)

accretion   The gradual accumulation of matter in one location, typically due to the action of gravity. (Chapter 7)

active optics   A technique for improving a telescopic image by altering the telescope's optics to compensate for variations in air temperature or flexing of the telescope mount. (Chapter 6)

adaptive optics   A technique for improving a telescopic image by altering the telescope's optics in a way that compensates for distortion caused by the Earth's atmosphere. (Chapter 6)

aerosol   Tiny droplets of liquid dispersed in a gas. (Chapter 15)

albedo   The fraction of sunlight that a planet, asteroid, or satellite reflects. (Chapter 8)

amino acids   The chemical building blocks of proteins. (Chapter 17)

andesite   Low-density, silicon-rich volcanic rocks found in the young northern lowlands of Mars. (Chapter 12)

angle   The opening between two lines that meet at a point. (Chapter 1)

angular diameter   The angle subtended by the diameter of an object. (Chapter 1)

angular distance   The angle between two points in the sky. (Chapter 1)

angular measure   The size of an angle, usually expressed in degrees, arcminutes, and arcseconds. (Chapter 1)

angular momentum   See *conservation of angular momentum*.

angular resolution   The angular size of the smallest feature that can be distinguished with a telescope. (Chapter 6)

angular size   See *angular diameter*.

annular eclipse   An eclipse of the Sun in which the Moon is too distant to cover the Sun completely, so that a ring of sunlight is seen around the Moon at mid-eclipse. (Chapter 3)

anorthosite   Rock commonly found in ancient, cratered highlands on the Moon. (Chapter 9)

Antarctic Circle   A circle of latitude 23½° north of the Earth's South Pole. (Chapter 2)

aphelion   The point in its orbit where a planet is farthest from the Sun. (Chapter 4)

apogee   The point in its orbit where a satellite or the Moon is farthest from the Earth. (Chapter 3)

apparent solar day   The interval between two successive transits of the Sun's center across the local meridian. (Chapter 2)

apparent solar time   Time reckoned by the position of the Sun in the sky. (Chapter 2)

arcminute   One-sixtieth (1/60) of a degree, designated by the symbol ′. (Chapter 1)

arcsecond   One-sixtieth (1/60) of an arcminute or 1/3600 of a degree, designated by the symbol ″. (Chapter 1)

Arctic Circle   A circle of latitude 23½° south of the Earth's North Pole. (Chapter 2)

asteroid   One of tens of thousands of small, rocky, planetlike objects in orbit about the Sun. Also called "minor planet." (Chapter 7, Chapter 17)

asteroid belt   A region between the orbits of Mars and Jupiter that encompasses the orbits of many asteroids. (Chapter 7, Chapter 17)

asthenosphere   A warm, plastic layer of the mantle beneath the lithosphere of the Earth. (Chapter 8)

astrobiology   The study of life in the universe. (Chapter 30)

astrometric method   A technique for detecting extrasolar planets by looking for stars that "wobble" periodically. (Chapter 7)

astronomical unit (AU)   The semimajor axis of the Earth's orbit; the average distance between the Earth and the Sun. (Chapter 1)

atmosphere (atm)   A unit of atmospheric pressure. (Chapter 8)

atmospheric pressure   The force per unit area exerted by a planet's atmosphere. (Chapter 8)

atom   The smallest particle of an element that has the properties characterizing that element. (Chapter 5)

atomic number The number of protons in the nucleus of an atom of a particular element. (Chapter 5, Chapter 7)

AU See *astronomical unit.*

aurora (plural **aurorae**) Light radiated by atoms and ions in the Earth's upper atmosphere, mostly in the polar regions. (Chapter 8)

aurora australis Aurorae seen from southern latitudes; the southern lights. (Chapter 8)

aurora borealis Aurorae seen from northern latitudes; the northern lights. (Chapter 8)

autumnal equinox The intersection of the ecliptic and the celestial equator where the Sun crosses the equator from north to south. Also used to refer to the date on which the Sun passes through this intersection. (Chapter 2)

average density The mass of an object divided by its volume. (Chapter 7)

B ring One of three prominent rings encircling Saturn. (Chapter 15)

Balmer line An emission or absorption line in the spectrum of hydrogen caused by an electron transition between the second and higher energy levels. (Chapter 5)

Balmer series The entire pattern of Balmer lines. (Chapter 5)

baseline In interferometry, the distance between two telescopes whose signals are combined to give a higher-resolution image. (Chapter 6)

belt A dark band in Jupiter's atmosphere. (Chapter 13)

Big Bang An explosion of all space that took place roughly 15 billion years ago and from which the universe emerged. (Chapter 1)

biosphere The layer of soil, water, and air surrounding the Earth in which living organisms thrive. (Chapter 8)

black hole An object whose gravity is so strong that the escape speed exceeds the speed of light. (Chapter 1)

blackbody A hypothetical perfect radiator that absorbs and re-emits all radiation falling upon it. (Chapter 5)

blackbody curve The intensity of radiation emitted by a blackbody plotted as a function of wavelength or frequency. (Chapter 5)

blackbody radiation The radiation emitted by a perfect blackbody. (Chapter 5)

blueshift A decrease in the wavelength of photons emitted by an approaching source of light. (Chapter 5)

Bohr orbits In the model of the atom described by Niels Bohr, the only orbits in which electrons are allowed to move about the nucleus. (Chapter 5)

bright terrain (Ganymede) Young, reflective, relatively crater-free terrain on the surface of Ganymede. (Chapter 14)

brown dwarf A starlike object that is not massive enough to sustain hydrogen burning in its core. (Chapter 7)

brown ovals Elongated, brownish features usually seen in Jupiter's northern hemisphere. (Chapter 13)

C ring One of three prominent rings encircling Saturn. (Chapter 15)

capture theory The hypothesis that the Moon was gravitationally captured by the Earth. (Chapter 9)

carbonaceous chondrite A type of meteorite that has a high abundance of carbon and volatile compounds. (Chapter 17, Chapter 30)

Cassegrain focus An optical arrangement in a reflecting telescope in which light rays are reflected by a secondary mirror to a focus behind the primary mirror. (Chapter 6)

Cassini division An apparent gap between Saturn's A and B rings. (Chapter 15)

CCD See *charge-coupled device.*

celestial equator A great circle on the celestial sphere 90° from the celestial poles. (Chapter 2)

celestial poles The points about which the celestial sphere appears to rotate. (Chapter 2) See also *north celestial pole; south celestial pole.*

celestial sphere An imaginary sphere of very large radius centered on an observer; the apparent sphere of the sky. (Chapter 2)

center of mass The point between a star and a planet, or between two stars, around which both objects orbit. (Chapter 7)

charge-coupled device (CCD) A type of solid-state device designed to detect photons. (Chapter 6)

chemical composition A description of which chemical substances make up a given object. (Chapter 7)

chemical differentiation The process by which the heavier elements in a planet sink toward its center while lighter elements rise toward its surface. (Chapter 7)

chemical element See *element.*

chondrule A glassy, roughly spherical blob found within meteorites. (Chapter 7)

chromatic aberration An optical defect whereby different colors of light passing through a lens are focused at different locations. (Chapter 6)

chromosphere A layer in the atmosphere of the Sun between the photosphere and the corona. (Chapter 18)

circumpolar A term describing a star that neither rises nor sets but appears to rotate around one of the celestial poles. (Chapter 2)

CNO cycle A series of nuclear reactions in which carbon is used as a catalyst to transform hydrogen into helium. (Chapter 18)

co-creation theory The hypothesis that the Earth and the Moon formed at the same time from the same material. (Chapter 9)

collisional ejection theory The hypothesis that the Moon formed from material ejected from the Earth by the impact of a large asteroid. (Chapter 9)

coma (of a comet) The diffuse gaseous component of the head of a comet. (Chapter 17)

coma (optical) The distortion of off-axis images formed by a parabolic mirror. (Chapter 6)

comet A small body of ice and dust in orbit about the Sun. While passing near the Sun, a comet's vaporized ices give rise to a coma and tail. (Chapter 7, Chapter 17)

condensation temperature The temperature at which a particular substance in a low-pressure gas condenses into a solid. (Chapter 7)

conduction The transfer of heat by directly passing energy from atom to atom. (Chapter 18)

conic section The curve of intersection between a circular cone and a plane; this curve can be a circle, ellipse, parabola, or hyperbola. (Chapter 4)

conjunction The geometric arrangement of a planet in the same part of the sky as the Sun, so that the planet is at an elongation of 0°. (Chapter 4)

conservation of angular momentum A law of physics stating that in an isolated system, the total amount of angular momentum—a measure of the amount of rotation—remains constant. (Chapter 7)

constellation A configuration of stars in the same region of the sky. (Chapter 2)

continuous spectrum A spectrum of light over a range of wavelengths without any spectral lines. (Chapter 5)

convection The transfer of energy by moving currents of fluid or gas containing that energy. (Chapter 8, Chapter 18)

convection cell A circulating loop of gas or liquid that transports heat from a warm region to a cool region. (Chapter 8)

convection current The pattern of motion in a gas or liquid in which convection is taking place. (Chapter 8)

convective zone The region in a star where convection is the dominant means of energy transport. (Chapter 18)

core (of the Earth) The iron-rich inner region of the Earth's interior. (Chapter 8)

corona (of the Sun) The Sun's outer atmosphere, which has a high temperature and a low density. (Chapter 18)

coronal hole A region in the Sun's corona that is deficient in hot gases. (Chapter 18)

coronal mass ejection An event in which billions of tons of gas from the Sun's corona is suddenly blasted into space at high speed. (Chapter 8, Chapter 18)

coudé focus An optical arrangement with a reflecting telescope. A series of mirrors is used to direct light to a remote focus away from the moving parts of the telescope. (Chapter 6)

crater A circular depression on a planet or satellite caused by the impact of a meteoroid. (Chapter 9)

crust (of a planet) The surface layer of a terrestrial planet. (Chapter 8)

crustal dichotomy (Mars) The contrast between the young northern lowlands and older southern highlands on Mars. (Chapter 12)

crystal A material in which atoms are arranged in orderly rows. (Chapter 8)

current sheet A broad, flat region in Jupiter's magnetosphere that contains an abundance of charged particles. (Chapter 13)

D ring One of several faint rings encircling Saturn. (Chapter 15)

dark terrain (Ganymede) Older, heavily cratered, dark-colored terrain on the surface of Ganymede. (Chapter 14)

decametric radiation Radiation from Jupiter whose wavelength is about 10 meters. (Chapter 13)

decimetric radiation Radiation from Jupiter whose wavelength is about a tenth of a meter. (Chapter 13)

declination Angular distance of a celestial object north or south of the celestial equator. (Chapter 2)

deferent A stationary circle in the Ptolemaic system along which another circle (an epicycle) moves, carrying a planet, the Sun, or the Moon. (Chapter 4)

degree A basic unit of angular measure, designated by the symbol °. (Chapter 1)

degree Celsius A basic unit of temperature, designated by the symbol °C and used on a scale where water freezes at 0° and boils at 100°. (Chapter 5)

degree Fahrenheit A basic unit of temperature, designated by the symbol °F and used on a scale where water freezes at 32° and boils at 212°. (Chapter 5)

density See *average density*.

differential rotation The rotation of a nonrigid object in which parts adjacent to each other at a given time do not always stay close together. (Chapter 13, Chapter 18)

differentiated asteroid An asteroid in which chemical differentiation has taken place, so that denser material is toward the asteroid's center. (Chapter 17)

differentiation See *chemical differentiation*.

diffraction The spreading out of light passing through an aperture or opening in an opaque object. (Chapter 6)

diffraction grating An optical device, consisting of thousands of closely spaced lines etched in glass or metal, that disperses light into a spectrum. (Chapter 6)

direct motion The apparent eastward movement of a planet seen against the background stars. (Chapter 4)

diurnal motion Any apparent motion in the sky that repeats on a daily basis, such as the rising and setting of stars. (Chapter 2)

Doppler effect The apparent change in wavelength of radiation due to relative motion between the source and the observer along the line of sight. (Chapter 5)

Drake equation An equation used to estimate the number of intelligent civilizations in the Galaxy with which we might communicate. (Chapter 30)

dust devil Whirlwind found in dry or desert areas on both Earth and Mars. (Chapter 12)

dust tail The tail of a comet that is composed primarily of dust grains. (Chapter 17)

E ring A very broad, faint ring encircling Saturn. (Chapter 15)

earthquake A sudden vibratory motion of the Earth's surface. (Chapter 8)

eccentricity A number between 0 and 1 that describes the shape of an ellipse. (Chapter 4)

eclipse The cutting off of part of or all the light from one celestial object by another. (Chapter 3)

eclipse path The track of the tip of the Moon's shadow along the Earth's surface during a total or annular solar eclipse. (Chapter 3)

eclipse year The interval between successive passages of the Sun through the same node of the Moon's orbit. (Chapter 3)

ecliptic The apparent annual path of the Sun on the celestial sphere. (Chapter 2)

electromagnetic radiation Radiation consisting of oscillating electric and magnetic fields. Examples include gamma rays, X rays, visible light, ultraviolet and infrared radiation, radio waves, and microwaves. (Chapter 5)

electromagnetic spectrum The entire array of electromagnetic radiation. (Chapter 5)

electromagnetism Electric and magnetic phenomena, including electromagnetic radiation. (Chapter 5)

electron A subatomic particle with a negative charge and a small mass, usually found in orbits about the nuclei of atoms. (Chapter 5)

electron volt (eV) The energy acquired by an electron accelerated through an electric potential of one volt. (Chapter 5)

element A chemical that cannot be broken down into more basic chemicals. (Chapter 5)

ellipse A conic section obtained by cutting completely through a circular cone with a plane. (Chapter 4)

elongation The angular distance between a planet and the Sun as viewed from Earth. (Chapter 4)

emission line spectrum A spectrum that contains bright emission lines. (Chapter 5)

Encke gap A narrow gap in Saturn's A ring. (Chapter 15)

energy flux The rate of energy flow, usually measured in joules per square meter per second. (Chapter 5)

energy level In an atom, a particular amount of energy possessed by an atom above the atom's least energetic state. (Chapter 5)

energy-level diagram A diagram showing the arrangement of an atom's energy levels. (Chapter 5)

epicenter The location on the Earth's surface directly over the focus of an earthquake. (Chapter 8)

epicycle A moving circle in the Ptolemaic system about which a planet revolves. (Chapter 4)

epoch The date used to define the coordinate system for objects on the sky. (Chapter 2)

equal areas, law of See *Kepler's second law.*

equilibrium resurfacing hypothesis (Venus) A proposed explanation for the young age of Venus's surface. It hypothesizes that volcanic eruptions on Venus are continually covering up craters at about the same rate that craters are formed by impact. (Chapter 11)

equinox One of the intersections of the ecliptic and the celestial equator. Also used to refer to the date on which the Sun passes through such an intersection. (Chapter 2) See also *autumnal equinox; vernal equinox.*

escape speed The speed needed by an object (such as a spaceship) to leave a second object (such as a planet or star) permanently and to escape into interplanetary space. (Chapter 7)

eV See *electron volt.*

excited state A state of an atom, ion, or molecule with a higher energy than the ground state. (Chapter 5)

exobiology The biology of life on other planets. (Chapter 30)

exponent A number placed above and after another number to denote the power to which the latter is to be raised, as n in 10^n. (Chapter 1)

extrasolar planet A planet orbiting a star other the Sun. (Chapter 7)

eyepiece lens A magnifying lens used to view the image produced at the focus of a telescope. (Chapter 6)

F ring A thin, faint ring encircling Saturn just beyond the A ring. (Chapter 15)

far side (of the Moon) The side of the Moon that faces perpetually away from the Earth. (Chapter 9)

favorable opposition An opposition of Mars that affords good Earth-based views of the planet. (Chapter 12)

filament A portion of the Sun's chromosphere that arches to high altitudes. (Chapter 18)

first quarter moon The phase of the Moon that occurs when the Moon is 90° east of the Sun. (Chapter 3)

fission theory The hypothesis that the Moon was pulled out of a rapidly rotating proto-Earth. (Chapter 9)

flare See *solar flare*.

focal length The distance from a lens or mirror to the point where converging light rays meet. (Chapter 6)

focal plane The plane in which a lens or mirror forms an image of a distant object. (Chapter 6)

focal point The point at which a lens or mirror forms an image of a distant point of light. (Chapter 6)

focus (of an ellipse) (plural foci) One of two points inside an ellipse such that the combined distance from the two foci to any point on the ellipse is a constant. (Chapter 4)

focus (of a lens or mirror) The point to which light rays converge after passing through a lens or being reflected from a mirror. (Chapter 6)

force A push or pull that acts on an object. (Chapter 4)

frequency The number of crests or troughs of a wave that cross a given point per unit time. Also, the number of vibrations per unit time. (Chapter 5)

full moon A phase of the Moon during which its full daylight hemisphere can be seen from Earth. (Chapter 3)

fusion crust The coating on a stony meteorite caused by the heating of the meteorite as it descended through the Earth's atmosphere. (Chapter 17)

G ring A thin, faint ring encircling Saturn. (Chapter 15)

galaxy A large assemblage of stars, nebulae, and interstellar gas and dust. (Chapter 1)

Galilean satellites The four large moons of Jupiter. (Chapter 14)

gamma rays The most energetic form of electromagnetic radiation. (Chapter 5)

geocentric model An Earth-centered theory of the universe. (Chapter 4)

global catastrophe hypothesis (Venus) A proposed explanation for the young age of Venus's surface. It hypothesizes that the entire planet was resurfaced over a short period with fresh lava, covering up any older craters. (Chapter 11)

global warming The upward trend of the Earth's average temperature caused by increased amounts of greenhouse gases in the atmosphere. (Chapter 8)

granulation The rice grain–like structure found in the solar photosphere. (Chapter 18)

granule A convective cell in the solar photosphere. (Chapter 18)

grating See *diffraction grating*.

gravitational force See *gravity*.

gravity The force with which all matter attracts all other matter. (Chapter 4)

Great Dark Spot A prominent high-pressure system that once existed in Neptune's southern hemisphere. (Chapter 16)

Great Red Spot A prominent high-pressure system in Jupiter's southern hemisphere. (Chapter 13)

greatest eastern elongation The configuration of an inferior planet at its greatest angular distance east of the Sun. (Chapter 4, Chapter 10)

greatest western elongation The configuration of an inferior planet at its greatest angular distance west of the Sun. (Chapter 4, Chapter 10)

greenhouse effect The trapping of infrared radiation near a planet's surface by the planet's atmosphere. (Chapter 8)

greenhouse gas A substance whose presence in a planet's atmosphere enhances the greenhouse effect. (Chapter 8)

ground state The state of an atom, ion, or molecule with the least possible energy. (Chapter 5)

heliocentric model A Sun-centered theory of the universe. (Chapter 4)

helioseismology The study of the vibrations of the Sun as a whole. (Chapter 18)

highlands (on Mars) See *southern highlands*.

highlands (on the Moon) See *lunar highlands*.

Hirayama family A group of asteroids with nearly identical orbits about the Sun. (Chapter 17)

hot spot (on Jupiter) An unusually warm and cloud-free part of Jupiter's atmosphere. (Chapter 13)

hot-spot volcanism Volcanic activity that occurs over a hot region buried deep within a planet. (Chapter 11)

hydrocarbon Any one of a variety of chemical compounds composed of hydrogen and carbon. (Chapter 15)

hydrogen burning The thermonuclear conversion of hydrogen into helium. (Chapter 18)

hydrogen envelope A huge, tenuous sphere of gas surrounding the head of a comet. (Chapter 17)

hydrostatic equilibrium A balance between the weight of a layer in a star and the pressure that supports it. (Chapter 18)

hyperbola A conic section formed by cutting a circular cone with a plane at an angle steeper than the side of the cone. (Chapter 4)

hypothesis An idea or collection of ideas that seems to explain a specified phenomenon; a conjecture. (Chapter 1)

ice rafts (Europa) Segments of Europa's icy crust that have been moved by tectonic disturbances. (Chapter 14)

ices Solid materials with low condensation temperatures, including ices of water, methane, and ammonia. (Chapter 7)

igneous rock A rock that formed from the solidification of molten lava or magma. (Chapter 8)

imaging The process of recording the image made by a telescope of a distant object. (Chapter 6)

impact breccia A type of rock formed from other rocks that were broken apart, mixed, and fused together by a series of meteoritic impacts. (Chapter 9)

impact crater A crater formed by the impact of a meteoroid. (Chapter 9)

inertia, law of See *Newton's first law of motion.*

inferior conjunction The configuration when an inferior planet is between the Sun and Earth. (Chapter 4)

inferior planet A planet that is closer to the Sun than the Earth is. (Chapter 4)

infrared radiation Electromagnetic radiation of wavelength longer than visible light but shorter than radio waves. (Chapter 5)

inner core (of the Earth) The solid innermost portion of the Earth's iron-rich core. (Chapter 8)

interferometry A technique of combining the observations of two or more telescopes to produce images better than one telescope alone could make. (Chapter 6)

intermediate-period comet A comet with an orbital period between 20 and 200 years. (Chapter 17)

internal rotation period The period with which the core of a Jovian planet rotates. (Chapter 13)

interstellar medium Gas and dust in interstellar space. (Chapter 7)

Io torus A doughnut-shaped ring of gas circling Jupiter at the distance of Io's orbit. (Chapter 14)

ion tail The relatively straight tail of a comet produced by the solar wind acting on ions. (Chapter 17)

ionization The process by which a neutral atom becomes an electrically charged ion through the loss or gain of electrons. (Chapter 5)

iron meteorite A meteorite composed primarily of iron. (Chapter 17)

irregular cluster (of galaxies) A sprawling collection of galaxies whose overall distribution in space does not exhibit any noticeable spherical symmetry. (Chapter 17)

isotope Any of several forms for the same chemical element whose nuclei all have the same number of protons but different numbers of neutrons. (Chapter 5)

jet An extended line of fast-moving gas ejected from the vicinity of a star or a black hole. (Chapter 7)

joule (J) A unit of energy. (Chapter 5)

Jovian planet Any of the four largest planets: Jupiter, Saturn, Uranus, or Neptune. (Chapter 7)

Jupiter-family comet A comet with an orbital period of less than 20 years. (Chapter 17)

kelvin (K) A unit of temperature on the Kelvin temperature scale, equivalent to a degree Celsius. (Chapter 5)

Kelvin-Helmholtz contraction The contraction of a gaseous body, such as a star or nebula, during which gravitational energy is transformed into thermal energy. (Chapter 7)

Kepler's first law The statement that each planet moves around the Sun in an elliptical orbit with the Sun at one focus of the ellipse. (Chapter 4)

Kepler's second law The statement that a planet sweeps out equal areas in equal times as it orbits the Sun; also called the law of equal areas. (Chapter 4)

Kepler's third law A relationship between the period of an orbiting object and the semimajor axis of its elliptical orbit. (Chapter 4)

kiloparsec (kpc) One thousand parsecs; about 3260 light-years. (Chapter 1)

kinetic energy The energy possessed by an object because of its motion. (Chapter 7)

Kirchhoff's laws Three statements about circumstances that produce absorption lines, emission lines, and continuous spectra. (Chapter 5)

Kirkwood gaps Gaps in the spacing of asteroid orbits, discovered by Daniel Kirkwood. (Chapter 17)

Kuiper belt A region that extends from around the orbit of Pluto to about 500 AU from the Sun, where many icy objects orbit the Sun. (Chapter 17)

last quarter moon The phase of the Moon that occurs when the Moon is 90° west of the Sun. (Chapter 3)

lava Molten rock flowing on the surface of a planet. (Chapter 8)

law of equal areas See *Kepler's second law.*

law of inertia See *Newton's first law of motion.*

law of universal gravitation A formula deduced by Isaac Newton that expresses the strength of the force of gravity that two masses exert on each other. (Chapter 4)

laws of physics   A set of physical principles with which we can understand natural phenomena and the nature of the universe. (Chapter 1)

libration   An apparent rocking of the Moon whereby an Earth-based observer can, over time, see slightly more than one-half the Moon's surface. (Chapter 9)

light-gathering power   A measure of the amount of radiation brought to a focus by a telescope. (Chapter 6)

light pollution   Light from cities and towns that degrades telescope images. (Chapter 6)

light scattering   The process by which light bounces off particles in its path. (Chapter 5, Chapter 15)

light-year (ly)   The distance light travels in a vacuum in one year. (Chapter 1)

limb darkening   The phenomenon whereby the Sun looks darker near its apparent edge, or limb, than near the center of its disk. (Chapter 18)

line of nodes   The line where the plane of the Earth's orbit intersects the plane of the Moon's orbit. (Chapter 3)

liquid metallic hydrogen   Hydrogen compressed to such a density that it behaves like a liquid metal. (Chapter 13)

lithosphere   The solid, upper layer of the Earth; essentially the Earth's crust. (Chapter 8)

local meridian   See *meridian.*

long-period comet   A comet that takes hundreds of thousands of years or more to complete one orbit of the Sun. (Chapter 17)

lower meridian   The half of the meridian that lies below the horizon. (Chapter 2)

lowlands (on Mars)   See *northern lowlands.*

luminosity   The rate at which electromagnetic radiation is emitted from a star or other object. (Chapter 5, Chapter 18)

lunar eclipse   An eclipse of the Moon by the Earth; a passage of the Moon through the Earth's shadow. (Chapter 3)

lunar highlands   Ancient, high-elevation, heavily cratered terrain on the Moon. (Chapter 9)

lunar month   See *synodic month.*

lunar phase   The appearance of the illuminated area of the Moon as seen from Earth. (Chapter 3)

Lyman series   A series of spectral lines of hydrogen produced by electron transitions to and from the lowest energy state of the hydrogen atom. (Chapter 5)

magma   Molten rock beneath a planet's surface. (Chapter 8)

magnetic axis   A line connecting the north and south magnetic poles of a planet or star possessing a magnetic field. (Chapter 16)

magnetic-dynamo model   A theory that explains the solar cycle as a result of the Sun's differential rotation acting on the Sun's magnetic field. (Chapter 18)

magnetogram   An artificial picture of the Sun that shows regions of different magnetic polarity. (Chapter 18)

magnetopause   That region of a planet's magnetosphere where the magnetic field counterbalances the pressure from the solar wind. (Chapter 8)

magnetosphere   The region around a planet occupied by its magnetic field. (Chapter 8)

magnification   The factor by which the apparent angular size of an object is increased when viewed through a telescope. (Chapter 6)

magnifying power   See *magnification.*

major axis (of an ellipse)   The longest diameter of an ellipse. (Chapter 4)

mantle (of a planet)   That portion of a terrestrial planet located between its crust and core. (Chapter 8)

mare (plural maria)   Latin for "sea"; a large, relatively crater-free plain on the Moon. (Chapter 9)

mare basalt   A type of lunar rock commonly found in the mare basins. (Chapter 9)

mass   A measure of the total amount of material in an object. (Chapter 4)

mean solar day   The interval between successive meridian passages of the mean Sun; the average length of a solar day. (Chapter 2)

mean Sun   A fictitious object that moves eastward at a constant speed along the celestial equator, completing one circuit of the sky with respect to the vernal equinox in one tropical year. (Chapter 2)

medium (plural media)   A material through which light travels. (Chapter 6)

medium, interstellar   See *interstellar medium.*

megaparsec (Mpc)   One million parsecs. (Chapter 1)

melting point   The temperature at which a substance changes from solid to liquid. (Chapter 8)

meridian (or local meridian)   The great circle on the celestial sphere that passes through an observer's zenith and the north and south celestial poles. (Chapter 2)

meridian transit   The crossing of the meridian by any astronomical object. (Chapter 2)

mesosphere   A layer in a planet's atmosphere above the stratosphere. (Chapter 8)

metamorphic rock   A rock whose properties and appearance have been transformed by the action of pressure and heat beneath the Earth's surface. (Chapter 8)

meteor   The luminous phenomenon seen when a meteoroid enters the Earth's atmosphere; a "shooting star." (Chapter 17)

meteor shower Many meteors that seem to radiate from a common point in the sky. (Chapter 17)

meteorite A fragment of a meteoroid that has survived passage through the Earth's atmosphere. (Chapter 7, Chapter 17)

meteoritic swarm A collection of meteoroids moving together along an orbit about the Sun. (Chapter 17)

meteoroid A small rock in interplanetary space. (Chapter 17)

microwaves Short-wavelength radio waves. (Chapter 5)

mineral A naturally occurring solid composed of a single element or chemical combination of elements, often in the form of crystals. (Chapter 8)

minor planet See *asteroid*.

minute of arc See *arcminute*.

model A hypothesis that has withstood experimental or observational tests; the results of a theoretical calculation that gives the values of temperature, pressure, density, and so forth throughout the interior of an object such as a planet or star. (Chapter 1)

molecule A combination of two or more atoms. (Chapter 7)

moonquake Sudden, vibratory motion of the Moon's surface. (Chapter 9)

nanometer (nm) One billionth of a meter: 1 nm = 10^{-9} meter = 10^{-6} millimeter = 10^{-3} μm. (Chapter 5)

neap tide An ocean tide that occurs when the Moon is near first-quarter or third-quarter phase. (Chapter 4)

near-Earth object (NEO) An asteroid whose orbit lies wholly or partly within the orbit of Mars. (Chapter 17)

nebula A cloud of interstellar gas and dust. (Chapter 1)

nebulosity See *nebula*.

negative hydrogen ion A hydrogen atom that has acquired a second electron. (Chapter 18)

NEO See *near-Earth object*.

neutrino A subatomic particle with no electric charge and little or no mass, yet one that is important in many nuclear reactions. (Chapter 18)

neutron A subatomic particle with no electric charge and with a mass nearly equal to that of the proton. (Chapter 5)

new moon The phase of the Moon when the dark hemisphere of the Moon faces the Earth. (Chapter 3)

Newtonian mechanics The branch of physics based on Newton's laws of motion. (Chapter 1, Chapter 4)

Newtonian reflector A reflecting telescope that uses a small mirror to deflect the image to one side of the telescope tube. (Chapter 6)

Newton's first law of motion The statement that a body remains at rest, or moves in a straight line at a constant speed, unless acted upon by a net outside force; the law of inertia. (Chapter 4)

Newton's form of Kepler's third law A relationship between the period of two objects orbiting each other, the semimajor axis of their orbit, and the masses of the objects. (Chapter 4)

Newton's second law of motion A relationship between the acceleration of an object, the object's mass, and the net outside force acting on the mass. (Chapter 4)

Newton's third law of motion The statement that whenever one body exerts a force on a second body, the second body exerts an equal and opposite force on the first body. (Chapter 4)

noble gas An element whose atoms do not combine into molecules. (Chapter 13)

node See *line of nodes*.

nonthermal radiation Radiation other than that emitted by a heated body. (Chapter 13)

north celestial pole The point directly above the Earth's North Pole where the Earth's axis of rotation, if extended, would intersect the celestial sphere. (Chapter 2)

northern lights See *aurora borealis*.

northern lowlands (on Mars) Relatively young and crater-free terrain in the Martian northern hemisphere. (Chapter 12)

nucleus (of an atom) The massive part of an atom, composed of protons and neutrons, about which electrons revolve. (Chapter 5)

nucleus (of a comet) A collection of ices and dust that constitutes the solid part of a comet. (Chapter 17)

objective lens The principal lens of a refracting telescope. (Chapter 6)

objective mirror The principal mirror of a reflecting telescope. (Chapter 6)

oblate Flattened at the poles. (Chapter 13)

oblateness A measure of how much a flattened sphere (or spheroid) differs from a perfect sphere. (Chapter 13)

Occam's razor The notion that a straightforward explanation of a phenomenon is more likely to be correct than a convoluted one. (Chapter 4)

occultation The eclipsing of an astronomical object by the Moon or a planet. (Chapter 14, Chapter 16)

oceanic rift A crack in the ocean floor that exudes lava. (Chapter 8)

1-to-1 spin-orbit coupling See *synchronous rotation*.

Oort cloud A presumed accumulation of comets and cometary material surrounding the Sun at distances of roughly 50,000 to 100,000 AU. (Chapter 17)

opposition The configuration of a planet when it is at an elongation of 180° and thus appears opposite the Sun in the sky. (Chapter 4)

optical telescope A telescope designed to detect visible light. (Chapter 6)

optical window The range of visible wavelengths to which the Earth's atmosphere is transparent. (Chapter 6)

organic molecules Molecules containing carbon, some of which are the molecules of which living organisms are made. (Chapter 30)

outer core (of the Earth) The outer, molten portion of the Earth's iron-rich core. (Chapter 8)

outgassing The release of gases into a planet's atmosphere by volcanic activity. (Chapter 8)

ozone A type of oxygen whose molecules contain three oxygen atoms. (Chapter 8)

ozone hole A region of the Earth's atmosphere over Antarctica where the concentration of ozone is abnormally low. (Chapter 8)

ozone layer A layer in the Earth's upper atmosphere where the concentration of ozone is high enough to prevent much ultraviolet light from reaching the surface. (Chapter 8)

P wave One of three kinds of seismic waves produced by an earthquake; a primary wave. (Chapter 8)

parabola A conic section formed by cutting a circular cone at an angle parallel to the side of the cone. (Chapter 4)

parallax The apparent displacement of an object due to the motion of the observer. (Chapter 4)

parsec (pc) A unit of distance; 3.26 light-years. (Chapter 1)

partial lunar eclipse A lunar eclipse in which the Moon does not appear completely covered. (Chapter 3)

partial solar eclipse A solar eclipse in which the Sun does not appear completely covered. (Chapter 3)

Paschen series A series of spectral lines of hydrogen produced by electron transitions between the third and higher energy levels. (Chapter 5)

penumbra (of a shadow) (plural **penumbrae**) The portion of a shadow in which only part of the light source is covered by an opaque body. (Chapter 3)

penumbral eclipse A lunar eclipse in which the Moon passes only through the Earth's penumbra. (Chapter 3)

perigee The point in its orbit where a satellite or the Moon is nearest the Earth. (Chapter 3)

perihelion The point in its orbit where a planet or comet is nearest the Sun. (Chapter 4)

period (of a planet) The interval of time between successive geometric arrangements of a planet and an astronomical object, such as the Sun. (Chapter 4)

periodic table A listing of the chemical elements according to their properties, invented by Dmitri Mendeleev. (Chapter 5)

permafrost Frozen soil that lies just underneath the surface. (Chapter 12)

photoelectric effect The phenomenon whereby certain metals emit electrons when exposed to short-wavelength light. (Chapter 5)

photometry The measurement of light intensities. (Chapter 6)

photon A discrete unit of electromagnetic energy. (Chapter 5)

photosphere The region in the solar atmosphere from which most of the visible light escapes into space. (Chapter 18)

photosynthesis A biochemical process in which solar energy is converted into chemical energy, carbon dioxide and water are absorbed, and oxygen is released. (Chapter 8)

physics, laws of See *laws of physics.*

pixel A picture element. (Chapter 6)

plage A bright region in the solar atmosphere as observed in the monochromatic light of a spectral line. (Chapter 18)

Planck's law The relationship between the energy of a photon and its wavelength or frequency; $E = hc/\lambda = h\nu$. (Chapter 5)

plane of the ecliptic The plane in which the Earth orbits the Sun. (Chapter 3)

planetesimal A small body of primordial dust and ice from which the planets formed. (Chapter 7)

plasma A hot ionized gas. (Chapter 13, Chapter 18)

plastic The attribute of being nearly solid yet able to flow. (Chapter 8)

plate A large section of the Earth's lithosphere that moves as a single unit. (Chapter 8)

plate tectonics The motions of large segments (plates) of the Earth's surface over the underlying mantle. (Chapter 8)

polymer A long molecule consisting of many smaller molecules joined together. (Chapter 15)

positional astronomy The study of the apparent positions of the planets and stars and how those positions change. (Chapter 2)

positron An electron with a positive rather than negative electric charge; the antiparticle of the electron. (Chapter 18)

power of ten The exponent n in 10^n. (Chapter 1)

powers-of-ten notation A shorthand method of writing numbers, involving 10 followed by an exponent. (Chapter 1)

precession (of the Earth) A slow, conical motion of the Earth's axis of rotation caused by the gravitational pull of the Moon and Sun on the Earth's equatorial bulge. (Chapter 2)

precession of the equinoxes The slow westward motion of the equinoxes along the ecliptic due to precession of the Earth. (Chapter 2)

primary mirror See *objective mirror.*

prime focus The point in a telescope where the objective focuses light. (Chapter 6)

primitive asteroid See *undifferentiated asteroid.*

prograde orbit An orbit of a satellite around a planet that is in the same direction as the rotation of the planet. (Chapter 14)

prograde rotation A situation in which an object (such as a planet) rotates in the same direction that it orbits around another object (such as the Sun). (Chapter 11)

prominence Flamelike protrusions seen near the limb of the Sun and extending into the solar corona. (Chapter 18)

proplyd See *protoplanetary disk.*

proton A heavy, positively charged subatomic particle that is one of two principal constituents of atomic nuclei. (Chapter 5)

proton-proton chain A sequence of thermonuclear reactions by which hydrogen nuclei are built up into helium nuclei. (Chapter 18)

protoplanet A Moon-sized object formed by the coalescence of planetesimals. (Chapter 7)

protoplanetary disk (proplyd) A disk of material encircling a protostar or a newborn star. (Chapter 7)

protosun The part of the solar nebula that eventually developed into the Sun. (Chapter 7)

Ptolemaic system The definitive version of the geocentric cosmology of ancient Greece. (Chapter 4)

pulsar A pulsating radio source thought to be associated with a rapidly rotating neutron star. (Chapter 1)

quantum mechanics The branch of physics dealing with the structure and behavior of atoms and their constituents as well as their interaction with light. (Chapter 5)

quasar A starlike object with a very large redshift. (Chapter 1)

radial velocity That portion of an object's velocity parallel to the line of sight. (Chapter 5)

radial velocity method A technique used to detect extrasolar planets. (Chapter 7)

radiant (of a meteor shower) The point in the sky from which meteors of a particular shower seem to originate. (Chapter 17)

radiation darkening The darkening of methane ice by electron impacts. (Chapter 16)

radiation pressure Pressure exerted on an object by radiation falling on the object. (Chapter 17)

radiative diffusion The random migration of photons from a star's center toward its surface. (Chapter 18)

radiative zone A region within a star where radiative diffusion is the dominant mode of energy transport. (Chapter 18)

radio telescope A telescope designed to detect radio waves. (Chapter 6)

radio waves The longest-wavelength electromagnetic radiation. (Chapter 5)

radio window The range of radio wavelengths to which the Earth's atmosphere is transparent. (Chapter 6)

redshift The Doppler shift of light from a receding source. (Chapter 5)

reflecting telescope A telescope in which the principal optical component is a concave mirror. (Chapter 6)

reflection The return of light rays by a surface. (Chapter 6)

reflector A reflecting telescope. (Chapter 6)

refracting telescope A telescope in which the principal optical component is a lens. (Chapter 6)

refraction The bending of light rays when they pass from one transparent medium to another. (Chapter 6)

refractor A refracting telescope. (Chapter 6)

refractory element An element with high melting and boiling points. (Chapter 9)

regolith The layer of rock fragments covering the surface of the Moon. (Chapter 9)

residual polar cap An ice-covered polar region on Mars that does not completely evaporate during the Martian summer. (Chapter 12)

respiration A biological process that produces energy by consuming oxygen and releasing carbon dioxide. (Chapter 8)

retrograde motion The apparent westward motion of a planet with respect to background stars. (Chapter 4)

retrograde orbit An orbit of a satellite around a planet that is in the direction opposite to which the planet rotates. (Chapter 14)

retrograde rotation A situation in which an object (such as a planet) rotates in the direction opposite to which it orbits around another object (such as the Sun). (Chapter 11)

rift valley A feature created when a planet's crust breaks apart along a line. (Chapter 12)

right ascension A coordinate for measuring the east-west positions of objects on the celestial sphere. (Chapter 2)

ring particles Small particles that constitute a planetary ring. (Chapter 15)

ringlet One of many narrow bands of particles of which Saturn's ring system is composed. (Chapter 15)

Roche limit The smallest distance from a planet or other object at which a second object can be held together by purely gravitational forces. (Chapter 15)

rock A mineral or combination of minerals. (Chapter 8)

runaway greenhouse effect A greenhouse effect in which the temperature continues to increase. (Chapter 11)

runaway icehouse effect A situation in which a decrease in atmospheric temperature causes a further decrease in temperature. (Chapter 12)

S wave One of three kinds of seismic waves produced by an earthquake; a secondary wave. (Chapter 8)

saros A particular cycle of similar eclipses that recur about every 18 years. (Chapter 3)

scarp A line of cliffs formed by the faulting or fracturing of a planet's surface. (Chapter 10)

scattering of light See *light scattering*.

scientific method The basic procedure used by scientists to investigate phenomena. (Chapter 1)

seafloor spreading The separation of plates under the ocean due to lava emerging in an oceanic rift. (Chapter 8)

search for extraterrestrial intelligence (SETI) The scientific search for evidence of intelligent life on other planets. (Chapter 30)

second of arc See *arcsecond*.

sedimentary rock A rock that is formed from material deposited on land by rain or winds, or on the ocean floor. (Chapter 8)

seeing disk The angular diameter of a star's image. (Chapter 6)

seismic wave A vibration traveling through a terrestrial planet, usually associated with earthquake-like phenomena. (Chapter 8)

seismograph A device used to record and measure seismic waves, such as those produced by earthquakes. (Chapter 8)

semimajor axis One-half of the major axis of an ellipse. (Chapter 4)

SETI See *search for extraterrestrial intelligence*.

shepherd satellite A satellite whose gravity restricts the motions of particles in a planetary ring, preventing them from dispersing. (Chapter 15)

shield volcano A volcano with long, gently sloping sides. (Chapter 11)

shock wave An abrupt, localized region of compressed gas caused by an object traveling through the gas at a speed greater than the speed of sound. (Chapter 8)

SI units The International System of Units (*Système international des unités*), based on the meter (m), the second (s), and the kilogram (kg). (Chapter 1)

sidereal clock A clock that measures sidereal time. (Chapter 2)

sidereal day The interval between successive meridian passages of the vernal equinox. (Chapter 2)

sidereal month The period of the Moon's revolution about the Earth with respect to the stars. (Chapter 3)

sidereal period The orbital period of one object about another as measured with respect to the stars. (Chapter 4)

sidereal time Time reckoned by the location of the vernal equinox. (Chapter 2)

sidereal year The orbital period of the Earth about the Sun with respect to the stars. (Chapter 2)

small-angle formula A relationship between the angular and linear sizes of a distant object. (Chapter 1)

SNC meteorite A meteorite that came to Earth from Mars. (Chapter 30)

solar constant The average amount of energy received from the Sun per square meter per second, measured just above the Earth's atmosphere. (Chapter 5)

solar corona Hot, faintly glowing gases seen around the Sun during a total solar eclipse; the uppermost regions of the solar atmosphere. (Chapter 3)

solar cycle See *22-year solar cycle*.

solar eclipse An eclipse of the Sun by the Moon; a passage of the Earth through the Moon's shadow. (Chapter 3)

solar flare A sudden, temporary outburst of light from an extended region of the solar surface. (Chapter 18)

solar nebula The cloud of gas and dust from which the Sun and solar system formed. (Chapter 7)

solar neutrino A neutrino emitted from the core of the Sun. (Chapter 18)

solar neutrino problem The discrepancy between the predicted and observed numbers of solar neutrinos. (Chapter 18)

solar system The Sun, planets and their satellites, asteroids, comets, and related objects that orbit the Sun. (Chapter 1)

solar transit The passage of an object in front of the Sun. (Chapter 10)

solar wind An outward flow of particles (mostly electrons and protons) from the Sun. (Chapter 7, Chapter 18)

south celestial pole The point directly above the Earth's South Pole where the Earth's axis of rotation, if extended, would intersect the celestial sphere. (Chapter 2)

southern highlands (on Mars) Older, cratered terrain in the Martian southern hemisphere. (Chapter 12)

southern lights See *aurora australis*.

spectral analysis The identification of chemical substances from the patterns of lines in their spectra. (Chapter 5)

spectral line In a spectrum, an absorption or emission feature that is at a particular wavelength. (Chapter 5)

spectrograph An instrument for photographing a spectrum. (Chapter 6)

spectroscopy The study of spectra and spectral lines. (Chapter 5, Chapter 6, Chapter 7)

spectrum (plural spectra) The result of dispersing a beam of electromagnetic radiation so that components with different wavelengths are separated in space. (Chapter 5)

speed Distance traveled divided by the time elapsed to cover that distance. (Chapter 4)

spherical aberration The distortion of an image formed by a telescope due to differing focal lengths of the optical system. (Chapter 6)

spicule A narrow jet of rising gas in the solar chromosphere. (Chapter 18)

spin-orbit coupling See *1-to-1 spin-orbit coupling*; *3-to-2 spin-orbit coupling*.

spring tide An ocean tide that occurs at new moon and full moon phases. (Chapter 4)

stable Lagrange points Locations along Jupiter's orbit where the combined gravitational effects of the Sun and Jupiter cause asteroids to collect. (Chapter 17)

Stefan-Boltzmann law A relationship between the temperature of a blackbody and the rate at which it radiates energy. (Chapter 5)

stony iron meteorite A meteorite composed of both stone and iron. (Chapter 17)

stony meteorite A meteorite composed of stone. (Chapter 17)

stratosphere A layer in the atmosphere of a planet directly above the troposphere. (Chapter 8)

subduction zone A location where colliding tectonic plates cause the Earth's crust to be pulled down into the mantle. (Chapter 8)

subtend To extend over an angle. (Chapter 1)

summer solstice The point on the ecliptic where the Sun is farthest north of the celestial equator. Also used to refer to the date on which the Sun passes through this point. (Chapter 2)

sunspot A temporary cool region in the solar photosphere. (Chapter 18)

sunspot cycle The semiregular 11-year period with which the number of sunspots fluctuates. (Chapter 18)

sunspot maximum/minimum That time during the sunspot cycle when the number of sunspots is highest/lowest. (Chapter 18)

supergranule A large convective feature in the solar atmosphere, usually outlined by spicules. (Chapter 18)

superior conjunction The configuration of a planet being behind the Sun as viewed from the Earth. (Chapter 4)

superior planet A planet that is more distant from the Sun than the Earth is. (Chapter 4)

supernova (plural supernovae) A stellar outburst during which a star suddenly increases its brightness roughly a millionfold. (Chapter 1, Chapter 17)

surface wave A type of seismic wave that travels only over the Earth's surface. (Chapter 8)

synchrotron radiation A type of nonthermal radiation emitted by charged particles moving through a magnetic field. (Chapter 13)

synchronous rotation The rotation of a body with a period equal to its orbital period; also called 1-to-1 spin-orbit coupling. (Chapter 3, Chapter 9)

synodic month The period of revolution of the Moon with respect to the Sun; the length of one cycle of lunar phases. Also called the lunar month. (Chapter 3)

synodic period The interval between successive occurrences of the same configuration of a planet. (Chapter 4)

T Tauri wind A flow of particles away from a young star. (Chapter 7)

tail (of a comet) Gas and dust particles from a comet's nucleus that have been swept away from the comet's head by the radiation pressure of sunlight and the solar wind. (Chapter 17)

temperature See *degree Celsius*; *degree Fahrenheit*; *kelvin*.

terminator The line dividing day and night on the surface of the Moon or a planet; the line of sunset or sunrise. (Chapter 9)

terrae Cratered lunar highlands. (Chapter 9)

terrestrial planet Mercury, Venus, Earth, or Mars; the classification sometimes also includes the Galilean satellites and Pluto. (Chapter 7)

theory A hypothesis that has withstood experimental or observational tests. (Chapter 1)

thermal equilibrium A balance between the input and outflow of heat in a system. (Chapter 18)

thermal radiation The radiation naturally emitted by any object that is not at absolute zero. Blackbody radiation is an idealized case of thermal radiation. (Chapter 13)

thermonuclear fusion The combining of nuclei under conditions of high temperature in a process that releases substantial energy. (Chapter 18)

thermosphere A region in the Earth's atmosphere between the mesosphere and the exosphere. (Chapter 8)

3-to-2 spin-orbit coupling The rotation of Mercury, which makes three complete rotations on its axis for every two complete orbits around the Sun. (Chapter 10)

tidal force A gravitational force whose strength and/or direction varies over a body and thus tends to deform the body. (Chapter 4, Chapter 15)

tidal heating The heating of the interior of a satellite by continually varying tidal stresses. (Chapter 14)

time zone A region on the Earth where, by agreement, all clocks have the same time. (Chapter 2)

total lunar eclipse A lunar eclipse during which the Moon is completely immersed in the Earth's umbra. (Chapter 3)

total solar eclipse A solar eclipse during which the Sun is completely hidden by the Moon. (Chapter 3)

totality (lunar eclipse) The period during a total lunar eclipse when the Moon is entirely within the Earth's umbra. (Chapter 3)

totality (solar eclipse) The period during a total solar eclipse when the disk of the Sun is completely hidden. (Chapter 3)

transit See *meridian transit; solar transit.*

Trojan asteroid One of several asteroids that share Jupiter's orbit about the Sun. (Chapter 17)

Tropic of Cancer A circle of latitude 23½° north of the Earth's equator. (Chapter 2)

Tropic of Capricorn A circle of latitude 23½° south of the Earth's equator. (Chapter 2)

tropical year The period of revolution of the Earth about the Sun with respect to the vernal equinox. (Chapter 2)

troposphere The lowest level in the Earth's atmosphere. (Chapter 8)

22-year solar cycle The semiregular 22-year interval between successive appearances of sunspots at the same latitude and with the same magnetic polarity. (Chapter 18)

ultramafic lava A type of lava enriched in magnesium and iron. These give the lava a higher melting temperature. (Chapter 14)

ultraviolet radiation Electromagnetic radiation of wavelengths shorter than those of visible light but longer than those of X rays. (Chapter 5)

umbra (of a shadow) (plural **umbrae**) The central, completely dark portion of a shadow. (Chapter 3)

undifferentiated asteroid An asteroid within which chemical differentiation did not occur. (Chapter 17)

universal constant of gravitation (G) The constant of proportionality in Newton's law of gravitation. (Chapter 4)

upper meridian The half of the meridian that lies above the horizon. (Chapter 2)

Van Allen belts Two doughnut-shaped regions around the Earth where many charged particles (protons and electrons) are trapped by the Earth's magnetic field. (Chapter 8)

velocity The speed and direction of an object's motion. (Chapter 4)

vernal equinox The point on the ecliptic where the Sun crosses the celestial equator from south to north. Also used to refer to the date on which the Sun passes through this intersection. (Chapter 2)

very-long-baseline interferometry (VLBI) A method of connecting widely separated radio telescopes to make very high-resolution observations. (Chapter 6)

visible light Photons detectable by the human eye. (Chapter 5)

VLBI See *very-long-baseline interferometry.*

volatile element An element with low melting and boiling points. (Chapter 9)

waning crescent moon The phase of the Moon that occurs between third quarter and new moon. (Chapter 3)

waning gibbous moon The phase of the Moon that occurs between full moon and third quarter. (Chapter 3)

water hole A region in the microwave spectrum suitable for interstellar radio communication. (Chapter 30)

watt A unit of power, equal to one joule of energy per second. (Chapter 5)

wavelength The distance between two successive wave crests. (Chapter 5)

wavelength of maximum emission The wavelength at which a heated object emits the greatest intensity of radiation. (Chapter 5)

waxing crescent moon The phase of the Moon that occurs between new moon and first quarter. (Chapter 3)

waxing gibbous moon The phase of the Moon that occurs between first quarter and full moon. (Chapter 3)

weight The force with which gravity acts on a body. (Chapter 4)

white ovals Round, whitish feature usually seen in Jupiter's southern hemisphere. (Chapter 13)

Widmanstätten patterns Crystalline structure seen in certain types of meteorites. (Chapter 17)

Wien's law A relationship between the temperature of a blackbody and the wavelength at which it emits the greatest intensity of radiation. (Chapter 5)

winter solstice The point on the ecliptic where the Sun reaches its greatest distance south of the celestial equator. Also used to refer to the date on which the Sun passes through this point. (Chapter 2)

X-ray burster A nonperiodic X-ray source that emits powerful bursts of X rays. (Chapter 1)

X rays Electromagnetic radiation whose wavelength is between that of ultraviolet light and gamma rays. (Chapter 5)

Zeeman effect A splitting or broadening of spectral lines due to a magnetic field. (Chapter 18)

zenith The point on the celestial sphere directly overhead an observer. (Chapter 2)

zodiac A band of 12 constellations around the sky centered on the ecliptic. (Chapter 2)

zonal winds The pattern of alternating eastward and westward winds found in the atmospheres of Jupiter and Saturn. (Chapter 13)

zone A light-colored band in Jupiter's atmosphere. (Chapter 13)

Answers to Selected Questions

Quiz: How to Get the Most from UNIVERSE
1. Paragraphs labeled by the "caution" icon **2.** Paragraphs labeled by the "analogy" icon **3.** These indicate references to sections, figures, or tables in previous chapters **4.** "Tools of the Astronomer's Trade," "The Heavens on the Earth" **5.** On the *Universe* web site; in addition, all but the "Web Links" can be found on the CD-ROM that accompanies this book **6.** A planetarium program; on the CD-ROM that accompanies this book; it comes with the book at no additional cost **7.** It shows whether the image was made with **R**adio waves, **I**nfrared radiation, **V**isible light, **U**ltraviolet light, **X** rays, or **G**amma rays **8.** On the *Universe* web site or on the CD-ROM that accompanies this book **9.** (a) page A-6 (b) page A-1 (c) page A-3 (d) page A-4 **10.** (a) page 403 (b) page 123 (c) page 84 (d) page 75 **11.** 656.9 nm **12.** At the back of the book, following the Index

 Only mathematical answers are given in the following, not answers that require interpretation or discussion. Your instructor will expect you to show the steps required to calculate each mathematical answer.

Part I
Chapter 1
19. About 3×10^{36} times larger **20.** 8.94×10^{56} hydrogen atoms **21.** 8.7×10^3 km **22.** 2.9×10^7 Suns **23.** (a) 1.58×10^{-5} ly (b) 4.84×10^{-6} pc **24.** 4.99×10^2 s **25.** 4.3×10^9 km **26.** (a) 9.95×10^{15} km (b) 10.5 years **27.** 4.7×10^{17} s **29.** (a) 1.1 m (b) 69 m (c) 4100 m **30.** 3.4×10^3 km **31.** 3.7 km **32.** 6.9 m **33.** 0.292 arcmin

Chapter 2
23. Around 11:28 P.M. **28.** (b) About 10 hours **32.** 26½° **36.** October 25, 1917 **39.** 50° **41.** (a) 6:00 P.M. (b) September 21 **42.** 3:50 A.M. local time

Chapter 3
23. (a) 0.91 hour (b) 11° **24.** 49 arcsec **35.** (a) July 2, 2019; southern Pacific and South America (b) July 24, 2055

Chapter 4
15. Semimajor axis = 0.25 AU, period = 0.125 year **16.** Average = 25 AU, farthest = 50 AU **18.** 12 N, 6 m/s² **21.** 16 times stronger **22.** Earth exerts 1.98×10^{20} N on the Moon **29.** 87.97 days **33.** (a) 4 AU (b) 8 years **34.** (a) 9.0 AU (b) 0.5 AU **35.** (a) 0.86 AU (b) 146 days **36.** ¹/₃ as much as on Earth **37.** Forces are approximately the same, Earth's acceleration is 100 times greater **38.** 67 newtons, 0.14 **39.** (a) 24 hours (b) 42,300 km **40.** 0.5 year **41.** 119 minutes **43.** (a) 5.00 years (b) 2.92 AU **44.** (a) 3.43×10^{-5} N (b) 3.21×10^{-5} N (c) 2.2×10^{-6} N

Chapter 5
1. 1.28 s **5.** 2.83×10^{14} Hz **6.** 0.336 m **11.** 753 nm **12.** 24,000 K **20.** 9980°F **21.** 9400 nm **22.** About 10 µm **23.** 5.1 times as much **24.** (a) 4890 nm = 4.89 µm (b) 540 times as much **25.** (a) 5.74×10^8 W/m² (b) 10,000 K **26.** 2.9 nm **28.** 2.43×10^{-3} nm **29.** (a) 1005 nm **30.** 372.2 nm **33.** Possible transitions: energy = 1 eV, λ = 1240 nm; energy = 2 eV, λ = 620 nm; energy = 3 eV, λ = 414 nm **34.** Coming toward us at 13.0 km/s **35.** 656.9 nm **36.** 8,600 km/s

Chapter 6
21. ¹/₂₅ = 0.04 **22.** Subaru has 12 times the light-gathering power **23.** (a) 222× (b) 100× (c) 36×; angular resolution = 0.75 arcsec **25.** 300 km (Hubble Space Telescope, Jupiter's moons); 110 km (human eye, our Moon) **26.** (a) 34 ly (b) 37 km **28.** (a) 0.18 m (b) 5.9×10^{-4} arcsec **29.** (a) 5.39×10^{-4} m (c) 0.56 m

Part II
Chapter 7
19. Mass = 6.4×10^{23} kg, average density = 3900 kg/m³ **22.** (a) 3.3×10^{21} J (b) Equivalent to 40 million Hiroshima-type weapons **23.** (a) 1.2 m/s **24.** 12 km/s **25.** (a) 617 km/s **26.** (a) 2.02 km/s (b) 19.4 km/s **27.** (a) About 50 times Earth's actual mass **29.** (a) 1.76 years **31.** (a) About 180 AU **32.** 2.9×10^7 AU = 140 parsecs = 460 light-years **33.** (a) About 600 AU (b) About 5×10^{40} cubic meters (c) About 10^{55} atoms (d) About 3×10^{14} atoms per cubic meter **35.** 2.03×10^{30} kg = 1.02 times the mass of the Sun **36.** (a) 12 m/s (b) 1.3×10^{-3} arcsec (c) 9.0×10^{-5} arcsec

Chapter 8
20. (a) 6.8×10^{16} W (b) 1.07×10^{17} W (c) 209 W (d) 246 K = –27°C **21.** 4 km **22.** Core: 17%; mantle: 82%; crust: 1% **23.** 0.020 (2% of the total mass) **24.** About 15,000 kg/m³ **25.** 2.2×10^8 years ago (based on the present-day separation of 6600 km) **29.** (a) 2.225 eV **33.** Pressure is 0.001 of sea-level value at 55 km altitude

Chapter 9
20. 130 N on the Moon, 780 N on Earth **21.** (a) 5.78×10^{-5} newton (b) 4.15×10^{-5} newton (c) 1.39 **26.** In 100 years, probability is 8.3×10^{-8} (one chance in 12 million); in 10^6 years, probability is 8.3×10^{-4} (one chance in 1200) **27.** 2.51 s **28.** 5.5×10^5 km **30.** (b) 10^6 times greater

Chapter 10
18. 730 km **21.** For T = 430°C, λ_{max} = 4.1 µm **23.** (a) 3.03 m/s (b) 1.26 nm **24.** (a) About 30 **25.** 300 N on Mercury, 130 N on the Moon, 780 N on Earth

Chapter 11

24. 0.23 nm **25.** For $T = 460°C$, $\lambda_{max} = 4.0$ µm **32.** 920 m

Chapter 12

25. Earth completes 2.14 orbits, Mars completes 1.14 orbits **27.** (a) 0.16 AU **28.** (a) 270 km (b) 27 km **29.** 2.0 cm **32.** (a) 3770 km (b) 370 km **46.** (a) Radius = 20,400 km, altitude = 17,000 km

Chapter 13

24. 12.7 km/s **26.** 8.5×10^{53} hydrogen atoms, 7.1×10^{52} helium atoms **27.** Roughly 600 km/h **28.** 127 K **29.** (a) About 160 K = $-113°C$ = $-171°F$ (b) About 6 atm (c) About 50 m **30.** 59.5 km/s, or 5.3 times the escape speed from the Earth **31.** 8300 newtons **32.** 71,000 kg/m³

Chapter 14

28. 760 km **29.** 0.046 arcsec **32.** 2.8×10^{11} (280 billion) years **33.** About 8.1 minutes **34.** (b) 159,000 km

Chapter 15

21. (a) 14.4 hours for inner edge of A ring, 7.9 hours for inner edge of B ring **25.** Escape speed = 2.6 km/s; mass of molecule = 2.0×10^{-26} kg, corresponding to a molecular weight of 12 **28.** 0.09 arcsec **29.** (a) 65.7 hours (b) 27.6 arcsec

Chapter 16

23. Sun-Uranus force = 1.39×10^{21} N, Neptune-Uranus force = 2.24×10^{17} N; Neptune reduces the sunward gravitational pull on Uranus by 1.61×10^{-4}, or 0.0161% **24.** (a) 2900 kg/m³ **28.** (a) 1.13×10^{-3} as bright as on Earth (b) 4.10×10^{-4} as bright as on Earth (c) 2.77 times brighter **29.** About 7200 km **31.** 0.95 arcsec

Chapter 17

25. 2.82 AU **26.** 30 km **28.** (a) 3.6×10^{12} kg (b) 0.83 m/s **30.** About 4% **31.** (a) Period = 350 years, lifetime = 3.5×10^4 years (b) Period = 1.1×10^4 years, lifetime = 1.1×10^6 years (c) Period = 3.5×10^5 years, lifetime = 3.5×10^7 years (d) Period = 1.1×10^7 years, lifetime = 1.1×10^9 years **34.** (a) 10^{15} kg; (b) 10^{-16} kg/m³

Part III

Chapter 18

24. (a) 1.8×10^{-9} J (b) 9.0×10^{16} J (c) 5.4×10^{41} J **25.** (a) 4.6×10^{-36} s (b) 2.3×10^{-10} s (c) 1.4×10^{15} s = 4.4×10^7 years **26.** 0.048 (4.8%) of the Sun's mass will be converted from hydrogen to helium; chemical composition of Sun (by mass) will be 69% hydrogen, 30% helium **27.** (a) 8.89×10^{28} kg of hydrogen consumed, mass lost = 6.23×10^{26} kg **28.** (a) 1.64×10^{-13} J (b) 2.43×10^{-3} nm **29.** 1.4×10^{13} kilograms per second **33.** 98,600 kg/m³ **34.** 1.9×10^{-7} nm **35.** (a) 1700 nm **37.** For the photosphere, 500 nm; for the chromosphere, 58 nm; for the corona, 1.9 nm **39.** For the umbra, 670 nm; for the penumbra, 580 nm **40.** (a) (Flux from patch of penumbra)/(flux from patch of photosphere) = 0.55 (b) (Flux from patch of penumbra)/(flux from patch of umbra) = 1.8

Chapter 19

24. Average Sun-Pluto distance = 1.92×10^{-4} pc; Proxima Centauri is 6780 times farther away **25.** (a) 9.7 pc (b) 0.10 arcsec **26.** 3.92 pc **27.** (a) 161 km/s (b) 294 km/s **28.** Distance = 105 pc; tangential velocity would have to be about 4340 km/s **29.** 110 km/s **30.** (a) +59.4 km/s (c) 486.23 nm **32.** The Sun is 1.11×10^{-3} as bright on Neptune as on Earth **33.** Star A is 16 times more luminous than star B **34.** Star D appears 9 times brighter than star C **35.** The farther star is 144 times more luminous than the closer star **36.** 37.0 AU **37.** 6.8 $L_\odot$ **38.** 0.38 pc = 7.9×10^4 AU **39.** (a) +13.8 (b) (Luminosity of HIP 72509)/(luminosity of Sun) = 2.4×10^{-4} **40.** +17 **41.** 6300 pc **42.** (a) Brightest: $M = -2.37$; dimmest: $M = +4.32$ (b) $M = +0.79$ **45.** (b) $m_B - m_V = -0.23$ (Bellatrix), +0.68 (Sun), +1.86 (Betelgeuse) **46.** 99 $R_\odot$ **47.** Radius increases by a factor of 2, luminosity increases by a factor of 64 **50.** $T = 10,000$ K, $R = 3.8$ $R_\odot$ **51.** $L = 35$ $L_\odot$, so distance = 14 pc **52.** (b) 0.26 $R_\odot$ = 1.8×10^5 km **53.** (a) 5.0 pc (b) 22.5 AU (c) 1.5 $M_\odot$ **54.** (a) 40 $M_\odot$ (b) $M_1 = 32$ $M_\odot$, $M_2 = 8$ $M_\odot$ **56.** 2500 pc, 125 times greater volume

Chapter 20

25. 0.34% **26.** 3.4×10^4 atoms per cm³ **30.** 1100 $R_\odot$ = 7.4×10^8 km = 4.9 AU **31.** (a) 250 AU = 3.7×10^{10} km (b) 4000 years (c) 1000 AU; 24 years **32.** 1.3×10^{31} m³ **33.** (a) 2.9° (b) 1300 km/s (c) 180 years

Chapter 21

26. 657,000 km **27.** (a) 618 km/s (b) 61.8 km/s **28.** (a) 12.0 km/s (b) 9.3 km/s **29.** 1.9×10^{21} kg, or 12% of the original mass of hydrogen **30.** 4.1×10^7 years (b) 3.2×10^{11} years **32.** About 1900 K (1600°C, or 2900°F) **33.** 3.4×10^7 years **36.** About 650 pc **37.** About 370 pc **39.** (a) 1.65×10^6 km

Chapter 22

22. 0.11 $R_\odot$ **24.** 7000 years ago **25.** (a) 97 nm **26.** (a) 1.8×10^9 kg/m³ (b) 6400 km/s **28.** (a) 1.7×10^{30} kg = 0.84 $M_\odot$ (b) 1.1×10^{12} newtons (c) 1.5×10^8 m/s, or 0.5 of the speed of light **29.** (a) 6.4×10^5 m/s **30.** 3.24×10^9 km = 21.7 AU **31.** (Maximum luminosity of SN 1993J)/(maximum luminosity of SN 1987A) = 4.5 **33.** (a) 1.3×10^{-7} $b_\odot$ (b) It would be 130 times brighter than Venus **34.** 1.3×10^8 pc = 130 Mpc **35.** 1100 km/s, or 3.8×10^{-3} of the speed of light

Chapter 23

21. About 5400 B.C. **23.** It will take about 3100 years, or until about A.D. 5100 (note that "expansion rate" is the rate at which the surface of the nebula moves) **24.** (a) Distance = 2200 pc = 7200 ly; estimate that explosion occurred around 6100 B.C. **25.** (b) Maximum correction = 10^{-4} of the pulsar period **26.** (a) Density of matter in a neutron = 4×10^{17} kg/m³ **29.** (a) 0.097 nm (b) Luminosity of neutron star = 5.8×10^{31} W = 1.5×10^5 $L_\odot$ **30.** (b) 14 km

Chapter 24

19. 0.98 of the speed of light **20.** 0.99995 of the speed of light **21.** (a) 25 years (b) 20 light-years as measured by an Earth

observer, 12 light-years as measured by the astronaut
23. 2.8 $M_\odot$ **24.** 5.7×10^8 years   **25. (a)** 2.01×10^6 km
26. 0.20 year = 70 days   **27.** 2.8 m   **28. (a)** R_{Sch} = 8.86 mm,
density = 2.05×10^{30} kg/m^3 **(b)** R_{Sch} = 2.95 km, density =
1.85×10^{19} kg/m^3 **(c)** R_{Sch} = 8.85×10^9 km = 59.1 AU,
density = 2.06 kg/m^3 **29.** 1.9×10^{17} kg/m^3 **30.** 2.7×10^{38} kg =
1.4×10^8 $M_\odot$

Part IV

Chapter 25
22. (a) 1.2×10^{12} cubic parsecs **(b)** 1.1×10^8 cubic parsecs
(c) Probability = 9.6×10^{-5}; we can expect to see a supernova
within 300 pc once every 350,000 years   **24.** 9.5×10^{-25} J; it takes
3.2×10^5 such photons to equal the energy of one H_α photon
25. (b) 5700 pc   **26.** 21 times   **28. (a)** 3.1×10^8 years
(b) 7.4×10^{11} $M_\odot$ **29.** A 10% error in radius results in a
10% error in mass; a 10% error in velocity results in a 20%
error in mass   **31.** 2.7×10^{11} $M_\odot$ **34. (a)** 8.9×10^6 km =
0.059 AU **(b)** 7.4×10^{-6} arcsec   **35. (a)** 2300 AU for S0-1,
950 AU for S0-2 **(b)** 0.29 arcsec for S0-1, 0.12 arcsec for S0-2
36. 2.5×10^6 $M_\odot$

Chapter 26
21. (a) 6.9 Mpc **(b)** 8.3 Mpc **(c)** 1.4 Mpc = 1400 kpc   **22.** 9.5 Mpc
23. (a) 1.1×10^{10} km = 70 AU **(b)** 7.0 Mpc   **24. (a)** 154 Mpc =
5.03×10^8 light-years   **25.** 54 km/s/Mpc   **26. (a)** 1.2×10^{70} atoms
(b) 3.6×10^{-6} atoms per cm^3 **27.** 1.2×10^{12} $M_\odot$ **28.** Period =
4.4×10^8 years; mass = 3.6×10^{11} $M_\odot$

Chapter 27
22. 13.1   **23.** 2.87×10^5 km/s = 0.958 of the speed of light
24. 9.7×10^4 km/s = 3.5×10^8 km/h = 0.32 of the speed
of light   **25.** 2.93×10^5 km/s = 0.976 of the speed of light
28. (a) 1.80×10^5 km/s = 0.600 of the speed of light **(b)** 134 hours
(c) 970 AU   **29. (a)** 30 years **(b)** 2006 **(c)** 5/3 of the speed of light
30. (a) Maximum luminosity = 9×10^{13} $L_\odot$ **31.** 1.5×10^8 $M_\odot$
32. 2.9×10^9 km = 20 AU

Chapter 28
25. 771 nm   **26. (a)** 20 billion years **(b)** 13 billion years
(c) 9.7 billion years   **27.** 1.6×10^8 km/s/Mpc   **28.** 65 times denser
29. 23.6 K   **30. (a)** 4.6×10^{-23} kg/m^3 **31. (a)** 9.5×10^{-18} kg/m^3
(b) 4.8×10^{-4} kg/m^3 **(c)** 1.3×10^{-7} kg/m^3 **32. (a)** $4.7 \times$
10^{-27} kg/m^3 **(b)** 1.9×10^{-26} kg/m^3 **33. (a)** 79 Mpc
(b) 5×10^{-25} kg/m^3 **(c)** T = 17.3 K, ρ_{rad} = 7.5×10^{-28} kg/m^3
34. Ω_Λ = $^1/_3$ **35. (a)** –0.11 **(b)** 0.16   **36.** 1.8×10^{-52} m^{-2}

Chapter 29
19. 1.1×10^{-26} J   **20.** 3.5×10^{-25} s   **21. (a)** 80.5 GeV
(b) 9.34×10^{14} K   **23.** For H_0 = 65 km/s/Mpc, mass =
7.9×10^{-35} kg = 8.7×10^{-5} of the mass of the electron; for
H_0 = 85 km/s/Mpc, mass = 1.4×10^{-34} kg = 1.5×10^{-4} of the
mass of the electron   **24. (a)** 1.1×10^{14} m = 0.011 light-year

Chapter 30
8. (a) 3.7 cm   **9. (a)** 10,000 years **(b)** 10 million years
11. (a) 1430 MHz   **12. (a)** 29.8 km/s **(b)** 0.10 m **(c)** 9.9×10^{-9} m,
0.0000099% of unshifted wavelength

Page numbers in *italic* indicate figures.

A ring, 323, *324*, 327, 330
A stars, 472, 486
A0620-00 X-ray source, 551
Abell, George O., *604*
Abell 2142, *604*
absolute magnitude, 426–429
absolute zero, 99, 100
absorption line spectrum, 107, 114
    of planetary nebula, 504
    of stars, 432
    of Sun, 109, 400–403, 432
absorption lines
    of gamma-ray bursters, 634
    of protostars, 463
    in spectrum, *136, 137*
    spectrograms, *136*
    stationary, 456
acceleration, 78–80
accretion, 166, 167, 464–465
accretion disks, 495, 541, *550*
    of supermassive black hole, 631–634
    see also circumstellar accretion disks
active galactic nuclei, 628, 630
active galaxies, 616, 623–629
    and black holes, 629–631
    unified model of, 631–633
active optics, 133
active sun, 405
actuators, 133
Adams, John Couch, 84, 343, 344
Adams, Walter S., 244
adaptive optics, 133–134
Adrastea, *316*, 317
aerosols, 334
Africa, 186, 191
    movement of, 188, *190*
    rain forests, 199
AGB stars, *see* asymptotic giant branch
    stars
Age of Aquarius, 33
Airy, George, 343
Alaskan legends, 18
albedo, 180
Aldebaran (α Tauri), 431, 435, 438–440,
    498
Aldrin, Edwin "Buzz," 211, *213*
Alexandria, Egypt, 56, *57*
Alexandria library, 66
Alexandria school, 57, 58
Algol (β Persei), *439*, 494–495
ALH 84001 (meteorite), 694–695
alien civilizations
    search for, 696–697
alien planets, *see* extrasolar planets
Allen Telescope Array, 697
Allende meteorite, 374, *690*

all-sky survey, 569
Almach (γ Andromedae), 498
Almagest (Ptolemy), 66
Alnitak, *456*
α Boötis, *see* Arcturus
Alpha Centauri, 20, 426
Alpha Herculis, 452
α Orionis, *see* Betelgeuse
alpha particle, 110–111, 484
Alpher, Ralph, 647
Alphonsus crater, 211
Altair (α Aquilae), *26*, 431
altitude, and temperature, 196
aluminum, 162, 166
Alvarez, Luis, 371
Alvarez, Walter, 371
AM radio transmissions, 97
Amalthea, *316*, 317
Amazon rain forest, 199
amino acids, 374
ammonia (NH_3), 158, 166, 334
Ananke, 317
Anasazi Indians, *525*
Anders, William, 62
andesites, 275
Andromeda (constellation), 24
Andromeda "Nebula," 589, 590
Andromeda Galaxy (M31), 126,
    587, 589–591, 600, 604, 606,
    615, 630
angle of incidence (i), 128, *129*
angle of reflection (r), 128, *129*
angles, 7
Anglo-Australian Telescope, *27*
angular diameter, 7–8
angular distance, 7, 28
angular measure, 7
angular momentum, 164, 631
    of a black hole, 554–555
    law of conservation of, 164, 631
    see also spin
angular resolution (θ), 132
    and radio telescopes, 138, 139
angular size, 7
    small-angle formula, 8
    see also angular diameter
anisotropic stellar motion, 594
annihilation, 674–676
annular solar eclipse, 54
anorthosite, 216–217
Antarctic Circle, *31*, 32
Antarctica, 694
    carbon dioxide concentrations in,
    199
    and ozone hole, 200–201
Antares (α Scorpii), 14, 435, 439, 498,
    509

antigravity, 656
antimatter, 393, 673, 675
antineutrinos, 675–677
antineutrons, 675
antiparticles, 556, 673–675
antiprotons, 673, 675
antitail (of a comet), 378
Antlia (the Air Pump), 601
Antlia galaxy, 601, *602*
Antoniadi, Eugène, 264
Antu telescope, 131, *134*
Apennine Mountains (Earth), 371
Apennine Mountains (Moon), *212*
aphelion, 75
Aphrodite Terra, *252, 253*
apogee, 54
Apollo (asteroid), *367*
Apollo missions, 211–215, *217, 218*
apparent brightness, 424–426, 428–429,
    431, 436, 437
apparent magnitude, 427–429
apparent motion
    of stars, 422
apparent solar day, 34–36
apparent solar time, 34, 35
Aquarius (the Water Bearer), 32, 33, 343,
    344, *505*, 648
Aquila (the Eagle), 25
Ara (the Altar), 453, 467
Arabia (Earth), *190*
Arabia (Mars), *264*
arcminutes, 8
arcseconds, 8
Arctic Circle, 31–32
Arcturus (α Boötis), 25, 435, 438, 439,
    498
Arecibo Observatory, Puerto Rico, 138,
    229, 248
Ares Valles, *268*, 275
argon, 162, 294–295, 399
Argyre Planitia, *274*
Ariel, 352–354
Ariel 5 satellite, 551
Aries (the Ram), 32, 33
Aristarchus of Samos, 57–58, 66, *67*, 76
Aristotle, 63, 71, 665
arithmetic, and powers-of-ten
    notation, 11
Armstrong, Neil, 211
Arnett, W., David, 579
Arp, Halton C., 622
asteroid belt, 77, 158, 366–368
asteroids, *152*, 158, 160, 168, 364,
    365–375, 379, 382–383
    see also Ceres; Juno; meteorites;
    meteoroids; Pallas; Vesta
asthenosphere, 189
Astro-1 mission, 585

astrobiology, 690
astrology, 33, 42
astrometric method, 170, 172
astronauts
    on Moon, 207
astronomical measures, ancient
        and modern, 58
astronomical unit (AU), 12, 68–70
astronomy, 18–19
    ancient, 2, 63–66
    and astrology, 42
    positional (naked-eye), 20, 21, 26
    X-ray, 143
asymptotic giant branch, 502
asymptotic giant branch (AGB) stars,
        502–504
Atkinson, Robert, 392
Atlantic Ocean, 187–189
    see also Mid-Atlantic Ridge
Atlantis (space shuttle), 144
Atlas, 335
atmosphere (atm), 196
atmosphere
    and biosphere, 198
    pressure, 196
    and telescope resolution, 133
atomic number, 112, 162
atoms, 99, 112
    energy levels in, 114–115
    models of, 2, 110
    movement of, 99
    nucleus of, 110–111
    size of, 10
    structure of, 110–111, 667
AU, see astronomical unit
Auriga (the Charioteer), 25, 482
aurora australis, see southern lights
aurora borealis, see northern lights
aurorae, 194
autumn, 29, 31
autumnal equinox, 30, 36
average density, 154, 670–671
average density of matter, 649–650
average mass density of matter, 659
average mass density of radiation, 659
Aztec civilization, 21

B ring, 323, 324, 327–329
B stars, 455, 466, 467, 469–472, 482,
        486, 568, 571, 578–580, 609–610
    see also OB associations
B1757-24 (pulsar), 530
Baade, Walter, 525, 617–618
Babcock, Horace, 410
Babylonian astronomers, 56
Bahcall, John, 418
Balch, Emily, 255
Balmer, Johann Jakob, 111
Balmer lines of hydrogen, 111, 114,
        432–434, 439, 622
Balmer series, 111, 114
Baltis Vallis, 255
bar magnet, 192
bar-code scanners, 105
Barnard, Edward E., 264, 421, 460
Barnard 33, see Horsehead Nebula

Barnard 72, 476
Barnard 86 nebula, 456, 457, 460, 476
Barnard 133, 476
Barnard 142, 476
Barnard 143, 476
Barnard objects, 460
Barnard's star, 421–423
Barnes, Joshua, 606
barred spiral (SB) galaxies, 591–593
Barringer Crater, 371
basalt, 190, 191
baseline (of radio telescopes), 138
Beagle 2, 692–693, 702
Bean, Alan, 212
Becklin, Eric, 581
Bekenstein, Jacob, 556
Belinda, 354
Bell, Jocelyn, 526
Bell Telephone Laboratories, 647
Bellatrix (γ Orionis), 23, 430, 431
Belopol'skii, Aristarkh, 491
belts
    Jupiter, 284, 285, 287–291
    Neptune, 348, 349
    Saturn, 330
BepiColombo mission, 236
BeppoSAX satellite, 634
beryllium, 483, 676
Bessel, Friedrich Wilhelm, 421–422, 506
Beta Centauri, 20
β Cygni, 452
β Lyrae variables, 494–495
β Persei, see Algol
Beta Regio, 255
Betelgeuse (α Orionis), 23, 420, 430,
        431, 438, 439, 436, 498, 509
Bianca, 354
Biermann, Ludwig, 377
Biesbrock, George van, 128
Big Bang, 6–7, 161, 489, 640, 645–649,
        666–667
Big Bang singularity, 666–668
Big Crunch, 660, 666, 671
Big Dipper, 7, 22, 25
Billion-channel ExtraTerrestrial Assay
        (BETA), 697
binary star systems, 444–445, 493–495,
        515–517, 533
    and black holes, 549–551
    neutron stars in, 534
    and pulsars, 534
    white dwarfs in, 506
    see also close binary systems; eclipsing
        binary star systems
binary stars (binaries), 441–443
    see also Sirius
biosphere, 198
Biot, Jean-Baptiste, 372
bipolar outflow, 464–465
BL Lac objects, see blazars
BL Lacertae, 627
black hole evaporation, 558, 661, 668
black holes, 5, 6, 519, 541, 542,
        548–558, 562, 581, 661, 667–668
    and active galaxies, 629–631
    electric charge of, 554

formation of, 549
    mass of, 554
    nonrotating, 552–554
    and quasars, 6
    supermassive, 630–631
Black Stone, 372
Black Widow Pulsar (PSR 1957+20), 533
blackbodies, 101–104, 406
blackbody curves, 99, 101, 104
blackbody radiation, 101–104
blackbody spectrum, 625
blazars, 627–628, 632, 633
blue light, 108
blue main-sequence stars, 571
Blue Mountains, Australia, 108
Blue Ridge Mountains, 108
blue supergiant stars, 513, 596
blueshift, 116, 170
Bode, Johann, 365
Bohr, Niels, 111–112, 432
Bohr formula for hydrogen wavelengths,
        114
Bohr model of the hydrogen atom,
        112–114
Bohr orbits, 112–114
Bok, Bart, 460
Bok globules, 460
Boltzmann, Ludwig, 102
Boltzmann constant, 159
Bondi, Herman, 667
BOOMERANG, 652, 653, 656, 677,
        688
Boötes (the Shepherd), 25
Bradbury, Ray, 264
Brahe, Tycho, 62, 73–74, 76, 518, 529
Braille (asteroid), 369
Brazil rain forest, 199
bright terrain, 313–315
brightness
    and apparent magnitude, 427
    fluctuation of in light sources, 628,
        629
    of stars, 424–429
    see also apparent brightness
brown dwarfs, 171, 172, 177, 434, 435,
        439, 443, 463
brown ovals, 289
Buie, Marc, 358
Bullock, Mark, A., 255
Bunsen, Robert, 105–106
Bunsen burner, 105
Burroughs, Edgar Rice, 264
bursters, see gamma-ray bursters; X-ray
        bursters
Butcher, Harvey, 609
Butler, Paul, 171

C ring, 323, 324, 327–329
Caesar, Julius, 35–37
Calar Alto Observatory (Spain), 293
calcium, 162, 166
calculators, and powers-of-ten notation,
        9–10
caldera, 266, 309
calendar, 35–37, 48
Caliban, 354

Callisto, 155, 302–306, 313, 315–317
Caloris Basin, 233
Calypso, *329*, 335
camera lens, in prism spectrograph, 136
Cameron, Alastair, 219
canali (Mars), 263–264
Cancer (the Crab), 32
Canes Venatici (the Hunting Dogs), *589*
Canis Major (the Large Dog), 25, *87*, 508
Canis Major R1 association, *472*
Canis Minor (the Small Dog), 25
Cannon, Annie Jump, 432
Canopus (α Carinae), 435
Capella (α Aurigae), 435
Capricornus (the Sea Goat), 32, 344, *32*
capture theory (of Moon's origin), 218, 219
Caracol (Chichén Itzá), 21
carbon, 195
    in living organisms, 690
    in solar system, 162
    in star cores, 508
carbon burning, 394, 508, 509
carbon dioxide (CO_2), 158, 195, 196, 468–469
    on Earth, 199, 250, 271–272
    on Mars, 271–273
    on Venus, 245, 250–251
carbon isotopes, 483–484
carbon monoxide, *503*, 690
carbon monoxide radiation, 468–469
carbon star, 503–504
carbonaceous chondrites, 374, 690
carbonate rocks, 272
carbonates, 195, 694–695
carbon-based chemicals, 690
Carlson, Shawn, 42
Carme, 317
Carr, Michael, 269
Carrington, Richard, 406
Carswell, Robert, 608
Casa Grande, Arizona, 21
Cassegrain focus, 129, 130, *132*
Cassen, Patrick, 307
Cassini division, 323, *324*, 327–329
Cassini spacecraft, 283, 287, 289, 337
Cassini, Gian Domenico, 263, 286, 323
Cassiopeia (constellation), 617
Cassiopeia A, 518, *519*, 617
Catalina Observatory, *323*
causality, *557*
celestial coordinates, 28
celestial equator, *26–28*, 30, 33, 34
celestial poles, *26–28*, 33
Celestial Police, 365
celestial sphere, *26–28*, 30–34, 44
    and geocentric model of universe, 63–64
    views of at different wavelengths, *145*
Celsius, Anders, 100
Celsius temperature scale, 99, 100
Centaurus (the Centaur), *460*
Centaurus A (NGC 5128), 626
Centaurus arm, 572

Centaurus X-3, 534, 535
center of mass, 170, 442
centimeter (cm), 11, 12
Central America, 199
central bulge, 566–568, *572*, 576, 591, 592
Cepheid variable stars, 490–492, 565, 589–590, 596, 597, 600
Cepheus constellation, *442*
Cerenkov, Pavel A., 514
Cerenkov radiation, 514
Ceres (asteroid), 158, 365–368
Cerro Pachón, Chile, 130, 131, 359
Cerro Paranal, Chile, 131, 134
Cerro Tololo, Chile, 133
cesium, 106
Cetus (constellation), 638
Chadwick, James, 525
Challis, James, 343
Chandra X-ray Observatory, 143, *144*, 507, *519*, 524, *552*, *604*, 630
Chandrasekhar, Subrahmanyan, 143, 507
Chandrasekhar limit, *506*, 507, 537
charge-coupled device (CCD), 134–137
Charon, 358–359
chemical differentiation, 154, 168, 183
chemical elements, 92, 161–163
    artificially produced, 112
    periodic table of, 112–113
    spectra of, 105–106
    and spectral analysis, 106
Chi (χ) Persei (NGC 884), 499
Chichén Itzá, Yucatán Peninsula, 21
Chicxulub Crater, 371–372, 382
Chihuahua, Mexico, 374
China plate, *190*
Chinese astronomers, 406, 525
Chinook winds, 480
chlorofluorocarbons (CFCs), 200
chondrules, 166–167
Christy, James W., 358
chromatic aberration, 126–129
chromosphere, 400, 402–405, 412
Chryse Planitia, 273
circle, 83, *84*
circumstellar accretion disks, 464–465
Clark, Alvan, 506
Clavius (moon crater), 208
Clementine spacecraft, 150, 212–213, 219
close binary systems, 493, 515–517, 533
    and black holes, 549–551
    and novae, 535–536
    see also binary star systems
closed universe, 653–655, 660
clouds
    and electromagnetic radiation, 244
    of Jupiter, 288–290–292, 294
    of Neptune, 347–348
    of Saturn, 330–331
    over Venus, 245–247
    see also nebula
clusters of galaxies, 600–604, 649
    and gravitational lensing, *609*
clusters of stars, *see* star clusters

COBE, *see* Cosmic Background Explorer
cocoon nebula, 462, 466
co-creation theory (of Moon's formation), 218, 219
cold dark matter, 678–680
collimating lens, 136, *137*
collisional ejection theory (of Moon's formation), 219
Colombo, Giuseppe, 229–230
color index B–V, 429
color-magnitude diagrams, 488, 502, *503*
color ratios (for stars), 431
coma (of a comet), 375–378
coma (optical defect), 130
Coma Berenices (Berenice's Hair), *143*, *568*, *627*, *683*
Coma cluster of galaxies, *600*, 602, *604*, 607
Coma star cluster (Mel 111), 499
combined average mass density, 653–656
Comet Borrelly, 378
Comet Hale-Bopp, *161*, 364, *378*
Comet Halley, 84, 364, 376, 378–380
Comet Hyakutake, *375*, 378
Comet Kohoutek, 377
Comet LINEAR, *381*
comet of 1577, 73
Comet Shoemaker-Levy 9, 292–294, 381
Comet Tempel 1, 378
Comet Wild 2, 378
Comet Wirtanen, 379
comets, 160–161, 168, 364, 375–382, 690–691
    formation of, 379
    orbits of, 83
comparison spectrum, 136, 137
compounds, 190
Compton, Arthur Holly, *144*
Compton Gamma Ray Observatory (CGRO), 144, *145*, *634*
Comte, August, 92
condensation temperature, 165
conduction, 395
Cone Nebula, *470*
confinement period, 682
conic sections, 83
conjunction, 67
conservation of angular momentum, 164, 631
constellations, 22–26
    and astrology, 42
    see also zodiac
contact binary, 494
continental drift, 186–188
    see also plate tectonics
continental shelves, *187*
continuous spectrum, 107
CONTOUR (Comet Nucleus Tour) spacecraft, 378
convection cells, 197, 247, 402
convection currents, 180, 196–197
    and plate movements, 189
convection, 180, *182*, 183, 189, 196–197, 247, 290, 395–397, *401*, 503
    in photosphere, 402, 410
convective zone, 397, 398, *411*, 503

cooling curves, 507
Coordinated Universal Time (UTC, UT), 35
Copernican model of the universe, 70, 73, 665
Copernicus, Nicolaus, 62, 66–70, 71, 76, 644
copper, 190
Cordelia, 354
core bounce, 509
core collapse, 509
core helium burning, 483–485, 488–489, 501–503, 508
core hydrogen burning, 467, 479–483, 485–488
corona (Sun), 400, 403–405, 412
Corona Australis (the Southern Crown), 458
coronal holes, 405
coronal mass ejection, 193, 413
Corvus (the Crow), 607
Cosmic Background Explorer (COBE), 566, 567, 648, 652
cosmic background radiation, see cosmic microwave background
cosmic censorship, law of, 554
cosmic microwave background, 647–650, 653, 655, 659, 666, 670, 675, 677, 688, 697
cosmic particle horizon, 645, 646, 671
cosmic singularity, 646
    see also Big Bang
cosmic string, 683–684
cosmological constant (Λ), 642, 656, 659, 660, 666, 671, 683
cosmological principle, 644
cosmological redshift, 644
cosmology, 640–661, 688
cosmology, relativistic, 658
coudé focus, 130
Cox, John, 492
Crab Nebula, 5, 524, 526, 528, 530–531
Crab pulsar, 526–529, 533
craters, 4, 151, 210
    on asteroids, 369
    Barringer Crater, 371
    on Callisto, 313, 315
    on Ganymede, 313, 314
    on Mars, 264, 268, 270
    on Mercury, 225, 232–233
    on Moon, 232
    see also impact craters
crepe ring, see C ring
Cressida, 354
critical density, 654, 670–671
crustal dichotomy, 266
Crux, see Southern Cross
crystal, 186
Cunitz, Maria, 254
Cunitz crater, 254
current sheet, 297
Curtis, Heber D., 589, 624
Cygnus (the Swan), 24, 25, 145, 428, 505, 536, 550, 551, 617
Cygnus A (3C 405), 617–618

Cygnus arm, 572
Cygnus Loop, 471, 518
Cygnus X-1, 549–550, 556
Cyrano de Bergerac, 210

D ring, 328, 329
Dactyl, 386
dark energy, 656, 658–660
dark energy density parameter (Ω_Λ), 656, 658–660
dark-energy-dominated universe, 659
dark matter, 117, 562, 573–576, 593, 606–609, 650, 678–680
dark matter halo, 576
dark-matter problem, 607, 608–609
dark nebulae, 456, 457, 465, 466, 468, 574
    and star formation, 469
    protostar formation in, 460–463
dark side of the Moon, see far side of the Moon
dark terrain, 313–315
Darwin, George, 218
Daspoort Tunnel, 203
Davis, Donald, 219
Davis, Marc, 597
Davis, Raymond, 399, 418
De revolutionibus orbium coelestium (On the Revolutions of the Celestial Spheres), 70
decametric (ten-meter) radiation, 295–296, 310
deceleration parameter, 660–661
decimals, and powers-of-ten notation, 10
decimetric (tenth-meter) radiation, 296
declination, 28
Deep Impact, 378
Deep Space 1, 369, 378
deferent, 65–66
deforestation, 199
degeneracy, 484, 485
degenerate-electron pressure, 484, 485, 508
degenerate neutron pressure, 525
degree (°), 7, 100
degrees Celsius (°C), 100
degrees Fahrenheit (°F), 100
Deimos, 277–278
Delphinus (the Dolphin), 26
Delta Aquarids, 383
Delta Cephei, 425–426, 452, 490, 491
Deneb, 26
density fluctuations, 677–678
density parameter (Ω_0), 654–656, 670
density waves, 577–579
density-wave model of spiral arms, 578–579
Desdemona, 354
Despina, 355
detached binary, 493
deuterium, 676
deuterium bottleneck, 676
diamonds, 190

diatom, 10
Diaz, Francisco Garcia, 515
Dicke, Robert, 647, 678
Dietz, Robert, 187
differential rotation, 286, 406–407, 410
differentiated asteroid, 374
diffraction, 132
diffraction grating, 136–137
dinosaurs, extinction of, 371
Dione, 335–337
direct motion, 65
dirty ice, see comets
dirty snowballs, see comets
Discovery (space shuttle), 3, 142
disk (of galaxy), see galactic disk
distance ladder, 597, 598
    and redshift, 621
distance indicators, 590
distance modulus, 429, 590
diurnal motion, 23–24, 26
Djorgovski, George, 597
Doppler, Christian, 115
Doppler effect, 116–117, 170, 244, 457–458, 633
    applications of, 117
    and binary systems, 444–445
    and red giants, 483
    in Saturn's rings, 326
    in stars that wobble, 177
Doppler shift, 547, 570–571, 573, 574, 601, 644, 666
    and cosmic microwave background, 648, 649
    in protostars, 463
    and Sun's vibrations, 397
Doppler shift equation, 116
double radio sources, 625, 627, 628, 632, 633
double stars, 441, 452
double-line spectroscopic binaries, 445, 447
double-slit experiment (Young), 95, 96
Draco (the Dragon), 33
Drake, Frank, 695
Drake equation, 695–696
Draper, Henry, 432
dredge-ups, 503
Dressler, Alan, 609
driving cells, 247
dry-ice snow, on Mars, 276
Dunham, Theodore, Jr., 244
dust
    on Mars, 274
    see also interstellar gas and dust
dust devils, 276
dust grains (interstellar medium), 456
    in solar nebula, 165, 166
dust storms, on Mars, 274, 277
dust tail, 377
dwarf elliptical galaxies, 595, 601–603
Dyce, Rolf B., 229
Dynamics Explorer 1 spacecraft, 194

E ring, 328, *329, 336*
E0 to E7 galaxies, 593
Eagle (lunar module), 211
Eagle Nebula, 466
Earth, 151, 179–201
    albedo of, 180, 181
    asthenosphere of, 253
    atmosphere of, 140, 180, 182,
        194–197, 199, 250, 271–272, 691
    average density of, 153, 154, 181, 183
    axis of rotation of, 29–33
    carbon dioxide on, 250, 271–272
    circumference of, 55–58
    climate of and solar activity, 411
    core of, 183–186, 234, 235
    crust of, 186–189
    curvature of, *10*
    data table, 181
    day length, 218
    diameter of, *10, 32, 57–58,* 152, 153,
        181
    equatorial plane of, 33
    escape speed of, 160, 181
    formation of, 183
    geologic processes on, 253, *254*
    and global warming, 200
    and gravitational force of Moon, 85,
        86
    and gravitational force of the Sun, 81
    heliocentric model of universe, 66
    human population of, 198, 199
    inclination of, 181
    interior structure of, 183–186
    life on, 390
    and line of nodes, 49
    location of in Galaxy, 131, 565–566
    and lunar eclipses, 49–51
    magnetic field of, 191–193, 350, *351,*
        235
    magnetosphere of, 193, 297
    mass of, 153, 181
    and meteoroids, 371
    and night sky, 23–24
    oblateness of, 295
    orbit of, 24, 29, 34, 151, 153, 181,
        198, *262*
    ozone layer of, 200–201
    plant life on, *198*
    and precession, 32–33
    radiation emitted from, 180–181
    relative size of, 57–58
    revolution of about Sun, 29
    rotation of, 23–24, 26, 32, 181
    rotation axis of, 198
    seasons on, 29–32
    semimajor axis of, 76
    shadow of, 47, 49–52
    sidereal period of, 68, 76
    surface of, 180, 182
    surface temperature of, 180–182, 198
    and tidal effects of Sun and Moon,
        85–86
    ultramafic lavas on, 309
    umbra of, 50–52
    volcanoes on, 250
    water on, 180, *181*

    weather patterns on, 290
Earth-Jupiter distance, 285
Earth-Mars distance, 261, 263
Earth-Moon distance, 44, 58, 206, 207,
    214–216, 218
Earth-Saturn distance, 325
Earth-Sun distance, *4, 9, 10,* 12, 29, 58,
    68, 153, 181, 391
earthquakes, 184, 185, 213–214
eccentricity (of an ellipse), 75
Eckart, Andreas, 581
eclipse path, 53–55
eclipse series, 56
eclipse year, 56
eclipses, 49–50
    Galilean satellites, 304–305
    predicting, 50, 55, 56
    see also lunar eclipses; solar eclipses;
        total lunar eclipse; total solar
        eclipse
eclipsing binary star systems, 445–447,
    494, 526, 534
ecliptic, 30–34, 44, 66
Eddington, Sir Arthur, 392, 492, 629
Eddington limit, 629–630
Egyptian pyramids, 21
Egyptians, ancient, 18
Einstein Observatory, 143, *624, 626*
Einstein, Albert, 84, 104, 113, 392, 394,
    531, 542, 545–546, 557, 641–642,
    644, 656, 666, 667, 671
    see also relativity, general theory of;
        relativity, special theory of
Einstein-Rosen bridge, 557
Eistla Regio (Venus), *254*
El Niño phenomenon, 198
Elara, 317
electric field, 95–97
electricity, 95
    see also electromagnetism
electromagnetic radiation, 96–101, 244,
    635
electromagnetic spectrum, 97
electromagnetism, 2, 95–96, 680–682,
    684
    Maxwell's theory of, 109–110
electron-positron pairs, 675
electron transitions, 114–115
electron-type neutrinos, 418
electron volt (eV), 105
electrons, 104, 110–111, 393, 568–569
    and Bohr orbits, 112–114
    and spin, 570
    in early universe, 675–676
    see also positrons
elementary particle physics, 669, 672
elements, *see* chemical elements; heavy
        elements; radioactive elements;
        refractory elements; volatile elements
eleven-dimensional theory, 684, 685
Elliot, James L., 351
ellipse, 83, *84*
elliptical (E) galaxies, 591, 593–595,
    597–599, 603, 610, 615
    see also blazars; giant elliptical
        galaxies; radio galaxies

Elmegreen, Debra and Bruce, 579
elongation (of planets), 67
Eltanin (γ Draconis), 498
Elysium Mons, 267
emission line spectrum, 107, 113–114
    of chromosphere, 402–403
    of corona, 404
    of iron vapor, 109
    of NGC 2363 nebula, 109
    of planetary nebula, 504–505
emission lines
    from hydrogen, 621–622
    of quasars, 618, 619
    of protostars, 463
    on spectrogram, *136*
emission nebula, 455, *456, 472*
empty space, 673, 674, 682
Enceladus, 328, 335, 336
Encke, Johann Franz, 327
Encke gap, 327, 330
energy flux (*F*), 102
energy-level diagram, 114, 115
energy levels (in atom), 114–115
energy transport, 394–396, 503
Enif (ε Pegasi), 498
epicenter (of earthquake), 184
epicycle, 65–66
Epimetheus, *329, 335*
epoch, 33–34
Epsilon Eridani, 425, 695
Epsilon Indi, 429
equatorial bulge, 32–33
equilibrium resurfacing hypothesis, 255
equinox, 30, 31
equinoxes, precession of the, 33
equivalence principle, 546
era of recombination, 651–652, 677, 678
Eratosthenes, 55–57, 58
ergoregion, 555
Eridanus (constellation), 426
Eros (asteroid), *158,* 369
error bars, 599
escape speed (of a planet), 160
Esposito, Larry W., 247
Eta Aquarids, 383
ethane, 334
Eurasia, 187
Eurasian plate, 188, *189*
Europa, 155, *157,* 302–305, 307,
    311–313, 317, 692
    data table, 304
European Southern Observatory, 134
European Space Agency, 140, 143, 172,
    236, 278, 337, *376,* 380, 405, 422,
    693, 698
evening star
    Mercury as, 226
    Venus as, 241
event horizon, 553–556
evolutionary tracks, 492–493
    of high-mass stars, 485, 508
    of post–main-sequence stars, 501–502
    of protostars, 460–461
    of white dwarfs, 507
Ewen, Harold, 569
excited states, 114

exobiology, *see* astrobiology
Exosat, 143
exponents, 9–10
extinction, interstellar, *see* interstellar extinction
extrasolar planets, 169–172, 177
extraterrestrial intelligence, search for (SETI), 689–697
eyepiece lens, 125, 126, *128*

F ring, 327, 328, 329–330
F stars, 472
Faber, Sandra M., 18
Fabricius, David, 490
Fahrenheit, Gabriel, 100
Fahrenheit temperature scale, 99, 100
false-color images, 139
false vacuum, 682–683
Fanale, Fraser, 306
far-infrared wavelengths, 566, 567
far side of the Moon, 48, 206
far-ultraviolet light, 141
Far Ultraviolet Spectroscopic Explorer (FUSE), 141–142
favorable opposition, 261
feet (ft), 12
feldspar, 190
Feynman, Richard P., 680
field, concept of a, 95
51 Pegasi, 171
55 Piscium, 452
59 Serpentis, 452
filaments (Sun), 412, 413
fireballs, 635
    meteors, *372*
    primordial, 651, 675
first quarter moon, 45
Fisher, Richard, 596
fission theory (of Moon's origin), 218–220
five-dimensional spacetime, 684
Fizeau, Armand-Hippolyte, 94
Fizeau-Foucault method (for measuring speed of light), 94
flat universe, 653–656, 660, 670–671
flatness problem, 670–672
Fleming, Williamina, 432
flocculent spiral galaxies, 579
floods, on Mars, 267, *268*
fluorescence, 457
fluorescent lights, 457
fluorosulfuric acid (HSO_3F), 246
FM radio transmissions, 97
focal length (of a lens), 124–126
focal length (of a mirror), 129
focal plane, 124, *125*
focal point (of a mirror), 129
focal point (of a lens), 124
focus (of an ellipse), 74
focus (of a lens), 124
force, 78
    and acceleration, 79
forces, physical, 681–682
    long-range, 681
    unification of, 684

see also electromagnetism; gravity; strong nuclear force; weak force
Foucault, Jean, 94
Fraunhofer, Joseph von, 105, Joseph, 432
Frederik II of Denmark, 73, *74*
frequency (of a wave) (ν), 97–98
Friedman, Alexander, 658, 666–667
Friedmann models of universe, 666–667
full moon, 45, *47*, 49, *50*, *52*, 85
fundamental plane, 597
fusion crust, 373

G ring, 328, *329*
G stars, 472
G5.4–1.2 supernova remnant, *530*
galactic cannibalism, 6, 606
galactic center, *see* galactic nucleus
galactic cluster, *see* open cluster
galactic collisions, 609–610
galactic disk, 566, 568, 570, 571, 573, 574, 577
galactic evolution, 609–611
galactic mergers, 609–610
galactic nucleus, 566, *568*, 571, 573, 574, 580–582
Galatea, 355
galaxies, 6, 562, 563, 587–611
    collisions and mergers of, 604–607, 609–610, 634
    distances to, 595–601
    distribution of, *683*
    formation of, 609–611, 677–680
    and quasars, 622–623
    recessional velocity of, 643
    rotation of, 595
    tidal interactions between, 87
    see also active galaxies; blazars; dwarf elliptical galaxies; elliptical galaxies; Milky Way Galaxy; radio galaxies; Seyfert galaxies; spiral arms; spiral galaxies
Galilean satellites, 71, 72, *302*, 302–318
    data table, 304
    see also Callisto; Europa; Ganymede; Io
Galilei, Galileo, 62, 70–72, 77, 93, 123, 126, *228*, 302, 303, 323, 344, 406, *407*, 562, 617, 665
Galileo Probe, 294
Galileo Regio, *313*, 314
Galileo (spacecraft), *44*, *206*, *210*, 212, 247, 287–289, 292, 294, 297, 302, 305, *308–316*, 333, 366, 367, 369, 386
Galle, Johann Gottfried, 343
GALLEX detector, 399
gallium, 399
Gamma Andromedae, 452
Gamma Leonis, 452
gamma-ray bursters, 537, 616, 634–636
gamma-ray photons, *392*, *393*, 483, 674
gamma-ray telescopes, 144, 512
gamma rays, 3, 97, 144
Ganymede, 155, 283, *293*, 302–306, 307, 313–317
    data table, 304

gas atoms, 159–160
gas tails (of comets), 160, *161*
gases, 158, 165, 474
    atoms in, 115
    compressing and expanding of, 480
    ideal, 484
    intracluster, 604
    see also interstellar gas and dust
Gaspra, 369
Gatewood, George, 172
gauss (G), 527
Gauss, Karl Friedrich, 365, 527
Gemini (the Twins), *25*, 32
Gemini North and Gemini South telescopes, 130, 131, 359
Geminids, 383
General Electric Company, 530–531
general theory of relativity, *see* relativity, general theory of
Genzel, Reinhard, 581
geocentric models of universe: 63–66
Georgi, Howard, 681
Georgium Sidus, *see* Uranus
germanium, 399
Ghez, Andrea, 581
giant elliptical galaxies, 593–594, 601–603, 606
giant molecular clouds, 469, 470, 472
giants (stars), 438, 440, 443, 454
Gibson, Everett, 694–695
Gingerich, Owen, 42
Giotto spacecraft, *376*
Glashow, Sheldon, 681
glass, and lenses, 128
Gliese 229 and 229B, *172*, 435
glitch (of pulsars), 531–532
global catastrophe hypothesis, 255
Global Positioning System (GPS), 14
global warming, 200, 488, *536*, 565–566, 568, 574, 625, 678
gluons, 681
gold, 162, 163, 190, 669
Gold, Thomas, 527, 667
Goldreich, Peter, 528
Golombek, Matthew, 702
Gondwanaland, 186–187, 191
Goodricke, John, 490, 494
grams (g), 13
grand-design spiral galaxies, 579
grand unified theories (GUTs), 681–682
granite, 190
granulation (Sun), 401, 402
granules (Sun), 401, 402
grating spectrograph, 136–137
gravitation, Newton's law of universal, 80–82, 84–85
gravitation, universal constant of, 81, 160
gravitational force, *see* gravity
gravitational lenses and lensing, 608, *609*

gravitational radiation, 548
gravitational red shift, 547–547, 555, 556
gravitational waves, *see* gravitational radiation
gravitons, 681
gravity (or gravitational force), 32, 62, 80, 680, 681
  and acceleration, 79, 80
  and black holes, 548, 549
  and density fluctuations, 677
  and general theory of relativity, 546–548
  Newton's theory of, 546, 547
  and tidal forces, 84–87
  and unification of physical forces, 684
Great Attractor, 18, 649
Great Dark Spot, 347, 348
Great Mosque, Mecca, 372
Great Red Spot, 157, 284, 285, 287–288, 291, 292, 305
Great Wall (of galaxies), 604
greatest eastern elongation, 67
  of Mercury, 226–228
  of Venus, 241, 242
greatest western elongation, 67
  of Mercury, 226–228
  of Venus, 241, 242
Greece, ancient: astronomy, 2, 55–58, 63–65, 665
greenhouse effect, 182–183, 195, 200, 244
  on Mars, 271–273
  on Venus, 196, 240, 245
greenhouse gases, 182, 195, 196, 198, 200
Greenstein, Jesse, 618
Greenwich Mean Time, 35
Gregory, James, 128
Gregory XIII, Pope, 37
Grinspoon, David J., 255
ground state, 114
groups (of galaxies), 600
Gruithuisen, Franz, 208
Gula Mons (Venus), 254
Gulf of Aqaba, 190
Gulf of Mexico, 689
Gulf of Suez, 190
Gum, Colin, 518
Gum Nebula, 518
Gunn, James, 609
Guth, Alan, 671, 682

H I hydrogen atoms, 456, 568, 569, 571
H II hydrogen ions, 456, 568
H II regions, 456, 457, 459, 460, 466, 467, 469–470, 476, 511, 512, 571, 572, 578, 595, 607
h Persei (NGC 869), 499
HALCA spacecraft, 139
Hale 5-meter telescope, 291
Hale, George Ellery, 408–410
Hall, Asaph, 277
Halley, Edmund, 84, 379
  see also Comet Halley
halo (of Milky Way), 567, 568, 574–576

Haro, Guillermo, 464
Hartmann, William, 219
Hawaiian Islands, 258, 266
Hawking, Stephen W., 552, 556, 557, 661, 665
Hayashi, C., 460
HD 171978, 444, 445
HD 193182, 111
HD 39801, *see* Betelgeuse
HDE 226868, 550
Head, James, 269
head-tail sources, 625–626
heavy elements, 161–163
  in early universe, 676
  in high-mass stars, 508–510
  origin of, 647
  in spiral galaxies, 591
  in supernovae, 511
heavy hydrogen, *see* deuterium
Heezen, Bruce C., 187
Heisenberg, Werner, 672
Heisenberg uncertainty principle, 556, 667, 668, 672–673
Helene, 335
heliocentric (Sun-centered) model of the universe, 66–72, 80
helioseismology, 397–398, 410
helium, 106, 158
  and Earth's formation, 194
  and hydrogen burning, 392–393, 394
  in chromosphere, 402
  in early universe, 647, 676, 682
  in Jupiter's atmosphere, 157, 286
  in neutron stars, 537
  in solar nebula, 165–166
  in solar system, 161, 162
  in stars, 492
  on Saturn, 331–332
  spectral lines of, 434
  in Sun's core, 479
  see also hydrogen burning
helium absorption line, 516
helium burning, 394, 483–485, 490, 508, 509
  see also core helium burning; shell helium burning
helium flash, 484–486, 488, 501, 502
helium ionization, 492
helium isotopes, 113, 392, 393
helium nuclei, 676
helium "rainfall," 333
helium shell flash, 504
Helix Nebula, 505
Hellas Planitia, 264, 266, 267
Helmholtz, Hermann von, 390
  see also Kelvin-Helmholtz contraction
Henry Draper Catalogue, 432
Henyey, Louis, 460
Herbig, George, 464
Herbig-Haro objects, 464
Hercules cluster, 603
Hercules constellation, 611
Hercules X-1, 534, 535
Hermann, Robert, 647
Hermes (near-earth object), 370
Herrnstein, Jim, 597

Herschel, William, 84, 96, 263, 343, 441, 563–564, 588
hertz (Hz), 97
Hertz, Heinrich, 97
Hertzsprung, Ejnar, 437
Hertzsprung-Russell (H-R) diagrams, 419, 437–439
  for dying stars, 507
  for luminosity classes, 440
  for mass-luminosity relation, 442–443, 444
  for Pleiades, 467, 468
  for protostars, 460–462
  for stellar evolution, 485–491, 501–502
  for young star cluster, 467
Hess, Harry, 187
Hewish, Anthony, 526
High Resolution Microwave Survey (HRMS), 696
high-velocity stars, 567
High-Z Supernova Search Team, 658
Himalayas, 190
Himalia, 317
Hipparchus, 33, 65, 422, 426
Hipparcos (High Precision Parallax Collecting Satellite), 422, 486
Hirayama, Kiyotsugu, 370
Hirayama families of asteroids, 370
Hobby-Eberly Telescope, 130, 131, 136
hogback ridges, 182, 190
Homestead Gold Mine, South Dakota, 399
Hooke, Robert, 285
horary astrology, 42
horizon problem, *see* isotropy problem
horizontal branch, 502, 503
horizontal-branch stars, 488, 502
Horsehead Nebula (Barnard 33), 456, 460, 469
hot dark matter, 678, 680
hot-spot reflection, 446, 447
hot-spot volcanism, 248, 253, 258, 266–267
hot spots, 294, 655
hourglass, 34
Howard, W. E., 229
Hoyle, Fred, 667
Hubble, Edwin, 589, 591–595, 598–599, 642, 666
Hubble classification, 591–595
Hubble constant (H_0), 599–601, 603, 618, 642, 644, 645, 688
Hubble flow, 598–599, 600, 618
Hubble law, 587, 599–601, 618, 622, 634, 642–644, 657
Hubble Space Telescope (HST), 3, 122, 142, 143, 157, 165, 260, 261, 264, 265, 287, 292, 293, 322, 330, 331, 333, 347, 349, 353, 358, 366, 368, 425, 427, 464, 465, 471, 513, 551, 590, 596, 600, 608–611, 624, 625, 632, 633, 635
Hulse, Russell, 548
Hulst, Henrik van de, 569
human population, 198, 199

Humason, Milton, 598
Huygens, Christiaan, 95, 263, 323, 333
Huygens probe, 337
Hyades star cluster (Mel 25), 499
Hyakutake, Yuji, *375*
Hydra-Centaurus supercluster, 649
hydrocarbons, 334
hydrogen, 109, 111, 112, 158, 568
    absorption lines of, *434*
    and Earth's formation, 194–195
    Balmer absorption lines of, 432–434, 439
    nearly universe, 676
    in interstellar space, *145*
    in Jupiter's atmosphere, *157*, 286
    in neutron stars, 536–537
    on Saturn, 331–332
    in solar nebula, 165–166
    in solar system, 161, 162
    in Sun, 394, 479–480
    see also liquid metallic hydrogen
hydrogen, ionized (H II), 568
    emission nebulae, 455–456
hydrogen, neutral (H I), 568, 569, 571
hydrogen atoms, 111–112, 651
    Bohr model of, 112–114
    diameter of, 10
    electron transitions in, 114–115
    magnetic interactions in, *569*
    mass of, 80
    and radio waves, 568–569
hydrogen bomb, 4, *5*
hydrogen burning, 392–394, 438, 508, 509
    see also core hydrogen burning; shell hydrogen burning
hydrogen cyanide (HCN), 334
hydrogen emission lines, 515, 517
hydrogen envelope, 376–377
hydrogen ions, negative, 402
hydrogen isotopes, 113, *392*
hydrogen molecules, 160, 468–469
hydrostatic equilibrium, 394, *395*, 479
hyperbola, 83, *84*
hyperbolic space, 653
Hyperion, 335
hypernova, 635
hypothesis, 2
HZ Herculis, 534

Iapetus, 335, 337
IC 434, *456*
IC 2163 galaxy, *87*
IC 2944, *460*
IC 4182 galaxy, 590
Icarus, *367*
ice, 158
    on Callisto, 315
    on Charon, 359
    in comets, 375, 376, 378, 381
    on Europa, 311–313, 692
    on Galilean satellites, 305
    on Ganymede, 313–314
    on Jupiter, 292, 295, 296
    on Mars, 268
    on Mercury, 234

on Moon, 212–213
on Pluto, 359
in Saturn's rings, 326
on Saturn's satellites, 335
in solar nebula, 165, 166
ice, liquid, 332
ice, methane, 347, *348*, 352
ice age, 198
ice rafts, 312, *313*
Ida (asteroid), 369, 386
ideal gas, 484
igneous rocks, 190–192
imaging, astronomical, 134–135
IMB (Irvine-Michigan-Brookhaven) neutrino detector, 514, 515
impact breccias, 217
impact craters, 208, 253
    on Earth, 210
    on Moon, 218
    on Venus, 253–255
Incan civilization, 21
inches (in.), 12
India, 199
India plate, *190*
Indonesia, 199
Industrial Revolution, 3
inertia, law of (Newton's first law of motion), 78, 79
inferior conjunction, 67
inferior planets, 67, 69
infinity (∞), 111
inflation (of universe), 671–672, 682
inflationary epoch, 671–672, 674
Infrared Astronomical Satellite (IRAS), 140, *141*, *145*, *459*, *567*, 580
infrared light, *566*
infrared radiation, 3, 96, 97, 99, 290
    and Earth's atmosphere, 140, 182, *183*
    and orbiting telescopes, 140–142
Infrared Space Observatory (ISO), 140, *141*
infrared telescopes, 122, 697–698
inner Lagrangian point, 493, 533
inner planets, *see* Earth; Mars; Mercury; terrestrial planets; Venus
instability strip, 491, 492
interacting galaxies, 615
interferometry, 138
intergalactic distances, 590
intermediate vector bosons, 681
intermediate-period comets, 380–381
internal rotation period, 297
International Astronomical Union, 366
International System of Units (SI), 12–13
interstellar clouds, and star formation, 471
    see also nebula
interstellar distances, 12–13
interstellar extinction, 457, 468, 564, 566, 571
interstellar gas and dust, *145*, 455–459, 462, 563, 564, 566, *567*, 591, 595, 604
interstellar medium, 162, 455–459, *464*, 472, 490, 506, 568
    and star formation, 460

and supernovae, 517
interstellar molecules, 468
interstellar reddening, 457–458
interstellar space, 5–6
intracluster gas, 604
inverse square law, 424–426, 429, 589, 635
Io, 83, 155, 296, 302, *305–311*, 317
    data table, 304
Io torus, 310
ion tail, 377
ionization, 115
iridium, 371
iron
    in Earth's interior, 183–186
    isotope of, 112
    in Mercury's interior, 235
    in solar nebula, 166
    in solar system, 162
iron atom, 112
iron meteorites, 373, *374*
iron vapor, 109
Irr I galaxies, 595
Irr II galaxies, 595
irregular cluster (of galaxies), 602, 603
irregular (Irr) galaxies, 591, 595
Irwin, James, *212*
Ishtar Terra (Venus), 252
island universe theory, 588–589
isotopes, 112–113
isotropic stellar motion, 595
isotropy, 670
isotropy problem, 670, 671
Italian Space Agency, 337

Jansky, Karl, 137
Janus, *329*, 335
Japanese Institute of Space and Astronautical Science (ISAS), 236, 278
Japanese National Large Telescope, 359
Jeans, James, 677
Jeans length, 677–678
Jefferson, Thomas, 372
Jesuit observers, 71, 72
Jet Propulsion Laboratory, 307
Jewitt, David, 359
Jiminy Cricket (Mars rock), *275*
Joule, James, 102
joule-seconds (J s), 104
joules (J), 102
Jovian nebula, 306
Jovian planets, 151–154
    atmospheres of, 158
    chemical composition of, 157–158
    formation of, 168, 349
    satellites of, 168
judiciary astrology, 42
Juliet, 354
Juno (asteroid), 366, *367*
Jupiter, 3, 64, 151, 212, 283–297, 330–333
    albedo of, 285
    and asteroid belt, 367–368
    and Io (moons of), 83
    angular diameter of, 285

atmosphere of, 158, 284–295, *331, 333*

average density of, 153, 285, 286

chemical composition of, *157*, 161

cloud layers, 290–292, 294

clouds of, *157*, 288–290, 292

comet impact, 292–294

data table, 285

diameter of, 9, 152, 153, 284, 285

escape speed of, 160, 285

formation of, 168, 283, 290, 294–295

interior of, 295–296, *332*

and Lagrange points, 370

liquid ice on, *332*

magnetic axis of, 297

magnetic field of, 296, 310, 312, *351*

magnetosphere of, 296–297, 310, 315, 332

mass of, 83, 153, 154, 284, 285

oppositions of, 284, 285

orbit of, 67, 69, 151, 153, 170, 285

plasma of, 297

radiation emitted from, 295–296, 310, 332–333

rings of, 317, 326

rotation of, 285–286

shape of, 295

sidereal period and semimajor axis, 68, 69, 76

small satellites of, 316–317

solar wind and, 297

surface gravity of, 285

surface temperature, 158

synodic period, 68, 69

temperature of, 285

tidal forces from, 307, 312

water on, 294

winds on, 288, 290–291, *293*, 294

see also Callisto; Europa; Galilean satellites; Ganymede; Io; Jovian planets

Jupiter-Earth distance, 285

Jupiter-family comets, 380, 381

Jupiter-Sun distance, 12, 68, 153, 285

K line of calcium, 601

K stars, 472

Ka'aba, Great Mosque, Mecca, 372

Kaluza, Theodor, 684

Kaluza-Klein theory, 684

Kamchatka Peninsula, *189*

Kamiokande detector, 399, 514, 515

Kant, Immanuel, 163, 588

Kappa Arietis, 444, *445*

Kappa Pegasi, *133*

Kapteyn, Jacobus, 564

Keck Observatory and telescopes (Mauna Kea, Hawaii), 127, 130–132, 140, 427, *635*

Keeler, James, 326

Keenan, P. C., 439

Kelvin, Lord, 390

Kelvin-Helmholtz contraction, 163, 166, 390, 434, 463

Kelvin temperature scale, 99, 100

kelvins (K), 99, 100

Kepler, Johannes, 42, 62, 73–77, 210–211, 518, 529, 641

Kepler's first law, 75, 82

Kepler's laws, 74–77

Newton's proof of, 82

Kepler's second law, 75–76, 82

Kepler's third law, 76, 77, 117, 574, 575, 631

for binary star systems, 441–442

see also Newton's form of Kepler's third law

Kerr black hole, 554–555

Kieffer, Susan, 308

kilograms (kg), 11–13

kilometer (km), 11–13

kiloparsec (kpc), 13

kinetic energy, 159

Kirchhoff, Gustav, 105–106, 113

Kirchhoff-Bunsen experiment, 105–106

Kirkwood, Daniel, 367

Kirkwood gaps, 367–368, 379

Kitt Peak National Observatory, *23*, 133, 134

Klein, Oskar, 684

Knapp, Michelle, 372

Kraft, Robert, 535

Krüger 60, *442*

krypton, 294–295

Kueyen telescope, 131, *134*

Kuiper, Gerard P., 326, 333, 380

Kuiper belt, 379–380

Lacerta (the Lizard), 627

Lagoon Nebula, 176, 476

Lagrange, Joseph Louis, 370

Lalande 21185, 172

Lamb, Willis, 673

Lamb shift, 673–674

Laplace, Pierre-Simon de, 163

Large Binocular Telescope, 131

Large Magellanic Cloud (LMC), 472, 511, *512, 595,* 600, 605, 607

Larissa, 355

Larson, Stephen, *323*

last quarter moon, 47

Laurasia, 186–187, 191

lava, 190, 192

on Io, 309

on Mars, 275

on Mercury, 232–233

on Moon, 208–210, *217*

ultramafic, 309

on Venus, 248, 249, 252, *254*

Lavinia Planitia, *254*

law of conservation of angular momentum, 164, 631

law of cosmic censorship, 554

law of equal areas (Kepler's second law), 76, 82

Le Verrier, Urbain Jean Joseph, 84, 343–344

leap years, 35, 37

Leavitt, Henrietta, *565*

Leda, 317

Lehmann, Inge, 185

Leighton, Robert, 397

Lemaître, Georges, 658

length contraction, 542–545

length, 11, 12

Lorentz transformation for, 545

lenses

diameter of in telescopes, 125–126, 127

glass in, 128

and refraction, 124

telescopes, 123–128

lenticular galaxies, 594, 603, 610

Leo (the Lion), 25, 32, 117, 440, 648

Leo I galaxy, *594*

Leonids, 383

Lepus (the Hare), *172*

Leviathan telescope, *588*

Levy, David, 292

Libra, 32

libration, 206

Lick Observatory and telescopes (Mount Hamilton, California), 125–126, 171

life, chemical building blocks of, 690–691

light, 92

as electromagnetic radiation, 96–98

gravitational bending of, 546, *547,* 607–608

invisible forms of, 96

laws governing, 3

nonvisible, 3

and photons, 104–105

reflection of, 44

refraction of 123–125

speed of, 12, 93–94, 542, *543,* 544–545

visible, 3, 96–98, 104

wave nature of, 95

wavelength of (λ), 96

see also electromagnetic radiation; gamma rays; infrared radiation; microwaves; near-infrared light; radio waves; ultraviolet radiation; X rays

light curves, 446–447, 517

light-gathering power (of a telescope), 125–127

light pollution, 134

light scattering, 107, 108, 328

light-year (ly), 12

lightbulbs, 98

lightning, 294

limb (of Sun), 400, *401*

limb darkening, 400, 401

limestone, 190, *191,* 195

Lin, Chia Chiao, 577

Lincoln Near Earth Asteroid Research (LINEAR) project, *381*

Lindblad, Bertil, 577

line of nodes, 49, *50, 56*

linear size, 8

liquid metallic hydrogen, 296, 332

liquids, atoms in, 115

lithium, 676

lithosphere, 189, 191

Little Ice Age, 411
Local Bubble, 570
Local Group, 601, 602, 649
logic, 2
long-period comets, 380–381
long-period variable stars, 490
long-range forces, 681
longitudinal waves, 184
lookback time, 619, 644
Lorentz, Hendrik Antoon, 544
Lorentz transformations, 544–545
Lowell, Percival, 263–264, 356, 357
low-mass black holes, 558
lower meridian, 34
luminosity, 427
   of stars, 420, 424–426
   of Sun, 103, 390–396
   see also absolute magnitude, 427
luminosity classes, 439–440
luminosity function, 426
Luna missions, 211, 212, 216
lunar colony, 221
lunar day, 48
lunar eclipses, 49–52
   2001–2005, 52
lunar highlands, 207–208, 216, 217
lunar missions, 211–213
lunar month, see synodic month
lunar mountains, 209–210
Lunar Orbiter program, 211–212, 234
lunar phases, 44–47
Lunar Prospector, 213
Lunar Rover, 212
Lunchbox (Mars rock), 275
Luu, Jane, 359
Lyman-alpha forest, 619
Lyman series, 114
Lyman spectral lines of hydrogen, 619, 620
Lynden-Bell, Donald, 629
Lyra (the Harp), 25, 33, 117, 428, 494
Lyrids, 383
Lysithea, 317

M stars, 488
M-theory, 685
M3 (NGC 5272), 499
M4 (NGC 6121), 499
M5 (NGC 5904), 499
M8 nebula, see Lagoon Nebula
M10 star cluster, 488
M11 cluster, 499
M12 (NGC 6218), 499
M13 (NGC 6205), 499
M15 (NGC 7078), 499, 505
M16, 465, 466
M17, see Omega Nebula
M20, see Trifid Nebula
M22 (NGC 6656), 499
M28 (NGC 6626), 499
M31 galaxy, see Andromeda Galaxy
M32 (NGC 221), 587, 639
M42, see Orion Nebula
M43 nebula, 476
M49 (NGC 4472), 615

M51 (NGC 5194), see Whirlpool Galaxy
M55 (NGC 6809), 488, 489, 499, 502, 503, 566
M58 (NGC 4579), 615
M59 (NGC 4621), 615
M60 (NGC 4649), 615
M61 (NGC 4303), 615
M63 (NGC 5055), 615
M64 (NGC 4826), 615
M65 (NGC 3623), 615
M66 (NGC 3627), 615
M74 (NGC 628), 585, 615
M81 (NGC 3031), 515, 605, 615
M82 (NGC 3034), 552, 605, 615
M83 (NGC 5236), 459, 469, 572, 615
M84 (NGC 4374), 593, 594, 615, 639
M86 (NGC 4406), 593, 594, 615, 639
M87 (NGC (4486), 551, 624–625, 639
M88 (NGC 4501), 615
M89 (NGC 4552), 615
M90 (NGC 4569), 615
M91 (NGC 4548), 615
M94 (NGC 4736), 615
M95 (NGC 3351), 615
M96 (NGC 3368), 615
M98 (NGC 4192), 615
M99 (NGC 4254), 615
M100 (NGC 4321), 6, 143, 589, 601, 615
M101 (NGC 5457), 615
M104 (NGC 4594), 615
M105 (NGC 3379), 615
M106 galaxy, 597
M108 (NGC 3556), 615
M110 (NGC 205), 587, 615, 639
Maat Mons, 249, 252
MACHOs, see massive compact halo objects
Magellan spacecraft, 248, 249, 252–255
magma, 190
magnesium, 162, 166, 185
magnetic-dynamo model, 410
magnetic fields, 95–97, 408, 409
   of dark nebula, 465
   of Earth, 191–193
   of Jupiter, 296, 297
   of neutron stars, 527–529
   of Sun, 408–412
   reversal of, 192
magnetic north pole, 192
magnetic poles, of neutron stars, 529
magnetic resonance imaging, (MRI), 570
magnetic south pole, 192
magnetism, 95, 191
   see also electromagnetism
magnetogram, 409
magnetopause, 193
magnetosphere, 193
magnification, 126, 127
magnitude, see absolute magnitude; apparent magnitude; color magnitude diagrams

magnitude scale, 426–429
main sequence, 437–438, 463, 466, 467, 486–488, 501, 503
   see also zero-age main sequence
main-sequence lifetime, 479–483, 489, 568, 578
main-sequence stars, 438–440, 442–444, 460, 461–466, 478–483, 485, 486, 494, 568, 571
major axis (of an ellipse), 74, 75
Malaysia, 199
mantle, of Earth, 183–186, 189
marble, 191, 195
Marcy, Geoff, 171, 177
mare, see maria
mare basalt, 216, 217
mare basins, 218
Mare Imbrium (Sea of Showers), 208, 209, 212
Mare Nectaris (Sea of Nectar), 208
Mare Nubium (Sea of Clouds), 208
Mare Orientale, 210, 234
Mare Serenitatis (Sea of Serenity), 208, 211
Mare Tranquillitatis (Sea of Tranquillity), 208, 210, 211
marginally bounded universe, 660
maria, 207–209
Mariner missions, 230–233, 235, 236, 245, 264–268, 271, 277, 377
Mars, 3, 64, 150, 260–278
   albedo of, 262
   angular diameter of, 261, 263
   atmosphere of, 157, 196, 264–265, 268, 270–273, 276–277
   average density of, 153, 262
   axis of rotation of, 263
   carbon dioxide on, 271–273
   chemical composition of, 157
   climate on, 271
   core of, 267
   craters on, 264, 268, 270
   crust of, 266, 267
   data tables, 262, 304
   day on, 263
   diameter of, 12–13, 153, 262
   dust devils on, 276
   dust storms on, 277
   escape speed of, 160, 262
   exploration of, 702
   floods on, 267, 268
   hot-spot volcanism on, 266–267
   lava flows on, 275
   life on, 274, 692–695, 702
   liquid water on, 260
   magnetic field of, 267
   mass of, 153, 262
   meteorites from, 694–695
   microorganisms on, 692–694
   motion of, 64, 66, 67
   oppositions of, 261–263
   orbit of, 67, 151, 153, 262
   path of, 2005–2006, 64
   permafrost on, 269
   polar ice caps, 263, 268, 269, 277

rainfall on, 272
regolith of, 274
rotation period of, 262, 263
seasons on, 263, 276–277
semimajor axis, 76
surface gravity of, 262
surface of, 264–270
surface temperature of, 158, 262, 268
synodic and sidereal periods, 68, 76
volcanoes on, 266, 271
water on, 267–271
see also Deimos; inner planets; Phobos
Mars 2, 265, 273
Mars 3, 265
Mars-Earth distance, 261, 263
Mars Exploration Rovers, 702
Mars Express, 693, 702
Mars Global Surveyor, 265–267, 269, 270, 275, 276, 278, 693, 702
Mars Pathfinder, 265, 273–276, 278, 702
Mars-Sun distance, 68, 153, 262
masers, 597–598
mass, 80
    and acceleration, 79
    and Heisenberg uncertainty principle, 673
    and main-sequence lifetime, 480–482
    inside Sun's orbit, 575
    measurement of, 11, 12
mass, center of, 170, 442
mass density of radiation, 649, 650
mass density parameter, 658, 659
mass ejection, 463–465, 507
mass-energy equation, 392, 393
mass loss, 483, *508*
mass-luminosity relation, 442–443, 463
mass-radius relation, 506
mass transfer, 493–495
    in binary systems, 536
massive compact halo objects (MACHOs), 575–576
mathematics, 2
Mathilde (asteroid), 369, 374, 375
matter
    average density of, 649–650
    decoupling of, 677, 678
    in early universe, 675
matter density parameter (Ω_m), 656
matter-dominated universe, 650
Matthews, Thomas, 618
Mauna Kea, Hawaii, 122, 130–134, 140, 359
Mauna Loa, 266
Maury, Antonia, 432
MAXIMA, 652, 656, 677
Maxwell, James Clerk, 95–96, 326
Maxwell Montes, 252–253
Mayan civilization, 21, 55
Mayor, Michel, 170–171
McDonald Observatory, Texas, 131
McKay, David, 694–695
mean solar day, 36
mean solar time, 35, 36
mean sun, 35

measurements, astronomical, ancient and modern, 58
Medicean Stars, *see* Galilean satellites
Medicine Wheel, Wyoming, 21
medicine, spin-flip transitions in, 570
Mediterranean Sea, 187
medium, media (for light), 123
megaparsec (Mpc), 13
megaton, 11
Megrez (δ Ursae Majoris), 431
Melipal telescope, 131, *134*
melting point (of rocks), 185–186
Mercury, 3, 64, 150, 225–236, 416
    albedo of, 226
    average density of, 153, 227, 234
    craters on, 225, 232–233
    data tables, 227, 304
    diameter of, 153, 227
    escape speed of, 160, 227
    iron core of, 234–235
    magnetic field of, 235–236
    magnetosphere, 235–236
    mantle of, *235*
    mass of, 153, 227
    orbit of, 67, 70, 151, 153, 226-228, 230–231
    phases of, *228*
    precession of orbit of, 546–547
    rotation of, 225, 227–231
    semimajor axis of, 76
    sidereal period of, 68, 76, 228
    and solar wind, 235–236
    and tidal forces, 230
    surface of, 225, 232–234
    surface gravity of, 227
    surface temperature of, 158, 227
    synodic period, 68, 228
    transits of, 228, *229*, 243
    see also inner planets
Mercury-Sun distance, 68, 153, 226, 227
meridian, 34
meridian transit, 34
mesosphere, *196*, 197
MESSENGER (Mercury Surface, Space Environment, Geochemistry, and Ranging), 236
metal-poor stars, *see* Population II stars
metal-rich stars, *see* Population I stars
metals (in astronomy), 434
metamorphic rock, 190, *191*
meteor showers, 382, 383
meteorites, 166, *167*, 372–375, 471, 689–690
    from Mars, 694–695
    see also carbonaceous chondrites
meteoritic swarm, 381–382
meteoroids, 208, 209, 214–218, 233–234, 364, 371, 372, 690
    see also asteroids
meteors, 372, 382
meter (m), 11, 12
methane (CH_4), 156, 158, 166, 200, 334, 345 346, 693
Metis, *316*, 317

Metzger, Mark, 634
Mexico, asteroid impact, 371–372
microlensing, 576
micrometer (or micron) (μm), 13, 97
micron, 13
    see also micrometer
Micronesian people, 20
microorganisms, in Martian meteorite, 694–695
microsecond, 11
Microwave Anisotropy Probe (MAP), 688
microwaves, 3, 97
Mid-Atlantic Ridge, 187–188, 189
Middle Ages, 66
mid-mass black hole, 552
midnight sun, 32
miles (mi), 12
Milky Way Galaxy, 1, 6, *145*, 458, *459*, 562–582, 600, 604, 605
    and Andromeda Galaxy, 606
    and black holes, 630
    center of and distance to Earth, 13
    chemical elements in, 162
    and dark matter, 117
    diameter of, *10*
    and extraterrestrial intelligence, 695–697
    halo of gas, 142
    location of, *604*
    luminosity of, 626
    mapping of, *571*
    mass of, 573, 574, 607
    and radio waves, 137
    rotation of, 573–574, *575*
    shape of, 566–568
    spiral arms, 571–572, 607
    and supernovae, 511–512
Miller, Joseph, 627
Miller, Stanley, 691
Miller-Urey experiment, 691
milliarcsecond, 11
millimeters (mm), 12
millisecond pulsars, 532–533
Mimas, 329, 335, 336
minerals, 190
Minkowski, Rudolph, 617–618
Minor Planet Center of the Smithsonian Astrophysical Observatory, 366
minor planets, *see* asteroids
Mintaka (δ Orionis), 23, 435
Mira, 490
Miranda, 352–*355*
mirrors
    and active optics, 133
    and adaptive optics, 133
    and reflection, 128–129
    concave, 128–130
    objective, 129–132
    parabolic, 130, *131*
    primary, *see* mirrors: objective
    in reflecting telescopes, 128–132
    secondary, 129–*132*
    spherical, 130

models, 2
of celestial sphere, 26
density-wave model of spiral arms,
578–579
of neutron stars, 527–529, 531–532
of stellar evolution, 454
Moe (Mars rock), *275*
molecular clouds, 468–470, 571–572
molecules, 156
temperature and speed of, 159
Monoceros (the Unicorn), *467, 469, 470,*
472, 551
month, 32, 48
Moon, 43, 150, 155, 205–221, 225
albedo of, 207, 216
angular diameter of, 7, 52, 206
apogee, 54
average density of, 207
core of, 213
craters on, *70,* 207–210, 211, 218,
232, 234, 264
crust of, 217
dark side, 48
data tables, 207, 304
in daytime, 47
diameter of, 58, 206–*208*
escape speed of, 160, 207
face of, 47–48
far side and near side of, 48, 206, 209,
210
formation of, 205
and geocentric model, 63–64
geologic history of, 219–221
gravitational force of on Earth, 32–33,
85, *86*
ice on, 212–213
inclination of equator, 207
interior of, 213–214
libration of, 206
and line of nodes, 49, *50*
lithosphere of, 210, 214
manned landing on, 211
maria, 232
mass of, 207
meteoric bombardment of, 208, 209,
214–218
missions to, 211–213
motion of, 32–33, 44, 64
orbit of, 32, 48, 53, 206, 207
origin of, 4, 218–*220*
path on celestial sphere, 32–33
penumbra of, *53*
phases of, 44–48
relative size of, 57–58
rotation of, 47–48, 86, 206, 214
shadow of and solar eclipses 49, *50*
sidereal period of, 207, 218
surface of, 206–210
surface composition of, *212*
surface gravity of, 207
surface temperature of, 207
synodic period of, 207
terminator of, 206
tidal effects of on Earth, 85–86,
214–216

and total solar eclipses, *55*
umbra of, 53, 54
waning of, 47
see also entries under lunar
Moon-Earth distance, 9, 58, 206, 207,
214–216, 218
Moon missions, 62
Moon rocks, 4, 62, 211, 213, 215–217,
219
moonlight, 44
moonquakes, 213
Morabito, Linda, 307
Morgan, W. W., 439
morning stars
Mercury as, 226
Venus as, 241
Morris, Mark, 581
Morris, Michael, 557
motion
and theory of relativity, 542–545
direct, 65
diurnal, 23–24, 26
retrograde, 65
superluminal, 628
see also acceleration; apparent motion;
Newton's laws of motion; proper
motion; speed; stellar motions;
tangential motion; velocity
Mount Everest, *190*
Mount Graham, Arizona, 131
Mount St. Helens, 247, *248, 250*
mountain building, 191
mountain ranges, 189
Mueller, M. W., 579
muon-type neutrinos, 418
muons, 545
Musca (the Fly), *505*
MyCn18 nebula, *505*

Naiad, 355
naked-eye astronomy, 20, 21
nanometer (nm), 11, 13, 96
Naos (ζ Puppis), 435
Nash Metropolitan, drawing of, *564*
National Aeronautics and Space
Administration (NASA),143, 172,
211, 273, 278, 337, 378, 405, *411,*
696, 697, 702
navigation by stars, 20
Nazca plate, 189
neap tides, 86
NEAR Shoemaker spacecraft, *158, 369*
near-Earth objects (NEOs), 370
near-infrared light, *3*
near-infrared wavelength, 140, 566, 567
near-ultraviolet light, 141
nebula, nebulae, 455, 456, 588
origin of, 4–5
see also cocoon nebula; dark nebula;
emission nebula; planetary nebula;
reflection nebulae; solar nebula;
spiral nebulae
nebulosity, 162, 455
negative curvature, 653–655
negative hydrogen ions, 402

neon burning, 508, 509
neon gas, 98, 162
neon signs, 98
Neptune, 4, 342–344, 347–351
atmosphere of, 347
average density of, 153, 345, 349
clouds on, 347–348
data table, 345
diameter of, 153, 345
discovery of, 64, 84
escape speed of, 160, 345
formation of, 349
internal structure of, 349–350
magnetic axis of, 350
magnetic field of, 350–351
mass of, 153, 345
orbit of, 67, 151, 153, 345
rings of, 351–*353*
rotation period of, 345
satellites of, 355–356
see also Triton
semimajor axis, 76
storms on, 347–348
surface gravity of, 345
synodic and sidereal periods, 68, 76
temperature of, 345
see also Triton
Neptune-Sun distance, 68, 345
Nereid, 355, 356
Netlanders, 702
neutrino-antineutrino background,
676–677
neutrino detectors, 399, 418, 514
neutrino flux, 399
neutrino oscillation, 400
neutrinos, 389, 392, 393, 398–400, 510,
514–515, 576, 678, 680, 682
in early universe, 675
missing, 418
neutron capture, 508–509
neutron stars, 519, 524–537, 554, 635
and black holes, 548, *549,* 551
and gravitational radiation, 548
magnetic poles of, *529*
mass of, 537
properties of, 531–532
see also pulsars
neutrons, 110–112, 525, 675–676
see also degenerate neutron pressure
new moon, 45, 49, *50,* 85
Newton, Isaac, 3, 62, 77–80, 82, 93–95,
129, 441, 641, 665, 677
Newton's first law of motion (law of
inertia), 78, 79
Newton's form of Kepler's third law,
82–83, 152, 526, 551, 554, 574, 575,
581, 630
Newton's law of universal gravitation,
80–82, 84–85
Newton's laws of motion, 78–80, 83,
84
Newton's second law of motion, 78–80
Newton's theory of gravity, 546, 547,
665, 666
Newton's third law of motion, 79, 80

Newtonian mechanics, 3, 83–84, 343, 546–548
Newtonian physics, *543*
Newtonian reflector, 129, 130
Newtonian space, 641
Next Generation Space Telescope (NGST), 142
NGC 628, *579*
NGC 891, *459*
NGC 1265, 626
NGC 1275, 623, *624*
NGC 1275, 634
NGC 1973-1975-1977, *162*
NGC 2024, *456*
NGC 2207, *87*
NGC 2264, 466, 467
NGC 2359, *508*
NGC 2363, 109, 115
NGC 3077, 605
NGC 3351, 425–426
NGC 3576, *458*
NGC 3603, *458*
NGC 3628, 615
NGC 3840, 117
NGC 4038, *607*
NGC 4039, *607*
NGC 4261, 632, *633*
NGC 4526, *596*
NGC 4565, *568*
NGC 4874, *600*
NGC 4889, *600*
NGC 5195, *589*, 615
NGC 6520, 465
NGC 6624 star cluster, *536*
NGC 6726-27-29, *458*
NGC 7027, *505*
NGC 7742, *624*, 633
NGC 7793, *579*
Nicholson, Seth B., 409
nickel, 162, 166
night sky, 23–24, 640, 645
Nimbus 7 spacecraft, *198*
1980 JE (asteroid), 366
1986 U10, 354
1992 QB$_1$, 359, 380
1993 SC, *380*
1994 XM$_1$, 370
nitrates, 196
nitrogen, 162, 196
nitrous oxide (N$_2$O)
NN Serpens, *446*
noble gases, 294–295
no-hair theorem, 555
nonthermal radiation, 295, 624–625
North America, 187
North America Nebula, *378*
north celestial pole, 27, 28, 33
North Star, Thuban and Vega as, 33
    see also Polaris
northern lights (aurora borealis), 193, *194*
northern lowlands (Mars), 265, 266, 269
Nova Cygni 1975 (V1500 Cyg), *536*
Nova Herculis 1934, *535*

novae, 428, 535–536
Nozomi orbiter, 702
Nu Boötis, 452
nuclear density, 510
nuclear fission, 392–393
nucleons, 510
nucleosynthesis, 676
nucleus (of an atom), 110–112
nucleus (of a comet), 376, 381

O stars, 455, 466, 467, 469–472, 482, 486, 568, 571, 578–580, 609–610
OB associations, 470, 472, 571, 595
    see also stellar association
OBAFGKM sequence, 432–433, 435
OBAFGKMLT sequence, 434, 435
Oberon, 352–354
objective lens, 125–126, 127, 128
oblateness, 295, 332
observable universe, 645, *646*
observation, 2, 3
Occam's razor, 66, 70, 77, 667
occultation, 304
    and Uranus rings, 351, *352*
    and Galilean satellites, 304, 305
ocean temperatures, *14*
ocean tides, 85–86
oceanic rifts, 189
oceans
    and biosphere, 198
    on Callisto, 315
    on Europa, 313
    variations in heights of, *14*
    on Venus, 250–251
Oceanus Procellarum (Ocean of Storms), 211
Oemler, Augustus, 609
Olbers, Heinrich, 365, 641
Olbers's paradox, 641, 645
Oldham, Richard Dixon, 185
Oliver, Bernard, 696
Olympus Mons, *157*, 266, *271*
Omega (M17) Nebula, 176, 476
Omicron Ceti, *see* Mira
1-to-1 spin-orbit coupling, 228, *230*
open cluster, 467
open universe, 653, 660
Ophelia, 354
Ophiuchi, *442*
Ophiuchus constellation, *422*, *442*, *488*
opposition, 67
optical double stars, 441
optical telescope, 123
optical window, 140
orbital eccentricities, 151, 171
orbiting telescopes, 140–145
orbits, 74–77, *82*
    prograde, 317
    retrograde, 318
organic molecules, 690
Orion (the Hunter), 22, 25, *141*, *145*, 162, 430, 455, 464, 469, 509
    see also Betelgeuse
Orion arm, 572

Orion Nebula (M42), 4, *5*, 22, 164, *165*, 176, 455, 462–463, 465, 469–470, 476, 572
Orionids, 383
Ostriker, Jeremiah, 593
outer planets, *see* Jovian planets; Jupiter; Neptune; Pluto; Saturn; Uranus
outgassing, 195
oxides, 166
oxygen, 195–196
    in solar system, 162
    in star cores, 508
oxygen burning, 394, 508, 509
oxygen isotopes, 484
ozone (O$_3$), 197, 200–201
ozone hole, 200–201

P waves, 184–185
Pacific plate, 188, *189*
Pacini, Franco, 530
pair production, 528, 674, 675
Pallas, 158, 365, 366, *367*, 368
pallasite, *373*
Palomar telescope, 617
Pan, 330, 335
Pandora, 329, 330, 335
Pangaea, 186, *187*, 191
Papua New Guinea, 199
parabola, 83, *84*
Paradis, Jan van, 634
parallax, 72, 420
    see also spectroscopic parallax; stellar parallax
parallax angles, 421–422, 436
parallax method, 595
Parkes Observatory, New South Wales, *138*
parsec (pc), 12–14, 421
Parsons, William, third Earl of Rosse, 588
partial lunar eclipse, 50, *51*
partial solar eclipse, 53
particle accelerators, 530–531, 682
particle-antiparticle pair, 673
Pasiphae, 317
Pati, Jogesh, 681
Pauli, Wolfgang, 484
Pauli exclusion principle, 484, *485*, 525
Payne, Cecilia, 433
PC 1247+3406, 619
Peale, Stanton, 307
Peebles, P. J. E., 593, 647, 651, 678
Pegasus (the Winged Horse), *133*, 171, *505*, *624*
Pele (volcano), *308*, *309*
Penrose, Roger, 554, 555
Penrose process, 555
penumbra
    of Earth's shadow, 50, *51*
    of sunspots, 406
penumbral eclipse, 50–52
perigee, 53
perihelion, 75
period (of a planet), 68

periodic table of the elements, 112–113
period-luminosity relation, 492, *565*, 589
Perlmutter, Saul, 658
permafrost, 269, 272
peroxides, 274
Perseids, 383
Perseus (constellation), 25, *439*
Perseus arm, 572
Perseus cluster of galaxies, *624*, 626
Peru-Chile Trench, 189–190
Pettengill, Gordon H., 229
PG 0052+251, *623*
PG 0316-346, *623*
PG 1012+008, *623*
"Phenomenon of Colours," 94
Phillips, Roger J., 255
Phobos, 277–278
Phobos 2, 265
Phoebe, 335, 337
photodisintegration, 510
photoelectric effect, 104
photographic film, 104, 134, *135*
photometry, 135, 424
photon energy, 112–115
photons, 104–105, 510, 674, 681, 682
    and black holes, *549*
    and cosmic microwave background,
        649–652
    in early universe, 647
    energy of (frequency), 105
    energy of (wavelength), 104, 105
    in H II regions, 456
    and light scattering, 108
    polarization of, 688
    and radiative diffusion, 395, 396
    redshift of, 644
    spin-flip transition, 569
    ultraviolet, 274
    virtual, 680
photosphere, 400–403, 405–411
photosynthesis, 195
physics, 3, 115
phytoplankton, 198
Piazzi, Giuseppe, 365
Pickering, Edward C., 432
Pickering, William, 356
Pilcher, Carl, 326
Pillan Patera, *308, 309*
Pioneer 10 and *11*, 287, 296, 305, 327,
    700
Pioneer Venus Multiprobe, 247
Pioneer Venus Orbiter, 240, 246,
    247
Pisces (the Fishes), 32, 33
pixels, 134–135
PKS 0405-123, *622*
PKS 2000-330, *619, 620*
plages, 412, 413
Plains Indians, 21
Planck, Max, 104, 113
Planck formula, 115
Planck time, 646–647, 671, 682, 684
Planck's constant, 104
Planck's law, 105, 510
plane of the ecliptic, 49

Planet Imager, 698
Planet X, *see* Pluto
planetary motions
    and Kepler's laws, 74–77
planetary nebula, 500, *501*, 504–505
planetary orbits: and eccentricities, 75
Planetary Society, 697
planetesimals, 166–168, 367
planets, 64
    atmospheres of, 155–156
    average densities of, 155, 306
    characteristics of, 153
    diameters of, 151–152
    distances from the Sun, 68–70
    elliptical orbits of , 74–75
    elongation of, 67
    escape speed of, 160
    formation of, 18, 163–169, 306
    inferior, 67, 69
    masses of, 152–154
    motions of in geocentric model,
        64–66
    motions of in heliocentric model, 70
    oblateness of, 295
    orbital period of, 68
    orbital speed variations of, 75–76
    orbits of, 67–70, 80, *81*, 151, *152,*
        365
    prograde rotation of, 243
    relative sizes of, *4*
    satellites of, 326
    semimajor axis of orbits of, 76
    sidereal periods of, 68, 69, 76
    and Sun to scale, *152*
    superior, 67, 69
    surface temperatures of, 158
    synodic periods of, 68, 69
    terrestrial, 151–154
    see also extrasolar planets; inner
        planets; Jovian planets; *individual
        entries, e.g.,* Earth, Mars, Pluto
plant life, *198*
plasma, 297, 651
    from supernovae, 635
    solar atmosphere, 408, 411, 413
plastic material, 186
plate boundaries, 188–190
plate tectonics, 186–191, 253
plates, 186
Plato, 71, 665
Pleiades (M45) cluster, *427*, 467, 499
Pluto, 151, 342, 356–359
    albedo, 357
    average density of, 153, 154, 357, 359
    data table, 357
    diameter of, *152*, 153, 357, 359
    discovery of, 64
    escape speed of, 160, 357
    mass of, 153, 357
    orbit of, 67, 75, 151, 153, 357
    polar ice caps of, 358
    rotation of, 244, 357
    semimajor axis of orbit of, 76
    surface gravity of, 357
    surface temperature of, 158, 357

synodic and sidereal periods, 68, 76
    see also Charon
Pluto-Sun distance, 68, 153, 357
polar ice caps
    on Mars, 263, 268, 269, 272, 277
    on Pluto, 358
Polaris (North Star), 7, 25, 27–29, 33,
    420
polarization, of photons, 688
polarized radiation, 625
Pollack, James B., 272, 306
polymers, 334
Polynesian people, 20
Pooh Bear (Mars rock), *275*
poor clusters, 600–*602*
Population I (metal-rich) stars, 489–490,
    492, *567*, 568, 578, 591
Population II (metal-poor) stars,
    489–490, 492, 493, *567*, 568, 591,
    593
Portia, 354
positional astronomy, 20, 21, 26
positive curvature, 653
positrons, *392*, 393, 514, 673, 674
pound (weight), 13
Pound, Robert, 548
powers of ten, 9
    see also exponents
powers-of-ten notation, 9–11
Praesepe (M44), 499
precession, 28, 32–33
    and calendar, 35
    of Mercury's orbit, 546–547
precession of the equinoxes, 33
primary mirrors, *see* mirrors: objective
prime focus (of reflecting telescope), 129,
    130
primitive asteroids, 374
primordial black holes, 552, 558
primordial fireball, 651, 675
prism, and spectrum, 94–95
prism spectrograph, 136
Procyon (α Canis Minoris), 426, 435
progenitor star, 512–513
prograde rotation, 243
Project Phoenix, 697
Project Ranger, 211
projection effect, 622
Prometheus (Io volcano), *308*, 329–330,
    335
prominences, solar, 412
proper distance, *see* proper length
proper length, 545
proper time, 544
proplyds, *see* protoplanetary disks
Prospero, 354
Proteus, 355
proton-antiproton annihilation, 675
proton-antiproton pair production,
    675
proton-proton chain, 394, 400
protons, *10*, 110–111, 392, 510,
    568–569, 675–676
    in isotopes, 112
    in neutron star, 532

in nucleus of an atom, 112
and spin, 570
protoplanetary disks, 164, *165*, 169, 465
protoplanets, 166, 167, 168, 306
Jupiter, 283
protostars, 453, 460–463, 486, *632*
evolution of, 463–464, 466
evolutionary track of, 460–463
mass gain and loss in, 463–465
see also T Tauri stars
protosun, 163–164, 166, 167, 169
Proxima Centauri, 12, 13, 422
pseudoscience, 42
PSR 1937+21, 532
Ptolemaic system, 65, 66, 70, 72
Ptolemy, 65, 66
Puck, 354
pulsars, 5, 526–533
slowdown of, 531
see also neutron stars
pulsating variable stars, 490–493
pulsating X-ray sources, 533–535
Purcell, Edward, 569
Pythagoras, 63
Pythagorean theorem, 423

Quadrantids, 383
quantum electrodynamics, 680–681
quantum gravity, theory of, 667
quantum mechanics, 84, 95, 114,
667–668, 672–673, 684
quantum state, 484
quarks, 680–682
quasars, 6, 582, 608, *609*, 616–623,
631–633
redshift of, 644
quasi-stellar objects, *see* quasars
quasi-stellar radio sources, *see* quasars
Queloz, Didier, 170–171
quiet Sun, 405

radar, 138
radial velocity, 116–117, 423, 444
radial velocity curve, 444, 445
radial velocity method, 170
radiants, 382
radiation, 96, 295–296
mass density of, 649, 650
see also blackbody radiation;
Cerenkov radiation; cosmic
microwave background;
electromagnetic radiation;
gravitational radiation; infrared
radiation; nonthermal radiation;
polarized radiation; synchrotron
radiation; thermal radiation
radiation darkening, 352
radiation-dominated universe, 650
radiation pressure, 377, 629–630
radiative diffusion, 395–397, 503
radiative zone, 396, 398, *411*
radio galaxies, 623–628, 633
radio lobes, 625, 626
radio sources, *see* quasars
radio stations, 97

radio telescopes, 122, 137–139,
617
see also Arecibo Observatory
radio transmissions, 97
alien, 695
radio waves, 3, 6, 97
and astronomy, 137
and hydrogen atoms, 568
and interstellar medium, 568
from Jupiter, 295
radio window, 140
radioactive age-dating, 694
radioactive elements, 110
radio-loud quasars, 619, 623, 627, 632,
633
radio-quiet quasars, 619, 623, 627
rain forests, 199
rainbow, 94, 95
see also spectrum
rainfall, on Mars, 272
Randi, James, 33
Ranger spacecraft, 211
Reber, Grote, 137, 617
Rebka, Glen, 548
recessional velocity, *599*, 601, 618, 619,
644
and expansion of universe, 657, 658
of galaxies, 643
and redshift, 620
recombination, 456
recombination, era of, 651–652, 677,
678
red emission nebulae, 571
red-giant branch, 501, *503*
red-giant stages, 501–502, 508
red giants, 438–439, 463, 479, 482–489,
493, 494, 507, 533
red light, 108
see also interstellar reddening
Red Sea, *190*, 267
red supergiants, 513, 515
redshift, 116, 170, 666
cosmological, 644
and distance, 621
and expanding universe, 645–646,
657, 658
of galaxies, 598–601, *617*
of gamma-ray bursters, 635
of photons, 644
of quasars, 618–623, 644
relativistic, 620
reflecting antenna, 137
reflecting telescopes, 128–132
reflection, 44, 128
reflection nebulae, 456, *458*
reflector, *see* reflecting telescope
refracting telescope, 124–128
refraction, 123–125
refractor, *see* refracting telescope
refractory elements, 219
regolith, 216
regular cluster (of galaxies), 602–603
Regulus (α Leonis), 25, 36, 425, 431,
440
relativistic cosmology, 658

relativity, general theory of, 84, 94, 531,
541, 546–549, 557, 641–641, 653,
658, 666–668, 684
relativity, special theory of, 94, 392,
542–546, 641, 673, 684
residual polar caps, on Mars, 268
Retherford, R. C., 673
retrograde motion, 65, 70, 244
and heliocentric model, 66, 67
geocentric explanation of, *65*
retrograde orbits, Jovian satellites,
318
retrograde rotation, 243–244
Reynolds, Ray, 307
Rhea, 336, 337
Rhine River valley, 267
rich clusters of galaxies, 600, 601, 603,
607, 609
rift valley, 267
Rigel (β Orionis), 28, 435, *438*, *439*,
509
right ascension, 28
and sidereal time, 36
ring particles, 326
ringlets, 327
rings, 151
Roche, Edouard, 326, 493
Roche limit, 326
Roche lobes, 493–495, 533
rocks, 190–191
andesites, 275
melting point of, 185–186
see also Moon rocks
Rocky Mountains, *182*
Roman Catholic Church
and calendar reform, 37
and Galileo, 71
Rømer, Olaus, 93
Röntgen, Wilhelm, 97
Rosalind, 354
ROSAT spacecraft, 143, *145*, 471
Rosen, Nathan, 557
Rosetta mission, 380
Rosette Nebula, 471, *472*
Rossi X-ray Timing Explorer, 535
Roswell, New Mexico, 695
rotating (Kerr) black hole, 554–555
rotation curves, 574
of Andromeda Galaxy, *630*
of galaxies, 607, *608*
rotation
differential, 286, 406–407, 410
internal rotation period, 297
prograde, 243
retrograde, 243–244
synchronous, 47, 86, 206, 303
RR Lyrae variable stars, 428–429, 493,
565, *566*, 596
light curve of, *565*
rubidium, 106
Rubin, Vera, 18
Ruhl, John, 688
runaway greenhouse effect, 251
runaway icehouse effect, 272
Russell, Henry Norris, 437

Rutherford, Ernest, 110, 432
Rutherford's model of the atom, 110
Rydberg, Johannes, 111
Rydberg constant, 111
Ryle, Martin, 667

S waves, 184–185
S/1999 J1, 317
S0 and SB0 galaxies, *see* lenticular
    galaxies, 594
Sa galaxies, 591–592, *595*
SAGE (Soviet-American Gallium
    Experiment) detector, 399
Sagitta (the Arrow), *26*
Sagittarius (the Archer), 32, 137, *145,
    169, 457, 488, 563, 564, 566,* 617
Sagittarius A, 580–581, 617
Sagittarius A*, 580–*582*
Sagittarius arm, 572
Sagittarius Dwarf, *595,* 601, *602*
Saha, Meghnad, 433
Salam, Abdus, 681
Sandage, Allan, 618
sandstone, 190, *191*
Santa Ana winds, 480
Santa Maria Rupes (Mercury), *233*
saros interval, *56*
satellites (of planets), 154–155
Saturn, 4, 64, 150, 322–337
    albedo of, 324
    angular diameter of, 8, 325
    atmosphere of, 330, 331, 333
    average density of, 153, 154, 324, 331,
        332
    clouds and storms on, 330–331
    core of, 332
    data table, 324
    diameter of, 153, 324
    escape speed of, 160, 324
    internal structure of, 331–332
    liquid ices within, 332
    magnetic field of, 322, *351*
    magnetosphere of, 332
    mass of, 153, 324, 332
    oppositions of, 2001–2005, 325
    optical and radio views of, *139*
    orbit of, 67, 151, 153, 324, 325
    radiation emitted from, 332–333
    rings of, 70, 317, 323–330, 339
    Roche limit of, 326
    rotation of, 324, 332
    satellites of, 322, 328–331, 334–337
    semimajor axis of, 76
    surface gravity of, 324
    synodic and sidereal periods, 68, 76
    temperature of, 324
    wind on, 331
    see also Titan
Saturn-Earth distance, 325
Saturn-Sun distance, 68, 153, 324
Saturn V rocket, 12
SAX J1808.4–3658, 535
Sb galaxies, 591–592
SBa galaxies, 592, *595*
SBb galaxies, 592

SBc galaxies, *592, 595*
Sc galaxies, 591–592, *595*
scarps, 232
    on Mercury, 232–233
Schiaparelli, Giovanni, 228, 263, 264
schist, *191*
Schmidt, Brian, 658
Schmidt, Maarten, 618
Schraber, Gerald G., 255
Schwabe, Heinrich, 407
Schwarzschild, Karl, 553
Schwarzschild radius, 553
Schwinger, Julian S., 680
Sciama, Dennis W., 665
scientific method, 2
Scorpius constellation, 32, 509
Scorpius X-1, 535
Scoville, Nicholas, 469
sea lions, 198
seafloor spreading, 188, 190, 253
search for extraterrestrial intelligence
    (SETI), 689–697
seasons, 29–32
    on Mars, 263, 276–277
    on Uranus, 346
seconds (s), 11–13
sedimentary rocks, 190, *191*
seeing disk, 133
seismic waves, 184–185
seismographs, 184
self-propagating star formation, 579
semidetached binary, 494
semimajor axis (of a planet's orbit), 76
semimajor axis (of an ellipse), 74, 75
SERENDIP IV (Search for
    Extraterrestrial Radio Emissions
    from Nearby, Developed, Intelligent
    Populations), 697
Setebos, 354
SETI, *see* search for extraterrestrial intel-
    ligence
SETI Institute, 697
SETI@home, 697
Seven Samurai (astronomers), 18
Seyfert, Carl, 623
Seyfert galaxies, *623, 624, 626–627,*
    633
shadow zone (of Earth), 185
Shapley, Harlow, *564–566,* 589
Shapley-Curtis debate, 589, 591, 598
shell helium burning, 502
shell hydrogen burning, 482, 483, 485,
    489, 501, 508
shepherd satellites, 330
shield volcano, 248
shock wave, 193
    and black holes, 631
    and star birth, 470–472
    from supernova, 511, 513
Shoemaker, Carolyn, 292
Shoemaker, Eugene, 292, 308
shooting stars, *see* meteors
short-range forces, 680
Showalter, Mark, *330*
Shu, Frank, 577

SI system, *see* International System of
    Units
sidereal clock, 36
sidereal day, 36
sidereal month, 48, *49*
sidereal period, 68
    and Kepler's third law, 76
    and synodic period, 69
sidereal time, 35, 36
sidereal year, 35
Siding Spring Mountain, Australia,
    27
silicon
    in solar nebula, 166
    in solar system, 162
silicon burning, 508, 509
silver, 190
single-line spectroscopic binaries, 445,
    550
singularities, 666
    of black hole, 553–556
    see also cosmic singularity
Sinope, 317
Sirius (α Canis Majoris), 103, 426, 427,
    431, 435, 436, 506
Sirius B, 436, *438,* 506, 548
Sitter, Willem de, 658
16 Cygni B, 171
61 Cygni, 422
skepticism, 2–3
sky, blue color of, 108, 456
slash-and-burn process, 199
Slipher, Vesto M., 598, 666
Small Magellanic Cloud, *565, 595,* 600
small-angle formula, 8–9
Smith, Bradford, 308
Smith, David, 266
SN 1987A, *see* supernova 1987A
SNC meteorites, 694
sodium, 162
*SOHO (Solar and Heliospheric
    Observatory),* 405
Sojourner, 275
solar constant, 103
solar corona, 43, *52, 53*
solar eclipses, 43, 49, 50, 52–55, 404
    during 1997–2020, 54, 55
    during 2001–2005, 54
solar energy, 180, 182, 183, 198,
    390–398
    flux of, 103
    on Jupiter, 290
solar flares, 389, 412–413
solar interior, 400
solar masses (M☉), 13
Solar Maximum Mission spacecraft,
    413
solar models, 399–400
solar nebula, 163–169, 294–295, 333,
    349, 390, 490, 554
    temperature distribution in, 166
solar neutrino problem, 399–400
solar neutrinos, 399–400, 418, 510
solar prominences, 412
solar radius (R☉), 395, 396

solar system, 3
    origin and evolution of, 4, 163–169
    see also Milky Way Galaxy
solar transits, 228
    of Mercury, 228, *229*, 243
    of Venus, 243
solar wind, 169, 192–193, 377, 405
    on Jupiter, 297
    and Mercury, 235–236
solids, 165
    atoms in, 115
Solomon, Philip, 469
sound waves, and Doppler effect, 116
South America, 186–188
South American plate, 189
south celestial pole, 27, 28, 33
South Pole–Aitken Basin, *210*, 212, 219
Southern Cross (Crux), 7, 20, 23
southern highlands (Mars), 265, 266, *268*, *269*
southern lights (aurora australis), 193, *194*
Southern Wall (of galaxies), 604
Soviet Union, space missions, 211–212, 245
Space Infrared Telescope Facility (SIRTF), 141
space, and relativity, 542, 666
space velocity, 423
spacecraft
    orbits of, 82
    see also individual entries, e.g., Pioneer spacecraft, *Voyager* spacecraft
spacetime, 542, 546, *547*, 666, 668, 684
    and black holes, 548, *549*, 554
    five-dimensional, 684
special theory of relativity, *see* relativity, special theory of
spectra, Kirchhoff's laws for, 106–107
spectral analysis, 106
spectral classes, 432–434, 435
spectral lines, 105–106, 109
    Lyman series, 114
    splitting of, 408–409
    wavelengths of, and relative motion 115–117
spectral types, 432, 439, 440
    and Hertzsprung-Russell diagram, 437
spectrogram, *136*
spectrograph, 136–137
spectroscopic binaries, 444–445
spectroscopic parallax, 425, 440, 441, 595
spectroscopy, 109, 136–136, 155–156, 444
    see also stellar spectroscopy
spectrum, spectra, 94–95
    see also absorption line spectrum; continuous spectrum; emission line spectrum
spectrum binary, 444
speed, 78, 542
    measurement of, 13

see also light, speed of
spherical aberration, 130, *131*
    Hubble Space Telescope, 142, *143*
Spica (α Virginis), 25, 435
spicules, 403, 411, 412
spin, 568–569
spin-flip transition, 569, 570
spin-orbit coupling, 228–231
spiral arms (of galaxies), 6, 568, 571, 572, 576–579, 592
    and star formation, 469
    density-wave model of, 578–579
    formation of, 606–607
spiral density waves, 579
spiral galaxies, 459, 572, 579, 591–592, 594–595, 603, 609–610, 615, 668
    and tidal forces, 86–87
    see also Seyfert galaxies
spiral nebulae, 588–589
spokes (in Saturn's rings), 339
spontaneous symmetry breaking, 682–683
spring tides, 85
spring (season), 29–31
stable Lagrange points, 370
stades, *56*
standard candles, 596
star catalogs, 34
star charts, 34
star clusters, 465–467, 485–489, *607*
    H-R diagram, *467*
    see also globular clusters
star death, 5, 161, 500, 504, 506–507, 510–511, 537
    and black holes, 548, 554
    supernovae, 510–511
star formation, 4–5, 162, 453–472, 610
    in elliptical galaxies, 593
    in spiral arms, 571, 578
    in spiral galaxies, 592
    self-propagating, 579
star trails, *27*
starburst galaxies, 605
Stardust spacecraft, 378
starlight, 92
Starobinsky, Alexei, 671
stars, 419–447
    absorption lines of, 434
    apparent motion of, 422
    atmospheres of, 447
    and Balmer absorption lines, 111
    blackbody spectrum of, 435
    chemical composition of, 434
    circumpolar, 27
    colors of, 430–431
    determining surface temperature of, 102
    ejected matter from, 161–162
    estimating distances to, 440
    evolution of, 6
    heavy elements in, 161–162, 163
    on Hertzsprung-Russell diagram, 437–439
    high-mass, 508–511
    high-velocity, 567

in galactic center, *582*
    light from, 44
    low-mass, 501–502, 506–507
    luminosity of, 420, 435–437
    metal poor, Population II, 489–490
    metal rich, Population I, 489–490
    naming of, 23
    navigation by, 20
    progenitor, 512–513
    pulsating, 490–493
    radii of, 435, 436, 447
    shining of, 4
    sizes and types of, 434–436
    surface temperatures of, 430–431, 433–437
    wobble of, 170, 172
    see also binary stars; double stars; giants; globular clusters; horizontal-branch stars; main-sequence stars; Population I stars; Population II stars; red giants; supergiant stars; T Tauri stars; variable stars; white dwarfs; *and entries under* stellar
stationary absorption lines, 456
steady-state model of universe, 667
Stefan, Josef, 102
Stefan-Boltzmann law, 102–104, 406, 435, 436
stellar association, 467, 471
stellar corpses, 524, 526, 536
stellar distances, 420–422
    and luminosity and brightness, 424–426
stellar evolution, 454, 478–495, 500–519
    and young star clusters, 465–467
stellar-mass black holes, 552, 556
stellar masses, 440–443, *444*
    and star formation, 463–465
stellar motions, 423
stellar nurseries, *5*, 460, 461, 471, 472
stellar parallax, 73, 420–422, 436, 437, 440
stellar spectra, 432–434, 436, 437
    and star types, 439–440
stellar spectroscopy, 432
stellar winds, 470, 471, 503
Stephano, 354
Stern, Alan, 358
Stjerneborg observatory, Denmark, 73
Stonehenge, 21
stones, *see* stony meteorites
stony asteroid, 382, *383*
stony iron meteorites, 373, 374
stony meteorites, 373
storms
    on Neptune, 347–348
    on Saturn, 330
    on Uranus, 346
stratosphere, *196*, 197
Strom, Robert G., 255
Strominger, Andrew, 685
strong nuclear force, 111, 537, 680–682
Subaru telescope, 131, 134, 359

subduction zone, 189
Sudbury Neutrino Observatory, 400
sulfur
    on Io, 308
    in solar nebula, 166
    in solar system, 162
    in Venus atmosphere, 246, 247
sulfur dioxide
    on Io, 308
    on Venus, 251
    and volcanoes, 250
sulfuric acid, on Venus, 246, *251*
summer, 29
    sunlight in, *30, 32*
summer solstice, 30–32, 56
summer triangle, 25, 26
Sun, 3, 150, 389–413, 471
    absorption line spectrum of, 400–403, 432
    angular diameter of, 52
    apparent brightness of, 424
    apparent magnitude of, 427
    atmosphere of, 109, 159, 400–405
    as a blackbody, 101
    and celestial sphere, 44
    central density of, 395
    changes in, 480
    chemical composition of, 391, 479–480
    color of, 431
    data table, 391
    diameter of, *4, 10, 58,* 391
    distances of planets from, 12, 68–70, 153
    and Earth's energy, 180
    and ecliptic, 30–34
    energy of, 390–397
    energy flux of, 103, 104
    formation of, 18, 163–164, 169
    future of, 502, 507
    and geocentric model, 63–64
    gravitational pull on Earth, 32–33, 81
    hydrogen in, 394
    interior of, 2, 394–397
    lifetime of, 18
    location of in Galaxy, 563–564, 569–570
    luminosity of, 103, 390–396, 424, 626
    and lunar eclipses, 49–52
    magnetic field of, 408–412
    magnetic poles of, 409–411
    main-sequence lifetime, 479
    mass inside orbit of, 575
    mass of, 80, 391, 395
    mean density of, 391
    motion of on celestial sphere, 64
    orbit of, 170, 391, 574
    and planets to scale, *152*
    polarity of, 409–411
    position on celestial sphere, 31
    radius of, 103, 170, 391
    as red giant, *482, 483*
    relative size of, *4,* 57–58
    rotation of, 406–407, 410, *411*
    and seasons on Earth, 29–32
    in spectral sequence, 435
    spectrum of, 105, 106, 109
    and summer solstice, 31
    surface area of, 103
    surface temperature of, *4,* 103
    temperatures of, *4,* 391
    theoretical model of, 394–396
    thermonuclear reactions in, *4, 5*
    tidal effects of on Earth, 85–86
    as timekeeper, 34–35
    vibrations of, 397–398
    and winter solstice, 31
    wobble of, 170
    see also corona; photosphere; proto-
        sun; *and entries under* solar
Sun-Earth distance, *4, 9, 10,* 58, 153, 181, 391
    astronomical unit, 12
    variation of, 29
Sun-Jupiter distance, 285
Sun-Mercury distance, 226, 227
Sun-Neptune distance, 345
Sun-Pluto distance, 357
Sun-Saturn distance, 324
Sun-Uranus distance, 344
Sun-Venus distance, *241,* 242
sunlight, 44, 390
    in summer and winter, *30*
    reflected from Titan, 156
sunquakes, 397
sunrises, 108
sunsets, 108
sunspot cycle, 407–408, 410, 411
sunspot maximum, 407
sunspot minimum, 407
sunspots, 70, 389, *400,* 405–412
suntans and sunburns, 104
superclusters (of galaxies), 603
superconductivity, 532
supercontinents, 186–187, 191
superfluidity, 531–532
supergiant stars, 436, *438*–440, 443, 454, *508*–510, 512–513, 550
supergrand unified theory, 682
supergranules, 402, 403
superior conjunction, 67
superior planets, 67, 69
Super-Kamiokande neutrino observatory, 400
superluminal motion, 628
supermassive black holes, 551–552, 582, 630–633
Supernova 1987A (SN 1987A), 511–515, 530
Supernova 1993J, 515
Supernova Cosmology Project, 658
supernova remnants, 471, 518–519, 525, 530
    see also neutron stars
supernovae, 5, *145,* 375, 500, 509–511, 515–518, 525, 529–531, *635, 636*
    as distance indicators, 590
    in 1572, 73
    in 1937, 590
    and star birth, 470–472
superoxides, 274
supersonic solar wind, 193
supersonic winds, 470
surface waves, 184
Surveyor program, 211, *212*
Sycorax, 354
Syene, Egypt, 56, *57*
symmetry, 673
symmetry breaking, 675, 681–683
symmetry restoration, 681
synchronous rotation, 47, 86, 206
    of Galilean satellites, 303
synchrotron, 530–531
synchrotron radiation, 296, 531, 580–581, 624, *625,* 632, 633
    from Jupiter, 297
synodic month, 48, *49*
synodic periods, 68
    and sidereal periods, 69
Syrtis Major, 263, *264*

T Coronae Borealis, 536
T Tauri stars, 463–464, *467,* 471
T Tauri wind, 169
tail (of comet), 375–378
Taj Mahal, *10*
tangential velocity, 423
Tarantula Nebula (30 Doradus), 511, *595*
Tau Cancri, 452
Tau Ceti, 695
tau-type neutrinos, 418
Taurids, 383
Taurus (the Bull), 25, *32,* 169, 468
Taylor, Joseph, 548
Teide 1, 435
telescopes, 70, 122–145
    infrared, 697–698
    Leviathan, 588
    and nonvisible forms of light, 3
Telesto, *329,* 335
temperature, 99
    and altitude, 196
    and electromagnetic radiation, 98–101
temperature scales, 99, 100
Temple of the Sun (Tiahuanaco, Bolivia), 21
terminator (of Moon), 206
terrae, *see* lunar highlands
Terrestrial Planet Finder, 697–698
terrestrial planets, 151–154
    atmospheres of, 158
    chemical composition of, 157–158
    formation of, 167–168
    surface temperatures of, 158
    see also Earth; Galilean satellites;
        Mars; Mercury; Venus
Tethys, *329, 335,* 336, 337
Tethys Sea, 186–187
Thalassa, 355
Thales of Miletus, 55
Tharsis rise, 266–267, *271*

Thebe, *316*, 317
Theia Mons, 248
theories, 2, 3
theory of everything (TOE), 682, 685
thermal equilibrium, 394, 395, 675
thermal pulse, 504
thermal radiation, 295, 624, 625
thermonuclear fusion, 392–394, 398, 483
thermonuclear reactions, 508–510
  in Big Bang, 647
  and special theory of relativity, 544
  in stars, 4
  on Sun, 4, *5*
  supernovae, 517
  in white dwarfs, 536
thermonuclear weapons, 4
thermosphere, *196*, 197
Third Cambridge Catalogue, 618
Thorne, Kip, 557
32 Eridani, 452
3-to-2 spin-orbit coupling, 230–231
3C 48, 618, 619
3C 75, 638
3C 273 (quasar), 616, 618, 619, 621,
  628, 630
3C 279 (blazar), 627–628
threshold temperature, 675
Thuban, 33
thunderstorms, 180, I82, 402
Tiahuanaco, Bolivia, 21
tidal bulge
  on Moon, 215–216, 218, *219*
  on Neptune, 356
tidal distortion
  and eclipsing binary, 446, 447
tidal forces, 84–87, 183, 326
  and black holes, 555, 558
  calculating, 215–216
  and colliding galaxies, 606
  and galaxies, 595
  from Jupiter, 307, 312
tidal heating, 307
time
  gravitational slowing of, 547, *548*
  and Heisenberg uncertainty principle,
    673
  Lorentz transformation of, 544
  measurement of, 11, 12
  and relativity, 542, 666
  units of, 13
  see also proper time; spacetime
time dilation, 543–545
time machines, *557*
time zones, 35
timekeeping, 34
Titan, 155–156, 322, 333–335, 345
Titania, 352–354
titanium oxide, 432, *433*, 434
Titius-Bode Law, 365
Tombaugh, Clyde W., 356
Tomonaga, Sin-Itiro, 680
TOPEX/Poseidon satellite, *14*
total lunar eclipse, 50, *51*
total solar eclipse, 43, 52–55, *404*
totality (of eclipses), 52, 53

TRACE (*Transition Region and Coronal
  Explorer*) spacecraft, *411*, 416
transits, of Galilean satellites, 304
  see also solar transits
transverse velocities, 667
transverse waves, 184
Trapezium, *455*, *462*, 470
trenches, deep-oceanic, 189
Triangulum, *618*
Trifid (M20) Nebula, *169*, 176,
  474–476
triple alpha process, 483–484,
  504
Triton, 155, 326, 351, 355–356
Trojan asteroids, *368*, 370
Tropic of Cancer, *31*, 32
Tropic of Capricorn, *31*, 32
tropical year, 35, 37
troposphere, *196*, 197
Trumpler, Robert J., 564, 457
TT Cygni, *503*, *504*
Tully, Brent, 596
Tully-Fisher relation, 596–597
tungsten atoms, *10*
Tunguska event, Siberia, 382, *383*
tuning fork diagram, 594–595
turbulence, 511
turnoff point, 489
21-centimeter emission line, 117,
  569–571, 573, 574, 596
22-year solar cycle, 410
24 Comae Berenices, 452
2-to-1 resonance, 367
TX Piscium, 452
Type I Cepheid variable, 492, 590
Type 1 Seyfert galaxies, 633
Type I supernova, 516
Type Ia supernova, 516–517, 529, 536,
  590, 596, 597, 657–658
Type Ib supernova, 516, 517, 518
Type Ic supernova, 516, 517, 519
Type 2 Seyfert galaxies, 633
Type II Cepheid, 492
Type II supernovae, 515, 517, 518, 529,
  551
Tyson, J. Anthony, 609

UBV photometry, 430–431
Uhuru orbiting X-ray observatory, *534*,
  549
Ulrich, Roger, 397
ultramafic lava, 309
Ultraviolet Imaging Telescope, 585
ultraviolet light, 104, 141–142
ultraviolet telescopes, 141–142
ultraviolet photons, 274
ultraviolet radiation, 3, 97
  and star formation, 470
Ulysses spacecraft, 405
umbra
  of Earth, 50–52
  of Moon, 53, 54
  of sunspots, 406
Umbriel, 352, 353, 354
unbounded universe, 660

uncertainty principle, *see* Heisenberg
  uncertainty principle
undifferentiated asteroids, 374
unidentified flying objects (UFOs), 695
unification of physical forces, 681–682
unified model of active galaxies, 631–633
universal constant of gravitation, 81, 160
universe, 591
  age of, 645
  birth of, 6
  boundary conditions of, 668
  closed, 660
  Copernican model of, 665
  critical density of, 654
  curvature of, 653–656
  dark-energy-dominated, 659
  density of, 670–671
  early, 669–685
  Earth-centered model of, 665
  expansion of, 117, 640, 642–646, 666,
    668, 671, 672
  expansion rate of, 656–658
  flat, 660, 670–671
  Friedmann models of, 666–667
  future evolution of, 660–661
  homogeneous, 644
  isotropic, 644
  marginally bounded, 660
  matter-dominated, 650
  nature of, 640–641
  observable, 645, *646*
  open, 660
  radiation-dominated, 650
  shape of, 653–656
  steady-state model of, 667
  unbounded, 660
  see also geocentric model of the
    universe; heliocentric model of the
    universe; Ptolemaic system
upper meridian, 34–36
upper meridian transit, 34
Upsilon Andromedae, 172
Uraniborg observatory, Denmark, 73, *74*
Uranus, 4, 64, 84, 342–351
  atmosphere of, 344–345, 346, 347
  average density of, 153, 344, 349
  data table, 344
  diameter of, 153, 344
  escape speed of, 160, 344
  formation of, 349
  internal structure of, 349–350
  magnetic axis of, 350
  magnetic field of, 350–351
  mass of, 153, 344
  orbit of, 67, 151, 153, 344, 346
  rings of, *347*, 351–353
  rotation of, 244, 344, 346
  satellites on, 352–354
  seasons of, 346
  semimajor axis, 76
  storms on, 346
  surface gravity of, 344
  synodic and sidereal periods, 68, 76
  temperature of, 344
  winds on, 347

Uranus-Sun distance, 68, 153, 344
Urey, Harold, 691
Ursa Major (the Great Bear), 22–23, 515, 552
Ursae Majoris, 171
Ursids, 383
UT, UTC, *see* Coordinated Universal Time
Utopia Planitia, 274

V404 Cygni, *551*
V616 Monocerotis, 551
Valles Marineris, 267, *270*
Van Allen, James, 193
Van Allen belts, 193, *194*
variable stars, 490–493
Vega (α Lyrae), 435
  spectrum of, 117
Veil Nebula, 518
Vela (the Sails), *169*, 518
Vela pulsar, 531–533
Vela satellites, 634
Vela supernova remnant, *666*
velocity, 78
  see also radial velocity; recessional velocity; space velocity; tangential velocity; transverse velocities
Venera spacecraft, 240, 245, 247, 249, *250*, 252
Venus, 3, 64, 151, 240–256
  albedo, 241, 242
  apparent size of, 70
  appearance of, 70–72
  atmosphere of, 196, 240, 243–247, 250–251, 255
  average density of, 153, 242
  climate of, 250–251
  cloud cover of, 240, 243–247
  crust of, 252, 253
  data table, 242
  diameter of, 152, 153, 242
  escape speed of, 160, 242
  geologic processes on, 253, *254*
  greatest elongations of, 242
  greenhouse effect on, 240, 251
  impact craters on, 253–255
  inferior conjunction of, *242*, 243
  interior of, 252
  lithosphere of, 253, 255
  magnitude of, 428
  mass of, 153, 242
  and Mayan civilization, *21*
  oceans on, 250–251
  orbit of, 67, 75, 151, 153, *241*–243
  phases of, 70–71, *72*
  rotation of, 242–244
  semimajor axis of, 76
  sidereal period, 68, 76, 77
  solar day on, 244
  solar transits of 243
  sunlight on, 244–247
  surface gravity of, 242
  surface of, 246–250, 252–256

surface temperature of, 242, 245, 250–251
synodic period, 68
volcanism on, 252, 255
volcanoes, 247–250, 266
  see also inner planets
Venus-Sun distance, 68, 153, *241*, 242
vernal equinox, 28, 30, 33
  and sidereal time, 36
  and tropical year, 35
Verne, Jules, 14, 211
Very Large Array (VLA), Soccoro, New Mexico, 138, *139, 297, 519, 581, 605, 609, 617, 628*, 638
Very Large Telescope (VLT), 134, *381, 446*
Very Long Baseline Array (VLBA), 138–139, 597
very-long-baseline interferometry (VLBI), 138
Vesta, 158, 366, 368
*Viking Lander*s, *265*, 273–277, 692, 694
*Viking Orbiter*s, 264–269, *270*, 274, 277, 693–694
virga (rain), 246
Virgo (the Virgin), 25, 32, 551
Virgo cluster, *593*, 601–603, *625*
virtual pairs, 556–557, 673, 674
virtual photons, 680
virus, *10*
visible light, 3, 96, 97, 666
  from hot objects, 98
  photons of, 104
  wavelength and frequency of, 98
visual binaries, 441–442, 444, 445
  center of mass of, 442, *443*
VLT UT 1–Antu, 131
VLT UT 2–Kueyen, 131
VLT UT 3–Melipal, 131
VLT UT 4–Yepun, 131
voids, 603
volatile elements, 219
volcanism, 248
  on Io, 307–309
  on Venus, 252, 255
volcanoes, 151, 248
  on Mars, 266, 271
  and plate boundaries, 188, *189*
  on Venus, 247–249, 266
Voyager 1, 366
  Galilean satellites, 305, 307, *308*
  Jovian system, 309
  Jupiter, *284*, 287–289, *291*
  Jupiter's satellites, 317
  Saturn, *324*, 326–328
  Saturn's satellites, *336*
  Titan, 334
Voyager 2, 366
  Galilean satellites, 305
  Jovian system, 309
  Jupiter, 287–289
  Jupiter's satellites, 317
  Miranda, *355*
  Neptune, 347, *348*, 350
  Neptune's rings, *353*

  Saturn's moons, *330*
  Saturn's rings, *329*, 339
  Saturn's satellites, 326–328, 330, 336
  Triton, 356
  Uranian satellites, 352
  Uranus, 344, *345*, 346, 350
  Uranus's rings, 351, 352
Voyager spacecraft
  Galilean satellites, 302
  Ganymede, 314, 315
  Jupiter, 296, 297
  Saturn, 322, 333
  Saturn's satellites, 329, *330*, 335–337
  Titan, 334

W Ursae Majoris stars, *494, 495*
Walker, Merle, 535
Walsh, Dennis, 608
waning crescent moon, 47
waning gibbous moon, 47
Ward, William, 219
water (H_2O), 158
  in solar nebula, 166
  on Europa, 311–313, 692
  on Jupiter, 294
  on Mars, 260, 267–271
water, frozen
  on Europa, 311
  in Saturn's ring, 328
water, liquid
  and life, 691–692
water hole, 696, *697*
water vapor, 195, 196, 268
  and Earth's atmosphere, 194–195
  and infrared radiation, 140
  on Venus, 250–251
watt (W), 102
wave frequency (ν), 97–98
wavelength (λ), 96
  relationship with frequency, 97–98
  of spectral lines, and relative motion, 115–117
wavelength of maximum emission, 100–101, 102
wavelength shift (Δλ), 116–117
waxing crescent moon, 45
waxing gibbous moon, 45
weak force, 680–682
weakly interacting massive particles (WIMPs), 576, 678
weather
  early warnings, *14*
  and telescopes, 134
Weaver, Thomas, 509
Wegener, Alfred, 186–187, 188
weight, 13, 80
Weinberg, Steven, 681
Weinberg-Glashow-Salam theory, 681
Wells, H. G., 14, 264, 278, 557
Weymann, Ray, 608
Wheeler, John A., 530
Whipple, Fred, 375

Whirlpool Galaxy (NGC 5914), 588, *589*, 591, 607, 615
white dwarfs, 436, *438–440*, 443, 454, 500, 506–507, 510, 515–516, 526, 536, 537
and black holes, 551
spectra of, *548*
white hole, 661
white light, *95, 98*
white ovals, Jupiter, *288–290*
Widmanstätten, Count Alois von, 373
Widmanstätten patterns, 373, *374*
Wien, Wilhelm, 102
Wien's law, 102, 406, 510, 566, 650
Wild Duck cluster (M11), 499
Wildt, Rupert, 286
William of Occam, 66
WIMPs, *see* weakly interacting massive particles
winding dilemma, 577
winds
on Earth's surface, *197*
on Jupiter, 290–291, *293, 294*
on Neptune, 348
on Uranus, 347

see also solar wind; zonal winds
winter, 29
on Mars, 276
sunlight in, *30*
winter solstice, *30,* 31, 32
winter triangle, 25
Witten, Edward, 684
Wolf, Max, 366
Woosley, Stanford, 509
wormholes, 556, 557

X-ray astronomy, 143
X-ray binary pulsars, 534–535
X-ray bursters, 5, 143, *536–537*
X-ray emissions
from galaxy clusters, *604*
X-ray flickering, 551
X-ray pulses, 534–535
X-ray sources, 552
black holes, 551
pulsating, 533–535
X-ray telescopes and observatories, 122, 142–144
X rays, 3
from strange galaxies, 6
xenon, 294–295
XMM-Newton (X-ray Multi-mirror Mission), 143–144

Yang Wei-T'e, 525
year, 35
Yepun telescope, 131, *134*
Yerkes Observatory, *128*
Young, Thomas, *95, 96*
Yucatán Peninsula, 21, 371
Yurtsever, Ulvi, 557
Yuty (crater), *270*

Zamama, *308*
ZAMS, *see* zero-age main sequence
Zappa, Frank, 366
Zappafrank (asteroid), 366
Zeeman, Pieter, 408
Zeeman effect, 408
zenith, 27
zero-age main sequence (ZAMS), 479, 485
zero curvature, 653
0024 + 1654 cluster, *609*
zodiac, 32, 64
zonal winds, 290–291
zones
on Jupiter, *284, 285,* 287–291
on Neptune, 348, *349*
on Saturn, 330
Zuber, Maria, 266
Zwicky, Fritz, 525

Northern Hemisphere Star Charts

The following set of star charts, one for each month of the year, is from *Griffith Observer* magazine. It is useful in the northern hemisphere only. For a set of star charts suitable for use in the southern hemisphere, see the *Universe* CD-ROM or web site.

To use these charts, first select the chart that best corresponds to the date and time of your observations. Hold the chart vertically and turn it so that the direction you are facing shows at the bottom.

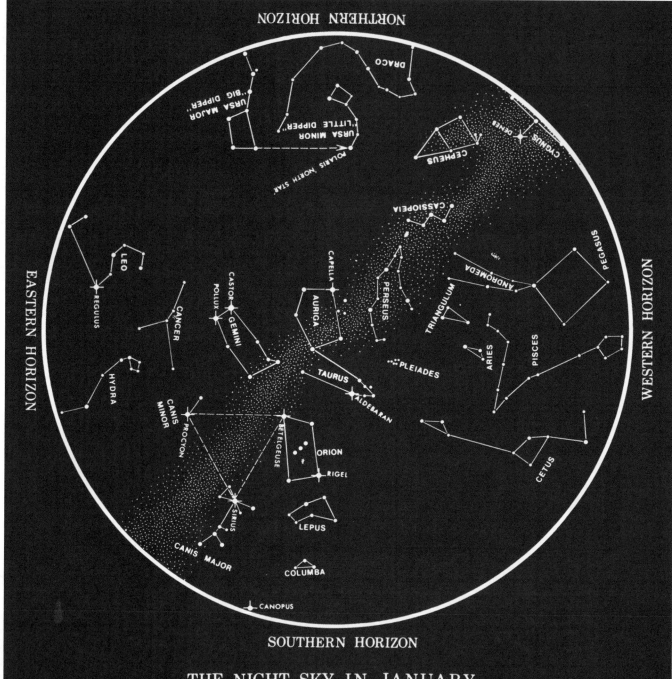

THE NIGHT SKY IN JANUARY

Chart time (Local Standard Time):

10 pm...First of January
9 pm...Middle of January
8 pm...Last of January

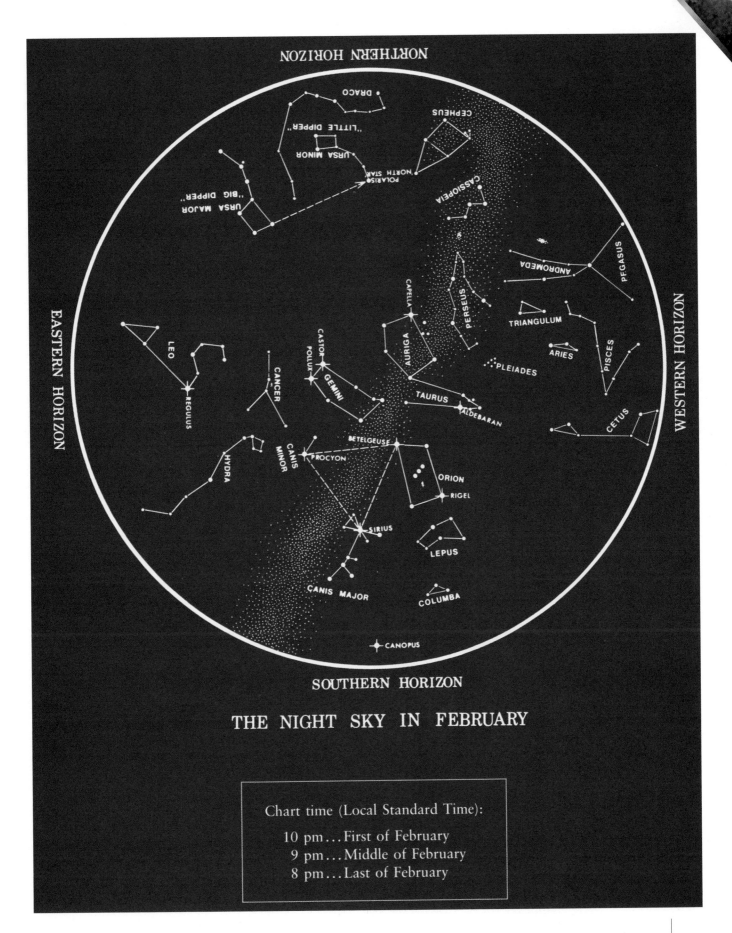

THE NIGHT SKY IN FEBRUARY

Chart time (Local Standard Time):

10 pm...First of February
9 pm...Middle of February
8 pm...Last of February

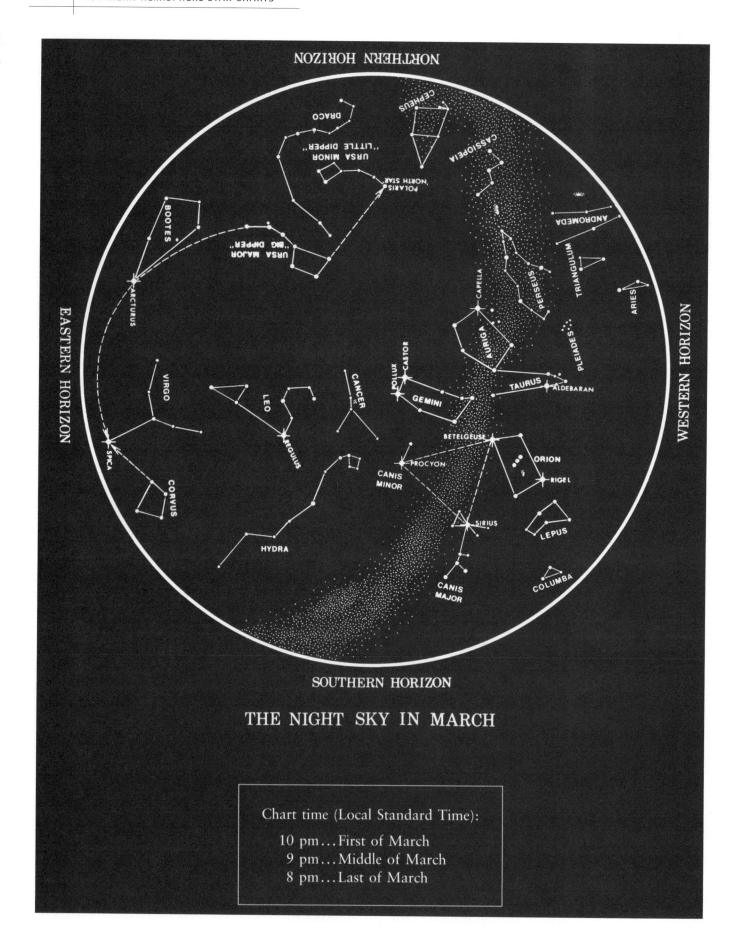

THE NIGHT SKY IN MARCH

Chart time (Local Standard Time):

10 pm...First of March
9 pm...Middle of March
8 pm...Last of March

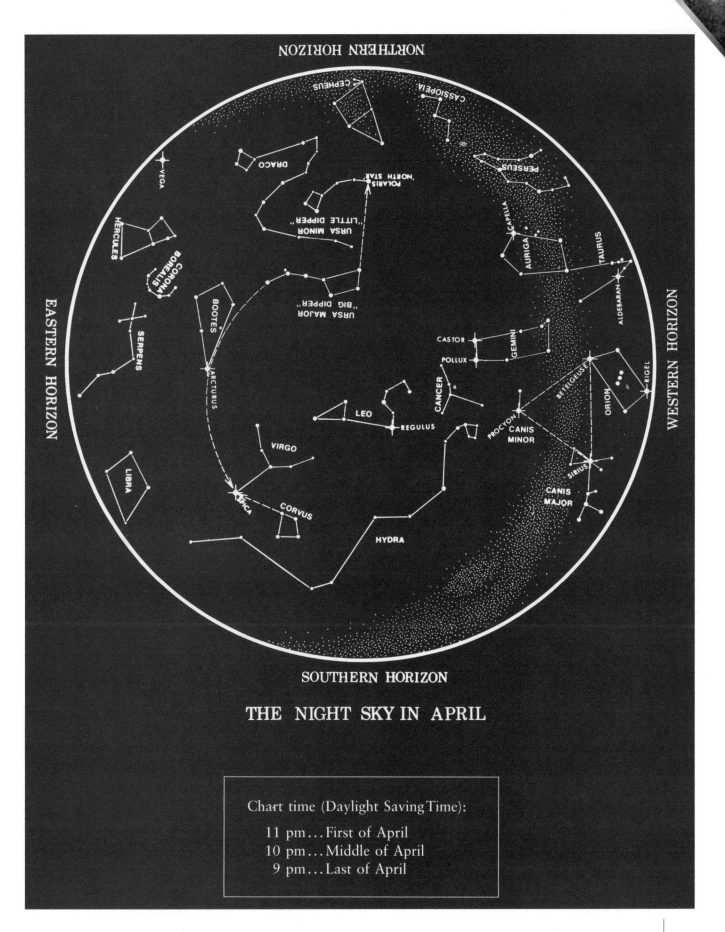

THE NIGHT SKY IN APRIL

Chart time (Daylight Saving Time):

11 pm...First of April
10 pm...Middle of April
9 pm...Last of April

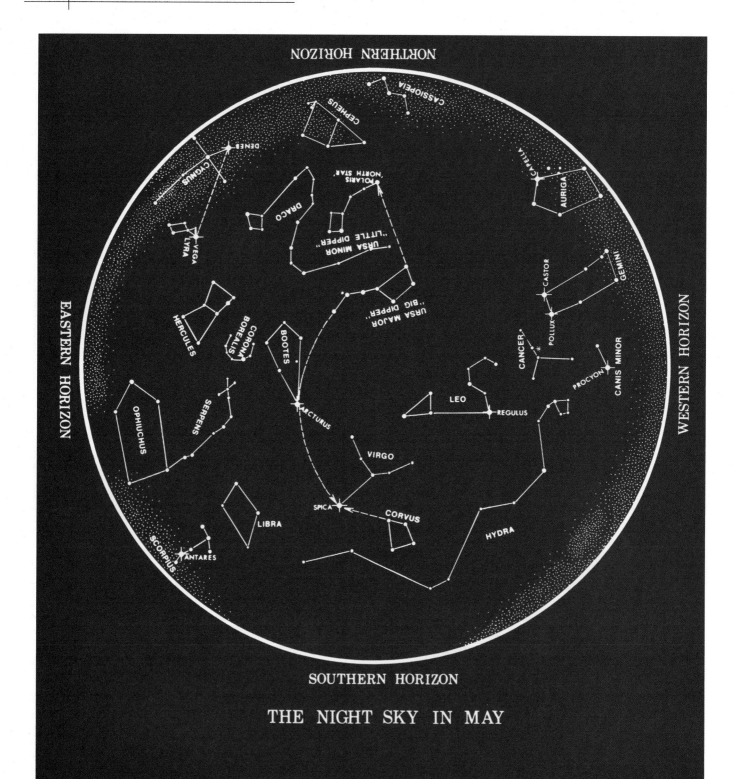

THE NIGHT SKY IN MAY

Chart time (Daylight Saving Time):

11 pm...First of May
10 pm...Middle of May
9 pm...Last of May

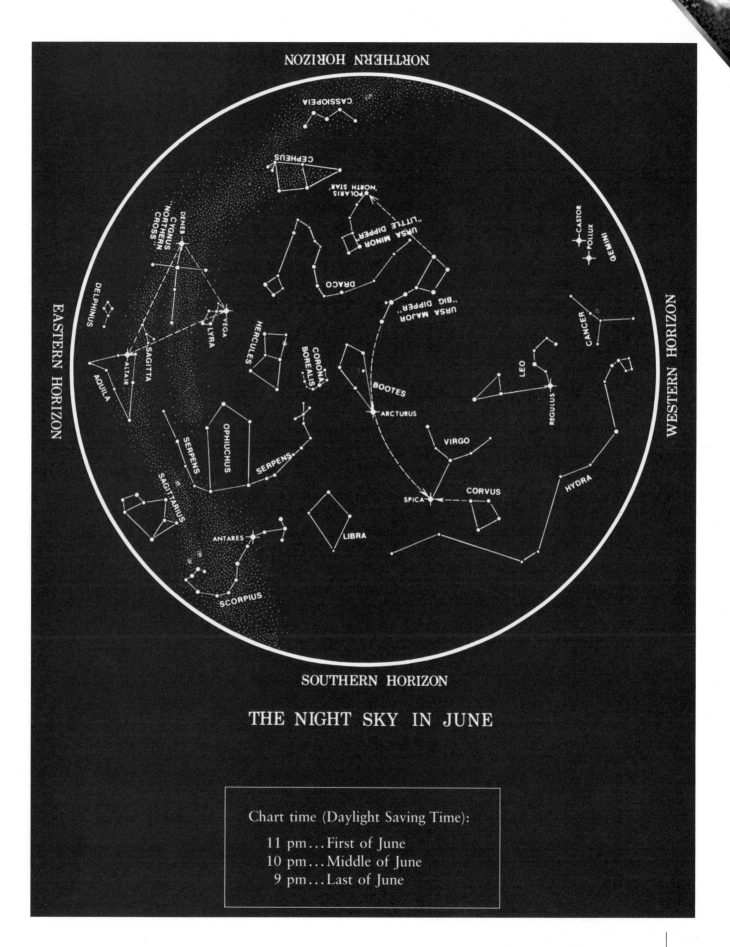

THE NIGHT SKY IN JUNE

Chart time (Daylight Saving Time):

11 pm...First of June
10 pm...Middle of June
9 pm...Last of June

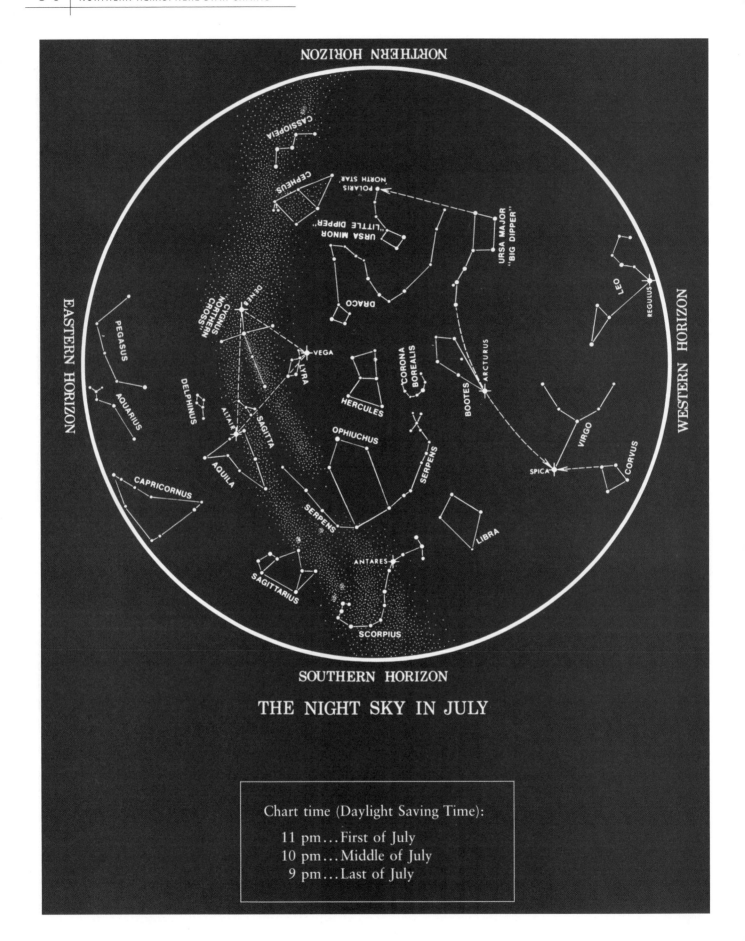

THE NIGHT SKY IN JULY

Chart time (Daylight Saving Time):

11 pm...First of July
10 pm...Middle of July
9 pm...Last of July

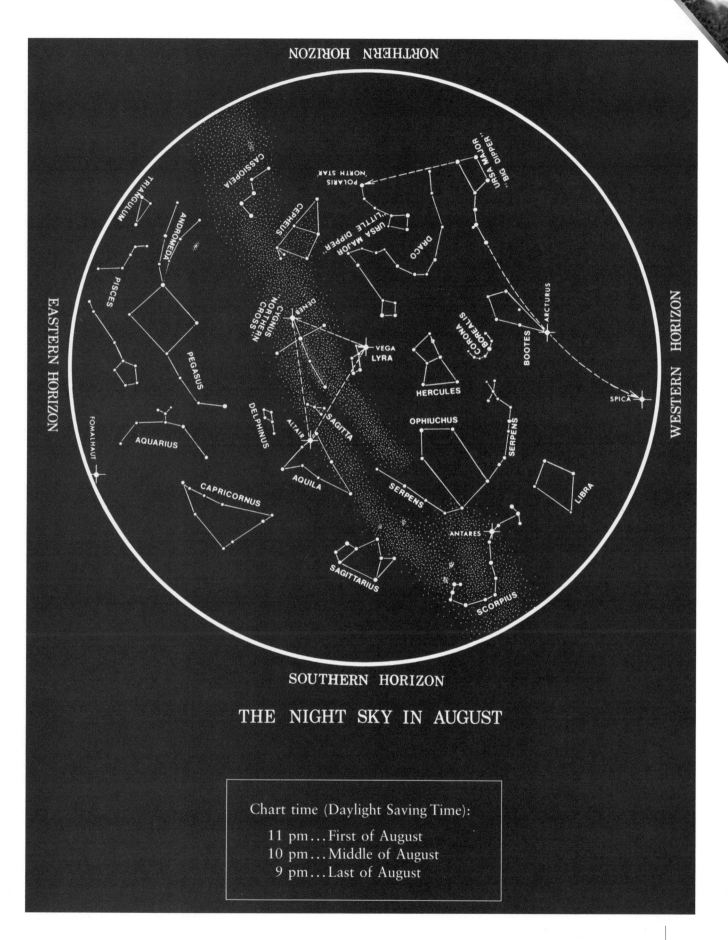

THE NIGHT SKY IN AUGUST

Chart time (Daylight Saving Time):

11 pm...First of August
10 pm...Middle of August
9 pm...Last of August

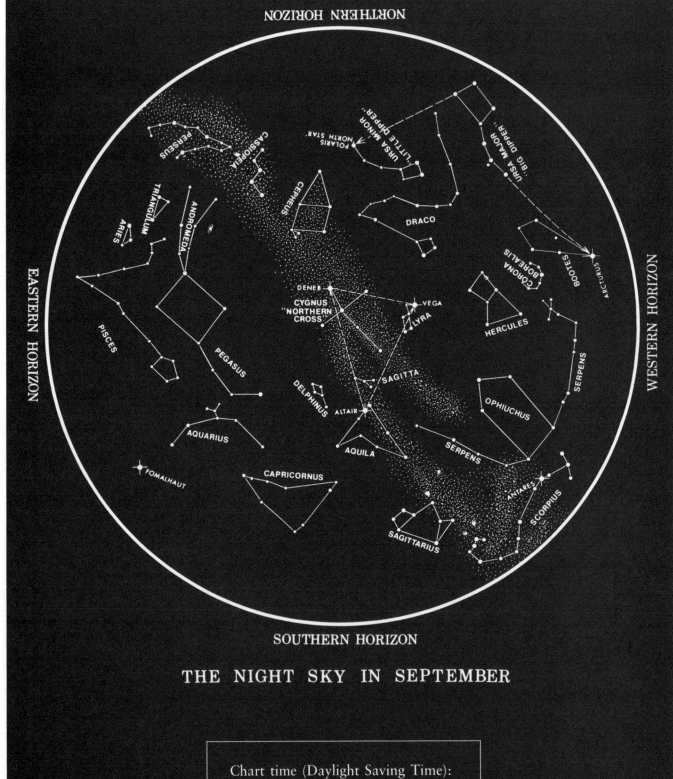

THE NIGHT SKY IN SEPTEMBER

Chart time (Daylight Saving Time):

11 pm...First of September
10 pm...Middle of September
9 pm...Last of September

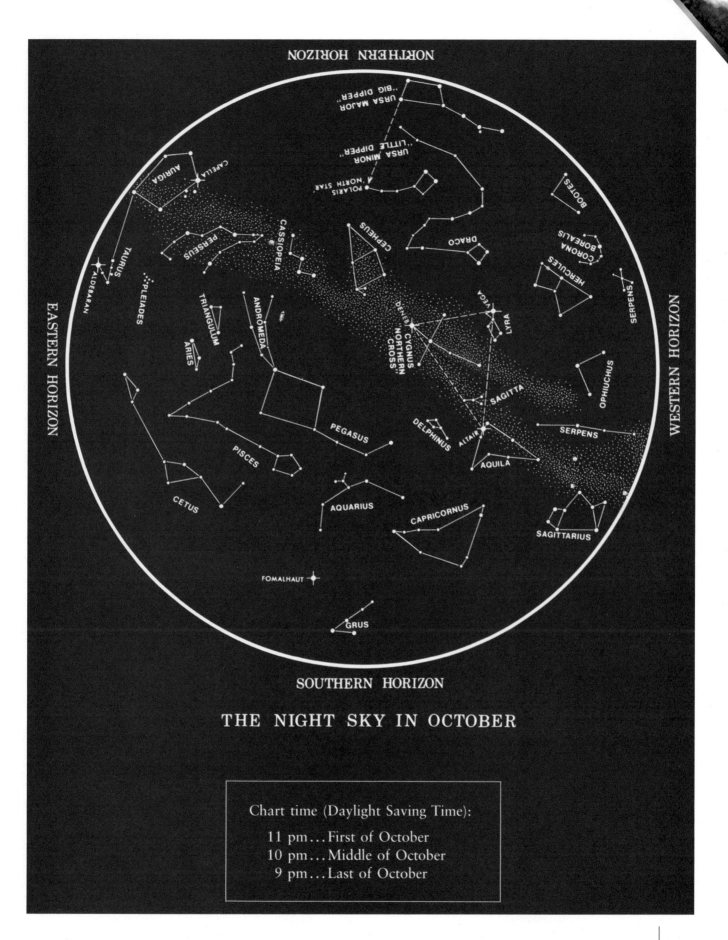

THE NIGHT SKY IN OCTOBER

Chart time (Daylight Saving Time):

11 pm...First of October
10 pm...Middle of October
9 pm...Last of October

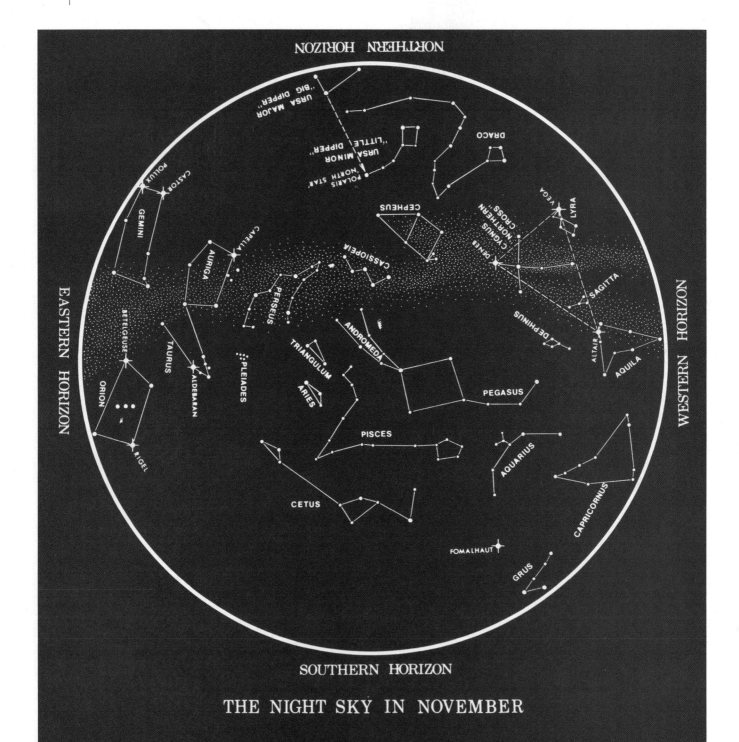

THE NIGHT SKY IN NOVEMBER

Chart time (Local Standard Time):

10 pm...First of November
9 pm...Middle of November
8 pm...Last of November

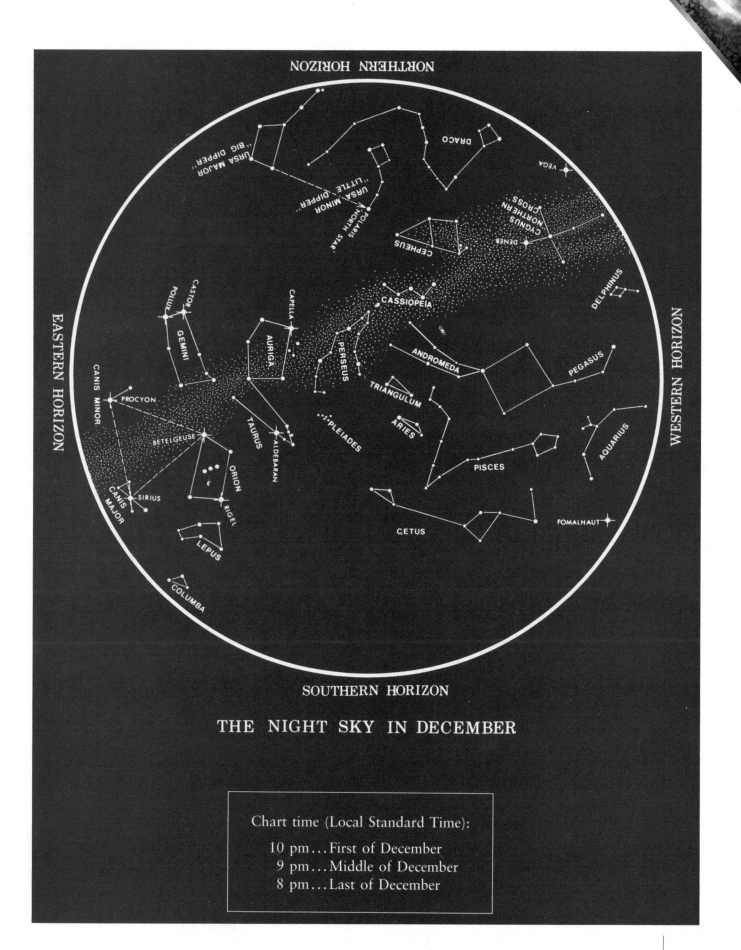

THE NIGHT SKY IN DECEMBER

Chart time (Local Standard Time):

10 pm...First of December
9 pm...Middle of December
8 pm...Last of December